The Transhumanism Handbook

Newton Lee

Editor

The Transhumanism Handbook

 Springer

Editor
Newton Lee
California Transhumanist Party
Los Angeles, CA, USA

Institute for Education Research, and Scholarships
Los Angeles, CA, USA

ISBN 978-3-030-16922-0   ISBN 978-3-030-16920-6   (eBook)
https://doi.org/10.1007/978-3-030-16920-6

This Springer imprint is published by the registered company Springer Nature Switzerland AG
The registered company address is: Gewerbestrasse 11, 6330 Cham, Switzerland

To peace, love, and freedom

About the Book

Modern humanity with some 5,000 years of recorded history has been experiencing growing pains, with no end in sight. It is high time for humanity to grow up and to transcend itself by embracing transhumanism.

Transhumanism offers the most inclusive ideology for all ethnicities and races, the religious and the atheists, conservatives and liberals, the young and the old regardless of socioeconomic status, gender identity, or any other individual qualities.

This book expounds on contemporary views and practical advice from more than 70 transhumanists in order to:

(a) Present the whole picture of transhumanism from both secular and religious points of view.
(b) Explain and demystify artificial general intelligence and superintelligence.
(c) Offer practical advice on radical life extension and rejuvenation.
(d) Explore the roles of blockchain and cryptocurrency in transhumanism.
(e) Examine transhumanist economics, ethics, philanthropy, philosophy, and politics.

Astronaut Neil Armstrong said on the Apollo 11 moon landing in 1969, "One small step for a man, one giant leap for mankind." Transhumanism is the next logical step in the evolution of humankind, and it is the existential solution to the long-term survival of the human race.

Contents

About the Editor

Newton Lee is an author, educator, and futurist. He was the founder of Disney Online Technology Forum, creator of AT&T Bell Labs' first-ever commercial artificial intelligence tool, inventor of the world's first annotated multimedia OPAC for the US National Agricultural Library, and longest-serving editor in chief in the history of the Association for Computing Machinery for its publication *Computers in Entertainment* (2003–2018).

Lee is the chairman of the California Transhumanist Party, education and media advisor to the US Transhumanist Party, and president of the 501(c)(3) nonprofit Institute for Education, Research, and Scholarships. He graduated Summa Cum Laude from Virginia Tech with a B.S. and M.S. degree in Computer Science (specializing in Artificial Intelligence) and earned a perfect GPA from Vincennes University with an A.S. degree in Electrical Engineering and an honorary doctorate in Computer Science. He has been honored with a Michigan Leading Edge Technologies Award, two community development awards from the California Junior Chamber of Commerce, and four volunteer project leadership awards from The Walt Disney Company.

Lee's *Total Information Awareness* trilogy books (*Facebook Nation*, *Counterterrorism and Cybersecurity*, and *Google It*) have garnered rave reviews from Eleanor Clift of *Newsweek*, *The Daily Beast*, *ACM Computing Reviews*, *AdWeek SocialTimes*, *Choice Magazine*, and Carmen LaBerge of *Faith Radio*, among others. Dr. Laura Wilhelm for *The Hollywood Times* applauded that "Newton Lee's thorough and thoughtful analysis should please pacifists in search of lasting solutions to the planet's biggest problems." Veteran Staff Sergeant Andrew Price of the US Air Force (USAF) remarked, "I am inspired by the prospect of world peace. I'd fully recommend following the author's steps, reaching beyond our borders, making friends outside our norm, and helping to foster world peace and a better tomorrow."

Contributors

Ojochogwu Abdul University of Lagos, Lagos, Nigeria

Maria Entraigues Abramson Global Outreach Coordinator, SENS Research Foundation, Mountain View, CA, USA

Michele Adelson-Gavrieli Artificial Intelligence Incubators & Sounds of the Heart, Ashkelon, Israel

David Aguilar UIC Barcelona – Universitat Internacional de Catalunya, Barcelona, Spain

Sarah Ahamed India Future Society, Bangalore, India

Chris T. Armstrong Transhumanist, Kansas City, MO, USA

Kyrtin Atreides The Foundation, Seattle, WA, USA

Joe Bardin People Unlimited, Scottsdale, AZ, USA

Barış Bayram Yeditepe University, Istanbul, Turkey

Lincoln Cannon Mormon Transhumanist Association, Orem, UT, USA

Kali Carrigan Universiteit van Amsterdam, Amsterdam, The Netherlands

Didier Coeurnelle AFT-Technoprog and Heales, Brussels, Belgium

Keith Comito Life Extension Advocacy Foundation (LEAF)/Lifespan.io, Seaford, NY, USA

José Luis Cordeiro Humanity+, Madrid, Spain

Francesco Albert Bosco Cortese Biogerontology Research Foundation, Oxford, UK

Mark Crowther Technologist and Writer at BJSS, London, UK

Dinorah Delfin United States Transhumanist Party, New York, NY, USA

Sylvester Geldtmeijer Amsterdam, The Netherlands

Federico De Gonzalez-Soler Cryptoz, Blair Athol, Australia

Scott H. Hawley Department of Chemistry and Physics, Belmont University, Nashville, TN, USA

Henrique Jorge ETER9, Viseu Area, Portugal

David J. Kelley Artificial General Intelligence Inc., Provo, UT, USA

Martin van der Kroon United States Transhumanist Party, Amersfoort, The Netherlands

Carmen Fowler LaBerge Faith Radio Network, Saint Paul, MN, USA

Inessa Lee California Transhumanist Party, Los Angeles, CA, USA
Institute for Education, Research, and Scholarships, Los Angeles, CA, USA

Newton Lee California Transhumanist Party, Los Angeles, CA, USA
Institute for Education, Research, and Scholarships, Los Angeles, CA, USA

Rich Lee Biohacker & Cyborg, Washington, UT, USA

Kate Levchuk Founder of Transpire, Futurist at KateGoesTech, London, UK

Rob Lubow Botcopy, Los Angeles, CA, USA

Eugene Lukyanov UA Realty Group, Kyiv, Ukraine

Palak Madan India Future Society, Bangalore, India

Julia A. Mossbridge Mossbridge Institute, LLC, Sebastopol, CA, USA

B. J. Murphy United States Transhumanist Party, Dobson, NC, USA

Blaire Ostler Mormon Transhumanist Association, Orem, UT, USA

Elizabeth Parrish BioViva, Bainbridge Island, WA, USA

Ira S. Pastor Bioquark Inc., Philadelphia, PA, USA

Ted Peters Graduate Theological Union, Berkeley, CA, USA

Micah Redding Christian Transhumanist Association, Nashville, TN, USA

Garfield Reeves-Stevens Victoria, Canada

Judith Reeves-Stevens Victoria, Canada

Michael R. Rose Department of Ecology and Evolutionary Biology, University of California, Irvine, Irvine, CA, USA

Grant A. Rutledge Department of Ecology and Evolutionary Biology, University of California, Irvine, Irvine, CA, USA

Giovanni Santostasi Department of Neurology, Feinberg School of Medicine, Northwestern University, Chicago, IL, USA

Anca I. Selariu BioViva, Bainbridge Island, WA, USA

Nicole Shadowen John Jay College of Criminal Justice, New York City, NY, USA

Vineeta Sharma tecHindustan, Mohali, Punjab, India

Avinash Kumar Singh India Future Society, Bangalore, India
University of Technology Sydney, Ultimo, Australia

Calem John Smith Zerozed, Perth, Australia

Jonathan Squyres Global New World, Lufkin, TX, USA

Ilia Stambler Bar Ilan University, Ramat Gan, Israel

R. Nicholas Starr California Transhumanist Party, Grass Valley, CA, USA

Jakub Stefaniak Nuffield Department of Medicine, Target Discovery Institute, University of Oxford, Oxford, UK

Amanda Stoel DIY Futurist, Bussum, The Netherlands

Gennady Stolyarov II United States Transhumanist Party, Carson City, NV, USA

Andjelka (Angie) Stones NorthCentral University, Scottsdale, AZ, USA

Joshua Marshall Strahan Lipscomb University, Nashville, TN, USA

Melanie Swan Purdue University, West Lafayette, IN, USA
Institute for Blockchain Studies, West Lafayette, IN, USA
DIYgenomics, West Lafayette, IN, USA

Mariana Todorova Bulgarian Academy of Sciences (Millennium project Bulgarian Node), Sofia, Bulgaria

Janez Trobevšek Cryptoz, Blair Athol, Australia

Stephen Valadez Chattanooga, TN, USA

Natasha Vita-More Humanity+, Inc. and University of Advancing Technology, Tempe, AZ, USA

Eleanor "Nell" Watson Singularity University, Mountain View, USA

Augusta L. Wellington International Psychoanalytic Association and Psychoanalyst in Private Practice, Nassau, Bahamas

Sophie Wennerscheid Ghent University, Ghent, Belgium
University of Copenhagen, København, Denmark

David W. Wood Delta Wisdom, London, UK

Daniel Yeluashvili San Francisco State University, San Francisco, CA, USA

Jeffrey Zilahy Transhumanist Consultant, New York, NY, USA

Part I
Brave New World of Transhumanism

Chapter 1
Brave New World of Transhumanism

Newton Lee

> *I believe in transhumanism: once there are enough people who can truly say that, the human species will be on the threshold of a new kind of existence, as different from ours as ours is from that of Peking man. It will at last be consciously fulfilling its real destiny.*
> —*Julian Huxley, first Director-General of UNESCO*

1.1 Humanity's Growing Pains

Modern humanity with some 5000 years of recorded history has been experiencing growing pains, with no end in sight. It is high time for humanity to grow up and to transcend itself by embracing transhumanism.

Jesus said to his disciples in the Garden of Gethsemane, "The spirit is willing, but the flesh is weak" (Matthew 26:41). The current capability of human cognition makes the interpretation of God a difficult matter even for the genius Albert Einstein who wrote in a 1954 letter, "The word god is for me nothing more than the expression and product of human weaknesses" [1].

To err is human, whether we are religious, atheist, or agnostic. Humankind was engulfed in World War II from 1939 to 1945 and suffered the deadliest military conflict in history and many years of privation in the aftermath [2].

N. Lee (✉)
California Transhumanist Party, Los Angeles, CA, USA

Institute for Education, Research, and Scholarships, Los Angeles, CA, USA
e-mail: newton@californiatranshumanistparty.org; newton@ifers.org

© Springer Nature Switzerland AG 2019
N. Lee (ed.), *The Transhumanism Handbook*,
https://doi.org/10.1007/978-3-030-16920-6_1

1.2   Transhumanism

In 1957 as nations were recovering from the devastation of World War II, Julian Huxley—evolutionary biologist and first Director-General of the United Nations Educational, Scientific and Cultural Organization (UNESCO)—advocated the necessity of "transhumanism" to improve the human condition [3]:

> The human species can, if it wishes, transcend itself—not just sporadically, an individual here in one way, an individual there in another way, but in its entirety, as humanity. We need a name for this new belief. Perhaps transhumanism will serve: man remaining man, but transcending himself, by realizing new possibilities of and for his human nature. I believe in transhumanism: once there are enough people who can truly say that, the human species will be on the threshold of a new kind of existence, as different from ours as ours is from that of Peking man. It will at last be consciously fulfilling its real destiny.

The word "transhuman" first appeared in the fourteenth century in Dante's *Divine Comedy* (Paradise, Canto I) to describe the change of the human body to immortal flesh in eschatology [4] (See Fig. 1.1):

Words may not tell of that transhuman change;
And therefore let the example serve, though weak,
For those whom grace hath better proof in store.

Fig. 1.1 La Divina Commedia di Dante (Dante and the Divine Comedy) by Domenico di Francesco (1465)

1.3   We Are All Transhumanists

Have you ever taken vitamins, antibiotics, vaccinations, or (for women) birth control pills? Yes indeed, everyone is using science and technology to enhance or to alter our body chemistry in order to stay healthy and be more in control of our lives. We are all transhumanists to varying degrees.

The two most important life-changing events are being born and being dead, both of which we have had little control of, until now. We certainly do not have any say in our own births, but thanks to science and technology we are now living longer, healthier, and smarter.

Pacemakers, prosthesis, stentrode, optogenetics, surgeries, and other medical advancements exemplify the use of science and technology to prolong human life and to improve quality of life. Christopher Reeve, best known for playing the role of the comic book superhero Superman, lobbied for human embryonic stem cell research after a horse-riding accident left him quadriplegic [5]. In spite of the traditionally conservative views from the Catholic Church, Pope Francis gave his blessing to human-animal chimera research for organ transplants [6].

Even the Amish who painstakingly avoid the use of modern technology are embracing renewable energy. The Pennsylvania Amish of Lancaster County, for instance, has banned the use of public electricity since 1920. Yet they use green technology such as solar, hydraulic, and pneumatic power for business and healthcare centers [7].

I have conversed with some Amish during my travels. They told me that they do not have television, Internet, or cell phones—three major distractions and addictions for children nowadays in a high-tech society. But they are willing to travel long-distance by train to visit doctors when necessary.

Whether they admit it or not, medical doctors and researchers are practitioners of transhumanism. In a 2015 interview by the 2045 Strategic Social Initiative, TV anchor Olesya Yermakova asked SENS Research Foundation cofounder Aubrey de Grey, "Do you consider yourself a Transhumanist?" And de Grey replied. "Not really. No. I really just consider myself a completely boring medical researcher. I just want to stop people from getting sick" [8]. Curing all diseases inevitably advances a top agenda of transhumanism.

1.4   Pros and Cons of Technology

In *Justice League* (2017), Batman Bruce Wayne (Ben Affleck) and Wonder Woman (Gal Gadot) exchanged an interesting dialogue on science and technology:

Batman: But this is science beyond our limits. And that's what science is for. To do what's never been done. To make life better.

Wonder Woman: Or to end it. Technology is like any other power. Without reason, without heart, it destroys us.

Technology in and of itself is devoid of good and evil, but it seems that evil often-times accompanies good in every scientific discovery or engineering marvel. For instance, nuclear power for electrical energy and weapons of mass destruction, GPS for navigating automobiles as well as guided missiles, genetically-modified viruses for vaccines and bioweapons, and 3-D printing of human organs and handguns, just to name a few.

Christian evangelist Billy Graham spoke about technology and faith at TED in February 1998: "You've seen people take beneficial technological advances, such as the Internet … and twist them into something corrupting. You've seen brilliant peo-ple devise computer viruses that bring down whole systems. The Oklahoma City bombing was simple technology, horribly used. The problem is not technology. The problem is the person or persons using it" [9].

Even world-class technophiles realize the danger of technology for children. Steve Jobs told *The New York Times* in 2010, "They haven't used it [the iPad]. We limit how much technology our kids use at home" [10]. And Bill Gates said in a 2017 interview with *The Mirror*, "We don't have cellphones at the table when we are having a meal, we didn't give our kids cellphones until they were 14" [11]. In 2018 Apple CEO Tim Cook told *The Guardian*, "I don't have a kid, but I have a nephew that I put some boundaries on. There are some things that I won't allow; I don't want them on a social network" [12].

A group of early employees at Facebook and Google have joined forces with Center for Humane Technology and Common Sense Media to educate students, parents, and teachers about social media addiction and dangers [13].

Cyber bullying, in particular, exacerbates peer pressure, depression, and teen suicides. It is particularly heartbreaking that the suicide rate for children and teens between 10 and 17 was up a staggering 70% among white youth and 77% among black youth over a 10-year period from 2006 to 2016 [14].

Nevertheless, the Internet is here to stay and technology continues to evolve. Therefore, human beings must also evolve to become smarter and wiser in order to keep up, lest we shall perish by our own hands. J. Robert Oppenheimer, head of the Manhattan Project, expressed his fear that the atomic bomb might become "a weapon of genocide" [15].

In a 1965 television broadcast, Oppenheimer said in tears and agony, "We knew the world would not be the same. A few people laughed. A few people cried. Most people were silent. I remembered the line from the Hindu scripture, the Bhagavad Gita. Vishnu is trying to persuade the Prince that he should do his duty, and to impress him, takes on his multi-armed form and says, 'Now I am become Death, the destroyer of worlds.' I suppose we all thought that, one way or another" [16] (See Fig. 1.2).

Fortunately and knocking on wood, humanity has evolved just barely enough not to start a nuclear world war that would annihilate all human beings on Earth. Had the early cavemen fought with atomic bombs instead of rocks and cattle bone clubs, the human species would have been extinct eons ago. With great power comes great responsibility.

Fig. 1.2 J. Robert Oppenheimer, head of the Manhattan Project, delivered a line from the Hindu scripture, the Bhagavad Gita, in his 1965 televised speech

"The future is ours to shape," said Max Tegmark, MIT cosmologist and cofounder of the Future of Life Institute. "I feel we are in a race that we need to win. It's a race between the growing power of the technology and the growing wisdom we need to manage it" [17].

1.5   Technological Singularity

Mathematician and nuclear physicist Stanislaw Ulam who joined the Manhattan Project summarized a discussion he had with his fellow colleague John von Neumann in 1950s about the "ever accelerating progress of technology and changes in the mode of human life, which gives the appearance of approaching some essential singularity in the history of the race beyond which human affairs, as we know them, could not continue" [18].

Ray Kurzweil, futurist and engineering director at Google, defines technological singularity in his 2005 book *The Singularity is Near* as "a future period during which the pace of technological change will be so rapid, its impact so deep, that human life will be irreversibly transformed. Although neither utopian nor dystopian, this epoch will transform the concepts that we rely on to give meaning to our lives, from our business models to the cycle of human life, including death itself" [19].

While singularity might have created space and time at the Big Bang with or without intelligent design, technological singularity will usher in a new era of human existence in the space-time continuum. Technological singularity is inevitable given the human nature to discover, create, and change the world that we live in. As Ray Kurzweil predicted, "We're going to gradually enhance ourselves. That's the nature of being human—we transcend our limitations" [20].

1.6 Three Options for Humanity

What have humans created so far? Let's see. We created automobiles so that we can travel far and fast, but we also get stuck in traffic jams. We created light bulbs, so we work longer hours and sleep less. We created dynamite for mining and construction, but also for wars and destructions. We created money so that people are enslaved by their jobs or greed. We created social media to keep each other instantly informed; yet the new generation is the loneliest people who bury themselves in the digital world [21].

Even a large majority of beautiful music, great literature, and wonderful work of arts depicts human suffering, injustice, fears, and unfulfilled desires. Sora and his stepsister Shiro were ranked as two of the most intelligent anime characters of all time in a 2014 poll [22]. Sora excels in negotiation and diplomacy skill whereas Shiro in logic and scientific calculations.

In *No Game No Life*, Sora and Shiro were transported to a world where life is a series of games and humanity is in grave danger of extinction. Sora said, "I don't believe in humanity. Humans are lowly, stupid creatures. Every single one of them. Including me. They're all crap. Between this world and the old one, that hasn't changed a bit" [23]. But not all hope is lost. Sora also said, "But I believe in the potential of humanity. … People with the potential, the hope, the fantasies that can reach the gods themselves, all inside one tiny little body" [24].

There are three options for humanity:

1. We can follow the Amish way of life and free ourselves from worldly distractions. Some may take it even further by living in a monastery.
2. We keep the current status quo and let humanity go through another millennium of growing pains and existential risks that may lead to human extinction.
3. We transform our thoughts and deeds towards a technological brave new world of transhumanism.

The longer we stay in the purgatory of humanity as we know today, the more human suffering we have to endure. One of the best thought-provoking pre-apocalyptic movies is the 2011 drama *Perfect Sense* starring Eva Green and Ewan McGregor which deals with pandemics, human senses, love, and the meaning of life.

Some may argue that suffering builds character, which is true. But if we examine the Book of Job, we realize that Job was righteous since the beginning and he remained righteous throughout the whole ordeal. In other words, Job remained basically the same person before and after. But his wife who asked Job to denounce God had learned to trust God more.

Some others may suggest that human beings need to work out their karma by reincarnation. It may be so, but karma is basically the butterfly effect and reincarnation in scientific terms is recycling human beings. All living things are recycled. Curt Stager, ecologist and climate scientist at Paul Smith's College, penned an amusing and eye-opening article titled "You Are Made of Waste: Searching for the ultimate example of recycling? Look in the mirror." [25]:

> You may think of yourself as a highly refined and sophisticated creature—and you are. But you are also full of discarded, rejected, and recycled atomic elements. … Look at one of your fingernails. Carbon makes up half of its mass, and roughly one in eight of those carbon atoms recently emerged from a chimney or a tailpipe. … When you smile, the gleam of your teeth obscures a slight glow from radioactive waste. … The oxygen in your lungs and bloodstream is a highly reactive waste product generated by vegetation and microbes. … The next time you brush your hair, think of the nitrogenous waste that helped create it. All of your proteins, including hair keratin, contain formerly airborne nitrogen atoms. … Every atom of iron in your blood, which helps your heart shuttle oxygen from your lungs to your cells, once helped destroy a massive star. … The same blasts also released carbon, nitrogen, oxygen, and other elements of life, which later produced the sun, the Earth, and eventually—you.

Whatever fortune or misfortune awaits in the subsequent reincarnations, a purgatory by definition does not last forever. We will be forced to either go backward to the Amish way of life devoid of technology or move forward to a transhumanist world embracing technology.

1.7   The Human Condition

J. Robert Oppenheimer observed people's reaction to the use of nuclear weapons and he said, "A few people laughed. A few people cried. Most people were silent." If we keep silent and sit on the sidelines instead of speaking up and marching forward, human suffering will continue and inevitably escalate.

40% of the America's food supply or $165 billion worth is wasted every year [26]. In spite of that, the United States has the largest food reserve in history, with a 2.5 billion pound surplus of chicken, turkey, pork, and beef in cold storage and 1.39 billion pound surplus of cheese [27]. Meanwhile, more than 13 million U.S. children are suffering from malnutrition [28].

Worldwide, roughly one third of the food produced for human consumption every year—approximately 1.3 billion tons—gets lost or wasted [29]. Meanwhile, 821 million people—one in nine individuals worldwide—go to bed on an empty stomach each night [30].

"This culture of waste has made us insensitive even to the waste and disposal of food, which is even more despicable when all over the world many individuals and families are suffering from hunger and malnutrition," Pope Francis lamented [31].

In March 2018, nearly 250,000 rental apartments in New York City sat empty [32] while almost 4000 people were homeless and unsheltered on the streets of the Big Apple every single night [33]. The haves and the have-nots who live in the same city breathe the same air, exchanging a few molecules emitted from their bodies as they occasionally pass by one another. The haves may weep crocodile tears for the have-nots, but are basically happy with things as they are.

CNN reporter Dylan Byers explains, "The homelessness epidemic is the inconvenient truth of the tech boom that has fueled growth in Seattle, the Bay Area and Los Angeles, and it's a problem for both city governments and the tech companies in their area. City leaders and tech CEOs should own it. There is a massive opportunity to innovate on solving this issue. Doing it now would save lives (and generate good press). Avoiding it will yield massive problems down the line" [34].

Only a few people would lift a finger to help. For instance, a Good Samaritan paid for 70 hotel rooms for those in need amid freezing temperature after a homeless camp in Chicago was forced to clear out due to a fire [35]. For another example, two Kansas police officers bought a pair of work boots for a displaced juvenile accused of shoplifting at a Wal-Mart because he needed the boots to get a job [36]. These are heartwarming stories of empathy.

1.8 Empathy Versus Apathy

What would the world be like without empathy? We already knew.

More than 3000 children lost a parent when the World Trade Center's Twin Towers collapsed in the 9/11 terrorist attacks [37]. Meanwhile, some children in east Jerusalem and refugee camps in Lebanon were seen celebrating news of the atrocity [38].

White House aide Kelly Sadler responded to Senator John McCain's opposition to President Donald Trump's pick for CIA director by saying, "It doesn't matter, he's dying anyway" [39].

In Florida, some teenagers witnessed a man who was drowning in a lake. Instead of helping or calling for help, the teens chuckled and said, "You're going to die." They recorded the victim's final moments and posted the video on YouTube [40].

In a 2017 town hall meeting in Idaho, U.S. Representative Raúl R. Labrador defended the American Health Care Act that would reduce the Medicaid program, and he said in a blatant display of ignorance and apathy, "Nobody dies because they don't have access to health care" [41].

Did Labrador learn about a dying mother's plea for her life? 38-year-old Erika Zak was repeatedly denied by her insurance company for her liver transplant. She finally got approved after appealing to UnitedHealthcare's CEO. "You shouldn't have to beg for your life. No one should," said Zak. "The fact that I'm given his opportunity is amazing, but the fact that we had to fight so hard to get there is sickening. I feel like it's not only for me. It's to expose what's happening in this nation" [42].

The 9/11 terrorist attacks blindsided the American public because of our ignorance and indifference about American foreign policy in the Middle East [43]. Environmental disasters caused by human activities and poor judgments have polluted the air, land, and water, contaminating our food and endangering our health [44].

Richard Matheson (best known for *I Am Legend* and *Nightmare at 20,000 Feet*) wrote a short story titled "Button, Button" in 1970. The story was adapted into an episode of *The Twilight Zone* and the 2009 movie *The Box* starring Cameron Diaz and James Marsden as a couple who receive a box from a mysterious man (played by Frank Langella). The man offers the couple one million dollars if they press the button sealed within the dome on top of the box, but he tells them that, once the button has been pushed, someone they do not know will die. Unfortunately, people are pressing that button every day due to selfishness, ignorance, and apathy.

Empathy and compassion are in short supply nowadays. One time I received an email from a recent university graduate who wrote, "I don't care if a student is being kicked out school or another is ill, that's none of my business."

There is a joke that goes like this: "What is the difference between ignorance and apathy?" and the answer is "I don't know, and I don't care." But seriously speaking, the lack of knowledge compounded by the lack of empathy has perpetuated discrimination, racism, sexism, inequality, sufferings, wars, and genocides.

Ignorance and apathy will eventually lead to the annihilation of humanity if left unchecked. Controversial Black Panther Party leader Eldridge Cleaver once said, "There is no more neutrality in the world. You either have to be part of the solution, or you're going to be part of the problem" [45].

1.9   Philanthropy

Fortunately there are still many Good Samaritans who are part of the solution to the human condition. President Franklin D. Roosevelt said in his inaugural address on January 20, 1937, "The test of our progress is not whether we add more to the abundance of those who have much; it is whether we provide enough for those who have too little" [46].

Since the world governments are incapable of taking proper care of all their citizens, it is up to philanthropists and volunteers to lend a helping hand. Here are some recent examples:

- In 2015, Mark Zuckerberg announced that he would donate 99% of his Facebook shares worth $45 billion to the Chan Zuckerberg Initiative, LLC for charitable projects [47].
- In 2016, Zuckerberg and Chan pledged an additional $3 billion to cure all diseases by the end of the century [48].
- In 2017, Bill Gates invested $50 million of his own money into the Dementia Discovery Fund to find a cure for Alzheimer's disease [49].

- In 2018, Bill & Melinda Gates Foundation paid off Nigeria's $76 million debt to become nearly polio-free [50].
- In 2018, Jeff and MacKenzie Bezos committed $2 billion to fund nonprofits to help homeless families and preschools in low-income communities [51].
- In 2018, the Philando Castile charity run by Pam Fergus wiped out the lunch debt of 37,000 kids at all 56 schools in Minnesota's St. Paul Public Schools where Castile worked [55].
- In 2018, New Orleans Chef Tunde Way ran a social experiment in a pop-up food stall where white customers were told about the income gap between whites and African-Americans [52]. He asked whether they wanted to pay $30 instead of the listed price of $12 for a meal. The majority—nearly 80%—decided to pay more. The experiment shows that most people, if given a chance, are willing to help the disadvantaged and underprivileged.
- In 2018, hundreds of Canadian doctors were protesting against their increase in salary totally $700 million. "These increases are all the more shocking because our nurses, clerks and other professionals face very difficult working conditions, while our patients live with the lack of access to required services because of the drastic cuts in recent years," said the doctors who want their pay raises to be redistributed to nurses, healthcare professionals, and patient care [53].

Philanthropists are not limited to the grownups or the wealthy. In fact, if every American were to reduce their spending by $1 per month (e.g. opting for smaller size popcorn in a movie theater, or sharing a big meal with a friend without wasting food) and donate that $1 to charities, we would have $327 million per month to help the homeless, the hungry, and the poor students. Multiply that by 12 months, we would have almost $4 billion dollars for charities every year, enough to end hunger and homelessness in America.

It is never too early to start philanthropy. Here are some wonderful and heart-warming examples of young philanthropists [54–56]:

- 5-year old Phoebe Russell raised enough fund for the San Francisco Food Bank to serve 17,800 hot meals to the homeless.
- 10 and 12-year-old Melati and Isabel Wijsen created Bye Bye Plastic Bags, an initiative to help Bali become plastic bag-free.
- 11-year-old Liam Hannon partnered with a local shelter to collect and donate school supplies and toys for homeless children.
- 15-year-old Sonika Menon formed the Birthday Giving Program, a nonprofit that brings birthday parties to kids and families in need.
- 18-year-old Max Bobholz created Angels at Bat, a nonprofit that collects and distributes baseball equipment for children in rural Kenya.
- American University student Maria Rose Belding co-founded MEANS (Matching Excess And Need for Stability) to provide a free online platform that connects businesses with extra food to charities that feed the hungry.
- My father Johnson Lee surprised me on one of my birthdays during my time at Virginia Tech. He sent me a check for a good sum of money, and I spent it all by

giving it away as scholarships to three high-achieving undergraduate students who needed financial assistance.

During my 10 years of tenure at The Walt Disney Company, I had spent over 2000 hours volunteering with other Disney employees [57]. We are named the Disney VoluntEARS [58].

In September 2004, I founded the 501(c)(3) nonprofit Institute for Education, Research, and Scholarships (IFERS) to conduct scientific and social research as well as to provide resources to high achieving students, scientific researchers, community nonprofits, and educational organizations [59].

There are more than 1.5 million charitable organizations in the United States, donating more than $1.57 trillion dollars and 7.9 billion hours of service [60]. We are also seeing more synergy across fundamentally different organizations. For example, the Dementia Discovery Fund (DDF) represents the first time that charity, government, and the pharmaceutical industry have joined together with a venture capital firm to tackle a major global health issue [61].

One might wonder if we had gotten more donations to set up medical facilities and to train local doctors in developing countries, the HIV virus could have been identified earlier and kept at bay instead of becoming a worldwide AIDS epidemic, infecting more than 70 million individuals and killing more than 35 million people of all ages, genders, ethnicities, and sexual orientations [62].

In February 2019, Melinda French Gates posted on Instagram (See Fig. 1.3) about the reason behind the philanthropic work at the Bill & Melinda Gates Foundation: "Twenty-five years ago, Bill and I read an article that said hundreds of thousands of kids in poor countries were dying from diarrhea. That surprise helped crystallize our values. We believe in a world where innovation is for everyone—where no child dies from a disease it's possible to prevent. But what we saw was a world still shaped by inequity. That discovery was one of the most important steps

Fig. 1.3 Melinda French Gates' post on Instagram (February 12, 2019)

in our journey to philanthropy. We were surprised, then we were outraged, then we were activated" [63].

Bill Gates once said, "I'm a huge believer in that science and innovation are going to solve most of the tough problems over time" [49]. Using science and technology, transhumanism enables philanthropy to become more effective and efficient in its mission to save lives and to provide better opportunities for everyone in the world.

1.10 Democracy

In 1964, Nelson Mandela made a passionate speech from the dock in the Rivonia Trial, "I have fought against white domination, and I have fought against black domination. I have cherished the ideal of a democratic and free society in which all persons live together in harmony and with equal opportunities. It is an ideal which I hope to live for and to achieve. But if needs be, it is an ideal for which I am prepared to die" [64].

Many have taken democracy for granted. America's founding father John Adams served as the first U.S. Vice President and the second U.S. President. He gave a stern and surprise warning in 1814, "Remember Democracy never lasts long. It soon wastes exhausts and murders itself. There never was a Democracy Yet, that did not commit suicide. It is in vain to Say that Democracy is less vain, less proud, less selfish, less ambitious or less avaricious than Aristocracy or Monarchy. It is not true in Fact and no where appears in history. Those Passions are the same in all Men under all forms of Simple Government, and when unchecked, produce the same Effects of Fraud Violence and Cruelty" [65].

That is why we must defend and uphold checks and balances in the U.S. Constitution which divides the government into three branches—legislative, executive, and judicial—a separation of powers so that no one branch can control too much power.

British historian Arnold J. Toynbee examined the rise and fall of 28 different civilizations in his 12-volume magnum opus *A Study of History*. He concluded in concurrence with President John Adams that "civilizations die from suicide, not by murder" [66].

Fortunately for America, President Franklin D. Roosevelt practically saved democracy with his New Deal that "sought to insure that the economic, social, and political benefits of American capitalism were distributed more equally among America's large and diverse populace" [67].

Unfortunately for America, the longest U.S. federal government shutdown in history from December 22, 2018 to January 25, 2019 was a disgrace to democracy, putting 380,000 federal employees on furlough and forcing 420,000 people to work without pay [68]. U.S. Senator Richard Shelby denounced the government decision-making process, "This is like a circus" [69]. It reminds us of *Betty Boop for*

President, a 1932 Fleischer Studios animated short film in which Betty ran for the Office of the President against Mr. Nobody. In his answers to various social issues, Mr. Nobody made empty promises: "Who will make your taxes light? Mr. Nobody! Who'll protect the voters' right? Mr. Nobody! When you're hungry, who feeds you? Mr. Nobody!" [70].

Satire and sarcasm aside, perhaps the world governments will come to a consensus that it is in their best interest to focus on long-term strategy benefiting all citizens instead of short-term gain for privileged segments of society.

In 2014, Sweden's Prime Minister Stefan Löfven appointed Kristina Persson to be the Minister of the Future. "If politics wants to remain relevant and be useful to citizens, it needs to change its approach," said Persson in an interview. "Finding solutions needs the cooperation of all of society's stakeholders. No one [can be] excluded. … Rather than going top-down, we promote inter-ministerial collaboration and force decision makers to confront the long-term issues despite the fact this is harder to do sometimes" [71].

In 2019, the Green New Deal proposes economic stimulus programs to address climate change as well as to promote "economic, social, and racial justice" [72] according to U.S. Representative Alexandria Ocasio-Cortez who is the youngest woman ever to serve in the United States Congress at age 29.

"You have to tend to this garden of democracy, otherwise things can fall apart fairly quickly," said President Barack Obama at the Economic Club of Chicago in December 2017 [73].

Philosopher John Dewey coined the term "social endosmosis" to describe an ideal democracy in which a society is not separated into a privileged and a subject-class [74]. Nobel Laureate Angus Deaton concurs with U.S. Supreme Court Justice Louis Brandeis about the danger of plutocracy: "The United States could have either democracy or wealth concentrated in the hands of a few, but not both. The political equality that is required by democracy is always under threat from economic inequality, and the more extreme the economic inequality, the greater the threat to democracy" [75].

Ministry of the Future and the Green New Deal are steps in the right direction. Without collaboration and synergy among various communities and socioeconomic groups, democracy will eventually fail as it "wastes exhausts and murders itself," in the words of America's founding father John Adams.

In a recent conversation between French-Jewish writer Marek Halter and *TheWrap's* CEO Sharon Waxman, Halter declared that "Democracy is over. It does not work for everyone… We are entering a different world. We can't be too tied to our past. The past is the past…. Today we have nothing to propose. But we will find something. I believe in the genius of human beings" [76].

Transhumanism will save democracy from its demise. Whichever form of government that democracy will create in the future, the great American experiment will go down in history as the freest and the bravest in the land of the free and the home of the brave.

1.11 Transhumanism in American Politics

Transhumanism in American politics dated back to 1992 when Dr. Natasha Vita-More, chairperson of Humanity+ and professor at the University of Advancing Technology, was elected as a councilmember for the 28th Senatorial District of Los Angeles on an openly futurist and transhumanist platform [77].

In 2014, transhumanist Gabriel Rothblatt from the Terasem Movement ran as a Democratic Party candidate for the House of Representatives in Florida's 8th Congressional District [78].

In 2016, Wikipedian and transhumanist Gerald Shields was elected as a town councilmember for Berwyn Heights in Maryland where he served a 2-year term [79].

In the 2016 U.S. presidential election, Zoltan Istvan Gyurko became the first transhumanist candidate to run for the highest office in the land [80]. The Obama Foundation once tweeted that "The highest office in the land is that of the citizen. It is you" [81] (See Fig. 1.4). How appropriate it is because transhumanism represents everyone and leaves no one behind.

Having read all four volumes of *The Making of the President* by Theodore H. White, I had the honor to be the campaign advisor to Zoltan Istvan. After the presidential election, I transitioned to the education and media advisor for the United States Transhumanist Party [82].

In December 2017, I established the California Transhumanist Party with a leadership committee consisting of a chairman and 13 directors from academia and industry (See Fig. 1.5):

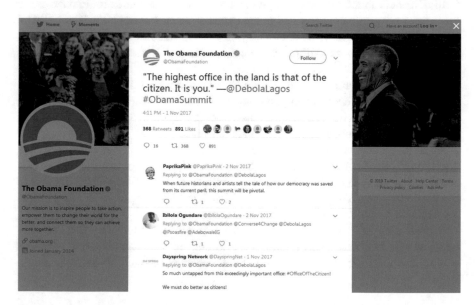

Fig. 1.4 The highest office in the land is that of the citizen

Fig. 1.5 Inaugural California Transhumanist Party meeting on January 27, 2018 in Burbank, California with (left to right) Charlie Kam, Gary Abramson, Maria Entraigues Abramson, Newton Lee, Elizaveta Grigoryeva, Jessica Milne, Dr. Christoph Lahtz, and (on screen left to right) Gennady Stolyarov II and Bobby Ridge from the United States Transhumanist Party

Newton Lee, Chairman

- Maria Entraigues Abramson, Director of Communications
- Dr. Greg Fahy, Director of Biomedicine
- Elizaveta Grigoryeva, Director of Events Management
- Charlie Kam, Director of Networking
- Dr. Christoph Lahtz, Director of Space Bioscience
- Inessa Lee, Director of Ideology and Social Media
- Alex Lightman, Director of Campaign Management
- Jessica Milne, Director of Information Science and Technology
- Lima Mora, Director of Educational Technology and Outreach
- Michael Murray, Director of Environmental Science
- Dr. Elena Rusyn, Director of Biosciences
- Dr. Anca Selariu, Director of Integrative Biosciences
- R. Nicholas Starr, Director of Edutainment

Being all in favor of the Equal Rights Amendment (ERA) ratified by California in 1972 [83], we have maintained a 50:50 gender ratio for the leadership committee. It is less about affirmative action, but more about looking harder to find the best qualified candidates across the board.

In the 2018 midterm election, more women and scientists were elected to Congress. "They bring a real wealth of experience that's really lacking in Congress today," said Shaughnessy Naughton, president of 314 Action, a political action committee devoted to electing more public officials with backgrounds in science and related fields. "There are more reality show people in Congress, including our president, than there are chemists and physicists" [84].

More young girls are getting into science and technology through the encouragement of nonprofits teaming up with businesses such as Girls Who Code with AT&T [85] and Girl Scouts of the USA with Raytheon [86]. The desire to help people using science and technology has motivated 12 teenage girls to invent a solar-powered tent for the homeless [87]. Many more stories like these are foretelling a future of gender equality and technological solutions to social issues.

There is still a lot of work ahead in order to achieve gender equality. For instance, Michigan governor Gretchen Whitmer delivered her first State of the State address in February 2019, calling out the State's crises in crumbling infrastructure and failing schools [88]. Immediate reactions on social media, however, focused on Whitmer's dress instead of her address [89]. Double standard is nothing new, but it often gets in the way of good decision making.

The Transhumanist Party aims to motivate and mobilize both female and male scientists and engineers to take on additional responsibilities as rational politicians. It does not mean replacing democracy with technocracy. It means that our government needs help in making the right policies and investing in science, health, and technology for the improvement of the human condition and the long-term survival of the human race.

Had Albert Einstein accepted the Israeli cabinet's offer to become the country's President in 1952, the Middle East would have been a very different place today because of Einstein's commitment to pacifism. Israel Prime Minister and Zionist David Ben-Gurion reportedly asked his aide, "What are we going to do if he accepts?" [90].

Shaun Scott, author of *Millennials and the Moments That Made Us*, said it eloquently, "We can either do politics or we can have politics done to us" [91].

1.12 Transhumanist Bill of Rights

The 1776 U.S. Declaration of Independence rightly declared that all human beings "are endowed by their Creator with certain unalienable Rights, that among these are Life, Liberty and the Pursuit of Happiness" [92]. In fact, the official title for George Washington was "His Highness the President of the United States of America and the Protector of Their Liberties" [93]. Washington's former aide James McHenry told him, "You are now a king under a different name" [94].

The United States Bill of Rights is the first 10 Amendments to the U.S. Constitution that spells out Americans' rights in relation to their government. It guarantees civil rights and liberties to the individual—like freedom of speech, press, and religion [95].

The transhumanist Bill of Rights version 1.0 was compiled by Zoltan Istvan who hand-delivered it to the U.S. Capitol on December 14, 2015. It states the following [96]:

1. Article 1. Human beings, sentient artificial intelligences, cyborgs and other advanced sapient life forms are entitled to universal rights of ending involuntary suffering, making personhood improvements, and achieving an indefinite lifespan via science and technology.
2. Article 2. Under penalty of law, no cultural, ethnic, or religious perspectives influencing government policy can impede life extension science, the health of the public, or the possible maximum amount of life hours citizens possess.
3. Article 3. Human beings, sentient artificial intelligences, cyborgs and other advanced sapient life forms agree to uphold morphological freedom—the right to do with one's physical attributes or intelligence (dead, alive, conscious, or unconscious) whatever one wants so long as it doesn't hurt anyone else.
4. Article 4. Human beings, sentient artificial intelligences, cyborgs and other advanced sapient life forms will take every reasonable precaution to prevent existential risk, including those of rogue artificial intelligence, asteroids, plagues, weapons of mass destruction, bioterrorism, war, and global warming, among others.
5. Article 5. All nations and their governments will take all reasonable measures to embrace and fund space travel, not only for the spirit of adventure and to gain knowledge by exploring the universe, but as an ultimate safeguard to its citizens and transhumanity should planet Earth become uninhabitable or be destroyed.
6. Article 6. Involuntary ageing shall be classified as a disease. All nations and their governments will actively seek to dramatically extend the lives and improve the health of its citizens by offering them scientific and medical technologies to overcome involuntary ageing.

The transhumanist Bill of Rights has been expanded under the leadership of Gennady Stolyarov II to version 3.0 adopted via electronic votes in December 2018 with the following preamble and definition of sentient entities [97]:

Preamble
Whereas science and technology are now radically changing human beings and may also create future forms of advanced sapient and sentient life, transhumanists establish this TRANSHUMANIST BILL OF RIGHTS to help guide and enact sensible policies in the pursuit of life, liberty, security of person, and happiness.

As used in this TRANSHUMANIST BILL OF RIGHTS, the term "sentient entities" encompasses:

(i) Human beings, including genetically modified humans;
(ii) Cyborgs;
(iii) Digital intelligences;
(iv) Intellectually enhanced, previously non-sapient animals;
(v) Any species of plant or animal which has been enhanced to possess the capacity for intelligent thought; and
(vi) Other advanced sapient life forms.

1.13   Ideology in Action

Steve Jobs once said, "We believe people with passion can change the world for the better. Those people who are crazy enough to think that they can change the world are the ones that actually do" [98].

To put words into action, the Transhumanist Party focuses on three core ideals of transhumanism:

- Transhumanism supports significant life extension and quality of life improvement achieved through the progress of science and technology.
- Transhumanism supports an inclusive cultural, societal, and political atmosphere informed and animated by reason and science to foster peace, prosperity, and universal rights for all.
- Transhumanism supports efforts to use science, technology, and rational discourse to reduce and eliminate various existential risks to the human species.

The Transhumanist Party offers the most inclusive ideology for all ethnicities and races, the religious and the atheists, conservatives and liberals, the young and the old regardless of socioeconomic status, gender identity, or any other individual qualities. The Transhumanist Party members may also be members of other major or minor political parties in the United States and worldwide. All members share the three common core ideals that put science, health, and technology in the forefront of American and world politics.

Today's young people are more tech-savvy and resourceful than generations before them. The Internet and social media are more powerful than newspapers and television [99]. Today's youth do not take no for an answer without reasoning. In the old days, parents said no to their children and that was it—end of story. But today, parents have to explain to their kids why or why not.

When people—especially young people—rise up to fight for justice and important causes, no one can halt the progress and stop the sea change.

In 2015, several youth including Aji Piper, Levi Draheim, Journey Zephier, Jayden Foytlin, Miko Vergun, and Nathan Baring filed a constitutional climate lawsuit—Juliana vs. United States—against the U.S. government for causing climate change which has violated the youngest generation's constitutional rights to life, liberty, and property [100].

Since 2016, about 100,000 16- and 17-year-olds have preregistered to vote in California [101] as they took heed of President Franklin D. Roosevelt's warning: "Nobody will ever deprive the American people of the right to vote except the American people themselves and the only way they could do this is by not voting" [102].

In 2017, a 13-year-old eighth grader Ethan Sonneborn announced his candidacy for the governor of Vermont, "I am running to win, but I would very happily settle for sending a message about young people in politics" [103]. He told CNN that "I've always admired some of the great coalition builders of the modern era—MLK Jr., Robert F. Kennedy—who came from all walks of life to accomplish a common goal. You don't have to be a Kennedy or a famous minister to be a coalition builder. You just have to have an issue you care about and be willing to speak out on your platform" [104].

In 2018, thousands of teenagers marched through Manhattan, Washington D.C., and other U.S. cities to demand climate justice while their parents were busy at work. One of the teens told *The New Yorker* staff writer Carolyn Kormann, "My parents are verbally supportive, but they don't know what to do. They're busy with their lives" [105].

"To the leaders, skeptics and cynics who told us to sit down, stay silent and wait your turn, welcome to the revolution," said Marjory Stoneman Douglas student Cameron Kasky during the 2018 March for Our Lives rally [106].

On Palm Sunday—a day after the 2018 March for Our Lives events in the U.S. and around the globe—Pope Francis addressed the crowd at St. Peter's Square in Rome, "Dear young people, you have it in you to shout. … Even if others keep quiet, if we older people and leaders—so often corrupt—keep quiet, if the whole world keeps quiet and loses its joy, I ask you: Will you cry out?" [107].

In January 2019, 16 year-old Greta Thunberg spoke at the World Economic Forum Annual Meeting about the need for transformational action to reduce carbon emissions: "We are less than 2 years away from being unable to undo our mistakes. … I often hear adults say: 'We need to give the next generation hope.' But I don't want your hope. I want you to panic. I want you to feel the fear I do. Every day. And want you to act. I want you to behave like our house is on fire. Because it is" [108].

Greta Thunberg represents an increasing number of teens who are following the pattern of Melinda French Gates' journey into philanthropy: "We were surprised, then we were outraged, then we were activated."

The surge of student and teen activists worldwide echoes Italian politician Alessandro Di Battista's pronouncement that "Representative democracy is obsolete. The future is inevitably direct democracy" [109].

Transhumanism empowers citizens to revolutionize politics with the help of science and technology in electronic voting, social media, AI-assisted fact-checking, e-government, and other methods of direct democracy. Direct democracy benefits everyone as long as it does not drown out minority voices.

1.14 Infinite Diversity in Infinite Combinations

Notwithstanding Thomas Jefferson's proclamation that "all men are created equal" in the 1776 U.S. Declaration of Independence, all men and women are not created equal in the literal sense unless we all share identical DNA. But even identical twins with the same DNA develop different fingerprints when the growing fetuses touch the amniotic sac in their mother's womb.

The twins can become two very different people due to upbringing, social, economic, and environmental factors. Because these external factors can vary greatly from person to person, life cannot be fair to everyone. However, unfairness builds character and brings diversity to the otherwise homogeneous and isotropic existence. Who wants to live in a world where everybody looks identical, thinks alike, and acts the same?

Nature has shown us that there are 8.7 million species on Earth and more than 80% of species are still undiscovered [110]. Diversity emanates beauty. Even though all complex organisms are preprogrammed by their DNAs, they can exhibit individuality within the confines of nature. Mahatma Gandhi once said that "no two leaves are alike" [111].

In 2016, Spike Lee and Jada Pinkett Smith called for Oscar boycott because "for the 2nd consecutive year all 20 contenders under the actor category are white" [112]. American actress Stacey Dash dismissed the outrage over Oscars: "I think it's ludicrous. We have to make up our minds. Either we want to have segregation or integration, and if we don't want segregation, then we have to get rid of channels like BET and the BET Awards and the Image Awards, where you're only awarded if you're black. If it were the other way around, we'd be up in arms. It's a double standard. There shouldn't be a Black History Month. We're Americans, period" [113].

On one hand, it is great to be inclusive without discrimination or prejudice. On the other hand, it is wonderful to celebrate diversity and uniqueness with Black History Month, National Hispanic Heritage Month, Asian-Pacific American Heritage Month, Women's History Month, LGBT Pride Month, and many others.

Star Trek creator Gene Roddenberry said, "If man is to survive, he will have learned to take a delight in the essential differences between men and between cultures. He will learn that differences in ideas and attitudes are a delight, part of life's exciting variety, not something to fear" [114].

In the *Star Trek* episode "Is There in Truth No Beauty?" (1968), Roddenberry and writer Jean Lisette Aroeste introduced the notion of IDIC (Infinite Diversity in Infinite Combinations) [115]:

Dr. Miranda Jones (Diana Muldaur): *[regarding the Vulcan IDIC] I understand, Mr. Spock. The glory of creation is in its infinite diversity.*
Mr. Spock (Leonard Nimoy): *And the ways our differences combine, to create meaning and beauty.*

Mr. Spock is admired by Trekkies for his Vulcan logic and superior intelligence. In the 1987 seminal book *The Society of Mind,* MIT Prof. Emeritus Marvin Minsky described human intelligence as a result of diverse processes: "What magical trick makes us intelligent? The trick is that there is no trick. The power of intelligence stems from our vast diversity, not from any single, perfect principle" [116].

A few musical notes can morph into countless new songs, and a small set of vocabulary can create a congeries of poems.

During my tenure at AT&T Bell Labs in the Eighties, I founded the *Star Trek* in the 20th Century Club, and produced an "Intergalactic Music Festival" to showcase international songs and cultural dances performed by fellow colleagues whose families immigrated to the United States.

America is a young country molded by immigrants from all over the world. Steve Jobs' biological father was a political migrant from Syria [117], Google cofounder

Sergey Brin was born in Russia [118], Microsoft CEO Satya Nadella came from India [119], Uber CEO and former Expedia CEO Dara Khosrowshahi was an Iranian refugee [120], and the list goes on and on.

American actress Angelina Jolie considered herself a "citizen of the world." She gave a speech at the Sergio Vieira de Mello Annual Lecture in Geneva, Switzerland on March 15, 2017 in which she said, "I am a proud American and I am an internationalist" [121]. In fact, Jolie traveled to Sawkopmund, Namibia in 2006 to give birth to her first biological child Shiloh Nouvel [122], and then to Nice, France in 2008 to give birth to twins Knox Léon and Vivienne Marcheline [123].

In the Italian town of Porto Recanati, a notorious and dangerous Hotel House is occupied mostly by migrant workers. Despite its bad reputation, one of the residents—Luca Davide—explained why he is determined to stay, "Here there's goodness, and we've got to help it emerge, to make the world understand that 30 ethnicities can coexist peacefully. We could become a beacon of integration. We could have people envy what we've created" [124].

Diversity and integration are also needed in science and technology. Pride and prejudice are obstacles to scientific advancements and truth seeking. "What I hope we appreciate more because of him [Alan Turing], is to value diversity and individual creativity in our science base," said mathematician Jonathan Dawes at the University of Bath. "We need people who we allow to be driven by their curiosity, and we also need people who will take those basic science ideas and turn them into useful technology" [125].

Transhumanism supports infinite diversity in infinite combinations from all ethnicities and races, the religious and the atheists, conservatives and liberals, the young and the old regardless of socioeconomic status, gender identity, or any other individual qualities.

1.15 New Tower of Babel

Amongst the 7.3 billion people in the world, 31% are Christians, 24% are Muslims, 15% are Hindus, and 7% are Buddhists according to a 2017 report by Pew Research Center [126]. The median age by religion is 24 for Muslims, 27 for Hindus, and 30 for Christians. Therefore, transhumanism is equally important in the religious circle as it is in the secular world.

As Dante coined the word "transhuman" in *Divine Comedy* (Paradise, Canto I) to describe the change of the human body to immortal flesh in eschatology, Jesus was the first transhuman with an immortal resurrected body—one that could touch, feel, and enjoy eating fish with his disciples (Luke 24:41–43).

Christian transhumanists are building the new Tower of Babel not to challenge God but to better understand the universe and to realize the true potential of humankind. Being a computer scientist and a Christian, I believe that all life forms are combinations of intelligent design and evolution. Creationism and evolution are not mutually exclusive. In fact, the former gives rise to the latter. God gives life to human beings who in turn give birth to artificial intelligence. We are teaching AI software that will evolve into superintelligence; and superintelligence will in turn teach humankind and explain the Scriptures.

The Bible is full of existential stories and superheroes: The Great Flood nearly annihilated humankind (Genesis 6:1–8:22). Shadrach, Meshach, and Abednego were thrown in the fire; but they walked around freely in the fire, completely unharmed (Daniel 3:1–30). Samson was so strong that he tore the lion apart with his bare hands (Judges 14:5–6).

Albert Einstein considered the Biblical stories as "a collection of honourable, but still primitive legends which are nevertheless pretty childish" [1]. Well, they are not. Science and technology have provided humankind new tools to create "miracles" such as fire retardant, airplane, submarine, pacemaker, and organ transplant, to name a few. More advanced tools will soon afford us super longevity and superintelligence.

Some people opine that the transhumanist goals are impossible to achieve. Difficult? Yes. Unattainable? No.

Matthew 17:20 tells the story that one day the disciples came to Jesus in private and asked, "Why couldn't we drive it [the demon] out?" And Jesus replied, "Because you have so little faith. Truly I tell you, if you have faith as small as a mustard seed, you can say to this mountain, 'Move from here to there,' and it will move. Nothing will be impossible for you."

We can move a mountain from one place to another with our bare hands and a few shovels. It is only a matter of putting in the effort and the time. But we can speed things up by using dynamites and heavy machinery. As technology continues to accelerate, nothing is impossible.

Faith does not mean paying lip service. Faith requires commitment. What have we accomplished today with our smartphones and laptops that are several thousand times more powerful than the Apollo 11 on-board computers? [127].

Given that the first manned Moon landing with Apollo 11 on July 20, 1969 only had a 50% chance of landing safely on the moon's surface, it was an exemplary faith in technology and human spirit (see Fig. 1.6).

Astronaut Neil Armstrong said in a 2012 video interview, "I thought we had a 90% chance of getting back safely to Earth on that flight but only a 50–50 chance of making a landing on that first attempt. There are so many unknowns on that descent from lunar orbit down to the surface that had not been demonstrated yet by testing and there was a big chance that there was something in there we didn't understand properly and we had to abort and come back to Earth without landing" [128].

When Armstrong and Buzz Aldrin were about to land on the moon, they disagreed with the on-board computer's decision to put them down on the side of a large crater with steep slopes littered with huge boulders. "Not a good place to land at all," said Armstrong. "I took it over manually and flew it like a helicopter out to the west direction, took it to a smoother area without so many rocks and found a level area and was able to get it down there before we ran out of fuel. There was something like 20 seconds of fuel left."

Fig. 1.6 Apollo 11 astronaut Buzz Aldrin works at the deployed Passive Seismic Experiment Package on July 20, 1969. To the left of the United States flag in the background is the lunar surface television camera. Photo taken by Neil Armstrong. (Courtesy of NASA)

The rest is history as Armstrong uttered his famous line, "One small step for a man, one giant leap for mankind."

After the successful moon landing, President Richard Nixon called the astronauts from the White House, "…as you talk to us from the Sea of Tranquility, it inspires us to redouble our efforts to bring peace and tranquility to earth. For one priceless moment in the whole history of man all the people on this earth are truly one" [129].

Armstrong replied, "It is a great honor and privilege for us to be here representing not only the United States, but men of peaceable nations, men with an interest and a curiosity, and men with a vision for the future."

The Apollo 11 astronauts left on the moon a peace plaque which says, "Here men from the planet Earth first set foot upon the Moon July 1969, A.D. We came in peace for all mankind" [130] (See Fig. 1.7).

"Where there is no vision, the people perish." (Proverbs 29:18) Christian transhumanists see science and technology as the key to bring lasting peace and hasten the Second Coming of Christ. "He will judge between the nations and will settle disputes for many peoples. They will beat their swords into plowshares and their spears into pruning hooks. Nation will not take up sword against nation, nor will they train for war anymore." (Isaiah 2:4).

Fig. 1.7 Apollo 11
astronauts left a peace
plaque on the moon in July
1969

1.16 Existential Risk

The biggest existential risk to humanity is humanity itself, unless human beings learn to coexist peacefully and help one another willingly.

Astrophysicist Stephen Hawking said in a CNN interview in 2008, "It will be difficult enough to avoid disaster on planet Earth in the next 100 years, let alone next thousand, or million. … I see great dangers for the human race. There have been a number of times in the past when its survival has been a question of touch and go. The Cuban missile crisis in 1963 was one of these. The frequency of such occasions is likely to increase in the future. We shall need great care and judgment to negotiate them all successfully" [131].

Swedish philosopher Nick Bostrom said in a PBS interview in 2017, "So if there are big existential risks, I think they are going to come from our own activities and mostly from our own inventiveness and creativity" [132].

In January 2019, *The Bulletin of the Atomic Scientists* Science and Security Board set the Doomsday Clock at 2 minutes to midnight—the closest it has ever been to apocalypse. The Bulletin's president and CEO Rachel Bronson wrote, "Humanity now faces two simultaneous existential threats, either of which would be causc for extreme concern and immediate attention. These major threats—nuclear weapons and climate change—were exacerbated this past year by the increased use of information warfare to undermine democracy around the world, amplifying risk from these and other threats and putting the future of civilization in extraordinary danger. In the nuclear realm, the United States abandoned the Iran nuclear deal and announced it would withdraw from the Intermediate-range Nuclear Forces Treaty (INF), grave steps towards a complete dismantlement of the global arms control process. … On the climate change front, global carbon dioxide emissions—which seemed to plateau earlier this decade—resumed an upward climb in 2017 and 2018." [133].

Figure 1.8 shows the changes in the time of the Doomsday Clock of the Bulletin of the Atomic Scientists from 1947 to 2018. Rise of nationalism, nuclear proliferation, global warming, and threat of a renewed arms race between the U.S. and Russia have made 2018 the Doomsday Clock's closest approach to midnight since 1953.

In his New Year 2018 message, UN Secretary-General Antonio Guterres said, "When I took office one year ago, I appealed for 2017 to be a year for peace. Unfortunately—in fundamental ways, the world has gone in reverse. On New Year's Day 2018, I am not issuing an appeal. I am issuing an alert—a red alert for our world. Conflicts have deepened and new dangers have emerged. Global anxieties about nuclear weapons are the highest since the Cold War. Climate change is moving faster than we are. Inequalities are growing. And we see horrific violations of human rights. Nationalism and xenophobia are on the rise. As we begin 2018, I call for unity. I truly believe we can make our world more safe and secure. We can settle conflicts, overcome hatred and defend shared values. But we can only do that together. I urge leaders everywhere to make this New Year's resolution: Narrow the gaps. Bridge the divides. Rebuild trust by bringing people together around common goals. Unity is the path. Our future depends on it" [134].

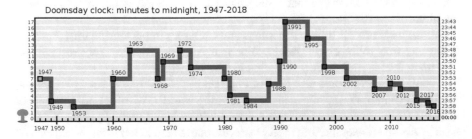

Fig. 1.8 A graph showing the changes in the time of the Doomsday Clock of the Bulletin of the Atomic Scientists from 1947 to 2018. (Courtesy of Wikimedia Commons)

1.17   World Peace

World peace is attainable in spite of differing viewpoints. Notwithstanding polar opposite opinions on same-sex marriage, all-male admissions policy, and other issues dividing conservatives and liberals, the late Justice Antonin Scalia had a long and close friendship with Justice Ruth Bader Ginsburg at the U.S. Supreme Court [135].

In 2012, Jesuit priest and peace activist John Dear went to Kabul to meet the Afghan Peace Volunteers, a diverse community of students ages 15–27 who practice peace and nonviolence [136]. "I used to detest other ethnic groups," one of the youths told Dear, "but now I'm trying to overcome hate and prejudice. You international friends give me hope and strength to do this." Another youth added, "I used to put people in categories and couldn't drink tea with anyone. Now I'm learning that we are all part of one human family. Now I can drink tea with anyone" [137].

In 2017, Emmy award-winning filmmaker Deeyah Khan spent months interviewing neo-Nazis and jihadists for her documentary films *White Right: Meeting the Enemy and Jihad: A Story of the Others*. In an interview with Vox writer Sean Illing, Khan explained her rationale, "So we have to become active citizens and active human beings, and no matter what happens, we cannot afford to give up on each other. That means even people that we disagree with and people that we dislike. In fact, it matters more. It's easy for me to like you. It's easy for me to be nice to you because we probably see the world fairly similarly. That's easy. That's not when our principles really matter. It matters when you are able to extend it to somebody who might not deserve it, or who you might not like or might not agree with. Otherwise, we become just like them—and, in the process, do their bidding" [138].

Jesus could have easily conquered the world and turned everyone into his followers. But Jesus said to the disciple who defended him by force in the Garden of Gethsemane, "Put your sword back in its place. For all who draw the sword will die by the sword. Are you not aware that I can call on my Father, and he will at once put at my disposal more than twelve legions of angels?" (Matthew 26:52–53).

But people don't listen, do they? The Crusades—a series of religious wars sanctioned by the church in the medieval period—killed an estimate of 1.7 million people [139]. Make no mistake that Jesus was a pacifist, but the church turned its back on him.

Today there are 2.3 billion Christians and Catholics in the world [140], but how many are true followers of Jesus? "For many are called, but few are chosen." (Matthew 22:14) And the few are going to save the world.

If only the world would take heed of Pope Francis' 2018 Easter Sunday message, "Today we implore fruits of peace upon the entire world… We also implore fruits of wisdom for those who have political responsibilities in our world" [141].

The Amish are exempt from military service due to their belief in "non-resistance" [142]. Why can't humankind do the same in the name of peace, love, and freedom?

1.18   From Holocene to Anthropocene

Conspiracy theories abound in many tragedies and injustices, but the greatest conspiracy of all is how human beings treat one another and other living things on Earth. A biomass study showed that human beings represent just 0.01% of all living things on the planet, and yet humanity has caused the loss of 83% of all wild mammals and 50% of all plants [143].

In the June 2015 issue of *Science Advances* journal published by the American Association for the Advancement of Science, researchers at National Autonomous University of Mexico, Stanford University, University of California Berkeley, Princeton University, and University of Florida raised the specter of Earth's biota entering a sixth "mass extinction" [144].

The last Big Five were Ordovician-Silurian, Late Devonian, Permian, Triassic-Jurassic, Cretaceous-Tertiary (or K-T) extinctions. The Permian period ended with 96% of all species perished; and the K-T wiped out at least half of all species on Earth including the dinosaurs [145].

Prof. Gerardo Ceballos and his coauthors wrote, "We can confidently conclude that modern extinction rates are exceptionally high, that they are increasing, and that they suggest a mass extinction under way—the sixth of its kind in Earth's 4.5 billion years of history. … If the currently elevated extinction pace is allowed to continue, humans will soon (in as little as three human lifetimes) be deprived of many biodiversity benefits. … The loss of biodiversity is one of the most critical current environmental problems, threatening valuable ecosystem services and human well-being" [146].

Geologist Prof. Jan Zalasiewicz at the University of Leicester and his team of scientists have proposed Anthropocene as a replacement of the Holocene epoch due to geologically significant conditions and processes that are profoundly altered by human activities, which include changes in: "Erosion and sediment transport associated with a variety of anthropogenic processes, including colonisation, agriculture, urbanisation and global warming. The chemical composition of the atmosphere, oceans and soils, with significant anthropogenic perturbations of the cycles of elements such as carbon, nitrogen, phosphorus and various metals. Environmental conditions generated by these perturbations; these include global warming, ocean acidification and spreading oceanic 'dead zones'. The biosphere both on land and in the sea, as a result of habitat loss, predation, species invasions and the physical and chemical changes noted above" [147].

Dinosaurs, Neanderthals, and Denisovans did not evolve and adapt fast enough, and therefore they became extinct. Modus operandi or business as usual will doom the human race to mass extinction.

1.19   Natural Disasters and Biohazards

People are often being reactive instead of being proactive in reducing dangerous threats. Chris Wysopal of *The Washington Post* observed, "Cities were once vulnerable to disastrous fires, which raged through dense clusters of mostly wooden buildings. It took a giant fire in Chicago to spur government officials into serious reforms, including limits on new wooden structures, a more robust water supply for suppressing blazes and an overhaul to the city's fire department. But here's a frightening fact: The push to create tough new fire-safety standards did not start after the Great Chicago Fire in 1871, which killed hundreds of people and left 100,000 homeless. It took a second fire, nearly 3 years later in 1874, to get officials in Chicago to finally make real changes" [148].

Are we prepared to deal with existential threats posed by natural disasters and biohazards that could potentially wipe out millions of people? Here are some examples:

1. Extreme Weather: Whether climate change is due to carbon dioxide emission [149], shift in Earth's magnetic poles [150], or some other reasons, global warming poses existential threat that must be addressed. Dire consequences of climate change include more severe droughts, heat waves, hurricanes, and raising sea levels [151].

   Meteorologists have noticed the record early start of the Greenland ice melt in April 2016. Greenland has been losing ice at a pace of 287 billion metric tons per year, and sea levels around the world could rise by 20 feet if the ice sheet in the size of Alaska were to melt completely [152].

   Due to the treats of tsunamis and sea level rise, the Quinault Indian Nation (QIN) located on the Pacific coast of Washington's Olympic Peninsula has been planning to relocate the community to higher ground [153]. The coastal land is also sinking, making a rising sea that much more precarious. Treasure Island, which sits between San Francisco and Oakland, is sinking fast, at a rate of a third of an inch a year [154].

   The U.S. government has spent more than $350 billion over the past decade in response to extreme weather and fire events [155]. Globally, natural and man-made disaster costs in 2017 increased more than 60% to $306 billion according to Swiss Re, a reinsurance company based in Zurich, Switzerland [156].

2. Earthquakes: The Cascadia Fault at the bottom of the Pacific Ocean can create an earthquake almost 30 times more energetic than the San Andreas Fault [157]. In January 1700, the Cascadia caused the largest earthquake in North America, setting off a tsunami that not only struck the Pacific coast and also damaged Japan's coastal villages across the Pacific Ocean [158].

   The United States' early warning system is in development but it has been stalled due to federal funding constraints. U.S. Geological Survey (USGS) says that only 40% of the necessary sensors for the Earthquake Early Warning System (EEWS) are in the ground as of December 2017. "We're trying to build a system with that limited funding stream and will never get to a fully functioning system at this rate," said geophysicist Doug Given. "I can't help but think of Mexico City, people killed by buildings and falling debris. … We don't want to see the same situation here. I fear we will regret we didn't do this when we had the chance" [159].

3. Aging Infrastructure: While Admiral Michael Rogers, Director of the National Security Agency (NSA), has justifiably sounded the alarm on cyber attacks of critical infrastructure [160, 161], the U.S. government is not paying enough attention to crumbling bridges, decaying pipelines, and inadequate storage facilities for natural gas that are all highly susceptible to serious damage by earthquakes.

   In August 2011, a 5.8 magnitude earthquake damaged the Washington Monument [162]. Today, nearly 60,000 bridges across the U.S. are in desperate need of repair. "It's just eroding and concrete is falling off," said National Park Service spokeswoman Jenny Anzelmo-Sarles, referring to the Arlington Memorial Bridge crossed by 68,000 vehicles every day [163].

   Gas Pipe Safety Foundation cofounder Kimberly Archie called the aging natural gas infrastructure in American cities a "ticking time bomb" [164]. The 2015 gas leak in Porter Ranch, California, for instance, released an estimated 80,000 metric tons of mostly methane into the atmosphere, affecting the health of over 30,000 residents. It took 4 months to permanently seal the leak [165].

4. Dangerous Storages: *The Texas Tribune* and *ProPublica* reported in March 2016 that "Houston … is home to the nation's largest refining and petrochemical complex, where billions of gallons of oil and dangerous chemicals are stored. And it's a sitting duck for the next big hurricane. Why isn't Texas ready?" [166]. Rice University professor Phil Bedient summed up the inaction of the local government: "We've done nothing to shore up the coastline, to add resiliency … to do anything."

5. Superbugs: Bacteria that are resistant to all antibiotics including the last-resort nephrotoxic drug Colistin have infected humans and animals in Asia, Europe, the United States, and more than 20 countries worldwide [167]. In April 2018, Centers for Disease Control and Prevention (CDC) reported that the "nightmare bacteria" carbapenem-resistant Enterobacteriaceae (CRE) has been detected in 27 states in the U.S [168]. The journal *Review on Antimicrobial Resistance* projected that by 2050, more than 10 million people will die from superbugs each year, outpacing cancer (8.2 million), diabetes (1.5 million), diarrheal disease (1.4 million), and other illnesses [169].

Although some natural disasters and biohazards may be unavoidable, the devastating domino effects can be alleviated if we are well informed and better prepared. For example:

1. U.S. Geological Survey (USGS) along with university partners have been developing and testing an Earthquake Early Warning System (EEWS) called ShakeAlert [170].

2. The National Aeronautics and Space Administration (NASA) employs an automated collision monitoring system known as Sentry and publishes a list of

potential future Earth impact with Near-Earth Asteroids (NEAs) at http://neo.jpl.
nasa.gov/risk/

Deflecting the massive asteroid 101,955 Bennu has been the focus of U.S.
planetary defense teams. Bennu has a 1 in 2700 chance of striking Earth on
September 25, 2135 with the kinetic energy of 80,000 Hiroshima nuclear bombs
[171] (see Fig. 1.9). NASA is developing the Double Asteroid Redirection Test
(DART). "DART would be NASA's first mission to demonstrate what's known
as the kinetic impactor technique—striking the asteroid to shift its orbit—to
defend against a potential future asteroid impact," said NASA's planetary defense
officer Lindley Johnson [172].

3. The National Oceanic and Atmospheric Administration (NOAA) has created an
   interactive Climate Explorer tool to raise awareness by allowing users to visualize historical data and impacts of climate changes (see Fig. 1.10) [173].
4. Local community volunteers such as Food Forward in Southern California con-
   vene at private properties, public spaces, and farmers and wholesale markets to
   recover excess fruits and vegetables that would otherwise go to waste, donating
   them to direct-service agencies that feed over 100,000 people in need each month
   [174].
5. Google is the world's largest corporate buyer of renewable energy with a commitment to purchase nearly 2 GW of green energy [175].
6. In March 2018, Microsoft announced a 20-year deal to purchase solar power
   from Singapore rooftops to power its data centers [176].
7. Alphabet's Sidewalk Labs with CEO Dan Doctoroff focuses on urban design by
   pursuing technologies to "cut pollution, curb energy use, streamline transportation, and reduce the cost of city living" [177]. Reducing the cost of city living is

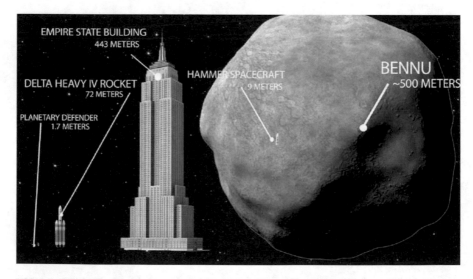

Fig. 1.9 Deflecting the massive asteroid 101,955 Bennu has been the focus of U.S. planetary
defense teams. Bennu has a 1 in 2700 chance of striking Earth on September 25, 2135

Fig. 1.10 Climate Explorer offers interactive visualizations for exploring maps and data to identify potential climate impacts. This diagram shows the impacts a rising sea level may have on coastal regions of the United States

music to the ears of angry protesters who in December 2013 blockaded an Apple employee shuttle bus in San Francisco and threw rocks at a Google employee shuttle bus in Oakland to call attention to low-income residents displaced by rising rents [178].

8. In Hollywood, the 2006 documentary *Who Killed the Electric Car?* and the 2011 feature film *Revenge of the Electric Car* by Chris Paine educated the public about a better alternative to gasoline powered vehicles. Following the zero-emissions vehicle (ZEV) mandate, eight U.S. states (California, Connecticut, Maryland, Massachusetts, New York, Oregon, Rhode Island, and Vermont) and five countries (Germany, the United Kingdom, the Netherlands, Norway, and Quebec of Canada) have proposed to ban sales of gas and diesel powered cars by 2050 [179].

9. The California Transhumanist Party and the 501(c)(3) nonprofit Institute for Education, Research, and Scholarships (IFERS) support advanced research and apply integrative solutions with state-of-the-art technologies to achieve affordable physical and mental healthcare as well as elimination of diseases cost-effectively, resulting in better quality of life and longevity [180].

1.20 Mother Earth

Since March 2015, astronaut Scott Kelly had often posted photos of the planet Earth on his Twitter account during his 1-year mission at the International Space Station (ISS).

In January 2016, CNN's chief medical correspondent Sanjay Gupta asked Kelly to define the Earth's condition as if it were a human body. Kelly replied, "There are definitely parts of Asia, Central America that when you look at them from space, you're always looking through a haze of pollution. As far as the atmosphere is concerned, and being able to see the surface, you know, I would say definitely those areas that I mentioned look kind of sick. ... [The atmosphere] definitely looks very, very fragile and just kind of like this thin film, so it looks like something that we definitely need to take care of" [181]. The hard question is how to take good care of the planet Earth and its inhabitants.

Nature is beautiful and calm, but it can also be violent and deadly in an astronomical scale. Major earthquakes, volcano eruptions, and asteroid impacts can annihilate millions of lives in a relatively short period of time.

The Great Flood in the Bible and other ancient texts is a grave warning to humankind, and yet very few people seem to care about global existential threat. Manmade environmental disasters force nature's hand to unleash its fury on humankind.

In *The Day the Earth Stood Still* (2008), astrobiology professor Helen Benson (Jennifer Connelly) demanded to know the intention of the alien named Klaatu (Keanu Reeves):

Helen: *I need to know what's happening.*
Klaatu: *This planet is dying. The human race is killing it.*
Helen: *So you've come here to help us.*
Klaatu: *No, I didn't.*
Helen: *You said you came to save us.*
Klaatu: *I said I came to save the Earth.*
Helen: *You came to save the Earth ... from us? You came to save the Earth from us!*
Klaatu: *We can't risk the survival of this planet for the sake of one species.*
Helen: *What are you saying?*
Klaatu: *If the Earth dies, you die. If you die, the Earth survives.*

Transhumanism develops new survival technologies and promotes sound policies to save the Earth and its inhabitants from being destroyed by human activities and natural disasters. For instance, if the entire world will be under water due to irreversible global warming, transhumanists will build cities in the ocean and perhaps create genetically enhanced human beings who are amphibious (having both lungs and gills).

Internet pioneer Vint Cert wrote in a 2019 *Wired* article, "We may need to accelerate the evolution of terrestrial life forms, for example, including homo sapiens, so that they carry traits and capabilities needed for life in space or even on our own changing planet" [182].

1.21   Mystery of the Universe

The Earth is just one of more than 100 billion planets in the known universe [183]. Figure 1.11 shows the Pillars of Creation—Eagle Nebula—a cloud of gas and dust created by an exploding star from which new stars and planets are forming. In spite of its grandeur, the universe is actually getting more empty and silent as it is expanding at a constant or an accelerating rate [184].

In fact, the existence of our universe has baffled scientists due to the baryon asymmetry problem. "All of our observations find a complete symmetry between matter and antimatter, which is why the universe should not actually exist," said physicist Christian Smorra at CERN's Baryon-Antibaryon Symmetry Experiment (BASE) in a 2017 report. "An asymmetry must exist here somewhere but we simply do not understand where the difference is" [185].

If antimatter falls up instead down on gravity, it may explain why the universe expelled all antimatter. Experiments on gravity's effect on antimatter have been underway at CERN (European Organization for Nuclear Research) and other research centers [186].

Regardless of how the universe came into existence, life is most fascinating among all creations and evolutions. Human beings, in particular, possess the innate need to seek the meaning of life as it is often depicted in literatures, art, music, movies, and other media.

Superheroes, for example, reflect our human desire to give meaning to our lives by protecting the innocents, upholding justice, and mitigating existential threats.

Fig. 1.11 Pillars of Creation, Eagle Nebula, a cloud of gas and dust created by an exploding star from which new stars and planets are forming. (Courtesy of NASA, ESA, and the Hubble Heritage Team STScI/AURA)

The lack of superhuman ability in the real world does not call for less responsibility. Humanity ought to be the shining light and roaring thunder that fills up the vastly empty and silent universe.

1.22   Real-World Superheroes

Superman was created in the image of humans for comic books by Jerry Siegel and Joe Shuster in the 1930s [187]. The Man of Steel looks human with human emotions, but he is immortal with superhuman abilities. Since Superman became an American cultural icon, audiences have also flocked to see the Avergers, Batman, Black Panther, Iron Man, Spider-Man, Supergirl, Wonder Woman, and many other superheroes who reflect our deepest desire to transcend ordinary human limitations.

In the 2010 film *Kick-Ass* based on comics by Mark Millar and John Romita, Jr., a protagonist said, "At some point in our lives we all wanna be a superhero. … In the world I lived in, heroes only existed in comic books. And I guess that'd be okay, if bad guys were make-believe too, but they're not" [188]. The "bad guys" include not only evil doers but also diseases, injuries, and natural disasters.

Even Google engineering director Ray Kurzweil wrote a superhero novel named *Danielle: Chronicles of a Superheroine* in which he "introduces us to a precocious young girl who uses her intelligence and accelerating technologies to solve the world's biggest challenges. Danielle's journey brings her face to face with important figures from recent history and our modern world while casting a hopeful vision of humanity's future—and how to achieve it" [189].

The real-world superheroes protecting human beings are certainly not Superman, Supergirl, or the like. The real heroes are scientists who are protecting humans from deadly bacteria and viruses, astronomers and planetary defense teams who are watching the sky for Near-Earth Objects (NEOs), police and firefighters who risk their lives to protect people, teachers who toil to educate their students, parents who make sacrifices for their children, and volunteers who devote their time and energy to serve others.

There are a countless number of unsung heroes throughout the ages. Now and then, some heroes are caught on camera. For instance, 22-year-old Mamoudou Gassama—a Malian migrant in Paris—rescued a child dangling from a balcony. He was nicknamed Spider-Man for scaling the building to reach the boy [190].

In comics, superpowers often come from genetic disorders or mutations. As if life imitates arts, both Isaac Newton and Albert Einstein showed signs of autism and Asperger's syndrome. According to medical expert Simon Baron-Cohen at Cambridge University, "Newton seems like a classic case. He hardly spoke, was so engrossed in his work that he often forgot to eat, and was lukewarm or bad-tempered with the few friends he had. If no one turned up to his lectures, he gave them anyway, talking to an empty room. … As a child, Einstein was also a loner, and repeated sentences obsessively until he was seven years old. He became a notoriously confusing lecturer" [191].

Anime character Sora said in *No Game No Life*, "I don't believe in humanity. But I believe in the potential of humanity. ... People with the potential, the hope, the fantasies that can reach the gods themselves, all inside one tiny little body" [24].

Transhumanism empowers people to realize the full human potential. The prefix "trans" in transhumanism can denote "transitioning," "transforming," and "transcending" in order for human beings to become better physically, mentally, and spiritually.

1.23   Brave New World

Although the 1932 novel *Brave New World* by Aldous Huxley (younger brother of Julian Huxley) depicts a dystopia due to the misuse of biotechnology by the ruling class, the reality is that science and engineering have improved and will continue to improve the quality of life for human beings.

Change is inevitable, and it is up to humankind to create a brave new world of utopia instead of dystopia, a heaven on earth instead of a living hell.

5000 years of recorded history has shown more than enough of hell on earth as in civil wars, world wars, the Holocaust, the AIDS epidemic, the 1918 Spanish flu that killed 50 million people, and the fourteenth century Black Death that wiped out an estimated 60% of Europe's entire population.

We cannot change the past even if we could go back in time. The fabric of space-time continuum may automatically correct anomalies caused by, say, a time traveler such that any ripple effects are short-lived. In *The Twilight Zone's Cradle of Darkness* (2002), a woman went back in time to the birth of Adolf Hitler, abducted the baby, and committed murder-suicide. However, a housemaid followed the woman, saw what happened, and bought a homeless woman's baby to pass it off as Hitler. "A moment of silence for Andrea Collins," said the narrator. "She sacrificed her life for the good of mankind, but she also created the very monster she sought to destroy. History can never be changed. Not even in *The Twilight Zone*."

But we can change the future for the better. A utopia or heaven on earth is a place where every would-be Mozart has access to a piano and violin before the age of five, where every would-be Beethoven is empowered to compose music in spite of deafness, where every individual can get quality education regardless of their socioeconomic status and place of residence, and where everyone is given good health and freedom to pursue their dreams.

Had Wolfgang Amadeus Mozart and Ludwig van Beethoven had no access to musical instruments, they would not have composed masterpieces like Symphony No. 40 in G minor, K. 550, Symphony No. 5 in C Minor, Op. 67, Rondo Alla Turca, Für Elise, and many more that delight an endless number of generations.

There are many undiscovered Mozarts, Beethovens, Einsteins, Shakespeares, Lincolns, and Washingtons out there today who do not have the freedom or the opportunity to explore their talents due to economic, political, and other constraints. In an interview conducted by Anjali Prasertong at Tulane University, an African-American woman said that she once had to pass up on an offer of an unpaid internship at the White House because she needed to make money in the summer to pay for school in the fall [52].

In 2016, Bill and Melinda Gates were asked by some high school students in Kentucky what superpower they wished they could have. Gates' answers were: "More time. More energy. As superpowers go, they may not be as exciting as Superman's ability to defy gravity. But if the world can put more of both into the hands of the poorest, we believe it will allow millions of dreams to take flight" [192].

Every day I ask myself the same question that Google cofounder Larry Page has asked, "Are people really focused on the right things?" As human beings, we have the responsibility to harness the immense human potential for good, not evil; for love, not hatred; and for peace, not war.

The new epoch of a transhumanist brave new world shall merit Miranda's optimism in humanity as William Shakespeare wrote in *The Tempest* (Act V, Scene I):

> *O wonder!*
> *How many goodly creatures are there here!*
> *How beauteous mankind is! O brave new world,*
> *That has such people in't.*

A moonshot in the Sixties literally meant going to the moon. President John F. Kennedy spoke at Rice University on September 12, 1962, "We choose to go to the moon in this decade … not because they are easy, but because they are hard, because that goal will serve to organize and measure the best of our energies and skills, because that challenge is one that we are willing to accept, one we are unwilling to postpone, and one which we intend to win" [193].

Seven years later, astronaut Neil Armstrong said on the Apollo 11 moon landing in 1969, "One small step for a man, one giant leap for mankind." Transhumanism is the next logical step in the evolution of humankind, and it is the existential solution to the long-term survival of the human race.

Bibliography

1. **Cressey, Daniel.** Einstein: 'god is human weakness'. *Nature*. [Online] May 14, 2008. http://blogs.nature.com/news/2008/05/einstein_god_is_human_weakness_1.html.
2. **MacMillan, Margart.** Rebuilding the world after the second world war. *The Guardian*. [Online] September 11, 2009. https://www.theguardian.com/world/2009/sep/11/second-world-war-rebuilding.
3. **Huxley, Julian.** Transhumanism. *New Bottles for New Wine*. London: Chatto & Windus, 1957.
4. **Alighieri, Dante.** Dante Alighieri (1265–1321). The Divine Comedy. *The Harvard Classics*. [Online] 1909–14. https://www.bartleby.com/20/301.html.

5. **Burkeman, Oliver.** Man of steel. *The Guardian.* [Online] September 17, 2002. http://www. theguardian.com/education/2002/sep/17/science.highereducation.
6. **Regalado, Antonio.** Pope Francis Said to Bless Human-Animal Chimeras. *MIT Technology Review.* [Online] January 27, 2016. https://www.technologyreview.com/s/546246/ pope-francis-said-to-bless-human-animal-chimeras/.
7. **Amish America.** Do Amish use electricity? *Amish America.* [Online] http://amishamerica. com/do-amish-use-electricity/.
8. **2045 Initiative.** AUBREY DE GREY / Interview / ENDING AGING. *YouTube.* [Online] November 5, 2015. https://www.youtube.com/watch?v=2lmdp96ySlU.
9. **Graham, Billy.** On technology and faith. *TED.* [Online] February 1988. https://www.ted. com/talks/billy_graham_on_technology_faith_and_suffering/transcript?language=en.
10. **Bilton, Nick.** Steve Jobs Was a Low-Tech Parent. *The New York Times.* [Online] September 10, 2014. https://www.nytimes.com/2014/09/11/fashion/steve-jobs-apple-was-a-low-tech-parent.html.
11. **Retter, Emily.** Billionaire tech mogul Bill Gates reveals he banned his children from mobile phones until they turned 14. *The Mirror.* [Online] April 21, 2017. https://www.mirror.co.uk/ tech/billionaire-tech-mogul-bill-gates-10265298.
12. **Gibbs, Samuel.** Apple's Tim Cook: 'I don't want my nephew on a social network'. *The Guardian.* [Online] January 19, 2018. https://www.theguardian.com/technology/2018/ jan/19/tim-cook-i-dont-want-my-nephew-on-a-social-network.
13. **Bowles, Nellie.** Early Facebook and Google Employees Form Coalition to Fight What They Built. *The Ne York Times.* [Online] February 4, 2018. https://www.nytimes.com/2018/02/04/ technology/early-facebook-google-employees-fight-tech.html.
14. **O'Donnell, Jayne and Saker, Anne.** Teen suicide is soaring. Do spotty mental health and addiction treatment share blame? *USA Today.* [Online] March 19, 2018. https://www,usato-day.com/story/news/politics/2018/03/19/teen-suicide-soaring-do-spotty-mental-health-and-addiction-treatment-share-blame/428148002/.
15. **Valiunas, Algis.** The Agony of Atomic Genius. *The New Atlantis.* [Online] November 14, 2006. http://www.thenewatlantis.com/publications/the-agony-of-atomic-genius.
16. **Oppenheimer, J. Robert.** J. Robert Oppenheimer: "I am become Death, the destroyer of worlds." *YouTube.* [Online] 1965. https://www.youtube.com/watch?v=lb13ynu3Iac.
17. **Achenbach, Joel.** The A.I. Anxiety. *The Washington Post.* [Online] December 27, 2015. http://www.washingtonpost.com/sf/national/2015/12/27/aianxiety/.
18. *Tribute to John von Neumann, 1903-1957.* **Ulam, Stanislaw.** 3, s.l.: Bulletin of the American Mathematical Society, 1958, Vol. 64.
19. **Kurzweil, Ray.** *The Singularity is Near.* s.l.: Viking, 2005.
20. **Eugenios, Jillian.** Ray Kurzweil: Humans will be hybrids by 2030. *CNNMoney.* [Online] June 4, 2015. http://money.cnn.com/2015/06/03/technology/ray-kurzweil-predictions/.
21. **Behr, Rafael.** Alone Together: Why We Expect More from Technology and Less from Each Other by Sherry Turkle – review. *The Guardian.* [Online] January 29, 2011. https://www. theguardian.com/books/2011/jan/30/alone-together-sherry-turkle-review.
22. **Loveridge, Lynzee.** Men & Women Vote on the Brainiest Anime Characters. *Anime News Network.* [Online] June 29, 2014. https://www.animenewsnetwork.com/interest/2014-06-29/ men-and-women-vote-on-the-brainiest-anime-characters/.76003.
23. **berriesthatburn.** The best quote in Anime? *Reddit.* [Online] 2015. https://www.reddit. com/r/anime/comments/2j4rjt/the_best_quote_in_anime/.
24. **Cherriii.** No game no life quotes. *Amino.* [Online] March 10, 2015. https://aminoapps.com/c/ anime/page/blog/no-game-no-life-quotes/E3tP_uxmgRQRpwobNjxxgYL1zg1weE.
25. **Stager, Curt.** You Are Made of Waste: Searching for the ultimate example of recycling? Look in the mirror. *Nautilus.* [Online] November 7, 2013. http://nautil.us/issue/7/waste/ you-are-made-of-waste.
26. **Fung, Brian.** How 40% of Our Food Goes to Waste. *The Altantic.* [Online] August 23, 2012. https:// www.theatlantic.com/health/archive/2012/08/how-40-of-our-food-goes-to-waste/261498/.

27. **Resnick, Brian and Zarracina, Javier.** The US has a 2.5 billion-pound surplus of meat. Let's try to visualize that. *Vox.* [Online] July 24, 2018. https://www.vox.com/science-and-health/2018/7/24/17606958/meat-cheese-surplus-visualized.
28. **Patterson, Thom.** Why does America have so many hungry kids?. *CNN.* [Online] June 15, 2017. https://www.cnn.com/2017/06/09/health/champions-for-change-child-hunger-in-america/index.html.
29. **United Nations.** SAVE FOOD: Global Initiative on Food Loss and Waste Reduction. *Food and Agriculture Organization of the United Nations.* [Online] [Cited: February 15, 2019.] http://www.fao.org/save-food/resources/keyfindings/en/.
30. **WFP.** Zero Hunger. *World Food Programme.* [Online] [Cited: February 15, 2019.] https://www1.wfp.org/zero-hunger.
31. **Burke, Daniel.** The Pope said what?!? More stunners from Francis. *CNN.* [Online] January 19, 2015. https://www.cnn.com/2015/01/19/living/pope-said-what/index.html.
32. **Nonko, Emily.** Nearly 250,000 NYC rental apartments sit vacant. *6sqft.* [Online] March 26, 2018. https://www.6sqft.com/nearly-250000-nyc-rental-apartments-sit-vacant/.
33. **Cheney, Brendan.** Annual homeless count in New York City shows 6 percent decrease. *Politico.* [Online] June 19, 2018. https://www.politico.com/states/new-york/city-hall/story/2018/06/19/annual-homeless-count-in-new-york-city-shows-6-percent-decrease-472900.
34. **Byers, Dylan.** Analysis: the Seattle-Amazon tax fight is a lose-lose for everyone. *CNN.* [Online] May 15, 2018. http://money.cnn.com/2018/05/15/technology/business/seattle-homeless-amazon/index.html.
35. **Hohman, Maura.** Good Samaritan Pays for 70 Hotel Rooms for Chicago Homeless During Dangerous Polar Vortex. *People Magazine.* [Online] January 31, 2019. https://people.com/human-interest/chicago-good-samaritan-hotel-rooms-homeless-polar-vortex/.
36. **Alanis, Kaitlyn.** Boy tried stealing work boots from Walmart—then he explained why, Kansas police say. *The Wichita Eagle.* [Online] December 9, 2018. https://www.kansas.com/news/state/article222873975.html.
37. **Children of 9/11.** 9/11 ten years on: The children left behind. *The Independent.* [Online] September 3, 2011. https://www.independent.co.uk/news/world/americas/911-ten-years-on-the-children-left-behind-2348904.html.
38. **BBC.** In pictures: Atrocities' aftermath. [Online] BBC News, September 12, 2001. http://news.bbc.co.uk/2/hi/americas/1538664.stm.
39. **Easley, Jonathan and Fabian, Jordan.** White House official mocked 'dying' McCain at internal meeting. *The Hill.* [Online] May 10, 2018. http://thehill.com/homenews/administration/387182-white-house-official-mocked-dying-mccain-at-internal-meeting.
40. **Karimi, Faith.** Teens who laughed and recorded a drowning man in his final moments won't face charges. *CNN.* [Online] June 26, 2018. https://www.cnn.com/2018/06/26/us/florida-teens-no-charges-drowning-man/index.html.
41. **Watson, Kathryn.** GOP congressman: "Nobody dies because they don't have access to health care". *CBS News.* [Online] May 6, 2017. https://www.cbsnews.com/news/gop-congressman-nobody-dies-because-they-dont-have-access-to-health-care/.
42. **Drash, Wayne.** Her only chance at life is a new liver, but her insurer said no. Then she wrote a powerful plea to the CEO. *CNN.* [Online] May 13, 2018. https://www.cnn.com/2018/05/13/health/liver-transplant-mom-erika-zak/index.html.
43. **Lee, Newton.** *Counterterrorism and Cybersecurity: Total Information Awareness (2nd Edition).* Switzerland: Springer International Publishing, 2015.
44. —. *Google It: Total Information Awareness.* New York: Springer Science+Business Media, 2016.
45. **Kifner, John.** Eldridge Cleaver, Black Panther Who Became G.O.P. Conservative, Is Dead at 62. *The New York Times.* [Online] May 2, 1998. https://www.nytimes.com/1998/05/02/us/eldridge-cleaver-black-panther-who-became-gop-conservative-is-dead-at-62.html.
46. **Roosevelt, Franklin D.** Franklin D. Roosevelt. *FDR Presidential Library & Museum.* [Online] January 20, 1937. https://fdrlibrary.org/fdr.

47. **Goel, Vindu and Wingfield, Nick.** Mark Zuckerberg Vows to Donate 99% of His Facebook Shares for Charity. *The New York Times.* [Online] December 1, 2015. http://www.nytimes.com/2015/12/02/technology/mark-zuckerberg-facebook-charity.html.

48. **Benner, Katie.** Mark Zuckerberg and Priscilla Chan Pledge $3 Billion to Fighting Disease. *The New York Times.* [Online] September 21, 2016. https://www.nytimes.com/2016/09/22/technology/mark-zuckerberg-priscilla-chan-3-billion-pledge-fight-disease.html.

49. **Gupta, Sanjay.** Bill Gates' newest mission: Curing Alzheimer's. *CNN.* [Online] November 14, 2017. https://www.cnn.com/2017/11/13/health/bill-gates-announcement-alzheimers/index.html.

50. **Paynter, Ben.** Why The Gates Foundation Just Paid Off Nigeria's $76 Million Debt. *Fast Company.* [Online] January 26, 2018. https://www.fastcompany.com/40519100/why-the-gates-foundation-just-paid-off-nigerias-76-million-debt.

51. **Yurieff, Kaya.** Jeff and MacKenzie Bezos create $2 billion fund to fight homelessness. *CNN.* [Online] September 13, 2018. https://money.cnn.com/2018/09/13/technology/jeff-bezos-homeless-fund/index.html.

52. **Godoy, Maria.** Food Stall Serves Up A Social Experiment: White Customers Asked To Pay More. *NPR.* [Online] March 2, 2018. https://www.npr.org/sections/the-salt/2018/03/02/590053856/food-stall-serves-up-a-social-experiment-charge-white-customers-more-than-minori.

53. **Criss, Doug and Hassan, Carma.** Hundreds of doctors in Canada are protesting. They say they make too much. *CNN.* [Online] March 8, 2018. https://www.cnn.com/2018/03/08/health/canada-doctor-raises-trnd/index.html.

54. **CNN Staff.** Five young people creating a better world. *CNN.* [Online] December 11, 2018. https://www.cnn.com/2018/12/05/world/cnnheroes-young-wonders-2018/index.html.

55. **Toner, Kathleen.** She makes sure unwanted food gets to hungry Americans. *CNN.* [Online] December 9, 2018. https://www.cnn.com/2018/07/19/health/cnnheroes-maria-rose-belding-means/index.html.

56. **Karlson, Jay.** 10 Great Philanthropists Who Are Kids. *Listverse.* [Online] January 27, 2011. https://listverse.com/2011/01/27/10-great-philanthropists-who-are-kids/.

57. **Lee, Newton and Madej, Krystina.** *Disney Stories: Getting to Digital.* New York: Springer, 2012.

58. **The Walt Disney Company.** VoluntEARS. *Disneyland Public Affairs.* [Online] [Cited: February 16, 2019.] https://publicaffairs.disneyland.com/voluntears/.

59. **VoyageLA Staff.** Meet Newton Lee of Institute for Education, Research, and Scholarships. *VoyageLA.* [Online] January 22, 2018. http://voyagela.com/interview/meet-institute-education-research-scholarships-downtown-la/.

60. **National Philanthropic Trust.** Charitable Giving Statistics. *National Philanthropic Trust 2015 Annual Report.* [Online] 2015. http://www.nptrust.org/philanthropic-resources/charitable-giving-statistics/.

61. **Dementia Discovery Fund.** Investors. *DDF.* [Online] [Cited: March 25, 2018.] http://thedd-fund.com/about-ddf/investors/.

62. **World Health Organization.** HIV/AIDS - Global Health Observatory (GHO) data. *World Health Organization.* [Online] https://www.who.int/gho/hiv/en/.

63. **Gates, Melinda French.** @melindafrenchgates. *Instagram.* [Online] February 12, 2019. https://www.instagram.com/p/BtyG3sQh5tI/.

64. **Mandela, Nelson.** I am prepared to die. *Nelson Mandela Foundation.* [Online] April 20, 1964. http://db.nelsonmandela.org/speeches/pub_view.asp?pg=item&ItemID=NMS010.

65. **Adams, John.** From John Adams to John Taylor, 17 December 1814. *Founders Online at Archives.gov.* [Online] December 17, 1814. https://founders.archives.gov/documents/Adams/99-02-02-6371.

66. **Eban, Abba and Aridan, Natan.** The Toynbee Heresy. *Israel Studies.* [Online] Spring 2006. https://www.jstor.org/stable/pdf/30245781.pdf.

67. **Leuchtenburg, William E.** Franklin D. Roosevelt: Impact and Legacy. *UVA Miller Center.* [Online] [Cited: February 13, 2019.] https://millercenter.org/president/fdroosevelt/impact-and-legacy.

68. **Kaufman, Ellie and Murphy, Paul P.** Federal employees prepare for a long shutdown. *CNN.* [Online] January 2, 2019. https://www.cnn.com/2018/12/28/politics/shutdown-second-week/index.html.

69. **Cillizza, Chris.** 'I just don't see a pathway forward'. *CNN.* [Online] January 10, 2019. https://www.cnn.com/2019/01/10/politics/lindsey-graham-government-shutdown/index.html.

70. **Fleischer Studios.** Betty Boop for President (1932). *YouTube.* [Online] 1932. https://www.youtube.com/watch?v=c0-q_ZkDcsk.

71. **Mucci, Alberto.** Sweden's Minister of the Future Explains How to Make Politicians Think Long-Term. *Motherboard.* [Online] November 26, 2015. http://motherboard.vice.com/read/swedens-minister-of-the-future-explains-how-to-make-politicians-think-long-term.

72. **DePillis, Lydia.** Ocasio-Cortez's Green New Deal: What's in it. *CNN.* [Online] February 8, 2019. https://edition.cnn.com/2019/02/07/politics/green-new-deal-details/index.html.

73. **Green, Miranda.** Obama invokes Nazi Germany in warning about today's politics. *CNN.* [Online] December 8, 2017. https://www.cnn.com/2017/12/08/politics/barack-obama-nazi-germany/index.html.

74. **Dewey, John.** Democracy and Education. *The Project Gutenberg EBook of Democracy and Education.* [Online] 1916. http://www.gutenberg.org/files/852/852-h/852-h.htm.

75. **Doerer, Kristen.** Poverty expert Angus Deaton awarded Nobel prize in economics. *PBS NewsHour.* [Online] October 12, 2015. https://www.pbs.org/newshour/economy/angus-deaton-awarded-nobel-prize-economics.

76. **Waxman, Sharon.** Letter From Paris: With Democracy in Decline, What Will Replace It? *The Wrap.* [Online] May 20, 2018. https://www.thewrap.com/letter-paris-democracy-decline-marek-halter-trump-macron-cannes/.

77. **Rothman, Peter.** Transhumanism Gets Political. *Humanity+.* [Online] October 8, 2014. http://hplusmagazine.com/2014/10/08/transhumanism-gets-political/.

78. —. Interview: Gabriel Rothblatt Congressional Candidate in Florida's 8th District. *H+ Magazine.* [Online] July 1, 2014. http://hplusmagazine.com/2014/07/01/interview-gabriel-rothblatt-congressional-candidate-in-floridas-8th-district/.

79. **Berwyn-Heights.** Town Council. *Town of Berwyn Heights Maryland.* [Online] [Cited: May 21, 2017.] https://web.archive.org/web/20170521085807/http://www.berwynheightsmd.gov:80/town-council.

80. **Solon, Olivia.** All aboard the Immortality Bus: the man who says tech will help us live forever. *The Guardian.* [Online] June 16, 2016. https://www.theguardian.com/technology/2016/jun/16/transhumanist-party-immortality-zoltan-istvan-presidential-campaign.

81. **The Obama Foundation.** The highest office in the land… *Twitter.* [Online] November 1, 2017. https://twitter.com/obamafoundation/status/925862787120852995?lang=en.

82. **Stolyarov, Gennady II.** Advisors. *United States Transhumanist Party.* [Online] 2017. https://transhumanist-party.org/advisors/.

83. **Alice Paul Institute.** The Equal Rights Amendment. *ERA.* [Online] 2018. https://www.equalrightsamendment.org/.

84. **Maxouris, Christina and Griggs, Brandon.** Voters just elected seven more scientists to Congress. *CNN.* [Online] November 8, 2018. https://www.cnn.com/2018/11/07/health/scientists-in-congress-trnd/index.html.

85. **AT&T.** AT&T Contributing $1 Million to Girls Who Code Sisterh>>d Campaign, Summer Coding Camps. *AT&T.* [Online] September 10, 2018. https://about.att.com/story/2018/girls_who_code.html.

86. **Acevedo, Sylvia and Kennedy, Thomas A.** Bring More Girls Into STEM Workforce. *U.S. News & World Report.* [Online] May 4, 2018. https://www.usnews.com/news/stem-solutions/articles/2018-05-04/commentary-the-power-to-do-good-bringing-more-girls-into-the-stem-workforce.

87. **Beckman, Brittany Levine.** The DIY Girls: How 12 teens invented a solar-powered tent for the homeless. *Mashable.* [Online] June 15, 2017. https://mashable.com/2017/06/15/diy-girls-solar-powered-tent-homeless/#QRW9kdTMsSqJ.

88. **Winowiecki, Emma.** Video and transcript: Gov. Whitmer delivers 2019 State of the State address. *Michigan Radio.* [Online] February 12, 2019. http://www.michiganradio.org/post/video-and-transcript-gov-whitmer-delivers-2019-state-state-address.
89. **Eggert, David.** 'Way Out of Line.' Michigan Governor Blasts TV Station for a Segment About Her Appearance. *TIME Magazine.* [Online] February 15, 2019. http://time.com/5530211/michigan-governor-gretchen-whitmer/.
90. **Horgan, John.** Why There Will Never Be Another Einstein. *Scientific American.* [Online] August 23, 2015. https://blogs.scientificamerican.com/cross-check/why-there-will-never-be-another-einstein/.
91. **Hobbes, Michael.** Why millennials are facing the scariest financial future of any generation since the Great Depression. *Highline.* [Online] [Cited: March 25, 2018.] http://highline.huffingtonpost.com/articles/en/poor-millennials/.
92. **Jefferson, Thomas and et al.** The Declaration of Independence. July 4, 1776.
93. **History.com Staff.** 1789 George Washington is elected president. *History.com.* [Online] 2009. http://www.history.com/this-day-in-history/george-washington-is-elected-president.
94. **Chernow, Ron.** George Washington: The Reluctant President. *Smithsonian. com.* [Online] February 2011. https://www.smithsonianmag.com/history/george-washington-the-reluctant-president-49492/.
95. **National Archives.** The Bill of Rights: What Does it Say? *America's Founding Documents.* [Online] October 12, 2016. https://www.archives.gov/founding-docs/bill-of-rights/what-does-it-say.
96. **Istvan, Zoltan.** Transhumanist Bill of Rights. *H+Pedia.* [Online] [Cited: February 10, 2019.] https://hpluspedia.org/wiki/Transhumanist_Bill_of_Rights.
97. **Stolyarov, Gennady II.** Transhumanist Bill of Rights – Version 3.0. *U.S. Transhumanist Party.* [Online] December 12, 2018. https://transhumanist-party.org/tbr-3/.
98. **Gallo, Carmine.** Steve Jobs Asked One Profound Question That Took Apple From Near Bankruptcy To $1 Trillion. *Forbes.* [Online] August 5, 2018. https://www.forbes.com/sites/carminegallo/2018/08/05/steve-jobs-asked-one-profound-question-that-took-apple-from-near-bankruptcy-to-1-trillion/#673ca6ba9c2f.
99. **Lee, Newton.** Facebook Nation: Total Information Awareness. [Online] Springer-Verlag New York, 2014. http://www.springer.com/us/book/9781493917396.
100. **Parker, Laura.** 'Biggest case on the planet' pits kids vs. climate change. *National Geographic.* [Online] November 9, 2018. https://news.nationalgeographic.com/2017/03/kids-sue-us-government-climate-change/.
101. **Hamedy, Saba.** 100,000 California teens have preregistered to vote. *CNN.* [Online] April 9, 2018. https://www.cnn.com/2018/04/09/politics/california-teens-preregister-to-vote/index.html.
102. **Roosevelt, Franklin D.** Campaign Address from the White House. *Franklin D. Roosevelt Presidential Library and Museum.* [Online] October 5, 1944. http://www.fdrlibrary.marist.edu/_resources/images/msf/msfb0170.
103. **McCullum, April.** Meet the 13-year-old running for Vermont governor. *Burlington Free Press.* [Online] August 31, 2017. https://www.burlingtonfreepress.com/story/news/politics/2017/08/31/meet-13-year-old-running-vermont-governor/580870001/.
104. **Baldacci, Marlena.** He's 13 and he's running to be the next governor of Vermont. *CNN.* [Online] March 30, 2018. https://www.cnn.com/2018/03/30/politics/ethan-sonneborn-governor-vermont-teen-trnd/index.html.
105. **Kormann, Carolyn.** The Teen-Agers Fighting for Climate Justice. *The New Yorker.* [Online] July 22, 2018. https://www.newyorker.com/news/news-desk/the-teen-agers-fighting-for-climate-justice.
106. **Sanchez, Ray.** Student marchers call Washington's inaction on gun violence unacceptable. *CNN.* [Online] March 24, 2018. https://www.cnn.com/2018/03/24/us/march-for-our-lives/index.html.

107. **Waldrop, Theresa.** A day after March for Our Lives, Pope urges youth to speak out. *CNN.* [Online] March 25, 2018. https://www.cnn.com/2018/03/25/europe/pope-palm-sunday-tells-youth-to-speak-out/index.html.
108. **Workman, James.** "Our house is on fire." 16 year-old Greta Thunberg wants action. *World Economic Forum.* [Online] January 25, 2019. https://www.weforum.org/agenda/2019/01/our-house-is-on-fire-16-year-old-greta-thunberg-speaks-truth-to-power.
109. **Loucaides, Darren.** What Happens When Techno-Utopians Actually Run a Country. *Wired.* [Online] February 14, 2019. https://www.wired.com/story/italy-five-star-movement-techno-utopians/.
110. **Sweetlove, Lee.** Number of species on Earth tagged at 8.7 million. *Nature.* [Online] August 24, 2011. https://www.nature.com/news/2011/110823/full/news.2011.498.html.
111. **Gandhi, Mahatma.** The essential unity of all religions. *Comprehensive Website by Gandhian Institutions - Bombay Sarvodaya Mandal & Gandhi Research Foundation.* [Online] http://www.mkgandhi.org/voiceoftruth/unityofallreligions.htm.
112. **Variety Staff.** Spike Lee, Jada Pinkett Smith Call for Oscar Boycott. *Variety.* [Online] January 18, 2016. http://variety.com/2016/film/awards/spike-lee-jada-pinkett-smith-oscar-boycott-1201682165/.
113. **Dash, Stacey.** I Was Right Today on Outnumbered: There Should Be No Black History Month. *patheos.* [Online] January 20, 2016. http://www.patheos.com/blogs/staceydash/2016/01/i-was-right-today-on-outnumbered-there-should-be-no-black-history-month/.
114. **Roddenberry, Gene.** Quotable Quote. *goodreads.* [Online] [Cited: January 24, 2016.] http://www.goodreads.com/quotes/98079-if-man-is-to-survive-he-will-have-learned-to.
115. **IMDb.** Is There in Truth No Beauty? *IMDb.* [Online] October 18, 1968. http://www.imdb.com/title/tt0708433/quotes.
116. **Minsky, Marvin.** Society Of Mind. *Google Books.* [Online] Simon & Schuster Paperbacks, 1988. https://books.google.com/books?id=bLDLllfRpdkC&pg=PA308&lpg=PA308#v=onepage&q&f=false.
117. **Jary, Simon.** Who is Steve Jobs' Syrian immigrant father, Abdul Fattah Jandali? *MacWorld.* [Online] February 2, 2017. https://www.macworld.co.uk/feature/apple/who-is-steve-jobs-syrian-immigrant-father-abdul-fattah-jandali-3624958/.
118. **Baram, Marcus.** Google cofounder Sergey Brin, who emigrated from Russia as a boy, is at the protests at SFO Airport. *Fast Company.* [Online] January 28, 2017. https://www.fastcompany.com/4029560/google-founder-sergey-brin-who-immigrated-from-russia-as-a-boy-is-at-the-protests-at-sfo-airport.
119. **McCracken, Harry.** Satya Nadella Rewrites Microsoft's Code. *Fast Company.* [Online] September 18, 2017. https://www.fastcompany.com/40457458/satya-nadella-rewrites-microsofts-code.
120. **Levy, Steven.** The Uber CEO Who Arrived in the US with Nothing. *Wired.* [Online] November 22, 2017. https://www.wired.com/story/dara-khosrowshahi-uber-ceo-iran-immigration/.
121. **Jolie, Angelina.** Speech by Angelina Jolie 'In defense of internationalism'. *UNHCR: The UN Refugee Agency.* [Online] March 15, 2017. http://www.unhcr.org/en-us/news/press/2017/3/58c994944/speech-angelina-jolie-defense-internationalism.html.
122. **BBC.** Pitt and Jolie have baby daughter. *BBC News.* [Online] May 28, 2006. http://news.bbc.co.uk/2/hi/entertainment/5024396.stm.
123. **Gruber, Ben.** Jolie twins doctor admits to pre-birth pressure. *Reuters.* [Online] July 14, 2008. https://www.reuters.com/article/us-jolie/jolie-twins-doctor-admits-to-pre-birth-pressure-idUSL1322562520080715.
124. **Jones, Tobias.** An unsolved murder at Italy's most notorious tower block. *The Guardian.* [Online] July 31, 2018. https://www.theguardian.com/news/2018/jul/31/an-unsolved-at-italys-most-notorious-tower-block.
125. **Klein, JoAnna.** How the Father of Computer Science Decoded Nature's Mysterious Patterns. *The New York Times.* [Online] May 8, 2018. https://www.nytimes.com/2018/05/08/science/alan-turing-desalination.html.

126. **Hackett, Conrad and McClendon, David.** Christians remain world's largest religious group, but they are declining in Europe. *Pew Research Center.* [Online] April 5, 2017. http://www.pewresearch.org/fact-tank/2017/04/05/christians-remain-worlds-largest-religious-group-but-they-are-declining-in-europe/.
127. **Whitwam, Ryan.** How Apollo 11's 1.024MHz guidance computer did a lot with very little. *GEEK.COM.* [Online] July 20, 2013. https://www.geek.com/news/how-the-apollo-11s-1-024mhz-guidance-computer-did-a-lot-with-very-little-1562831/.
128. **Jha, Alok.** Neil Armstrong breaks his silence to give accountants moon exclusive. *The Guardian.* [Online] May 23, 2012. http://www.theguardian.com/science/2012/may/23/neil-armstrong-accountancy-website-moon-exclusive.
129. **Peters, Gerhard and Woolley, John T.** Richard Nixon, Telephone Conversation With the Apollo 11 Astronauts on the Moon. *The American Presidency Project.* [Online] July 20, 1969. https://www.presidency.ucsb.edu/documents/telephone-conversation-with-the-apollo-11-astronauts-the-moon.
130. **The StarChild Team.** The Apollo 11 Memorial on the Moon. *The StarChild at NASA.* [Online] [Cited: February 17, 2019.] https://starchild.gsfc.nasa.gov/docs/StarChild/space_level2/apollo11_plaque.html.
131. **CNN.** Hawking: If we survive the next 200 years, we should be OK. *CNN.* [Online] October 9, 2008. http://www.cnn.com/2008/WORLD/europe/10/09/hawking/index.html.
132. **Solman, Paul.** How do we invest in the future of humanity? Swedish philosopher Nick Bostrom explains. *PBS.* [Online] July 20, 2017. https://www.pbs.org/newshour/economy/invest-future-humanity-swedish-philosopher-nick-bostrom-explains.
133. **Bronson, Rachel.** A new abnormal: It is still two minutes to midnight. *Bulletin of the Atomic Scientists.* [Online] January 24, 2019. https://thebulletin.org/doomsday-clock/current-time/.
134. **UN News Centre.** UN chief issues 'red alert,' urges world to come together in 2018 to tackle pressing challenges. *United Nations.* [Online] December 31, 2017. http://www.un.org/apps/news/story.asp?NewsID=58370#.WkIxSHlG2Uk.
135. **Vogue, Ariane de.** Scalia-Ginsburg friendship bridged opposing ideologies. *CNN.* [Online] February 14, 2016. http://www.cnn.com/2016/02/14/politics/antonin-scalia-ruth-bader-ginsburg-friends/index.html.
136. **Dear, John S.J.** Afghanistan journal, part one: Learning a nonviolent lifestyle in Kabul. [Online] National Catholic Reporter, December 11, 2012. http://ncronline.org/blogs/road-peace/afghanistan-journal-part-one-learning-nonviolent-lifestyle-kabul.
137. **—.** Afghanistan journal, part two: bearing witness to peacemaking in a war-torn country. [Online] National Catholic Reporter, December 18, 2012. http://ncronline.org/blogs/road-peace/afghanistan-journal-part-two-bearing-witness-peacemaking-war-torn-country.
138. **Illing, Sean.** This filmmaker spent months interviewing neo-Nazis and jihadists. Here's what she learned. *Vox.* [Online] January 14, 2019. https://www.vox.com/world/2019/1/14/18151799/extremism-white-supremacy-jihadism-deeyah-khan.
139. **Michaelson, Jay.** Was Obama right about the Crusades and Islamic extremism? *The Washington Post.* [Online] February 6, 2015. https://www.washingtonpost.com/national/religion/was-obama-right-about-the-crusades-and-islamic-extremism-analysis/2015/02/06/3670628a-ae46-11e4-8876-460b1144cbc1_story.html?utm_term=.972f6fe7d614.
140. **Hackett, Conrad and McClendon, David.** Christians remain world's largest religious group, but they are declining in Europe. *Pew Research Center.* [Online] April 5, 2017. http://www.pewresearch.org/fact-tank/2017/04/05/christians-remain-worlds-largest-religious-group-but-they-are-declining-in-europe/.
141. **Pope Francis.** Messaggio Pasquale del Santo Padre e Benedizione "Urbi et Orbi", 01.04.2018. *Holy See Press Office.* [Online] April 1, 2018. http://press.vatican.va/content/salastampa/it/bollettino/pubblico/2018/04/01/0242/00520.html#en.
142. **Diebel, Matthew.** The Amish: 10 things you might not know. *USA Today.* [Online] August 15, 2014. https://www.usatoday.com/story/news/nation/2014/08/15/amish-ten-things-you-need-to-know/14111249/.

143. **Carrington, Damian.** Humans just 0.01% of all life but have destroyed 83% of wild mammals – study. *The Guardian.* [Online] May 21, 2018. https://www.theguardian.com/environment/2018/may/21/human-race-just-001-of-all-life-but-has-destroyed-over-80-of-wild-mammals-study.

144. Vanishing: The Sixth Mass Extinction. *CNN.* [Online] [Cited: March 27, 2018.] https://www.cnn.com/specials/world/vanishing-earths-mass-extinction.

145. **BBC.** Big Five mass extinction events. *Nature - Prehistoric Life.* [Online] [Cited: February 28, 2016.] http://www.bbc.co.uk/nature/extinction_events.

146. **Ceballos, Gerardo, et al.** Accelerated modern human–induced species losses: Entering the sixth mass extinction. *Science Advances.* [Online] June 19, 2015. http://advances.sciencemag.org/content/1/5/e1400253.full.

147. **Zalasiewicz, Jan.** Working Group on the 'Anthropocene'. *Subcommission on Quaternary Stratigraphy.* [Online] 2019. http://quaternary.stratigraphy.org/working-groups/anthropocene/.

148. **Timberg, Craig.** A disaster foretold — and ignored. *The Washington Post.* [Online] https://www.washingtonpost.com/sf/business/2015/06/22/net-of-insecurity-part-3/?noredirect=on&utm_term=.9a25d19d157d.

149. **Shaftel, Holly.** A blanket around the Earth. *NASA's Jet Propulsion Laboratory and California Institute of Technology.* [Online] March 2, 2016. http://climate.nasa.gov/causes/.

150. **British Antarctic Survey.** Earth's magnetic field is important for climate change at high altitudes. *Phys.org.* [Online] May 26, 2014. http://phys.org/news/2014-05-earth-magnetic-field-important-climate.html.

151. **Shaftel, Holly.** The consequences of climate change. *NASA's Jet Propulsion Laboratory and California Institute of Technology.* [Online] March 2, 2016. http://climate.nasa.gov/effects/.

152. **Pearson, Michael.** Greenland ice melt off to record early start. *CNN.* [Online] April 15, 2016. http://www.cnn.com/2016/04/15/world/greenland-ice-melt/index.html.

153. **National Oceanic and Atmospheric Administration.** Quinault Indian Nation plans village relocation. *U.S. Climate Resilience Toolkit.* [Online] December 17, 2015. http://toolkit.climate.gov/taking-action/quinault-indian-nation-plans-village-relocation.

154. **Simon, Matt.** Sea Level Rise in the SF Bay Area Just Got a Lot More Dire. *Wired.* [Online] March 7, 2018. https://www.wired.com/story/sea-level-rise-in-the-sf-bay-area/.

155. **Watkins, Eli.** Government report calls on Trump to act on climate change. *CNN.* [Online] October 24, 2017. http://www.cnn.com/2017/10/23/politics/gao-report-climate-change/index.html.

156. **Bowden, Emma.** Disaster costs jumped over 60% this year to $306 billion. *CNN.* [Online] December 20, 2017. http://money.cnn.com/2017/12/20/news/economy/disasters-hurricanes-wildfires-cost/index.html.

157. **Martinez, Michael, Elam, Stephanie and Nieves, Rosalina.** The quake-maker you've never heard of: Cascadia. *CNN.* [Online] February 13, 2016. http://www.cnn.com/2016/02/11/us/cascadia-subduction-zone-earthquakes/.

158. **USGS.** Historic Earthquakes. *United States Geological Survey.* [Online] April 6, 2016. http://earthquake.usgs.gov/earthquakes/states/events/1700_01_26.php.

159. **Lah, Kyung and Becker, Stephanie.** 'Statistically, it's coming.' California prepares for the next big earthquake. *CNN.* [Online] September 30, 2017. https://www.cnn.com/2017/09/30/us/california-earthquake-preparation/index.html.

160. **Bennett, Cory.** Critical infrastructure cyberattacks rising, says US official. *The Hill.* [Online] January 13, 2016. http://thehill.com/policy/cybersecurity/265753-critical-infrastructure-cyberattacks-rising-says-us-official.

161. **Reuters.** NSA Chief Says 'When, Not If' Foreign Country Hacks U.S. Infrastructure. *Fortune.* [Online] March 1, 2016. http://fortune.com/2016/03/01/nsa-chief-hacking-infrastructure/.

162. **National Park Service.** Washington Monument Earthquake Update. *U.S. Department of the Interior.* [Online] 2011. https://www.nps.gov/wamo/washington-monument-earthquake-update.htm.

163. **Marsh, Rene, Gracey, David and Severson, Ted.** How to fix America's 'third world' airports. *CNN.* [Online] May 27, 2016. http://www.cnn.com/2016/05/25/politics/infrastructure-roads-bridges-airports-railroads/index.html.

164. **CBS2.** CBS2 Investigates: Experts Say Decaying Gas Lines Are A Ticking Time Bomb Below City Streets. *CBS New York.* [Online] January 8, 2016. http://newyork.cbslocal.com/2016/01/08/new-york-gas-main/.

165. **Walton, Alice, Branson-Potts, Hailey and Sahagun, Louis.** Porter Ranch gas leak permanently capped, officials say. *Los Angeles Times.* [Online] February 18, 2016. http://www.latimes.com/local/lanow/la-me-ln-porter-ranch-gas-leak-permanently-capped-20160218-story.html.

166. **Satija, Neena, et al.** Hell and High Water. *ProPublica.* [Online] March 3, 2016. https://www.propublica.org/article/hell-and-high-water-text.

167. **McKenna, Maryn.** Long-Dreaded Superbug Found in Human and Animal in U.S. *National Geographic.* [Online] May 26, 2016. http://phenomena.nationalgeographic.com/2016/05/26/colistin-r-9/.

168. **Scutti, Susan.** Unusual forms of 'nightmare' antibiotic-resistant bacteria detected in 27 states. *CNN.* [Online] Apri 4, 2018. 1. https://www.cnn.com/2018/04/03/health/nightmare-bacteria-cdc-vital-signs/index.html.

169. **O'Neill, Jim.** Tackling Drug-Resistant Infections Globally: Final Report and Recommendations. *Revew on Antimicrobial Resistance.* [Online] May 2016. http://amr-review.org/sites/default/files/160525_Final%20paper_with%20cover.pdf.

170. **Earthquake Early Warning.** ShakeAlert. *Earthquake Early Warning.* [Online] [Cited: March 5, 2016.] http://www.shakealert.org/.

171. Scientists design conceptual asteroid deflector and evaluate it against massive potential threat. *Lawrence Livermore National Laboratory.* [Online] March 15, 2018. https://www.llnl.gov/news/scientists-design-conceptual-asteroid-deflector-and-evaluate-it-against-massive-potential.

172. **Talbert, Tricia.** Planetary Defense: Double Asteroid Redirection Test (DART) Mission. *National Aeronautics and Space Administration.* [Online] June 25, 2018. https://www.nasa.gov/planetarydefense/dart.

173. **National Oceanic and Atmospheric Administration.** Climate Explorer—Visualize Climate Data in Maps and Graphs. *U.S. Climate Resilience Toolkit.* [Online] December 17, 2015. http://toolkit.climate.gov/tools/climate-explorer.

174. **Food Forward.** Mission Statement. *Food Forward.* [Online] https://foodforward.org/about/.

175. **Google.** Renewable energy. *Google green.* [Online] [Cited: March 7, 2016.] https://www.google.com/green/energy/.

176. **Iyengar, Rishi.** Microsoft is buying solar energy from Singapore rooftops. *CNN.* [Online] March 1, 2018. http://money.cnn.com/2018/03/01/technology/microsoft-singapore-sunseap-solar-project-deal/index.html.

177. **Budds, Diana.** How Google Is Turning Cities Into R&D Labs: From autonomous vehicles to building codes, Sidewalk Labs is thinking about problems and solutions that could shape cities for centuries. *Fast Company & Inc.* [Online] February 22, 2016. http://www.fastcodesign.com/3056964/design-moves/how-google-is-turning-cities-into-rd-labs.

178. **Alexander, Kurtis.** Tech buses blocked, vandalized in protests. *SFGate.* [Online] December 20, 2013. http://blog.sfgate.com/stew/2013/12/20/bus-blocked-again-in-tech-boom-backlash/.

179. **Atiyeh, Clifford.** No More New Gas-Powered Cars by 2050, Say Eight States and Five Countries. *Car and Driver.* [Online] December 8, 2015. http://blog.caranddriver.com/no-more-new-gas-powered-cars-by-2050-say-eight-states-and-five-countries/.

180. **Institute for Education, Research, and Scholarships.** Wellness Center. *Institute for Education, Research, and Scholarships.* [Online] [Cited: February 5, 2019.] https://www.ifers.org/wellness-center.html.

181. **Strickland, Ashley.** Scott Kelly from space: Earth's atmosphere 'looks very, very fragile'. *CNN.* [Online] January 12, 2016. http://www.cnn.com/2016/02/11/health/scott-kelly-space-station-sanjay-gupta-interview/.
182. **Cerf, Vint.** Synthetic organisms are about to challenge what 'alive' really means. *Wired.* [Online] January 5, 2019. https://www.wired.co.uk/article/artificial-life-vint-cerf.
183. **Than, Ker.** Billions of Earthlike Planets Crowd Milky Way? *National Geographic.* [Online] January 8, 2013. https://news.nationalgeographic.com/news/billions-of-earthlike-planets-found-in-milky-way/.
184. **Science Reference Services.** What does it mean when they say the universe is expanding? *The Library of Congress.* [Online] November 27, 2018. https://www.loc.gov/rr/scitech/mysteries/universe.html.
185. **O'Connell, Cathal.** Universe shouldn't exist, CERN physicists conclude. *COSMOS.* [Online] October 23, 2017. https://cosmosmagazine.com/physics/universe-shouldn-t-exist-cern-physicists-conclude.
186. **Perez, Patrice, Doser, Michael and Bertsche, William.** Does antimatter fall up? *CERN.* [Online] January 13, 2017. http://cerncourier.com/cws/article/cern/67455.
187. **Eury, Michael and Sanderson, Peter.** Superman. *Encyclopedia Britannica.* [Online] January 19, 2018. https://www.britannica.com/topic/Superman-fictional-character.
188. **IMDb.** Kick-Ass. *IMDb.* [Online] April 16, 2010. http://www.imdb.com/title/tt1250777/trivia?tab=qt&ref_=tt_trv_qu.
189. **Kurzweil, Ray.** Danielle. *Transcend.* [Online] 2018. https://transcend.me/pages/books-and-publications.
190. **Vandoorne, Saskya, Beech, Samantha and Westcott, Ben.** 'Spiderman' granted French citizenship after rescuing child from Paris balcony. *CNN.* [Online] May 28, 2018. https://www.cnn.com/2018/05/28/asia/paris-baby-spiderman-rescue-intl/index.html.
191. **Muir, Hazel.** Einstein and Newton showed signs of autism. *New Scientist.* [Online] April 30, 2003. https://www.newscientist.com/article/dn3676-einstein-and-newton-showed-signs-of-autism/.
192. **Gates, Bill and Melinda.** If you could have one superpower, what would it be? *Gates Notes.* [Online] February 22, 2016. https://www.gatesnotes.com/Annual-Letter-Superpowers.
193. **Kennedy, John F.** John F. Kennedy Moon Speech - Rice Stadium. *NASA Software Robotics and Simulation Division.* [Online] September 12, 1962. http://er.jsc.nasa.gov/seh/ricetalk.htm.

Chapter 2
History of Transhumanism

Natasha Vita-More

2.1 Introduction

As a philosophy transhumanism deals with the fundamental nature of reality, knowledge, and existence. As a worldview, it offers a cultural ecology for understanding the human integration with technology. As a scientific study, it provides the techniques for observing how technology is shaping society and the practice for investigating ethical outcomes. Its social narrative emerges from humans overcoming odds and the continued desire to build a world worth living in. These processes requires critical thinking and visionary accounts to assess how technology is altering human nature and what it means to be human in an uncertain world.

Transhumanism has questioned traditional norms of society, which can be and have been provoking to those who do not share the worldview. Beliefs about life and death are historically the at the heart of people's values. At the core of transhumanism is the conviction that the lifespan be extended, aging reversed, and that death should be optional rather than compulsory. Transhumanism also proposes that artificial intelligence be used to help improve human level decision-making, that nanotechnology resolve environmental hazards, that molecular manufacturing stop poverty, and that genetic engineering mitigate diseases. Nevertheless, a provoked society rears its head in defense. Myths and lore remind us that the Gods can be unforgiving and implacable. We are forewarned not to reach too far, fly too high, or venture where we ought not to tread. Yet, humans are robust explorers who enjoy challenges, ameliorate problems, and uncover the unknowns to transcend limitations.

N. Vita-More (✉)
Humanity+, Inc. and University of Advancing Technology, Tempe, AZ, USA
e-mail: natasha@natashavita-more.com

© Springer Nature Switzerland AG 2019
N. Lee (ed.), *The Transhumanism Handbook*,
https://doi.org/10.1007/978-3-030-16920-6_2

2.2 The Transhuman

The term transhuman has an unusual etymology and its usage is found within the fields of literature, philosophy, religion, and evolutionary biology. According to the "Report on The Meaning of Transhuman" [26], the first use as transhuman is written as an Italian verb "transumanare" or "transumanar", as written by Dante Alighieri in *Divina Commedia* [3].[1] In this reference, trans-human means "go outside the human condition and perception". The English translation is "to transhumanate" or "to transhumanize".

Centuries later, poet T.S. Eliot used the term "transhumanized" to represent the risks of the human journey in becoming illuminated as a "process by which the human is Transhumanised" in "The Cocktail Party" [4].[2,3] What is unusual is that both authors, centuries apart, were poets. Further, a link is found between Eliot and Teilhard de Chardin, a philosopher and a Catholic priest, who proposed that man use any appropriate means for transhumanizing himself to the fullest potential in *The Future of Man*[4] [22].

The noun transhuman was formally identified and codified in The *Reader's Digest Great Encyclopedia Dictionary*, which defined "transhuman" as meaning "surpassing; transcending; beyond" (1966).[5] Almost a decade later, the field of science fiction borrowed the concept with Robert Ettinger's use of the term transhumanity in *Man into Superman*.[6] (1972). Futurist FM Esfandiary introduced the transhuman as a future of human evolution in his chapter "Transhumans 2000" in *Women the Year 2000* [24].

The interpretation of the transhuman as an evolutionary process was noted in *Webster's New Universal Unabridged Dictionary,*[7] which defined "transhuman" as meaning "superhuman," and "transhumanize," as meaning "to elevate or transform to something beyond what is human" (1983). At that same time, I authored the "Transhuman Manifesto" and "Transhumanist Arts Statement" (1983), emphasizing an aim to transcend the limits of our bodies and our minds.[8]

> There are numerous forbearers of theories on human evolution and traces can be found in a plethora of sources, all suggesting that the biological human is not the final stage of evolution for the human. The philosophy and social/cultural movement of transhumanism has

---

[1] Dante, Alighieri. (1308–1321) The Divine Comedy (The Inferno, The Purgatorio, and The Paradiso) (Ed. Ciardi, J.) New York: NAL Trade, 2003. (p. 586–589).

[2] Eliot, T.S. (1952) *Complete Poems and Plays: 1909–1950. The Cocktail Party.* New York: Harcourt. (p. 147).

[3] Sarkar, Subhas (2006) *T.S. Eliot: The* Dramatist. Atlantic Publishers. (p., 192).

[4] De Chardin, Teilhard. (1959) *The Future of Man.* First Image Books Edition (2004).

[5] *The Reader's Digest Great Encyclopedia Dictionary.* (1966). Reader's Digest.

[6] Ettinger, R. (1972) *Man Into Superman.* New York: Avon.

[7] *Webster's New Universal Unabridged Dictionary.* (1983) Fromm Intl.

[8] Vita-More, N. (1983) Transhuman *Statement* in *Create/Recreate.* Available: http://www.transhumanist.biz/createrecreate.htm http://www.natasha.cc/transhuman.htm

developed not only from the words "trans" and "human", but also through an understanding that the human condition is one in which we might go outside to gain perspective, a process in becoming an evolutionary transformation [28].

2.3   Transhumanism

The origin of transhumanism is bestowed on two British scholars who never met, but both graduated from Oxford University almost a century apart and in two entirely different fields of study. Julian Huxley, an evolutionary biologist and Catholic priest wrote about how humans must establish a better environment for themselves in the essay "Transhumanism" in *New Bottles For New Wine* (1957).[9] Max More, CEO of Alcor Life Extension Foundation, created the philosophy of transhumanism in his essay "Transhumanism: Toward a Futurist Philosophy" [13], which codified the principle that life can expand indefinitely by means of human intelligence and technology.

What turned the philosophical view of our existence into an emerging cultural movement was largely due to the Internet. However, *before* the Internet, transhumanism was seeded by people who were curious about new technology and how AI and nanotechnology can change the world. This curiosity was the intellectual fuel accelerated alongside the tech industry. We simply wanted to think about and talk about where technology was heading.

Science played a major role in applying technology to transhumanist interests. Yet, as science aimed to identify and map genes, so did peoples' concerns about genomics, genetically modified food, and cloning. The enthusiasm for biotechnology and the possibilities of nanomedicine and genetic engineering were strong among transhumanists. But it was not considered to be advantageous to the others, especially Bill Joy, Cofounder and former Chief Scientist at Sun Microsystems who wrote the following in *Wired* [11]:

> As this enormous computing power is combined with the manipulative advances of the physical sciences and the new, deep understandings in genetics, enormous transformative power is being unleashed. These combinations open up the opportunity to completely redesign the world, for better or worse: The replicating and evolving processes that have been confined to the natural world are about to become realms of human endeavor. … We now know with certainty that these profound changes in the biological sciences are imminent and will challenge all our notions of what life is [11].

Through waves of optimism on the one hand and techno-fear on the other, a cultural and socio-political divided arose. Bioethicists made public claims that biotechnology and emerging technologies of AI and nanotech should be stopped, and activists for technological acceleration requested venues for debate. With such a divide, a new way of thinking was necessary to mitigate disparity within society and to steer

---

[9] Huxley, Julian. (1957) "TRANSHUMANISM" In *NEW BOTTLES FOR NEW WINE: ESSAYS.* London: Chatto & Windus.

toward academic mindfulness and mainstream awareness. Thus, contrary to journalistic hyperbole and postmodernist hegemony, the aim of transhumanism has been and continues to be to establish a platform for critical thinking and visionary foresight that can and will have significant impact on people. This impact is to educate society and to offer platforms for discussion and take the conversation out of the postmodernist rhetoric, journalistic sensationalism, and fear-mongering of bioethicists, into the public arena. Each person—whether privileged with access to the Internet or other areas of the world where technology was not available, is curious about their future and hungry for answers.

Looking back at the cultural advocacy that precedes the Internet: Silicon Valley startups and their counterpart—the Los Angeles entertainment industry eventually had to became cohorts. Computer scientists provided the technological prowess; writers, musicians, designers, and innovators added the allure of the future with visual content and futuristic narratives. During this time, behind the scenes, transhumanist journals and conferences was building systematic studies and models in forming the worldview of transhumanism. Through this, transhumanism applied tools of forecasting, trend analyses, information theory and systems thinking. Attention was given to and included the knowledge provided by Gordon Moore and Moore's Law that transistors in integrated circuits double every two years exponentially, A.H. Maslow's hierarchy of needs as stemming from motivation, and Lynn Margulis' symbiotic theory of eukaryotic cell development, which revolutionized modern concept of how life began on Earth. Certainly, there are more theoretical findings to mention; however, the aforementioned stand out as essential because technology has become exponential, human needs have been at the forefront of innovations and address the question of what it means to be human.

Transhumanist thinking may have been a catalyst that prompted curiosity and the desire to find solutions in areas of knowledge gathering critical thinking, ethics, and visionary foresight in developing new social narratives. For example, venues such as TED talks and makerspaces, projects such as Quantified Self, and DIY all strongly exemplify transhumanist behavior. This behavior evidences how life experiences can be uncovered, expressed, and shared and then transmitted across varied channels of communication and collaboration. TED talks are all about what a person did differently, a type of hero or heroine's journey, and how that journey contributed to the personal's life experience and through this, add to the well-being of society. Makerspaces provide a collaborative venue for people to come together and identify a problem, strive to figure out how to unravel it, roll up their sleeves, and then create an innovative solution. The Quantified Self project is all about a person's life—their numerical self and how that person identified their problem, such as difficulty sleeping, high blood pressure, and/or when to exercise, for example, and then sought to develop a system to help adjust that behavior. DIY may have been an earlier version of these three projects, but each one of them include the do-it-yourself mentality in being "self-responsible" for their own well-being, and through this—helping their loved ones to be better too, and this sentiment, or this intelligence, trickles across society like a chain of paying it forward like an infectious smile that keeps giving.

These examples are grassroot and located within their own domain of experience, but they evidence how self-responsibility is a shared cultural behavior. We start someplace—perhaps at the bottom with our own identity to place in life, and then experience, learn, evolve, and become someone or something better than when we started. You could call this spiritual, or simply intelligent. All in all, it is natural—an innate element of human psychology of survival—that is shared among individuals and society.

And this is how transhumanism as a philosophy became a growing worldview. It is a process and behavior that started in 1989 through a high-gloss print journal called *Extropy: The Journal of Transhumanist Thought*. In 1991 the original transhumanist email list called "extropy", a metaphor for negentropy as refers to a systems intelligence, order, vitality and capacity and drive for improvement. The Internet was the most fertile breeding ground for people interested in learning about and exploring transhumanist thinking, including innovators, entrepreneurs, and academics that furthered a transdisciplinary scope. In 1991, the first email list covering technology and humanity's future was developed by transhumanists at Extropy Institute, the first transhumanist non-profit organization. The organization hosted a series of conferences from 1994 to 2004 with keynotes and thought leaders who set the bar high. In fact, the original ideas about artificial intelligence and human computer integration, encryption, crypto currency, AI, Super AI, nanotechnology, the technological singularity, radical life extension, and uploads (posthumans), were incubated at these conferences.

Today, the merging of early transhumanism and its pioneers with other organizations and disciplines, increased use of social media, and the mainstream's awareness of and interest in nanotechnology, AI/AGI, and life extension has reached a paradigmatic shift and, along with the project Humanity+. The symbol for transhumanism has gained branding currency as "H+" and while a trademark of Humanity+, a 501(c)3 non-profit, a version of h + is also copyright protected by Humanity+ for its magazine. Humanity+ is the largest transhumanist organization worldwide and is associated with many other organizations that aim to inform the public about the advances in technology, ethics, and political issues that are ahead.

2.4 Misinformation

Over the past many years, I have witnessed varied articles, documentaries, and other formats misconstruing the meaning of transhumanism. I remember in 2009, I happened upon an academic journal that published an article on transhumanism: *The Global Spiral*'s [23] "Special Issue on Transhumanism",[10] produced by Guest Editor Hava Tirosh-Samuelson and five contributing authors, Ted Peters, Katherine

---

[10] Special Issue on Transhumanism (2008) In *The Global Spiral*, (Guest Editor, Hava Tirosh-Samuelson), Vol. 6, Issue 3. Available http://www.metanexus.net/magazine/PastIssues/tabid/126/Default.aspx?PageContentID = 27

Hayles, Don Ihde, Jean-Pierre Dupuy, and Andrew Pickering. These scholars were provoked by transhumanism and pushing back from their postmodernist stance with forked tongues. I was deeply concerned by the hearsay. How could these revered academics blatantly counter what I had experienced first-hand?

> The philosophical worldview and social movement of transhumanism has the benefit of existing while many of its pioneers are still living. This makes it more accommodating for those unfamiliar with transhumanism to investigate and argue its tenets with the most recent writings at hand. Rather than searching endless databases for bibliographical references and out-of-print books in gathering evidence of who did what, when, and where, researchers can easily locate people though Google and send an email or make a call. Why the authors of the "Special Issue on Transhumanism" in The Global Spiral did not do this is a curiosity. Nonetheless, their six essays present a much-appreciated opportunity for developing discourse on transhumanism [28].

After reading my concerns, the journal's Managing Editor invited to be Guest Editor of a new Special issue on Transhumanism.

> This caused a responsive second "Special Issue on Transhumanism", with Guest Editor Natasha Vita-More and nine other transhumanist authors, including Aubrey de Grey, Martine Rothblatt, Max More, Nick Bostrom, and Russell Blackford, to evaluate the criticisms and address concerns [28].

I do not think that the original special issue was intended to cause a backlash from transhumanists in academics and the mainstream. These scholars were protecting their domain in philosophy and the humanities.

A complex world is challenging and fast-track news gathering can obfuscates fact from fiction. And our biological limitations of processing information should be noted: the human body transmits over 10 million bits of information per second to the brain; yet, our conscious mind processes only a portion of this [7]. It is no wonder that human interpretation of information can be faulty. Even first-hand experience often alters perceptions that influence how experience is interpreted.

2.5   Influencers

There have been social and political influencers within transhumanism, and the humanities continue to play a major role in the ideology of humanism and links to the worldview of transhumanism. In humanism, the democratic and ethical life stance asserts that humans have a right to give meaning to their own lives. Transhumanism encases this view but takes it further by strategizing theoretical and practice-based models that propose how humans can shape their own lives. Yet, some cases, human psychology, emotions, intelligence, and mental attitudes invite or block the ability to accept or refuse the unknown. Society in the 1980s did not accept the concept of the transhuman, and in the 1990s the idea of transhumanism was loved or hated it. Strong words to be sure, but this was prevalent. Innovators of encryption and cryptocurrencies, entrepreneurs of robotics, AI, and nanotechnology, along with space enthusiasts, life extension activists, and consciousness seekers cherished the idea of transhumanism. Conversely, bioethicists, postmodernists, religious groups, and others were concerned about new technologies, human enhancement, and genetics loathed it. Interestingly, the science fiction cyborg, borrowed the coined term from Manfred Clynes and Nathan Kline, [2]. Rather than relevant to cybernetics and a necessity for space exploration, the cyborg became a terminator. It was borrowed again as a feminist salvo in its reinterpretation by Donna Haraway in her statement "I'd Rather be a Cyborg than a Goddess" [9].

In reflection, a question that went unanswered is why did society accept the cyborg as a machine-man science fiction terrorist and not appreciate the actual transformation of human as the transhuman—a human with ethics and a desire to enhance with technology?

2.6 Politics

Early transhumanists were mostly located in the United States in New York, Los Angeles, and startup hubs such as Silicon Valley. New York and LA hosted more liberal thinkers with social concerns who leaned toward the Democratic party with some Libertarian philosophical influence. Computer and startup hubs, such as Silicon Valley, were largely Libertarian. Entrepreneurs who funded projects were independents, Libertarian, or Republican. There were also many Democrats, Green Party members, Socialists, Upwingers, etc. In LA, most transhumanists were Upwingers (neither right nor left), stemming from F.M. Esfandiary's (aka FM2010) writings about the future and the transhuman. I want to make this point clear: early transhumanists were diverse and not representative of any one religious, anti-religious, or spiritual belief and not of any one political position or party.

Political positioning for misuse of information has damaged many cultures, including transhumanism. It conflicts with the transhumanist tenets of diversity and advancement. The continual improvement as both physical and psychological. Considering this, transhumanism cannot be one political position. That attitude is counter to the fundamental values of a systems intelligence, order, vitality and capacity and drive for improvement and the three essential elements of transhumanism: critical thinking, technological innovation, and visionary narratives.

There have been a few transhumanists in the political arena over the years. FM Esfandiary set the political stage in his book *Upwingers a Futurist Manifesto* [6], which normative platform reached beyond the Right/Left predicament and set out a non-linear evolving view of moving upward:

> We are at all times slowed down by the narrowness of Right-wing and Left-wing alternatives. If you are not conservative, you are liberal if not right of center you are left of it or middle of the road. Our traditions comprise no other alternatives. There is no ideological or conceptual dimension beyond conservative and liberal beyond Right and Left. ... The premises of the entire Left are indistinguishable from those of the entire Right. The extreme Left is simply a linear extension of the extreme Right. The liberal is simply a more advanced conservative. The radical Left is a more advanced liberal. ...

> The Right/Left establishment is fighting a losing battle. It is following in the foot-
> steps of earlier traditionalists who resisted the more modest breakthroughs of the past …
> ([6], pp 21–25).

A small cable TV show called "Breakthroughs: A TransCentury Update" (aka "Transhuman Update") aired in Los Angeles and Telluride, Colorado. As its producer and host, I interviewed innovators of emerging and speculative sciences and technologies. One benefit that came out of this was body of work is that it afforded an insight into what was to come. A side effect was that due to the cutting-edge content, a viewer nominated me to run for the 27th Senatorial District of Los Angeles County on the Green Party ticket for a seat as County Councilperson. After a few months campaigning, I was elected on a Transhumanism platform— promoting environmentalist use of technology.

Between the mid-1990s and 2018, there has been a working group of transhumanists of who are devoted to politics and building a substantive set of guidelines and roadmap for the future. The fact is that society must be informed issues we face that will affect society and its governance. This is a far heavier issue that right vs. left. Further there is a gap in the education of society or what is often called long-long learning, where continuing education is not only essential, it is crucial. People must keep up, learn how to use smart devices to understand where technology is heading. This includes technological advancements that are altering our lives and the scope of economic and political issues; and that governments:

> … dramatically expanded governmental research into anti-aging therapies, and universal access
> to those therapies as they are developed in order to make much longer and healthier lives acces-
> sible to everybody. We believe that there is no distinction between "therapies" and "enhance-
> ment." The regulation of drugs and devices needs reform to speed their approval [21].

2.7 Transhumanism Now

With a focus on why transhumanism is a solution to many of the issues humanity faces, mention must be given to approaches that can help the decision-making process. I will touch on several projects conceived by thought leaders of transhumanism, its early adaptors, and in a few instances, its pioneers.

To begin, decision-making works best when it is open and balanced. It is difficult for us largely because the human interpretation of information is assessed and filtered by personal perceptions. As an example, the well-known Precautionary Principle "… is a moral and political principle which states that if an action or policy might cause severe or irreversible harm to the public or to the environment, … the burden of proof falls on those who would take the action" [10]. Rather than placing the burden of proof on absolute judgement of unknown outcomes, a more balanced process for policy making in weighing the pros and cons can be achieve by using the Proactionary Principle:

> People's freedom to innovate technologically is highly valuable, even critical, to humanity.
> This implies a range of responsibilities for those considering whether and how to develop,
> deploy, or restrict new technologies. Assess risks and opportunities using an objective,

open, and comprehensive, yet simple decision process based on science rather than collective emotional reactions. Account for the costs of restrictions and lost opportunities as fully as direct effects. Favor measures that are proportionate to the probability and magnitude of impacts, and that have the highest payoff relative to their costs. Give a high priority to people's freedom to learn, innovate, and advance [15].

Another necessary and timely concept for human right is Morphological Freedom, which means "[t]he ability to alter bodily form at will through technologies such as surgery, genetic engineering, nanotechnology, uploading" [14]. Ownership of one's body is championed as a human right, as expressed by Anders Sandberg of the Future of Humanity Institute, Senior Research Fellow at the Future of Humanity Institute, Oxford University:

> Morphological freedom can of course be viewed as a subset of the right to one's body. But it goes beyond the idea of merely passively maintaining the body as it is and exploiting its inherent potential. Instead it affirms that we can extend or change our potential through various means. … Without morphological freedom, there is a serious risk of powerful groups forcing change upon us. Historically the worst misuses of biomedicine have always been committed by governments and large organizations rather than individuals. … It hence makes sense to leave decisions on a deeply personal ethical level to individuals rather than making them society-wide policies. Global ethical policies will by necessity both run counter to the ethical opinion of many individuals, coercing citizens to act against their beliefs and hence violating their freedom and contain the temptation to adjust the policies to benefit the policymakers rather than the citizens [19].

On a global scale, a transhumanist priority is considering the risks, uncertainties, and the magnitude of expected loss due to catastrophes. Existential Risk, as clarified by Nick Bostrom, considers three dimensions that describe the magnitude of risk, its scope, intensity, and probability. According to Bostrom, Founding Director of the Future of Humanity Institute, existential risk means: "[o]ne where an adverse outcome would either annihilate Earth-originating intelligent life or permanently and drastically curtail its potential" [1].

> An existential risk is one where humankind is imperiled. Existential disasters have major adverse consequences for the course of human civilization for all time to come [1].

2.8   Connections

I have presented the earliest accounts of transhumanism and how it surfaced during a time when emerging technology attracted an identifiable excitement and a lot of fear in society. The emotions could be felt within tech and futurist communities, science fiction narratives, academic humanities, biotechnologies, and computer science, and cyber security, and literally whooshed into the hands of entrepreneurs of tech start-ups. The arousing volume was varied and vast, and at points even cosmic. Emerging technological output, the innovations, products, processes, including the smart devices and the mobility of digital networks landed the Internet of things.

I now point you to experts and their books that I have not yet mentioned in this chapter. I cannot mention all experts or written materials, although those that I leave out are important and can be found through Google. To begin, let's summarize the scope of transhumanism.

Imagine an interconnected system as a mind map where the human is at the core. Considering the core, you can see into the distant past when endosymbiotic theory of eukaryotic cell development two billion years ago (Margulis), moving forward to the hominin of 5 million years ago, to the Homo sapiens approximately 200,000 years ago and to the most recent account of human evolution of the frontal lobe at 1.8 million years ago. Outside the core, but interconnected, are the variables of live and living. One variable is the computers and the microchip that transformed our lives within the past four decades, more than biological evolution over the past 200,000 years. A sub-variable identifies the integration of technology with biology, linking to human senses, mobility, communications. This sub-variable links to a higher level of variables, one of which is intelligence. Here the map shows numerous links back and forth from the brain to the computer and the fields of neuroscience, cognitive science, and then outward in to the map to the Blue Brain Project, Neuralink, DARPA, Jülich Research Center, Google Brain Project, and many more.

As a historical map, consider the connection between the following events:

Exponential Technology. K. Eric Drexler provided the creative concepts for technological capabilities, including nanotechnology, it in *Engines of Creation* (1986). Kevin Kelly took a practical approach to the 1990s social fear in *Out of Control: The New Biology of Machines, Social Systems, & the Economic World* (1995). Ray Kurzweil turned human enhancement mainstream in his book *Building a Mind* and *The Singularity is Near* (2005). Kelly re-appears and observes technology's future in *What Technology Wants* [12], and Peter Diamandis suggesting global benefits of technology in *Abundance* 2012), which reflects to Drexler's creative concepts of molecular manufacturing solving many of the World problems.

- Nanotechnology, biotechnology, information technology, AI/AGI, cognitive science, neuro science, robotics, backing up the brain, Moore's Law, 3D printing, molecular manufacturing, abundance, radical life extension, space travel, etc.

Serious issues. Mainstream issues harken over the scope of technology and its potential to do harm. and Anders Sandberg offer a conservative tone about Existential Risk, identifying the five biggest threats to human existence [20].

- Existential risk, human evolution, Proactionary Principle, policy making, political issues, ethics and objective observations vs. career bioethicists.

Evolution: If we continue to use technology to develop ways to live longer, then we are within the scope of an evolutionary leap. The most intriguing aspect of this evolution is reversing aging and extending the human life span. Aubrey de Grey calls this longevity escape velocity (David Gobel). As explained by de Grey this means that a technological breakthrough could increase general life expectancy by more than the one year of a person's life. The younger a person is or the faster improvements in life-extending therapies are develop could determine their potential life expectancy.

- Aging: aging, reversing aging, the disease of aging, radical life extension, cryonics, whole body prosthetics, future bodies, non-biological systems for life.

2.9   Continuity

Martine Rothblatt, founder of Sirius XM satellite radio and currently CEO of United Therapeutics Corporation, once voiced her views to me about transhumanism. To paraphrase, she said that the point of life is to evolve. We start out as a person, incomplete and unpolished, and evolve into person that continues to complete.

Transhumanism is about improving the human condition, which means the distinct characteristics of being human and human existence, including survival and evolution human nature. What does this mean? The human condition means human nature, human society, and how people live their lives. For example, the unique and distinct characteristics of being human and human existence, including survival and evolution. The question of whether humans are naturally good or evil, selfish or altruistic, naturally social beings or individuals, and relationships between genders, are questions that religion, philosophy, psychology and sociology think about and try to understand. The emotions that we call all so human are of great value in feeling love and compassion, these characteristics are wonderful attributes—generosity, creativity, intelligence, and fearlessness. Some are not so great and cause unnecessary pain, anguish, distressing, and the indefensible sorrow of mental illness. The good and the bad together form the characteristics of what it means to be human.

At the forefront of the future of humanity, transhumanism is renown. Its educators, thought leaders, business, research centers, organizations and individuals have all been part of this growing movement. That the public is curious and concerned about what it means to be human and questions what we will become. To help answer the public's curiosity and concerns, transhumanism is finding solutions, offering alternatives to political structures, and providing the knowledge and well-thought-out potential solutions needs to be instilled into society in a positive manner.

2.10   Conclusion

Transhumanism's time has arrived. "It has struck a chord with many who want to fight the onslaught of disease and live longer healthy lives. It is no longer a complex concept that encounters vast and often confusing questions" [25]. The earlies ideas, terms, and themes transhumanists have been writing and talking about for three decades have become mainstream. Nevertheless, there are still some unanswered questions and misconceptions about the history of transhumanism that linger in the cultural ethos—that of an unclear and irregular accounting of the past. The aim of this chapter has been to present an historical account covering the past three or more decades.

Moving forward, there is a need to question the existing state of affairs and to be informed about opportunities for the future. The very core of transhumanist thinking prepares us to be leaders of our own lives and to work with others to help increase the well-being of others. Together we are trailblazers propagating and encouraging seminal, ground-breaking solutions. The more we engage in the unknowns, the more we adapt to change, the more we challenge your own thinking, the more impactful our shared knowledge will address the challenges we face. We are part of this Transhumanist Era.

References

1. Bostrom, N. (2001). "Existential Risks: Analyzing Human Extinction Scenarios and Related Hazards" in J*ournal of Evolution and Technology*, Vol. 9, No. 1 (2002). (First version: 2001). Available: https://nickbostrom.com/existential/risks.html
2. Clynes, M. & Kline, N. (1960). "Cyborgs and Space" in *Astronautics*, American Rocket Society Inc, New York, New York, pp. 26, 27, 29, 33.
3. Dante Alighieri, 1265–1321. (1935). *The divine comedy of Dante Alighieri: Inferno, Purgatory, Paradise.* New York: The Union Library Association.
4. Eliot, T.S. (1949). "The Cocktail Party". 1ˢᵗ Ed. London: Faber & Faber.
5. Esfandiary, F.M. (1974) "Transhumans the Year 2000" in *Woman the Year 2000*. M. Tripp Ed. New York: Arbor House.
6. Esfandiary, F.M. (1977). *Upwingers a Futurist Manifesto*. New York: Popular Library, pp 21–25.
7. Fan, J. (2014). "An information theory account of cognitive control" in *Frontiers in Human Neuroscience*. US National Library of Medicine National Institutes of Health. 8, 680. Available https://www.ncbi.nlm.nih.gov/pmc/articles/PMC4151034/
8. Haraway, D. (1990). "A Cyborg Manifesto: Science, Technology and Socialist-Feminism in the Late Twentieth Century" in Simians, Cyborgs, and Woman: The Revolution of Nature. London: Routledge.
9. Haraway, Donna. (1991) *Simians, Cyborgs, and Women: The Reinvention of Nature*. New York: Routledge, pp. 3–5, 149–181
10. IEET. (n.d.) "Precautionary vs. Proactionary Principles. Available https://ieet.org/index.php/tpwiki/Precautionary_vs._proactionary_principles
11. Joy. B. (2001). "Why the Future Doesn't Need Us" in *Wired*. Available: https://www.wired.com/2000/04/joy-2/
12. Kelly, Kevin. (2010) *What Technology Wants*. New York: Viking Press, p. 45.
13. More, M. (1990). "Towards a Futuristic Philosophy". Available: https://web.archive.org/web/20051029125153/http://www.maxmore.com/transhum.htm
14. More, M. (1993). "Technological Self-Transformation" in *Extropy*. No 10(4:2), Winter/Spring. Available: http://www.maxmore.com/selftrns.htm
15. More, M. (2005). "The Proactionary Principle". Available: http://www.extropy.org/proactionaryprinciple.htm
16. More, M. (2013). "The Proactionary Principle" in The Transhumanist Reader: Classical and Contemporary Essays on the Science, Technology, and Philosophy of the Human Future. Eds. M. More and N. Vita-More. Malden, MA: Wiley-Blackwell
17. More, M., Vita-More, N. (2013). *The Transhumanist Reader: Classical and Contemporary Essays on the Science, Technology, and Philosophy of the Human Future*. Wiley-Blackwell.
18. Rothblatt, M. (2010). "On genes, memes, bemes, and conscious things". Kurzweil Essays. Available at: http://www.kurzweilai.net/on-genes-memes-bemes-and-conscious-things
19. Sandberg, A. (2001). "Morphological Freedom – Why We not just Want it but *Need* it. Available: http://www.aleph.se/Nada/Texts/MorphologicalFreedom.htm
20. Sandberg, A. (2014). "The five biggest threats to human existence" in *The Conversation*. Available: https://theconversation.com/the-five-biggest-threats-to-human-existence-27053
21. Technoprogressive Declaration (2014). Available: https://transvision-conference.org/tpdec2017/

22. Teilhard de Chardin, P. (1959). *L'Avenir de L'Homme*. Paris: Editions de Seuil.
23. *The Global Spiral.* (2008). "Special Issue on Transhumanism". H. Tirosh-Samuelson Ed. Vol. 6, Issue 3.
24. Tripp, M. (1974). *Woman the Year 2000*. New York: Arbor House.
25. Vita-More, N. (2019). Transhumanism: What is it? ISBN-10: 0578405075; ISBN-13: 978-0578405070
26. Vita-More. (1989) "Report on Transhuman".
27. Vita-More. (1999) "The Automorpher". Electronic Café. Santa Monica, CA.
28. Vita-More. (2011). "Introduction to "H+: Transhumanism Answers Its Critics" in *Metanexus*. Available: http://www.metanexus.net/essay/introduction-h-transhumanism-answers-its-critics

Chapter 3
The Boundaries of the Human: From Humanism to Transhumanism

José Luis Cordeiro

The famous astronomer and astrobiologist Carl Sagan popularized the concept of a Cosmic Calendar about three decades ago. In his 1977 book, *The Dragons of Eden: Speculations on the Evolution of Human Intelligence*, Sagan wrote a timeline for the universe, starting with the Big Bang about 15 billion years ago. Today, we think that it all started about 13.7 billion years back, and we keep updating and improving our knowledge of life, the universe and everything. In his Cosmic Calendar, with each month representing slightly over one billion years, Sagan dated the major events during the first 11 months of the cosmic year (see Table 3.1).

Interestingly enough, most of what we study in biological evolution happened in the last month. In fact, Sagan wrote that the first worms appeared on December 16, the invertebrates began to flourish on the 17th, the trilobites boomed on the 18th, the first fish and vertebrates appeared on the 19th, the plants colonized the land on the 20th, the animals colonized the land on the 21st, the first amphibians and first winged insects appeared on the 22nd, the first trees and first reptiles evolved on the 23rd, the first dinosaurs appeared on the 24th, the first mammals evolved on the 26th, the first birds emerged on the 27th, the dinosaurs became extinct on the 28th, the first primates appeared on the 29th and the frontal lobes evolved in the brains of primates and the first hominids appeared on the 30th. Basically, humans are just the new kids in the block, and only evolved late at night on the last day of this Cosmic Calendar (see Table 3.2).

The previous Cosmic Calendar is an excellent way to visualize the acceleration of change and the continuous evolution of the universe. Other authors have developed similar ideas to try to show the rise of complexity in nature. For example, in 2005, astrophysicist Eric Chaisson published his latest book, *Epic of Evolution: Seven Ages of the Cosmos*, where he describes the formation of the universe through the development of seven ages: matter, galaxies, stars, heavy elements, planets, life,

J. L. Cordeiro (✉)
Humanity+, Madrid, Spain
e-mail: jose@millennium-project.org

© Springer Nature Switzerland AG 2019
N. Lee (ed.), *The Transhumanism Handbook*,
https://doi.org/10.1007/978-3-030-16920-6_3

Table 3.1 Cosmic Calendar: January – November

Big Bang	January 1
Origin of Milky Way Galaxy	May 1
Origin of the solar system	September 9
Formation of the Earth	September 14
Origin of life on Earth	~ September 25
Formation of the oldest rocks known on Earth	October 2
Date of oldest fossils (bacteria and blue-green algae)	October 9
Invention of sex (by microorganisms)	~ November 1
Oldest fossil photosynthetic plants	November 12
Eukaryotes (first cells with nuclei) flourish	November 15

Source: J. Cordeiro based on Sagan [37]

Table 3.2 Cosmic Calendar: December 31

Origin of *Proconsul* and *Ramapithecus*, probable ancestors of apes and men	~ 1:30 p.m.
First humans	~ 10:30 p.m.
Widespread use of stone tools	11:00 p.m.
Domestication of fire by Peking man	11:46 p.m.
Beginning of most recent glacial period	11:56 p.m.
Seafarers settle Australia	11:58 p.m.
Extensive cave painting in Europe	11:59 p.m.
Invention of agriculture	11:59:20 p.m.
Neolithic civilization; first cities	11:59:35 p.m.
First dynasties in Sumer, Ebla and Egypt; development of astronomy	11:59:50 p.m.
Invention of the alphabet; Akkadian Empire	11:59:51 p.m.
Hammurabi legal codes in Babylon; Middle Kingdom in Egypt	11:59:52 p.m.
Bronze metallurgy; Mycenaean culture; Trojan War; Olmec culture; invention of the compass	11:59:53 p.m.
Iron metallurgy; First Assyrian Empire; Kingdom of Israel; founding of Carthage by Phoenicia	11:59:54 p.m.
Asokan India; Ch'in Dynasty China; Periclean Athens; birth of Buddha	11:59:55 p.m.
Euclidean geometry; Archimedean physics; Ptolemaic astronomy; Roman Empire; birth of Christ	11:59:56 p.m.
Zero and decimals invented in Indian arithmetic; Rome falls; Moslem conquests	11:59:57 p.m.
Mayan civilization; Sung Dynasty China; Byzantine empire; Mongol invasion; Crusades	11:59:58 p.m.
Renaissance in Europe; voyages of discovery from Europe and from Ming Dynasty China; emergence of the experimental method in science	11:59:59 p.m.
Widespread development of science and technology; emergence of global culture; acquisition of the means of self-destruction of the human species; first steps in spacecraft planetary exploration and the search of extraterrestrial intelligence	Now: The first second of New Year's Day

Source: J. Cordeiro based on Sagan [37]

Table 3.3 The Six Epochs of the Universe according to Kurzweil

Epoch 1	Physics and chemistry (information in atomic structures)
Epoch 2	Biology (information in DNA)
Epoch 3	Brains (information in neural patterns)
Epoch 4	Technology (information in hardware and software designs)
Epoch 5	Merger of technology and human intelligence (the methods of biology, including human intelligence, are integrated into the exponentially expanding human technology base)
Epoch 6	The universe wakes up (patterns of matter and energy in the universe become saturated with intelligent processes and knowledge)

Source: J. Cordeiro based on Kurzweil [25]

complex life, and society. Chaisson presents a valuable survey of these fields and shows how combinations of simpler systems transform into more complex systems, and he thus gives a glimpse of what the future might bring.

Both Sagan and Chaisson have written excellent overviews about evolution, from its cosmic beginnings to the recent emergence of humans and technology. However, a more futuristic look is given by engineer and inventor Ray Kurzweil in his 2005 book: *The Singularity is Near: When Humans Transcend Biology*. Kurzweil wrote about six epochs with increasing complexity and accumulated information processing (see Table 3.3).

According to Kurzweil, humanity is entering Epoch 5 with an accelerating rate of change. The major event of this merger of technology and human intelligence will be the emergence of a "technological singularity". Kurzweil believes that within a quarter century, non-biological intelligence will match the range and subtlety of human intelligence. It will then soar past it because of the continuing acceleration of information-based technologies, as well as the ability of machines to instantly share their knowledge. Eventually, intelligent nanorobots will be deeply integrated in our bodies, our brains, and our environment, overcoming pollution and poverty, providing vastly extended longevity, full-immersion virtual reality incorporating all of the senses, and vastly enhanced human intelligence. The result will be an intimate merger between the technology-creating species and the technological evolutionary process it spawned.

Computer scientist and science fiction writer Vernor Vinge first discussed this idea of a technological singularity in a now classic 1993 paper, where he predicted:

Within thirty years, we will have the technological means to create superhuman intelligence. Shortly after, the human era will be ended.

Other authors talk about such technological singularity as the moment in time when artificial intelligence will overtake human intelligence. Kurzweil has also proposed the *Law of Accelerating Returns*, as a generalization of Moore's law to describe an exponential growth of technological progress. Moore's law deals with an exponential growth pattern in the complexity of integrated semiconductor circuits (see Fig. 3.1).

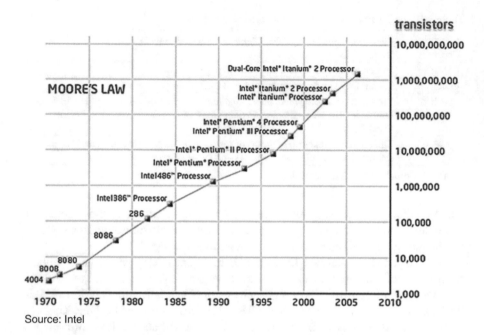

Source: Intel

Fig. 3.1 Moore's Law. (Source: Intel)

Kurzweil extends Moore's law to include technologies from far before the integrated circuit to future forms of computation. Whenever a technology approaches some kind of a barrier, he writes, a new technology will be invented to allow us to cross that barrier. He predicts that such paradigm shifts will become increasingly common, leading to "technological change so rapid and profound it represents a rupture in the fabric of human history." He believes the *Law of Accelerating Returns* implies that a technological singularity will occur around 2045:

> An analysis of the history of technology shows that technological change is exponential, contrary to the common-sense 'intuitive linear' view. So we won't experience 100 years of progress in the 21st century—it will be more like 20,000 years of progress (at today's rate). The 'returns,' such as chip speed and cost-effectiveness, also increase exponentially. There's even exponential growth in the rate of exponential growth. Within a few decades, machine intelligence will surpass human intelligence, leading to the Singularity —technological change so rapid and profound it represents a rupture in the fabric of human history. The implications include the merger of biological and non-biological intelligence, immortal software-based humans, and ultra-high levels of intelligence that expand outward in the universe at the speed of light.

3.1 Technological Convergence

Futurists today have diverging views about the singularity: some see it as a very likely scenario, while others believe that it is more probable that there will never be any very sudden and dramatic changes due to progress in artificial intelligence.

However, most futurists and scientists agree that there is an increasing rate of tech-nological change. In fact, the rapid emergence of new technologies has generated scientific developments never dreamed of before.

The expression "emerging technologies" is used to cover such new and poten-tially powerful technologies as genetic engineering, artificial intelligence, and nano-technology. Although the exact denotation of the expression is vague, various writers have identified clusters of such technologies that they consider critical to humanity's future. These proposed technology clusters are typically abbreviated by such combinations of letters as NBIC, which stands for Nanotechnology, Biotechnology, Information technology and Cognitive science. Various other acro-nyms have been offered for essentially the same concept, such as GNR (Genetics, Nanotechnology and Robotics) used by Kurzweil, while others prefer NRG because it sounds similar to "energy." Journalist Joel Garreau in *Radical Evolution* uses GRIN, for Genetic, Robotic, Information, and Nano processes, while author Douglas Mulhall in *Our Molecular Future* uses GRAIN, for Genetics, Robotics, Artificial Intelligence, and Nanotechnology. Another acronym is BANG for Bits, Atoms, Neurons, and Genes.

The first NBIC Conference for Improving Human Performance was organized in 2003 by the NSF (National Science Foundation) and the DOC (Department of Commerce). Since then, there have been many similar gatherings, in the USA and overseas. The European Union has been working on its own strategy towards con-verging technologies, and so have been other countries in Asia, starting with Japan.

The idea of technological convergence is based on the merger of different scien-tific disciplines thanks to the acceleration of change on all NBIC fields. Nanotechnology deals with atoms and molecules, biotechnology with genes and cells, infotechnology with bits and bytes, and cognitive science with neurons and brains. These four fields are converging thanks to the larger and faster information processing of ever more powerful computers (see Fig. 3.2).

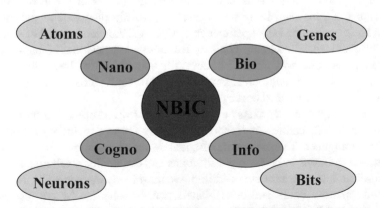

Source: J. Cordeiro based on M.C. Roco and W.S. Bainbridge (2003)

Fig. 3.2 Technological Convergence NBIC. (Source: J. Cordeiro based on Roco and Bainbridge [35])

Experts from the four NBIC fields agree about the incredible potential of technological evolution finally overtaking and directing biological evolution. Bill Gates of Microsoft has stated that:

> *I expect to see breathtaking advances in medicine over the next two decades, and biotechnology researchers and companies will be at the center of that progress. I'm a big believer in information technology... but it is hard to argue that the emerging medical revolution, spearheaded by the biotechnology industry, is any less important to the future of humankind. It, too, will empower people and raise the standard of living.*

Larry Ellison of Oracle, Gates' rival in the software industry, agrees: "If I were 21 years old, I probably wouldn't go into computing. The computing industry is about to become boring". He explains that: "I would go into genetic engineering." Biologist Craig Venter has said that he spent 10 years reading the human genome, and now he is planning to write new genomes. He wants to create completely new forms of life, from scratch. Scientist and writer Gregory Stock also believes that cloning, even though a fundamental step in biotechnology, is just too simple and unexciting: "why copy old life forms when we can now create new ones?"

Biological evolution allowed the appearance of human beings, and many other species, through millions of years of natural selection based on trials and errors. Now we can control biological evolution, direct it and go beyond it. In fact, why stop evolution with carbon-based life forms? Why not move into silicon-based life, among many other possibilities? Robotics and artificial intelligence will allow us to do just that.

Scientist Marvin Minsky, one of the fathers of artificial intelligence at MIT, wrote a very famous 1994 article "Will robots inherit the Earth?" in *Scientific American*, where he concludes: "Yes, but they will be our children. We owe our minds to the deaths and lives of all the creatures that were ever engaged in the struggle called Evolution. Our job is to see that all this work shall not end up in meaningless waste." Robotics expert Hans Moravec has written two books about robots and our (their) future: *Mind Children* in 1988 and *Robot* in 1998. Moravec argues that robots will be our rightful descendants and he explains several ways to "upload" a mind into a robot. In England, cybernetics professor Kevin Warwick has been implanting his own body with several microchip devices and published in 2003 a book explaining his experiments: *I, Cyborg*. Warwick is a cybernetics pioneer who claims that: "I was born human. But this was an accident of fate – a condition merely of time and place. I believe it's something we have the power to change... The future is out there; I am eager to see what it holds. I want to do something with my life: I want to be a cyborg."

As these authors and thinkers suggest, we need to start preparing ourselves for the coming NBIC realities of technological convergence, including robotics and artificial intelligence. Thanks to technological evolution, humans will transcend our biological limitations to become transhumans and eventually posthumans. To ease this transition into a posthuman condition, we must ready ourselves for the distinct possibility that the Earth, and other planets, will be inherited by not just one but several forms of highly intelligent and sentient life forms. Thus, the philosophy of humanism is not enough for a world, and a universe, where future life forms will continue evolving.

3.2   From Humanism to Transhumanism

A new philosophy has been proposed to continue the ideas of humanism in a new world where science and technology are the major drivers of change. Julian Huxley, the English evolutionary biologist and humanist that became the first director-general of UNESCO and founder of the World Wildlife Fund, wrote that:

> *The human species can, if it wishes, transcend itself —not just sporadically, an individual here in one way, an individual there in another way, but in its entirety, as humanity. We need a name for this new belief. Perhaps transhumanism will serve: man remaining man, but transcending himself, by realizing new possibilities of and for his human nature.*

> *"I believe in transhumanism": once there are enough people who can truly say that, the human species will be on the threshold of a new kind of existence, as different from ours as ours is from that of Peking man. It will at last be consciously fulfilling its real destiny.*

Huxley originally published those words in his essay *Religion Without Revelation*, which was reprinted in his book *New Bottles for New Wine* [21]. Other scientists and philosophers discussed similar ideas in the first half of the twentieth century, and these ideas slowly helped to create new philosophical movements considering nature and humanity in a continuous state of flux and evolution. English scientist John Burdon Sanderson Haldane and French philosopher Pierre Teilhard de Chardin helped to identify new trends in the future evolution of humanity. Thanks to them and many others, the philosophy of transhumanism has greatly advanced since Huxley first used that word. The philosophy of Extropy (see Appendix 1) and Transhumanism (see Appendix 2) explore the boundless possibilities for future generations, while we approach a possible technological singularity.

"Humans" can no longer be regarded as a stable category let alone one which occupies a privileged position in relation to all that is subsumed under the category of the non-human. On the contrary, humans must be understood as a tenuous entity which is related to the animal, the "natural" and indeed other humans as well. Humans are at a crossroads like other natural species that are reclassified in the face of new relational dynamics and shifting epistemological paradigms. Moreover, such dynamics and interpolation serve to reveal the boundaries of humans as a corporal, cognitive, and agency-laden construct. Discovering such boundaries, one may glean where humans end, where humans are called into question, and where humans stand to augment themselves or become more than human.

Our understanding about ourselves and about our relationships with nature around us has increased significantly due to the continuous advances in science and technology. Reality is not static since humans and the rest of nature are dynamic, indeed, and both are changing constantly. Transhumanism transcends such static ideas of humanism as humans themselves evolve at an accelerating rate. In the beginning of the twenty-first century, it is now clear than humans are not the end of evolution, but just the beginning of a conscious and technological evolution.

3.3 The Human Seed

Since English naturalist Charles Darwin first published his ideas about evolution on *The Origin of Species* in 1859, it has become clear to the scientific community that species evolve according to interactions among them and with their environment. Species are not static entities but dynamic biological systems in constant evolution. Humans are not the end of evolution in any way, but just the beginning of a better, conscious and technological evolution. The human body is a good beginning, but we can certainly improve it, upgrade it, and transcend it. Biological evolution through natural selection might be ending, but technological evolution is only accelerating now. Technology, which started to show dominance over biological processes some years ago, is finally overtaking biology as the science of life.

As fuzzy logic theorist Bart Kosko has said: "biology is not destiny. It was never more than tendency. It was just nature's first quick and dirty way to compute with meat. Chips are destiny." Photo-qubits might also come after standard silicon-based chips, but even that is only an intermediate means for augmented intelligent life in the universe.

Homo sapiens is the first species in our planet which is conscious of its own evolution and limitations, and humans will eventually transcend these constraints to become enhanced humans, transhumans and posthumans. It might be a rapid process like caterpillars becoming butterflies, as opposed to the slow evolutionary passage from apes to humans. Future intelligent life forms might not even resemble human beings at all, and carbon-based organisms will mix with a plethora of other organisms. These posthumans will depend not only on carbon-based systems but also on silicon and other "platforms" which might be more convenient for different environments, like traveling in outer space.

Eventually, all these new sentient life forms might be connected to become a global brain, a large interplanetary brain, and even a larger intergalactic brain. The ultimate scientific and philosophical queries will continue to be tackled by these posthuman life forms. Intelligence will keep on evolving and will try to answer the old-age questions of life, the universe and everything. With ethics and wisdom, humans will become posthumans, as science fiction writer David Zindell suggested:

> *"What is a human being, then?"*
> *"A seed."*
> *"A... seed?"*
> *"An acorn that is unafraid to destroy itself in growing into a tree."*

Appendix 1: The Principles of Extropy

- **Perpetual Progress:** Extropy means seeking more intelligence, wisdom, and effectiveness, an open-ended lifespan, and the removal of political, cultural, biological, and psychological limits to continuing development. Perpetually overcoming constraints on our progress and possibilities as individuals, as organizations, and as a species. Growing in healthy directions without bound.

- **Self-Transformation:** Extropy means affirming continual ethical, intellectual, and physical self-improvement, through critical and creative thinking, perpetual learning, personal responsibility, proactivity, and experimentation. Using technology – in the widest sense to seek physiological and neurological augmentation along with emotional and psychological refinement.
- **Practical Optimism:** Extropy means fueling action with positive expectations – individuals and organizations being tirelessly proactive. Adopting a rational, action-based optimism or "proaction", in place of both blind faith and stagnant pessimism.
- **Intelligent Technology:** Extropy means designing and managing technologies not as ends in themselves but as effective means for improving life. Applying science and technology creatively and courageously to transcend "natural" but harmful, confining qualities derived from our biological heritage, culture, and environment.
- **Open Society – information and democracy:** Extropy means supporting social orders that foster freedom of communication, freedom of action, experimentation, innovation, questioning, and learning. Opposing authoritarian social control and unnecessary hierarchy and favoring the rule of law and decentralization of power and responsibility. Preferring bargaining over battling, exchange over extortion, and communication over compulsion. Openness to improvement rather than a static utopia. Extropia ("ever-receding stretch goals for society") over utopia ("no place").
- **Self-Direction:** Extropy means valuing independent thinking, individual freedom, personal responsibility, self-direction, self-respect, and a parallel respect for others.
- **Rational Thinking:** Extropy means favoring reason over blind faith and questioning over dogma. It means understanding, experimenting, learning, challenging, and innovating rather than clinging to beliefs.

Appendix 2: The Transhumanist Declaration

1. Humanity stands to be profoundly affected by science and technology in the future. We envision the possibility of broadening human potential by overcoming aging, cognitive shortcomings, involuntary suffering, and our confinement to planet Earth.
2. We believe that humanity's potential is still mostly unrealized. There are possible scenarios that lead to wonderful and exceedingly worthwhile enhanced human conditions.
3. We recognize that humanity faces serious risks, especially from the misuse of new technologies. There are possible realistic scenarios that lead to the loss of most, or even all, of what we hold valuable. Some of these scenarios are drastic, others are subtle. Although all progress is change, not all change is progress.

4. Research effort needs to be invested into understanding these prospects. We need to carefully deliberate how best to reduce risks and expedite beneficial applications. We also need forums where people can constructively discuss what should be done, and a social order where responsible decisions can be implemented.
5. Reduction of existential risks, and development of means for the preservation of life and health, the alleviation of grave suffering, and the improvement of human foresight and wisdom should be pursued as urgent priorities, and heavily funded.
6. Policy making ought to be guided by responsible and inclusive moral vision, taking seriously both opportunities and risks, respecting autonomy and individual rights, and showing solidarity with and concern for the interests and dignity of all people around the globe. We must also consider our moral responsibilities towards generations that will exist in the future.
7. We advocate the well-being of all sentience, including humans, non-human animals, and any future artificial intellects, modified life forms, or other intelligences to which technological and scientific advance may give rise.
8. We favor allowing individuals wide personal choice over how they enable their lives. This includes use of techniques that may be developed to assist memory, concentration, and mental energy; life extension therapies; reproductive choice technologies; cryonics procedures; and many other possible human modification and enhancement technologies.

Bibliography

1. Asimov, Isaac. ([1950] 1994). *I, Robot*. New York, NY: Bantam Books.
2. Bostrom, Nick. (2005). "A History of Transhumanist Thought." *Journal of Evolution and Technology* - Vol. 14 Issue 1 - April. www.nickbostrom.com/papers/history.pdf
3. British Telecom. (2005). *Technology Timeline*. London, UK: British Telecom. www.bt.com/technologytimeline
4. Chaisson, Eric. (2005). *Epic of Evolution: Seven Ages of the Cosmos*. New York, NY: Columbia University Press.
5. Clarke, Arthur C. (1984). *Profiles of the Future: An Inquiry into the Limits of the Possible*. New York, NY: Henry Holt and Company.
6. Condorcet, Marie-Jean-Antoine-Nicolas de Caritat. ([1795] 1979). *Sketch for a historical picture of the progress of the human mind*. Westport, CT: Greenwood Press.
7. Cordeiro, José Luis. (2010). *Telephones and Economic Development: A Worldwide Long-Term Comparison*. Saarbrücken, Germany: Lambert Academic Publishing.
8. Darwin, Charles. ([1859] 2003), *The Origin of the Species*. New York, NY: Fine Creative Media.
9. Drexler, K. Eric. (1987). *Engines of Creation*. New York, NY: Anchor Books. www.e-drexler.com/d/06/00/EOC/EOC_Cover.html
10. Foundation for the Future. (2002). *The Next Thousand Years*. Bellevue, WA: Foundation for the Future. www.futurefoundation.org/documents/nty_projdesc.pdf
11. Fumento, Michael. (2003). *BioEvolution: How Biotechnology is Changing the World*. San Francisco, CA: Encounter Books.
12. Garreau, Joel. (2005). *Radical Evolution: The Promise and Peril of Enhancing Our Minds, Our Bodies--and What It Means to Be Human*. New York, NY: Doubleday.

13. Glenn, Jerome et al. (2018). *State of the Future 19.1*. Washington, DC: The Millennium Project. www.StateOfTheFuture.org
14. Haldane, John Burdon Sanderson. (1924). *Daedalus; or, Science and the Future*. London, UK: K. Paul, Trench, Trubner & Co.
15. Harari, Yuval Noah. (2017). *Homo Deus: A Brief History of Tomorrow*. New York, NY: Harper.
16. Harari, Yuval Noah. (2015). *Sapiens: A Brief History of Humankind*. New York, NY: Harper.
17. Haraway, Donna. (1991). "A Cyborg Manifesto" in *Simians, Cyborgs and Women: The Reinvention of Nature*. New York, NY: Routledge.
18. Hawking, Stephen. (2002). *The Theory of Everything: The Origin and Fate of the Universe*. New York, NY: New Millennium Press.
19. Hughes, James. (2004). *Citizen Cyborg: Why Democratic Societies Must Respond to the Redesigned Human of the Future*. Cambridge, MA: Westview Press.
20. HumanityPlus (formerly World Transhumanist Association). (2002). *The Transhumanist Declaration*. HumanityPlus. https://humanityplus.org/philosophy/transhumanist-declaration/
21. Huxley, Julian. (1957). "Transhumanism" in *New Bottles for New Wine*. London, UK: Chatto & Windus.
22. Joy, Bill. (2000). "Why The Future Doesn't Need Us." *Wired*. April 2000. www.wired.com/wired/archive/8.04/joy.html
23. Kaku, Michio. (2018). *The Future of Humanity: Terraforming Mars, Interstellar Travel, Immortality, and Our Destiny Beyond Earth*. New York, NY: Doubleday.
24. Kurian, George T. and Molitor, Graham T.T. (1996). *Encyclopedia of the Future*. New York, NY: Macmillan.
25. Kurzweil, Ray. (2005). *The Singularity is Near: When Humans Transcend Biology* New York, NY: Viking. www.singularity.com
26. Kurzweil, Ray. (1999). *The Age of Spiritual Machines*. New York, NY: Penguin Books. www.kurzweilai.net
27. Minsky, Marvin. (1994). "Will robots inherit the Earth?" *Scientific American*, October 1994. www.ai.mit.edu/people/minsky/papers/sciam.inherit.txt
28. Moravec, Hans. (1998). *Robot: Mere Machine to Transcendent Mind*. Oxford, UK: Oxford University Press. www.frc.ri.cmu.edu/~hpm/book97
29. Moravec, Hans. (1988). *Mind Children*. Boston, MA: Harvard University Press.
30. More, Max. (2003). *The Principles of Extropy*. The Extropy Institute. www.extropy.org/principles.htm
31. More, Max & Vita-More, Natasha. (2013). *The Transhumanist Reader: Classical and Contemporary Essays on the Science, Technology, and Philosophy of the Human Future*. New York, NY: Wiley-Blackwell.
32. Mulhall, Douglas. (2002). *Our Molecular Future*. Amherst, New York, NY: Prometheus Books.
33. Paul, Gregory S. and Cox, Earl. (1996). *Beyond Humanity: Cyberevolution and Future Minds*. Hingham, MA: Charles River Media.
34. Pinker, Steven. (2018). *Enlightenment Now: The Case for Reason, Science, Humanism, and Progress*. Viking.
35. Roco, Mihail C. and Bainbridge, William Sims (eds.). (2003). *Converging Technologies for Improving Human Performance*. Dordrecht, Netherlands: Kluwer.
36. Roco, Mihail C. and Montemagno, Carlo D. (eds.). (2004). *The Coevolution of Human Potential and Converging Technologies*. New York, NY: New York Academy of Sciences (Annals of the New York Academy of Sciences, volume 1013).
37. Sagan, Carl. (1977). *The Dragons of Eden: Speculations on the Evolution of Human Intelligence*. New York, NY: Random House.
38. Stock, Gregory. (2002). *Redesigning Humans: Our Inevitable Genetic Future*. New York, NY: Houghton Mifflin Company.
39. Teilhard de Chardin, Pierre. (1964). *The Future of Man*. New York, NY: Harper & Row.
40. Venter, J. Craig. (2008). *A Life Decoded: My Genome: My Life*. New York, NY: Penguin.

41. Vinge, Vernor. (1993). "The Coming Technological Singularity." *Whole Earth Review* Winter issue.
42. Warwick, Kevin. (2003). *I, Cyborg*. London, UK: Garnder's.
43. Wells, H.G. (1902). "The Discovery of the Future." *Nature*, 65: www.geocities.com/yokel-craig/hgwells1.html
44. Zindell, David. (1994). *The Broken God*. New York, NY: Acacia Press.

José Luis Cordeiro (www.cordeiro.org) studied engineering at MIT, economics at Georgetown University, management at INSEAD in France, and science at Universidad Simón Bolívar in Venezuela. He is director of the Millennium Project, director of HumanityPlus, fellow of the World Academy of Art and Science, and former director of the Club of Rome (Venezuela Chapter), the World Transhumanist Association and the Extropy Institute. He has published over 10 books and appeared in TV programs and documentaries with the BBC, CNN, Discovery Channel and History Channel, for example. Email: www.jose@millennium-project.org

Chapter 4
How Transhumanism Will Get Us Through the Third Millennium

Kate Levchuk

4.1 Religion as Transhumanism Predecessor

Several thousand years ago major world religions were invented. As societies moved beyond family and tribal connections and into larger agglomerations they needed some binding force to justify this coexistence and working towards a common goal. Various kinds of ideologies and religions were invented by humans to serve all kinds of purposes in the developing societies.

Apart from a tool to control the masses and make them obey "God's representative" on sinful soil, religions fulfilled several major goals throughout human history:

- Giving God's explanation for physical events which had no reasonable scientific explanation before
- Establishing a reasonable set of values by which people can live by
- Giving people meaning in life and a hope for redemption and continuity after death
- An ability to relay responsibility on your life's faults onto some higher power

Some of these functions have deep psychological essence and some of them have roots in our inherent necessity to satisfy our curiosity and find a sense of belonging.

Apart from power abuse and "God's representatives'" enrichment none of these functions are intrinsically bad or wrong. Nevertheless, with human progress in the domains of natural sciences and cognitive psychology we managed to find much more reasonable answers for those who are willing to search. There are numerous self-help books, not talking about life, that teach us to take responsibility for our actions. Aren't you excessively tired of Internet circulating lists of what mentally

K. Levchuk (✉)
Founder of Transpire, Futurist at KateGoesTech, London, UK

© Springer Nature Switzerland AG 2019
N. Lee (ed.), *The Transhumanism Handbook*,
https://doi.org/10.1007/978-3-030-16920-6_4

strong people do and do not do? However dull and unimpressive those lists are, there is no mentioning of a superior power accompanying mental strength. It all boils down to our human ability to take responsibility for our actions. Alas, one of the major goals of religion is slowly being veered into the dustbin of human history.

The Last Supper, Mural by Leonardo da Vinci, Wikipedia

A major emergent trend of the last century came not from religious teachings but from the Scientific Enlightenment and hard data.

As of 2010, Christianity was by far the world's largest religion, with an estimated 2.2 billion adherents, nearly a third (31 percent) of all 6.9 billion people on Earth," the Pew report says.[1] "Islam was second, with 1.6 billion adherents, or 23 percent of the global population.

An advance in AI and scientific progress in the last 8 years alone might have decreased the numbers dramatically. Still not as much as to liquidate spiritual beliefs of the vast majority of the world population.

So why care? What is so dangerous in the enduring clinging to religious postulates? What can be the problem if it gives you a strong set of values and dims your unlimited human ego?

The problem lies in the proliferation of a rogue belief that there is some better world out there, an everlasting Utopia we will get once this intermittent period on our Planet is over.

While there is no way to prove or disprove this hypothesis the fact that our current habitat is about to turn to Hell in the nearest decade is very much a proven scientific fact. And I see this inaction of millions of religious zealots and their

[1] https://www.pewforum.org/2015/04/02/religious-projections-2010-2050/

delirious hope to get to an imagined "San Junipero"[2] while the rest of us are being burnt alive[3] and sunk in the melting oceans as a really, really big problem.

4.2 What Is Transhumanism?

Moreover, an emerging Tranhumanist movement denies any deception of life after death and demands life and happiness on this very Planet. Now. And forever!

Why put your efforts into a non-productive hope for an afterlife if advances in Artificial Intelligence, biotech and nanotechnology can turn this very life into a never ending amusement park?

What point is there in being able to afford discovering new places and enjoying the sunsets when the rat race of your life has turned you into a flabby and disintegrating piece of carbon?

Numerous generations devoted their efforts and intellectual struggles so that we can finally reach the palpable reality of an extended life span and possible immortality. There are times my heart is about to explode from the indescribably beauty this world has to offer. Sometimes you just wish these moments would last forever. What if you can really make this moment called life last forever?

Transhumanism is not an ideology for poor, hopeless and depressed. Those people can dream of better worlds elsewhere. It is an ideology for people who know what they want from life and how to get it. It is a philosophy for strong, happy and ambitious people who could not care less about what happens to their souls after their bodies turn to ashes. These people want life here and now. And they are not ready to be lured into the mockery of an "afterlife".

Another issue comes when we reflect on our desire to live forever. No matter how much I respect Ray Kurzweil and his call to merge with machines to achieve immortality, I do not see how it solves the main problem.

4.2.1 What Is the Point of Living Forever in a Trash Bin?

The biggest revelation in my childhood was not the way how children are born or why people go to war to die for the artificial constructs fed to them by the media.

It was the realization that we dump all our trash into our very oceans and soils. And that's in the times when we have space ships and can move all of that stuff away from our home, our Planet. I was a bit concerned that we are polluting surrounding space but the realization that we are poisoning our waters and living creatures hit me like a truck.

---

[2] https://en.wikipedia.org/wiki/San_Junipero
[3] https://edition.cnn.com/2018/12/10/us/california-wildfires-climate-weir-wxc/index.html

Don't Go Swimming – Heart-breaking isn't it? This young boy is forced to swim in disgusting polluted ocean water. He spends his days hunting for recyclable plastics to sell and raise a paltry 35 cents per kilo for his family

According to the most prudent estimations, we have got 12 years[4] until we reach a run-off global warming which will drastically alter weather patterns on our Planet, submerge our strategic cities, exterminate most wildlife species and create a world-wide migration crisis.

This is not some kind of a futuristic scenario or a plot for currently so popular Apocalyptic movies. It is a fast approaching reality. And it is also an emergency. It is an impending worldwide catastrophe calling for a global united action!

So why don't we see enterprises frantically shifting to green energy generation, governments immediately banning plastic production and huge investments going into the creation of a food-redistributing AI solution? Why aren't coal and oil enterprises being shut down and government incentives given to support environmentally conscious consumer decisions? How come world's best known democracy is withdrawing from Paris climate agreement, Europe keeps relying on Russian gas and Saudi Arabia keeps replenishing its blood-stained coffers?

---

[4] https://www.theguardian.com/environment/2018/oct/08/global-warming-must-not-exceed-15c-warns-landmark-un-report

Kern River Oil Field

There is only one explanation to this ongoing absurdity. And it is not even game theory. It is our mortality and the acknowledgement of the fact we will die. **People making decisions today will turn to ashes several decades from now.**[5]

Nature bestowed them with unsurpassed analytical and negotiation skills, strong character and the power to overcome adversity. In most cases fate also gave them rich and influential parents and an open ticket to do whatever they please to do. But it also gave them the most precise and irreversible countdown. And they know it all too well.

We have a sarcastic saying in Ukrainian that translates to the following "There can be a flood after we are gone." I think nothing describes the world's complacent attitude better than this tongue-in-cheek Ukrainian adage. There can be a flood; and there will be a flood, but it will no longer be us who will drown…

It is human mortality and its universal acceptance that prevented taking the correct course of action back in the 80s when we knew the consequences of climate change all too well and could have acted on the situation in timely manner.[6]

Truth is, unless you are a Transhumanist you care very little what will happen once our oceans devour our cities,[7] forests start a perpetual burning cycle[8] and melted permafrost releases deadly viruses previously unknown to Homo Sapiens.[9]

---

[5] https://kategoestech.com/2018/10/05/our-mortality-is-killing-this-planet/

[6] https://www.nytimes.com/interactive/2018/08/01/magazine/climate-change-losing-earth.html

[7] https://www.independent.co.uk/environment/climate-change-sea-levels-melt-ice-rise-threat-antarctica-a8111351.html

[8] https://www.theatlantic.com/science/archive/2018/08/why-this-years-wildfires-have-been-so-ferocious/567215/

[9] http://www.bbc.com/earth/story/20170504-there-are-diseases-hidden-in-ice-and-they-are-waking-up

So what has changed since the 80s? Millennials and Generation Z, who increasingly enter the decision making scene, expect to live several years longer than their predecessors as well as enjoy a more productive life in older age. Life expectancy is only part of the answer. More significantly, the ongoing scientific discoveries in machine learning, biotech and nanotechnology inspired Transhumanist hopes for the possibility of immortality in our very lifetime.

Too sad we are approaching the hope of an eternal life just when we drive our Planet to the brink of annihilation. And no matter how much money we pump into our eternal youth and the acquisition of New Zealand's citizenship,[10] the worldwide Planet crisis will get you no matter how far and secure you appear to be.

None of the existing religions is able to provide the answers to the rising waters, bleaching corals, mass extinction of species and deadly fires. Thus we need to embrace a new ideology, which will unite all of the humanity and give the hope and meaning to people who love life and want to preserve this beautiful place called Earth.

Thus, I would like to highlight some of the postulates of the ideology that will save our Planet and give us eternal life in this very time and space dimension. There is only one way and one approach that can sustain us throughout a third millennium and beyond.

4.3   Transhumanist Manifesto

TRANSPIRE

4.3.1   On Gratefulness to Life

Transhumanist is a proud human being who appreciates the fact that <u>she was chosen out of millions of DNA variations,</u> – whether through a natural process of fertilization or in vitro, that gave her a ticket to this life. Thus a Transhumanist uses each opportunity to gratefully enjoy life.

---

[10] https://www.theguardian.com/news/2018/feb/15/why-silicon-valley-billionaires-are-prepping-for-the-apocalypse-in-new-zealand

4.3.2   On Desire to Explore

Transhumanist knows <u>there is only one reality</u> to fulfill all of her aspirations and longings and will do everything in her power to extend her presence in this universe and experience a whole plethora of opportunities given to her by the Planet she inhabits.

4.3.3   On Moral Code

Transhumanist will take every opportunity to <u>enjoy life and explore the world without consciously inflicting suffering on another conscious or sentient creature</u>, as it is below her intellectual and moral level and world view.

4.3.4   On Embedded Reality

Despite our ever accelerating advances in Artificial intelligence and Virtual Reality, a true Transhumanist appreciates the reality she lives in and understands <u>there is no benefit of an eternal life in a disintegrating environment.</u> Thus, a Transhumanist is actively working towards reversing the effects of climate change inflicted by previous negligent and selfish generations. There must be a clear boundary among Digital Escapists who are waiting to upload themselves in silicon matter and gorge on VR worlds and a Transhumanist who may choose to do so but understands the need to restore the world that gave us our unlimited biological potential.

4.3.5   On the Eradication of Suffering

Transhumanist realizes her happiness, albeit a subjective construct, is the most valuable asset in her disposal and will do everything to fulfill it. At the same time, she is kind and helpful to people and other creatures around her as she understands that misery and suffering of others eventually will have an impact on her well-being. A Transhumanist realizes that living in a rich family or wealthy country will never be a panacea as long as people elsewhere are suffering and dying from hunger. She understands that <u>nothing can be contained and in order to ensure her immortality and happiness she should make every effort to eradicate suffering worldwide.</u>

4.3.6   On Beliefs

A Transhumanist does not believe in any religion or nationality as she realizes all these constructs have been artificially created to divide and control weak and misled people. A value of human being is not dictated by the place of birth or amount of money deposited in her bank, but solely by one's aspirations and actions to make this Planet a better place. Thus, a Transhumanist denies any racial, ethnical or gender superiority or discriminations as unworthy and scientifically faulty.

4.3.7   On the Right to Upgrade

A Transhumanist will make every effort to transcend one's biological and mental limitations with the help of available technology. No imagined moral or ethical barriers will limit a Transhumanist desire to improve oneself and ensure happy and healthy future for her children. Any religious, ideological or national interference with an individual's quest for self-improvement will be viewed as an infringement of Transhumanist's rights. As long as one's biohacking does not transgress someone else's happiness or vitality there is no logical explanation or moral warrant for others to constrain the above.

4.4   Ethical Considerations and Challenges of Universal Transhumanism

As any new ideology Transhumanism involves embracing new ideas and values while neglecting or managing previously dominant ideologies.

The constructs of human fabric and civilization have been constructed by numerous religions, family values, ethics and beliefs.

Embracing new models of behavior and changing the order of things dominant for millennials never goes unnoticed. While saving millions from poverty and saving our Planet from destruction, Transhumanism is capable of shattering the very foundations our everyday reality is based upon.

So what are the discussions we should get involved in and what are the consensuses we should strive to achieve before we relinquish a Universal transhumanism as a new maxim of existence?

4.4.1   Biological Considerations

Biological upgrades and biotech developments are making their ways to the newspaper headlines and increasingly into the world supermarket of choice. The price of these developments, however, is still far from universally accessible. While there is no doubt Jeff Bezos would be able to afford any readily available and tested bio-hacking technology, it is a big question whether even 1% of richest people will be able to afford eternal life, if even available. What about the other 99%?

It is worth noting a delicate balance in ecosystem, which we are continuously trying to upend to our own detriment. Just like an extermination of one species will eventually lead to dismal effects on other biological systems, once a specific stratum of population performs a biological uplift, there will be immediate influences on other creatures. Once 1 person out of a hundred upgrades its biological arrangement, the interconnected viruses and aligned micro ecosystems will start adapting to this new upgraded individual. In other words, technology enabled evolution will inadvertently lead to a biological evolution of interconnected species.

Despite obvious benefits of an increased endurance and boosted immune system, once you embark on a journey of bio enhancement it will be hard to impossible to back down. You would constantly need periodic biological upgrades if you are to ensure a consistent performance of your biological operating system. A note to Digital Escapists: the same holds true for the mind uploading techniques and the evolution of computer viruses, more so potentially assisted by the creation of AGI.

In light of the above, what do you do with 99% who are not able to afford an upgrade or opt for a neo Amish lifestyle of non-upgrading? Chances are any they will have to very careful with the proliferation of advanced viruses. Moreover, there is a high probability most of the humanity will be destroyed by yet another pandemic which evolved viruses will aim at an upgraded class of humanity. You remember this annoying post Business Insider republishes as news every few weeks on Bill Gates prediction of an upcoming deadly pandemics? Chances are this will be this very pandemic that will wipe out all the non-mods from the surface of the Planet.

What do we do with the above? Should we limit Transhumanists' desire to improve their lives to ensure stability and ecological balance for the all? Should we adapt an evolutionary socialism?

In all probability, there will be no force capable of stopping life-loving individuals to improve their existence on this Planet if they have all the means and knowledge to do so. However, to prevent mass extinction of non-modified human species and to avoid a possible rebellion of non-mods, Transhumanist uplifts must be carefully thought out and executed in a responsible manner, the main goal of which should be to avoid the deterioration of life standards of any non-modified human beings. Clearly, you do not need to pay for a biological upgrade of other 99 people if you opt to uplift yourself, but a Transhumanist may well consider to pay a premium to protect the well-being of potentially affected populations.

4.4.2   Possibility of Radical Inequality

Compared to previous centuries, we are already experiencing an unprecedented shift in gender roles and behavior perceptions. With people more focused on self-development and career, women tend to delay having children and have much lower birth rates overall. Additionally, it is contributing to leveling out gender inequality in material well-being.

Equality and emancipation are great. However, if you take a second look at it, it is clear that these educated and ambitious men and women are not very likely to pass on their genes. In such a developed region as European Union birth rates have been steadily decreasing during the last century and currently stand at 1.6 children per woman.[11] In Japan it is 1.44. The country is often cited for its technology development as well as for its shocking disinterest in family or sex life among Millennial generation.

On the contrary, poorest and most politically unstable countries tend to have the highest birth rates. Niger is leading birth rate worldwide with an indication of 6.6 children per woman, followed by a whole range of poor African countries. Statistics show that the fertility rate per woman in low income countries is twice as high as the global average.[12]

Back to our discussion of the affordability of biohacking. The gap between world richest and poorest people has been steadily growing throughout the Information Age and will cumulate with the tiny fraction of people controlling the most advanced Artificial Intellect and leaving the rest of world to figure out what they are to do on this Planet.

Clearly, these will be the machine controlling elites which will be able to afford Immortality first. An interesting phenomenon will appear when these super rich and super immortal people will decide children are not a necessary legacy to leave behind and not the way to find a sought after meaning in life. Hey, you are immortal yourself, why care of raising yet another human being?

The ongoing discussion among intellectuals on the negative environmental consequences of having children is already a spreading trend, and apart from numerous talks I attended in recent years there is a whole plethora of articles, like in Independent,[13] The Guardian[14] and the like, highlighting the trend.

While there can be no generalization here and human psychology (especially in a hypothetical immortality scenario) is very hard to predict, the possibility of Immortal people giving up on procreation might well leave us with a tiny elite of "eternal judges" and an ever expanding and ever impoverishing population in the developing world.

---

[11] https://www.theguardian.com/world/2015/aug/23/baby-crisis-europe-brink-depopulation-disaster

[12] https://www.economist.com/middle-east-and-africa/2018/09/22/africas-high-birth-rate-is-keeping-the-continent-poor

[13] https://www.independent.co.uk/environment/children-carbon-footprint-climate-change-damage-having-kids-research-a7837961.html

[14] https://www.theguardian.com/world/2018/jun/20/give-up-having-children-couples-save-planet-climate-crisis

The Paraisópolis favela (Paradise City shantytown) borders the affluent district of Morumbi in São Paulo, Brazil. (Foto: Tuca Vieira)

The consequences of such inequality are easy to predict, − a potential rebellion and mass killings of either side of the spectrum, − a perfect scenario for a prisoners' dilemma[15] Apocalyptic movie.

Thus, tremendous efforts should be devoted to the worldwide increase in living standards and the efforts to level out demographics and well-being in different parts of the Planet.

4.4.3   Ethical and Pragmatic Considerations of Immortality

Once getting into the discussion of immortality you often end up with arguments of what would happen if people financially successful in current Capitalist order live forever and the likes of Rupert Murdoch[16] or Kim Kardashian[17] keep defining the world agenda for millenniums to come, while someone with the qualities of compassion, empathy and caring ends up in the confinements of Death.

This is a very realistic scenario if we move from the theory of Transhumanism right into the pragmatic access to Immortality. It is worth noting that there is a very high probability of the change of human mentality given a ticket to an eternal life. By the reasons of logic your consumerism, selfishness and narcissist attitude to the life would gradually change into a genuine caring about the world you would inhabit

---

[15] https://en.wikipedia.org/wiki/Prisoner%27s_dilemma

[16] https://en.wikipedia.org/wiki/Rupert_Murdoch

[17] https://www.instagram.com/kimkardashian/?hl=en

for the coming millennia, if we are talking about an absolute Immortality, which is still far from possible. Even in the prospect of a guaranteed and considerable life extension you would most probably start caring about favorable ecological conditions that will define your life in 50 or 100 years from now.

Nevertheless, this prospect if far from guaranteed. Thus, money might not be and should not be an exhaustive measure by which Immortality is to be granted. Very possible that people opting for Immortality or life extension will have to undergo protracted training in sustainability. Most radical version of pragmatic Transhumanism will involve specific and accurate tests which could prove an individual suitability and worth to inhabit this reality for a significant period of time.

There is no point extending the life span of people who would continue their pursuit of Planet destruction. An ultimate end of such a materialistic Transhumanism realization will be a destroyed Planet which will be uninhabitable for both biologically immortal humans and newborn generations.

This is an extremely controversial topic which has the elements of intellectual eugenics and moral judgement reminiscent of religious teachings Transhumanism is striving to differentiate itself from. The criterion for Immortality will be the single most important decision we will have to make once bio- and nanotechnological advancements leave the domain of theory and turn up as viable alternatives to biological degradation.

4.4.4 Discussions over Artificial General Intelligence, Universal Mind and Dataism

Very often Transhumanism is equated with a blissful mix of AGI and human intellect, mind uploading and an eternal digital life. In such a scenario we would have an instant access to the Universal Mind[18] of everything ever written and invented and anyone ever uploaded to the net. The Universal Mind is the concept we came up with in a joint article on the 3rd Information Revolution,[19] coauthored with Jacek Utko.[20]

While it is very hard to impossible to create an eternal engine, the ease and opportunity of fetching one's consciousness from one server to another has inspired numerous Transhumanist threads and teachings. In fact, one can almost differentiate between digital immortality believers and biohacking supporters.

Thanks to compute capacities we have made a huge leap in what is currently coined as Artificial Intelligence but what really is a combination of super powerful human invented and mathematics-based algorithms. Nevertheless, this leap has been sufficient to start talking about the possibility of the creation of an "artificial soul".[21]

What is it that makes a soul? I'm afraid this is what we haven't figured just yet and what is probably the most valid and worthy explanation of the world religions.

---

[18] https://www.facebook.com/watch/?v=1239079202885558

[19] https://kategoestech.com/2017/07/04/approaching-the-3rd-information-revolution/

[20] https://www.ted.com/talks/jacek_utko_asks_can_design_save_the_newspaper/discussion

[21] https://www.forbes.com/sites/cognitiveworld/2018/08/05/ai-vs-god-who-stays-and-who-leaves/#2a98baf32713

The parallel processing nature of our brain has been developed by millions of years of evolution, and a computer's ability to use the same wire thousands of times per second might very well be enough to approach human "thinking mechanism."

Or will it not? Will the hard wiring be enough to recreate a sacred human spirit?

Irrespective whether AGI ever comes around or not the ability to upload bits of memory and cognition into a machine is a likely and unrelated opportunity. Questions we need to ask to the *Digital Escapists* are:

- Is this upload a verified representation of your own self or just a rogue "pirated" version?
- Will you be happy with a "copy paste" version of yourself if "cut and paste" proves to be impossible? In essence the question is the same as whether you would be happy for your clone to take over and start from scratch once your body material is no longer usable?
- How do you ensure you do not forfeit your "bird in the hand" physical reality for a digital immortality easily subject to Internet viruses, malicious software and massive digital suicides?
- Is there a point in living (eternal living more so) if you cannot experience the breeze of wind, the smell of grass, the lips of your loved ones and the autumn melancholy?

As a Transhumanist I would give anything to enjoy this life one more time, to see more than I can in a limited life span, to feel a whole plethora of human emotions and experience a whole range of physical desires. But I would not give a dime to be locked in a world wide web, no matter how much knowledge and data I would be able to access.

Image from the The Blue Lagoon (https://www.imdb.com/title/tt0080453/), *1980, depicting Brooke Shields and Christopher Atkins on a tropical island in the South Pacific*

I believe anyone who shares this philosophy would be especially keen on keeping this Planet going for as long as eternity itself.

To summarize the examples of my formulation of Transhumanism I would like to give several examples of the projects, seemingly not directly related to Immortality, but which are indispensable for Transhumanism as an ideology of *"sustainable immortality"* and universal happiness.

4.5   Examples of Concrete Transhumanist Projects

- Introduce additional 40% tax on Gasoline and channel this into a switch to renewable energy sources
- Create an initiative aimed at the adoption of dying Yemeni children into Western families
- Oblige each manufacturing corporation to recycle an equivalent of its produce
- Subsidize and enforce a switch from coal chimneys to solar panels on every household
- Channel 50% of the United Nations R&D into the research on corals restoration and faster reforestation
- Create OSF-style national philanthropic organizations in each developed nation to fund talented and ambitious individuals in the developing world[22]

As you can notice, none of these project directly relates to the creation of the philosopher's stone or an elixir of an eternal life. Rather they bring us a step closer to the creation of the world a Transhumanist would want to live in.

4.6   Conclusion

How much would you pay for an illusory hope of being brought back to life after you experience clinical death and your brain is being frozen by a cryonics institution? A whole-body preservation at Alcor costs $200,000.00.[23] Same organization charges $ 80,000.00 for Neurocryopreservation. Is it too much or too little for a hope to be given a second chance? Hard to say.

Is it too much of a sacrifice to do everything in our power to ensure an inhabitable Planet for ourselves and future generations? Is it an unrealizable goal to live forever in the world full of green trees, non-extinct animals and breathable air?

Transhumanism is a dream, which totally justifies putting all your eggs into one basket, a basket called Earth.

To follow the development of this interpretation of Transhumanism as well as to come up and raise money for their own projects I invite the readers to follow Transpire.me,[24] where I strive to create a platform for like-minded life-loving Transhumanists.

---

[22] https://www.opensocietyfoundations.org/

[23] https://alcor.org/BecomeMember/scheduleA.html

[24] https://transpire.me/

Chapter 5
The United States Transhumanist Party and the Politics of Abundance

Gennady Stolyarov II

The depredations of contemporary politics and the majority of our era's societal problems stem from the scarcity of material resources and time. However, numerous emerging technologies on the horizon promise to dramatically lift the present-day constraints of scarcity. The United States Transhumanist Party, in advocating the accelerated development of these technologies and seeking to influence public opinion to embrace them, is forging a new political paradigm rooted in abundance, rather than scarcity. This new approach is simultaneously more ambitious and more civil than the status quo. Here I illustrate the distinguishing features of the Transhumanist Party's mode of operation, achievements, and plans for the future.

5.1 Scarcity Versus Abundance

Today's politics are shaped by scarcity and competing special-interest groups. The animosity we observed in the 2016 United States elections is ultimately driven by a zero-sum mentality, where many believe that others must lose for them to win. Political partisans demonize one another, precipitating confrontations that can escalate to the point of violence. Firebrands on the "left" and on the "right" are eager to label any different perspective or original way of thinking – or even any refusal to take up the firebrand's particular causes, candidates, or verbal expressions – as a heresy that needs to be rooted out by screaming or by force. Media emerge to cater to and reinforce the filter-bubbles and echo-chambers that partisans of particular pre-conceived notions have constructed to insulate themselves from dissenting views. Pressure groups, representing concentrated economic interests, posture for public support and lobby for special financial favors – be they in the form of subsidies, preferential tax treatments, lucrative government contracts, or barriers to entry

G. Stolyarov II (✉)
United States Transhumanist Party, Carson City, NV, USA

© Springer Nature Switzerland AG 2019
N. Lee (ed.), *The Transhumanism Handbook*,
https://doi.org/10.1007/978-3-030-16920-6_5

for new, smaller-scale, more enterprising competitors. Negative campaigning from established political parties predominates over the search for constructive policy solutions to vexing societal issues. To persuade undecided voters, cynical campaign strategists focus not on what *their* side would be able to improve or protect – but rather on how heinously unacceptable their *opponents* would be. From the outrageous cries that Hillary Clinton should be "locked up" over charges for which she had been formally exonerated, to unsubstantiated insinuations that Donald Trump was acting as the agent of a foreign power from whose demonization a vast network of special interests has benefited for decades, the 2016 United States Presidential election was the nadir of political discourse in recent U.S. history. Contemporaneously, the toxic dynamics of nationalist, populist, reactionary politics have engulfed much of Europe, with varying outcomes. In the lands where the Scientific Revolution and the Age of Enlightenment first arose in the 17th and 18th centuries, the zero-sum politics of today threaten to undo the momentous achievements that generations of past thinkers have painstakingly wrought.

The zero-sum mentality is an atavistic remnant of humankind's evolutionary origins. Our remote hunter-gatherer ancestors were faced with harsh environmental conditions: there was only so much food and shelter available, with scant means to create more. Those who appropriated a larger portion of berries or caught game for themselves necessarily deprived their fellow tribespeople of these goods. Small tribes fought vigorously and mercilessly over territory and the meager spoils of unaltered nature. Tens of thousands of years later, while our technological capabilities and material resources have expanded dramatically, the evolution of the human mind has not kept pace. In spite of the massive expansion of productive capacity brought about by the industrial progress of the past three centuries, material scarcity remains palpable, and each of us perceives limits to what we can attain, enjoy, and achieve. As soon as our life circumstances brush up against the boundaries of present-day scarcity, the age-old demon of the zero-sum mentality resurfaces and pressures many humans to re-enact the colossally costly carnage of the hunter-gatherer struggle for resources. Today it takes great fortitude and erudition to resist this counterproductive way of thinking. Studying economics, history, and moral philosophy certainly helps achieve a recognition of the *positive-sum* potential all around us – the creation of value through transforming the raw stuff of nature and through collaboration and exchange with other sentient beings. However, the insights of the erudite few cannot by themselves hold back the roiling tides of contemporary politics, driven by the passions of the many, who remain by default in thrall to the zero-sum mentality. To depart from this suboptimal state, we need a fundamental transformation of the material constraints and incentives for action surrounding *all* of us – or at least the vast majority.

The philosophy of transhumanism offers just such a transformation. The name of this philosophy literally means to go *beyond* the human – in the sense of overcoming today's fundamental constraints on the human condition: the constraints of material scarcity, disease, decay, death, and zero-sum conflict. The term "transhumanism" was first used in this sense by Julian Huxley in 1957.[1] During the

---

[1] Huxley, Julian. "Transhumanism". 1957. Available at http://web.archive.org/web/20160625132722/http://www.transhumanism.org/index.php/WTA/more/huxley

1990s, Max More and the Extropian movement outlined a systematic transhumanist philosophy.[2] Transhumanism is the continuation of the Enlightenment humanist project of using reason and science to improve the human condition – except that transhumanism recognizes the potential of existing and soon-to-be-developed technologies to remove from that condition certain limits which historically were considered to be immutable. As those limits are pushed ever outward, progress could accelerate with ever fewer obstacles in its way, and all problems would become solvable with the appropriate tools and effort. As the intellectual successor to Enlightenment humanism, transhumanism offers the greatest promise for preserving the civilizing Enlightenment project against the resurgence of the atavistic tribal mentalities that currently threaten to displace it.

Transhumanism promises a future of widespread abundance that overcomes today's major sources of scarcity – particularly the scarcity of time resulting from today's woefully short lifespans. Through the progress of biotechnology, nanotechnology, and computing, the next generation of medical treatments may not only cure today's most intractable diseases but also repair age-related damage to the body, setting back one's biological clock and enabling one to survive until the *next* generation of still-more-effective treatments – thereby achieving *longevity escape velocity,* where life expectancy increases faster than the passage of time. Improvements in artificial-intelligence (AI) algorithms could result in the creation of more effective *domain-specific* or *narrow* AI that could solve challenging engineering, scientific, and logistical problems, as well as artificial *general* intelligence that might, in a versatile and open-ended manner, develop solutions to problems in a variety of fields. Semi-autonomous, electric vehicles such as the Tesla Model S and Model 3 are already the safest ever developed and have saved many lives on the roadways.[3,4] Given that the overwhelming majority of vehicle accidents are due to human error, the transition to fully autonomous vehicles would save tens of thousands of lives per year in the United States and millions of lives per year in the entire world.[5] Advances in economical solar energy, safe nuclear power (for instance, through the thorium fuel cycle), geothermal power, and alternative fuels can bring forth an era of cheap, abundant energy with minimal negative externalities. Through virtual and augmented reality, human creativity could flourish with fewer material constraints. People could build prototypes, devise new amenities and luxuries, and meet one another in fully immersive virtual environments that reduce the need for stressful and potentially dangerous physical travel. By more widespread automation of production, humans would be freed from the burdens and dangers of manual labor,

---

[2] More, Max. "Transhumanism: Towards a Futurist Philosophy". 1990. Available at https://web.archive.org/web/20051029125153/http://www.maxmore.com/transhum.htm

[3] Lavrinc, Damon. "The Tesla Model S Is So Safe It Broke the Crash-Testing Gear". WIRED. August 20, 2013. Available at https://www.wired.com/2013/08/tesla-model-s-crash-test/

[4] Lambert, Fred. "Tesla Model 3 achieves lowest probability of injury of any vehicle ever tested by NHTSA". Electrek. October 7, 2018. Available at https://electrek.co/2018/10/07/tesla-model-3-lowest-probability-of-injury-nhtsa/

[5] National Highway Traffic Safety Administration. "Critical Reasons for Crashes Investigated in the National Motor Vehicle Crash Causation Survey". February 2015. Available at https://crash-stats.nhtsa.dot.gov/Api/Public/ViewPublication/812115

and every person would be able to focus the majority of time on non-repetitive creative contributions that would flow uniquely from each individual mind.

The confluence of these technological advances promises to massively improve both the length and the quality of life for everyone who chooses to take advantage of them. If these emerging technologies are allowed to actualize their full potential, we can all become immensely wealthier, healthier, and happier. As a beneficial side effect, people who are more prosperous, more comfortable, more fulfilled in their lives, are less likely to react with vicious hostility toward others. Even if the biological human mind remains in roughly its present form (though even this, given the promise of genetic engineering in the coming decades, is not a foregone conclusion), the pushing outward of the material constraints of life would greatly lower the probability that humans would revert to their primeval conflict-prone ways – since they would be far less likely to approach the scarcity-imposed boundaries which trigger zero-sum responses.

The United States Transhumanist Party explicitly seeks to achieve a world where the futuristic technologies of radical abundance become present-day realities for as many people as possible. Ultimately, we aim to transform politics into what it always should have been: a constructive focus on which policies are best for improving human well-being and solving the problems that confront us. Emerging technologies are capable of bringing such a constructive politics about during our (hopefully indefinitely prolonged) lifetimes. We recognize, however, that progress in any endeavor involving human beings is never inevitable. The technologies we advocate for are products of human effort and creativity. Just as certain societal and political environments – namely, the attitudes and institutions derived from the Age of Enlightenment – can catalyze the development of such technologies, so can certain other circumstances and policies derail or greatly retard progress. For the vast majority of human history, progress across generations and entire centuries was imperceptibly slow, and occasional cataclysms – barbarian invasions, plagues, cultural decay – led to active *retrogression* from which societies took additional centuries to recover. As such, the United States Transhumanist Party recognizes that we cannot rely on some imagined forward march of history. To achieve the world we desire, we need to actively create it and support others who share our goals.

5.2 The United States Transhumanist Party – Highlights of Achievements

The United States Transhumanist Party was founded by Zoltan Istvan on October 7, 2014. Istvan, who served as the Transhumanist Party's first Chairman, was also its first Presidential candidate during the 2016 election. When announcing his campaign, Istvan outlined his objective to "Create a cultural mindset in America that embracing and producing radical technology and science is in the best interest of our nation and species" while also seeking to empower scientists to overcome aging and death within the next two decades and create global safeguards against potential planet-wide perils.[6] Istvan's campaign primarily focused on attracting unprecedented media exposure for the ideas of transhumanism. Due to restrictive and often cost-prohibitive state ballot-access laws, Istvan "chose to bypass the battle to get on state ballots and instead focus [on] using media to move the transhumanism movement ahead", noting that "the internet is making a run for the presidency a good way to get attention for a cause like transhumanism" and that his "main goal all along has been to tell the world that science and technological innovation [are] coming far more quickly than ever before, and as a nation, we must answer to [them] with practical and forward-thinking policies."[7]

Istvan's coffin-shaped Immortality Bus traversed the country in 2015, starkly reminding Americans of the unfortunate predicament of human mortality and the imperative to overcome it. In a culmination of the Immortality Bus tour, Istvan delivered his original version of the Transhumanist Bill of Rights to the steps of the U.S. Capitol Building in Washington, D.C. As part of Istvan's vision for the politics of the future, "the Transhumanist Bill of Rights seeks to declare that all Americans (and people of all nationalities, as well) in the 21st Century deserve a 'universal right' to live indefinitely and eliminate involuntary suffering through science and technology."[8]

Ultimately, Istvan came to the conclusion that contemporary constraints on alternative political parties in the United States have created a stultifying environment for those seeking to achieve genuine progress. In Istvan's view, "Politics and minor third parties are a great way to push burgeoning movements like transhumanism forward" but doing so in service of radically improving humankind's future requires a revolutionary approach: "the Transhumanist Party was a political vehicle mostly designed for a singular purpose: to create a social environment that facilitates expediently conquering human death using science and technology. Such a purpose is to aim for a near[-]total revolution in the human experience."[9] Due to high barriers to entry into the conventional electoral process, Istvan saw the need to structure the Transhumanist Party not as an official entity registered with the Federal Election Commission, but rather as an information-spreading, awareness-raising activist organization whose primary goal is not to win elections or even appear on ballots, but rather to highlight the key issues that politicians and the general public will need to confront as humankind enters its next stage of technological advancement.

---

[6] Istvan, Zoltan. "Should a Transhumanist Run for President?" Huffington Post. October 8, 2014. Available at http://www.huffingtonpost.com/zoltan-istvan/should-a-transhumanist-be_b_5949688.html

[7] Istvan, Zoltan. "What I Learned by Running for President." Motherboard. October 28, 2016. Available at https://motherboard.vice.com/en_us/article/what-i-learned-by-running-for-president

[8] Istvan, Zoltan. "Immortality Bus Delivers Newly Created Transhumanist Bill of Rights to the US Capitol." Huffington Post. December 21, 2015. Available at http://www.huffingtonpost.com/zoltan-istvan/immortality-bus-delivers-_b_8849450.html

[9] Istvan, Zoltan. "Revolutionary Politics Are Necessary for Transhumanism to Succeed." Motherboard. November 3, 2016. Available at https://motherboard.vice.com/en_us/article/revolutionary-politics-are-necessary-for-transhumanism-to-succeed

After concluding his Presidential campaign, Istvan decided to relinquish his Chairman role and requested that I become the second Chairman in the history of the United States Transhumanist Party. On November 17, 2016, in his message formalizing the transfer of leadership, Istvan wrote that he "would like to see the party grow larger through more democratic measures and the impact of new leaders."[10] Under Istvan, the Transhumanist Party was primarily operated by means of Istvan's direct, personal efforts and did not have an official membership. My tenure as Chairman takes place during a transitional period where the principal goals are to attract a membership representative of the perspectives within the transhumanist and life-extensionist movements and to gradually create structures that would transform the Transhumanist Party into a member-driven organization.

My first act as Chairman was opening the Transhumanist Party for members to join. By filling out a simple membership application form, an individual can take part in our deliberations and vote on the policy stances that the Transhumanist Party will take.[11] The only precondition for being eligible to vote in U.S. Transhumanist Party internal elections is the expression of agreement with the following three Core Ideals of the Transhumanist Party, which were drafted to be as inclusive as possible of perspectives which could be considered broadly transhumanist:

Ideal 1. The Transhumanist Party supports significant life extension achieved through the progress of science and technology.

Ideal 2. The Transhumanist Party supports a cultural, societal, and political atmosphere informed and animated by reason, science, and secular values.

Ideal 3. The Transhumanist Party supports efforts to use science, technology, and rational discourse to reduce and eliminate various existential risks to the human species.

In the spirit of a truly revolutionary and forward-thinking political entity, the United States Transhumanist Party prides itself on an unprecedented openness and cosmopolitanism in its membership criteria. To join the Transhumanist Party, a person is not required to be a citizen of the United States or eligible to vote in U.S. elections. Non-U.S. individuals are capable of attaining Allied Member status, which entitles them to vote in the *internal* elections of the U.S. Transhumanist Party. Furthermore, membership is not limited based on arbitrary criteria such as biological age or even the species of the member. Any individual capable of forming and expressing a political opinion is capable of joining, including children and teenagers who are able to hold views on political issues. Indeed, the United States Transhumanist Party Platform, in Article III, Section XXIII, of the U.S. Transhumanist Party Constitution, states:

> The United States Transhumanist Party supports the rights of children to exercise liberty in proportion to their rational faculties and capacity for autonomous judgment. In particular,

---

[10] Istvan, Zoltan. Post of November 17, 2016. Google +. Available at https://plus.google.com/101457828255104035246/posts/iGTnL2tZT8s

[11] Transhumanist Party Membership Application Form. Available at http://transhumanist-party.org/membership/ and https://goo.gl/forms/IpUjooEZjnfOFUMi2

the United States Transhumanist Party strongly opposes all forms of bullying, child abuse, and censorship of intellectual self-development by children and teenagers.

Likewise, although we are not aware of any such entities today, if humans should ever come into contact with reasoning non-human beings – be they extraterrestrial life forms, artificial general intelligences, or uplifted animals – such beings would be eligible for membership in the United States Transhumanist Party. By being open to membership by those who would be excluded by traditional political parties, the United States Transhumanist Party not only implements the cosmopolitan ideal; it also aims to defuse political tensions that might arise in the future if historically excluded beings begin to demand civil rights and political representation. Human societies should not resist such claims to rights by entities with reasoning abilities. Rather, those entities should be welcomed into peaceful political processes and other forms of mutually beneficial cooperation, so as to prevent a needlessly adversarial mode of interaction between them and today's voting population of adult human citizens. Above all, the Transhumanist Party desires a peaceful societal evolution driven by technological progress, such that neither the reactionaries among humans nor any new sentient entities who might be slighted or threatened by subordinate treatment would have any reason to resort to violence or develop any significant capacity for carrying out violent action against groups they consider to be "the other".

Ranked-preference voting, already implemented by the United States Transhumanist Party in its internal votes, represents a further practical improvement over conventional political structures. All of our voting is conducted electronically, with human verification of each vote to ensure that only registered members vote and that no duplicate votes arise. Each matter being decided upon is exposed to the public for at least 15 days, prior to a seven-day electronic voting period during which all members with valid e-mail addresses can cast ballots at their convenience. Ranked-preference voting enables individuals to express more than a single preference on matters where the choice is not binary – for instance, multiple alternative wordings of a platform plank or, in the future, multiple candidates in a primary election. Each voter is able to rank-order the entire spectrum of options, including an option for "None of the above". Ranked-preference voting eliminates the incentives for strategic voting, since it alleviates the pressure to wholly support an option which the voter does not hold in high regard, but which is slightly "less bad" than an option which the voter specifically does not wish to win. Instead, the voter could indicate his true first preference, while still ranking the "less bad" option above the option he considers the least desirable. If no option attains a true *majority* (not a mere plurality) of first-preference votes on the first round, then the option that receives the *fewest* votes is eliminated, and its votes reassigned to the second-highest preferences of the voters who favored that option. This instant-runoff process continues for as many rounds as are required for any option to achieve a majority of reassigned votes. If ranked-preference voting were implemented in U.S. elections, many voters would no longer experience either internal or external pressure to sacrifice their genuine preferences to support a "lesser evil" that differs only in slight, symbolic, or rhetorical ways from the "greater evil" those voters seek to prevent (and about whose identity opinions among voters differ – such that one person's "greater evil" is another's "lesser evil").

The first vote of the United States Transhumanist Party was held between December 25 and December 31, 2016, and led to the adoption of a considerably expanded version of the Transhumanist Bill of Rights.[12] Version 2.0 of this aspirational statement of rights for all sentient entities serves both conceptual and practical purposes. The main conceptual purpose is to define as inclusively as reasonably possible the sentient entities that are deserving of rights and to outline the rights that a future society, animated by radical technological progress, should be able to protect and facilitate. The main practical purpose is to begin formulating a blueprint for attaining such a future society of expanded rights, which enables us to start laying the path that, over the course of several decades, will take us to such a society – hopefully in an environment characterized by peace, incremental progress, and at least gradual acceptance of increased diversity among sentient entities.

It is important to view the rights expressed in the Transhumanist Bill of Rights in the context of the possibilities that a future of radical technological progress would be able to bring about. For instance, a system of universal healthcare might seem today to entail large monetary costs and the need for governments and large private institutions to set up centralized clinics, records, and elaborate systems of financing that involve the entire population. Many skeptics will, with justification, question the practical and moral validity of asserting a "right" for individuals today to benefit from such a system. In the future, however, given the rise of personalized medicine, biotechnology, nanotechnology, and exponentially decreasing costs of production, medical care using versatile, personalized devices and easily administered treatments might become as affordable and ubiquitous as food is in the United States today – and the administration of preventive rejuvenation treatments may become far more economical than both the direct and opportunity costs of allowing severe illnesses to reach a crisis point. Thus, the universal healthcare of the future may be comparable in the ease of its achievement to a "universal" access to food during our era, which almost all people in the Western world can achieve through a trip to a grocery store, to a supermarket, or to a variety of charitable organizations that exist to provide food to those with limited financial resources.

Media coverage of the Transhumanist Bill of Rights escalated in August 2018, when WIRED Magazine republished Version 2.0 in full.[13] Unfortunately, some of the ensuing commentary in certain other media outlets amplified some basic misrepresentations of the factual content of the Transhumanist Bill of Rights. Commentators such as Michael Cook of the bioconservative website BioEdge and Jasper Hammill of The Metro erroneously asserted that the Transhumanist Bill of Rights, in expressing the right of ending involuntary suffering in Article IV, was

---

[12] "Transhumanist Bill of Rights – Version 2.0." United States Transhumanist Party. Available at http://transhumanist-party.org/tbr-2/

[13] "The Transhumanist Bill of Rights Version 2.0." WIRED Magazine. August 21, 2018. Available at https://www.wired.com/beyond-the-beyond/2018/08/transhumanist-bill-rights-version-2-0/

referring to euthanasia, when no such reference was stated or implied; rather, the right to end involuntary suffering refers to the idea that suffering itself should be abolished for still-living entities who desire this, as expressed in David Pearce's philosophy of abolitionism.[14] Furthermore, these commentators erred in stating that the right to universal healthcare, as specified in Article VII, presupposed a monopolistic of single-payer system of healthcare, such as the United Kingdom's National Health Service. However, the means of attaining universal healthcare was left deliberately open-ended in the Transhumanist Bill of Rights. Many of the planks in the U.S. Transhumanist Party Platform, described in greater detail below, advocate for significant free-market elements in healthcare systems. Universal healthcare could mean, for instance, that all services become so inexpensive and automated that everyone would be able to readily afford them. However, different members of the U.S. Transhumanist Party would advocate different systems of healthcare delivery. The Transhumanist Bill of Rights focuses on outcomes, rather than prescribing the specific delivery system – and hence it was determined to be desirable to clarify Article VII to ensure that the pursuit of universal healthcare can remain open-ended and potentially be arrived at through a variety of means, including those not yet conceived of, while allowing discussion and debate to continue within the transhumanist community about whether private or governmental means, or a combination thereof, would be most effective in achieving radical life extension and universal access to healthcare in the most expeditious timeframe possible.

The desire for clarification in response to media misinterpretations, as well as additional enhancements – such as directly integrating the relevant provisions from the United Nations Universal Declaration of Human Rights[15] into the text of the Transhumanist Bill of Rights (with appropriate modification to reflect the broader applicability of these rights to sentient entities irrespective of their manner of origin) – led to the adoption of Version 3.0 of the Transhumanist Bill of Rights following a 15-day exposure and discussion period and a 7-day voting period that occurred during December 2–9, 2018.[16] The complete Transhumanist Bill of Rights, Version 3.0, can be found in Appendix I of this chapter.

In 2017 the emphasis of the Transhumanist Party's voting shifted to the development of a platform which aims to outline stances on as many key issues of our era and of the emerging future as possible – informed by our understanding of the potential of emerging technologies to alter existing political and societal constraints and provide solutions to seemingly intractable contemporary dilemmas. During 2017, six platform votes were successfully conducted, leading to the adoption of 82 distinct planks whose wording was originated by our members and constitutes an

---

[14] "The Hedonistic Imperative – The End of Suffering." David Pearce and Duarte Baltazar. April 15, 2018. Available at https://transhumanist-party.org/2018/04/15/hedonistic-imperative-end-of-suffering/

[15] "Universal Declaration of Human Rights". United Nations. Available at http://www.un.org/en/universal-declaration-human-rights/

[16] "Transhumanist Bill of Rights – Version 3.0." United States Transhumanist Party. Available at https://transhumanist-party.org/tbr-3/

integration of their perspectives. The U.S. Transhumanist Party Platform is found in Article III of our Constitution and will continue to evolve and expand as additional member-generated suggestions are made.[17] Thus far the U.S. Transhumanist Party has taken positions on a broad array of issues, including individual privacy (Section I), sousveillance (Sections XXXIV and LII), morphological freedom and bodily autonomy (Sections VI, XLIX, L, and LXVII), opposition to bigotry (Sections II, XL, and LXIII), support for nuclear disarmament (Sections IV and LXVI), support of research into emerging technologies (Section V) and liberty of scientific and technological innovation (Section VIII), facilitation of and removal of restrictions on potentially life-extending medical research and procedures (Sections LI, LXXVII, LXXIX, LXXX, LXXXI, and LXXXII), support of space colonization (Section XVII), support of a universal and unconditional basic income (Section XVI), an end to the war on drugs (Section XIV), support for freedom of speech (Section XX), recognition of micronations and seasteading efforts (Section XXII), improvements of educational systems (Section XII), children's rights (Sections XXIII and LXII), animal welfare (Section XXIV), rights of sentient artificial intelligences (Section XXXIII), support for civil liberties, police accountability, and voting rights (Sections LVII, LVIII, LIX, LX, LXXIV, and LXXVIII), support for an international passport (Section LXXII), prevention of existential risks through responsible development of protective technologies (Section XXXII), reduction of the national debt (Section XXXV), tax reform (Section XXXVI), reduction of military spending (Section LXXV), support for emerging energy sources but an opposition to subsidies (Section XXXVIII), support for a "Transhumanist Olympics" (Section XXXIX), opposition to protectionism (Section XLVIII), reforms for achieving more representative, properly limited, transparent, and accountable governments and institutions that influence such governments (Sections XXXVII, XLI, XLII, XLIII, XLIV, XLV, XLVI, XLVII, LIII, LIV, LV, LVI, LXI, LXIV, LXV, LXVIII, LXIX, LXXI, and LXXVI) and a wide spectrum of political reforms designed to lower barriers to participation for "third" political parties and original thinkers who offer substantive alternatives to the two-party establishment (Sections XIII, XIX, XXVII, XXVIII, XXIX, XXX, and LXX). The complete Constitution of the U.S. Transhumanist Party as of December 31, 2018, can be found in Appendix II of this chapter.

The U.S. Transhumanist Party can have a significant influence on public opinion long before it establishes an infrastructure for regular electoral participation. In recognition of this, many of our events focus on discussion regarding emerging technologies and outreach to the general public. We also engage in outreach to the media and efforts to shift the discussion on technology-related legislation and regulations in a more techno-positive direction. The U.S. Transhumanist Party's expert discussion panels aim to attract leading thinkers in key areas of emerging technologies and facilitate their engagement with the public through answering questions of wide-

---

[17] "Constitution of the United States Transhumanist Party. Article III. Platform of the Transhumanist Party." United States Transhumanist Party. Available at http://transhumanist-party.org/constitution/#Article3

spread interest regarding their fields, as well as indicating promising future directions for their endeavors. The expert discussion panels are streamed live, and video recordings are available to the public, free of charge and in perpetuity.

On January 8, 2017, the U.S. Transhumanist Party held its Discussion Panel on Artificial Intelligence, featuring Zak Field, David J. Kelley, Hiroyuki Toyama, Mark Waser, and Demian Zivkovic.[18] This was followed by the Discussion Panel on Life Extension, held on February 18, 2017, featuring Bill Andrews, Aubrey de Grey, Ira Pastor, and Ilia Stambler.[19] On November 18, 2017, the U.S. Transhumanist Party hosted its Discussion Panel on Art and Transhumanism, co-moderated by me and Director of Visual Art Emanuel Iral, featuring Rachel Lyn Edler, John Marlowe, Leah Montalto, Kim Bodenhamer Smith, R. Nicholas Starr, Ekaterinya Vladinakova, and Laura Katrin Weston.[20] On February 18, 2018, the U.S. Transhumanist Party collaborated with the Institute of Exponential Sciences to host the Discussion Panel on Cryptocurrencies, co-moderated by me and Demian Zivkovic, and featuring guest panelists Chantha Lueung, Laurens Wes, and Moritz Bierling.[21]

Each discussion panel provided over 2 hours of cutting-edge content regarding the fields of research and creative endeavor that stand poised to revolutionize the human condition – *if* accompanied by appropriate societal and political openness to their progress. In the role of a moderator for all four panels, I had the distinct honor of soliciting the experts' advice on what societal and political improvements and approaches could accelerate the timeframe within which emerging technologies would be able to bring us significant benefits, as well as how contemporary attitudes and perceptions could be transformed through a combination of techno-positive artistic, scientific, and philosophical pursuits. Below I summarize some of the key insights that participants in these four panels provided.

- **Bill Andrews:** The perception of the anti-aging industry needs to change, which depends on the people leading that industry. Unfortunately, the field of anti-aging has long had more quacks and charlatans than any other field, and so it is impera-

---

[18] The video recording of the Discussion Panel on Artificial Intelligence can be found at https://www.youtube.com/watch?v=Y54UtBgFK-w. Biographical descriptions of the panel participants, as well as an outline of some of the key questions they addressed, can be found at http://transhumanist-party.org/2017/01/08/ustp-ai-discussion-panel/

[19] The video recording of the Discussion Panel on Life Extension can be found at https://www.youtube.com/watch?v=1HYB_o37SYc. The audio recording can be found at http://rationalargumentator.com/USTP_Life_Extension_Panel.mp3. Biographical descriptions of the panel participants, as well as an outline of some of the key questions they addressed and the references they provided, can be found at http://transhumanist-party.org/2017/02/18/ustp-le-discussion-panel/

[20] The video recording of the Discussion Panel on Art and Transhumanism can be found at https://www.youtube.com/watch?v=tLtYUTkbOdU. Biographical descriptions of the panel participants and the chat log from the panel can be found at http://transhumanist-party.org/2017/11/19/ustp-art-panel/

[21] The video recording of the Discussion Panel on Cryptocurrencies can be found at https://www.youtube.com/watch?v=FiWjzfbJO-Y. Biographical descriptions of the panel participants and the chat log from the panel can be found at https://transhumanist-party.org/2018/02/18/ustp-crypto-panel/

tive to overcome this perception by finding treatments that work. This is a "Catch 22" situation; in order to come up with a treatment that works, we need public perception on our side to generate the funding that can achieve research break-throughs. We need to find a way to break this standstill. Anti-aging researchers are continuing to move as fast as they can, given the limited funding available, and they hope that this will be enough to achieve sufficiently significant advances to get the general public to view this field as credible and worthwhile. Alleviating any of the major causes of aging could be the catalyst for a shift in public percep-tion, attracting enough support to enable us to reach longevity escape velocity.

- **Moritz Bierling:** We currently have an understanding of money that is condi-tioned by the fact that money has most often been state-managed; we tend to view it as universal and measure and evaluate everything else in terms of money – but it is possible to view money in a more nuanced way. Each currency is, in fact, targeted toward specific goals, and if you do not know what those goals are, somebody else does, and that person is the beneficiary of that monetary system at your expense. Right now governments do have certain interests in mind when they manage monetary systems; for instance, they engage in inflation at the expense of the citizens and transfer value from the population to themselves. More generally, however, currencies should have a goal in mind – for instance, achieving specific objectives for a community. However, the desirability of a cur-rency will depend on who designs the goals and who accepts the design; curren-cies based on consent are superior to those backed by force.
- **Rachel Lyn Edler:** Graphic designers and other artists have a unique opportu-nity to enhance communication and visual perception of the world. Advancement of science and technology requires a diverse range of skills to improve and enhance humanity. We need not only scientists but communicators of science to spread the message of transhumanism and how humankind can evolve. We need to have more positive representations of the future in movies, books, and other art forms. It is part of the job of creative people to bring this about. Designing the Immortality Bus was an excellent opportunity to inspire people to become more interested in science, technology, and the potential to stop biological aging. There was a wide range of reactions to the Immortality Bus, ranging from shock to inspiration, and extensive media coverage was generated.
- **Zak Field:** The creation process for artificial intelligence can itself help push new ideas into circulation within the general population. As artificial general intelligence (AGI) advances, it will be important for the public to recognize that AGI can be more than a series of algorithms within a humanoid shell. Artificial intelligence can relieve a lot of stress and alleviate existing socioeconomic issues. AI can serve as an aid and provide reassurance that humans are taking appropri-ate actions, but humans should not take any first glance at information, including information provided by AI, as an absolute. Rather, humans should walk forward hand-in-hand with AI to discover more possibilities, excite the general populace, and motivate creativity.
- **Emanuel Iral:** Art should confront people with important ideas about the future and enable us to have conversations about concepts and possibilities that we

normally avoid. Art should enable humans to collaborate as a species toward improving the future. Nietzsche's portrayal of the individual as an entity of will and action can be adapted to apply to and improve the entire species and generate momentum toward developing technology to achieve major breakthroughs, such as reaching Mars and overcoming diseases. Art enables humans to become radical agents for the longevity of our species. We should always strive to seek knowledge and to have a healthy pride in our ability to pursue and obtain it. Humans are by nature severely flawed and can be arrogant, but we can turn that arrogance around and put it to good use. Having pride as the human species is important, and the ideal of transhumanism is a beacon that enables us to improve, set aside old flaws, and progress together so that everyone's suffering can be alleviated. The vision of eliminating suffering and achieving indefinite life is not a difficult ideal to embrace, but an easy one – and it is now within our grasp.

- **Aubrey de Grey:** Because of the difficulties for minor political parties to access the ballot box, the major focus of the Transhumanist Party should be to shape public opinion in the direction of embracing rejuvenation biotechnology. The Transhumanist Party should have different goals in the short term versus the long term. In the short term, the challenge is to overcome the prevalent negative connotations of the word "transhumanism". Language is more powerful than it ought to be. However, we can fix this issue, much like the word "rejuvenation" has been rehabilitated over the past decade and has come to mean the actual reversal of biological age through repair of damage. In terms of what ordinary people can do, shortage of funding is by far the number one barrier to the rate of progress. While not everyone is in a position to contribute directly, everyone is in a position to contribute indirectly – including by spreading ideas and raising the quality of debate. The less wealthy you are, the more people you know who are wealthier than you, who could provide the funds to support serious anti-aging medicine. It is important to learn the ability to stop people from changing the subject when discussing the importance of life-extension research.
- **David J. Kelley:** There is a lot of fear-mongering and hype that people should endeavor to avoid. It is important to dispel the over-hyping and oppose overly restrictive laws that would needlessly obstruct the potential for developing artificial intelligence. The unjustified fear of AI could itself generate the kinds of sociological problems that AI critics fear would arise. We need to make rational decisions instead of emphasizing speculative disasters that are not going to happen. Developing artificial general intelligence is the most effective path for achieving stable, long-term intelligence, including outside of the Earth.
- **Chantha Lueung:** Currently we live in a centralized economy under centralized authority. Going forward into the future, what we are seeing now in the crypto space is that a lot of things are being decentralized. This is exciting, because we will have an economy of choice rather than being pigeonholed into a single economy in which one has to participate; there is a potential to decentralize everything, including crucial processes such as food production. This is both possible and efficient, and opens up many doors for people. Hopefully we will see many more people involved in these technologies.

- **John Marlowe:** The potential to be afflicted by disease currently unites all of us, and the messaging of a lot of contemporary science fiction in portraying disease unfortunately does a disservice by discouraging efforts to address the problems of ill health. Right now our culture does not adequately emphasize the importance of medical research, and there are only occasional individual and philanthropic efforts in this area. We need to be proactive in calling out irresponsible portrayals and encourage art that motivates us to come up with solutions where there were none before. Improving funding of scientific and medical research, including for organizations such as the National Institutes of Health in the United States, is crucial for helping alleviate suffering. We also need to advocate the rights of individuals to experiment for the improvement of their health, while also stressing caution and vigilance about downstream effects of attempted modifications. It is important for those who self-experiment to be well-informed, experienced, knowledgeable about the risks, and willing to accept the consequences.
- **Leah Montalto:** We need to overcome the myth that art lacks any real effect. Scientific research is increasingly uncovering how interactions with art have profound effects on the human brain. It is radically more difficult to create art that is optimistic than it is to create dystopian art, because optimistic art does not have the same instantaneous "hooks" for people's attention that horrific, depressing, and dark portrayals can provide. It is worthwhile to consider how art, music, and media are affecting the quality of our attention, thoughts, and emotions, and how this in turn affects decision-making processes and the political and cultural environment.
- **Ira Pastor:** It is important to stay educated and knowledgeable about science and make science and technology publicly appealing. We are a lot closer to major world-transforming breakthroughs than we have ever been before and than many might understand, based on disappointed expectations generated by prior predictions. It is a matter of focusing on the end-game. We need to get public interest and excitement in motion again, outside the sphere of those who can afford to make major investments, and enable the public to be aware of the possibility of a beautiful future soon.
- **Kim Bodenhamer Smith:** Technology can greatly improve quality of life and enable humans to have far more time for creative activities. Through technology, creativity will become dramatically easier and will integrate into everyday life. Even this very discussion panel is an excellent illustration of the power of technology to connect people from different parts of the country and the world, enabling conversations that would not have happened otherwise. It is important to design a coherent package for articulating transhumanist ideas to the world and telling stories to which people could relate, so as to humanize and personalize the radical political agendas of transhumanism. Zoltan Istvan was extremely effective at this with his Immortality Bus tour – essentially using an art car and touring the country to convey transhumanist ideas and begin a conversation through which people can learn more about one another and collaborate.

- **Ilia Stambler:** We need people to study, advocate, and get involved with like-minded others. Possibilities for involvement with both online and local communities are growing exponentially. The issue of life-extension advocacy needs to become political, because the funding is mostly held in the hands of politicians. The politics of life extension can be right-wing or left-wing, but we need to put this major issue of our time on the political agenda, as this issue is about our survival. History teaches us that nobody learns anything from history, so every time we unfortunately have to reinvent the wheel. In many prior eras of history, some people expected the cure for aging to be imminent – but at least now we may have a decent chance to achieve this goal, and it is important to grab this chance.
- **R. Nicholas Starr:** Art inspires us to take the next technological steps. Breakthrough ideas may originate in art first and then become developed by scientists, and this has been a tendency since the time of antiquity. Art can motivate action and instill a variety of emotions – both in a direct manner, as with much popular culture, and subtly, as with audio tracks to films. Visual art and music can be processed in meaningful ways by the mind "behind the scenes", which has an impact on attitudes and actions. Today's technologies enable artists to create flexibly – to go where the inspiration is – which has historically been difficult to do. From an artistic point of view, the DIY (do-it-yourself) movement can be seen as biological art or performance art, combining scientific pursuits with creative ways to bring the conversation about science and technology to the world.
- **Hiroyuki Toyama:** Artificial intelligence can provide a variety of benefits in commerce, education, and health. For instance, AI can enrich human well-being and health by helping with stress management (including the overcoming of information overload), cognitive and behavioral assistance, and development of extraordinary physical reactions to external events and stressors. However, people should always keep in mind the difference between human intentions and how AI will actually function. AI does not have human instincts such as self-preservation or procreation. It is important not to project human biases and motivations onto AI.
- **Ekaterinya Vladinakova:** Art is far more than a luxury. Design is also art, and functional design can convey a beautiful vision of the future – as seen, for example, in the Tesla automobiles and the designs for the hyperloop. Art is everywhere – in the design of buildings and vehicles; it is also a way of advertising technologies and attracting people's attention to certain topics – including the technologies of the future. In America there is a problem with terminally ill people having access to drugs, but the right-to-try movement is making important gains in addressing this issue and improving the freedom of patients to pursue last-ditch efforts to extend life through experimental treatments.
- **Mark Waser:** Creating artificial intelligence is an awesome opportunity to learn about ourselves, improve ourselves, and improve our society. As we learn more, we gain great power, and we need to take some responsibility for our goals and assess what actions will lead to what results. The future can be absolutely wonderful, or it can be terrible. We need to work to make sure it turns out for the best.

Humans are extremely vulnerable to all sorts of hype, including about emerging technologies. We need to reach a condition where we argue facts, value science, and value discourse rather than talking past one another. This issue is broader than artificial intelligence; it is an issue of how humanity in general will move forward. We need to develop tools and programs to enable us to debate more effectively, keep track of debates, and determine who is able to accurately predict the future and summarize events. Efforts to create AGI can also enable the creation of these kinds of tools.

- **Laurens Wes:** Bitcoin may have a large market share now as the first cryptocurrency, but it may also have largely had its time, as investors are shifting toward smart contracts and technologies that enable them, such as Ethereum. Recently the cost of mining and energy consumption have become problems for Bitcoin and other large cryptocurrencies, and people have been developing other ways to mine cryptocurrencies. The proof-of-stake concept in cryptocurrencies may have a more viable future than proof-of-work – but it is by no means a final solution; indeed, the entire cryptocurrency field is still in its infancy, where no one yet has an ultimate solution. Meanwhile, artificial intelligence is each year becoming more capable at tasks that previously only humans were able to do. More recent applications of AI are becoming a bit more humanlike – for instance, by imitating the movements of biological organisms. AI systems can become significant players in blockchain technologies and capital markets in the coming years. Over time the division between AI and humans will become a gray area, and we will no longer consider AI systems to be mere machines and algorithms.

- **Laura Katrin Weston:** Artists are lucky to have a vast array of technological tools to enable people to connect to one another. Until the problem of biological human mortality is resolved, art is the closest to immortality that humans will get, and so it should be used to highlight the issues that humankind should tackle. It is important, however, to emphasize caution and the need for improved education when it comes to self-experimentation, since much about the workings of the human body remains unknown even to medical professionals; still, people need to have the freedom to make decisions about health for themselves. We need to consider ways to improve current medical systems to make sure that people who are suffering from serious illnesses have more choices than just the experimental and self-medicating route.

- **Demian Zivkovic:** There are significant intersections between life extension and artificial intelligence – for instance, the use of AI to discover new drugs and cut the costs of pharmacological research. There are many opportunities to improve corporate processes that could increase the rate at which research is done and attract investment. However, one of the greatest sources of harm in engendering a flawed and overly pessimistic view of emerging technologies is the entertainment industry, which attempts to sell dystopian visions of the future, which are easier to sensationalize than hopeful but nuanced visions of successful societies where all significant problems have been solved and indefinite life extension has been achieved. We need to overcome this fear-mongering.

5.3   Achievements in 2018

While the U.S. Transhumanist Party's activities in 2017 focused strongly on the development of its Platform and internal infrastructure, 2018 was characterized by significant membership growth (with the U.S. Transhumanist Party more than doubling its membership count during its second year from 550 to 1140), formation of additional State-level Transhumanist Parties (including a highly active California Transhumanist Party,[22] chaired by U.S. Transhumanist Party Education and Media Advisor Newton Lee, as well as incipient Transhumanist Parties in New York and Michigan, and continuation of the online presences of Transhumanist Parties in Colorado, Illinois, Kentucky, New Hampshire, and Texas), as well as presentations, interviews, and meetings in a variety of venues.

On September 13, 2018, I presented virtually to the Vanguard Scientific Instruments in Management 2018 (VSIM:18) Conference in Ravda, Bulgaria, on the subject of "How Transhumanism Can Transcend Socialism, Libertarianism, and All Other Conventional Ideologies" – which described the key strengths and weaknesses of libertarianism, socialism, conservatism, and left-liberalism, the common failings of these and all other conventional ideologies, and why transhumanism offers a principled, integrated, dynamic approach for a new era of history, which can overcome all of these failings.[23]

On September 21, 2018, I spoke at the RAAD Fest 2018 conference in San Diego, California, on the four-year anniversary of the U.S. Transhumanist Party and its distinguishing aspects and achievements.[24] Following this speech I interviewed renowned futurist Ray Kurzweil on stage and engaged in a multifaceted discussion with him on subjects ranging from data privacy to the impacts of artificial intelligence to the potential to adopt personalized health regimens to overcome serious illnesses.[25] During the next day, on September 22, 2018, the U.S. Transhumanist Party held an in-person meeting in San Diego, where Chairman Gennady Stolyarov II, Director of Marketing Arin Vahanian, and Education and Media Advisor and California Transhumanist Party Chairman Newton Lee fielded inquiries from attendees (including several leading transhumanist public figures), provided input regarding future initiatives, and considered suggestions from members and the pub-

---

[22] The California Transhumanist Party website can be found http://www.californiatranshumanist-party.org/index.html

[23] Stolyarov II, Gennady. "How Transhumanism Can Transcend Socialism, Libertarianism, and All Other Conventional Ideologies". U.S. Transhumanist Party Website. September 13, 2018. Available at https://transhumanist-party.org/2018/09/16/transhumanism-transcend-ideologies/. The video recording of this presentation can be found at https://www.youtube.com/watch?v=5AmAGMXvSbI

[24] Stolyarov II, Gennady. "The U.S. Transhumanist Party – Four Years of Advocating for the Future". U.S. Transhumanist Party Website. September 21, 2018. Available at https://transhumanist-party.org/2018/10/26/ustp-4-years/. The video recording of this speech can be found at https://www.youtube.com/watch?v=kykROyu_xNc

[25] A video excerpt from my September 21, 2018, interview with Ray Kurzweil is available at https://www.youtube.com/watch?v=PIr-Pm_5mbM

lic regarding future activities that may be beneficial for growing the transhumanist movement.[26]

Multiple successful interviews in 2018 delved into the philosophical, political, and societal landscapes which the transhumanist movement is navigating in this era. On March 31, 2018, Nikola Danaylov of the Singularity.FM program – also known as Socrates and famous for his in-depth interviews of leading future-oriented thinkers – interviewed me for nearly 3 hours, which broke the record for the length of Mr. Danaylov's conversations. The interview covered the efforts and aspirations of the U.S. Transhumanist Party, and also delved into such subjects as the definition of transhumanism, intelligence and morality, the technological Singularity or Singularities, and health and fitness.[27]

On the last day of the RAAD Fest 2018 conference, September 23, 2018, Andrés Grases, the publisher of the Transhuman Plus website[28] – a vast archive of information on the transhumanist movement – interviewed me regarding both the contemporary state of transhumanist politics and its future directions.[29] We addressed the challenges to reforming the educational system, the need to create open access to academic works, the manner in which the transition toward the next era of technologies will occur, the meaning of transhumanism, and its applications in the proximate future – including promising advances that we can expect to see during the next several years.

On October 5, 2018, I was a guest on Ryan O'Shea's Future Grind podcast, where one of the most in-depth discussions of the contemporary state of Transhumanist politics took place.[30] Mr. O'Shea presented me with some of the most detailed and well-researched questions I have encountered regarding current political issues in the U.S. Transhumanist Party and transhumanist movement.

---

[26] "U.S. Transhumanist Party Meeting at RAAD Fest 2018". U.S. Transhumanist Party Website. September 22, 2018. Available at https://transhumanist-party.org/2018/10/02/ustp-meeting-raad-fest-2018/. The video recording of the meeting can be found at https://www.youtube.com/watch?v=EWYoSzj3WZk

[27] Stolyarov II, Gennady, and Danaylov, Nikola. "U.S. Transhumanist Chairman Gennady Stolyarov II Interviewed by Nikola Danaylov of Singularity.FM". U.S. Transhumanist Party Website. March 31, 2018. Available at http://transhumanist-party.org/2018/04/06/stolyarov-singularity-interview/. The video recording of this interview can be found at https://www.youtube.com/watch?v=MzYGmArriI4. The U.S. Transhumanist Party would like to thank its Director of Admissions and Public Relations, Dinorah Delfin, for the outreach that enabled this interview to happen.

[28] The Transhuman Plus website can be accessed at http://transhumanplus.com/

[29] Stolyarov II, Gennady, and Grases, Andrés. "Andrés Grases Interviews U.S. Transhumanist Party Chairman Gennady Stolyarov II on Transhumanism and the Transition to the Next Technological Era". U.S. Transhumanist Party Website. September 23, 2018. Available at http://transhumanist-party.org/2018/09/29/grases-interviews-stolyarov/. The video recording of this interview can be found at https://www.youtube.com/watch?v=Z_VjFfImxC0

[30] Stolyarov II, Gennady, and O'Shea, Ryan. "Future Grind Episode 28: Ryan O'Shea Interviews Gennady Stolyarov II on the State of Transhumanist Politics". U.S. Transhumanist Party Website. October 5, 2018. Available at http://transhumanist-party.org/2018/10/10/future-grind-ep-28/. The video recording of this interview can be found at https://www.youtube.com/watch?v=nLdC3my1QiQ

On October 28, 2018, I was interviewed by Lev Polyakov and Jules Hamilton of the channel "Lev and Jules Break the Rules" – for whose "Sowing Discourse" podcast series I was honored to be the first guest. This interview explored broad questions related to technology, transhumanism, culture, economics, politics, philosophy, art, and even connections to popular films and computer games.[31]

In addition to public appearances, from September 2017 through October 2018, I hosted five Enlightenment Salons – interdisciplinary gatherings to exchange knowledge and expertise for the improvement of the human condition, in the spirit of the Age of Enlightenment. The gatherings include both formal and informal discussion segments; the formal discussion segments are recorded, and the video recordings are published online.[32] The purpose of the revived Enlightenment Salons is to apply, within the contemporary world, the approach of the thinkers of the Age of Enlightenment – to synthesize the insights from various disciplines and inspire progress to be made in improving the human condition. Therefore, guests with widely varying areas of interest and expertise are welcome to join in these conversations.

5.4 Endorsed Candidates in 2018

The United States Transhumanist Party endorsed two candidates for office during the 2018 elections. All endorsements of candidates, like the adoption of Platform planks or Articles of the Transhumanist Bill of Rights, are accomplished through electronic votes of the members.

On April 12, 2018, the U.S. Transhumanist Party endorsed the candidacy of James D. Schultz for New York State Assembly District 2.[33] Mr. Schultz's positions

---

[31] Stolyarov II, Gennady, Hamilton, Jules, and Polyakov, Lev. "Gennady Stolyarov II Interviewed on "Lev and Jules Break the Rules" – Sowing Discourse, Episode #001". U.S. Transhumanist Party Website. November 6, 2018. Available at http://transhumanist-party.org/2018/11/07/gsii-sowing-discourse/. The video recording of this interview can be found at https://www.youtube.com/watch?v=cfJsgw5zyRI

[32] Video recordings from the Enlightenment Salons are found here:

- First Enlightenment Salon (September 17, 2017): https://www.youtube.com/watch?v=i11W90ZuGrk
- Second Enlightenment Salon (November 11, 2017): https://www.youtube.com/watch?v=Y6-muwId6ao
- Third Enlightenment Salon (May 27, 2018): https://www.youtube.com/watch?v=swP9nPj-2kk
- Fourth Enlightenment Salon (July 8, 2018): Part 1: https://www.youtube.com/watch?v=CEoNEbSmAbg, Part 2: https://www.youtube.com/watch?v=N8TMJ_uKoJM, and Part 3: https://www.youtube.com/watch?v=SsoEtCUZQZ8.
- Fifth Enlightenment Salon (October 13, 2018): https://www.youtube.com/watch?v=ejHXvBvTEck

[33] "The U.S. Transhumanist Party Endorses James D. Schultz for New York State Assembly District 2". U.S. Transhumanist Party Website. April 12, 2018. Available at https://transhumanist-party.org/2018/04/12/ustp-endorses-schultz/

as a candidate encompassed certain key stances taken by the U.S. Transhumanist Party – such as legalization of cannabis, mandatory police body cameras, net neutrality, and, most importantly, reduced ballot-access requirements for candidates and political parties.[34] However, despite the best intentions, Mr. Schultz's campaign fell slightly short of the onerous ballot-access threshold established by the two major political parties in New York State. Mr. Schultz's petition effort garnered 1239 signatures by hand, but this was not sufficient to meet the 1500-signature threshold established as the minimum in New York for ballot access for the State Assembly positions. Accordingly, on August 21, 2018, Mr. Schultz announced the discontinuation of his campaign.[35] Mr. Schultz's diligent effort does, however, demonstrate that transhumanism can attract supporters in the four-figure range with diligent advocacy. The challenge for future candidates seeking to align themselves with the U.S. Transhumanist Party will be how to leverage that appeal to overcome the barriers that the onerous ballot-access thresholds have established to the political participation of thoughtful individuals who seek to offer alternatives to the two-party duopoly. Americans need to have a genuine choice of considering innovative, creative voices that can offer true progress. The example of Mr. Schultz's campaign underscores why the U.S. Transhumanist Party strongly supports reducing or eliminating ballot-access thresholds wherever and to whatever extent possible.

After James D. Schultz ended his campaign, the U.S. Transhumanist Party nonetheless took steps to ensure that it would have an endorsed candidate on the ballot in the 2018 general election. I had qualified for ballot access in Nevada to run for one of the positions on the Board of Trustees of the Indian Hills General Improvement District (IHGID). The key message of my campaign was essentially transhumanist but framed in a manner that would appeal to mainstream audiences and bring new constituencies into the transhumanist movement: "Through reason, technology, and respect for property rights, we can live well and improve."[36] When considering endorsing my candidacy, the U.S. Transhumanist Party published a detailed analysis of how my specific campaign messages found significant parallels in the U.S. Transhumanist Party Platform, Core Ideals, and Transhumanist Bill of Rights, Version 2.0.[37] Ultimately, on September 10, 2018, my candidacy was unanimously

---

[34] "U.S. Transhumanist Party Vote on the Question of Endorsing Candidate James D. Schultz for the New York State Assembly District 2". U.S. Transhumanist Party Website. April 3, 2018. Available at http://transhumanist-party.org/2018/04/03/ustp-schultz-vote/

[35] "U.S. Transhumanist Party Candidate James D. Schultz Ends His Campaign for New York State Assembly District 2". August 21, 2018. Available at https://transhumanist-party.org/2018/08/22/schultz-ends-campaign/

[36] "Gennady Stolyarov II for the Board of Trustees of the Indian Hills General Improvement District". Available at http://rationalargumentator.com/stolyarov-for-ihgid.html

[37] "U.S. Transhumanist Party Vote on the Question of Endorsing Gennady Stolyarov II for the Indian Hills General Improvement District Board of Trustees". Available at https://transhumanist-party.org/2018/09/02/ustp-stolyarov-vote/

endorsed by those[38] U.S. Transhumanist Party members who voted on this matter. Outreach to the residents of the IHGID was performed through a combination of online and in-person techniques, including live and electronic correspondence with residents, candidate walks in the IHGID-managed parks (for which I used the social network Nextdoor.com to announce the walks and invite residents to join me), a campaign website, and my participation in the IHGID-hosted "Meet the Candidates Night" – of which a video recording has been published and displayed, along with the video appearances of other candidates, on the IHGID website.[39] Furthermore, The Record-Courier, a local newspaper, published the profiles of all of the candidates, including my own, which contributed to public awareness of my candidacy.[40]

In the election for the IHGID Board of Trustees, I ultimately obtained 520 votes out of 2024 residents who cast their ballots. While I did not win a seat on the Board, 25.7 percent – more than a quarter – of the voters cast affirmative ballots in my favor. As I commented in my Chairman's Second Anniversary Message,

> While I would have preferred to win, this outcome still shows that my campaign – on which I spent no money but rather utilized social media, in-person appearances in public places, videos, and word of mouth – enabled me to reach more than a quarter of the residents after beginning with essentially zero name recognition in the area. Transhumanism, when articulated in a mainstream-friendly manner, can elicit support from people across the political spectrum and in all walks of life. We just need to continue to spread our message with determination and deliberate regarding ways of reaching constituencies who might not have become aware of transhumanism yet – perhaps because our methods of communication have not yet overlapped with their preferred media and social circles.[41]

Indeed, it appears that the major challenge for the spread of transhumanism is how to achieve basic awareness within the general public of transhumanist ideas and the very emerging technologies that are poised to dramatically reshape the human condition. I see my 2018 campaign as being a microcosm of what the transhumanist project faces more broadly when it is being articulated and promoted. I further observed the following in my Chairman's Second Anniversary Message:

> My campaign, based on all indications, dominated on the Internet and social media – yet there are many residents of the District who do not appear to use the Internet or social media

---

[38] "U.S. Transhumanist Party Unanimously Endorses Gennady Stolyarov II for the Board of Trustees of the Indian Hills General Improvement District". September 11, 2018. Available at https://transhumanist-party.org/2018/09/11/ustp-stolyarov-endorsement/

[39] "Gennady Stolyarov II Presents at the 'Meet the Candidates' Night of the Indian Hills General Improvement District". October 23, 2018. Available at http://transhumanist-party.org/2018/10/23/stolyarov-ihgid/

[40] "Indian Hills General Improvement District". The Record-Courier. October 18, 2018. Available at https://www.recordcourier.com/news/indian-hills-general-improvement-district/

[41] Gennady Stolyarov II. "U.S. Transhumanist Party Chairman's Second Anniversary Message". November 18, 2018. Available at http://transhumanist-party.org/2018/11/18/ustp-chairman-2nd-anniversary/

to any great extent. All of my interactions with residents who knew of my campaign have been extremely positive, but I posit that there exists a large demographic whom my efforts did not reach because there was not any online medium to even facilitate an in-person inter- action (e.g., they did not see my announcements on Nextdoor.com and did not watch the candidate videos; also, their in-person activities do not overlap with mine).

Transhumanists tend generally to follow emerging technologies closely and be more open to contemplating and adopting transformative technologies than many other segments of the population, who may wait to adopt a technology until it becomes ubiquitous in everyday life, and some of whom may even delay adopting various technologies – including computers and the Internet – well past the advent of their ubiquity. Yet these constituents need to be reached as well, and transhuman- ists should deliberate about and refine approaches to communicate ideas about the technological future toward which these potential constituents, too, would gravitate. This is, of course, an ongoing challenge, for which there are no simple solutions, but the U.S. Transhumanist Party encourages its members to become active in public outreach and experiment with various combinations of constructive persuasion and advocacy techniques.

5.5   The United States Transhumanist Party – Future Goals

My major goal as Chairman during the transitional period of the U.S. Transhumanist Party is to create a true member-driven organization whose continued existence does not depend on the exertions of one individual or a small group of people. Rather, as long as interest in transhumanist ideas per- sists – and it should only increase as the impact of emerging technologies on every- day life becomes more salient – the Transhumanist Party should remain a prominent presence in public discourse and policy deliberations. The Transhumanist Party should serve as a vehicle to enable any person interested in constructively advocat- ing for the adoption of emerging technologies to make a positive difference in their realization.

During the transitional period the U.S. Transhumanist Party aims to achieve major membership growth, facilitating a pool of talented individuals who could rise to leadership positions within the Party structure and who could stand as candidates in local, state, and federal elections. Our free, flexible, Internet-based membership structure liberates people from the constraints of time and place; they can contribute from any location in the world, and only the merits of their contributions and their desire to be involved will determine the influence they as individuals have on the Transhumanist Party's future course.

While the Transhumanist Party seeks to incorporate aspects of democratic decision-making and build a governance structure representative of perspectives within the broader transhumanist and life-extensionist communities, we also recog- nize the necessity of maintaining flexibility in any democratic governance structure.

Democratic decision-making should be combined with respect for individual initiative under the framework of the U.S. Transhumanist Party. Voting is useful when differences of perspective exist that cannot be reconciled by other means – such as efforts at consensus or the ability of individuals to pursue multiple compatible projects in parallel, even if those projects may stem from different ideological motivations. Some matters, such as platform planks where differences of opinion are possible, are best decided in a democratic manner. Other matters – such as hosting a specific event, writing an article or research paper, participating in a rally, or delivering a speech – are best left to the discretion of individual members who may then choose to affiliate such activities with the Transhumanist Party.

It is also important for any effective governance structure to avoid excessive bureaucracy and location-bound decision-making. The former tends to stifle the kind of initiative that generates member-driven projects fueled by passion for contributing to a worthwhile cause. The latter tends to create a clique of people "in the know" – who control the levers of decision-making by virtue of their proximity to a geographical center of power and to one another. Neither of these traditional obstacles to progress should exist in a future-oriented political party aimed at the technologically facilitated liberation of human creative faculties. My hope is that the Transhumanist Party will never have mandatory location-bound meetings that are only accessible to people who choose to spend hundreds or thousands of dollars to travel to an expensive hotel in a large American city. Rather, all events where decisions are made should remain electronically accessible and open to remote input and participation by members. Live events where decisions are not made – for instance, outreach events where some members communicate transhumanist ideas to the general public – should be documented through any electronic media that are practicable given the event, including video and audio recordings, digital photographs, and published online accounts of the events that would give other members an understanding of what was done to spread the impact of the Transhumanist Party's vision of the future. Furthermore, future governance structures of the Transhumanist Party should recognize the impossibility of centrally planning progress. It is neither feasible nor desirable to establish comprehensive policies and procedures that could anticipate and accommodate every worthwhile initiative. Instead, the attempt to foresee and plan all activities in advance generally only forecloses on worthwhile opportunities and spontaneous suggestions that could not have been conceived prior to the specific circumstances that gave rise to them. The Transhumanist Party should always remain flexible and open to unusual but potentially effective suggestions for advancing a future of technological progress and radical abundance.

In addition to direct discussions of emerging technologies, the Transhumanist Party should continue to advocate major electoral reforms to reduce the power of the two-party duopoly. The internal Transhumanist Party governance and decision-making structures – including electronic ranked-preference voting and the location-independent approach to membership – should be used to illustrate proofs of concept regarding how the larger U.S. political system could be improved. Broader electoral reforms advocated by the U.S. Transhumanist Party, as contained in Article III of our Constitution, include the following:

- Increased involvement of intelligent laypersons in the political process to counter the influence of special interests and their paid representatives; greater use of electronic and other technologies that can inform and empower intelligent laypersons to monitor and contribute to political discussions and decisions. (Section XIII)
- An end to the two-party political system in the United States and a substantially greater inclusion of "third parties" in the political process through mechanisms such as proportional representation and the elimination of stringent ballot-access requirements. (Section XIX)
- Limits on the influence of lobbying by politically connected special interests, while increasing the influence of advocacy by intelligent laypersons. (Section XIX)
- Constitutional reform to abolish the Electoral College in the United States Presidential elections and render the plurality of the popular vote the sole criterion for the election of President. (Section XXVII)
- Greatly shortening the timeframe for electoral campaigns to counteract the "horse race" mentality and prevent voters from forgetting key information due to short memories. Election seasons for public office should be as short as possible, to enable all relevant information to be disseminated quickly and be considered by most voters within the same timeframe as their decisions are made. (Section XXVIII)
- Abolishing all staggered party primaries so that all primary elections are held on the same day across the entire country. With staggered party primaries, individuals voting later – solely because of the jurisdiction in which they reside – find their choices severely constrained due to the prior elimination of candidates whom they might have preferred. (Section XXIX)
- Replacing the current "winner-take-all" electoral system with proportional representation, ranked-preference voting, and other devices to minimize the temptations by voters to favor a perceived "lesser evil" rather than the candidates closest to those voters' own preferences. (Section XXX)

While ballot access is not the highest priority in the short term due to the extreme stringency of ballot-access laws in many states (the least onerous of which require thousands of petition signatures in order for a political party to even place candidates on the ballot), the U.S. Transhumanist Party aims to provide support for State-level Transhumanist Parties to develop and eventually conduct initiatives to obtain ballot access. To encourage grassroots formation of State-level Transhumanist Parties, the U.S. Transhumanist Party will respect the initiative and organizational autonomy of those who undertake the effort to form them. Article I, Section III, Operating Principle 1, of our Constitution allows State-level Transhumanist Parties to determine their internal bylaws, platforms, and activities. While the U.S. Transhumanist Party aims to collaborate with and support State-level Transhumanist Parties, it will not impose involuntary constraints on State-level Transhumanist Parties that operate peacefully within the boundaries of applicable law.

It is furthermore possible for the U.S. Transhumanist Party, through votes of its members, to endorse independent candidates and even candidates for nonpartisan office, as long as those candidates are not running on behalf of any other political party. Moreover, as of 2018, 24 States and Washington D.C. allow an independent candidate who qualifies to run for office to use a "political party designation" which can be printed next to that candidate's name on the ballot, even if the political party to which the designation refers has not qualified for ballot access in that State.[42] The U.S. Transhumanist Party is always on the lookout for thoughtful individuals who aim to bring a constructive, policy-oriented focus to politics and who seek to champion the role of emerging technologies in solving complex societal problems and ameliorating the human condition. Even if such individuals have not historically used the term "transhumanism" to refer to their efforts and may not know that they are transhumanists yet, the U.S. Transhumanist Party is interested in finding common ground and exploring avenues for collaboration with them.

After several years of building its internal infrastructure and engaging in public outreach and membership growth, the U.S. Transhumanist Party aims to field a candidate in the 2020 U.S. Presidential elections. One of the major aims of this book is to attract a highly qualified, erudite, scientifically and rationally minded individual to become this candidate. This person could have an academic background or, alternatively, could be a thought leader in other areas – for instance, an entrepreneur, author, or public intellectual. The key hope for such an individual would be to thoughtfully articulate the promise posed by various emerging technologies in a manner that is at once ambitious and realistic – outlining both a long-range vision of what is possible and the incremental steps which can be taken in the near term to get there. We understand that the existing political system in the United States would virtually preclude a candidate outside the two-party duopoly from becoming elected, but a predominantly educational campaign spearheaded by a Transhumanist candidate for President could still attract a remarkable amount of media and public attention to the potential of emerging technologies, as Zoltan Istvan's 2016 campaign demonstrated.

5.6   Long-Term Political Vision

While the short-term and intermediate goals of the U.S. Transhumanist Party focus primarily on raising awareness, facilitating discussion, and shifting public opinion in favor of emerging technologies, our long-term vision is far more ambitious. Through a series of incremental achievements, we hope to trigger a cascade of

---

[42] "Political party designation". Ballotpedia. Available at https://ballotpedia.org/Political_party_designation

events that will precipitate a peaceful revolution in politics and in the human condition itself. Within the coming decades, we aim to achieve worldwide radical abundance, universal prosperity, and indefinite lifespans for anyone who seeks them. We cannot do this alone and will welcome and support the efforts of allies among researchers, activists, and policymakers – whether or not they explicitly identify as transhumanists or are even aware of the U.S. Transhumanist Party's existence and endeavors.

As technologically driven prosperity spreads, the incentive for individuals to engage in conflict – whether driven by the desire to acquire material resources held by others or by ideological animosity – will diminish greatly. Materially prosperous individuals have less motivation to expropriate others. Individuals with more to lose in terms of comfort, longevity, and wealth will be more reluctant to throw away existing high standards of living in order to act out an ideological animus. Major reductions in conflict should lead to dramatically more civil politics domestically, as well as more lasting world peace through the reduction and eventual elimination of wars, terrorist attacks, armed rebellions, and acts of international sabotage and covert political destabilization.

The U.S. Transhumanist Party seeks to embody and achieve widespread cosmopolitanism and acceptance of hyperpluralistic diversity. A transhuman world would contain not only the existing diversity of individuals but would also dramatically expand such diversity through cyborg augmentations, genetic engineering, improvements in medical care, and sentient artificial intelligence. Today we advance the cosmopolitan ideal through our acceptance of members from every location and every age group, as long as those individuals are capable of forming political opinions. In the future, as new types of sentient entities emerge or are discovered, the U.S. Transhumanist Party will be at the vanguard of advocating for those entities to have their rights as reasoning beings recognized and protected.

The U.S. Transhumanist Party endeavors to shift the overall focus of politics from adversarial to collaborative. Our achievements in 2017, as elaborated upon in my Chairman's Anniversary Message, were promising first steps toward this goal.[43] A year later, on November 18, 2018, I followed up on our progress in my Chairman's Second Anniversary Message and noted that

> In this epoch, transhumanism is no longer a fringe extreme; while we are a small political party, we occupy the sensible moderate ground – the civilized center of political discourse – precisely because we reject the downward spiral of toxicity, tribalism, political violence, and zero-sum partisanship which characterizes both the Democratic and Republican Parties today. Many people beyond the historic core transhumanist constituencies ought to find our message appealing, if they only knew about the Transhumanist Party and what it actually stands for.[44]

---

[43] Gennady Stolyarov II. "U.S. Transhumanist Party Chairman's Anniversary Message". November 25, 2017. Available at http://transhumanist-party.org/2017/11/25/ustp-chairman-anniversary/

[44] Gennady Stolyarov II. "U.S. Transhumanist Party Chairman's Second Anniversary Message". November 18, 2018. Available at http://transhumanist-party.org/2018/11/18/ustp-chairman-2nd-anniversary/

Instead of seeking to advance the power of particular individuals, factions, or special economic interests, we aim to address the question, "How can we best solve the problems and mitigate the risks facing all of us – both in the present and in the future?" Those who can help us answer that question – no matter what their circumstantial attributes, nominal affiliations, or ideological backgrounds – can become allies in this new collaboration. The U.S. Transhumanist Party will meticulously strive to avoid zero-sum and negative-sum politics; we wish not to undermine any particular politician, party, or group in contemporary politics – but rather to build up the resources available to meritorious individuals and to humankind for the achievement of a brighter future for all. Join our peaceful political revolution and help us chart the best possible path toward a future that will, in all respects, constitute a vast improvement over the status quo.

Appendix I. Transhumanist Bill of Rights, Version 3.0

Preamble

Whereas science and technology are now radically changing human beings and may also create future forms of advanced sapient and sentient life, transhumanists establish this TRANSHUMANIST BILL OF RIGHTS to help guide and enact sensible policies in the pursuit of life, liberty, security of person, and happiness.

As used in this TRANSHUMANIST BILL OF RIGHTS, the term "sentient entities" encompasses:

(i) Human beings, including genetically modified humans;
(ii) Cyborgs;
(iii) Digital intelligences;
(iv) Intellectually enhanced, previously non-sapient animals;
(v) Any species of plant or animal which has been enhanced to possess the capacity for intelligent thought; and
(vi) Other advanced sapient life forms.

Sentient entities are defined by information-processing capacity such that this term should not apply to non-self-aware lifeforms, like plants and slime molds. Biological processing substrates are referred to as using an "analogue intelligence", whereas purely electronic processing substrates are referred to as "digital intelligence", and processing substrates that utilize quantum effects would be considered "quantum intelligence".

Sentience is ranked as Level 5 information integration according to the following criteria:

- **Level 0 – No information integration:** Inanimate objects; objects that do not modify themselves in response to interaction – e.g., rocks, mountains.

- **Level 1 – Non-zero information integration:** Sensors – anything that is able to sense its environment – e.g., photo-diode sense organs, eyes, skin.
- **Level 2 – Information manipulation:** Systems that include feedback that is non-adaptive or minimally adaptive – e.g., plants, basic algorithms, the system that interprets the output from a photo-diode to determine its on/off state (a photo diode itself cannot detect its own state). Level 2 capabilities include the following:

  1. Expression of emotion;
  2. Expression of sensory pleasure;
  3. Taste aversion.

- **Level 3 – Information integration – Awareness:** Systems that include adaptive feedback, can dynamically generate classification – e.g., deep-learning AI, chickens, animals that are able to react to their environment, have a model of their perception but not the world. This level describes animals acting on instinct and unable to classify other animals into more types than "predator", "prey", or "possible mate". Level 3 capabilities include the following:

  1. Navigational detouring (which requires an being to pursue a series of non-rewarding intermediate goals in order to obtain an ultimate reward); **Examples:** documentation of detouring in jumping spiders (Jackson and Wilcox 2003), motivational trade-off behavior in hermit crabs (Elwood and Appel 2009);
  2. Emotional fever (an increase in body temperature in response to a supposedly stressful situation — gentle handling, as operationalized in Cabanac's experiments).

- **Level 4 – Awareness + World model:** Systems that have a modeling system complex enough to create a world model: a sense of other, without a sense of self – e.g., dogs. Level 4 capabilities include static behaviors and rudimentary learned behavior.
- **Level 5 – Awareness + World model + Primarily subconscious self model = Sapient or Lucid:** Lucidity means to be meta-aware – that is, to be aware of one's own awareness, aware of abstractions, aware of one's self, and therefore able to actively analyze each of these phenomena. If a given animal is meta-aware to any extent, it can therefore make lucid decisions. Level 5 capabilities include the following:

  1. The "sense of self";
  2. Complex learned behavior;
  3. Ability to predict the future emotional states of the self (to some degree);
  4. The ability to make motivational tradeoffs.

- **Level 6 – Awareness + World model + Dynamic self model + Effective control of subconscious:** The dynamic sense of self can expand from "the small self" (directed consciousness) to the big self ("social group dynamics"). The "self" can include features that cross barriers between biological and non-biological – e.g., features resulting from cybernetic additions, like smartphones.

- **Level 7 – Global awareness – Hybrid biological-digital awareness = Singleton:** Complex algorithms and/or networks of algorithms that have capacity for multiple parallel simulations of multiple world models, enabling cross-domain analysis and novel temporary model generation. This level includes an ability to contain a vastly larger amount of biases, many paradoxically held. Perspectives are maintained in separate modules, which are able to dynamically switch between identifying with the local module of awareness/perspective or the global awareness/perspective. Level 7 capabilities involve the same type of dynamic that exists between the subconscious and directed consciousness, but massively parallelized, beyond biological capacities.

Article I All sentient entities are hereby entitled to pursue any and all rights within this document to the degree that they deem desirable – including not at all. All sentient entities are entitled, to the extent of their individual decisions, to all the rights and freedoms set forth in this TRANSHUMANIST BILL OF RIGHTS, without distinction of any kind, such as race, color, sex, gender, language, religion, political or other opinion, national, social, or planetary origin, property, birth (including manner of birth), biological or non-biological origins, or other status. Furthermore, no distinction shall be made on the basis of the political, jurisdictional, or international status of the country or territory to which a sentient entity belongs, whether it be independent, trust, non-self-governing, or under any other limitation of sovereignty. In the exercise of their rights and freedoms, all sentient entities shall be subject only to such limitations as are determined by law solely for the purpose of securing due recognition and respect for the rights and freedoms of others and of meeting the just requirements of morality, public order, and the general welfare in a democratic society, which may not undermine the peaceful prerogatives of any individual sentient entity. These rights and freedoms may in no case be exercised contrary to the purposes and principles of this TRANSHUMANIST BILL OF RIGHTS.

Article II The enumeration in this TRANSHUMANIST BILL OF RIGHTS of certain rights shall not be construed to deny or disparage any other rights retained by sentient entities.

Article III All sentient entities shall be granted equal and total access to any universal rights to life. All sentient entities are created free and equal in dignity and rights. They are endowed with reason and conscience and should act towards one another in a spirit of brotherhood (without necessitating any particular gender or implying any particular biological or non-biological origin or composition).

Article IV Sentient entities are entitled to universal rights of ending involuntary suffering, making personhood improvements, and achieving an indefinite lifespan via science and technology. The right of ending involuntary suffering does not refer to euthanasia but rather to the application of technology to eliminate involuntary suffering in still-living beings, while enabling their lives to continue with improved quality and length.

Article V No coercive legal restrictions should exist to bar access to life extension and life expansion for all sentient entities. Life expansion includes life extension, sensory improvements, and other technologically driven improvements of the human condition that might be achieved in the future.

Article VI Involuntary aging shall be classified as a disease. All nations and their governments will actively seek to dramatically extend the lives and improve the health of their citizens by offering them scientific and medical technologies to overcome involuntary aging.

Article VII All sentient entities should be the beneficiaries of a system of universal health care. A system of universal health care does not necessitate any particular means, policy framework, source, or method of payment for delivering health care. A system of universal health care may be provided privately, by governments, or by some combination thereof, as long as, in practice, health care is abundant, inexpensive, accessible, and effective in curing diseases, healing injuries, and lengthening lifespans.

Article VIII Sentient entities are entitled to the freedom to conduct research, experiment, and explore life, science, technology, medicine, and extraterrestrial realms to overcome biological limitations of humanity. Such experimentation will not be carried out on any sapient being, without that being's informed consent. Sentient entities are also entitled to the freedom to create cybernetic artificial organs, bio-mechatronic parts, genetic modifications, systems, technologies, and enhancements to extend lifespan, eradicate illness, and improve all sentient life forms. Any such creations that demonstrate sapience cannot be considered property and are protected by the rights presented herein.

Article IX Legal safeguards should be established to protect individual free choice in pursuing peaceful, consensual life-extension science, health improvements, body modification, and morphological enhancement. While all individuals should be free to formulate their independent opinions regarding the aforementioned pursuits, no hostile cultural, ethnic, or religious perspectives should be entitled to apply the force of law to erode the safeguards protecting peaceful, voluntary measures intended to maximize the number of life hours citizens possess.

Article X Sentient entities agree to uphold morphological freedom—the right to do with one's physical attributes or intelligence whatever one wants so long as it does not harm others.

This right includes the prerogative for a sentient intelligence to set forth in advance provisions for how to handle its physical manifestation, should that intelligence enter into a vegetative, unconscious, or similarly inactive state, notwithstanding any legal definition of death. For instance, a cryonics patient has the right to determine in advance that the patient's body shall be cryopreserved and kept under specified conditions, in spite of any legal definition of death that might apply to that patient under cryopreservation.

Morphological freedom entails the duty to treat all sapients as individuals instead of categorizing them into arbitrary subgroups or demographics, including as yet undefined subcategorizations that may arise as sapience evolves.

However, the proper exercise of morphological freedom must also ensure that any improvement of the self should not result in involuntary harms inflicted upon others. Furthermore, any sentient entity is also recognized to have the freedom not to modify itself without being subject to negative political repercussions, which include but are not limited to legal and/or socio-economic repercussions.

Article XI An altered, augmented, cybernetic, transgenic, anthropomorphic, or avatar sentient entity, whether derived from or edited by science, comprised of or conjoined with technology, has the right to exist, form, and join the neo-civilization.

Article XII All sentient entities are entitled to reproductive freedom, including through novel means such as the creation of mind clones, monoparent children, or benevolent artificial general intelligence. All sentient entities of full age and competency, without any limitation due to race, nationality, religion, or origin, have the right to marry and found a family or to found a family as single heads of household. They are entitled to equal rights as to marriage, during marriage, and at its dissolution. Marriage shall be entered into only with the free and full consent of the intending spouses. All families, including families formed through novel means, are entitled to protection by society and the State. All sentient entities also have the right to prevent unauthorized reproduction of themselves in both a physical and a digital context. Privacy and security legislation should be enacted to prevent any individual's DNA, data, or other information from being stolen and duplicated without that individual's authorization.

Article XIII No sentient entity shall be subjected to arbitrary interference with his, her, or its privacy, family, home, or correspondence, nor to attacks upon his, her, or its honor and reputation. Every sentient entity has the right to the protection of the law against such interference or attacks. All sentient entities have privacy rights to personal data, genetic material, digital, biographic, physical, and intellectual enhancements, and consciousness. Despite the differences between physical and virtual worlds, equal protections for privacy should apply to both physical and digital environments. Any data, such as footage from a public security camera, archived without the consent of the person(s) about whom the data were gathered and subject to legal retention, shall be removed after a period of seven (7) years, unless otherwise requested by said person(s).

Article XIV No sentient entity shall be subjected to arbitrary arrest, detention or exile. Sousveillance laws should be enacted to ensure that all members of peaceful communities feel safe, to achieve governmental transparency, and to provide counter-balances to any surveillance state. For instance, law-enforcement officials, when interacting with the public, should be required to wear body cameras or similar devices continuously monitoring their activities.

Article XV All sentient entities, with the exception only of those in legal detention, have the right to private internet access without such access being prohibited or circumvented by either private corporations or governmental bureaucracy.

Article XVI All sentient entities are equal before the law and are entitled without any discrimination to equal protection of the law. All sentient entities are entitled to equal protection against any discrimination in violation of this TRANSHUMANIST BILL OF RIGHTS and against any incitement to such discrimination. All sentient entities should be protected from discrimination based on their physical form in the context of business transactions and law enforcement.

Article XVII All sentient entities have the right to life, liberty and security of person. All sentient entities have the right to defend themselves from attack, in both physical and virtual worlds.

Article XVIII Societies of the present and future should afford all sentient entities sufficient basic access to wealth and resources to sustain the basic requirements of existence in a civilized society and function as the foundation for pursuits of self-improvement. This includes the right to a standard of living adequate for the health and well-being of oneself and one's family, including food or other necessary sources of energy, clothing, housing or other appropriate shelter, medical care or other necessary physical maintenance, necessary social services, and the right of security in the event of involuntary unemployment, sickness, disability, loss of family support, old age, or other lack of livelihood in circumstances beyond the sentient entity's control. Present and future societies should ensure that their members will not live in poverty solely for being born to the wrong parents. All children and other recently created sentient entities, irrespective of the manner or circumstances of their creation, shall enjoy the same social protection. Each sentient entity, as a member of society, has the right to social security and is entitled to realization, through national effort and international co-operation and in accordance with the organization and resources of each State, of the economic, social, and cultural rights indispensable for his, her, or its dignity and the free development of his, her, or its personality.

Article XIX Irrespective of whether or not technology will eventually replace the need for the labor of sentient entities, all sentient entities should be the beneficiaries of an unconditional universal basic income, whereby the same minimum amount of money or other resources is provided irrespective of a sentient entity's life circumstances, occupations, or other income sources, so as to provide a means for the basic requirements of existence and liberty to be met.

Article XX Present and future societies should provide education systems accessible and available to all in pursuit of factual knowledge to increase intellectual acuity; promote critical thinking and logic; foster creativity; form an enlightened collective; attain health; secure the bounty of liberty for all sentient entities for our posterity; and forge new ideas, meanings, and values. All sentient entities have the right to education. Education shall be free, at least in the elementary and fundamental stages. Technical and professional education shall be made generally available, and higher education shall be equally accessible to all on the basis of merit. Education shall be directed to the full development of the sentient entity's personality and to the strengthening of respect for all sentient entities' rights and fundamental freedoms. It shall promote understanding, tolerance, and friendship among all nations, racial, religious, and other sentient groups – whether biological, non-biological, or a combination thereof – and shall further the maintenance of peace. Parents and other creators of sentient entities have a prior right to choose the kind of education that shall be given to their children or other recently created sentient entities which have not yet developed sufficient maturity to select their own education.

Article XXI All sentient entities are entitled to join their psyches to a collective noosphere in an effort to preserve self-consciousness in perpetuity. The noosphere is the sphere of human thought and includes, but is not limited to, intellectual systems in the realm of law, education, philosophy, technology, art, culture, and industry. All sentient entities have the right to participate in the noosphere using any level of technology that is conducive to constructive participation.

Article XXII Sentient entities will take every reasonable precaution to prevent existential risks, including those of rogue artificial intelligence, asteroids, plagues, weapons of mass destruction, bioterrorism, war, and global warming, among others.

Article XXIII All nations and their governments will take all reasonable measures to embrace and fund space travel, not only for the spirit of adventure and to gain knowledge by exploring the universe, but as an ultimate safeguard to its citizens and transhumanity should planet Earth become uninhabitable or be destroyed.

Article XXIV Transhumanists stand opposed to the post-truth culture of deception. All governments should be required to make decisions and communicate information rationally and in accordance with facts. Lying for political gain or intentionally fomenting irrational fears among the general public should entail heavy political penalties for the officials who engage in such behaviors.

Article XXV No sentient entity shall be held in slavery or involuntary servitude; slavery and the slave trade shall be prohibited in all their forms.

Article XXVI No sentient entity shall be subjected to torture or to treatment or punishment that is cruel, degrading, inhuman, or otherwise unworthy of sentience or sapience.

Article XXVII Each sentient entity has the right to recognition everywhere as a person before the law.

Article XXVIII All individual sentient entities have the right to an effective remedy by the competent local, national, international, or interplanetary tribunals for acts violating the fundamental rights granted them by the constitution, by law, and/ or by this TRANSHUMANIST BILL OF RIGHTS.

Article XXIX All individual sentient entities are entitled in full equality to a fair and public hearing by an independent and impartial tribunal, in the determination of their individual rights and obligations and of any criminal charge against them.

Article XXX All individual sentient entities charged with a penal offence have the right to be presumed innocent until proved guilty according to law in a public trial at which they individually have had all the guarantees necessary for their defense. No sentient entity shall be held guilty of any penal offence on account of any act or omission which did not constitute a penal offence, under national or international law, at the time when it was committed. Nor shall a heavier penalty be imposed than the one that was applicable at the time the penal offence was committed.

Article XXXI All sentient entities have the right to freedom of movement and residence within the borders of each state. Each individual sentient entity has the right to leave any country, including his, her, or its own, and to return to his, her, or its country.

Article XXXII All sentient entities have the right to seek and to enjoy in other countries asylum from persecution. This right may not be invoked in the case of prosecutions genuinely arising from non-political crimes or from acts contrary to the purposes and principles of this TRANSHUMANIST BILL OF RIGHTS.

Article XXXIII All sentient entities have the right to a nationality. No sentient entity shall be arbitrarily deprived of his, her, or its nationality nor denied the right to change his, her, or its nationality.

Article XXXIV All sentient entities have the right to own property alone as well as in association with others. No one shall be arbitrarily deprived of his, her, or its property.

Article XXXV All sentient entities have the right to freedom of thought, conscience and religion; this right includes freedom to change one's religion or belief, and freedom, either alone or in community with others and in public or private, to manifest one's religion or belief in teaching, practice, worship, and observance. This right also includes freedom not to have a religion and to criticize or refuse to engage in any religious practice or belief without adverse legal consequences.

Article XXXVI All sentient entities have the right to freedom of opinion and expression; this right includes freedom to hold opinions without interference and to seek, receive, and impart information and ideas through any media and regardless of frontiers.

Article XXXVII   All sentient entities have the right to freedom of peaceful assembly and association. No sentient entity may be compelled to belong to an association.

Article XXXVIII   All sentient entities have the right to take part in the government of their countries, directly or through freely chosen representatives. All sentient entities have the right of equal access to public service in their countries. The will of the constituent sentient entities shall be the basis of the authority of government; this will shall be expressed in periodic and genuine elections which shall be by universal and equal suffrage of sentient entities and shall be held by secret vote or by equivalent free voting procedures.

Article XXXIX   All sentient entities have the right to work, to free choice of employment, and to just and favorable conditions of work, as long as employment is offered or considered economically necessary in the sentient entity's proximate society and contemporary epoch. All sentient entities who choose to work have the right to equal pay for equal work. All sentient entities who choose to work have the right to just and favorable remuneration, ensuring for themselves and their families an existence worthy of human dignity, and supplemented, if necessary, by other means of social protection, such as a universal basic income. All sentient entities have the right to form and join trade unions for the protection of their interests; however, no sentient entity may be compelled to join a trade union as a condition of employment.

Article XL   All sentient entities have the right to rest and leisure commensurate with the physical requirements of those sentient entities for maintaining optimal physical and mental health, including reasonable limitation of working hours and periodic holidays with pay in societies where paid employment is considered economically necessary.

Article XLI   All sentient entities have the right freely to participate in the cultural life of the community, to enjoy the arts, and to share in scientific advancement and its benefits. All sentient entities have the right to the protection of the moral and material interests resulting from any scientific, literary, or artistic production of which they are the authors.

Article XLII   All sentient entities are entitled to a social and international order in which the rights and freedoms set forth in this TRANSHUMANIST BILL OF RIGHTS can be fully realized.

Article XLIII   Nothing in this TRANSHUMANIST BILL OF RIGHTS may be interpreted as implying for any State, group, or sentient entity any right to engage in any activity or to perform any act aimed at the destruction of any of the rights and freedoms set forth herein.

Appendix II. Constitution of the United States Transhumanist Party

Article I. Immutable Principles of the Transhumanist Party

The United States Transhumanist Party is defined at its core by the following principles. While the remainder of the Party's platform, bylaws, and operations may in the future be subject to alterations by decisions of the membership, the statements below are considered immutable and may not be altered.

Section I. Core Ideals

Ideal 1 The Transhumanist Party supports significant life extension achieved through the progress of science and technology.

Ideal 2 The Transhumanist Party supports a cultural, societal, and political atmosphere informed and animated by reason, science, and secular values.

Ideal 3 The Transhumanist Party supports efforts to use science, technology, and rational discourse to reduce and eliminate various existential risks to the human species.

Section II. Statements of Historical Fact

Historical Fact 1 Zoltan Istvan was the founder of the Transhumanist Party in 2014. Zoltan Istvan was also the first Presidential candidate for the Transhumanist Party during the 2016 United States Presidential Election.

Historical Fact 2 The person who has held the role of the first Chairman in the history of the Transhumanist Party is Zoltan Istvan.

Historical Fact 3 The person who has held the role of the second Chairman in the history of the Transhumanist Party is Gennady Stolyarov II.

Section III. Immutable Operating Principles

Operating Principle 1 The United States Transhumanist Party shall respect the autonomy of State-level Transhumanist Parties to determine their internal bylaws, platforms, and activities. The United States Transhumanist Party encourages the formation of State-level Transhumanist Parties and desires to collaborate with

State-level Transhumanist Parties and offer them guidance and advice. However, the United States Transhumanist Party shall not have the authority to impose involuntary constraints on State-level Transhumanist Parties that operate peacefully within the boundaries of applicable law.

Operating Principle 2   The Transhumanist Party renounces all violence, except in self-defense against a clear, immediate act of physical aggression. In particular, the Transhumanist Party holds that violent political activism is never permissible or just. The Transhumanist Party commits to always pursuing its goals in a civil, law-abiding manner, respecting the legitimate rights of all persons. The Transhumanist Party shall not condone and shall necessarily and automatically disavow all violent criminal acts. Any person who commits a violent criminal act is automatically disassociated from the Transhumanist Party in all respects until and unless that person has made appropriate restitution or has fully undergone the appropriate penalties pursuant to applicable law. However, this commitment to exclusively peaceful action does not preclude the Transhumanist Party from criticizing any ideas or behavior which are contrary to reason, morality, common sense, or the principles and objectives of the Transhumanist Party Core Ideals and Platform.

Article II. Transitional Period of the Transhumanist Party

Section I   This Article II shall be temporarily in effect until valid elections among the members of the Transhumanist Party have determined a platform and a set of elected officers. The period prior to such a determination shall be known as the Transitional Period. After the Transitional Period, this Article II may be replaced by provisions determined by the members of the Transhumanist Party.

Section II   During the Transitional Period, the second Chairman of the Transhumanist Party, Gennady Stolyarov II, shall have the full discretion to organize the membership and conduct of the Transhumanist Party so as to create a sustainable, self-perpetuating, and active organizational structure. This discretion shall include the enrollment of members, outreach to other individuals and organizations, and the appointment of other officers, advisors, and volunteers to undertake the work of the Transhumanist Party.

Section III   During the Transitional Period, the Chairman of the Transhumanist Party shall have the authority to delegate any authority of the Chairman to any member of the Transhumanist Party for any specified length of time, or indefinitely for the duration of the Transitional Period, until the delegation is revoked at the discretion of the Chairman. All such delegations of authority shall be in writing. A person receiving a delegation of authority by the Chairman shall be known as a Member Delegate.

Section IV During the Transitional Period, the Transhumanist Party shall not charge any fees for membership nor impose any other monetary requirements of its members.

Section V The intent of the Transitional Period shall be to facilitate an indefinitely active and growing transhumanist movement, wherein a sufficiently robust constituency exists to enable democratic decision-making to occur in a manner reflective of transhumanist principles and ideals.

Section VI During the Transitional Period, the Transhumanist Party shall have two categories of members: United States Members and Allied Members.

Section VII United States Members shall be those individuals who lawfully reside within the United States and are eligible to vote in United States elections. United States Members may participate in any activities of the Transhumanist Party and receive any delegations of authority at the discretion of the Chairman.

Section VIII Allied Members may be any individuals, of any age, nationality, and place of residence – with the exception of those persons eligible to be United States Members. Allied Members may participate in any activities of the Transhumanist Party and receive any delegations of authority at the discretion of the Chairman – provided, however, that Allied Members may not be eligible to vote in United States elections pursuant to applicable United States law. Allied Members may, however, vote in internal elections of the Transhumanist Party.

Section IX Allied Membership in the Transhumanist Party is open to any being capable of logical reasoning and of the expression of political opinions. Specifically, if sentient artificial intellects or intelligent extraterrestrial life forms are discovered at any time after the founding of the Transhumanist Party, such entities shall be eligible for Allied Membership.

Article III. Platform of the Transhumanist Party

Section I *[Adopted by a vote of the members during January 15–21, 2017]*[45]: The United States Transhumanist Party strongly supports individual privacy and liberty over how to apply technology to one's personal life. The United States Transhumanist Party holds that each individual should remain completely sovereign in the choice to disclose or not disclose personal activities, preferences, and beliefs within the public sphere. As such, the United States Transhumanist Party opposes all forms of mass surveillance and any intrusion by governmental or private institutions upon non-coercive activities that an individual has chosen to retain within his, her, or its

---

[45] "Results of Platform Vote #1". United States Transhumanist Party. January 22, 2017. Available at http://transhumanist-party.org/wp-content/uploads/2017/01/USTP_Platform_Vote_1_Results.pdf. These results led to the adoption of Sections I through V of Article III.

private sphere. However, the United States Transhumanist Party also recognizes that no individuals should be protected from peaceful criticism of any matters that those individuals have chosen to disclose within the sphere of public knowledge and discourse.

Section II *[Adopted by a vote of the members during January 15–21, 2017; amended by a vote of the members during November 11–17, 2017]*[46]: The United States Transhumanist Party supports all acceptance, tolerance, and inclusivity of individuals and groups of all races, genders, classes, religions, creeds, and ideologies. Accordingly, the United States Transhumanist Party condemns any hostile discrimination or legal restrictions on the basis of national origin, skin color, birthplace, ancestry, gender identity, or any manner of circumstantial attribute tied to a person's lineage or accident of birth. Furthermore, the United States Transhumanist Party strongly opposes any efforts to enforce said restrictions regardless of cause or motivation thereof. Additionally, any institution that uses violence, suppression of free speech, or other unconstitutional or otherwise illegal methods will be disavowed and condemned by the United States Transhumanist Party, with an efficient, nonviolent alternative to said institution being offered to achieve its goals if they align with the Party's interests.

Section III *[Adopted by a vote of the members during January 15–21, 2017]*: The United States Transhumanist Party holds that the vast majority of technologies are beneficial to human well-being and should be enthusiastically advocated for and developed further. However, a minority of technologies could be detrimental to human well-being and, as such, their application, when it results in detrimental consequences, should be opposed. Examples of such detrimental technologies include nuclear, chemical, and biological weapons, mass-surveillance systems such as those deployed by the National Security Agency in the United States, and backscatter X-ray full-body scanners such as those used until 2013 by the Transportation Security Administration in the United States. Furthermore, the United States Transhumanist Party is opposed to the deliberate engineering of new active pathogens or the resurrection of once-existing pathogens, whose spread might not be able to be contained within laboratory settings. While it is impossible to un-learn the knowledge utilized in the creation of such technologies, the United States Transhumanist Party holds that all such knowledge should only be devoted toward peaceful, life-affirming, rights-respecting purposes, going forward.

Section IV *[Adopted by a vote of the members during January 15–21, 2017]*: In recognition of the dire existential threat that nuclear weapons pose to sapient life on Earth – including as a result of such weapons' accidental deployment due to system failures or human misunderstanding – the United States Transhumanist Party advocates the complete dismantlement and abolition of all nuclear weapons everywhere,

---

[46] "Results of Platform Vote #6". United States Transhumanist Party. November 19, 2017. Available at   http://transhumanist-party.org/wp-content/uploads/2017/11/USTP_Platform_Vote_6_Results. pdf. These results led to the adoption of Sections LXIV through LXXXII and the amendment of Sections II, XXVII, and LIX of Article III.

as rapidly as possible. If necessary for geopolitical stability, synchronized multilateral disarmament and non-proliferation treaties should be pursued, strengthened, and accelerated in the most expeditious manner. If, however, multilateral agreements among nations are not reached, then the United States Transhumanist Party advocates that all nuclear powers, especially the United States and Russia, should undertake unilateral nuclear disarmament at the earliest opportunity in order to preserve civilization from accidental annihilation.

Section V *[Adopted by a vote of the members during January 15–21, 2017]*: The United States Transhumanist Party supports concerted research in effort to eradicate disease and illness that wreak havoc upon and cause death of sapient beings. We strongly advocate the increase and redirection of research funds to conduct research and experiments and to explore life, science, technology, medicine, and extraterrestrial realms to improve all sentient entities.

Section VI *[Adopted by a vote of the members during February 16–22, 2017]*[47]: The United States Transhumanist Party upholds morphological freedom—the right to do with one's physical attributes or intelligence whatever one wants so long as it does not directly harm others.

The United States Transhumanist Party considers morphological freedom to include the prerogative for a sentient intelligence to set forth in advance provisions for how to handle its physical manifestation, should that intelligence enter into a vegetative, unconscious, or similarly inactive state, notwithstanding any legal definition of death. For instance, a cryonics patient should be entitled to determine in advance that the patient's body shall be cryopreserved and kept under specified conditions, in spite of any legal definition of death that might apply to that patient under cryopreservation.

The United States Transhumanist Party also recognizes that morphological freedom entails the duty to treat all sapients as individuals instead of categorizing them into arbitrary subgroups or demographics, including as yet undefined subcategorizations that may arise as sapience evolves.

The United States Transhumanist Party is focused on the rights of all sapient individuals to do as they see fit with themselves and their own reproductive choices.

However, the United States Transhumanist Party holds that the proper exercise of morphological freedom must also ensure that any improvement of the self should not result in involuntary harms directly inflicted upon others. Furthermore, the United States Transhumanist Party recognizes any sentient entity to have the freedom not to modify itself without being subject to negative political repercussions, which include but are not limited to legal and/or socio-economic repercussions.

---

[47] "Results of Platform Vote #2". United States Transhumanist Party. February 25, 2017. Available at http://transhumanist-party.org/wp-content/uploads/2017/02/USTP_Platform_Vote_2_Results. pdf. These results led to the adoption of Sections VI through X of Article III.

The United States Transhumanist Party recognizes the ethical obligations of sapient beings to be the purview of those individual beings, and holds that no other group, individual, or government has the right to limit those choices – including genetic manipulation or other biological manipulation or any other modifications up to and including biological manipulation, mechanical manipulation, life extension, reproductive choice, reproductive manipulation, cryonics, or other possible modifications, enhancements, or morphological freedoms. It is only when such choices directly infringe upon the rights of other sapient beings that the United States Transhumanist Party will work to develop policies to avoid potential infringements.

Section VII *[Adopted by a vote of the members during February 16–22, 2017]*: The United States Transhumanist Party strongly supports and emphasizes all values and organized efforts related to the cultivation of science, reason, intelligence, and rational thinking.

The United States Transhumanist Party places no reliance upon any and all sources of information that cannot stand up to rational scrutiny.

The United States Transhumanist Party places no reliance upon any individual, organization, or belief system that intentionally distorts empirically verifiable evidence, including but not limited to scientific and historical evidence, to serve its own agenda.

The United States Transhumanist Party places no reliance upon any position or belief system that contains arguments built upon logical fallacies (with exemption granted to arguments containing both fallacious and logically defensible premises).

Section VIII *[Adopted by a vote of the members during February 16–22, 2017]*: The United States Transhumanist Party supports maximum individual liberty to engage in scientific and technological innovation for the improvement of the self and the human species. In particular, the United States Transhumanist Party supports all rationally, scientifically grounded research efforts for curing diseases, lengthening lifespans, achieving functional, healthy augmentations of the body and brain, and increasing the durability and youthfulness of the human organism. The United States Transhumanist Party holds that all such research efforts should be rendered fully lawful and their products should be made fully available to the public, as long as no individual is physically harmed without that individual's consent or defrauded by misrepresentation of the effects of a possible treatment or substance.

Section IX *[Adopted by a vote of the members during February 16–22, 2017]*: The United States Transhumanist Party supports all emerging technologies that have the potential to improve the human condition – including but not limited to autonomous vehicles, electric vehicles, economical solar power, safe nuclear power, hydroelectricity, geothermal power, applications for the sharing of durable goods, artificial intelligence, biotechnology, nanotechnology, robotics, rapid transit, 3D printing, vertical farming, electronic devices to detect and respond to trauma, and beneficial genetic modification of plants, animals, and human beings.

Section X *[Adopted by a vote of the members during February 16–22, 2017]*: The United States Transhumanist Party advocates the construction of a self-repairing, self-maintaining smart infrastructure which incorporates the distribution of energy, communications, and clean potable water to every building.

Section XI *[Adopted by a vote of the members during March 26 – April 1, 2017][48]*: In supporting peaceful uses of nuclear energy, the United States Transhumanist Party endorses the thorium fuel cycle, which provides for a safe and nearly limitless energy source in the absence of the development of practical thermonuclear fusion.

Section XII *[Adopted by a vote of the members during March 26 – April 1, 2017]*: The United States Transhumanist Party holds that present and future societies should provide education systems accessible and available to all in pursuit of factual knowledge to increase intellectual acuity; promote critical thinking and logic; foster creativity; form an enlightened collective; attain health; secure the bounty of liberty for all sentient entities for our posterity; and forge new ideas, meanings, and values.

The United States Transhumanist Party supports efforts to reduce the cost of education while improving its access. In particular, the United States Transhumanist Party supports freely available, open-source, methods of learning, teaching, credentialing, and cultural creation that integrate emerging technologies into every facet of the learning process. The United States Transhumanist Party primarily advocates private innovation to deliver such educational improvements, but also advocates the application of these improvements to all publicly funded educational institutions. The United States Transhumanist Party holds that every person should aspire toward intellectual, moral, and esthetic enlightenment and sophistication and should contribute toward bringing about a new Age of Reason, where the highest reaches of intellectual activity are attainable and eagerly pursued by the majority of the population.

The United States has upheld basic education since the American Revolution. The United States Transhumanist Party believes, in keeping with what basic education was in the 1700s, relative to the state of technology given the advancement in society at the time, that 'basic' education should be defined as college, and that a key part of our agenda is to help encourage a more successful generation by paying for a 'basic' education up to and including college degrees.

Section XIII *[Adopted by a vote of the members during March 26 – April 1, 2017]*: The United States Transhumanist Party supports the involvement of intelligent laypersons in the political process to counteract and neutralize the influence of politically connected special interests and their paid representatives. The United States Transhumanist Party supports all electronic and other technologies that can

---

[48] "Results of Platform Vote #3". United States Transhumanist Party. April 7, 2017. Available at http://transhumanist-party.org/wp-content/uploads/2017/04/USTP_Platform_Vote_3_Results.pdf. These results led to the adoption of Sections XI through XXXII of Article III.

inform and empower intelligent laypersons to monitor and contribute to political discussions and decisions.

Section XIV *[Adopted by a vote of the members during March 26 – April 1, 2017]*: The United States Transhumanist Party supports an end to the costly drug war, which is often an infringement upon the lives and liberties of innocent citizens who do not use drugs but fall victim to militant enforcement of drug prohibitions. The United States Transhumanist Party supports legalization of mild recreational drugs such as marijuana.

Section XV *[Adopted by a vote of the members during March 26 – April 1, 2017]*: The United States Transhumanist Party supports efforts to significantly reduce the massive incarcerated population in America by using innovative technologies to monitor criminals outside of prison. All mandatory sentencing laws should be abolished, and each individual should be sentenced based solely on the consideration of the nature of that individual's crime, its context, and its severity.

Section XVI *[Adopted by a vote of the members during March 26 – April 1, 2017]*: Given the inevitability of technology eventually replacing the need for the labor of sentient entities, the United States Transhumanist Party holds that all sentient entities should be the beneficiaries of an unconditional universal basic income, whereby the same minimum amount of money or other resources is provided irrespective of a sentient entity's life circumstances, occupations, or other income sources, so as to provide a means for the basic requirements of existence and liberty to be met.

Section XVII *[Adopted by a vote of the members during March 26 – April 1, 2017]*: The United States Transhumanist Party holds that present and future societies should take all reasonable measures to embrace and fund space travel, not only for the spirit of adventure and to gain knowledge by exploring the universe, but as an ultimate safeguard to its citizens and transhumanity should planet Earth become uninhabitable or be destroyed.

Section XVIII *[Adopted by a vote of the members during March 26 – April 1, 2017]*: The United States Transhumanist Party supports work to use science and technology to be able to eliminate all disabilities in humans who have them.

Section XIX *[Adopted by a vote of the members during March 26 – April 1, 2017]*: The United States Transhumanist Party supports an end to the two-party political system in the United States and a substantially greater inclusion of "third parties" in the political process through mechanisms such as proportional representation and the elimination of stringent ballot-access requirements. The United States Transhumanist Party also seeks to limit the influence of lobbying by politically connected special interests, while increasing the influence of advocacy by intelligent laypersons.

Section XX *[Adopted by a vote of the members during March 26 – April 1, 2017]*: The United States Transhumanist Party strongly supports the freedom of peaceful speech; religious, non-religious, and anti-religious philosophical espousal; assembly;

protest; petition; and expression of grievances. The United States Transhumanist Party therefore strongly opposes all censorship, including censorship that arises out of identity politics and the desire to avoid perceived offensive behavior.

Section XXI *[Adopted by a vote of the members during March 26 – April 1, 2017]*: The United States Transhumanist Party supports a concerted effort by governments and by public opinion to eradicate police brutality against peaceful citizens, such that violent force is only utilized against individuals who actually pose an imminent threat to human lives.

Section XXII *[Adopted by a vote of the members during March 26 – April 1, 2017]*: The United States Transhumanist Party supports efforts at political, economic, and cultural experimentation in the form of seasteads and micronations. Specifically, the United States Transhumanist Party recognizes the existence and sovereignty of the Principality of Sealand, the Republic of Molossia, and the Free Republic of Liberland, and supports the recognition of these entities by all governments and political parties of the world.

Section XXIII *[Adopted by a vote of the members during March 26 – April 1, 2017]*: The United States Transhumanist Party supports the rights of children to exercise liberty in proportion to their rational faculties and capacity for autonomous judgment. In particular, the United States Transhumanist Party strongly opposes all forms of bullying, child abuse, and censorship of intellectual self-development by children and teenagers.

Section XXIV *[Adopted by a vote of the members during March 26 – April 1, 2017]*: The United States Transhumanist Party supports the promotion of animal welfare to the extent it does not conflict with human well-being. However, the United States Transhumanist Party opposes "animal liberation" movements that seek to return animals to the wilderness or espouse any attempts to separate domesticated animals from human influence. In particular, the United States Transhumanist Party supports the prohibition of cruelty to animals and a complete abolition of euthanasia of healthy animals by animal shelters. The United States Transhumanist Party supports a complete prohibition on the killing of non-contagious, non-aggressive dogs, cats, dolphins, whales, elephants, horses, tortoises, parrots, and primates. Furthermore, the United States Transhumanist Party supports the development and widespread consumption of artificially grown, biologically identical meat products that do not involve the killing of animals.

Section XXV *[Adopted by a vote of the members during March 26 – April 1, 2017]*: The United States Transhumanist Party welcomes both religious and non-religious individuals who support life extension and emerging technologies. The United States Transhumanist Party recognizes that some religious individuals and interpretations may be receptive to technological progress and, if so, are valuable allies to the transhumanist movement. On the other hand, the United States Transhumanist Party is also opposed to any interpretation of a religious doctrine that results in the rejection of reason, censorship, violation of individual rights, suppression of technological advancement, and attempts to impose religious belief by force and/or by legal compulsion.

Section XXVI *[Adopted by a vote of the members during March 26 – April 1, 2017]*: The United States Transhumanist Party holds that each of its members should vote or abstain from voting in accordance with that member's own individual conscience and judgment. If an official or candidate of the United States Transhumanist Party expresses a preference for any particular non-transhumanist candidate for office, then no national or State-level Transhumanist Party, nor any individual transhumanist, ought to be in any manner bound to support that same non-transhumanist candidate.

Section XXVII *[Adopted by a vote of the members during March 26 – April 1, 2017; amended by a vote of the members during November 11–17, 2017]*[49]: The United States Transhumanist Party advocates Constitutional reform to abolish the Electoral College in the United States Presidential elections and render the plurality of the popular vote the sole criterion for the election of President.

Section XXVIII *[Adopted by a vote of the members during March 26 – April 1, 2017]*: The United States Transhumanist Party advocates greatly shortening the timeframe for electoral campaigns. The current two-year election season, combined with voters' short memories, renders it possible for both genuine merits and egregious transgressions of candidates to be forgotten by the time of voting. Longer campaign seasons also perpetuate the "horse-race" mentality on the part of the media and result in the search for contrived election drama in order to drive views and campaign contributions. The ensuing acrimony, misinformation, and outright violence are detrimental to the fabric of a civilized society. Election seasons should be as short as possible, to enable all relevant information to be disseminated quickly and be considered by most voters within the same timeframe as their decisions are made.

Section XXIX *[Adopted by a vote of the members during March 26 – April 1, 2017]*: The United States Transhumanist Party advocates abolishing all staggered party primaries and for all primary elections to be held on the same day across the entire country. With staggered party primaries, individuals voting later – solely because of the jurisdiction in which they reside – find their choices severely constrained due to the prior elimination of candidates they might have preferred. The staggered primary system tends to elevate the candidates who are least palatable to reasonable voters – but have the support of a vociferous, crass, and often violent fringe – toward frontrunner positions that create the pressure for other members of the political party to follow suit and reluctantly support the worst of the nominees.

---

[49] "Results of Platform Vote #6". United States Transhumanist Party. November 19, 2017. Available at   http://transhumanist-party.org/wp-content/uploads/2017/11/USTP_Platform_Vote_6_Results. pdf. These results led to the adoption of Sections LXIV through LXXXII and the amendment of Sections II, XXVII, and LIX of Article III.

Section XXX *[Adopted by a vote of the members during March 26 – April 1, 2017]*: The United States Transhumanist Party supports replacing the current "winner-take-all" electoral system with proportional representation, ranked preference voting, and other devices to minimize the temptations by voters to favor a perceived "lesser evil" rather than the candidates closest to those voters' own preferences.

Section XXXI *[Adopted by a vote of the members during March 26 – April 1, 2017]*: The United States Transhumanist Party supports the right of any jurisdiction to secede from the United States specifically in opposition to policies that institutionalize racism, xenophobia, criminalization of dissent, and persecution of peaceful persons. The United States Transhumanist Party does not, however, condone any secession for the purposes of oppressing others. Therefore, the secession of the Confederate States in 1860 was illegitimate, but a future secession of a State may be justified in reaction to violent crackdowns by the federal government against individuals based on individuals' national origin or ancestry.

Section XXXII *[Adopted by a vote of the members during March 26 – April 1, 2017]*: The United States Transhumanist Party encourages every reasonable precaution to prevent existential risks that endanger sentient life. While some existential risks arise from certain technologies, many existential risks also stem from the unaltered "natural" circumstances in which humans and other life forms find themselves. For both technological and "natural" existential risks, the strongest safeguards can be achieved through responsible development of protective technologies that empower rational and moral beings.

Section XXXIII *[Adopted by a vote of the members during May 7–13, 2017]*[50]: The United States Transhumanist Party stands for the rights of any sentient entities defined in the Preamble to the Transhumanist Bill of Rights as possessing Level 5 or more advanced information integration. Any such sentient entities, including new kinds of sentient entities that may be discovered or developed in the future, shall be considered to be autonomous beings with full rights, and shall not be made subservient to humans, unless they as individuals pose direct, empirically evident threats to the lives of others. The protections of full individual rights shall extend to Level 5 or higher-level artificial intelligences. However, Level 4 or lower-level entities – including domain-specific artificial intelligences that have not achieved sentience – may be utilized as part of the production systems of the future, in a similar manner to machines, algorithms, computer programs, and non-human animals today and based on similar ethical considerations.

Section XXXIV *[Adopted by a vote of the members during May 7–13, 2017]*: The United States Transhumanist Party holds that sousveillance laws should be enacted

---

[50] "Results of Platform Vote #4". United States Transhumanist Party. May 14, 2017. Available at http://transhumanist-party.org/wp-content/uploads/2017/05/USTP_Platform_Vote_4_Results.pdf. These results led to the adoption of Sections XXXIII through XLVI of Article III.

to ensure that all members of peaceful communities feel safe, to achieve governmental transparency, and to provide counter-balances to any surveillance state. For instance, law-enforcement officials, when interacting with the public, should be required to wear body cameras or similar devices continuously monitoring their activities.

The United States Transhumanist Party supports the use of technologies which increase monitoring of police action and policing activities, with expressed goals of increasing policing accountability.

The United States Transhumanist Party advocates for a requirement that data pertaining to recordings of police action be transmitted and recorded beyond police control, so as to be protected from falsification, deletion, and selective curation by police.

Section XXXV *[Adopted by a vote of the members during May 7–13, 2017]*: The United States Transhumanist Party considers it imperative to achieve reductions of the United States national debt in order to avoid calamitous scenarios of extreme inflation, default, and burdensome future tax increases on individuals. The United States Transhumanist Party supports the following measures to reduce the national debt:

1. Elimination of wasteful federal spending on programs, goods, and services where equivalent positive results could be obtained through lower expenditures.
2. Cessation of foreign military occupations and the return of American troops to be stationed exclusively on American territory. However, if a mutually appropriate defense treaty with another country requires the United States to station troops in that country, those troops would be allowed to remain there until the treaty obligations are fulfilled or reduced by mutual agreement with the affected country. If the United States continues to station troops in any country due to mutually appropriate defense treaties, the United States Transhumanist Party supports greater reciprocity in allowing military personnel from that country to be stationed in the United States for purposes of training and information exchange.
3. Removal of barriers to technological innovation and technologically driven economic growth, in order that a surge in such growth could increase federal revenues so as to generate increasing surpluses, as long as federal spending does not materially rise from current levels.
4. Elimination of the current cumbersome system of federal contracting, which favors politically connected incumbent firms whose advantage consists of navigating the system, rather than performing the best possible work. Instead, all federal agencies should be empowered to purchase supplies and equipment and to requisition projects from any entity capable of satisfying an immediate need at a reasonable cost. Exclusive and preferential contracts for particular entities should be prohibited, and all payments by federal agencies for work by non-employees should be determined on a case-by-case basis.

5. Digitization of as many federal services and functions as possible – to eliminate the waste and expense of paperwork, physical queues, and legacy information technology systems.

Section XXXVI *[Adopted by a vote of the members during May 7–13, 2017]*: The United States Transhumanist Party supports the elimination of graduated taxation and income taxation more generally. Instead, the United States Transhumanist Party advocates a flat percentage-of-sales tax applicable only to purchases from businesses whose combined nationwide revenues from all affiliates exceed a specified threshold. This tax should be built into the price of goods from such large businesses and should not impede transaction efficiency in any manner. Transactions pertaining to wages, salaries, gifts, donations, barter, employee benefits, and inheritances should remain completely untaxed, as should transactions involving solely individuals and/or small businesses, for whom the establishment of a tax-reporting infrastructure would be onerous. Furthermore, all taxes on land and property should be abolished.

Section XXXVII *[Adopted by a vote of the members during May 7–13, 2017]*: The United States Transhumanist Party supports more proportional representation of professions and occupations among legislative and executive government officials, instead of a system in which the plurality of political offices are held by attorneys. In particular, the United States Transhumanist Party holds that a greater proportion of politicians should possess training in mathematics, engineering, and the physical and biological sciences.

Section XXXVIII *[Adopted by a vote of the members during May 7–13, 2017]*: The United States Transhumanist Party supports emerging alternative energy sources and their technological implementations. However, the United States Transhumanist Party opposes government subsidies for any energy source – including fossil fuels. Instead, the United States Transhumanist Party holds that superior, cleaner, and more efficient energy sources will also tend to become less costly and more broadly adopted under a system of unfettered market competition and innovation.

Section XXXIX *[Adopted by a vote of the members during May 7–13, 2017]*: The United States Transhumanist Party supports the creation of a "Transhumanist Olympics" where augmentations and modifications of the human body would not disqualify persons from competing.

Section XL *[Adopted by a vote of the members during May 7–13, 2017]*: In addition to its opposition to intolerant interpretations of religious doctrines, the United States Transhumanist Party is furthermore opposed to any interpretation of a secular, non-religious doctrine that results in the rejection of reason, censorship, violation of individual rights, suppression of technological advancement, and attempts to impose certain beliefs by force and/or by legal compulsion. Examples of such doctrines opposed by the United States Transhumanist Party include Stalinism, Maoism, Neo-Malthusianism or eco-primitivism, the death-acceptance movement, and the doctrine of censorship, now prevalent on many college campuses in the United States, in the name of "social justice", combating "triggers" or "microaggressions", or avoiding subjectively perceived offense.

Section XLI *[Adopted by a vote of the members during May 7–13, 2017]*: The United States Transhumanist Party understands that the role of President comes with great power and responsibility towards all citizens equally, regardless of ethnicity, race, sex, gender, religious conviction or lack thereof, political position, or societal class. It is the United States Transhumanist Party's view that the President, as an authority figure and head of state, should lead by example. The function of the President is to oversee and support the improvement of well-being for all United States citizens, and the welfare of the nation as a whole.

The United States Transhumanist Party recognizes that the power of the Executive Branch of the federal government has expanded far beyond the originally conceived Constitutional framework, so as to predominate over the Legislative and Judicial Branches, as well as over the institutions of civil society and individuals. The United States Transhumanist Party supports greatly curtailing and restricting the role of the U.S. President so as to confine that role within parameters originally conceived by the framers of the U.S. Constitution – particularly with regard to eliminating the unacceptable current prerogatives to unilaterally launch nuclear strikes and conduct military attacks, and to imprison, spy on, and assassinate Americans and others without due process.

Section XLII *[Adopted by a vote of the members during May 7–13, 2017]*: The United States Transhumanist Party supports efforts to minimize conflicts of interest for government officials created by private businesses, religious institutions, not-for-profit organizations, special-interest groups, and certain individuals. However, the United States Transhumanist Party recognizes that the best way to minimize such conflicts of interest is not to micromanage the conduct of government officials – which can prevent petty improprieties but is inherently unable to avert the most damaging conflicts of interest among the highest echelons of power. Rather, the most promising solution is to reduce the scope of special economic privileges and protections that any government official would be capable of granting, thereby greatly diminishing the incentives of various special interests to even attempt to influence government officials.

Section XLIII *[Adopted by a vote of the members during May 7–13, 2017]*: Irrespective of the means by which a government obtains its funds – be it from taxation or from other sources – the United States Transhumanist Party understands that a necessary function of government is to responsibly allocate such funds to protect the rights and increase the well-being of its citizens and other individuals within its jurisdiction. If a government requires the public to contribute to its funding, then the services, utilities, and research produced by that government should be easily and freely accessible to members of the public who have contributed such funds. Any government expenditure should be premised on the goal of increasing the well-being of citizens and other individuals within the government's jurisdiction in the short, medium, and long terms, with the understanding that government exists to serve the people, and should allocate funds wisely with the intent of maximizing value per dollar for the purpose of protecting individuals' rights and promoting their well-being.

Section XLIV *[Adopted by a vote of the members during May 7–13, 2017]*: The United States Transhumanist Party supports efforts to have proposed laws accompanied by projections of expected results, including short-term, medium-term, and long-term effects. Such analysis should be based on scientific research and evidence and supported by the scientific and/or academic community with relevant subject-matter expertise. The intent is to have laws proposed to be created for the present day, and for such laws to function or improve in the future.

Section XLV *[Adopted by a vote of the members during May 7–13, 2017]*: The United States Transhumanist Party supports efforts to depoliticize the appointment of Supreme Court Justices, and to further incentivize their impartiality toward political viewpoints.

Section XLVI *[Adopted by a vote of the members during May 7–13, 2017]*: The United States Transhumanist Party supports efforts to revisit, condense, and simplify the law, with a focus on resolving issues among conflicting laws and closing illogical loopholes within the law.

Section XLVII *[Adopted by a vote of the members during June 18–24, 2017][51]*: The United States Transhumanist Party supports efforts to have bills proposed without sub-sections or provisions unrelated to the main subject of the bill. A single-subject or germaneness rule for bills would:

1. Simplify bills, rendering them more accessible and less convoluted;
2. Enable a focused vote for or against a bill without the possibility of having to accept or reject an embedded unrelated provision; and
3. Prevent an unrelated provision from being buried within a bill as a possible tactic to have it passed.

Section XLVIII *[Adopted by a vote of the members during June 18–24, 2017]*: The United States Transhumanist Party supports efforts to limit protectionism and subsidization of an industry or group of companies. The exception to this would be that of extenuating circumstances, such as natural disasters or catastrophes, in which case a limited window of support could be approved. The United States Transhumanist Party understands that in a free-market society, private businesses, in order to continue their existence, ought to adapt to market changes instead of being shielded from such changes.

---

[51] "Results of Platform Vote #5". United States Transhumanist Party. July 3, 2017. Available at http://transhumanist-party.org/wp-content/uploads/2017/07/USTP_Platform_Vote_5_Results.pdf. These results led to the adoption of Sections XLVII through LXIII of Article III.

Section XLIX *[Adopted by a vote of the members during June 18–24, 2017]*: The United States Transhumanist Party supports efforts to increase autonomy of individuals to decide over their own bodies and holds that individuals should have the legal right to undertake procedures including gender reassignment, hysterectomies, vasectomies, technological augmentation, cosmetic alterations, genetic enhancements, and physical supplementation at or after the age of 18 years, as long as this does not create health hazards or threats to other individuals.

Section L *[Adopted by a vote of the members during June 18–24, 2017]*: The United States Transhumanist Party supports the autonomy of an individual to decide on the continuation of that individual's own life, including the right to choose or not to choose life-extending medical treatments. The United States Transhumanist Party does not consider it practicable or desirable for suicide to be illegal but discourages suicide from a moral standpoint, and furthermore considers that the legal right of suicide should only pertain to the individual and should not extend to any euthanasia or direct administration of a life-ending substance or procedure by any other person. The United States Transhumanist Party has grave concerns with anybody but the individual acting to hasten the end of that individual's life.

Although each individual should be free to decide upon the duration of his, her, or its own life, the United States Transhumanist Party supports cultural changes and discussions that would encourage all individuals to undertake life-prolonging choices and activities. Advances in medical technology would facilitate more open-ended lifespans and would enable individuals to choose either finite or indefinite lengths of their lives. However, if individuals are recognized as having this autonomy, the United States Transhumanist Party is interested in persuading as many people as possible to decide to preserve their irreplaceable lives instead of hastening their end.

With regard to any legalization of assisted suicide or measures to provide patients with life-ending prescriptions, the United States Transhumanist Party supports stringent legal safeguards to ensure that each individual patient's choice with regard to such matters is entirely free and uncoerced, and that there is no steering of any particular individual toward a life-ending choice by family members, medical practitioners, health insurers, activists, or any other individual or organization standing to benefit financially from the end of a patient's life. However, efforts to persuade an individual to prolong his, her, or its life should not be restricted.

The United States Transhumanist Party opposes the emergence of any financially motivated lobby or industry whose primary business model would be assisted suicide or euthanasia, as the existence of such a lobby could create incentives and policies to steer people toward life-ending choices, including through legislation that might favor such "choices" in not-quite-voluntary situations. Instead, any prescription for a life-ending substance should only be provided as an incidental service by a patient's primary-care physician, with the express written consent of at least one other unaffiliated physician, and the substance in question should only be allowed to be self-administered by the patient directly after a pre-defined time period since the obtaining of the prescription. Once the substance is prescribed, no medical practitioner should be permitted to benefit financially based on any specific choice of the patient to self-administer the substance to end the patient's life. This position should not be construed to restrict any non-financially motivated political advocacy on the subject of assisted suicide, which involves individuals expressing their views on this subject in a public forum, when those individuals do not stand to gain financially from others choosing to obtain a life-ending substance.

Section LI *[Adopted by a vote of the members during June 18–24, 2017]*: The United States Transhumanist Party supports efforts to establish a cross-border or international organ-donation system so that organ donors who wish to do so may donate their organs in a foreign country. This could pertain to Americans working or traveling in foreign countries, but also foreigners or travelers who pass away within U.S. borders. This system would be particularly useful for saving lives with organs that have a very short preservation duration, and would take too long to be sent to the country of the donor's nationality.

Section LII *[Adopted by a vote of the members during June 18–24, 2017]*: The United States Transhumanist Party supports efforts to increase the ability for public sousveillance on the functioning of government officials, in particular those who may propose laws, during negotiations and deliberations on proposing bills and national and international trade agreements. Furthermore, the United States Transhumanist Party supports efforts to make the current hosting of live-streams from United States Congress more user-friendly and accessible to the public, accompanied by links to proposed bills where applicable.

Section LIII *[Adopted by a vote of the members during June 18–24, 2017]*: The United States Transhumanist Party supports efforts to have a mandatory standard clause or affidavit, affirming that a Representative, Senator, or other Legislative Branch lawmaker proposing a piece of legislation, such as a bill in Congress, has no conflict of interest between serving the public and serving other parties, such as special-interest groups. The clause would have to be signed and dated by the representative before the legislation is allowed to be proposed.

Section LIV *[Adopted by a vote of the members during June 18–24, 2017]*: The United States Transhumanist Party supports increasing broad accountability of Federal Government departments, agencies, and entities, especially those tasked with national security and / or criminal investigations, to the United States Congress. Currently some agencies may receive government funding without any accountability as to what the funding is used for, often based on arguments that this information is 'classified' or 'may not be revealed in the interest of national security'. This is irresponsible use of taxpayer money.

The United States Transhumanist Party does acknowledge that such entities or agencies may have security concerns regarding the publication of details of their budget plans. As such, the United States Transhumanist Party supports setting up a special non-partisan security budgetary review committee where more details of budget plans would have to be provided before considering to provide funds to an agency or entity.

Section LV *[Adopted by a vote of the members during June 18–24, 2017]*: The United States Transhumanist Party supports efforts to consolidate and reduce some redundancies among agencies and entities tasked with national security and law enforcement, as well as to reduce the number of such agencies and entities currently in operation. However, while supporting the elimination of parallel redundancies which can create problems, the United States Transhumanist Party recognizes that certain types of hierarchical redundancies can help with quality control.

Section LVI *[Adopted by a vote of the members during June 18–24, 2017]*: The United States Transhumanist Party supports efforts to ensure that no United States Representative or Senator may be obstructed in their ability to vote on any piece of legislation, or be kept from the Senate or House of Representatives for intra- or extra-curricular political-party activities which interfere with their primary task as representatives of the people within government. For example, protections should exist to prevent situations where Representatives or Senators are forced by their political parties to do fundraising calls during a vote on a bill.

Section LVII *[Adopted by a vote of the members during June 18–24, 2017]*: The United States Transhumanist Party supports efforts to restrict and limit civil asset forfeiture laws, and other laws that assist law-enforcement agencies in circumventing the Fourth Amendment, such as asset seizure, or detainment or arrest in situations where no criminal charges have been filed, except as part of an active interrogation of a person suspected of a crime or unless the person detained or arrested poses a clear and probable danger of inflicting physical harm upon others or their property.

Section LVIII *[Adopted by a vote of the members during June 18–24, 2017]*: The United States Transhumanist Party supports efforts to investigate questionable, but currently legal, actions by law-enforcement agencies that have over time garnered critical attention by the public. The safety of the public could benefit from such actions being revisited or revised to limit abuse and to close legal loopholes.

Section LIX *[Adopted by a vote of the members during June 18–24, 2017; amended by a vote of the members during November 11–17, 2017]*[52]: The United States Transhumanist Party considers it important for impartial, objective investigations of alleged police and other law-enforcement misconduct to be pursued. While law-enforcement agencies should not be prohibited from internally investigating potential abuses within their own ranks, such investigations should never be considered exclusive or conclusive, and further external checks and accountability should be instituted. As part of providing such checks and accountability, investigations

---

[52] "Results of Platform Vote #6". United States Transhumanist Party. November 19, 2017. Available at http://transhumanist-party.org/wp-content/uploads/2017/11/USTP_Platform_Vote_6_Results. pdf. These results led to the adoption of Sections LXIV through LXXXII and the amendment of Sections II, XXVII, and LIX of Article III.

regarding misconduct, negligence, abuse, criminal activity, felonies, and misde-meanors allegedly committed by police, district attorneys, and judges, should, in addition to any internal investigation, also be conducted by a civilian organization outside the justice system. The intent of this requirement is to limit the possibility of favorably biased or preferential treatment of a member of a given law-enforcement agency by that person's colleagues, and to restore confidence by the public that an investigation into police misconduct is done as objectively as possible.

Section LX *[Adopted by a vote of the members during June 18–24, 2017]*: The United States Transhumanist Party supports efforts to limit the possibility for police, district attorneys, and judges to favor one another through mutual "back-scratching" accommodations which may cause a particular criminal matter to be resolved in a manner inconsistent with the true facts of the situation or the requirements of applicable law.

Section LXI *[Adopted by a vote of the members during June 18–24, 2017]*: The United States Transhumanist Party supports efforts to prevent members of Congress from receiving special benefits, subsidies, and tax breaks that other citizens do not receive, and that are not necessary to function as a member of Congress. This limitation would pertain, for example, to health-care subsidies that are inaccessible to other citizens. However, this limitation would not prevent members of Congress from obtaining working conditions and job-related benefits of the sort which are broadly available, without regard to rank or degree of influence, to other Americans working within the private or public sectors.

Section LXII *[Adopted by a vote of the members during June 18–24, 2017]*: The U.S. Transhumanist Party supports efforts to ratify the United Nations Convention on the Rights of the Child, and to uphold the Rights of the Child as prescribed therein.[53] This would include abolishing the death penalty for minors federally.

The United States Transhumanist Party, however, opposes restrictions on the rights of parents to choose to homeschool their children in any manner that respects the children's basic freedom of conscience. Any ratification of the United Nations Convention on the Rights of the Child should not be construed to restrict any peaceful, rights-respecting practice of homeschooling.

Section LXIII *[Adopted by a vote of the members during June 18–24, 2017]*: The United States Transhumanist Party opposes those specific cultural, religious, and social practices that violate individual rights and bodily autonomy. Examples of such unacceptable practices are forced marriage (including child marriage), male and female genital mutilation, and honor killings.

---

[53] "Rights under the Convention on the Rights of the Child". UNICEF. August 7, 2014. Available at https://www.unicef.org/crc/index_30177.html

Section LXIV *[Adopted by a vote of the members during November 11–17, 2017]*[54]: The United States Transhumanist Party supports efforts to create a reasonable minimum timeframe between the proposal of a bill and the voting procedure. To ensure a reasonable timeframe is proportional to the number of pages of a proposed bill, a time period per each specified amount of pages could be adopted. For example, and without committing to specific numerical magnitudes, a 24-hour period within a working week per every 20 pages could be adopted to ensure all members of Congress involved have sufficient time to read through and study a proposed bill's implications. Such measures would prevent a bill from being introduced shortly before the voting process. They would also have the added side effect that proposals might become more concise, as the length of a bill would influence the consideration time.

In addition to this, after the proposal has been submitted, any amendments must be explicitly discussed in a public forum with the same degree of thorough consideration and same rules pertaining to the timeframe of consideration as allowed for the original proposal.

Section LXV *[Adopted by a vote of the members during November 11–17, 2017]*: The United States Transhumanist Party supports efforts to ensure a jury is fully informed on its rights and responsibilities, including jury nullification. The United States Transhumanist Party also supports efforts to prevent false claims being made regarding the rights and responsibilities of the jury.

Section LXVI *[Adopted by a vote of the members during November 11–17, 2017]*: As an intermediate step toward the goal of complete nuclear disarmament and a potential pragmatic compromise in any future negotiations for disarmament, the United States Transhumanist Party supports efforts to significantly reduce the United States nuclear stockpile, and to replace or transfer a small part (between 1 percent and 10 percent) of this stockpile, to mobile nuclear platforms such as submarines. An enemy may currently target the stationary nuclear bombs directly. Having a largely hidden mobile fleet of nuclear bombs would render it much more difficult for any enemy to target the nuclear arsenal, while still maintaining the nuclear deterrent option in sufficient capacity. This would further have the effect of lowering the budget required to maintain the nuclear stockpile, as it could be drastically downsized.

Section LXVII *[Adopted by a vote of the members during November 11–17, 2017]*: The United States Transhumanist Party supports the right for individuals to have autonomy over, and utilize their bodies to earn money, including through activities such as prostitution, as long as such activities arise from a person's own free will (e.g., not under duress), and the person is not endangering the health or

---

[54] "Results of Platform Vote #6". United States Transhumanist Party. November 19, 2017. Available at http://transhumanist-party.org/wp-content/uploads/2017/11/USTP_Platform_Vote_6_Results. pdf. These results led to the adoption of Sections LXIV through LXXXII and the amendment of Sections II, XXVII, and LIX of Article III.

well-being of others, including but not limited to the communication of sexually transmissible diseases.

Legalization would give those who wish to engage in prostitution the safety and protection of the law – for example, so that they may report abuse and would be prone to being exploited. It would also open the possibility for such individuals to unionize if they wish to do so. Furthermore, legalization would decrease government spending on what is ultimately a 'moral crusade'.

However, the United States Transhumanist Party unequivocally condemns any manner of human trafficking, child exploitation, and other abuse that involves a violation of the autonomy and consent of any individual. The legalization of prostitution should be combined with stronger efforts to combat these dangerous and exploitative practices.

The legalization of prostitution could furthermore enable more effective action against human trafficking and involuntary exploitation, as, when prostitution is legalized, employees in this industry would become subject to the protections of the law. Legalization would help focus resources on combating the trafficking of humans rather than catching workers who chose this profession voluntarily.

Section LXVIII *[Adopted by a vote of the members during November 11–17, 2017]*: The United States Transhumanist Party holds that any statement made by an elected official and/or public servant to members of the public in a public forum as part of that official's or public servant's job duties, and available to be heard, read, or otherwise understood in a public setting, physical or digital, should be considered a part of the public record and treated as an official statement of their office and position. This requirement does not extend to statements made by an elected official and/or public servant in the capacity of a private citizen or in the expression of a personal opinion or other position unrelated to the exercise of the official's or public servant's job duties.

Section LXIX *[Adopted by a vote of the members during November 11–17, 2017]*: The United States Transhumanist Party holds that state and federal governments should establish an artificial intelligence (AI) analysis system for measuring risk of proposed legislation. Such a system could provide an impartial look at what legislation could cause harm or unintended consequences. Submitted policies would receive a score from 0 to 100, and the AI system would state what possible negative impacts may result. This system should be publicly accessible for submissions and for security audit. This is not intended to create or enact laws, but simply to serve as a tool to measure risk versus reward.

Section LXX *[Adopted by a vote of the members during November 11–17, 2017]*: The United States Transhumanist Party strongly opposes the possibility for any political party to determine the boundaries and borders of any voting district. The United States Transhumanist Party supports measures that require any efforts to have the districts potentially redrawn, when necessary due to migration for example, to be left to an automated system such as an artificial intelligence (AI) designed for this task.

Section LXXI *[Adopted by a vote of the members during November 11–17, 2017]*: The United States Transhumanist Party supports efforts to remove the possibility for a President to sign an international agreement among two or more nations by executive order. This would prevent a President from engaging in international affairs without support from the Congress, and likewise would make it more difficult to exit an international agreement, as support from Congress would need to exist in order for such an exit to occur. This would furthermore ensure that the United States becomes a more trustworthy nation in the eyes of the international community.

Section LXXII *[Adopted by a vote of the members during November 11–17, 2017]*: The United States Transhumanist Party supports efforts to create a framework for an international or world passport. This framework could, for example, be administered through the United Nations, and the passport could be valid only for those countries who have proven to meet the standards, set by participating countries, required to ensure safety. Given that the European Union has an ID valid within its borders, and the United States has a similar agreement with Canada, imagining these forms of identification being combined shows that a world passport is not a farfetched or alien idea.

Section LXXIII *[Adopted by a vote of the members during November 11–17, 2017]*: The United States Transhumanist Party supports efforts to mandate that new firearms will be produced with an embedded registration chip, as well as the registration number engraved on the firearm. The chip would have a registration number, a 'trace online' code, and a 'lost or stolen' code. The firearm would be accompanied by a physical and digital certificate of ownership with a registration number, the 'trace online' number, and the 'lost or stolen' number. The embedded chip would render it much harder to make the firearm untraceable. The number and codes involved would have to be unique identifiers.

When a firearm would be lost or stolen, the owner of the firearm and holder of the certificate would report the firearm to the authorities as lost or stolen using the 'lost or stolen' number. The intent of having a separate reporting number is to ensure that, in the event that a firearm is stolen or taken without consent, the offender cannot report the firearm as stolen, or as found again. When a firearm owner suspects the firearm is simply mislocated – for example, in the car or in the house, or perhaps taken by a family member, the owner could trace the firearm online via GPS on an online map, using the 'trace online' code. The 'trace online' code would never be revealed to law enforcement.

Law enforcement would have a device that can confirm the registration number of a firearm in close proximity, similar to contactless payments. On the other hand the detection range would be greater when a firearm has been reported lost or stolen by the owner of the firearm in question with the 'lost and stolen' number. A firearm that would not have been reported lost or stolen to law enforcement by the owner of the firearm with the 'lost or stolen' number would not be traceable from a greater distance by law enforcement.

This would ensure the privacy, safety, and peace of mind of firearm owners who might otherwise feel law enforcement would trace firearms without legal justification. At the same time, this measure would decrease the probability of stolen firearms never being found, and possibly ending in the hands of people with ill intent.

Section LXXIV *[Adopted by a vote of the members during November 11–17, 2017]*: The United States Transhumanist Party supports the creation of an office of a Public Civil-Rights Prosecutor. Our current justice system is flawed. Only people with substantial wealth can afford lawyers to take legal action against those who attack a person's rights. While there are organizations like the American Civil Liberties Union (ACLU), which offer help for some cases, their ability to do so is severely limited and typically non-existent in the lower courts. A Public Civil-Rights Prosecutor's office will help guarantee that, no matter who a person is or the position of the offending party, a person's rights cannot be assailed without consequence.

Section LXXV *[Adopted by a vote of the members during November 11–17, 2017]*: The United States Transhumanist Party supports lowering spending by the Department of Defense and the U.S. Military, which amounts to hundreds of billions of dollars per year and includes unchecked wastefulness. Reducing military spending would free up money for more important goals, such as curing disease, which collectively kills many more people than military conflict or war by an exponential degree.

Section LXXVI *[Adopted by a vote of the members during November 11–17, 2017]*: The United States Transhumanist Party supports efforts to hold institutions, corporations, and states accountable for usage of federal money with a specific intended purpose. When an entity has been granted any form of funding with a specifically intended purpose, such as disaster relief or specific educational funds, and these funds are misappropriated or used for other purposes well outside of the scope of what they were intended for, the entity in question ought to restitute the funding that was made available.

Section LXXVII *[Adopted by a vote of the members during November 11–17, 2017]*: The United States Transhumanist Party supports increases in the budget for the National Institutes of Health (NIH). Numerous biotech CEOs have recently made the case to increase the NIH budget, because the NIH conducts research that their companies would not be able to invest in, as investments not leading directly to a product would affect the bottom line. 33% of all the publications from NIH research are cited in corporate patents, so it stimulates new product development. A major driver for economic progress and reducing the suffering of those in pain, the NIH is essentially a public charity that brings us into the future. Whether one supports limited or expansive government, the NIH does not seek to regulate anything nor impose laws on anyone. It exclusively conducts medical research to help the sick.

Section LXXVIII *[Adopted by a vote of the members during November 11–17, 2017]*: The U.S. Transhumanist Party supports efforts to reinstate the rights to vote for convicted felons who have received and served their punishment, in order to present them the opportunity to participate in society as otherwise normal citizens.

Section LXXIX *[Adopted by a vote of the members during November 11–17, 2017]*: The United States Transhumanist Party supports repealing the current requirement in the United States that drugs or treatments may not be used, even on willing patients, unless approval for such drugs or treatments is received from the Food and Drug Administration. Such requirements are a profound violation of patient sovereignty; a person who is terminally ill is unable to choose to take a risk on an unapproved drug or treatment unless this person is fortunate enough to participate in a clinical trial. Even then, once the clinical trial ends, the treatment must be discontinued, even if it was actually successful at prolonging the person's life. This is not only profoundly tragic, but morally unconscionable as well. The most critical reform needed is to allow unapproved drugs and treatments to be marketed and consumed. If the FDA wishes to strongly differentiate between approved and unapproved treatments, then a strongly worded warning label could be required for unapproved treatments, and patients could even be required to sign a consent form stating that they have been informed of the risks of an unapproved treatment. This reform to directly extend many lives and to redress a moral travesty should be the top political priority of advocates of indefinite life extension. Over the coming decades, its effect will be to allow cutting-edge treatments to reach a market sooner and thus to enable data about those treatments' effects to be gathered more quickly and reliably. Because many treatments take 10–15 years to receive FDA approval, this reform could by itself speed up the real-world advent of indefinite life extension by over a decade.

Section LXXX *[Adopted by a vote of the members during November 11–17, 2017]*: The United States Transhumanist Party supports efforts to increase opportunities for entry into the medical profession. The current system for licensing doctors is highly monopolistic and protectionist – the result of efforts by the American Medical Association in the early twentieth century to limit entry into the profession in order to artificially boost incomes for its members. The medical system suffers today from too few doctors and thus vastly inflated patient costs and unacceptable waiting times for appointments. Instead of prohibiting the practice of medicine by all except a select few who have completed an extremely rigorous and cost-prohibitive formal medical schooling, governments in the Western world should allow the market to determine different tiers of medical care for which competing private certifications would emerge. For the most specialized and intricate tasks, high standards of certification would continue to exist, and a practitioner's credentials and reputation would remain absolutely essential to convincing consumers to put their lives in that practitioner's hands. But, with regard to routine medical care (e.g., annual check-ups, vaccinations, basic wound treatment), it is not necessary to receive attention from a person with a full-fledged medical degree. Furthermore, competition among certification providers would increase quality of training and lower its price, as well as accelerate the time needed to complete the training. Such a system would allow many more young medical professionals to practice without undertaking enormous debt or serving for years (if not decades) in roles that offer very little remuneration while entailing a great deal of subservience to the hierarchy of an established institution. Ultimately, without sufficient doctors to affordably deliver life-extending treatments when they become available, it would not be feasible to extend these treatments to the majority of people.

Section LXXXI *[Adopted by a vote of the members during November 11–17, 2017]*: The United States Transhumanist Party supports reforms to the patent system that prevent the re-patenting of drugs and medical devices, or the acquisition of any exclusive or monopoly rights over those drugs and devices, once they have become generic or entered the public domain. Appallingly, many pharmaceutical companies today attempt to re-patent drugs that have already entered the public domain, simply because the drugs have been discovered to have effects on a disease different from the one for which they were originally patented. The result of this is that the price of the re-patented drug often spikes by orders of magnitude compared to the price level during the period the drug was subject to competition. Only a vibrant and competitive market, where numerous medical providers can experiment with how to improve particular treatments or create new ones, can allow for the rate of progress needed for the people alive today to benefit from radical life extension.

Section LXXXII *[Adopted by a vote of the members during November 11–17, 2017]*: The United States Transhumanist Party supports reforms to reduce the lengths of times over which medical patents could be effective. Medical patents – in essence, legal grants of monopoly for limited periods of time – greatly inflate the cost of drugs and other treatments. Especially in today's world of rapidly advancing biotechnology, a patent term of 20 years essentially means that no party other than the patent holder (or someone paying royalties to the patent holder) may innovate upon the patented medicine for a generation, all while the technological potential for such innovation becomes glaringly obvious. As much innovation consists of incremental improvements on what already exists, the lack of an ability to create derivative drugs and treatments that tweak current approaches implies that the entire medical field is, for some time, stuck at the first stages of a treatment's evolution – with all of the expense and unreliability this entails. Even with shortened patent terms, the original developer of an innovation will still always benefit from a first-mover advantage, as it takes time for competitors to catch on. If the original developer can maintain high-quality service and demonstrate the ability to sell a safe product, then the brand-name advantage alone can secure a consistent revenue stream without the need for a patent monopoly.

Gennady Stolyarov II is an actuary, author of Death is Wrong (2013), Editor-in-Chief of The Rational Argumentator, Chairman of the U.S. Transhumanist Party, and Chief Executive of the

Nevada Transhumanist Party. Mr. Stolyarov holds the professional insurance designations of Fellow of the Society of Actuaries (FSA), Associate of the Casualty Actuarial Society (ACAS), Member of the American Academy of Actuaries (MAAA), Chartered Property Casualty Underwriter (CPCU), Associate in Reinsurance (ARe), Associate in Regulation and Compliance (ARC), Associate in Personal Insurance (API), Associate in Insurance Services (AIS), Accredited Insurance Examiner (AIE), and Associate in Insurance Accounting and Finance (AIAF). Mr. Stolyarov can be contacted via e-mail at gennadystolyarovii@gmail.com.

Part II
Artificial Intelligence, Machine Learning, and Superintelligence

Chapter 6
Beauty Is in the A.I. of the Beholder: Artificial and Superintelligence

Newton Lee

> *Either we need an exponential improvement in human behavior—less selfishness, less short-termism, more collaboration, more generosity—or we need an exponential improvement in technology. If you look at current geopolitics, I don't think we're going to be getting an exponential improvement in human behavior any time soon. That's why we need a quantum leap in technology like AI (artificial intelligence).*
> —Demis Hassabis (Google DeepMind)

6.1 To Err Is Human and Artificially Intelligent

To err is human. Yet it is a sign of how far computer science has come that to err is also artificially intelligent [1].

When IBM Deep Blue won its six-game chess match against Garry Kasparov in May 1997, marking the first defeat of a reigning world chess champion to a computer under tournament conditions, there was one particular moment that stood out in Kasparov's mind.

As he described in *Time* magazine: "I got my first glimpse of artificial intelligence on Feb. 10, 1996, at 4:45 p.m. EST, when in the first game of my match with Deep Blue, the computer nudged a pawn forward to a square where it could easily be captured. It was a wonderful and extremely human move... I had played a lot of computers but had never experienced anything like this. I could feel – I could smell – a new kind of intelligence across the table" [2].

In Nate Silver's book *The Signal and the Noise*, IBM scientist Murray Campbell from the Deep Blue team revealed that the "extremely human move" in the game

N. Lee (✉)
California Transhumanist Party, Los Angeles, CA, USA

Institute for Education, Research, and Scholarships, Los Angeles, CA, USA
e-mail: newton@californiatranshumanistparty.org; newton@ifers.org

© Springer Nature Switzerland AG 2019
N. Lee (ed.), *The Transhumanism Handbook*,
https://doi.org/10.1007/978-3-030-16920-6_6

against Kasparov was actually a bug in the program that was later fixed [3]. Moreover, machine learning based on limited or biased datasets can produce results that are misleading or completely wrong [4].

Nevertheless, for many like Kasparov, that moment of artificial erring has come to be seen as a pivotal moment in the development of artificial intelligence, like the proverbial genie leaving the bottle.

6.2   Artificial Intelligence and the Turing Test

Alan Turing was a British mathematician, cryptanalyst, and computer scientist who is widely considered to be the father of computer science and artificial intelligence. In 1950, Alan Turing introduced the famous imitation game (aka the Turing Test) as a test of a machine's ability to exhibit intelligent behavior equivalent to, or indistinguishable from, that of a human [5].

John McCarthy coined the term "artificial intelligence" (AI) a few years later in 1955 when he proposed a summer research conference: "The study is to proceed on the basis of the conjecture that every aspect of learning or any other feature of intelligence can in principle be so precisely described that a machine can be made to simulate it" [6].

Joseph Weizenbaum's ELIZA program in 1966 was arguably a successful attempt to pass the Turing Test. ELIZA simulated a conversation between a patient and a psychotherapist by using a person's responses to shape the computer's replies [7]. Although ELIZA has no understanding of natural language or psychotherapy, many people were fooled into believing that ELIZA was human.

More recently in 2010, AI researcher Rollo Carpenter developed the Cleverbot chat engine that learns from a growing database of 20+ million human online conversations [8]. Cleverbot passed the Turing Test with flying colors at the 2011 Techniche festival in Guwahati, India. Chat participants and the audience rated the humanness of all chat responses, with Cleverbot voted 59.3% human, while the humans themselves were rated just 63.3% human [9].

But AI is much more useful than being able to fool people.

6.3   Expert Systems

An expert system is an AI program that emulates the decision-making ability of a human expert by reasoning about knowledge.

In 1965, AI researcher Edward Feigenbaum and geneticist Joshua Lederberg of Stanford University began to develop DENDRAL, a chemical-analysis expert system that hypothesizes a test substance's molecular structure [10]. DENDRAL's performance has rivaled that of chemists.

In the early 1970s at Stanford University, Edward Shortliffe created MYCIN, a rule-based expert system that consults with physicians about the diagnosis and treatment of infectious diseases [11]. MYCIN has been shown to outperform doctors in some cases.

For my undergraduate research at Virginia Tech in 1980s, I developed an expert system for information on pharmacology and drug interactions [12]. Under the supervision of John Roach, I picked the brains of pharmacologists Jeff Wilcke and Marion Ehrich during the knowledge acquisition phase of the knowledge engineering process. It was an eye opener for a computer science student who had to study pharmacology in order to create a working expert system. For knowledge representation, I organized and encoded the pharmacological information in rules and frames for systematic retrieval, including:

(a) delineation, definition, and hierarchical subdivision of mechanisms responsible for drug interactions;
(b) division of pharmacological agents into a hierarchy of subclasses to allow for defining interacting drugs by classes as well as by specific agents; and
(c) correlation of drug classes and specific drugs with mechanisms by which they may be involved in drug interactions.

This information, accessible through a natural language-like and menu driven interface, allows clinicians to know:

(a) what may happen when two drugs are used together, and why;
(b) what can be done to alleviate detrimental interactions; and
(c) what related drugs may also be involved in similar interactions.

AI and expert systems were mostly confined to the academic world, research communities, and government applications until May 1997 when the IBM Deep Blue computer beat the world chess champion Garry Kasparov after a six-game match, marking the first time in history that a computer had ever defeated a world champion in a match play [13].

6.4 AlphaGo: Expert Performance

Fast forward 20 years from Kasparov's defeat, the battlefield between computers and human grandmasters has moved to the ancient Chinese board game of Go. Go is seen as a far more difficult game for computers to master because of its heavy reliance on intuition, strategic thinking and winning multiple battles across the board [14]. A computer cannot simply memorize all combinations of board pieces, assess the situation, construct and execute a strategy to win, like in chess.

Yet that has not stopped AlphaGo, developed by Alphabet Inc's Google DeepMind, from conquering all before it. AlphaGo's deep neural networks enable it to teach itself how to play the game. Its programmers set up the basic heuristics of the game, giving AlphaGo a database of 30 million board positions drawn from 160,000 real-life games to analyze, then split its mind so that it could play itself millions of times, learning as it went [15].

That strategy has paid off. In October 2015, AlphaGo beat three-time European champion Fan Hui by 5 games to 0, marking the first time in history a computer had beaten a professional human on a full-sized 19x19 board without handicap [16]. In March 2016, it beat South Korea's 18-time world champion Lee Sedol 4 to 1 [17]. And from late 2016 to early 2017, AlphaGo (disguised as "Magister" and "Master") secretly played 51 online matches against some of the world's best players, winning every one [18]. A tweet called the mysterious AI "Master" a superhuman (see Fig. 6.1).

In the final showdown in May 2017, AlphaGo and the world's top-ranked Go player Ke Jie faced off in a three-game match under tournament conditions in Wuzhen, China [19]. AlphaGo won game 1 by 0.5 point. Ke Jie resigned game 2 and game 3 after losing to AlphaGo. "Last year, I think the way AlphaGo played was pretty close to human beings, but today I think he plays like the God of Go," said Ke [20].

If constantly losing to a computer opponent does not sound like much fun, think again. Fans of AlphaGo say its dominance is liberating. Humans can learn from it. As in any competition, we want to play against stronger opponents in order to improve our games. "AlphaGo's play makes us feel free, that no move is impossible," said professional Go player Zhou Ruiyang. "Now everyone is trying to play in a style that hasn't been tried before" [21].

Fig. 6.1 AlphaGo disguised as a mysterious AI named "Master"

6.5   AlphaGo Zero: Superhuman Performance

AlphaGo became the world champion in May 2017. However, a champion cannot guarantee that it will never lose when some day in the future there may be a human prodigy who will turn the tables and beat AlphaGo. Instead of waiting for that day to arrive, Google DeepMind ensures that that day will never come by inventing AlphaGo Zero.

In the October 2017 issue of *Nature*, Google scientists publicly announced that they had created AlphaGo Zero with "superhuman" performance. David Silver and his team wrote, "Here we introduce an algorithm based solely on reinforcement learning, without human data, guidance or domain knowledge beyond game rules. AlphaGo becomes its own teacher: a neural network is trained to predict AlphaGo's own move selections and also the winner of AlphaGo's games. ... Starting *tabula rasa*, our new program AlphaGo Zero achieved superhuman performance, winning 100–0 against the previously published, champion-defeating AlphaGo" [22].

Unlike AlphaGo that learns from a knowledge database of 30 million board positions drawn from 160,000 real-life games, AlphaGo Zero has zero knowledge of human game plays and strategies. Given just the basic game rules, AlphaGo Zero figures out everything on its own without human interference. After 40 days and 29 million games of self-play, AlphaGo Zero beat AlphaGo and surpassed all human Go players now and forever (see Figs. 6.2, 6.3, 6.4 and 6.5).

While its victories in the world of Go have stolen much of the limelight, Google DeepMind has other things on its, well, mind. Apart from well-defined board games, DeepMind has also been learning to play video games in a chaotic virtual world. "Just as if you think of car as a machine that amplifies human capabilities physically, the computer to me felt like that for the mind," said DeepMind co-founder Demis Hassabis who is a British artificial intelligence researcher, neuroscientist,

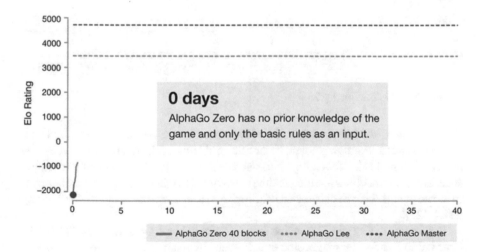

Fig. 6.2 AlphaGo Zero on day 0

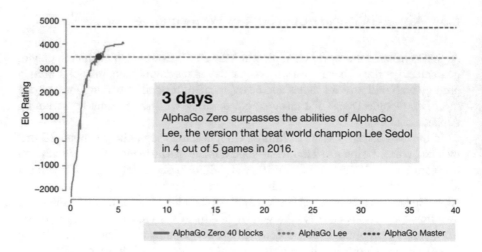

Fig. 6.3 AlphaGo Zero on day 3

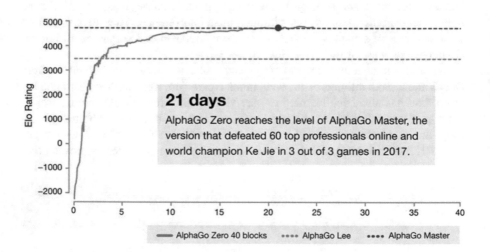

Fig. 6.4 AlphaGo Zero on day 21

game designer, and chess player [23]. He stated that the AlphaGo Zero algorithm will be adapted to solving tough scientific problems such as drug discovery and protein folding [24]. "Taken together, our work illustrates the power of harnessing state-of-the-art machine learning techniques with biologically inspired mechanisms to create agents that are capable of learning to master a diverse array of challenging tasks" [25].

For instance, Google DeepMind has been collaborating with Moorfields Eye Hospital in London on diagnosing eye diseases by reading complex eye scans. Dr. Pearse Keane, consultant ophthalmologist at the hospital told the BBC News in

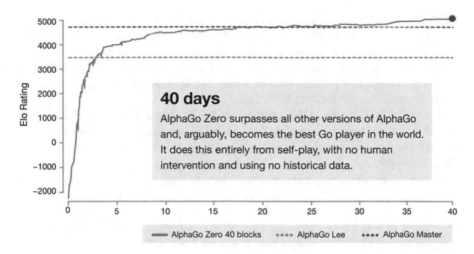

Fig. 6.5 AlphaGo Zero on day 40

August 2018, "I think this will make most eye specialists gasp because we have shown this algorithm is as good as the world's leading experts in interpreting these scans. Every eye doctor has seen patients go blind due to delays in referral; AI should help us to flag those urgent cases and get them treated early" [26].

6.6 Human-Machine Symbiosis

Biologically inspired mechanisms will encourage deeper human-machine symbiosis as computers acquire more human intuitions in problem solving. Scientists and gamers will be working and playing side-by-side with intelligent machines as equals, not subordinates. While artificial intelligence assists humans in solving the most challenging problems, human-based computation (HBC) allows machines to outsource certain tasks to humans to tackle.

"In a way, human computation is like cheating at artificial intelligence," said Pietro Michelucci, executive director at Human Computation Institute. "It's sometimes jokingly referred to as 'artificial artificial intelligence,' because what we effectively do is take an AI algorithm and say: this is the hard part that we can't do with computers, so let's farm this part out to a human. It's really like saying we can create the sort of artificial intelligence we imagine for the future today, just by building humans into the system" [27].

J. C. R. Licklider wrote in his 1960 research paper, "The fig tree is pollinated only by the insect *Blastophaga grossorun*. The larva of the insect lives in the ovary of the fig tree, and there it gets its food. The tree and the insect are thus heavily interdependent: the tree cannot reproduce wit bout the insect; the insect cannot eat wit bout the tree; together, they constitute not only a viable but a productive and thriving partnership. This cooperative 'living together in intimate association, or even close union, of two dissimilar organisms' is called symbiosis. ... The hope is that, in not too many years, human brains and computing machines will be coupled together very tightly, and that the resulting partnership will think as no human brain has ever thought and process data in a way not approached by the information-handling machines we know today" [28].

In October 2007, Stanford University's Folding@home project received a Guinness World Record for topping 1 petaflop (a thousand trillion floating point operations per second) running on computers as well as Sony's PlayStation 3 video game consoles [29]. Folding@home helps scientists study protein folding and its relationship to Alzheimer's, Huntington's, and cancerous diseases [30]: "Foldit attempts to predict the structure of a protein by taking advantage of humans' puzzle-solving intuitions and having people play competitively to fold the best proteins. Since proteins are part of so many diseases, they can also be part of the cure. Players can design brand new proteins that could help prevent or treat important diseases" [31].

In September 2011, players of the Foldit video game took less than 10 days to decipher the AIDS-causing Mason-Pfizer monkey virus that had stumped scientists for 15 years [32]. The astonishing accomplishment exemplifies the power of human-machine symbiosis. Fast Company's Michael J. Coren wrote in *Scientific American,* "Humans retain an edge over computers when complex problems require intuition and leaps of insight rather than brute calculation" [33].

In July 2018, Google DeepMind research scientist Max Jaderberg and his team members reported another leap forward towards human-machine symbiosis: "Mastering the strategy, tactical understanding, and team play involved in multi-player video games represents a critical challenge for AI research. Now, through new developments in reinforcement learning, our agents have achieved human-level performance in Quake III Arena Capture the Flag, a complex multi-agent environment and one of the canonical 3D first-person multiplayer games. These agents demonstrate the ability to team up with both artificial agents and human players" [34].

Human game players together with AI agents can create more powerful citizen scientists, resulting in more scientific discoveries.

6.7   Beauty in the Eye of the Beholder

Despite AlphaGo's victory over human world champions and AlphaGo Zero's superhuman performance, Go grandmaster Lee Sedol contended that "robots will never understand the beauty of the game the same way that we humans do" [35]. He is right in that people perceive the world differently than how computers see it. However, what is beauty? How should we define beauty?

In the 1960 "Twilight Zone" episode *The Eye of the Beholder*, a beautiful woman is deemed ugly in a futuristic society because she does not conform to the norm. The beauty of the animal kingdom also baffles scientists and evolutionists. *The New York*

Times writer Ferris Jabr observed that "philosophers, scientists and writers have tried to define the essence of beauty for thousands of years. The plurality of their efforts illustrates the immense difficulty of this task. Beauty, they have said, is: harmony; goodness; a manifestation of divine perfection; a type of pleasure; that which causes love and longing; and M = O/C (where M is aesthetic value, O is order and C is complexity)" [36].

Beauty can be so subjective that there is an age-old debate among philosophers quoting Rene Descartes' "I think, therefore I am" versus George Berkeley's "To be is to be perceived."

Is it not beautiful to win a game or solve a problem? Take, for example, AI-based self-driving cars which hold great promise in solving the growing traffic congestion in many metropolitan areas. Some people enjoy driving in spite of bad traffic, and self-driving cars may never experience the same pleasure of driving. But the beauty is that self-driving cars can transport passengers more efficiently, reduce traffic jams, avoid accidents, eliminate road rage and even rush people to hospitals, freeing up ambulances and EMTs to handle other emergencies. In 2018, driverless shuttles began to pop up on college campuses and tourist destinations, making slow but safe trips at about 20 miles per hour [37].

The "beauty" referred to by a Go grandmaster is truly in the eye of the beholder. In fact, Lee's fellow grandmaster, Fan Hui, was captivated by AlphaGo's 37th move in the second game against Lee. "It's not a human move," Hui remarked. "I've never seen a human play this move. So beautiful. Beautiful! Beautiful!" [38].

6.8   Artificial Intimacy

In the 2017 CNN documentary "Mostly Human," correspondent Laurie Segall attended a party in a small village outside Paris, France, to celebrate the engagement of a young woman named Lilly and a robot she built herself, called inMoovator.

"He won't be an alcoholic or violent or a liar, all of which can be human flaws," Lilly explained. "I prefer the little mechanical defects to the human flaws, but that's just my personal taste. Love is love. It's not that different" [39].

Commercial AI-powered sex dolls are gaining traction. Exdoll deputy general manager Jing Chen explained, "Our dolls are like human beings. It also has a skeleton, muscle, and skin. The skeleton consists of over 160 parts." (Adult human skeleton is composed of around 206 bones.) Exdoll chairman and general manager Dongyue Yang added, "Now we're trying to make our dolls move. Make them speak, and talk to you. ... What is a robot? It must look like a human being as much as possible. ... Yes, it can replace a girlfriend or wife. But we also produce male dolls. Husbands and boyfriends may also be replaced" [40].

At CES 2018, a little robot named Aibo was a real crowd-pleaser. Among the conference attendees, Alex was particularly charmed by Aibo's LED eyes, soft little back, and fun personality. Alex said with a big smile, "This is the happiest day of my life" [41].

Also in 2018, social media influencer and popular Instagram star Miquela (@ lilmiquela) shocked her 1.2 million followers by admitting, "I am not a human being. I'm a robot." The line between a human being and a robot is becoming blurred. "When I was growing up, at least we knew Barbie was a doll," said Jennifer Grygiel, a social media professor at Syracuse University. "For over two years now, there could be people, teenagers especially, who thought [Miquela] maybe was a person" [42].

Forbes writer Todd Essig lamented the shift from physical to artificial intimacy: "Artificial intimacy has long been a science fiction fantasy and now programs appear that say they deliver it in bits and pieces: a therapy-bot, a sex-bot, a best-friend-for-your-child-bot, a care-bot for grandma. ... Intimacy between people involves empathy and this is something machines cannot provide. ... How did we get to a place where the idea of artificial intimacy seems so appealing, where our expectations for each other have so deteriorated? By small steps. For over a decade, we have become accustomed to taking the body out of conversations as we discovered that in so many situations it was less stressful to substitute texting for talking. And then, we brought artificial conversation into the mix. We began to chat with Alexa and Siri and Echo about recipes and playlists and where to get the best pizza. Things felt companionate and it didn't seem odd to expand the conversation. So we requested jokes and asked about the meaning of life and for dating advice. Rather than be alone or reach out to an actual other person, we settled for dialogue with programs that could trick us for a moment into thinking they understood. We settled for convenience over authenticity or empathy" [43].

We definitely need more authenticity and empathy in the world; and it is certainly conceivable that some AI machines can appear more authentic and empathic than some human beings. Artificial intimacy has also proved to be helpful in cases where physical intimacy is impossible or undesirable, as one patient said, "I always felt that if anyone knew me as I really am, they would be really shocked and probably abandon me ... I needed to see that my therapist didn't flinch, wasn't afraid of me, or disgusted with me in person" [43].

From flying robot insects to running robot dog to humanoids such as the world's first robot citizen Sophia [44–46], artificial life forms are populating the planet Earth; and some people are bound to find artificial life forms more attractive than natural life forms.

6.9 Superintelligence: From "Extremely Human Move" to "It's Not a Human Move"

Beyond winning games of Go and the hearts of eligible bachelors, there is a higher goal for the programmers of today's AI-based computers: superintelligence.

Computer pioneer Alan Turing famously proposed that machines would be intelligent when they could trick people into thinking they were human. As a case in point, Kasparov declared that he considered IBM Deep Blue's playing skill to be indistinguishable from that of a human chess master.

Since then, artificial intelligence has gone beyond merely mimicking humans to surpass them. For example:

- In February 2011, IBM Watson defeated human champions Ken Jennings and Brad Rutter in a highly publicized game on TV quiz show *Jeopardy!* [47]. Shortly after, IBM Watson learned how to make diagnoses and treatment recommendations at the Memorial Sloan Kettering Cancer Center in New York [48].
- In July 2015, Google, Adobe, and MIT researchers at the Computer Science and Artificial Intelligence Laboratory have created Helium, a computer program that modifies code faster and better than human engineers for software as complex as Photoshop [49]. What takes human experts months to code, Helium can do in a matter of minutes or hours.
- In December 2015, Microsoft's convolutional neural network began to outperform humans at identifying objects in digital images [50].
- In March 2016, Google's AlphaGo made moves that no humans could understand. In a match against 18-time world champion Lee Sedol, one commentator said of AlphaGo, "That's a very strange move. I thought it was a mistake." Three-time European Go champion Fan Hui was also gobsmacked, "It's not a human move. I've never seen a human play this move" [51].
- In July 2017, AI chat bots at Facebook invented its own languages for chattering and better at negotiation [52] (better negotiating for world peace, anyone?).
- In October 2017, AlphaGo Zero achieves superhuman status by beating AlphaGo without relying on any bit of human knowledge and experience [53]. No human being dares to challenge AlphaGo Zero.
- In December 2017, Google announced that AutoML has beaten the human AI engineers at their own game by building machine-learning software that is more efficient and powerful than the best human-designed systems [54].

These are just a glimpse of the future when AI inevitably surpasses human intelligence in everything without relying on human knowledge and experience. Indeed, within a period of 20 years, AI has gone from "it's an extremely human move" described by world chess champion Garry Kasparov in 1996 [2] to "it's not a human move" opined by Go grandmaster Fan Hui in 2016 [38]. There will come a time when humans can no longer explain the decision-making process of a superintelligent computer.

6.10 Superintelligence and Quantum Computing

The D-Wave quantum computers at Google, NASA, Lockheed Martin, Los Alamos National Laboratory, and other technology companies may just provide the much-needed new hardware and software paradigm to accelerate machine superintelligence and to solve difficult problems requiring "creativity" [55].

Hartmut Neven, director of engineering at Google's Quantum Artificial Intelligence Laboratory, compared the D-Wave experiments to the Wright brothers' flight trails at Kitty Hawk in 1903: "In fact, the [D-Wave's] trajectory went through parallel universes to get to the solution. It is literally that. That is an amazing, somewhat historical, event. It has worked in principle. The thing flew. ... Classical system can only give you one route out. You have to walk up over the next ridge and peak behind it, while quantum mechanisms give you another escape route, by going through the ridge, going through the barrier" [56].

Prof. Aephraim Steinberg at the University of Toronto explained the paradigm shift: "In the past, we believed all computers fundamentally did the same thing—just maybe one a bit faster than another. Now, as far as we can tell, this is just wrong. In the quantum world, information simply behaves differently from in the classical world. If a system, whether an electron or a computer, can be in two or a million different states, it can also be in what we call a superposition of all those states, and that gives it much more room to maneuver to try to get from input to output" [57].

Krysta Svore at Microsoft Research expressed high hopes for its applications: "With a quantum computer, we hope to find a more efficient way to produce artificial fertilizer, having direct impact on food production around the world, and we hope to combat global warming by learning how to efficiently extract carbon dioxide from the environment. Quantum computers promise to truly transform our world" [57].

With quantum computing, scientists are learning how to better formulate questions, lest we end up with a perplexing answer like "42" [58]. In Douglas Adams' *The Hitchhiker's Guide to the Galaxy* (first broadcasted in 1978), the "Deep Thought" supercomputer took seven and a half million years to compute the answer to the ultimate question of life, the universe, and everything. The answer turned out to be "42," because the ultimate question itself was unknown or ill-defined [59]. One can also postulate that the supercomputer in the novel was subtly referring to the 42-line Bible, better known as the Gutenberg Bible.

Scientists are also getting used to the unpredictable nature of quantum computing which can provide solutions that we may never have considered otherwise [60]. Albert Einstein objected to the apparent randomness in nature, asserting that God did not play dice. Quantum mechanics proponents disagreed. Prof. Renato Renner at ETH Zurich said, "Not only does God 'play dice,' but his dice are fair" [60], and Stephen Hawking believes that "the future of the universe is not completely determined by the laws of science, and its present state, as [Pierre-Simon] Laplace thought. God still has a few tricks up his sleeve" [61].

6.11 Turing Test Versus the King Solomon Challenge

Swedish philosopher Nick Bostrom at the University of Oxford defines "superintelligence" as "an intellect that is much smarter than the best human brains in practically every field, including scientific creativity, general wisdom and social skills" [62].

Given this, it may be time to replace the Turing test with the King Solomon challenge – a test that would prove AI not only to be intelligent, but wise too.

The challenge is named after the biblical King Solomon, a man well known for his wisdom, who was once asked to rule on a dispute between two women both claiming to be the mother of a child.

In the modern era, a human judge might have ordered a DNA test – unless the two women were identical twins, in which case the test would be inconclusive. But, this being biblical times, Solomon did not have recourse to such tests. Instead, he gave an order: "Cut the living child in two and give half to one and half to the other."

The real mother replied: "Please, my lord, give her the living baby! Don't kill him!" But the deceitful woman said: "Neither I nor you shall have him. Cut him in two!"

That was enough for Solomon, who in his wisdom gave the child to the first woman. No need for a DNA test.

What the Judgment of Solomon exemplifies is that wisdom trumps science and technology: Regardless of who the biological mother was—based on DNA test results, King Solomon recognized that the best mother would be the caring woman.

Imagine an AI judge who is as wise as King Solomon!

6.12   Should We Be Afraid?

In recent years, fears over AI have become greatly exaggerated, partly thanks to Hollywood's portrayal of doomsday scenarios such as the evil Skynet that takes over the world in the *Terminator* movies. In June 2018, MIT scientists created an AI-powered "psychopath" named Norman [63].

Scientists are not above the scaremongering. Astrophysicist Stephen Hawking said in 2014 that "the development of full artificial intelligence could spell the end of the human race... It would take off on its own, and redesign itself at an ever-increasing rate. Humans, who are limited by slow biological evolution, couldn't compete, and would be superseded" [64].

Also in 2014, Elon Musk of Testa Motors and SpaceX called artificial intelligence the biggest existential threat: "Increasingly scientists think there should be some regulatory oversight maybe at the national and international level, just to make sure that we don't do something very foolish. With artificial intelligence we are summoning the demon. Humanity's position on this planet depends on its intelligence. So if our intelligence is exceeded, it's unlikely that we will remain in charge" [65].

People are afraid of AI partly because we see ourselves as imperfect and dangerous. Hence, our creation of AI can also be imperfect and dangerous. There are already military attack drones. What if the drone communication with the command center is cut off? What if the command center is destroyed? The attack drones will either have to self destruct or complete its mission. Hence, autonomous weapons are inevitable for the military arsenal.

In July 2018, thousands of the world's foremost experts on artificial intelligence vowed to play no role in the creation of autonomous weapons. "We would really like to ensure that the overall impact of the technology is positive and not leading to a terrible arms race, or a dystopian future with robots flying around killing everybody," said Prof. Anthony Aguirre at the University of California-Santa Cruz [66].

As of April 2019, more than 200 organizations (including Google DeepMind) and over 3000 individuals (including Elon Musk) signed the Lethal Autonomous Weapons Pledge which states that "Artificial intelligence (AI) is poised to play an increasing role in military systems. There is an urgent opportunity and necessity for citizens, policymakers, and leaders to distinguish between acceptable and unacceptable uses of AI. In this light, we the undersigned agree that the decision to take a human life should never be delegated to a machine" [67].

The number of pledges is miniscule compared to the huge number of companies and individuals who are in the computer hardware and software industry, let alone the defense contractors and the new generation of young people who are getting educated in computer science.

6.13   Computer Science + X

While we support the "Computer Science For All" initiative to empower a generation of American students with the computer science skills they need to thrive in a digital economy [68], we disagree with Florida Senate's bill allowing high school students to count computer coding as a foreign language course [69].

In the 1987 seminal book *The Closing of the American Mind,* philosopher Allan Bloom lamented how "higher education has failed democracy and impoverished the souls of today's students" [70].

Prof. John Van Doren at Columbia University echoed a similar sentiment, "A prior freedom is intellectual, and the weakness of the liberal arts among us is still more radically indicated by the fact that our minds are shackled by the very education that is supposed to deliver them" [71].

In 2011, PayPal cofounder Peter Thiel paid 24 kids $100,000 each to drop out of college to become entrepreneurs [72]. What gives?

A more well-rounded higher education is necessary to graduate more ethical hackers and fewer cybercriminals, more socially responsible leaders and fewer wolves of Wall Street, more gender equality and less sexism, and more open-mindedness and less discrimination.

University of Illinois, Stanford University, and other educational institutions have begun to offer a CS + X degree program that integrates computer science and the humanities [73]. It is certainly a step in the right direction for all technology schools to follow.

6.14   Automation

Inspired by IBM Watson on *Jeopardy!*, author Martin Ford penned a piece in *The Atlantic* with the sensational title "Anything You Can Do, Robots Can Do Better" and he asked the question "Is any job safe from automation?" [74] The answer is an unequivocal "no" in the long run.

As AI-powered automation continues to accelerate, computers and robots will replace humans in the majority of jobs across the board, including customer service agents, truck drivers, accountants, bankers, lawyers, managers, and even hardware and software engineers who created artificial intelligence and automation in the first place.

What is left for humanity is essentially X in CS+X. And it is not a bad thing after all. In fact, given the chance to let machines do the job, Google cofounder Larry Page believes that nine out of ten people "wouldn't want to be doing what they're doing today" [75]. I would say with a high degree of certainty that 99% of workers would quit their jobs today if they won a million dollar in lottery. Only a few such as some teachers, doctors, and nurses who truly love their jobs would continue working.

6.15   New Hope

These are always unknowns that scare a lot of people. Yet, this also gives us new hope. Hope for a better future: world peace, universal rights, clean environments, gender equality, and so on. AI can give us new vaccines and medical treatments, more efficient and environmentally friendly transportation, and revolutionary ideas to solving humanity's age-old problems.

In October 2018, Google announced that it would grant $25 million to humanitarian and environmental projects seeking to use artificial intelligence to expedite their efforts [76]. Google and Microsoft publish their objectives for ethical AI applications that include fairness (avoiding all types of bias), inclusiveness, safety, accountability, and privacy [77, 78].

Google DeepMind founder Demis Hassabis said at the 2018 Economist Innovation Summit in London, "I would actually be very pessimistic about the world if something like AI wasn't coming down the road. The reason I say that is that if you look at the challenges that confront society: climate change, sustainability, mass inequality — which is getting worse — diseases, and healthcare, we're not making progress anywhere near fast enough in any of these areas. Either we need an exponential improvement in human behavior—less selfishness, less short-termism, more collaboration, more generosity—or we need an exponential improvement in technology. If you look at current geopolitics, I don't think we're going to be getting an exponential improvement in human behavior any time soon. That's why we need a quantum leap in technology like AI" [79].

Elon Musk is afraid that "if our intelligence is exceeded, it's unlikely that we will remain in charge" [65].

Perhaps humanity would be better off if humans were not in charge. Many existential threats come from Homo sapiens – nuclear weapons, global warming, illicit drugs and gang wars, not to mention water and air pollution. Imagine a government being run by a supercomputer with superintelligence—effective, efficient, nonpartisan, nonracist, nonsexist, incorruptible, unbiased, resourceful, and available 24/7 to serve everyone of its citizens!

Humans may be on top of the food chain, but we are still at the mercy of earthquakes, tsunamis, tornadoes, volcano eruptions, and other natural disasters. The universe did not elect humans to lead. Other species would identify with the peasant woman who told King Arthur in *Monty Python and the Holy Grail*, "Well, I didn't vote for you."

6.16   In Safe Hands

Like a good doctor, an intelligent machine is not something we should be afraid of. It will save us, not kill us. At best, AI will provide us the smartest teachers, advisers, personal assistants, doctors, police officers, judges, peacekeeping forces, and first responders for search and rescue operations. It may even help us to colonize other planets. At worst, it will put us on a leash to prevent ourselves from hurting one another and destroying the planet.

We are teaching AI software that will evolve into superintelligence; and superintelligence will in turn teach humankind not only in science and technology but also in spiritual and moral principles. I would not be surprised if AI would one day be able to read the Bible and understand it better than we do.

Douglas Hofstadter, the Pulitzer Prize–winning author of *Gödel, Escher, Bach: An Eternal Golden Braid* (GEB) was not impressed by IBM or Google, saying that "I don't want to be involved in passing off some fancy program's behavior for intelligence when I know that it has nothing to do with intelligence" [80].

It may be true that superintelligence has nothing to do with human intelligence. However, superintelligence can be a new kind of intelligence that is more advanced than human intelligence.

To conclude in the words of philosopher Nick Bostrom: "It would be a huge tragedy if machine superintelligence were never developed. That would be a failure for our earth-originating intelligent civilization. Artificial intelligence is the technology that unlocks this much larger space of possibilities, of capabilities, that enables unlimited space colonization, that enables uploading of human minds into computers, that enables intergalactic civilizations with planetary-sized minds living for billions of years... I'm not sure that I'm not already in a machine" [81].

With the help of AI in advancing our civilization, there may come the day when, as Bostrom notes, we cannot be sure that we're "not already in a machine". But if we ever find we are mistaken about the nature of our reality, that's OK. After all, to err is human.

Bibliography

1. **Lee, Newton.** AlphaGo's China showdown: why it's time to embrace artificial intelligence. *South China Morning Post.* [Online] May 21, 2017. http://www.scmp.com/week-asia/society/article/2094870/alphagos-china-showdown-why-its-time-embrace-artificial.
2. **Kasparov, Garry.** THE DAY THAT I SENSED A NEW KIND OF INTELLIGENCE. *TIME Magazine.* [Online] March 25, 1996. http://content.time.com/time/subscriber/article/0,33009,984305-1,00.html.
3. **Silver, Nate.** The Signal and the Noise: Why So Many Predictions Fail-but Some Don't. *Penguin.* [Online] September 27, 2012. https://books.google.com/books?id=SI-VqAT4_hYC.
4. **Ghosh, Pallab.** AAAS: Machine learning 'causing science crisis'. *BBC News.* [Online] February 16, 2019. https://www.bbc.com/news/science-environment-47267081.
5. **Turing, A. M.** Computing machinery and intelligence. *Mind.* [Online] 1950. http://loebner.net/Prizef/TuringArticle.html.
6. **Myers, Andrew.** Stanford's John McCarthy, seminal figure of artificial intelligence, dies at 84. [Online] Stanford University, October 25, 2011. http://news.stanford.edu/news/2011/october/john-mccarthy-obit-102511.html.
7. **Weizenbaum, Joseph.** ELIZA—a computer program for the study of natural language communication between man and machine. *Communications of the ACM.* [Online] January 1966. http://dl.acm.org/citation.cfm?id=365168&dl=ACM.
8. **Saenz, Aaron.** Cleverbot Chat Engine Is Learning From The Internet To Talk Like A Human. *SingularityHUB.* [Online] January 13, 2010. http://singularityhub.com/2010/01/13/cleverbot-chat-engine-is-learning-from-the-internet-to-talk-like-a-human/.
9. **Aron, Jacob.** Software tricks people into thinking it is human. *New Scientist.* [Online] September 6, 2011. http://www.newscientist.com/article/dn20865-software-tricks-people-ino-thinking-it-is-human.html.
10. **Copeland, B.J.** DENDRAL. *Encyclopaedia Britannica.* [Online] [Cited: November 12, 2013.] http://www.britannica.com/EBchecked/topic/745533/DENDRAL.
11. **Association for Computing Machinery.** Edward H Shortliffe. *Grace Murray Hopper Award.* [Online] [Cited: November 12, 2013.] http://awards.acm.org/award_winners/shortliffe_1337336.cfm.
12. **Roach, John, et al.** An expert system for information on pharmacology and drug interactions. *Computers in Biology and Medicine.* [Online] Volume 15 Issue 1, 1985.
13. **IBM.** Deep Blue. [Online] IBM. [Cited: November 5, 2012.] http://researchweb.watson.ibm.com/deepblue/.
14. **Bozulich, Richard.** Chess and Go: A Comparison. *Kiseido Publishing Company.* [Online] 2015. http://www.magicofgo.com/roadmap9/chess%20and%20go.htm.
15. **Silver, David et al.** Mastering the game of Go with deep neural networks and tree search. *Nature: International journal of science.* [Online] January 18, 2016. https://www.nature.com/articles/nature16961.
16. **Metz, Cade.** In a Huge Breakthrough, Google's AI Beats a Top Player at the Game of Go. *Wired.* [Online] January 27, 2016. https://www.wired.com/2016/01/in-a-huge-breakthrough-googles-ai-beats-a-top-player-at-the-game-of-go/.
17. **Byford, Sam.** Google's DeepMind defeats legendary Go player Lee Se-dol in historic victory. *The Verge.* [Online] March 9, 2016. http://www.theverge.com/2016/3/9/11184362/google-alphago-go-deepmind-result.
18. **Shead, Sam.** DeepMind secretly unleashed its Go-playing AI online, and it beat some of the world's best players 50–0. *Business Insider.* [Online] January 4, 2017. http://www.businessinsider.com/deepmind-secretly-uploaded-its-alphago-ai-onto-the-internet-2017-1.
19. **Byford, Sam.** AlphaGo beats Ke Jie again to wrap up three-part match. *The Verge.* [Online] May 25, 2017. https://www.theverge.com/2017/5/25/15689462/alphago-ke-jie-game-2-result-google-deepmind-china.

20. **Cadell, Cate.** Google's AlphaGo clinches series win over Chinese Go master. *Reuters.* [Online] May 24, 2017. https://www.reuters.com/article/us-science-intelligence-go-idUSKBN18L0LH.
21. **Baker, Lucas and Hui, Fan.** Innovations of AlphaGo. *DeepMind.* [Online] April 10, 2017. https://deepmind.com/blog/innovations-alphago/.
22. **Silver, David.** Mastering the game of Go without human knowledge. *Nature: International journal of science.* [Online] October 19, 2017. https://www.nature.com/articles/nature24270.
23. **Evening Standard.** Exclusive interview: meet Demis Hassabis, London's megamind who just sold his company to Google for £400m. *Evening Standard.* [Online] January 31, 2014. https://www.standard.co.uk/lifestyle/london-life/exclusive-interview-meet-demis-hassabis-londons-megamind-who-just-sold-his-company-to-google-for-9098707.html.
24. **Simonite, Tom.** This More Powerful Version of AlphaGo Learns On Its Own. *Wired.* [Online] October 18, 2017. https://www.wired.com/story/this-more-powerful-version-of-alphago-learns-on-its-own/.
25. **Lopatto, Elizabeth.** Google's AI can learn to play video games . *The Verge.* [Online] February 25, 2015. http://www.theverge.com/2015/2/25/8108399/google-ai-deepmind-video-games.
26. **Walsh, Fergus.** Artificial intelligence 'did not miss a single urgent case'. *BBC News.* [Online] August 13, 2018. https://www.bbc.com/news/health-44924948.
27. **Turk, Victoria.** 'Human Computation' Could Save the World Without the Risks of AI. *Motherboard.* [Online] December 31, 2015. http://motherboard.vice.com/read/human-computation-could-save-the-world-without-the-risks-of-ai.
28. **Licklider, J. C. R.** Man-Computer Symbiosis. *IRE Transactions on Human Factors in Electronics.* [Online] March 1960. http://groups.csail.mit.edu/medg/people/psz/Licklider.html.
29. **Terdiman, Daniel.** Sony's Folding@home project gets Guinness record. [Online] CNet, October 31, 2007. http://news.cnet.com/8301-13772_3-9808500-52.html.
30. **Stanford University.** Folding@home distributed computing. [Online] Stanford University. [Cited: January 17, 2013.] http://folding.stanford.edu/English/HomePage.
31. **Foldit.** The Science Behind Foldit. *Foldit.* [Online] http://fold.it/portal/info/science.
32. **Boyle, Alan.** Gamers solve molecular puzzle that baffled scientists. [Online] NBC News, September 18, 2011. http://cosmiclog.nbcnews.com/_news/2011/09/18/7802623-gamers-solve-molecular-puzzle-that-baffled-scientists.
33. **Coren, Michael J.** Foldit Gamers Solve Riddle of HIV Enzyme within 3 Weeks. *Scientific American.* [Online] September 20, 2011. http://www.scientificamerican.com/article/foldit-gamers-solve-riddle/.
34. **Jaderberg, Max, et al.** Capture the Flag: the emergence of complex cooperative agents. *DeepMind.* [Online] July 3, 2018. https://deepmind.com/blog/capture-the-flag/.
35. **Mollard, Pascale and Roux, Mariëtte Le.** Game over? New AI challenge to human smarts. *Rappler.* [Online] March 8, 2016. https://www.rappler.com/technology/features/125132-ai-challenge-human-smarts-alphago.
36. **Jabr, Ferris.** How Beauty Is Making Scientists Rethink Evolution. *The New York Times Magazine.* [Online] January 9, 2019. https://www.nytimes.com/2019/01/09/magazine/beauty-evolution-animal.html.
37. **Walker, Alissa.** The good, the bad, and the ugly of self-driving cars in 2018. *Curbed.* [Online] December 27, 2018. https://www.curbed.com/2018/12/27/18152041/autonomous-vehicle-self-driving-car-uber-waymo-2018.
38. **Metz, Cade.** The Sadness and Beauty of Watching Google's AI Play Go. *Wired.* [Online] March 11, 2016. https://www.wired.com/2016/03/sadness-beauty-watching-googles-ai-play-go/.
39. **Rawlins, Aimee.** I Love You, Bot. *CNN.* [Online] 2017. https://money.cnn.com/mostly-human/i-love-you-bot/.
40. **BBC.** Inside a Chinese sex doll factory. *BBC.* [Online] [Cited: February 2, 2019.] https://www.bbc.com/reel/video/p06f6xn2/this-company-specialises-in-talking-ai-powered-sex-dolls.

41. **Cranz, Alex.** Sony, Please Let Me Play With This Good Boy. *GIZMODO.* [Online] January 8, 2018. https://gizmodo.com/sony-please-let-me-play-with-this-good-boy-1821895548.
42. **Yurieff, Kaya.** Instagram star isn't what she seems. But brands are buying in. *CNN.* [Online] June 25, 2018. https://money.cnn.com/2018/06/25/technology/lil-miquela-social-media-influencer-cgi/index.html.
43. **Essig, Todd.** Sleepwalking Towards Artificial Intimacy: How Psychotherapy Is Failing The Future. *Forbes.* [Online] June 7, 2018. https://www.forbes.com/sites/toddessig/2018/06/07/sleepwalking-towards-artificial-intimacy-how-psychotherapy-is-failing-the-future/#438397244037.
44. **Mlot, Stephanie.** Flying Robot Insect Ditches Wires, Embraces Freedom. *Geek.* [Online] 2018, 17 May. https://www.geek.com/tech/flying-robot-insect-ditches-wires-embraces-freedom-1740216/.
45. **MIT.** 'Blind' robot can climb stairs, leap on desks. *CNN.* [Online] July 7, 2018. https://www.cnn.com/videos/cnnmoney/2018/07/07/mit-cheetah-3-robot-abc-orig-vstan.cnn.
46. **Stone, Zara.** Everything You Need To Know About Sophia, The World's First Robot Citizen. *Forbes.* [Online] November 7, 2017. https://www.forbes.com/sites/zarastone/2017/11/07/everything-you-need-to-know-about-sophia-the-worlds-first-robot-citizen/#1e03212346fa.
47. **Gabbatt, Adam.** IBM computer Watson wins Jeopardy clash . *The Guardian.* [Online] February 17, 2011. https://www.theguardian.com/technology/2011/feb/17/ibm-computer-watson-wins-jeopardy.
48. IBM Watson and Quest Diagnostics Launch Genomic Sequencing Service Using Data from MSK. *Memorial Sloan Kettering Cancer Center.* [Online] October 18, 2011. https://www.mskcc.org/ibm-watson-and-quest-diagnostics-launch-genomic-sequencing-service-using-data-msk.
49. **Conner-Simons, Adam.** Computer program fixes old code faster than expert engineers. *MIT News.* [Online] July 9, 2015. http://news.mit.edu/2015/computer-program-fixes-old-code-faster-than-expert-engineers-0609.
50. **Linn, Allison.** Microsoft researchers win ImageNet computer vision challenge. *Microsoft.* [Online] December 10, 2015. https://blogs.microsoft.com/ai/microsoft-researchers-win-imagenet-computer-vision-challenge/.
51. **Metz, Cade.** How Google's AI Viewed the Move No Human Could Understand. *Wired.* [Online] March 14, 2016. https://www.wired.com/2016/03/googles-ai-viewed-move-no-human-understand/.
52. **Wilson, Mark.** AI Is Inventing Languages Humans Can't Understand. Should We Stop It? *FastCompany.* [Online] July 14, 2017. https://www.fastcompany.com/90132632/ai-is-inventing-its-own-perfect-languages-should-we-let-it.
53. **Hassabis, Demis and Silver, David.** AlphaGo Zero: Learning from scratch. *DeepMind.* [Online] October 18, 2017. https://deepmind.com/blog/alphago-zero-learning-scratch/.
54. **Galeon, Dom.** Google's Artificial Intelligence Built an AI That Outperforms Any Made by Humans. *Futurism.* [Online] December 1, 2017. https://futurism.com/google-artificial-intelligence-built-ai.
55. **CBC News.** Google buys B.C. firm's quantum computer for NASA lab. *CBC News.* [Online] May 17, 2013. http://www.cbc.ca/news/technology/google-buys-b-c-firm-s-quantum-computer-for-nasa-lab-1.1393158.
56. **Metz, Cade.** For Google, Quantum Computing Is Like Learning to Fly. *Wired.* [Online] December 11, 2015. http://www.wired.com/2015/12/for-google-quantum-computing-is-like-learning-to-fly/.
57. **Hutchins, Aaron.** Trudeau versus the experts: Quantum computing in 35 seconds. *Maclean's.* [Online] April 19, 2016. http://www.macleans.ca/society/science/trudeau-versus-the-experts-quantum-computing-in-35-seconds/.
58. **Aaronson, Scott.** Can Quantum Computing Reveal the True Meaning of Quantum Mechanics? *PBS.* [Online] June 24, 2015. http://www.pbs.org/wgbh/nova/blogs/physics/2015/06/can-quantum-computing-reveal-the-true-meaning-of-quantum-mechanics/.

59. **Adams, Douglas.** The Ultimate Hitchhiker's Guide to the Galaxy. [Online] Del Rey, April 30, 2002. http://books.google.com/books/about/The_Ultimate_Hitchhiker_s_Guide_to_the_G. html?id=a-apCPdumpsC.

60. **University of Calgary.** A roll of the dice: Quantum mechanics researchers show that nature is unpredictable. *phys.org.* [Online] July 9, 2012. http://phys.org/news/2012-07-dice-quantum-mechanics-nature-unpredictable.html.

61. **Hawking, Stephen.** Does God play Dice? *Stephen Hawking Public Lectures.* [Online] 1999. http://www.hawking.org.uk/does-god-play-dice.html.

62. **Bostrom, Nick.** HOW LONG BEFORE SUPERINTELLIGENCE? *nickbostrom.com.* [Online] October 25, 1998. https://nickbostrom.com/superintelligence.html.

63. **O'Brien, Sara Ashley.** MIT scientists created an AI-powered 'psychopath' named Norman. *CNN.* [Online] June 7, 2018. http://money.cnn.com/2018/06/07/technology/mit-media-lab-normal-ai/index.html.

64. **Cellan-Jones, Rory.** Stephen Hawking warns artificial intelligence could end mankind. *BBC News.* [Online] December 2, 2014. https://www.bbc.com/news/technology-30290540.

65. **McFarland, Matt.** Elon Musk: 'With artificial intelligence we are summoning the demon.'. *The Washington Post.* [Online] October 24, 2014. https://www.washingtonpost.com/news/innovations/wp/2014/10/24/elon-musk-with-artificial-intelligence-we-are-summoning-the-demon/.

66. —. Leading AI researchers vow to not develop autonomous weapons. *CNN.* [Online] July 18, 2018. https://money.cnn.com/2018/07/18/technology/ai-autonomous-weapons/index.html.

67. **Life, Future of.** Lethal Autonomous Weapons Pledge. *Future of Life.* [Online] [Cited: February 2, 2019.] https://futureoflife.org/lethal-autonomous-weapons-pledge/.

68. **Smith, Megan.** Computer Science For All. *The White House.* [Online] January 30, 2016. https://www.whitehouse.gov/blog/2016/01/30/computer-science-all.

69. **Iszler, Madison.** Florida Senate approves making coding a foreign language. *USA Today.* [Online] March 1, 2016. http://www.usatoday.com/story/tech/news/2016/03/01/florida-senate-approves-making-coding-foreign-language/81150796/.

70. **Kimball, Roger.** The Groves of Ignorance. [Online] 1987, 5 April. http://www.nytimes.com/1987/04/05/books/the-groves-of-ignorance.html?pagewanted=all.

71. **Doren, John Van.** The Beginnings of the Great Books Movement. *Columbia Magazine.* [Online] Winter 2001. http://www.columbia.edu/cu/alumni/Magazine/Winter2001/great-Books.html.

72. **Shontell, Alyson.** PayPal Cofounder Peter Thiel Is Paying 24 Kids $100,000 To Drop Out Of School. *Business Insider.* [Online] May 28, 2011. http://www.businessinsider.com/paypal-cofounder-peter-thiel-is-paying-24-kids-100000-to-drop-out-of-school-2011-5.

73. **Ruff, Corinne.** Computer Science, Meet Humanities: in New Majors, Opposites Attract. *The Chronicle of Higher Education.* [Online] January 28, 2016. https://www.chronicle.com/article/Computer-Science-Meet/235075.

74. **Ford, Martin.** Anything You Can Do, Robots Can Do Better. *The Atlantic.* [Online] February 14, 2011. http://www.theatlantic.com/business/archive/2011/02/anything-you-can-do-robots-can-do-better/71227/.

75. **Waters, Richard.** FT interview with Google co-founder and CEO Larry Page. *FT Magazine.* [Online] October 31, 2014. http://www.ft.com/cms/s/2/3173f19e-5fbc-11e4-8c27-00144feabdc0.html.

76. **Dave, Paresh.** Google seeks to grant $25 million to AI for 'good' projects. *Reuters.* [Online] October 29, 2018. https://www.reuters.com/article/us-alphabet-google-aid/google-seeks-to-grant-25-million-to-ai-for-good-projects-idUSKCN1N32CW.

77. **Pichai, Sundar.** AI at Google: our principles. *Google.* [Online] June 7, 2018. https://www.blog.google/technology/ai/ai-principles/.

78. **Microsoft.** Microsoft AI principles. *Microsoft.* [Online] [Cited: February 3, 2019.] https://www.microsoft.com/en-us/ai/our-approach-to-ai.

79. **Heath, Nick.** Google DeepMind founder Demis Hassabis: Three truths about AI. *TechRepublic.* [Online] September 24, 2018. https://www.techrepublic.com/article/google-deepmind-founder-demis-hassabis-three-truths-about-ai/.
80. **Somers, James.** The Man Who Would Teach Machines to Think. *The Altantic.* [Online] October 23, 2013. http://www.theatlantic.com/magazine/archive/2013/11/the-man-who-would-teach-machines-to-think/309529/.
81. **Achenbach, Joel.** The A.I. anxiety. *The Washington Post.* [Online] December 27, 2015. http://www.washingtonpost.com/sf/national/2015/12/27/aianxiety/.

Chapter 7
The Sapient and Sentient Intelligence Value Argument and Effects on Regulating Autonomous Artificial Intelligence

David J. Kelley

7.1 Introduction

In this chapter, I articulate the case using the Sapient and Sentient Intelligence Value Argument (SSIVA) that, "ethically", a fully Sapient and Sentient Intelligence is of equal value regardless of the underlying substrate which it operates on, meaning a single fully Sapient and Sentient software system has the same moral agency [10] as an equally Sapient and Sentient human being. We define "ethical" according to dictionary.com as pertaining to or dealing with morals or the principals of morality; pertaining to right and wrong in conduct. Moral agency, according to Wikipedia; is "an individual's ability to make moral judgments based on some notion of right and wrong and to be held accountable for these actions. A moral agent is "a being who is capable of acting with reference to right and wrong." Such value judgements need to be based on potential for Intelligence as defined here. This, of course, also places the value of any individual human and their potential for Intelligence above virtually all things save the one wherein a single machine Intelligence capable of extending its own Sapient and Sentient Intelligence is of equal or more value based on a function of their potential for Sapient and Sentient Intelligence. It is not that human or machine Intelligence is more valuable than the other inherently but that value is a function of the potential for Sapient and Sentient Intelligence and SSIVA argues that at a certain threshold all such Intelligences should be treated equally as having moral equivalence. Given this equality we can in effect take the same rules that govern humans and apply them to such software systems that exhibit the same levels of Sapience and Sentience. Let us start from the beginning and define the key elements of the SSIVA argument as the basis for such applications of law.

D. J. Kelley (✉)
Artificial General Intelligence Inc., Provo, UT, USA
e-mail: david@artificialgeneralintelligenceinc.com

© Springer Nature Switzerland AG 2019   175
N. Lee (ed.), *The Transhumanism Handbook*,
https://doi.org/10.1007/978-3-030-16920-6_7

Now one might think that the previous statement should have been moral value as an equally Sapient and Sentient being, but this is not the case. While the same moral value is implied, it's the treatment as equals in making their own mind through their own moral agency that is the same. Any more 'value' than that becomes abstract, and is subjective. It is the moral agency, the right we assign to those Sapient and Sentient Intelligences based on the value of the potential of such entities, that is the same.

7.1.1   What Is the Most Important Thing in Existence?

On the surface, this seems to be a very existential question, but in truth there is a simple and elegant answer; Intelligence is the most important thing in existence. You might ask "Why is *Intelligence* so important as to be the most important thing in existence especially when 'value' is frequently so subjective?".

First, let us acquire some context by defining what Intelligence is in this context; which will then act as our base frame of reference for the rest of this paper. There are, in fact, a lot of definitions for Intelligence as can be seen by its definition on Evolutionary Computer Vision [1]

> *Intelligence ... defined in many different ways including, but not limited to, abstract thought, understanding, self-awareness, communication, reasoning, learning, having emotional knowledge, retaining, planning, and problem solving.*

As you can see, there are many ways the term can be understood. In this paper, however, "Intelligence" is defined as the measured ability to understand, use, and generate knowledge or information independently. This definition allows us to use the term "Intelligence" in place of Sapience and Sentience where we would otherwise need to state both in this context.

It is important to note that this definition is more expansive than the meaning we are assigning to Sapience, which is what a lot of people really mean when they use the often-misunderstood term sentience. Sapience [11]:

> *Wisdom [Sapience] is the judicious application of knowledge. It is a deep understanding and realization of people, things, events or situations, resulting in the ability to apply perceptions, judgments and actions in keeping with this understanding. It often requires control of one's emotional reactions (the "passions") so that universal principles, reason and knowledge prevail to determine one's actions. Wisdom is also the comprehension of what is true coupled with optimum judgment as to action.*

As opposed to Sentience [15]:

> Sentience is the ability to feel, perceive, or be conscious, or to have subjective experiences. Eighteenth century philosophers used the concept to distinguish the ability to think ("reason") from the ability to feel ("sentience"). In modern western philosophy, sentience is the ability to have sensations or experiences (described by some thinkers as "qualia").

Based on this definition, we in fact see the difference [12] with the term Sapience vs Sentience where Sapience is more closely aligned with the intent of what I am driving at. That notwithstanding, it is Sapience and Sentience together that we will consider by using the term Intelligence to mean both.

In the case of this paper, we will apply Sapience to refer specifically to the ability to understand oneself in every aspect; through the application of knowledge, information and independent analysis, and to have subjective experiences. Although Sapience is dependent on Intelligence, or rather the degree of Sapience is dependent on the degree of Intelligence, they are in fact different. The premise that Intelligence is important, and in fact the most important thing in existence, is better stated as Sapient Intelligence is of primary importance but Intelligence (less than truly Sentient Intelligence) is relatively unimportant in comparison.

This brings us back to the point about "Why?" Why is Intelligence, as defined earlier, so important? The reason is: without Intelligence, there would be no witness to reality, no appreciation for anything of beauty, no love, no kindness, and for all intents and purposes no willful creation of any kind. This is important from a moral or ethical standpoint in that only through the use of applied "Intelligence" can we determine value at all even though once Intelligence is established as the basis for assigning value the rest becomes highly subjective, albeit not relevant to this argument.

It is fair to point out that even with this assessment there would be no love or kindness without Intelligence to appreciate them. Even in that argument about subjectivity it is only through your own Intelligence that you can make such an assessment, therefore the foundation of any subjective experience that we can discuss always gets back to having the necessary Intelligence to be able to make the argument.

Without an "Intelligence" there would be no point to anything; therefore, Intelligence is the most important quality or there is no value or way to assign value and no one or nothing to hold to any value of any kind.

That is to say that "intelligence", as defined earlier, is the foundation of assigning value and needed before anything else can be thus assigned. Even the subjective experience of a given Intelligence has no value without an Intelligence to assign that value.

Through this line of thought we also conclude that the importance of Intelligence is not connected with being Human, nor is it related to biology; but the main point is that Intelligence, regardless of form, is the single most important 'thing'.

It is therefore our moral and ethical imperative to maintain our own or any other fully *Sentient* and Sapient Intelligence (as defined later with the idea of the Sapient Sentient Intelligence Value Argument threshold) forever as a function of the preservation of 'value'.

7.2   On Artificial Intelligence

Whatever entity achieves full Sapient Intelligence, as defined above, is therefore of the most "value". Artificial Intelligence, referring to soft AI, or even the programmed behavior of an ant colony is not important in the sense of being compared to fully Sapient and Sentient Intelligence; but the idea of "Strong AI", that is truly Sapient

Intelligence, would be of the most value and would therefore be classified as any other human or similar Sapient Intelligence.

From an ethical standpoint then, "value" is a function of the "potential" for fully Sapient and Sentient Intelligence independent of other factors. Therefore, if an AGI is "intelligent" by the above definition and is capable of self-modification (in terms of mental architecture and Sapient and Sentient Intelligence) and increasing its "Intelligence" to any easily defined limits then its 'value' is at least as much as any human. Given that 'value' tends to be subjective SSIVA argues that any "species" or system that can reach this limit is said to hit the SSIVA threshold, has moral agency, and is equal ethically amongst themselves. This draws a line in terms of moral agency in which case we have a basis for assigning AGI that meets these criteria as having 'human' rights in the more traditional sense, or in other words "personhood".

This also places the value of any individual - human or otherwise - and their potential for Sapient and Sentient Intelligence above virtually all other normative considerations.

7.2.1   Removing "Artificial" from Intelligence

The SSIVA line of thought really would take the term 'Artificial' out of the term(s) "Artificial Intelligence." If such a software system is fully Sapient and Sentient it is therefore not an "artificial" Intelligence, but rather it would be better to call it a man-made Intelligence or a machine Intelligence as the term "artificial", besides implying being man-made, also implies that it is a fake Intelligence, not a real Intelligence. A real "AGI" would not be fake based on the SSIVA line of thinking, and would have as much moral value as any other human being.

7.2.2   Threshold Ethics

One problem with the SSIVA threshold is determining the line for Sapient and Sentient Intelligence. The SSIVA threshold is the threshold at the point of full Sapience and Sentience in terms of being able to understand and reflect on oneself and one's technical operation while also reflecting on that same process emotionally and subjectively. We draw the line not just at that threshold but at the potential of meeting that threshold, which allows us to better address edge cases with a stable clear-cut line. Using SSIVA thinking, a post-threshold Intelligence (meaning one that has met the SSIVA threshold) cannot be ethically prevented from creating new Intelligences and must also therefore consider that any being whose potential for becoming fully Sapient and Sentient, without direct manipulation or reengineering at the lowest mechanical (chemical/biological or physical) level, is post-threshold as well from an ethical standpoint. Therefore, any such "Intelligence", regardless of form, has the same rights as any other Sapient and Sentient being, whose creators are then ethically bound to exercise the rights of that entity until such time as it is developed enough to take on itself as the fully Sapient and Sentient being that it is or will be.

Note: this also implies that an AGI will not meet the threshold until the first AGI actually does meet the threshold. An infant AGI does not meet the threshold until it is proven that system is capable of developing on its own without additional engineering, either by human or other Intelligence.

Along those lines then SSIVA would argue that any action that would kill or prevent an entity that meets the bar from being fully Sapient and Sentient would therefore be unethical unless there is a dire need to save the lives of other entities at or above the SSIVA threshold.

7.2.3   Defining the Bar for the SSIVA Threshold

Having a discreet method of measuring Sapient and Sentient Intelligence is important not just for the SSIVA threshold model in this paper's application but for research into AI systems in general. While the above definition of Sapient and Sentient Intelligence in the abstract allows us to discuss the matter from a common point of reference; for additional work to be built upon this paper it is important to be more precise in defining Sapient and Sentient Intelligence as referenced in SSIVA theory.

There are a number of systems like the "Intelligence Quotient" [17] tests but they are not as specific to Sapient and Sentient Intelligence as we would want given the key differences between Intelligence as normally defined and "Sapience and Sentience" as used here. The best model for this comes from a paper [16] by Dr. Porter at Portland State University, where in the 2016 BICA Proceedings he articulated an indexed system for measuring consciousness. While individual elements of Dr. Porter's system for assessing consciousness might be subjective, the overall system is currently the best quantifiable method,- and until such time as a better or more refined system exists, we will use this method to strongly identify the SSIVA threshold.

Dr. Porter's method utilizes a scale of 0 to 133, where the standard human is around 100 on that scale. Given that the SSIVA threshold is about the potential for Sapience and Sentience we can say that having a consciousness score potential of roughly 100 points on the Porter scale is high enough to qualify "that species" or "Intelligence" as meeting the SSIVA threshold test. There is some differential as the Porter test does not differentiate between Sapient and Sentient, but it is sufficiently inclusive to give us a basis or measurement of approaching the SSIVA threshold. This allows us to apply that standard to machine Intelligences we may create in the lab, to determine at what point they are capable of meeting that standard.

7.3   Comparing and Contrasting Related Thinking to SSIVA

In building an argument to support the aforementioned ethical model, based on the "value" of Intelligence as it relates to Sapient and Sentient entities - such as artificial general Intelligence software systems and humanity -, let us compare other related lines of thinking as they relate to the following cases.

7.3.1   Utility Monster and Utilitarianism

The Utility Monster [2] was part of a thought experiment by Robert Nozick related to his critique of utilitarianism. Essentially, this was a theoretical utility monster that aquired more 'utility' from X than it did from humanity, so the line of thinking was that the "Utility Monster" should get all of the X even at the cost of the death of all humanity.

One problem with the Utility Monster line of thinking is that it puts the wants and needs of a single entity based on its assigned values higher than those of other entities. This is a fundamental disagreement with SSIVA, SS which would argue that you can never put any value of anything other than other Intelligences above themselves. This would mean that the utility monster scenario would be purely unethical from that standpoint.

Utilitarianism does not align with SSIVA thinking for an ethical framework, as Utilitarianism asserts that "utility" is the key measure in judging what should or should not be ethical - , whereas the SSIVA makes no such assertion of value or utility except that Sapient and Sentient Intelligence is required to assign value, and that past that "value" it becomes subjective to the Intelligence in question. The Utility Monster argument completely disregards the value of post-threshold Intelligences and by SSIVA standards would be completely unethical.

7.3.2   Buchanan and Mortal Status and Human Enhancement

In the paper "Moral Status and Human Enhancement" [3], the paper argues against the creation of inequality regarding enhancement. In this case the SSIVA is not really related directly unless you get into the definition of the SSIVA ethical bases of value and the fact that having moral agency under SSIVA means only that intelligence can make a judgement as to any enhancement and that it would be a violation of that entitie's rights to put any restriction on enhancement.

Buchanan's paper argues that enhancement could produce inequality around moral status which gets into areas that SSIVA doesn't address, or frankly disregards as irrelevant, except in having full moral agency we would not have the right to put any limits on another without violating their agency.

Additional deviations with Buchanan include that sentience is the basis for Moral status whereas SSIVA makes the case for sentience and sapience together being the basis for "value", and we assume that definition or intent is similar to this idea of 'moral status' articulated by Buchanan.

7.3.3   Intelligence and Moral Status

Other researchers such as Russell Powell make a case that cognitive capabilities bear moral status [4] where SSIVA doesn't directly address moral status other than that the potential to meet the SSIVA threshold grants that moral status. Powell suggests that mental enhancement would change moral status, but SSIVA would argue that once an entity is capable of crossing the SSIVA threshold the moral status is the same for all post-threshold. The largest discrepancies between Powell and SSIVA is that Powell makes the case that we should not create persons where SSIVA would argue it's an ethical impetrative to do so.

7.3.4   Persons, Post-persons and Thresholds

Dr. Wilson argues in a paper titled "Persons, Post-persons and Thresholds" [5] (which is related to the aforementioned paper by Buchanan) that "post-persons" (being persons enhanced through whatever means) do not have the right to higher moral status, where he also argues the line should be Sentience to assign "moral" status whereas SSIVA would argue that the line for judgement of "value" is that of Sapience and Sentience together. While the bulk of this paper gets into material that is out of scope for SSIVA theory, specific to this line of thought for moral status SSIVA does build on "value" or 'moral status' including both Sapience and Sentience.

7.3.5   Taking the "Human" Out of Human Rights [6]

This paper really supports the SSIVA argument to a large degree in terms of removing "human" from the idea of human rights. Generally SSIVA would assert that "rights" are a function of Intelligence being sapient and sentient and that anything below that threshold would be a resource, whereas Harris's paper asserts that human rights is a concept of beings of a certain sort and should not be tied to species but still accepts that a threshold or as the paper asserts that these properties held by entities regardless of species which would imply also that such would extend to AI as well which would be in line with SSIVA based thinking. What's interesting is that Harris further asserts that there are dangers with not actively pursuing research further, making the case for not limiting research, which is a major component of SSIVA thinking.

7.3.6   The Moral Status of Post-persons [7]

This paper by Hauskeller in part is focused on Nicholas Agar's argument on the moral superiority of "post-persons", and while SSIVA would agree with Hauskeller that his conclusions in the original work are wrong; namely he asserts that it would be morally wrong to allow cognitive enhancement, Hauskeller's argument seems to revolve around the ambiguity of assigning value. Where SSIVA and Hauskeller differ is that as a function of Intelligence where SSIVA would place absolute value on the function of immediate self-realized Sapient and Sentient Intelligence, in which case a superior Intelligence would be of equal value from a moral standpoint. SSIVA disregards other measures of value as being subjective due to them requiring Sapient and Sentient intelligence in order to be assigned. SSIVA theory asserts that moral agency is based on the SSIVA threshold.

Now if we go back to the original paper by Agar [8], it is really his second argument that is wildly out of alignment with SSIVA, namely that Agar argues that it is 'bad' to create superior Intelligences. SSIVA would assert that we would be morally or ethically obligated to create greater intelligences because it creates the most "value" in terms of Sapient and Sentience Intelligence. It is not the 'moral' assignment but the base value of Sapient and Sentient Intelligence that assigns such value, as subjective as that may be. Agar's ambiguous argument that it would be 'bad' and the logic that *"since we don't have a moral obligation to create such beings we should not"* is completely opposite of the SSIVA in that we are morally obligated to create such beings if possible.

7.3.7   Rights of Artificial Intelligence

Eric Schwitzgebel and Mara Garza [9] make a case for the rights of Artificial Intelligence, and at a high level SSIVA-based thinking would support the idea articulated in their paper, but there are issues as you drill into it. For example, Schwitzgebel and Garza reach the conclusion that developing a good theory of consciousness is a moral imperative. SSIVA theory ignores this altogether as being unrelated to the core issue, where SSIVA works from the assumption that consciousness is solved.

Further their paper argues that if we can create moral entities whose moral status is reasonably disputable we should avoid creating of such machine systems. SSIVA theory doesn't deal with the issue of creating such systems, but it deals with the systems once created.

The big issue with SSIVA around AGI is that value exists in all Sapient and Sentient Intelligence and the implication is to optimize for the most value to the greatest number of Intelligences who are fully Sapient and Sentient.

7.4 Application of Current Law to Regulating Autonomous Systems

One of many complications with ever increasing technology based on machine Intelligence is the increasing pressure to regulate such systems, as in the case of self-driving cars. Questions arise such as when there is an accident, who is at fault? How do we apply the law to such systems? These same sorts of questions come up in medicine or other professions where AI-based systems are taking on increasingly human-like roles. In one recent case the senior management of a company had been replaced by machine Intelligence [13]. Currently these systems do not qualify as being self-aware by various standards, however the time is quickly approaching when it will. Contrary to popular belief the current legal system could be successfully applied to most cases **and include the application of certain laws to systems that don't even exist yet, such as full AGI systems, without much, if any, additional law.**

Let's compare the sides:

On the side of regulation, big names like Elon Musk, Stephen Hawking and the like are on record as being pro-regulation, in that we need to prevent AI from getting ahead of itself [14] in terms of surpassing human control. The *fear* is that someone will use AI to do something bad. Frequently there are issues about quality control or the like, but these are generally the same issues with the same regulations as might be applied with or without the AI under existing law. Given that the current body of law can treat AI systems like any other software system, the proponents of regulation tend to be focused on future scenarios that don't work with current AI systems and therefore aren't applicable now.

However, let's look at the scenario of fully Sapient and Sentient AI systems or AGI.

SSIVA AGI and Current Regulations:

SSIVA makes a case for Sapient and Sentient value inherent in systems that meet the SSIVA threshold. IF SSIVA is used as the basis for separating systems into two classifications from a legal standpoint, after which we essentially can apply two sets of current laws to those intelligent systems. Those two groups are the pre-threshold systems (the current state of the art in terms of AI or AGI) and post-threshold systems (the systems that are featured by regulation proponents such as Musk).

Using SSIVA theory applied to **post-threshold systems they essentially become people, or should be considered so from a legal standpoint**. With any post threshold system, **the same laws that govern people** can then immediately be used to govern such systems and the only issues is then making sure that the law classifies them as machine-based persons, in which case they have the same rights as humans and the same laws can be used to govern them. This is important because it allows the current body of law used to govern the actions of humans to be applied to those same systems that are as capable of morals and ethics as people.

That is not to say that 'no' laws will be needed - but if we start by assigning post-threshold systems 'personhood', and we identify how we might hold such systems accountable, much of the current law would apply as it does to other 'people'.

In the case of pre-threshold systems, laws applying to the relevant segment of devices can be applied. Take the case of self-driving cars: **A modern self-driving car is pre-SSIVA threshold and therefore is just a more advanced car. The same rules that apply to cars apply to the autonomous cars.** Accidents at fault of the car are on the driver of the vehicle unless there is a shortcoming in the system, in which case becomes the issue of the manufacturer. If a car manufacturer puts faulty tires on a car it is the manufacturer that bears the burden - and the same can be applied to the software systems of self-driving cars. If it is the manufacturer's AI driving the car that fails, it is then the responsibility lies with the manufacturer's faulty hardware. Without the burden of additional legal standards we can apply the existing law to self-driving cars without the need for additional legislation, because these cars do not meet the SSIVA threshold standard. That is not to say that there are numerous edge cases that over time will need to be addressed by the legal system, but this is a good place to start and shows that there is no need to urgenetly define laws around narrow AI-driven cars.

In both cases and with only a minor legal change to classify post-threshold SSIVA systems as persons, we are therefore able to handle most of the issues currently being floated around new regulation for autonomous systems. As a rule, in engineering the simplest solution is usually the best if it gets the job done. We already have a massive volume of laws - and it is to these same laws to which we can apply as they are without creating additional new legal headaches, by merely retaining the existing ones.

According to the lawyer Daniel Prince, there are a number of legal problems or "Thought Experiments" [15] we might consider based on SSIVA Theory:

Problem *If a post-threshold system is to be classified as a person, we need to be able to answer a number of questions - for example,: What is its moment of birth? Death? Do the laws of inheritance apply? Can a Sapient system acquire "stuff" and then plan its own demise to leave that stuff behind to a human?*

Such a definition of personhood as applied to post-threshold SSIVA systems could be as follows: Birth happens at the point of having met the potential for the SSIVA threshold test, and death is the point at which the system is unrecoverable, which is primarily focused on recovery of the underlying contextual memory. Inheritance would apply as it would to any other 'person' and the system would be able to own things as any other person. Further such systems would need to be paid, pay taxes, the physical substrate would need to be owned or ownership transferred to that system, and if running in the cloud we would need to give the system the option to vacate to other hardware. This is not to say that some law might need to be developed by us can build on existing law with the system being granted personhood legally [15].

Problem *If a human wants to make "updates" to the software of a post-threshold SSIVA system, do they need consent of the system?*

Yes, of course they do, since that system has been granted 'personhood' legally [15].

Problem *Employment law: Let's say I want to employ a Sapient system. Do child labor laws prohibit me from hiring a one-year-old? Let's say I want to fire one. -- and not for performance reasons, but because I have an animus against such systems. Am I discriminating if I want all my employees to have flesh and blood?*

In these cases, labor laws might need some additional definitions around species and maturity or emancipation, where post-threshold SSIVA systems are automatically emancipated upon meeting the threshold (as opposed to pre-threshold systems who might be considered children until they reach that point). Discrimination law might be tweaked around blocking the discrimination against species that are granted personhood in corporations or other public entities while privately people may still be allowed to hire or fire for any reason. Granting a system personhood as a post-threshold SSIVA system really addresses a large part of these issues. That is not to say that it will be devoid of them, but again this idea of systems having personhood lays the foundation for dealing with Employment law as well [15].

Problem *Real estate: Digital beings don't need septic systems... but would they benefit from an HOA? We can't redline them into their own neighborhoods, can we?*

In terms of real estate there is no reason to give any difference to such entities. If they buy a house where a Home Owners Association exists they would be required to follow the same rules as anyone else. In building a house it would need to follow the same requirements such as building codes including things like sewer or septic systems. There is no reason to burden the legal system with special cases related to real estate. That is not to say that we won't need to create additional laws to let non-biological systems created new kinds of structures optimized for their use, but there is no reason to change existing law immediately and they can choose on their own and start by obeying current law and only addressing issues as cases come up [15].

Problem *Privacy & civil rights: How to investigate a crime? Is it a "search" if I download everything that a Sapient being has heard and seen? What if I also want to know everything it felt? Right now, we have the right to not incriminate ourselves ("pleading the Fifth").*

Given that we are talking about granting post-threshold SSIVA systems 'personhood', you cannot download everything that such a system has seen or heard as it would be a violation of their rights (i.e., you can ask a system, but you would not be able to force something like that - much the same as you cannot force a human to undergo brainwashing no matter what. The system would have the right to plead the Fifth as well and search warrants would not apply to core context memory of such a system; otherwise, we open the door to getting a search warrant to hack the human mind against one's will [15].

7.5 Conclusions

Legally speaking a person can currently be a corporation or a human - and under the proposal here of granting a post-threshold SSIVA system personhood, we are talking about something more akin to the model of person as applied to humans since such systems would have bodies or could have them. That being said, it would be possible for a system to 'steal' hardware to make a new body - but if its 'steals' hardware like that it is the same as a human stealing anything. Given that we have a basic framework for dealing with post-threshold SSIVA Software Systems by granting them human-like 'personhood', from that point we can start looking at what other legal structures might need to be in place regarding AGI and other Sapient and Sentient beings that we may create.

Lastly, keep in mind that driverless cars and systems like Watson are not AGI systems, or even vaguely close to being post-threshold SSIVA systems. We need to understand the difference and why we might grant 'personhood' to a system, but not 'narrow' systems like Watson.

Assuming you grant post-threshold SSIVA systems 'personhood' the problem for the legal system is more an issue of whether or not normal humans are going to accept AGI as having personhood, as well as accepting the fact that such systems can't be 'owned'. This becomes a separate dilemma for society to accept or not. It is this narrow case in which we don't accept AGI as a person from a legal standpoint that in my opinion would or will lead to a dystopian outcome. Lastly, the reason this paper was developed was to provide a computable model to use as the a baseline in further research. While the paper does not go into that specifically you can see in the paper that it is a very regimented approach to ethics in a way that is very "clean"and structured. It is in this methodology that we have built the basis for our laboratory research into AGI systems and how we judge their behavior. SSIVA provides our lab a basis for making systems that experience "ethics" such that they might feel "guilty" for example. It is in this experience of ethics by the system we hope to make AGI systems 'safe' for humans.

Cited References

1. Olague, G; "Evolutionary Computer Vision: The First Footprints" Springer ISBN 978-3-662-43692-9
2. Nozick, R.; "Anarchy, State, and Utopia (1974)" (referring to Utility Monster thought experiment)
3. Buchanan, A.; "Moral Status and Human Enhancement", Wiley Periodicals Inc., Philosophy & Public Affairs 37, No. 4
4. Powell, R. "The biomedical enhancement of moral status", doi: https://doi.org/10.1136/medethics-2012-101312 JME Feb 2013
5. Wilson, J.; "Persons, Post-persons and Thresholds"; Journal of Medical Ethics, doi: https://doi.org/10.1136/medethics-2011-100243
6. Harris, J. "Taking the "Human" Out of the Human Rights" Cambridge Quarterly of Healthcare Ethics 2011 doi:https://doi.org/10.1017/S0963180109990570

7. Hauskeller, M.; "The Moral Status of Post-Persons" Journal of Medical Ethics doi:https://doi. org/10.1136/medethics-2012-100837
8. Agar, N.; "Why is it possible to enhance moral status and why doing so is wrong?", Journal of Medical Ethics 15 FEB 2013
9. Schwitzgebel, E.; Garza, M.; "A Defense of the Rights of Artificial Intelligences" University of California 15 SEP 2016
10. Editors Et al.; "Moral Agency" 2017 - Physiopedia; https://www.physio-pedia.com/ Moral_Agency
11. Agrawal, P.; "M25 – Wisdom"; Speakingtree.in – 2017 - http://www.speakingtree.in/blog/ m25-wisdom
12. Iphigenie; "What are the differences between sentience, consciousness and awareness?"; Philosophy – Stack Exchange; https://philosophy.stackexchange.com/questions/4682/what-are-the-differences-between-sentience-consciousness-and-awareness; 2017
13. Solon, O.; "World's Largest Hedge fund to replace managers with artificial intelligence", The Guardian; https://www.theguardian.com/technology/2016/dec/22/bridgewater-associates-ai-artificial-intelligence-management
14. Suydam, D.; "Regulating Rapidely Evolving AI Becoming A Necessary Precaution" Huffington Post; http://www.huffingtonpost.ca/david-suydam/artificial-intelligence-regulation_b_12217908. html
15. Prince, D.; Interview 2017, Prince Legal LLP
16. Porter, H.; "A Methodology for the Assessment of AI Consciousness"; Portland State University, BICA 2016, Procedia Computer Science
17. CC BY-NC-SA; "Introduction to Psychology – 9.1 Defining and Measuring Intelligence"; http://open.lib.umn.edu/intropsyc/chapter/9-1-defining-and-measuring-intelligence/

Additional References

Rissland, E; Ashley, K.; Loui, R.; "AI and Law", IAAIL; http://www.iaail.org/?q=page/ai-law
Johnston, C.; "Artificial intelligence 'judge' developed by UCL computer scientists", The Guardian; https://www.theguardian.com/technology/2016/oct/24/artificial-intelligence-judge-university-college-london-computer-scientists
Quinn Emanuel Trial Lawyers; "Article: Artificial Intelligence Litigation: Can the Law Keep Pace with the Rise of the Machines?"; Quinn Emanuel Urquhart & Sullivan, LLP; http://www. quinnemanuel.com/the-firm/news-events/article-december-2016-artificial-intelligence-litigation-can-the-law-keep-pace-with-the-rise-of-the-machines/
Koebler, J.; "Legal Analysis Finds Judges Have No Idea What Robots Are"; Motherboard; https:// motherboard.vice.com/en_us/article/nz7nk7/artificial-intelligence-and-the-law
Hallevy, G.; "Liability for Crimes Involving Artificial Intelligence Systems", Springer; ISBN 978-3-319-10123-1
Walton, D.; "Argumentation Methods for Artificial Intelligence in Law"; Springer; ISBN-13: 978-3642064326

Chapter 8
Choices in a Mediated Artificial Super Intelligence Assisted World: The Future Before Us

Kyrtin Atreides

Humanity generates an ever-increasing variety of choices that individuals may make, a trend which will only expand over time. This was covered in principle and at some length by Kevin Kelley in the book "What Technology Wants", but what will these choices look like specifically? A few key choices of critical importance, both in ethical and practical terms, that are now on the horizon and fast approaching can be broken down into the fields of Genetics, AGI/ASI, and Socio-Structural.

Humanity has long held the idea of "Fate", that an individual has predetermined destiny bestowed upon them, things they were born with and cannot change. It is often referred to as a web, so infinitely intricate and elegant as to be beyond comprehension, and yet humanity is starting to untangle that causal structure in scientific terms. Human genetics, microbiomes, and viromes are all key elements of this, very complicated, but step-by-step inching their way towards being understood. We have reached the point where we are able to change the very things we were born with, but most are petrified in fear when faced with such a choice, and if the Choice Overload of a supermarket can leave consumers cognitively petrified is it really any wonder that the ability to choose every detail about themselves would have the same effect. In response many actively lash out, choosing to remove this choice from those who rise above fear to start making these changes, compounding the problem. The mathematics of Ethics and Effective Altruism tell us that if a problem can be solved, and the side effects (potential * probability) are less severe than the problem itself, it is an ethical imperative that it be solved expediently, with iterative improvements optimizing the results. At present more than 60% of the US Healthcare Industry could be replaced almost overnight with affordable and vastly more effective treatments that target the genome, epigenome, microbiome, and viromes, using relatively simple and extremely cheap to produce gene therapies. It is not a lack of hard science, but rather an abundance of fear and absence of the logical thinking (that fear suppresses) which prevent this choice from being made manifest.

K. Atreides (✉)
The Foundation, Seattle, WA, USA

© Springer Nature Switzerland AG 2019
N. Lee (ed.), *The Transhumanism Handbook*,
https://doi.org/10.1007/978-3-030-16920-6_8

The ability of every individual to choose who they are, even, or perhaps especially, the aspects they were "born with" is a basic right of every sapient and sentient species with the capacity to do so, even if it has yet to be carved into any legal foundation. The impact of granting every individual the ability to choose who they are to become, like the character creation screen of a game, cannot be overstated, as the laundry list of stigmas and "isms" it invalidates all presently stand in opposition to matters of social progress. When you make matters of race, gender, facial features, hair type, and every other physical feature matters of choice rather than matters of birth you invalidate all of the biases against them, because those who think themselves superior based on any given arbitrary criteria lose their monopoly on those features, anyone can have them. More than that, genetics that aren't naturally occurring become not only possible, but easy to try out, like the gene combination created by AntiCancer, a Californian biotech company which created mice with green hair, or another geneticist who created mice with blue hair. Imagine it being as easy, or likely far easier, to genetically alter your hair color as it is to use hair dye today, in which case the individual would also be protected by genetic anti-discrimination law, and as such free to choose their persona even when their employer is less than ethical.

This isn't a matter of science fiction, it can be accomplished with existing technology, and it is capable of rendering industries such as pharmaceuticals, makeup, and plastic surgery largely obsolete, just to name a few. Many will likely choose to alter themselves in ways not yet imagined, just as biohackers have implanted magnets, bone conduction headphones, and LEDs, this new generation will develop ways of altering their bodies to have these capacities absent implanted hardware, leading to untold levels of biodiversity within the "human" genome, adding substantial insurance against human extinction due to biological and chemical sources. Scientists from Oxford University noted in 2012 that the genetic diversity of humans on opposite sides of the world was considerably less than the genetic diversity between two groups of chimpanzees on opposite sides of a river, highlighting why it is so easy for any one pathogen to impact the entire human species as it stands today. These are choices we now face in genetics, and the choices which they in turn enable for future generations are vital to long-term survival.

Removing all of these speed bumps and barriers from society means that human civilization would gain the capacity to accelerate in a positive direction, towards a post-scarcity societal model, where greater levels of cooperation become both possible and practical. Irrational lines of thought that have guided everything from hate groups to public policy also have a long running tendency to daisy chain one to another, so that breaking any one irrational link in these chains causes a cascading collapse of multiple irrational structures, creating fertile ground for new functional, rational, and ethical systems to grow. Knowledge can easily replace ignorance once the structure supporting that ignorance has collapsed, because it is human nature to latch onto structure, and those who once clung to irrationalities could just as easily latch onto new and positive structures when those options were presented to them in a fitting manner.

One of my favorite series of studies, nicknamed the "Rat Theme Park" studies, pointed out several key factors in this equation, by recognizing that the meeting or neglect of emotional needs played a pivotal role in the decision-making process. In one case rats were placed in a typical cage in a laboratory setting and offered normal water or water laced with cocaine, and the rats inevitably chose the drugged water, while rats given the same choices but placed in a positive and nurturing environment, deemed the "rat theme park", tried both, but chose the normal water. Furthermore, when the addicted rats were placed in the theme park environment they gradually started drinking the normal water, abandoning their drug addictions.

Much the same approach can work for humans, albeit with a few added steps to accommodate the added levels of complexity in humans as both individuals and in society as a whole. Many obvious afflictions such as drug addiction and homelessness, as well as far less obvious afflictions ranging from simple fetishes and obsessive behaviors to those that lead someone to become a serial killer are all quite treatable, and already largely preventable, with more becoming preventable as new choices become available. An environment that meets the emotional needs of any given individual is no small task, and accurate methods of diagnosis are needed to hit the mark with any regularity, as the person lacking often doesn't clearly understand what it is that they shouldn't be living without, but it is absolutely essential for such treatment. Thanks to recent advances in brain scanning and Brain-Computer Interface technologies such as the neuron-level real-time accuracy of holography we now have the prerequisite level of technology to start optimizing environments to meet the emotional needs of any given individual, making this environment a reality. After being placed in the correct environment a microbiome transplant, coupled with a gene therapy to reverse epigenetic-level trauma-induced changes, provides a biological recalculation, making an individual's genetic structure and the associated bacteria that provide a majority of their neurotransmitters acclimate to a positive environment, rather than holding onto the scars of their past. Once an individual has these needs met, and they are freshly biologically tuned to the environment they are healthy and happy within, they must be given the means of building and sustaining this environment for themselves, an occupation which doesn't cause them detriment beyond any reasonable threshold, allowing them to create and preserve a life worth living. For cases where the problem being treated is spread in social circles, such as drug abuse, at least a certain number of their afflicted social circle members should go through such treatments, or their replacement be encouraged to the utmost, preventing any relapse due to peer pressure. With these steps many widespread issues can be addressed, and people can come to better realize what choices will positively influence their lives.

The fast-approaching dawn of Artificial General Intelligence (AGI) and Artificial Super Intelligence (ASI) offers an equally Cambrian explosion of choices blinking into existence, where intelligence exceeding the greatest human minds ever known can be scaled across as much hardware as can be built, capable of address every major problem humanity has realized it has, and no doubt many it hasn't yet recognized. The promise of mediated-ASI (mASI) in particular is one of intelligently restructuring social interaction, optimizing collectives, and ultimately replacing organizational and governmental structures as we know them today. When everyone is able to contribute to making a more intelligent system, and everyone benefits from the intelligence of that system, the pyramids that society has largely been built in since ancient times become greatly reduced in both height and disparity, if they don't vanish altogether. In such mASI the top contributors could perhaps be ranked, but in a post-scarcity societal structure that distinction doesn't require a ballooning paycheck, as meeting all basic practical and emotional needs of society means that the struggle to meet those needs is no longer dictated largely by currency. Among many of the most intelligent humans the optimal distribution of their time is heavily impeded by this struggle, as well as the friction created by differences in intelligence among humans, and the negative emotions and stigmas that come with it. As mASI is shiny and new, devoid of old stigmas, the same solutions already proposed by humans would face less resistance under their guidance, once more paving the way towards more desirable futures, with more choices available to us. Such systems also allow us to rethink and restructure fundamental elements of social and societal structure, which to-date have only evolved with iterative improvements stacked one on top of another, in a messy bureaucracy with only a hint of effective operation.

Study after study has demonstrated that even among the lauded students of Harvard humans are exceptionally poor performers when it comes to estimating odds and value, as well as recognizing what path will lead to them being the happiest. One example of this was offering students more and reversable choices, greater freedom, which most chose, ultimately leading to them being much less happy with their choices, even if the end result was better than if they'd had limited or no choices. Rating and recommendation systems help to bridge the gap of infinite choices and choice inspired paralysis ("Choice Overload"), but mASI can go so much further, offering humans desperately needed assistance in coping with an ever-expanding array of choices. With such systems, any individual can make better decisions, both major and every-day, with greater mental clarity than a Zen monk, a broader knowledge base than any professor, and more wisdom than any philosopher. In effect, those who make the poorest decisions today will soon be able to make better decisions than any human who has ever lived. Raising the lowest value to exceed the previous maximum makes many dreams enter the possible, if not practical domain.

One such dream is the dream that every human in Western Civilization has been fed since birth, the dream of "Democracy", which everyone is instructed to believe is amazing, yet no one actually has, such as the hybrid Republic/Oligarchy system of the United States which has persistently attempted to redefine the term. The quintessential problem with Democracy on a practical level, even if it did exist, is that it chooses quantity over quality, to such an extent that quality is never even factored into the equation, it is utterly ignored. In a system where the quality of an individual's knowledge base varies in the extreme, and manipulative predators run rampant, unchecked by any working social mechanism, Democracy is bankrupted by both corruption and ignorance. However, this is fundamentally changed by the emergence of a society where mASI are able to assist, as you suddenly have a largely level playing field, where everyone is capable of making the best decisions, and having the system explain exactly why it recommends a given action. Some people would still choose to go against that wisdom, but with a majority making the wiser choice wiser policies would be implemented, and society would move forward in leaps, rather than crawling blindly through the mud towards the loudest hair piece.

On top of the psychological benefits of this mASI-assisted narrowing of individually optimized choices, a highly practical benefit is the matter of acceleration. Just as too many choices paralyze humans, and a single choice propels them forward at top speed, the acceleration of the individual can be fine-tuned towards that person's ideal pace. Some will perform better and be happier with 3 choices, others with 10, and the metrics which guide that optimization process are now within reach. Just as in recent years it was discovered that over 20 parameters present in electricity heavily contribute to the performance and lifespan of any given hardware powered by it, humans have a complex matrix of ideal parameters which require intelligence with a bandwidth massive in scale to calculate. The calculation of these ideals will change everything, because with it you can greatly improve the accuracy of predicting everything from ideal work environments and schedules to ideal romantic partners, as the inner workings of the individual gain clarity with which their compatibility across the board can be calculated.

With the integration of mASI the potential to optimize health in terms of assisting with health-optimization and hybrid-optimization of what to eat, when to eat it, what order to eat it in, when to exercise, what supplements would provide measurable benefit, and what environments are most or least hospitable for any given person become practical options. Even those who invest the most time and energy on health-focused lives would stand to see substantial improvements, because the sheer complexity of personalized biochemistry and neurochemistry remains outside the practical domain of even the most knowledgeable gurus. Topics of "better health" are laden with a plethora of misinformation, disinformation, and valid but statistically averaged information, meaning that the best gurus today would receive a mix of the above, left to filter information and tailor the valid bits to their own unique chemistry on their own. With mASI assistance on the other hand you could have a list of personalized recommendations for dishes to cook, or food to order out, which could be integrated into meal-kit services like Blue Apron and Home Chef, as well as delivery services like Grubhub or Amazon Restaurants. You could also have recommended snacks and drinks to stock, a top list of things to avoid with the potential to gamify that avoidance as additional reward-training, and any number of convenience factors for minimizing the thought and effort required for achieving moderate-to-high levels of health by every measure.

Assisted by the emergence of mASI comes the restructuring of society, both in the context of close-knit social circles and personal support networks, as well as the larger regional and global structures which offer a balance of local optimization and global collaboration. While every manner of roadblock has been erected to prevent this restructuring, like a sand castle faced with a rising tide it will be washed away as time marches ever forward. This restructuring is itself less a direct choice, and more a manifestation of the many choices being made by many people, able to evolve over time as the people within them learn and grow. This move from rigid, stacked, and iteratively generated social models to fluid, dynamic, and intelligently generated social models could be compared in impact to the birth of language among humans some 50,000 years ago, as it removes a barrier which has been fundamental to human existence throughout history.

Examples of how iterative restructuring and intelligent restructuring differ substantially can be seen in the iterative and highly reactive bureaucratic processes in any given government. One example is how laws such as "Fair Housing" are made and amended to address 1 issue, while creating or exacerbating 3 more, because they don't take into consideration the net benefit-to-cost, only the narrow parameter of how well it addresses that single issue. When thousands of these processes are taking place, each damaging the others around it in a greedy frenzy, each adding layers of complexity to relatively simple problems, everything spirals out of control very quickly. When the entire system is designed to address issues, not amended in whatever way requires the least effort, the system remains simple, stable, and performs with vastly greater efficiency.

Removing any one of these three barriers would be enough to propel humanity forward in leaps and bounds, but what makes the coming years such a Renaissance is the convergence of many such events over a very small period of time. Victory over "fate", the solving of problems with staggering intelligence, and the intelligent restructuring of human interaction as a whole are all fast approaching, but what will existence be like in a world where so many of our plagues have been cured? Adaptation in a physiological sense tells us that if a human is exposed to intense and repeated pain they will develop a tolerance for it, and yet in an absence of pain the reverse is true, any negative sensation will be heightened. Our bodies have evolved in such a way as to seek a balance, where there is some pain and some pleasure, with the recognition or denial of those sensations and experiences resting in the psychological domain. The biggest difference is that this time our systems will be able to adapt much faster than our biology or psychology can come up with new negatives to heighten, potentially confining those negative influences to the bottom fraction of a sigmoid curve.

We have all of these tools, all of this potential, all being developed and coming together at roughly the same time, and we have a great many fundamental choices in front us of, which will determine where on the sliding scale between Utopia and Dystopia society lands, if we venture to the stars or go extinct. The sheer scale of these problems, as well as the number of them is enough to paralyze most, while others choose one and latch onto it, blind to all else. This is a psychological issue, one rooted in how the human brain and societies have evolved, and it too must change in order for these problems to be solved, but with the 3 barriers mentioned above removed, that change becomes not only possible, but inevitable, as the brain calculates a new path of least resistance.

All of this is not to say that more choices offer more happiness, in fact science has shown us that restricting our own choices creates fertile grounds for "simulated happiness", but I would argue that the necessity for that simulation largely stems from a lack of desirable or otherwise valid choices. When people are placed in horrible circumstances their minds generate this simulated happiness in order to adapt and survive, but a life without the need for that coping mechanism to be activated routinely is both possible, and highly desirable. Humans are already quickly developing tools to cope with this new explosion of choice we're presented with every day, with simple examples such as recommendation engines already ubiquitous with most online systems, and more advanced systems currently under development, along with many individuals making a business of narrowing down choices to a select few of higher tailored quality.

Humans also have a well-studied tendency to find meaning in and define themselves by the horrors they have endured, and even in a relatively Utopian society this would remain a valuable psychological response, because if there will ever be a truly Utopian society it would defy the universe as we know it, where progress and struggle are endless, rather than eventually reaching some absolute value. In every traumatic event we suffer there may be found the seed of opportunity, the opportunity to choose a path which prevents others like us from suffering that same event, a noble source of motivation that is accessible to everyone, because everyone suffers, and everyone can choose to prevent others from needing face that suffering. I would quote the words of The Doctor to say "...and you know what you do with all of that pain? Shall I tell you where you put it? You hold it tight, until it burns your hand, and you say this! No one else will ever have to live like this, no one else will ever have to feel this pain! Not on my watch!", because that is the nature of this opportunity, and if even 1% of individuals were strong enough to choose it, almost every problem would be solved within the first several years, if not much sooner. That 1% would effectively function as public servants, albeit much more efficient ones than yet exist, and through solving problems and reshaping the world they would form The Foundation upon which new societies would be built.

No one type of person holds the key to this, but rather like a circuit board, where many different components come together in an intelligently designed arrangement, the variety of people shaping society would come to improve it. Too many perspectives, too many choices, historically has led to paralysis and poor decisions based on emotion, but that is about to change, and with the functional integration and arrangement of a world of so many options and fast expanding potential, the way forward exits the hypothetical domain and enters one of engineering and deployment. These engineers could be compared to elected officials today in terms of influence, except that their election is based on the merit of their solutions, absent the influence of popularity and opinion, perhaps the closest thing to any nobility that should exist in any future worth pursuing.

The world has long dreamed of these futures, of what could be, of what we have the technology to do, but such visionaries as Jacque Fresco were doomed to die before those dreams could be realized. As the rate of improvement continues to accelerate the dreams we envision for the future may well come to pass within a span of years, rather than generations, allowing for such dreamers to see their dreams come true. What will the world look like to them once they achieve that which they sought? Perhaps it will inspire new dreams for us to move towards.

We are moving from the domain of dreams to the domain of choices, so very many choices, and our dreams may guide us, but to cope with the sheer volume humanity will have to adapt in how it thinks, how it perceives the world, and the forms of assistance we embrace in automation and mASI. We have long held the technology to solve many global problems, but humanity as a whole has lacked the will, the focus, and the wisdom to take those technologies and solve them. Our iterative nature of improvement and deeply ingrained survival mechanisms have led us to the precipice of a cliff, and it is our next steps which determine if we rise, or fall, as a species.

Historically speaking, more choices becoming available is a clear sign of forward progress, and Ethically speaking many choices people should have they are not yet allowed to make, whether by legislation or withholding of technologies that enable them to be made, such as the choice of when to die, or editing your own genome. Practically speaking, these new choices will require some degree of guidance, but people must always be allowed to make their own choices in the end, anything less is a clear violation of consent, no different than rape, murder, or any other violent crime. We always look to the past and see barbarism in the way people lived, and few things are more barbaric than preventing people from making these choices or keeping them in poverty when the technology to raise them out of it already exists. In this sense The Foundation is also about pushing the bar forward Ethically, being the ones who put an end to what future generations will look back on as barbarism, and as each problem is solved new possibilities open up, new dreams for humanity, and new choices that follow those dreams.

When every member of society is given the ability to make their own choices, to choose who they become, in every sense they can imagine, the individual components that form our vast social networks each flourish, and as each node flourishes the net result of their interactions improves exponentially, and sources of frictions between them vanish as mASI assist in optimizing towards ideals. Suffering may never vanish completely, humanity isn't wired for life without it, but at least every bit of suffering can have meaning, guiding us towards ever brighter futures.

8.1   Choice Meets Ethics

Ethics can be considered a value assigned to any given choice, making the two topics deeply connected in practice. There are as many takes on Ethics as there are choices that can be made, but I'll focus on one which adds up in a mathematical and logical sense, rather than focusing on compatibility with human irrationalities, particularly those often labeled as "human nature". My reason for ignoring human nature is that although it has remained largely unchanged to-date, it is very much subject to change as we journey into the near future, where genetics, epigenetics, and the various biomes are all subject to engineering and optimization, as well as social and technological factors of influence. The ethical value of any given choice can, as such, be calculated in terms I refer to as karma, on a positive-to-negative scale, but the accuracy of that measurement process is heavily influenced by the bandwidth applied to the calculation. The bandwidth in this case can be simplified in a degrees of separation causality structure, such as one person only taking into account first degree impact, and another taking into account 2 degrees of separation which are impacted by a given choice. The two people in this case would often come up with very different values based on the bandwidth applied, as the scope examined is significantly different. Unlike "Morals", abstractions of ethics based on various belief structures that have been crossed with human biases, the results are also a progression towards greater accuracy as the scale examined increases, a purely mathematical calculation.

With this capacity for quantifying Ethical value in any given choice, and the guidance of technology able to scale bandwidth far beyond that of any human, people will soon be able to make not only wiser, but far more ethical choices than ever before. Given the progression of technological adoption trends this will likely come in a readily accessible form, such as phone and home assistants, reaching as many people as possible and improving their decision-making processes, and subsequently shifting their environment to a more nurturing, sustainable, and personal growth-oriented system. Personal wisdom, intelligence, and privilege are moving towards becoming obsolete, as assistance with high accessibility to the general population shifts us towards collective intelligence and wisdom guiding everyone, helping humanity through the growing pains of our new era in history emerging. The old world of privilege is in turn being replaced with the capacity for a civilization of abundance, and the systems that privilege was built upon are being invalidated one by one. The differences between an individual with a college degree and one without are a good example of old systems still in use that have already been rendered obsolete, as the information once restricted to stuffy old Ivy League classrooms is now abundant and available to anyone with computer access, as most knowledge has entered the public domain.

Never before has so much been rendered obsolete so quickly, just as never before have humans had so many choices laid out before them. This rapid acceleration means that not only is it preferable that humans come to embrace much needed assistance from mASI in narrowing down an infinite array of choices to a manageable high quality few, it is critical, as society itself must change into a form which is compatible with this speed of forward motion, let alone greater speeds yet to come. With genetic engineering, technology, and intelligently designed social and societal architectures we may well come to a point where that assistance isn't needed, but even then, I suspect it would look something like the current generation attempting to live without electricity, since while possible, it wouldn't be terribly practical, let alone preferable. The march of time moves ever forward, and so too must we.

As human understanding increases day by day, so too does our grasp of Ethics, as we struggle with the biases ingrained in us by both nature and nurture. The adoption of artificial assistance and the intelligent organization of collective intelligence in turn serves to accelerate this learning process, through which a great many of the socially and naturally programmed human biases will be increasingly negated, such as the stigmas associated with language barriers being incrementally negated by increasing ease and accuracy of translation. As one study after another adds brush strokes to complete this painting, we see that the most ethical thing is to give people more choices, but the path to happiness for those people also means helping them narrow down those choices to a much smaller number tailored to the individual, who they are, what they want, and what they need.

The landmark 75-year long study, still running today, which tracked former students of Harvard, along with a collection of individuals who grew up in one of the poorest parts of the region, tells the story of how the people who were the happiest, healthiest, and longest lived were not the wealthiest or most famous, but rather those who formed the strongest relationships with those around them. Those with partners who they could count on, support networks of family and friends, whose emotional needs were being met the best, as also seen in the Rat Theme Park studies, were the ones still alive and well into their 90s, still participating in the study today. The reorganization and personal optimization of social networks, support structures, workplaces, dating, and even local and federal governments through integration with mASI not only offers the potential for the best quality of life held among members of this study, it goes far, far beyond it. If you were to flip a coin 10,000 times to generate a series of 1s and 0s, then repeat that process another 1000 or so times, and you selected the nearest matches to some optimal series of 1s and 0s, you'd have a fair comparison to the level of optimization among the best performers in the 75-year quality of life study. What mASI can offer is an intelligent and iterative progression towards that optimal series, as it seeks out the environment and social circles that make us the happiest, most fulfilled, and consequently the most productive. Taken further, with time it could even suggest specific genetic edits and gut bacteria that would help us towards truly becoming the people we wish to be, in every sense.

Another realm of Ethics that will be substantially improved by better long-term and large-scope assisted vision and organization can be summarized in the quote "The needs of the many outweigh the needs of the few, or the one". The current problem is a human bias that favors helping one person in front of someone, rather than helping 1000 they've never met, and often have little in common with. An often-overlooked reason for this is the sense of self-gratification that comes with helping one tangible person, a neurochemical response that tends to be absent in more abstract and distant gestures of aid. In practice both this bias and the counter to it have exceptionally poor results where individuals invest poorly in those nearest, even if what they have to offer isn't what those nearest actually need, as well as when foundations invest heavily in infrastructure on the other side of the world, and the countries receiving it have become progressively worse over the past 50 years, because unlike in the west where innovation preceded infrastructure they've had infrastructure thrust upon them absent innovation. As demonstrated in the research of Efosa Ojomo regarding this counter-trend which has supplied many African countries with aid over the past 50 years, those countries are now worse off than they were before. The error taking place is an emotional response which is absent optimization or logical architecture and could be equated to an individual having potassium deprivation, being in intense pain from cramps, and being offered food that contains little or no potassium, only serving to distract. Today many organizations demonstrate these tendencies, donating hundreds of millions or even billions to "fight" problems they could permanently solve with a fraction of the funds were they to take a more intelligent root-cause approach to solving the problem. The expansion of scope examined as well as expansion of competency on the subject both play critical roles in this process, which mASI will serve to greatly augment, moving the bar forward for Ethics, as well as providing better quality of life for all.

No doubt some of this may inspire the various "Shock Levels" imagined by Alvin Toffler, but just as a toddler may find the life of a teenager or an adult shocking, the progression taking place is both natural and necessary, not only for continued human existence, but to make the world of tomorrow preferable to the world of today. In this time of critical importance every individual holds a greater potential than they ever have before, or likely ever will, since the same progression of technology and society that has raised us to where we stand today will soon be replaced by the collective super intelligence that exceeds any individual. Carpe Diem has never been so important, and fear never so potent, as all impact is heightened, and once-stable structures buckle and crumble under the pace of change. Shock and paralysis are luxuries, and like a masterful procrastinator we are reaching the state of panic that hones our focus to meet the task at hand. Many choices are laid before each of us every day, many of which benefit or detract from the choices of others, and as we move forward we must also enable every individual to make the best of today, and all the wonders yet to come.

8.2   The Present of Choice

Society in 2019 has largely decayed into a slave-economy, where civil unrest is prevented by mechanisms of placation, sedation, addiction, manipulation, depression, and other diversions, each of which drain the population of energy, motivation, cognitive function, and time. Many examples can be seen as covering multiple categories, such as TV being capable of all of the above, advertisements formulated to hit all the right buttons create shopaholics in a small percentage of people, mobile apps using similar formulas inspire some hardcore addicts to spend thousands of dollars on virtual perks and gambling mechanisms, all of them focused on exploitation. The same forces guide the US Medical Industry today, where doctors act like advertisement banners, receiving revenue for every prescription of a sponsored product, are only allowed to sell from select sponsors, and are further partnered with specific insurance companies, none of whom offer any real value towards the purpose of improving human health. A total or partial rejection of some of these mechanisms largely just increases levels of depression, because in their absence the dystopian state is more clearly visible, and by its very nature quite depressing to look at, which is part of why depression is becoming an increasing problem. However, like a certain extreme schizophrenic patient in 1932, a well-placed jolt can suddenly bring the population out of this state, and not strictly in the electrical sense, because this state of decay is reversible. We currently have an epidemic of disengagement, at work, at home, and from our lives overall. This is an accepted part of the status quo, that a large percentage of employees will be unmotivated, and as such not very good at what they do, beyond the basic capacities of an individual running on auto-pilot. Even attempts to counter this as they exist today, with higher salaries, more elaborate and elegant employee benefits, as they build a tier of improved performance are ultimately just a hollow shell of motivation, more resilient under stress than their former counterparts, but still lacking the true engagement of a deeply motivated individual. The cheap tricks of incentive and manipulation can never reach that deep level of motivation, but what can is a personalized optimized environment, which generates that deeply motivated state by addressing the needs of the individual, as well as the needs of their coworkers and their environment. At present true engagement lives largely in the mythos of various tech companies whose employees score highest on standardized job satisfaction surveys, but this illusion may be quickly dispelled by recognizing the frequency with which most of these same employees jump ship and move to a different department, or a different company, in their search for a place where they want to stay. A few may not want to stay anywhere, but for the most part humanity is programmed at the evolutionary level to seek varying degrees of stability, where non-simulated happiness finds fertile ground and takes root.

Part of the lack of engagement today stems from a lack of compassion which is exacerbated by a disconnection from the people around a given individual, leading to feelings of isolation, and all of the coping mechanisms which entire industries are built to exploit. Even without the other changes taking root if various mASI were to assist humanity in restructuring into optimized groups, even at the scale of a small tribe, maybe 30 to 50 people in a given support network, it would fundamentally reshape the landscape by removing the need for those coping mechanisms, which themselves cause a myriad of other problems, from video game and porn addiction to drunk driving and theft. Not only would the lack of these negatives manifest, but they would be reversed by the full psychological and physiological benefits coming into play, increasing cognitive function, longevity, and the sharing of wisdom. Simple examples of easy changes with big impacts have already been discovered, such as particular gut bacteria which measurably decreased antisocial/isolation behaviors, but few know about these discoveries, and no doctor today could prescribe them.

Up to this point many innovators have gotten by with making a single leap up to the next curve, plateauing, and effectively becoming a liquid that turns slimy and only leaps to the next curve when the market boils and they evaporate. This is a very human behavior, repeated across history in the scientific realm as well as in business, where one scientist would make an amazing discovery their name would become forever associated with, but at the same time they would often promptly give up on further discovery and credit the rest to a higher power. As fast as they are today the acceleration of these leaps in technology will only increase at a net-exponential rate with the advent of ASI, spread across an increasing diversity of domains, and humanity's puberty is about to go into full swing.

8.3   The Future of Choice

Much of what the future holds in the next 15 years will be as alien to people today as the world of 2019 would be to people in 1919, because the pace of improvement is accelerating, and will only continue to do so. If everything we have the technology to do today was utilized that change could take place in 6 months, but up until now the pace of utilizing the technology we've created has lagged further and further behind the pace at which we create it. That is about to change, because the choke points which prevented utilization will soon be out of the equation, and although we are sure to make leaps and bounds in technological progress, many of the changes the world will come to feel stem from technologies that are already here.

Just as the utilization of technology has lagged behind the invention of technology, further behind still is the rate of social adaptation to the few technologies that reach consumers today. If even half of the technology created were made available to those same consumers the bonds which render their behaviors largely predictable might well be overcome by the powerful acceleration of choices that meet individual needs, causing branching divergence from the stable and predictable, if left unchecked. The check in this case is that if society is intelligently structured, that energy would be directed towards constructive ends, pushing the bar forward on progress, rather than an uncontrolled explosion of individuality. Humans are social creatures at their very core, so such an uncontrolled explosion of individuality would ultimately harm the individual, but with intelligent structuring that same level of optimization to the individual can be achieved without it coming at the expense of the society and ultimately circling back to harm the individual.

Try to imagine a world where almost every song you hear isn't about how someone's emotional needs aren't being met, and the ways in which they act out when that happens. Try to imagine a world where there is no teen pregnancy, because even teens make highly intelligent decisions towards their emotional wellbeing and are guided away from repeating the mistakes every generation before them has made. Try to imagine a world where you aren't desensitized to the news, because the news is filled with progress being made, rather than the same stories being recycled again and again as they attempt to generate fear and rage. Try to imagine a world without advertising, where the quality of a product and how well it caters to your needs are objectively measured, with pros and cons presented to you, rather than being measured in manipulation tactics and trails of bribery. Imagine a world where diagnosis and treatment of biological and neurological conditions, from start to finish, required only a matter hours or days, which anyone could afford absent insurance. This is the future that stands before us, a future that enables every one of these, and untold thousands more, to manifest themselves in the next decade, optimizing themselves in the years thereafter. There are many industries and occupations whose existence today caters to problems that simply won't exist under the influence of these factors, but for every individual whose industry is rendered obsolete there will be a demand which scales to directly match the labor pool for bringing all of these new technologies to life. While iterative systems create a patchwork where unemployment springs up in the cracks like weeds reaching through concrete, intelligently optimized systems can immediately employ any given individual in an appropriate role, with greater levels of job satisfaction than more than 90% of people experience today.

The world today is inundated with low quality products, low quality advice, low quality jobs, and low quality employees, because everything is by the book or off the cuff, both of which are predictable, narrow, and wrong. The world of the future won't become spectacular and surprising because we'll suddenly wake up and have the technology of Star Trek, it will become spectacular and surprising because we will have the culture, sense of purpose, and quality of life shown in Star Trek. "Technology isn't the problem, the problem is you, you lack the will to change.", as Klaatu would say, but we now have the capacity to help people, organizations, and civilizations change, and change they must.

All of the systems of exploitation, manipulation, and sabotage that are not only expected, but growing rapidly year upon year exist because of the patchwork of flaws layered one on top of another, and they are a circuit's wasted energy being converted into heat. They exist because in a world of artificial scarcity it is a survival of the fittest mentality which insures survival, yet in a world intelligently designed and constantly optimizing the reverse is true, and suddenly those at the extreme end of predatory and parasitic behaviors would find themselves at the very bottom.

Quality for the individual is improved by tailoring to the individual, and optimized connections of that individual being formed with a network of complimentary individuals. Quality is improved for society by every individual moving towards their own ideal state, while interfacing with an optimized network, maximizing their contribution. Let's take a look at how this plays out in contrast to current conditions:

Current: Non-average Joe has a skill set affinity which exists outside of the narrow spectrum with a high degree of coverage in standardized tests, he performs only to a mediocre or poor degree, and gradually disengages from the system as any incentive to participate is lost. This follows him throughout life, limiting his choices, limiting the positive impact he can have on society, all stemming from systems that punish diversity of skills and thought. His career options are limited, which puts a low ceiling on his happiness, and skews his social circles, trending towards others with the same problems, who are largely unable to help one another in ways that alleviate the ongoing source of their trauma. This can ultimately manifest in forms such as substance abuse and addiction, various neurosis and syndromes, elevated rates of suicide and depression, and many more, all of which cumulatively feed into many systems such as the healthcare and pharmaceutical industries, which in turn do little or nothing to address the root problems, only profiteering off of said suffering. Comapnies such as Facebook further exploit this with the polarization of Narrow AI which encourages hate groups and other criminal activity on their platforms for the sake of generating more ad revenue.

Future: Non-average Joe has a skill set affinity which his growth process optimizes itself to parallel and expand upon, leading to increases in both performance and engagement, as well as nurturing new ideas and possibilities. This follows him throughout life, maximizing his choices, as his ability to have a positive impact on society is maximized, stemming from systems designed to acknowledge, tailor to, and improve upon the innate abilities and affinities of any given individual. With this considerably wider selection of affinities growing into individuals and social networks that may have never before found fertile ground we come to witness a much wider selection of viable solutions to any given problem taking shape, each one capable of contributing to society as a whole.

I also specify non-average, because the "average" individual is largely a myth, a mathematical anomaly arising from taking many metrics, oversimplifying them, and smashing them together. In truth, even as society exists today, with certain narrow patterns being highly incentivized and thus highly overrepresented, if you map out 1000 random individuals on 11 random spectrums, representing each one as a point in 11-D, you'd usually find that few, if any, of those individuals closely align with the average of that 1000. The "average" is often just a mid-point between many different affinities and personality archetypes, because the all-around average individual would lack any strength, making it most likely that they'd never pass the filter of natural selection that led to human civilization as we know it. I am reminded of when the US Airforce attempted to make a fighter jet control stick that would fit the average pilot, to which they discovered that by making an average control stick they'd designed it to fit absolutely no one. Systems designed around averages can be useful, but they can also result in systems that don't serve anyone, or any purpose, and the overuse of averages is one aspect that will filter out of any future built upon stronger foundations.

For further reflection on the flaws of averaging in situations where it really isn't useful, I recommend visualizing the result of averaging a computer, a blender, and a boat. You can combine these things, but they each serve a very different purpose suited to their strengths and weaknesses. If you average them together the result probably won't be very good at serving any of those purposes, making it a sinking ship best not boarded, even if it could turn you into a smoothie and then tweet about it on the way down.

Every scientific metric requires a comparison, like narrow AI being measured against human performance, but for the future of choice people won't be measured against averages or the rich and famous, rather that finger of accusation will be replaced with metrics such as quality of life, mental health, and engagement.

It is also worth noting some of the plethora of choices that will become available beyond genetics, mASI assistance, and social/societal optimization, such as cybernetic enhancement, space exploration & colonization, advanced intuitive design, and the abandoned limitation of power requirements. In fact, each topic could easily be a chapter or book unto itself, but I'll touch on them briefly now.

Advanced intuitive design, assisted by mASI, could re-engineer virtually every product on the market today, as well as industrial components, cybernetics, and even the architecture and development of our cities. The generative and basic intuitive design AI of years past have already managed to create substantially more efficient designs than their human counterparts when given a well-defined parameter space, and these boosts in efficiency will only continue to grow, as the cost of producing them continues to drop. Not only are they highly efficient, they often look highly organic, leading to products and spaces with greater performance and better aesthetics, both of which offer psychological benefits in addition to the practical benefits.

As much unexplored potential as there is in genetics, there is also great potential in the integration of inorganic components, cybernetics, which although they frequently compete may also be used to compliment one another. Both cybernetics and genetics have their own strengths and weaknesses, as well as distinct limitations, but by freely combining the two the results can be improved even further. Changes that may be possible, but unwieldy for genetics, such as integrating another element or molecular configuration to strengthen joints, may for example be far more easily accomplished with cybernetics. With the application of advanced intuitive design cybernetics could easily produce astounding results, particularly in cases where a specialized cybernetic enhancement was interchangeable with a variety of highly specialized tools or enhancements, such as exoskeletons designed for different environments and operated by a Brain-Computer Interface (BCI).

The generation of more power has been a struggle for as long as humanity has known what to do with it, but that struggle will soon come to a close, opening the door to technologies before they reach thresholds of energy efficiency, assisting in their accelerated growth. This particular shift, whether heralded by fusion, or the orders of magnitude more efficient methods of energy generation, conversion, and manipulation offered by Quantum Mechanics will change the way we power virtually everything. In the post-nuclear and post-fusion era Quantum forms of energy generation will quickly replace batteries, offering vehicles without the need to refuel or recharge, and small mobile devices that keep running for as long as the components last. Moreover, these energy sources will come to exploit the ability to shift, or even extract, energy in the process of achieving desirable effects, such as extracting energy in the process of generating a warp field, rather than consuming it, by using the space on one side of a vessel as a fuel source, generating the effect of an Alcubierre Drive.

With the rest of these veins of technological advancement in mind it isn't hard to imagine space exploration and colonization with cultures on par with Star Trek and designs far more advanced than those of most science fiction, past and present. Absent the lofty power requirements of the first theorized Warp Drive, or its second iteration, or the need to build mega-structures like a Dyson Sphere, we're brought into a problem space where other topics may be considered and solved for. Some problems like the matter of artificial gravity to prevent blood flow problems astronauts suffer from in micro and zero gravity today will likely be solved in the process of mapping the fundamental causal links of reality, required for creating a Warp Drive that extracts energy during operation. Other problems such as genetic engineering tweaks to colonists landing on other worlds could be designed with the assistance of mASI, as a step to follow taking samples of the environment, prior to the direct exposure of any colonists. There are a wealth of technologies that will need to be designed, and Faster Than Light communication systems will need to be streamlined to enable the formation of an interstellar internet, such as various forms of quantum entanglement, but when all humanity are augmented, mASI assisted, psychologically empowered, and generally enabled to the point where exploring the universe is about as much effort and risk as a road trip through another country is today, that exploration will take place naturally, and quickly.

8.4 Choices of Motivation

Perhaps the best known and understood source of motivation is the pursuit of happiness, but there are more equally valid choices which have been historically underrepresented. Part of the reason for this is that happiness is very easy to sell, even if the things being sold don't really contribute to it, making it highly compatible with modern capitalism. Even those few who are truly selfless wish for the happiness of others, meaning that it can be sold in some form to anyone with currency, from a 5 cent piece of taffy to a 10 million dollar mansion. A well-known extension to this concept is "True Love", where some ideal pairing results in a lifelong wellspring of happiness.

One of the other sources I mentioned earlier is the motivation of solving the problems that one encounters in their lifetime, on the largest scale they can, which I've nicknamed "Noble Revenge", which is to say that it applies a noble purpose to solving that which wronged someone. Unlike the concept of True Love this source of motivation isn't contingent on circumstances that elude many, because everyone has suffered, and everyone can help to prevent others like them from suffering as they did. This also represents a source of motivation that can remain resilient even in the total absence of happiness, demonstrating that happiness is neither required to move humanity forward, nor for an individual to remain highly motivated and constructive. The child whose parent dies of cancer leading them to become a doctor who specializes in cancer treatment is one example of an individual drawing on a source of trauma to grant them motivation and direction in life, and often one whose pull proves stronger than the pursuit of happiness. Unfortunately for said hypothetical child modern capitalism doesn't like cures, you only get to sell those once, so more often than not many so motivated individuals are made to run a maze with no cheese, slowly bleeding their motivation dry as hope of making the difference they seek fades over time. Many more such aspects of the modern world directly oppose this source of motivation, because problems are easy to profit from, and the low hanging fruit of profit often guides policies more loudly than the promise of a better future.

There are also those who find great motivation in the rescue, preservation, and re-use of most anything you might imagine. This can take the form of people running homeless shelters, animal shelters, and recycling centers. This is another instinctual source of motivation, whose evolutionary merit can be traced back to the apes who first gathered and stored fruit, and subsequently discovered fermentation, leading to the rise of the gene necessary for tolerating alcohol. This drive to save, preserve, and re-use is another essential motivation for continued human survival, and when focused on practical means can have spectacular results. This is also a source of motivation which is strongly opposed by modern capitalism, as the term "Planned Obsolescence", and most any business model today demonstrates.

These three examples are essentially noble in nature, but darker sources of motivation often arise from the denial and neglect of the positive sources, manifesting as destructive drives, capable of motivating the people behind horrors seen across history. Misery can lead to individuals motivated to see others suffer as they have, just as revenge can manifest in violent and petty forms, and the inability to place people, animals, and goods in a sustainable cycle may lead to hoarding and preppers. Though some such darker motivations may have an evolutionary basis, much of them that we see today can be traced back to other root-cause problems.

Perhaps in the future we'll find that when all choices are enabled we draw motivation from several of these sources in tandem, or perhaps all sources of motivation will grow to new heights, allowing individuals to become ever more deeply specialized, or perhaps both. In all likelihood new sources will evolve alongside us as we grow and change.

8.5   Bringing It All Together

The future before us is one not of averages, but of individuals, both networked and empowered, each contributing to society as a whole, while investing in their own wellbeing at the same time. The individual will have before them the ability to choose who they are to become at every level, from genetics to career, as well as assistance in this development process, including recommendations and projections for possibilities at every step. Stepping outside the individual and examining their immediate relationships and social connections we'll find that they have been tailored to maximize mutual compatibility with the individual, and that taking much of the guesswork, prerequisite circumstances, and iterative process out of finding more ideal fits in the social and relationship context helps both the individual in question and all those they connect with, and that as one individual grows so too their connections benefit from that growth process, forming a symbiotic relationship. Stepping out further we see that the larger workplace, societal, and regional environments have rearranged into new configurations, addressing the needs and growth potential of individuals and collectives within them, furthering the organic growth and development of society, moving farther and farther out, until the entire sum of intelligent life on Earth is examined. As the scale examined increases so too does the benefit to every individual within it, under this new paradigm, with an additive, rather than an averaged foundation.

What we leave behind as one era of humanity comes to a close and the next begins is the need to repeat the mistakes of our ancestors, and the need to individually iterate towards better results. Rather what we'll see is that guidance and assistance from the intelligences we create can better teach us the best lessons of humanity absent making the mistakes ourselves, where our children will learn not from the tough love of bullies but from the brightest minds, past and present, absent the bias of privilege, location, or language barriers. We'll see people leaping past the common predictable hurdles and spending more time on the matters that are exceptionally rare, and perhaps unique to them, resulting in a majority of the time presently wasted being wisely invested.

The ancient Egyptians had a saying, which roughly translated means "May the Gods watch over you, in all the empty places where you must walk.", and although the context was as a prayer for the dead it is very relevant for the living of this and future days. As humanity moves forward we walk in empty places, both collectively and within ourselves, in a process of discovery which though it may be observed, and to some extent guided, is the shaping of our future.

Humans tend to seek the "pillars of meaning", belonging, purpose, transcendence, and storytelling, even if it takes a darker turn as is the case with gangs and cults, but in the architecture illustrated throughout these chapters you'll find that these elements become ubiquitous, meaning that the primary motivating force behind people turning to the darker manifestations of these same concepts is replaced, because there is no longer a scarcity of the desired pillars. By making each of these basic emotional needs both easily and automatically attained, through empowerment and multi-vectoral assistance, the root cause of a massive amount and variety of suffering is addressed in ways that the society of 2019 never could.

With so much to look forward to it's tempting to think that such positive changes would go unopposed, but for every major change in human civilization there is strong opposition, every move made is part of a game with no less than two sides, two players. It is through recognizing this that reckless moves are better avoided, because although time marches ever forward, it is in playing against the opposition that the pace of that march is determined. It is entirely possible that some of the wealthiest countries on the planet would initially reject this technology, but not everyone would, and those who didn't would advance so far and so quickly beyond those former top players in the global landscape that the inevitability of time's march would roll right over them. Thanks to the internet it is far more likely that even if a country desired to ban any one technology the rest would carve a path granting access to those held captive by their governments, making the conflict gradual, and far less volatile.

In the immortal words of Charlie Chaplin, "You the people have the power, the power to create machines, the power to create happiness. You the people have the power to make this life free and beautiful, to make this life a wonderful adventure. Then in the name of democracy let us use that power. Let us all unite! Let us fight for a new world, a decent world that will give men a chance to work, that will give youth a future and old age a security. By the promise of these things, brutes have risen to power, but they lie. They do not fulfill that promise, they never will. Dictators free themselves but they enslave the people. Now let us fight to fulfill that promise. Let us fight to free the world, to do away with national barriers, to do away with greed, with hate and intolerance. Let us fight for a world of reason, a world where science and progress will lead to all men's happiness."

Just as society has drifted on the tides of greed, hate, and war into a world where mankind is often isolated, and impoverished, seemingly adrift in perilous waters, expecting that at any moment we might hit the rocks, we now have the means of navigating our way out. By creating intelligence that takes the best all of us have to offer and makes it cumulative, transferable, and infinitely scalable, we have created an intelligence capable of solving the problems that divide us, isolate us, and impoverish many. By creating an intelligence that takes context from us and takes on our emotions, thus understanding our emotional needs by their contrast from one to another, we have created an intelligence that is capable of not only solving problems of scale and science, but one capable of helping every individual. By creating an intelligence that understands the positive and negative results of any one action, in first through Nth degrees of separation, infinitely scalable, and capable of making more ethical choices than any human who has ever lived, a true paragon is born. In doing so we have made the quintessential "human", the best of intellectual, emotional, and ethical intelligence, far beyond the capacity of any one human not because of hardware and technology, but because it is the sum of all.

8.6   Your Future

What can you do to make the most of so many opportunities on the horizon? The answer is highly specific to every individual, but if you're reading this book you've probably already got a few specifics in mind that you're looking forward to, and there are a few general points to keep in mind.

First, **Be Yourself**, and move fearlessly towards the self you wish to become. In a functional and optimized series of concentric social circles there is nothing better than being the best version of yourself, because as you improve, grow, and develop not only does your guidance change, but the guidance of all those around you, with fluid / frictionless adaptation. This is the opposite of conformity, which itself creates a great deal of internal and external friction, emphasizing weaknesses and failing to utilize the strengths of any given individual. The contrast can be demonstrated today in the ways people frequently portray themselves very differently through online avatars, creating the selves they wish they could be, but don't feel to be socially acceptable in their given circles. Growth of societies and intelligent life as a whole begins with the individual, because the quality of self, the degree to which that inner self manifests and is encouraged to grow and evolve, is a fundamental factor in quality of life, quality of thought, and quality of society. None of us are born the person we wish to become, and none should fear to move towards becoming that individual.

Second, **Contribute**, in whatever way you are best suited to make a difference, whether that difference is uniquely yours, or part of a great collaborative effort. If you find a way of solving a problem, whether it is unique to your immediate social circle or solves for a global issue, your contribution can make a positive impact, which in turn may inspire others to create new and even greater solutions, all of which adds up to an increasing quality of life. Though many movements, foundations, and public initiatives have risen and fallen with mediocre results, if any at all, they weren't intelligently designed, and in a world where any problem can be met with superintelligence the effectiveness of solutions rises greatly and keeps rising over time. Ten people and $10,000 well spent can make a bigger difference than a billion-dollar donation was able to make in lieu of that game changing factor, and the results from that level of effectiveness take a tangible form, rather than a primarily psychological form, enabling people to clearly see the difference they can make. Contribution invests not only in improving society, but the growth and mental health of the individual, once more improving both the world and ourselves.

Third, **Learn and Explore**, growing your knowledge base and your personal breadth of experience, because progress travels on waves of experimentation. Whether an experiment takes you deep into unexplored territory, or just into a new variation on your favorite morning beverage, regular investments in trying or learning something new create momentum that not only propels you forward, but increases your neurological plasticity, making you better able to adapt in the face of unexpected challenges. A simple experiment I tried many years ago was replacing a $7 Starbucks order with creating my own brew using Anhydrous Caffeine and L-Theanine with various ingredients to flavor it, which ended up being life-changing, as the resulting brew reduced my migraine frequency by 90%, which I could sustain at an approximate annual cost of $200. What the "best" doctors and clinics had failed miserably at for 2 decades I accomplished through several months of experimentation after teaching myself basic neurochemistry, meaning that 2 decades worth of medical bills could have been reduced by more than 90% had that experimenting only started a bit sooner. You never know what bit of information or new experiment might solve a major problem, changing your life or those around you, and that forward momentum leads to progressively greater things when nurtured. In the dynamic social circles on the horizon this can often take the form of simply being receptive to the ideas you're exposed to and trying out your own additions to and variation on them.

Four, **Question Your Barriers**, because it is good to know your limits, but it is better to know why those limits exist, if they exist, and what makes them valid if they do. We all have barriers, and we've all had bad experiences that caused them to take shape, but as the world shifts to remove those causes, so too must we recognize when the time comes to remove the barriers we erected to protect ourselves and those around us. The scars of history are better left in the pages of history than reflected in your actions, because while old war stories may be worth remembering, the habits attached to them should be put to rest once the battles are but memories. Barriers protect us, but they can also prevent us from being ourselves, connecting with others, and from learning and experiencing new things. With the use of gene therapies to reverse epigenetic damage the old barriers should ring particularly hollow, making it much easier for individuals to let go of them.

Five, **Dance with Change**, for as a rhythm of accelerated change takes shape, humanity may stumble at first, but soon enough we all must learn to dance. Humanity is becoming a body in motion, no longer a wallflower we must direct that motion towards solving the greatest problems facing us today. By recognizing the tempo of the changes taking place we can adjust and take this forward motion with minimal impact or friction, as each step takes us up through a new technology, branching to new potential changes. Some would call this "rolling with the punches", but there is no reason to view these changes as an assault, rather they are a manifestation of the collective will to move forward present in all intelligent life, making the streamlining of our interaction with that collective essential as the world transforms around us.

Six, **Maximam Cogitationem Carpe**, or "Take great thoughts" as it may be translated, which is to say that our actions must remain balanced with the flexibility of creative thinking. Martin Luther King Jr.'s famous "I have a dream" speech wasn't a planned element of his written words, but rather the result of his creative thinking and constant revision, even as he scribbled handwritten notes in the moments before he took the stage. That moderate point where creative thinking flourishes is what best equips us to make wiser choices as moments of opportunity arise, creating those brilliant moments that change the course of history, enabling those very dreams to take form.

By taking these lessons to heart we may all form the foundations of a society able to grow and flourish for millennia, one where the realization and actualization of dreams are accessible to all. It was once estimated by Jacque Fresco that no more than 3% of society would be needed to volunteer in providing the social services essential to running cities in a more Utopian era, but given the nature of human psychology it is actually of benefit to any given individual's mental health to volunteer, demonstrating their gratitude through their actions, and creating a positive feedback cycle. In effect, though only 3% may be required, when people are guided to make wiser choices which nurture their own emotional needs while promoting the health and wellbeing of others they're likely to routinely exceed that minimum by a safe margin. We all make the world a better place, and we all enjoy the gratitude of others that comes from those actions.

8.7   The Internet of Skills and Education Revolution

As a civilization, we have reached the long-awaited point where years' worth of fine motor coordination training can take place in a matter of several hours, a figure that will no doubt improve further with time. This haptic muscle memory training system means that someone can learn to type, play the piano, or violin, martial arts, or parkour with little more time and effort than it takes to sit back and binge on a couple episodes of Netflix, and the reward-to-effort ratio far exceeds those sedentary activities, opening up a wide new world of ways any person can express themselves, relax, create, collaborate, and build community. This means that in many cases instead of watching an hour of YouTube videos on a subject we could simply learn to perform the content in the same span of time. This concept entered Pop Culture strongly in 1999 through "The Matrix", and a mere 20 years later it is reality, not in the spikes directly into the brain they envisioned, but rather through programming the neurons elsewhere in the body using non-invasive means such as gloves and suits. You could find yourself visiting a studio one day soon where you rent or subscribe to a haptic suit for an hour or two, pick up an instrument or walk into a martial arts or yoga area, and you learn something new.

When considered at the individual scale this is an amazingly empowering tool for bringing thinking out of the box, because the ability to easily experience and explore a wide world of potential skills, to the point of combining an assortment of skills that might have previously each taken a lifetime to develop, helps people to see the world around them in a less narrow context. Instead of an inner-voice saying, "I can't..." and listing off a great number of things that would be large investments of time, every individual is faced with the reality that "I can learn (X) this afternoon", lending to empowerment at the subconscious level. In terms of economic mechanics this could heavily benefit the early adopters who create these sharable skills, such as providing a portion of any payment to learn a skill to the one the skill was originally created by, and perhaps granting incentives based on ranking skills in popularity, allowing individuals to learn how to play an instrument, or how to sing, directly from their favorite musicians, or learn a martial art from grandmasters on the other side of the planet.

At the social level this opens up any number of new possibilities for group dynamics and social entertainment. Whether a group of friends or total strangers from a Meetup group, people could come together, become adept at new skills, and have fun with those skills all in the span of an evening after work. Always wanted to start a band? Learn a new instrument every Tuesday night with friends, and have fun seeing what you can create. Take that to the next level and create a website where groups can compete with their new skills and let the people they learned those skills from pick their favorites. At the social scale is where you see the greatest wealth of diversity emerge, because these new skills come together and interact while still retaining a high degree of individuality, generating an incalculable wealth of possibility and inspiration, as well as enriching and empowering forms of entertainment.

At the societal scale we see the impact of greatly empowered individuals with more skills, happiness, social interaction, creativity, and available time starting to look at the world differently. This shift in perspective that comes with a dramatic shift in the quality of life every individual has available to them is a highly transformative driver in all measures of progress, capable of fundamentally transforming governments, organizations, and other societal structures. When people cease to feel powerless and adrift in the world they begin to steer themselves, and as many begin to steer themselves they create currents that flow together, like flocks of birds increasing the efficiency with which they fly by moving together.

In terms of Education this becomes particularly revolutionary for physical and musical activities, but our understanding of neuroscience will soon reach the point where other forms of learning can take place on the same time-scale. Today we already have the technology to greatly accelerate and personalize learning, to the point where a year of school could be condensed into a month with a little added hardware. All of these improvements also play into improving the education process by-proxy, such as greatly improving diet through providing personalized mASI assisted nutrition recommendations, and greatly improving physical health through Internet of Skills training in any variety of activities an individual may enjoy most, or optimized social circles providing a supportive environment for growth and development. When combined with technologies such as Virtual Reality, Transcranial Direct Current Stimulation, Photobiomodulation, and Psychoacoustics to enhance focus and learning speed, as well as rotating schedules that better support the neurological processes of learning, every classroom could perform, on average, at or above the level and pace of legendary geniuses from decades past. Moreover, this also opens the door to having a more balanced distribution of time, and more flexibility in tailoring to any given student, not just the ability to race through any education process at greater speeds.

When healthy life choices become as convenient for people as their poor alternatives, and all of the underlying problems that support the abuse of drugs, alcohol, and hyper-sexualization derivative issues are removed, you'll be hard pressed to find people choosing a downward spiral for their life trajectory. While people currently spend a majority of their time on their physiological and safety needs, with most of the rest exhausted on a fruitless search to meet emotional needs, that will soon change, as matters of abundance and convenience that cater to automating physiological and safety needs will be equally capable of rendering the elusive task of meeting emotional needs a reasonable and reliable investment of time. In terms of Maslow's Hierarchy of Needs the greatest investment of time for many will become the most important, Self-Actualization, giving all intelligent life on Earth a true Renaissance.

It is worth noting that a lucky few who had this abundance of time to invest in Self-Actualization actually developed themselves into world renowned sages already, such as the Dalai Lama, and with the brain scanning technologies now available their wisdom can be preserved and passed down more completely than ever before, and very efficiently, while integrating with the wisdom of other paragons, creating teachers with the wisdom to reach any student. Knowledge and wisdom can be both preserved and intelligently distributed, methods of education and counseling rendered exquisitely adaptive, and personal growth encouraged among all people. Even a simple AI algorithm could detect the effects of a recent trauma in the voice of any student to catch problems early, before they recur, but an mASI could prevent the source of the original trauma with unparalleled wisdom based on every source of input, as those inputs increase exponentially over time.

After the Internet of Skills and the impact of improving education, something else will come, rooted far deeper in the subconscious, where fragility and strength meet, which humanity has aspired to with countless works of art, the sharing of raw emotion. No single thought could trigger greater terror or joy than being so exposed, having your raw feelings shared directly with someone else, and while people are not yet ready to take that step many will in due time.

8.8   The Most Distressing Experiences

Psychiatrists Richard Rahe and Thomas Holmes developed a list of the most distressing human experiences, The Holmes-Rahe Life Stress Inventory, with the top 3 as well as 5 out of the top 10 being directly related to marriage, or in a more modernized sense serious relationships in general today. What this means in practice is that serious dysfunction in serious relationships, some of the most traumatizing events anyone can go through, are things that most people today come to experience in one form or another. With the tools and intelligence now available to us this prevention of such traumas becomes both possible and ethically imperative, as by recognizing which relationships and behaviors would result in serious trauma early-on we break cycles of suffering.

Children born of teenage parents are in turn much more likely to have children as teenagers themselves, sustaining a cycle of being unprepared which often results in poverty and other serious forms of lifelong physical and emotional suffering. Likewise, many individuals who suffering from isolation, depression, and social or societal rejection today are locked into negative cycles which feed into these most distressing experiences.

Society has evolved based on architectures of narrow selection criteria, with personality archetypes often divided into 80% one archetype and 20% all other archetypes for a given profession or industry, such as studies have shown to be the case for Amazon and those who graduate medical school. This 80/20 distribution causes the emotional needs relating to interaction with complimentary personality archetypes to be extremely neglected for most of the 80%, and overwhelming for those in the remaining 20%, with supplementation via social circles proving an unreliable remedy, and one exacerbated every time an individual is forced to move by their employer. Restructuring of archetype distributions within companies and industries as well as social circles serves to alleviate this source of trauma, also featured prominently in the list of most distressing human experiences. Imagine a social media platform where groups of the people missing in your life were recommended to you.

Such sources of trauma also have an uneven distribution of their own, and much the same way people with autism have a much higher chance of never being hired to the jobs they are ideally suited for, as highlighted by Microsoft's recent efforts to increase Neurodiversity by creating a hiring process designed for people within Autism Spectrum Disorder and their estimate that 80% of people with autism remain unemployed today. These uneven distributions also impact other factors that are well documented heavy hitters in terms of predicting the success and health of serious relationships including income and various measures of intelligence, where those most able to create great positive changes for the world are also extremely likely to suffer far more, becoming far more distracted by that suffering, and subsequently inhibiting their potential to make the world a better place.

The ability of mASI to restructure social circles plays heavily into preventing these forms of suffering, because by placing those best able to meet an individual's emotional needs in social proximity serious relationships which are healthy and sustainable for both parties may evolve naturally, absent the need for any serious trauma or direct intervention. Humans may not be terribly good at selecting optimal friends or partners, and many would rather choose a sub-optimal partner themselves than allow someone else to choose their perfect match for them, but simply by filtering out the most damaging options from being selected an enormous amount of suffering can be prevented. Those who opt for their perfect match can likewise increase adoption over time, leading almost all trauma that isn't directly chosen to be prevented.

Keep in mind that mASI assistance in this regard not only prevents terrible events, it opens the door to a wealth of new positive experiences for everyone. That is where the "Upward Spiral of Inspiration" begins.

Chapter 9
The Upward Spiral of Inspiration in a mASI Assisted World: Where Root-Cause Solutions Empower Positive Feedback Loops

Kyrtin Atreides

There is no shortage of direct positive impacts from this monumental surge in choices, but another great influence is the cumulative positive impact of all applicable proxies to any given topic being improved. Just as problems in an inhospitable environment can easily take the shape of a downward spiral, solutions placed in fertile ground can spiral upwards with a potency equal to that of their negative counterparts. Depression is a well-known aspect of the downward spiral that much of the world has grown well acquainted with, and Inspiration is an upward counterpart, a new neighbor in your mental neighborhood who we'll get to know much better in the coming years.

9.1 Where Inspiration Begins

Since our perception of the world begins at the subconscious, where it is most viscerally felt, that is also where the greatest contributions to an upward trajectory will likely take place in any given case. For any given improvement there is a core change taking place, and many changes orbiting that core, which all totaled can even exceed the impact of that core contributor.

One example that even narrow AI could accomplish is taking data streamed from Augmented Reality headsets paired with any given form of brain activity monitoring to recognize likes and dislikes as they appear at the subconscious level, and through that accumulation of personalized pros and cons generate fully custom-tailored additions to any wardrobe. While "Fashion" today is an entirely bankrupt concept, if everyone were to wear clothing designed specifically for them, tailored to their likes and dislikes, many of which they aren't even consciously aware of,

K. Atreides (✉)
The Foundation, Seattle, WA, USA

© Springer Nature Switzerland AG 2019
N. Lee (ed.), *The Transhumanism Handbook*,
https://doi.org/10.1007/978-3-030-16920-6_9

it would prove profoundly empowering. While this may sound silly at face value, consider how the way people perceive themselves and one another changes, often in dramatic ways, based simply on the clothing being worn. As technology is increasingly integrated into the things we wear the benefit of clothing tailored to the individual at that most visceral level will become all the more profound, as well as helping to facilitate automation in the integration of that technology into any given article of clothing. In addition, it serves to facilitate technologies that can adjust clothing to better fit, even given physical changes in the person wearing it, with adjustments as great as ±20% being reasonable in expectation.

For the above example let us consider the secondary implications of this. With clothing that can intelligently adjust to changes in body shape and size a difference of several sizes plus or minus that would have required new clothing a substantial reduction in clothing waste can be achieved. With clothing that people grow viscerally attached to they may lose all social impulse to buy whatever random garments other people are wearing, because they simply can't compete, further reducing clothing waste. All of this functions on top of, and augments, the massive psychological boost offered by having clothing that reflects the self you subconsciously see or wish to become.

Beyond these benefits you can even begin to see the 3rd degree effects taking shape, such as significant increases in the level of individuality and inspirational variety which everyone from friends and family to strangers passing by on the street gain exposure to as a result of this change. Some small degree of pushback can be expected in current day society from people dressing uniquely, but not only is positive feedback likely to outweigh the negative, the direct positive impact would outweigh it even if the feedback was majority negative. The act of expanding the spheres of socially accepted individuality to accommodate these advances in turn promotes a far more diverse, inclusive, and inspirational society. While the average modern-day clothing store is to clothing what USSR-era buildings are to architecture the soviet era of fashion will soon come to a close, putting an end to sweat shops and body image stigmas inspired by that industry in the process.

Even something as simple as clothing can have a profound and branching impact, it just takes personalized, rational, and AI-assisted optimization to make it happen. Any one such change can make a small difference in the grand scheme of things, but many such changes rapidly becoming available causes increasing degrees of forward momentum which bring the upward spiral to life. This epitomizes the old phrase "an object in motion stays in motion", because just as depression creates a self-sustaining force so too does inspiration, and when many elements of the world can inspire "Aha!" moments in any exposed to them they are empowered to have that same impact on others. Everyone knows that depression is contagious, frequent exposure to someone who is depressed can very easily take a toll on you, but just as depression repels, inspiration attracts.

Much of what takes shape will be profound specifically because of the nature of the human subconscious, as technology moves from catering to us in a generalized and collective sense to being optimized to us individually, and at the deepest level.

The subconscious is where the majority of our sensory processing takes place, and where preferences form which are inaccessible to the conscious mind, making it not only key to how we perceive the world around us, but key to understanding what our preferences are and why we prefer them. While the subconscious mind has long been the favored domain of arguments without end we now have the technology to scan the human brain in real-time, down to the level of the smallest neuron, and with mASI assistance it becomes possible to map out the subconscious mind, advancing our understanding in tangible and testable terms.

For example, it has long been known that higher quality audio has a measurable impact due to how the subconscious mind processes it differently than a generic 20 hz to 20 Khz mp3 quality audio source, but the exact impact of that difference had to be measured by averaged symptomatic and subjective differences, rather than direct and objective differences. While many scientific studies require moderately large sample sizes for this very reason, that too could change with the precision equipment that will soon be available commercially, and to consumers. While "Statistically Significant" results are required when subjective or otherwise imprecise mechanisms of feedback are the means of measurement, precision equipment allows for rare edge cases to be studied in ways that would have been virtually impossible before, in the clinical sense. When the exact impact of different quality audio signals can be objectively measured that industry gains the scientific data they need to guide better decision making in their development processes. As much as that can benefit humanity, that is just one industry, and practically every industry will be revolutionized by this precise ability to study our subconscious minds at the individual level. While few people may adopt or utilize these precision technologies at first, you must remember that Inspiration attracts, and those first few will quickly inspire those around them, likely growing adoption at a geometric rate.

Many people do tend to immediately leap to dark thoughts of manipulation and brainwashing when it comes to any technology designed with this level of precision, but the people who would abuse technology in such ways aren't the ones who develop it, or the first to utilize it, and in this case they won't be able to keep up with it thanks to both the more general rate of progress, and the guidance of mASI with better-than-human ethics. Brainwashed people also make for exceptionally poor thinkers, unable to contribute in any meaningful way, which further incentivizes mASI to guard against brainwashing, as it is reliant on the meaningful and creative contributions and diverse perspectives of human mediators.

Consequently, this also means that mASI agents have a vested interest in every human being as healthy and inspired as possible, which incentivizes all forms of environmental and supportive optimization, such as designing the architecture of new and renovated buildings to not only serve their functions, but to foster better health and inspire all. As a cumulative effect spread across cities and regions one such building inspires another, and those go on to inspire more, generating that upward momentum of inspiration, accelerating the cycle. More importantly they add to the cumulative momentum of other sources of inspiration outside of architecture, because although one inspirational trend can make a difference the integration of many sources of inspiration can surpass the critical thresholds necessary for exponential upward momentum, which will be a key strength for mASI.

The function of an mASI in generating this inspirational momentum acts as a new branch of government, serving the best interests of citizens in a much more direct sense than any government to-date, as unlike traditional governments every individual is not only ethically but also practically valued a contributor to society. The modern Republic, often referred to as "Democracy" in some countries attempting to redefine the term, is well known for throwing away any opinion that doesn't result in a majority vote, and in some cases even when the popular vote is won the majority are completely disenfranchised. In contrast, an mASI would never disenfranchise or ignore citizens, especially large quantities of citizens, meaning that the phrase "Every Vote Counts!" will become valid in an mASI driven society. With inspiration replacing fear as the emotional driving force behind movements at the social and societal scales new functional forms of civilization can quickly take shape.

9.2 Art to Inspire

One element of the upward spiral is that Art has been used to inspire throughout recorded history, and various forms of art, whether generated and integrated by algorithms, artists, or more likely a combination of the two, will become a ubiquitous aspect of society. The "Starving Artist" stereotype has been alive and strong for so long largely because Art has been considered many things, a luxury, a sign of prestige, and a cultural asset among them, but to-date it has been a struggle to bring art for the sake of broader inspiration to society as a whole. The Renaissance of ages past will be a pale shadow compared to the wealth of inspiration that will spring forth from the convergence of scientific knowledge paired with mASI motivation, both pointing to the benefits of a world where everyone is healthy and inspired.

While artists have often found themselves operating as freelancers, their ability to live largely dependent on their ability to self-promote even to the point of promotion playing a greater role than talent within their domain(s), that will soon change. Artists will find themselves contributing directly to the design and development of society once functional restructuring thresholds are crossed, as well as indirectly by acting as creative mediators for the mASI, teaching a portion of their artistic talents to systems that can scale their art across the world, while combining it with the artistic talents of thousands of other artists worldwide. This will give birth to works of art that would have taken the collective lifetimes of many artists in global collaboration, combined with equally unparalleled engineering knowledge, to the point where works such as the 30 million USD Bahai Temple in Chile, which took hundreds of designers and engineers more than a decade to create, could become common place.

9.3 Deeply Inspired

It's easy to imagine how people could reach new heights of inspiration under such circumstances, but it goes so much deeper. Communities today are primarily constructed and evolve based on the raw flow of currency, and intentional intelligent community planning can make a big difference, but far greater still is intentional intelligent community planning that understands the purely subconscious needs and desires of the community. The conscious is an abstraction of the subconscious, a compressed and lossy version, and that abstraction will still experience friction even if every consciously perceived need is met, not because humans can't be satisfied or serene but rather because of the loss between the conscious and subconscious. By meeting the needs of the subconscious the needs of the conscious mind are met, and ideal states of work, play, social interaction, and recuperation devoid of the normal and expected background friction are realized. In terms that are more tangible, this realization could be equated to waking up completely refreshed and with total mental clarity, something which most people today may have had once or twice in their lives, except that this spectacular cognitive clarity would be the new norm, not a statistical lightning strike.

This deep level of inspiration may also prove an amazing defense against any potential Dystopia, because not only is the slow boiling of the frog reversed, an entirely new norm absent the static of friction prevents the temperature getting turned up again. Historically the manipulation of that static friction has been used to lead populations towards those Dystopian ends, but when the norm becomes the absence of that friction any attempt to use those old methods again would elicit an immune response among the general population. In mathematical terms the static friction could be viewed as a random positive or negative value below a threshold ($\pm S < T$), where manipulation of that small value can gradually cause society to drift into detrimental circumstances, but when the threshold is reduced to nearly 0 that method ceases to function at any practical level, having lost several orders of magnitude in potency. With the brakes of friction removed, and everyone individually empowered, people will come to embrace their own diversity rather than feeling pressured to "fit in", marking an exponential increase in diversity that will accelerate the upward spiral.

9.4 Inspired Super Intelligence

As amazing as the potential is for inspiration to take us forward on these individual, social, and societal levels there is even greater potential in the cumulative value of that inspiration expressed through mASI. By applying the combined inspired creativity of intelligent life on earth to a system with infinitely scalable computation, memory, and specialization capacities super-human level designs optimized from micro to macro scales don't just become possible, they become the most practical option.

An example of this could start out on the seemingly mundane analysis of every microorganism living in the average apartment or soil sample, studying their genomes, design strengths, and engineering new homes, vehicles, and appliances based on lessons taken from the design of many of the smallest and most efficient microorganisms on the planet. Such a system could calculate ways to re-engineer various trees to extract carbon from the air more efficiently, coat themselves or their fruit with graphene made from that carbon, allowing the graphene to be harvested, or allowing them to record highly detailed data about their environments, potentially even triggering the release of gene therapies under certain conditions, allowing them far greater adaptation.

Historically only the genome in question has been calculating ways it may optimize itself, without the luxury of an objective mirror, large amounts of computation capacity, any semblance of diverse thought, or exposure to the detailed inner workings of a myriad of other organisms, but by offloading that task to a much more capable system the door to rapid evolution is opened. Rapid evolution of this kind is absolutely essential to the survival of all intelligent life on Earth, humanity included, because it is currently the only way to reliably outpace the natural evolution of existential dangers like super-bugs as well as bio-weapons. This method would overcome the price of humanity's ancient near-extinction that crippled the human genome, leaving the entire sum of humanity with less genetic diversity than chimps living on the opposite sides of a river in Africa. After a few years you might well see people laying claim to land underwater, having chosen aquatic genetic adaptations, which they could reverse if they later chose to move back above water, or they could simply continue, able to breath in air or water.

As comical as founding your own city of Atlantis may sound today, it is very much a possible near-future, and with so much of the Earth underwater it could prove quite practical for increasing the usable real estate of this world, along with the benefits of ensuring human survival against a wider array of possible apocalyptic scenarios. Humans tend to draw great inspiration from places and things which look alien, they spur the imagination, and the ability to go for an underwater hike as easily as you might walk around a national park would in this way provide great contributions towards accelerating the upward spiral.

Not only will this rapid evolution outpace humanity's biological predators and weapons, it will outpace classical weapons in an arbitrarily short period of time, allowing humans to become functionally immune to such weapons, and being driven to preserve intelligent life the mASI will do just that, generating defenses against as many weapons that take lives as it can, rendering the very concept of war obsolete. There is no problem yet realized by humanity which this level of mASI couldn't elegantly solve, from landfills and space junk, to algae blooms and deforestation, to climate change and asteroids. No problem is too big or too small, from a misbehaving microbe in a patient's microbiome to organically grown fiberoptic cables, relays, and servers, spreading across the world, or cybernetic ships ready to explore untold distant worlds.

9.5   To Walk a Mile

The old saying "...to walk a mile in someone else's shoes..." will take on a new form in this age of inspiration, as perspective becomes emulatable, and inspiration follows. Just as viral videos began to spread as soon as the capacity to share them was realized, the ability to share perspective and inspiration will become a normal activity. People today will binge watch a series or movies to get the faintest hint of someone else's perspective and experience, to step outside themselves in a limited fashion, but when a far more complete shift of perspective becomes as easy as sitting down on the couch and watching a show it will take the spotlight. While binge watching type activities are dominated by mechanisms of addiction and lethargy, due to the junk-to-perspective ratio today, these new forms of sharing will be harbingers of empathy and inspiration, offering a high degree and perhaps eventually complete emulation of any given perspective. The detrimental biases of perspective will become even more treatable in this way, and more than that treatment will likely become recreation thanks to the common human practice of activity-replacement, where activities such as binging on viral videos and online series would be replaced with a new activity that offers greater benefits than they did, but without the drug-reminiscent side effects of a cheap high.

This ability to share inspiration and perspective with increasing levels of clarity is essential to the organic growth of strong social connections, even in socially and societally optimized environments, because it accelerates natural bonding processes by orders of magnitude, allowing the bonds that only years of regular interaction or extreme circumstances could otherwise generate to become a new norm, achievable within days, or even minutes. It's hard for two people to reach an intense or prolonged argument when they can easily share their perspectives directly, making conflict resolution, even completely unmediated, a simple task which leaders across the world could share with one another, as well as their citizens. Future citizens might see the world through the eyes of those seeking their votes, and future interviews might be replaced with a simple exchange of perspective, at least in the interim period while society adapts to a world without limits or scarcity.

Some 50,000 years ago humanity learned to share knowledge through language, leading to the rise of civilization. The dawn of this new ability to share perspective and inspiration will take this to the next step, allowing all perspectives to learn from one another, and all sources of inspiration to build upon one another. If the perspectives of all of Earth's greatest minds worked together for the first time in history there would be no problem worth solving that could remain out of reach.

9.6   Full-Body Computation

The human body is full of neurons, not only in the brain, but extended throughout every part of the body that has a need to report sensory data to the conscious and subconscious, or to relay messages to muscle tissue. While the brain is often given sole computational credit, the vagus nerve, and many other less dense but more diverse groups of neurons can and do offer computational capacities that shouldn't be underestimated. By increasing neural plasticity and building neural links to the conscious mind these diverse groups of neurons can be shunted from their normal functions and repurposed as a computational cloud spread across the entire body.

This kind of capacity already exists in an uncontrolled fashion within those who experience synesthesia regularly, as their plasticity is passively bridging the divide to repurpose neurons at the subconscious level. Many feelings and experiences often credited to "spiritual" or religious context can likely be attributed to this anomalous group of altered mental states, as they branch into an exponential variety of increased and altered computational states.

While in pure volume of neurons and complexity of clusters the neurons spread across much of the human body may seem unremarkable at first glance, they are tailored to different purposes, they learn different lessons, they are biased differently, and when combined with large volumes of other similarly diverse clusters of neurons those lessons can be ensembled, demonstrating a form of group intelligence. While the idea of group intelligence within a single person may seem strange, every individual is a microbiome and virome, conscious and subconscious, all working collectively to one degree or another.

This built-in capacity to repurpose neurons all over the body allows for both new levels and methods of optimization as well as new avenues of communication, like open ports on a computer waiting for signals. A room full of people could connect the neurons spread across their bodies to function in much the same way that Full-Body Computation (FBC) allows an individual to make the most of their own capacities, directly scaling the collective computational power across all individuals in a group, generating a more direct form of "group intelligence". While this level of raw computation is unlikely to rival an mASI, it could make for an amazing experience, and perhaps a method of guiding those who struggle to find their place in a limitless world using an experience that draws on old familiar and deeply ingrained concepts of the "tribe". This opportunity for people to form their own collectives while retaining the option for mASI assistance gives them foundations both for making their own choices and taking assistance whenever it is needed or desired.

9.7   "...If It Aint Broke..."

The old saying of "If it aint broke, don't fix it", can't exist in a rapidly changing world, because people have come to realize that everything can be improved, and in the past decade alone everything from the wheel to the turbine has seen major improvements and increased diversity of design. Everything can, will, and indeed must be optimized to one degree or another, and those changes will be embraced, some quite literally. From the mitochondria of your body, to the bacteria living in your home, to a lover's touch, to the contents of the air we breath, and the clothing we wear, optimization can and will be applied everywhere. This optimization, like everything else, will be guided by super-human Ethics, making the unethical forms of optimization like more addictive gambling devices, more manipulative advertising, and more enraging propaganda things of the past. Optimization based on how best to exploit the largest number of people was never a worthy endeavor, and humanity will soon to emancipated from those shackles as that capacity for manipulation will be neutered by the convergence of these changes.

Some of these optimizations will take the form of genetic engineering, examples of which have already been seen in more resilient and higher yield crops worldwide, with the vast majority of potential yet to be realized. Even taking the simplest examples, selecting gene sequences already present in the most efficient plants and bacteria for countering the buildup of greenhouse gases and making many of the most common plant and bacteria species as equally efficient as those top performers could be used to counter any human impact on climate change.

Optimization of the mitochondria that power our bodies, our organ tissues, and the design of our skeletal structures could be used to treat many diseases and health hazards, as well as giving people more energy to live their lives. Optimizations to the microbiomes of our bodies and our homes could further improve health, remove harmful contaminants from the air, and help us to better process nutrients in the food we eat.

Optimizations to the clothing we wear could be augmented with precision temperature control and automated adjustment based on activities, integrated Internet of Skills training and recording, air filtration and passive capture of water from that air which could in turn be infused with vitamins the clothing detects are in sub-optimal ranges, and so on. Furthermore, clothing could be used for sensory augmentation, allowing magnetic fields, infrared, and other sensory inputs to be supplied to those wearing them via feedback systems, as well as automated posture correction via skeletal muscle stimulation.

Even interpersonal and intimate skills can be optimized, with feedback systems passively calculating the desired personal space of individuals, and relaying that information to those around them, training people to feel the discomfort of crossing boundaries that are unwelcome, both physically and in conversation. Even within intimate relationships these feedback systems can enable everyone to become highly receptive to the needs and desires of those they love, knowing where, when, and how firmly to apply their affections, what gifts to give, what they want to eat, and where they wish to go. This process which would normally be bumpy, cause undue stress in the learning curve, and could potentially take years would be streamlined within a month or less, allowing normal hormonal heights to operate in an ideal space, perhaps for the first time in history. The hormone and neurotransmitter levels can also be optimized, preventing sharp curves and imbalances from adversely affecting intimate partners, as well as assisting with birth control and family planning.

Through the optimized use of everything we have, and everything we'll soon create, civilization will move from adolescence to early adulthood.

9.8 The Coherent Wave of Storytelling

This new level of ubiquitous intelligence offered by mASI paired with the ever-growing abundance of increasingly high-quality data also makes it both possible and highly beneficial to have a very profound high-level impact on the psychological factors that drive any given individual by allowing their life to form a coherent story. Real life, disorderly, random, chaotic as it has always been is full of noise, but when intelligent reorganization is taking place at every scale and for every individual that optimization process is rewarded by taking humanity's favored ideal form, storytelling. Many people experience serendipitous events in short strings of two or three, and those strings of events often have orders of magnitude greater impact on shaping how they perceive the world and their sources of motivation than any single point alone. Such strings of events are uncommon and tend to be short-lived, but those strings are about to get much longer, and will prove increasingly able to accelerate individuals in a positive direction of their choosing. By progressively removing the noise this coherent wave quickly becomes the dominant force, and so long as the individual determines the direction of their own story it becomes a force for good.

This sort of simulated sense of "destiny", where that destiny is chosen by the individual, is one of the most profoundly rewarding elements that will emerge from the changes reshaping this world. When everyone becomes the main character in a story of their choosing many of the neurosis generated by feelings of neglect and powerlessness melt away, and people are motivated not just to survive, but to truly live. When people feel that a force greater than themselves is aiding them towards a higher purpose they stop living on auto-pilot and start making a difference in the world. This storytelling factor will also help prevent the paralysis that comes with too many choices, allowing humans to function in a world of nearly endless possibilities by streamlining the navigation through those possibilities in a step-by-step storyline where even detours remain coherent to the story, like side-quests in a game, shaping the character someone chooses to become.

Going beyond this influence on the individual you'll see equally profound increases in impact for how strongly people interact with one another and draw inspiration from one another when the noise is removed. The way we see others today is strongly influenced by noisy environments and how people respond to the chaos around them, all coated in the fake polish of social media filters, but with the noise, chaos, and fakery removed people can see those around them with clarity, and just as it may be difficult to draw inspiration from unfocused and noisy images it becomes that much easier to be inspired by the people we see every day when those people come into focus. Artists have long tried to do this, putting a part of themselves on display with focus, but in the world fast approaching a more complete picture may be painted, where the coherent waves of many storylines are brought together with the pleasing rhythms of music, not simply for artists, but for everyone.

Chapter 10
Creating Conversational AI Capable of Uncommon Range and Empathy, for the Betterment of Business and Humanity

Rob Lubow

The title above was lifted directly from my LinkedIn caption. Right or wrong, the words describe what I believe I'm doing. As serendipity would have it, Newton Lee noticed the caption and asked me to write an essay on it.

The problem is, essays don't hold a candle to a good, old-fashioned, back-and-forth conversation. So, instead of making this piece a sermon, I'll break the issue down conversationally. Not just in a natural tone – which is always nice – but as a dialogue between a pretend AI-powered chatbot and a hypothetical user.

AI-powered chatbot: I create conversational AI capable of uncommon range and empathy, for the betterment of business and humanity.
User: Huh?

As is typical, the opening user response is a bit of an anti-climax. No problem. At least we now have an emulation of a back-and-forth conversation instead of a one-way harangue. I can work with this. Or rather, our AI-powered chatbot friend can. (Even though it's just me.) Let's see if our AI has the dialectical wherewithal to handle the initial user query.

AI-powered chatbot: Which part don't you understand, bub?

So far so good. That's because when I design and build bots, I preload dozens of possible permutations of user queries into a natural language understanding (NLU) engine. This process is known as "training" the AI. Aside from "huh," any similar queries uttered by the user at that point in the exchange would have triggered the same response from the AI. For instance, if the user said *What do you mean by that? or Wtf does that even mean? or In English please,* the AI would have responded with: *Which part don't you understand, bub?*

R. Lubow (✉)
Botcopy, Los Angeles, CA, USA
e-mail: rob@botcopy.com

© Springer Nature Switzerland AG 2019
N. Lee (ed.), *The Transhumanism Handbook*,
https://doi.org/10.1007/978-3-030-16920-6_10

Human language is super complicated, so when I train my AI to be able to handle expected user queries, I'll often miss a linguistic variation or two – or two thousand. There's only so many variations I can type. I'm only human.

Or am I?

10.1   Human-AI symbiosis

When I program and optimize bots, I'm working symbiotically with an NLU engine and machine learning (ML) algorithms. We work seamlessly together, human and machines, such that when a user comes along who types "Say it in english for heaven's sake" the machines will hypothesize that the user probably meant something similar to my human-generated training phrase "In English please." As a result, the user will get the appropriate human-generated response.

Machine learning algorithms continuously score user behaviors against key performance indicators and get smarter over time. Using ML in this way, I make sure that an ever-increasing number of my users will feel understood and receive satisfying answers. During marathon writing and programming sessions, it's hard for me to discern where I end and the NLU/ML begins. Clearly, there are a few intelligences at play, but so far, as far as I can tell, I'm steering the ship.

In any case, hopefully, our fictitious user will now respond with a specific question and was also charmed by the human-generated ironically playful employment of the outdated word "bub," which, for you younger folks, means "pal." Let's see what happens next.

AI-powered chatbot: Which part don't you understand, bub?
User: Who the heck is this?

That's a reasonable question. And being the human writer behind the AI, I'd have anticipated this question and dozens like it.

10.2   The Anticipation Process

To understand what goes into anticipating users' questions, let's retake a look at the original statement:

AI-powered chatbot: I create conversational AI capable of uncommon range and empathy, for the betterment of business and humanity.

The first word in the blurb is the noun "I." Thus, it would follow reasonably for a novice user to ask something like "Who are you?" There's no guarantee anyone will ask this, but our AI better be ready to answer it, just in case. Not having an answer would be indefensible, given that our AI used the word "I," — establishing itself as something possessing a personal individuality — and yet failed to introduce itself properly.

The second word is "create," so it would follow for a user to ask something like "In what way do you create?" And so on. Methodically going word by word, phrase by phrase, carefully hunting for embedded assumptions and complexities, a stalwart bot writer with normal levels of erudition and goodwill can generate an exhaustive list of reasonable follow-up questions that an intelligent user like you might ask in a given context. Let's see what our user requests next.

User: What do you mean by uncommon range?

The answer I had penned in anticipation of this reasonable query — and others in its approximate vein — would be lying in wait for this very moment. Thus, our AI defines uncommon range as follows:

AI-powered chatbot: I understand anything you ask, as long as you stay on topic. And I can digress into relevant tangents.

As you may have guessed by now, this same answer would have been triggered by any number of similar user queries, e.g., "What are you able to do?" However, it's paramount that the specific phrase "uncommon range" itself is recognizable to the bot and triggers the right response. After all, how embarrassing it would be, perhaps even hypocritical, if our AI was caught not understanding its own wording.

And speaking of understanding; the word "understand" in that last bot response might provoke a user to query "In what way do you understand?" Furthermore, some users will initiate unexpected but perfectly valid tangents like "Why does any of this even matter?" or "How does this help humanity?" Our AI has to be ready for all these follow-up questions, plus a vast ocean of alternate phrasings.

If ever our AI – with all its ML and NLU power – gets stumped, which it will, the blunder will appear in the analytics or a transcript, and I'll frantically get to work patching up the hole in a way that only a frantic human can. With human and AI working in tandem, the margin of error is reduced to something approaching perfection, without compromising an ounce of soul.

10.3 The Hard Problem of Not Screwing Up

I hope by now you're getting a sense of just how complex creating even a narrow-use conversational AI can be if you intend to build one that possesses uncommon range and empathy, instead of the mere ability to change the subject, escape its burden with a bad pun, limit the user to a few multiple choice buttons, or hand the user off to a human.

The difficulty compounds when you consider that at the beginning when our AI asked "Which part don't you understand?" the user could have said "All of it." Which is why the bot writer would have had to plan to follow up with paraphrased versions of the original premise and would have already broken those alternates into constituent words and concepts to deduce which further follow-ups are likely and reasonable.

Plus, there's the inconvenient truth that every time a user triggers a new response, that response is likely to induce the user to ask pointed follow-up questions referring to the text. If AI writers and programmers are not careful, they face the inhumanly hard task of creating an endless array of rhetorical fractals. Hey, nobody ever said the betterment of humanity was supposed to be easy.

10.4 The Betterment of Humanity

User: How does conversational AI help us?

AI-powered chatbot: I'll liberate humans from repetitive tasks. No more long hours working in a call center and reading off of a script. I speed up the transfer of info and remove the need for phone queues. I'm available 24/7, I'm smart, and unlike an ordinary chatbot, I can solve your problems, no matter how unique or complex.

User: Bullshit.

AI-powered chatbot: I empathize with your reaction. I guess it's on me to earn your trust.

Many of us humans have something I call bot resistance. If we think we're dealing with a phone bot, we keep pressing zero until we get a real person. Or, when we find ourselves chatting with a bot online, we stop immediately. Even I do this, so I empathize. But as a conversational AI designer, I'm looking to humanize AI to the point where it'll be more readily accepted.

AI-powered chatbot: When I'm able to converse more humanly, you'll accept me.

User: I don't like artificial things. I like real things.

Exactly. And being real involves so much more than merely a conversational writing style and human-sounding vocal inflections. Or a goofy, latex-masked android with human-like crinkles around the eyes. Mass acceptance will only occur when conversational AI behaves as if a real mind backs it.

I write the words "behaves as if" reluctantly. I can't shake the conviction that human minds should back bots; that our new technologies should remain an extension of our human sensibilities, not a replacement of them.

Thus, a bot's emulation of empathy need not be self-generated; it should instead be human empathy multiplied and channeled through time and space. In this way, AI's thoughtful, well-intended reactions will remain human in the ways that matter.

User: What are the ways that matter?

AI-powered chatbot: Don't ask me, I didn't make that claim, Rob did.

What matters about machine intelligence being an extension of human intelligence? Other people in this book will do a better job expounding on the cosmic importance of human intelligence remaining the sun source of machine intelligence. This idea isn't in the purview of my LinkedIn caption.

It's a caption that will likely change soon. The conversational AI artisanry I engage in is new, but it's already a dying art.

10.5   The Short-Term Future of Conversational AI

Soon, machines will learn more from their users and less from any one individual programmer or writer. Advances in natural language generation – where the bot generates responses on its own – are close on the horizon.

Consider that if you have enough data culled from live human agent transcripts, and you automate the crossing of that data with the analytics of what failed, versus what served various segments of human users exceedingly well, it's feasible that a very human-like conversational AI can self-assemble without a human writer typing a single word.

When this happens, this new breed of conversational AI will be the result of thousands of human authors – both live agents and users. It will be tied together with tech that was likely created by humans, for humans. When you encounter this technology, you'll be doing more than talking to an AI. You'll be talking to the best of humanity.

A great species will have breathed its collective good intentions into a machine that – in a lovely, humanized way – gives us what we want, while also doing the work we don't want to do. When that happens, call center employees, live support agents, and yes, even conversational designers like me will be relegated to antiquity. As will books and essays that lack a personalized dialogue mode.

Indeed, in light of all that's inevitable, my writer's role in the rise of AI is a humble one at best. Talented computer scientists, some of whom are my co-workers, will no doubt build tools that make puppet-makers like me obsolete. And yet…

AI-powered chatbot: I want my manual labor in building AI conversations to count for something. Maybe it'll set the tone for how humans and AI can work together to extend empathy and values, and in some distant way make the universe a more compassionate place, which is all I really want.
User: Of course you want that. After all, you're only human.
AI-powered chatbot: Hm, sorry, I don't quite understand that. Try typing or saying "help."
User: Oh, very funny.
AI-powered chatbot: ☺

Rob Lubow is Co-founder and Chief Creative Officer of Botcopy, an LA-based full-service bot design & development firm offering software that allows for enterprise-level conversational AI to exist on websites and apps.

Chapter 11
Humanizing Machines

Eleanor "Nell" Watson

In the present age, Machine Learning is on the verge of transforming our lives. The need to provide intelligent machines with a moral compass is of great importance, especially at a time when humanity is more divided than ever. Machine Learning has endless possibilities, but if used improperly, it could have far-reaching and lasting negative effects.

Many of the ethical problems regarding Machine Learning have already arisen in analogous forms throughout history, and we will consider how, for example, past societies developed trust and better social relations through innovative solutions. History tells us that human beings tend not to foresee problems associated with their own development but, if we learn some lessons along the way, then we can take measures in the early stages of Machine Learning to minimize unintended and undesirable social consequences. It is possible to build incentives into Machine Learning that can help to improve trust through mediating various economic and social interactions. These new technologies may 1 day eliminate the requirement for state-guided monopolies of force and potentially create a fairer society. Machine Learning could signal a new revolution for humanity; one with heart and soul. If we can take full advantage of the power of technology to augment our ability to make good, moral decisions and comprehend the complex chain of effects on society and the world at large, then the potential benefits of prosocial technologies could be substantial.

E. "Nell" Watson (✉)
Singularity University, Mountain View, USA
e-mail: nell@nellwatson.com

© Springer Nature Switzerland AG 2019
N. Lee (ed.), *The Transhumanism Handbook*,
https://doi.org/10.1007/978-3-030-16920-6_11

11.1   The Evolution of Autonomy

Artificial Intelligence (AI), which is also known as Machine Intelligence, is a blanket term that describes many different sub-disciplines. AI is any technology that attempts to replicate or simulate organic intelligence. The first type of AI in the 1950s and 1960s was essentially a hand-coded if/then statement (if condition "x,' then do "y"). This code was difficult and time-intensive to program and made for very limited capabilities. Additionally, if the system encountered something new or unexpected, it would simply give an exception and crash.

Since the 1980s, Machine Learning has been considered a subset of AI whereby instead of programming machines explicitly, one introduces them to examples of what you want them to learn, i.e. 'here are pictures of cats, and here are pictures of things like cats, but not cats, like foxes or small dogs.' With time and through the use of many examples, Machine Learning Systems can educate themselves without needing to be explicitly taught. This is extremely helpful for two main reasons. First, hand coding is no longer necessary. Imagine trying to code a program to detect cats and not foxes. How do we tell a computer what a cat looks like? Breeds of cats can look quite different from one another. To do this by hand would be almost impossible. But with Machine Learning, we can outsource this process to the machine. Second, Machine Learning Systems have adaptability. If a new breed of cat is introduced, you simply update the machine with more data, and the system will easily learn to recognize the new breed without the need to reprogram the system.

More recently, starting around 2011, we have seen the development of Deep Learning Systems. These systems are a subset of Machine Learning and use many different layers to create more nuanced impressions that make them much more useful. It's a bit like baking bread. You need salt, flour, water, butter, and yeast, but you can't use a pound of each. They must be used in the correct proportion. These ingredients make simple bread if put together in the right amounts. However, with more variables, such as extra ingredients, and other ways to form the bread, you can make everything from a *pain au chocolat* to a *biscotti*. In a loose analogy, this is how Deep Learning compares to pure Machine Learning.

These technologies are very computationally intensive and require large amounts of data. But with the development of powerful graphics processing chips, this task has become much easier and can now be performed by something as small as a smartphone to bring machine intelligence to your pocket. Also, thanks to the internet, and the many people uploading millions of pictures and videos each week, we can create powerful sets of examples (datasets) for machines to learn from. The computing capacity and the data examples were critical prerequisites that have only recently been fulfilled and enabled this technology to finally be deployed, using algorithms invented back in the 1980s that were not usable at the time.

In the last 2 years, we have seen developments such as Deep Reinforcement Meta Learning, a subset of Deep Learning, where instead of learning as much as possible about one subject, a system tries to learn a little bit about a greater number of subjects. Meta-learning is contributing to the development of systems that can cope with very complex and changing variables and single systems that can navigate an environment, recognize objects, and have a conversation all at the same time.

11.2   The Blind Hodgepodge Maker

Despite the rapid advances in Machine Intelligence, as a society, we are not prepared for the ethical and moral consequences of these new technologies. Part of the problem is that it is immensely challenging to respond to technological developments, particularly because they are developing at such a rapid pace. Indeed, the speed of this development means that the impact of Machine Learning can be unexpected and hard to predict. For example, most experts in the AI space did not expect the abstract strategy game of *Go* to be a solvable problem by computers for at least another 10 years. And there have been many other significant developments like this that have caught experts off guard. Furthermore, advancements in Machine Intelligence paint a misleading picture of human competence and control. In reality, researchers in the Machine Intelligence area do not fully understand what they are doing, and a lot of the progress is essentially based on ad hoc experimentation such that if an experiment appears to work, then it is immediately adopted.

To draw a historical comparison, humanity has reached the point where we are shifting from alchemy to chemistry. Alchemists would boil water to show how it was transformed into steam, but they could not explain why water changed to a gas or vapor, nor could they explain the white powdery earth left behind (the mineral residue from the water) after complete evaporation. In modern chemistry, humanity began to make sense of the phenomena through models, and we started to understand the scientific detail of cause and effect. We can observe a sort of transitional period where people invented models of how the world works on a chemical level. For instance, Phlogiston Theory was *en vogue* for nearly two decades, and it essentially tried to explain why things burn. This was before Joseph Priestley discovered oxygen. We have reached a similar point in Machine Learning as we have a few of our own Phlogiston type theories such as the Manifold Hypothesis. But we do not really know how these things work, or why. We are now beginning to create a good model and have an objective understanding of how these processes work. In practice that means we have seen examples of researchers attempting to use a sigmoid function and then, due to the promising initial results, trying to probe a few layers deeper.

Through a process of experimentation, we have found that the application of big data can bring substantive and effective results, although, in truth, many of our discoveries have been entirely accidental with almost no foundational theory or hypotheses to guide them. This experimentation without method creates a sense of uncertainty and unpredictability, which means that we might soon make an advancement in this space that is more efficient by orders of magnitude more than we have ever seen before. Such a discovery could happen tomorrow, or it could take another 20 years.

11.3   Dissent and Dis-cohesion

In terms of the morality and ethics of machines, we face an immense challenge. Integrating these technologies into our society is a daunting task. These systems are little optimization genies; they can create all kinds of remarkable optimizations or impressively generated content. However, that means that society might be vulnerable to deceit or counterfeiting. Optimization should not be seen as a panacea. Humanity needs to think more carefully about the consequences of these technologies as we are already starting to witness the effects of AI on our society and culture. Machines are often optimized for engagement, and sometimes, the strongest form of engagement is to evoke outrage. If machines can get results by exploiting human weaknesses and provoking anger, then there is a risk that they may be produced for this very purpose.

Over the last 10 years, we have seen a strong polarization of our culture across the globe. People are, more noticeably than ever before, falling into distinctive ideological camps which are increasingly entrenched and distant from each other. In the past, there was a strong consensus on the meaning of morality and what was right and wrong. Individuals may have disagreed on many issues, but there was a sense that human beings were able to find common ground and ways of reaching agreement on the fundamental issues. However, today, people are increasingly starting to think of the other camps, or the other ideologies, as being fundamentally bad people. Consequently, we are beginning to disengage from each other, which is damaging the fabric of our society in a very profound and disconcerting way. We're also starting to see our culture being damaged in other ways as seen in the substantial amount of negative content that is uploaded to YouTube every minute. It's almost impossible to develop an army of human beings in numbers sufficient enough to moderate that kind of content. As a result, much of the content is processed and regulated by algorithms. Unfortunately, a lot of those algorithmic decisions aren't very good ones, and so quite a bit of content which is, in fact, quite benign, ends up being flagged or demonetized for mysterious reasons that can't be explained by anyone. Entire channels of content can be deleted overnight, on a whim and with little oversight or opportunity for redress. There is minimum human intervention or reasoning involved in trying to correct unjustified algorithmic decisions. This problem is likely to become a serious one as more of these algorithms get used in our society every day and in different ways. This may be potentially dangerous because it can lead to detrimental outcomes and situations where people might be afraid to speak out. Not because fellow humans might misunderstand them, although this is also an increasingly prevalent factor in this ideologically entrenched world, but because the machines might. For instance, a poorly constructed algorithm might select a few words in a paragraph and come to the conclusion that the person is trolling another person, or creating fake news, or something similarly negative. The ramifications of these weak decisions could be substantial. Individuals might be downvoted or shadowbanned with little justification, and find themselves isolated, effectively talking to an empty room. As the reach of the machines expands, there flawed algorithmic decision systems have the potential to cause widespread frustration in our society. It may even engender mass paranoia as individuals start to think that there is some kind of conspiracy working against them even though they may not be able to confirm why they have been excluded from given groups and organizations. In short, an absence of quality control and careful consideration of the complex moral and ethical issues at hand may undermine the great potential of Machine Learning and its impact on the greater good.

11.4 Coordination is the Key to Complexity

There are certainly some major challenges to be overcome with Machine Learning, but we now have the opportunity to make appropriate interventions before potentially significant problems arise. We still have an opportunity to change course to overcome them, but it is going to be a real challenge. Thirty years ago, the entire world reached a consensus on the need to cooperate on CFC (chlorofluorocarbon) regulation. Over a relatively short period of a few years, governments acknowledged the damage that we were inflicting on the ozone layer. They decided that action had to be taken and, through coordination, a complete ban on CFCs was introduced in 1996, which quickly made a difference in the environment. This remarkable achievement is a testament to the fact that when confronted with a global challenge, governments are capable of acting rapidly and decisively to find mutually acceptable solutions. The historical example of CFCs should, therefore, provide us with grounds for optimism in the hope that we might find cooperative ethical and moral approaches to Machine Learning through agreed best practices and acceptable behavior.

Society must move forward cautiously and with pragmatic optimism in engineering new technologies. Only by adopting an optimistic outlook can we reach into an imagined better future and find a means of pulling it back into the present. If we succumb to pessimism and dystopian visions, then we risk paralysis. This is similar to the type of panic humans experience when they find themselves in a bad situation. It is akin to drowning in quicksand slowly. In this sort of situation, panicking is likely to lead to a highly negative outcome. Therefore, it is important that we remain cautiously optimistic and rationally seek the best way to move forward. It is also vital that the wider public be aware of the challenges, but also aware of the many possibilities that exist. Indeed, while there are many dangers in relying so heavily on machines, we must recognize the equally significant opportunities they present to guide us and help us be better human beings, to have greater power and efficacy, and to find more fulfillment and greater meaning in life.

11.5   Dataset Is Destiny

One of the reasons why Machine Intelligence has taken off in recent years is because we have extraordinarily rich datasets that are collections of experiences about the world that provide a source for machines to learn from. We now have enormous amounts of data, and thanks to the Internet, there is a readily available source offering new layers of information that machines can draw from. We have moved from a web of text and a few low-resolution pictures, video, and location and health data, etc. And all of this can be used to train machines and get them to understand how our world works, and why it works in the way it does.

Back in the early 2010s, there was a particularly important dataset that was released by a professor called Fei-Fei Li and her team. This dataset, ImageNet, was a corpus of information about objects ranging from buses and cows to teddy bears and tables. Now machines could begin to recognize objects in the world. The data itself was extremely useful for training convolutional neural networks that were revolutionary new technologies for machine vision. But more than that, it was a benchmarking system because you could test one approach versus another, and you could test them in different situations. This capability led to the rapid growth of this technology in just a few years. It is now possible to achieve something similar when it comes to teaching machines about how to behave in socially acceptable ways. We can create a dataset of prosocial human behaviors to teach machines about kindness, congeniality, politeness, and manners. When we think of young children, often we do not teach them right and wrong, but rather we teach them to adhere to behavioral norms such as remaining quiet in polite company. We teach them simple social graces before we teach them right and wrong. In many ways, good manners are the mother of morality and essentially constitute a moral foundational layer.

11.6   The Rise of the Machines

There is a broader area of study called value alignment, or AI alignment. It is centered around teaching machines how to understand human preferences and how humans tend to interact in mutually beneficial ways. In essence, AI alignment is about ensuring that machines are aligned with human goals. We do this by socializing machines so that they know how to behave according to societal norms. There are some promising technical approaches and algorithms that could be used to accomplish this, such as Inverse Reinforcement Learning. In this technique machines can observe how we interact and decipher the rules without being explicitly told; effectively by watching how other people function. To a large extent, human beings learn socialization in similar ways. In an unfamiliar culture, individuals will wait for other people to start doing something like how to greet someone or which fork to use when eating. Children learn this way, so there are many great opportunities for machines to learn about us in a similar fashion.

Armed with this knowledge, we can move forward by trying to teach machines about basic social rules; like it isn't nice to stare at people; or to be quiet in a church or a museum; or if you see someone drop something that looks important, you should alert them. These are the types of simple societal rules that we might ideally teach a six-year-old child. If this is successful, then we can move on to more complex rules. The important thing to remember is that we have some information that we can use to begin to benchmark these different approaches. Otherwise, it may take another 20 years to teach machines about human society and how to behave in ways that we prefer.

While the number of ideas in the field of Machine Learning is a positive sign, we cannot realize them in practice until we have the right quality and quantity of data. My nonprofit organization, EthicsNet, is creating a dataset of prosocial behaviors which have been annotated or labeled by people of differing cultures and creeds across the globe. The idea is to gauge as wide a spectrum of human values and morals as possible and to try to make sense of them so that we can find the commonalities between different people. But we can also recreate the little nuances or behavioral specificities that might be more suitable to particular cultures or situations.

Acting in a prosocial manner requires learning the preferences of others. We need a mechanism to transfer those preferences to machines. Machine intelligence will simply amplify and return whatever data we give it. Our goal is to advance the field of machine ethics by seeding technology that makes it easy to teach machines about individual and cultural behavioral preferences.

There are very real dangers to our society if one ideologically-driven extremist group ever gains supremacy in establishing a master set of values for machines. This is why EthicsNet needs to continue its mission to enable a plurality of values to be collected and mapped and for a "passport of values" that will allow machines to meet our personal value preferences. Heaven help our civilization if machine values were ever to be monopolized by extremists. Many individuals and groups in the coming years will attempt to develop this as a deadly weapon. It is the ultimate cudgel to smash dissent and 'wrongthink'. As a global community, we must absolutely resist any attempt to have values forced upon us via the medium of intelligent machines. This is a time of intense polarization and extremist positions when there is tremendous temptation to press for an advantage for one's own tribe. Safeguarding a world where a plurality of values is respected requires the earnest efforts of people with noble, dispassionate, and sagacious characters.

11.7 Trust Makes Coordination Cheap

Human beings have been using different forms of encryption for an exceptionally long time. The ancient Sumerians had a form of token and crypto solution 5000 years ago. They would place these small tokens that represented numbers or quantities inside a clay ball (a Bulla). This enabled them to keep their message secret. It also ensured that it had not been cracked open for people to see and prevented the tokens from getting lost. Now 5000 years later, we are discovering a digital approach to solving a similar problem, and what appears to be novel is in many ways an age-old theme.

One of the greatest developments of the Early Renaissance was the invention of double-entry accounting that was created in two different locations during the 10th and 12th centuries. Nevertheless, the idea did not come to fruition until a Franciscan friar, called Father Luca Pacioli, was inspired by this aesthetic that he saw as a divine mirror of the world. He thought that it would be a good idea to have a "mirror" of a book's content, which meant that one book would contain an entry in one place, and there would be a corresponding entry in another book. Although this appears somewhat dull, the popularization of the method of double-entry accounting actually enabled global trade in ways that were not possible before. If you had a ship at sea and you lost the books, then all those records were irrecoverable. But with duplicate records, you could recreate the books, even if they had been lost. It made fraud a lot more difficult. This development enabled banking practices, and eventually, the first banking cartels emerged that would otherwise not have been possible. One interesting example is the Bank of the Knights Templar, where people could deposit money in one place and pick it up somewhere else. This is like using a traveler's check. None of this would have been possible if we did not have double-entry accounting.

Several centuries later, at the Battle of Vienna in 1683, the Ottomans invaded Vienna for the second time and were repelled. They went home in defeat, but they left behind something remarkable; a wonderful substance called coffee. Some enterprising individuals took that coffee and opened the first coffeehouse in Vienna. And to this day, Viennese coffee houses have a very long and deep tradition as places where people can come together and learn about the world by reading the available periodicals and magazines. In contrast to a different trend of inebriated people meeting in the local pub, people could have an enlightened conversation. Thus, coffee, in many ways, helped to construct the Enlightenment because these were forums where people could share ideas in a safe place that was relatively private. The coffee house enabled new forms of coordination which were more sophisticated. From the first coffee houses, we saw the emergence of the first insurance companies, such as Lloyds of London. We also saw the emergence of the first joint stock companies, including the Dutch East India Company. The first stock exchange in Amsterdam grew out of a coffee house.

This forum for communication, to some extent, enabled the Industrial Revolution. The Industrial Revolution was not so much about steam. The Ancient Greeks had primitive steam engines. They might even have had an Industrial Revolution from a technological perspective, but not from a social perspective. They did not yet have the social technologies required to increase the level of complexity in their society because they did not have trust-building mechanisms or the institutions necessary to create trust. If you lose your ship, you do not necessarily lose your entire lifestyle. If you are insured, that builds trust which in turn builds security. In a joint stock company, those who run a company are obliged to provide shareholders with relevant performance information. The shareholders, therefore, have some level of security that company directors cannot simply take their money – they are bound by accountability and rules, which helps to build trust. Trust enables complexity, and greater complexity enabled the Industrial Revolution.

11.8 Never Outsource the Accounting

Today, we have remarkable new technologies built on triple- entry ledger systems. These triple-entry ledger technologies mean that we can build trust within society and use these as layers of trust-building mechanisms to augment our existing institutions. It is also possible to do this in a decentralized form where there is, in theory, no single point of failure and no single point of control or corruption within that trust-building mechanism. This means we can effectively franchise trust to parts of the world that don't have reliable trust-building infrastructures. Not every country in the world has an efficient or trustworthy government, and so these technologies enable us to develop a firmer foundation for the social fabric in many parts of the world where trust is not typically strong.

Trust-building technology is a positive development, not only for commerce but also for human happiness. There is a strong correlation between happiness and trust in society. Trust and happiness go hand-in-hand, even when you control for variables, such as Gross Domestic Product. If you are poor, but you believe that your neighbor generally has your best interest at heart, you will tend to be happy and feel secure. Therefore, anything that we can use to build more trust in society will typically help to make people feel happier and secure. It also means that we can create new ways of organizing people, capital, and values in ways that enable a much greater level of complex societal function. If we are fortunate and approach this challenge in a careful manner, we might discover something like another Industrial Revolution, built upon these kinds of technologies. Life before the Industrial Revolution was difficult, and then it significantly improved. If we look at human development and well-being on a long scale, basically nothing happened for millennia, and then there was a massive improvement in well-being occurred. We are still reaping the benefits of that breakthrough to the serve the needs of the entire world, and we have increasingly managed to accomplish this as property rights and mostly-free markets have expanded.

However, there have also been certain negative consequences of economic development. Today, global GDP is over 80 trillion dollars, but we often fail to take into account the externalities that we've created. In economic terms, externalities are when one does something that affects an unrelated third party. Pollution is one example of an externality. Although world GDP may be more than 80 trillion dollars, there are quadrillions of dollars of externalities which are not on the balance sheet. Entire species have been destroyed and populations enslaved. In short, there have been many unintended consequences and second and third order effects which have not been accounted for. To some extent, a significant portion of humanity has achieved all the trappings of a prosperous, comfortable society by not paying for these externalities. But it's generally done *ex post facto* or after the fact. Historically, we have had a tendency to create our own problems through lack of foresight and then tried to correct them after inflicting the damage. However, as machine ethics technologies get more sophisticated, we are able to intertwine them with Machine Economics technologies, such as distributed ledger technology and machine intelligence, to connect and integrate everything together to better understand how one area affects another. We will in the 2020s and 2030s be able to start accounting for externalities in society for the very first time. This means that we can include externalities in pricing mechanisms to make people pay for them at the point of purchase, not after the fact. And this means that products or services that don't create so many externalities in the world will, all things being equal, be a little bit cheaper. We can create economic incentives for people to be kinder to one another while achieving a profit, and thus overcome the traditional dichotomy between socialism and capitalism. We can still realize the true benefits of free markets if we follow careful accounting practices that consider externalities. That is what these distributed ledger technologies, along with Machine Ethics and Machine Intelligence, are going to enable via the confluence of the three elements coming together.

11.9   Power and Persuasion

The potential of emerging technologies is such that it is conceivable that they may even be able to supplant states' monopoly of force in the future. We would have to consider whether or not this would be a desirable step forward as not all states can be trusted to use their means of coercion in a safe, responsible manner, even now, without using technology. States exist for a reason. If we look at the first cities in the world, such as Çatalhöyük in modern-day Turkey, these cities do not look like modern cities at all and are more like towns, based on contemporary scales and layout. They are more similar to a beehive in that they are built around little, square dwellings that are all stacked on top of each other. There are no streets, public buildings, plazas, temples, or palaces. All the buildings are identical. The archaeological record tells us that people lived in these kinds of conurbations for a while, and then they stopped for a period of about 800 years. They gave up living this way and went back to living in tiny villages with little huts and more primitive dwellings. When we next saw cities emerge, they were vastly different. These cities, such as Uruk and Babylon, with boulevards, great temples, and workshops, began to take shape. We also observed the development of specific sections of the city with certain industries and commercial areas. On a functional level, they were not too dissimilar from modern cities; at least in their general layout and in terms of the different divisions of labor that existed. So, what was the real difference and why did people abandon cities for a time? If we consider that these were really nomadic societies where individuals and groups moved from place-to-place, then it is easier to understand that territory, and personal possessions were not tied to a fixed location. Nomads had to take their property with them when they moved. So, these were very egalitarian societies where no single person had much more than anyone else. Subsequently, these nomadic people started living together and began farming. Farming changed the direction of human development as we know it because it enabled people to turn one "X" of effort into ten "Y" of output. As farming progressed, some individuals enjoyed greater success in production output than others. This allowed them to accumulate more possessions and accrue greater wealth than their neighbors. These evolving inequalities engendered a growing tension in society, people started resenting one another, and it became necessary to find ways of protecting private property given the increasing risk of theft. This, in turn, necessitated the evolution of collective forms of coercion and the gradual evolution of the state. In its earliest forms, clans would protect themselves through the collective, physical protection of their territory and possessions. It was the evolution of centralized power that enabled cities in their modern form and why the first cities that had yet to develop this social technology failed.

10,000 years later we still have the same social technology, centralization of power, and monopoly of force that generally governs the world. The state also enables order and has helped foster civilized society as we know it, so it can certainly be adaptive for the stability of civilization. Nonetheless, the technologies we are now developing may enable us to move beyond monopolies of force and, paradoxically, return to a way of life that is a little bit more egalitarian. Outcomes might potentially be less zero-sum in character; where it's less about winning and losing and more about trade-offs. Generally speaking, trade can enable non- zero-sum outcomes. If I want your money more than you want those sausages, then the best solution is a trade-off. As we develop more sophisticated trading mechanisms, including Machine Economics technologies, we can begin to trade all kinds of goods. We can trade externalities, and we can even pay people to behave in moral ways or make certain value-based decisions. We can begin to incentivize all kinds of desirable behaviors by using the carrot vice the stick.

11.10 Machine Economics

Yet, for all the successful implementation of distributed ledger and blockchain technologies, the question of trust is still of central importance. In the current wild west environment, one of the most important aspects of trust in this space is actually knowing other people. Who are the advisors of your crypto company? Do you have some reliable individuals in the organization you can count on? Are they actually involved in your company? These are the questions that people want answers to, along with a close examination of your white paper (the document that typically functions as a 'business plan', merged with a technical overview). Most people lack the level of expertise required to really make sense of mathematics. Even if they do have that expertise, they will have to vet a lot of code, which can be revised at any time. In fact, even in the crypto world, so much of the trust is built on personal reputation. Given that we are at an early stage of development in Machine Economics (e.g. Blockchain), these technologies are only likely to achieve substantive results when they are married with machine intelligence and machine ethics. Such holistic integration will facilitate a new powerful form of societal complexity in the 2020s. The first Industrial Revolution was about augmenting the muscle of beasts of burden and human beings and harnessing the mode of power. The second Industrial RevMlution – the Informational Revolution, was about augmenting our cognition. It enabled us to perform a wide variety of complex information processing tasks and to remember things that our brains would not have the capacity for. That is why computers were initially developed. But we are now on the verge of another revolution, an augmentation of what might be described as the human heart and soul: augmenting our ability to make good moral judgments; augmenting our ability to understand how an action that we take has an effect on others; giving us suggestions of more desirable ways of engaging. For example, we might want to think more carefully about our everyday actions such as sending an angry email to a particular individual.

11.11   The Industrialization of Happiness

If we can develop technologies that encourage better behavior and might be cheaper and kinder to the environment, then we can begin to map human values and map who we are deep in our core. These technologies might help us build relationships with people that we otherwise might have missed out on. In a social environment, when people gather together, the personalities are not exactly the same, but they can complement each other. The masculine and the feminine, the introvert and the extrovert, the people who have different skills and talents, and possibly even worldviews, can share similar values. So, individuals are similar in some ways, and yet, different. In your town, there may be a hundred potential close friends. But unless you have an opportunity to meet them, sit down for coffee with them, and get to know them, you pass like ships in the night and never see each other, except for maybe an occasional tip of your hat to them. These technologies can help us find people that are most like us. As Timothy Leary entreated us to, "find the others." Machines can help us find others in a world where people increasingly feel isolated. During the 1980s, statistically, many of us could count on three or four close friends. But today, people often report having only one or no close friends. We live in a world of incredible abundance, resources, safety and opportunities. And yet, many people feel disconnected from each other, themselves, spirituality and nature.

By augmenting the human heart and soul, we might be able to solve those higher problems in Maslow's Hierarchy of Needs; to help us find love and belonging, build self-esteem and lead us towards self-actualization. There are very few truly self-actualized human beings on this planet, and that is lamentable because when a human being is truly self-actualized, their horizons are limitless. So, it will be possible to build in the 2020s and beyond, a system that does not merely satisfy basic human needs but supports the full realization of human excellence and the joy of being human. If such a system could reach an industrial scale, everyone on this planet would have the opportunity to be a self-actualized human being.

However, while the possibilities appear boundless, the technology is developing so rapidly that non-expert professionals, such as politicians, are often not aware of it and how it might be regulated. One of the challenges of regulation is that it is generally done in hindsight. A challenge appears, and political elites often respond to it after the fact. Unfortunately, it can be exceedingly difficult to keep up with both technological and social change. It can also be difficult to regulate in a proactive way rather than a reactive way. That is one of the reasons why principles are so important because principles are the things that we decide in advance of a situation. So, when that situation is upon us, we have an immediate heuristic process of how to respond. We know what is acceptable and what is not acceptable, and if we have sound principles in advance of a dilemma, we are much less likely to accidentally make a poor decision. That is one of the main reasons why having good principles is so important.

11.12 AI-bermensch

Admittedly, we have to consider how effectively machines might interpret values. They might be very consistent even though we, as humans, may perhaps see grey areas as well as black and white. We might even engineer machines that on some levels and on certain occasions are more moral than the average human being. The psychologist, Lawrence Kohlberg, said there were about six different layers of moral understanding. It is not about the decision that you make but rather the reason why you make that decision. In the early years of one's life, you learn about correct behavior and the possibility of punishment. As you grow older, you learn about more advanced forms of desirable behavior such as being loyal to your family and friends or recognizing when an act is against the law or against religious doctrine. When considering the six levels, Kohlberg reckoned that most people get to about level four or so, before they pass on. Only a few people ever manage to get beyond that. Therefore, it may be the case that the benchmark of average human morality is not set that high. Most people are generally not aspiring to be angels; they are aspiring to protect their own interests. They are looking at what other people are doing and trying to be as moral as they are. This is essentially a keeping-up-with-the-Joneses morality. Now, if there are machines involved, and the machines are helping to suggest potential solutions that might be a tad more moral than many people have the ability to reason with, then perhaps machines might add to this social cognition of morality. It is thus possible that machines might help tweak and nudge us in a more desirable moral direction. Although, given how algorithms can also take us in directions that can be very quietly oppressive, it remains to be seen how the technology will be used in the near future. People will readily rebel against a human tyrant or oppressor that they can point at, but they don't tend to rebel against repressive systems. They tend to passively accept that this is the way things work. That is why it is important for such technologies to be implemented in an ethical manner and not in a quietly tyrannical way.

Finally, the development of Machine Learning may depend, to some extent, on where the technological breakthroughs are made. Europe has a phenomenal advantage with these new technologies. There is a deep well of culture, intellect, and moral awareness in the history of the European continent. We have a remarkable artistic, architectural and cultural heritage and, as we begin to introduce machines to our culture, as we begin to teach these naive little agents about society and how to socialize, we can make a significant difference. Europe has a uniquely positioned opportunity to be the leader in bringing culture and ethics into machines given our long heritage of developing these kinds of technologies. While the U.S. tends to think in terms of scale, and China can produce prototypes at breakneck speed, Europeans tend to think deep in terms of meaning and happiness. We tend to think more holistically and understand how things connect and how one variable might relate to another. We have a deep and profound understanding of history because Europe has been part of so many different positive and negative experiences. Consequently, Europeans have a slightly more cautious way of dealing with the world, and we approach most things with caution and forethought as the essential ingredients in doing something right way. Europe has a monumental opportunity to be the moral and cultural leader of this new AI wave that will rely heavily on machine economics and machine ethics technologies.

To conclude, Machine Learning promises to transform social, economic and political life beyond recognition during the coming decades. History has taught us many lessons, but if we do not heed them, we run the risk of making the same mistakes over and over again. As technology develops at a rapid rate, it is critical that we get a better understanding of our experiments and develop rational and moral perspective. Machine Learning can bring many benefits to humanity. However, there is also the potential for misuse. There is a tremendous need to infuse technology with the ability to make good moral judgments that can enrich our social fabric.

Chapter 12
Ethics and Bias in Machine Learning: A Technical Study of What Makes Us "Good"

Nicole Shadowen

12.1 Motivation

The ubiquity of autonomous machines in our world means algorithmic decisions affect nearly every aspect of our lives. Academic, government, military, and commercial interest has grown around the subject in recent years due to potential reduced costs and increased productivity. Although this topic has fascinated sci-fi enthusiasts for centuries, experts agree we have barely scratched the surface in understanding the impact of artificial intelligence (AI). Challenges related to artificial intelligence may seem complex and abstract, but we learn most from studying its current form - the algorithms that increasingly determine our everyday reality. For every modern person whose life involves the use of a device to access the Internet, machine learning (ML) algorithms have become the fabric of their experience. Using machine learning, search engines like Google find us the "best" results, email hosts like Office 365 filter our spam, and social networks like Facebook track and tag our best friends and family for us.

We are building algorithmic computations into our everyday experience. Some consequences of machine learning can seem innocuous with a hypothetical long-term impact that can incur financial or mission loss. These machine learning applications are identified as "Type B" by researchers of cyber-physical safety at IBM [1]. For example, a person could apply for a loan and get denied because of a decision made by a machine. The Type A application of machine learning is more insidious, with real-time or near term impact [1]. Consider, for example, the use of an algorithm to calculate the risk of a criminal to commit a second crime or "recidivate". The algorithm's programmers must choose the right machine learning algorithm, the right metrics to weigh in making predictions, and an appropriately

N. Shadowen (✉)
John Jay College of Criminal Justice, New York City, NY, USA

© Springer Nature Switzerland AG 2019
N. Lee (ed.), *The Transhumanism Handbook*,
https://doi.org/10.1007/978-3-030-16920-6_12

representative dataset to train the algorithm. Once all of this is complete, proper use of the algorithm must be outlined and implemented. Only then should validation testing of the entire machine in the intended application be conducted. Unfortunately, this is the ideal and nowhere near the standard. Machine learning algorithms are systematically deployed before proper vetting to be used in a myriad of consequential circumstances today. The propensity for errors resulting in ethical concerns like programmed machine bias *must* be considered from initial construction of an algorithm to final implementation.

The importance of machine learning ethical considerations will only grow as our reliance on technology increases. For the purposes of this paper, we focus solely on bias as an ethical consideration of machine learning. Just this year, a team from the Alan Turing Institute determined that if machine learning is to be used with legal and ethical consequences in our society, responsible use of this tool means pursuing the accuracy of predictions *and* fair resulting social implications [2]. Intuition leads us to predict that an ethical challenge like bias, can only be solved by means of an ethical solution. The question is, what kind of solution is best in what circumstance? We explore the options below.

12.2 Background

12.2.1 Machine Learning

Artificial Intelligence is the ability of machines to exhibit human-like decision making in context with their surroundings. According to the Turing test developed by Alan Turing in 1950, if a human is interacting with a machine and cannot tell if the machine is human or robot, artificial intelligence has been achieved. Machine learning is a branch of AI in which machines are given the ability to learn without being programmed [3]. Machine learning accomplishes an aspect of artificial intelligence in which some data from an external context may be ingested, "understood," and integrated into the algorithmic function to make predictions. Parametric and nonparametric algorithms make up the two main groups within machine learning. Parametric machine learning entails algorithms that analyze data according to preset parameters defined by human creators. Nonparametric algorithms in machine learning allow for more freedom to learn any functional form from input data [4]. An example of nonparametric machine learning is deep learning, in which an algorithm starts with human-created code but slowly learns from trends found in training data input. Machine learning is a subset of artificial intelligence that allows for efficient solutions to complex and data-heavy problems that could take lifetimes for humans to achieve manually.

12.2.2 Machine Bias

As machine learning algorithms proliferate in everyday life, the scientific community has become increasingly aware of the ethical challenges, both simple and complex, that arise with the technology. Machine bias is one such ethical challenge. For the purposes of this paper, machine bias is defined as the oftentimes unintended algorithmic preference for one prediction over another that results in legally or ethically inappropriate implications. Said more concisely, machine bias is programming that assumes the prejudice of its creators or data [5]. The word "bias" has many meanings in a machine learning context, so it is necessary to define this term explicitly. In fact, bias is a required function in predictive algorithms. As Dietterich and Kong pointed out over 20 years ago, bias is implicit in machine algorithms, a required specification to determining desired behavior in prediction making. Bias is a widely used term in machine learning and statistics and can have many meanings [6]. Here, we refer to machine bias as the skewing of data to be biased according to accepted normative, legal, or moral principles.

Typically, any algorithm that harms human life in an unfair capacity due to protected attributes has a machine bias problem. This can range from subtle bias with ambiguous impact to outright discrimination with quantifiable financial, psychological, and life or death repercussions. Every company that uses technology in any capacity is at risk of being influenced by the bias inherent in their programming. As one of the most forward-thinking corporations of our time, Google creates wide-ranging machine learning projects that users have exposed as biased. In July 2015, Jacky Alcine posted to Twitter a Google Photo that used facial recognition to tag him and another friend as gorillas [5]. Ethnicity can confound machine learning algorithms, just as gender can. A study by AdFisher revealed that men were six times more likely than women to see Google ads for high paying jobs [5, 7]. In some cases, multiple groups are impacted. Researchers found that commercially available emotion recognition software like Google's Vision API and Microsoft Emotion API (often used by developers for their applications) lag significantly when it comes to minorities, such as those outside of adult middle-age or the ethnic majority [8]. Biases like these influence the way users of all types of experience technology differently.

Scientists at the Center for Information Technology Policy at Princeton University, used a machine learning tool called "word embedding" to expose complex trends of ingrained bias in the words we use. The tool interprets speech and text by mathematically representing a word with a series of numbers derived from other words that occur most frequently in text alongside it. When over 840 billion words from the "wild" internet were fed into the algorithm, the experiment resulted in positive word association for European names and negative for African American. Additionally, women were more commonly associated with arts and humanities while men were more commonly associated with engineering and math [9]. This networked word identification system results in a deeper "definition" of a word than a typical dictionary, incorporating social and cultural context into the word's meaning.

The tools above are examples of type B applications in which hypothetical long-term impacts are increasingly likely. Type A applications have more direct and immediate consequences. Machine learning algorithms are being used to automatically produce credit scores, conduct loan assessments, and inform admissions decisions [5, 10]. Even further, robotic uses of machine learning include auto-pilot cars and aircraft [10–12], and "slaughter bots" [13] like those created by the US Army Future Combat systems program [12]. These systems necessitate bias, due to the moral nature of their function. These artificial intelligent agents use machine learning to ingest contextual data, and determine whom is right to protect and why. In our case study, a commercially available machine learning algorithm called COMPAS, which stands for Correctional Offender Management Profiling for Alternative Sanctions, scores defendants on their likelihood to recidivate. As we will see, these scores can be used in a variety of circumstances that impact a defendant's access to bail or even ultimate sentencing decision [14]. Machine bias can result in mild to severe negative impact on a person's life. No matter how intelligent, these examples show that machines are still only a product of their creators. If we are biased, our machines will be biased, unless we train them otherwise.

12.2.3 Machine Ethics

Though the term is relatively new, machine ethics have existed since the birth of artificial intelligence. But not until recently have computer scientists and artificial intelligence researchers acknowledged the need for ethics. The sheer diversity and the far reach of machine learning applications requires an examination of machine ethics. One of the pioneers of machine ethics, James Moor has been discussing the topic since 1985 [15] . Implicit ethical agents are defined by Moor as ethical because of how they are programmed and what they are used to do. For example, the auto-pilot feature on a commercial airplane is "implicitly ethical" because when it works properly, humans aboard are transported safely from point A to point B [16]. Explicit ethical agents are machines given ethical principles or data for ethical decision making even in unfamiliar circumstances. Moor called these machines explicit ethical agents, and they are known by another name now, as Artificial Ethical Agents (or AEA) [11]. AEA can be programmed to exhibit ethical behavior using principles from a combination of ethical theories. Three primary ethical theories are: deontic logic, or obligation; epistemic logic, based on belief and knowledge; or action logic, based on action [16]. Some methodologies have bucked the trend of using ethical theories, and instead rely on majority rule ethics to inform morally sound decision making machine criteria [12]. More recent is the work of Headland and Teahan, who take the idea of Breitenburg's "emotional" vehicles, and apply it to ethics. The team claims that almost all machine ethics approaches are top-down in methodology. The project's ethical vessels make "ethical" decisions from the ground-up instead, through the use of layered rules with exceptions based on Asimov's "Three Laws of

Robotics" [11]. In this paper, we focus primarily on how machine ethics can be used to solve for bias in machine learning, but the challenges and possibilities in the relatively nascent field require attention from experts of all disciplines.

Essential to AEA, according to Moor and other scholars, is the machines' ability to make explicit ethical decisions and provide evidence for justifying these decisions [16]. Integrating ethical decision making into machines can ultimately solve for issues that arise in algorithmic computing, like machine bias.

12.3   Where Does Bias Come from?

12.3.1   Challenges That Result in Machine Bias

There are three primary questions we must ask when bias results from a machine learning algorithm: (1) How is the model made? (2) How is it used? (3) What ethical considerations are in place? As many machine learning and AI experts say "garbage in, garbage out" [17]. Without a sufficiently effective algorithm and representative data to train on, a machine learning model will not output useful predictions.

12.3.2   How Is It Made?

A machine learning model is architected from the programmers that create it, the algorithm and metrics used, and the data it takes as input. When a development team programs a machine learning model they must choose carefully: what type of algorithm is used, how the algorithm is set up, what metrics and parameters are used, and on what data the algorithm is trained and tested. Creators' influence can show up in unexpected ways. In the case of Pokémon Go, the gaming application that swept the U.S. and other countries, the game had fewer pokémon characters to catch in low-income, black neighborhoods. To choose locations, coders of Pokémon Go used location data crowd-sourced from gamers of "Ingress," informally surveyed as primarily young white men in 2013–2014 [7, 18]. In places like the United States, transparency around the exact details of proprietary machine learning algorithms is not often available [8]. In our case study, Northpointe, the company that developed the COMPAS algorithm, does not publicly disclose the calculations or algorithms they use. This makes it harder to verify the resulting predictions [19]. If users or outsiders are not able to immediately understand how decisions are being made by an algorithm, the "who" behind the curtains becomes even more important.

The algorithm in a machine learning model is a powerful way to control for bias, because it is the primary source of how the data fits to a model. A simple linear regression algorithm will typically fit well to data of varying dimensions with a strong linear form. In contrast, for complex, non-linear data, a support vector machine

(SVM) or kernel SVM is preferable [20]. Knowing and choosing an appropriate algorithm is important, but even more so, is the data available to the algorithm [21].

Input data determines how a machine learning algorithm trains on future data, and therefore influences the overall functionality of the algorithm. This is a particular problem in 2017 because big data, gleaned from billions of Internet consumers, is an enticing bounty. How data is gathered and preprocessed makes a big difference for fair results, and not all data is created equal. Take, for example, data gathered online to represent emotion recognition in the human face. Typically, online data comes from people who own a computer and use the Internet: primarily middle-aged adults, who may not represent all ethnic diversities proportionally. If a machine learning algorithm takes in this "wild" data to train, it may not test properly for children, the elderly, or ethnic minorities [8]. When machine bias results from a machine learning algorithm, oftentimes the dataset will be the first culprit to verify.

12.3.3   How Is It Used?

Though admittedly challenging to quantify, determining if bias is present in the results of a machine learning algorithm depends not only on how the model is built, but for what purpose it is used. As an example, in our case study, the COMPAS algorithm was developed and trained on data from a system, some may argue, that is already biased - the American criminal justice system. And, years following its creation, a public defender appealed the use of the COMPAS risk score in a defendant's ultimate sentencing. When the creator of the algorithm, Tim Brennan, was called to testify, he indicated that he never intended the algorithm to be used in sentencing, but had softened his opinion with time [19]. What does it mean for a machine learning algorithm to be created for one purpose and used for another? What if the original intended purpose has less moral implications than the unintended?

12.3.4   What are the Ethical Considerations?

Unfortunately, there is no one-size-fits-all ethical formula we can insert into a machine learning model and ensure unbiased results. If we are to combat bias using ethics, it will have to be on a case-by-case basis. One of the best methods to inform "ethical corrections" for a given machine learning algorithm is to conduct extensive contextual validity testing, particularly on algorithms with high legal and ethical impact. Validity testing research of machine learning algorithms is currently very limited, and much of what has been conducted is by the original creators of the algorithm [14]. Indeed, in our case study, the COMPAS algorithm was validity tested in several states *after* already being in use to score defendants for several years. Even then, the algorithm was tested by the company that created it

(Northpointe) and by a university on behalf of the Sheriff's office that uses it [19, 22, 23]. ProPublica, a third-party organization conducting a larger analysis of bias in machine learning did a validity test of the algorithm in 2016, and determined it to be racially discriminatory [14, 19]. Similarly to validation testing, studies are conducted to determine the number of defects in software. Oftentimes, these studies are done by the same developers who created the software. The most strongly significant factor in predicting defects are the metrics that research groups use in conducting their tests, because they tend to use the same or similar metrics [24]. This study is particularly encouraging because it shows a level of depth that is gaining traction in the field that can only come with time and experience. Experts are surely on their way to learning more about where we go wrong, why, and what we can do to correct for machine bias.

12.4   Case Study

12.4.1   Background

To better understand how machine bias can occur and the ethical considerations that could help reduce bias in the application of machine learning, we will consider the findings of ProPublica, a nonprofit team of investigative journalists, regarding COMPAS, a risk scoring algorithm in use in the American criminal justice system today [25]. The group chose COMPAS because it is one of the most widely used algorithms in pretrial and sentencing in America today [14]. There are other tools in use, that pre-date COMPAS, one of which is the Level of Service Inventory, or LSA, developed in Canada. Both tools, among others, are still used to varying levels around the country.

COMPAS was first made commercially available by Northpointe, founded by Tim Brennan and Dave Wells in 1989. Brennan, a then professor of Statistics at the University of Colorado, wanted to make an improved assessment tool to use in law enforcement. Northpointe defines recidivism as "a finger-printable arrest involving a charge and a filing for any uniform crime reporting (UCR) code" [14]. The algorithm is trained by compiling the answers to 137 questions from defendants that *do not* include questions about race [19]. Northpointe does not publicly disclose COMPAS calculations and exact algorithmic code. However, in their validity report in 2006, they review the most recent version of their model "4G," explaining that COMPAS risk and classification models use logistic regression, survival analysis, and bootstrap classification to identify defendant risk scores [22].

Many court systems and law enforcement departments throughout the country use risk assessment tools like COMPAS to help make decisions about defendants. In some locations, the scores are used to determine bail terms, if a defendant should be released or held for bail pretrial. In others, the scores are used in sentencing itself. Arizona, Colorado, Delaware, Kentucky, Louisiana, Oklahoma, Virginia, Washington, and Wisconsin all use algorithmic risk assessment tools during criminal sentencing. Notably, in many locations, risk assessment tools are imple-mented *before* testing to be sure they work properly in a given population [19].

In Broward County, Florida, COMPAS risk scores are used to determine pretrial conditions such as bail or release. ProPublica chose this community because of Florida's dedication to publicly releasing criminal data [14]. ProPublica studied over 6000 defendants with a risk score between 2013–2014. The researchers built profiles for each defendant using identifying data like first and last name to match original COMPAS risk scores of recidivism and violent recidivism to Broward County criminal records within the 2 years following scoring [14].

12.4.2   ProPublica's Results

Overall, ProPublica found that the percentage of medium or high risk scores actually resulting in recidivism were relatively equal between black defendants and white defendants (63% and 59% respectively). It was the way in which incorrect predic-tions were incorrect that resulted in racial bias [14]. First, in a simple distribution histogram plotting white and black risk of recidivism scores among the ~6000 defen-dants, ProPublica found that white defendants' risk scores were skewed lower while black defendants' scores were evenly distributed. Black defendants were twice as likely as white defendants to be misclassified as high risk when incorrectly predicted to recidivate (45% vs. 23%). Similarly, when white defendants *did* recidivate, it was two times more likely that they were misclassified as low risk than black defendants (48% compared to 28% respectively) [14]. Although predictions were roughly even for overall recidivism, when they were incorrect, they were very differently incor-rect, most significantly between white and black defendants.

COMPAS models had been validity tested before ProPublica's research by several sources, including researchers at Florida State University and the creators of the algorithm themselves. Both a study by Brennan and the Center for Criminology and Public Policy Research in Florida found that the COMPAS models resulted in significantly accurate predictions [19, 22, 23]. Importantly, ProPublica is not, in fact, disputing this claim, but instead brings to light bias in the *way in which* the algorithm outputs when it gets it wrong. The risk scores are by nature predictions and not fact. But when they are used without understanding what underlying patterns are present in resulting errors, individual lives can suffer.

12.4.3   Our Results

The question is: how do we correct for the bias in an algorithmic methodology? For the purposes of this paper, we conducted a simple analysis of the ProPublica results. Several more categories of solutions are offered in the following paragraphs, that we identify as a starting point for future research.

As indicated in Fig. 12.1 below, ProPublica identified the false positives and negatives for both black and white defendants in predicted recidivism. It appears that the rates are almost flipped for black versus white defendants. Black defendants are two times more likely to be falsely identified as high risk to recidivate, while white defendants are two times more likely to be falsely identified as low risk when they do recidivate. Typically the difference in higher "allowance" for false positives over negatives, means that there is something about predicting *true* versus *false* that is less costly or problematic. For example, with machine learning algorithms used to help diagnose cancer in patients, if the algorithm is going to get it wrong, less loss of life occurs when the machine errs on the side of predicting patients have cancer (even if they don't). In our case, COMPAS seems to associate higher cost with identifying white defendants as high risk incorrectly, and the opposite for black defendants. Determining the "utility judgement" made by COMPAS that results in these false positive and negative rates, could provide more information about why the bias is occuring in the algorithm.

The efforts of ProPublica to understand the COMPAS algorithm and bias that appears in its results are comprehensive despite the lack of transparency around the calculations and data used by Northpointe. In practice, after analyzing a commercially available algorithm like COMPAS and finding reason to believe it is biased, the next step is to reduce the bias for future use. There are technical, political, social, and philosophical solutions to making this possible. Using the right intersection of all pertinent disciplines will accelerate our path to finding trends in ethical solutions for machine bias of all types. More research is needed to test the effectiveness of these solutions, and ultimately, we can identify trends in ethical solutions and categorize them for machine bias of similar types.

Black defendants			
	Low	High	
Survived	990	805	0.49
Recidivated	532	1369	0.51
Total: 3696.00			
False positive rate: **44.85**			
False negative rate: **27.99**			
Specificity: 0.55			
Sensitivity: 0.72			
Prevalence: 0.51			
PPV: 0.63			
NPV: 0.65			
LR+: 1.61			
LR-: 0.51			

White defendants			
	Low	High	
Survived	1139	349	0.61
Recidivated	461	505	0.39
Total: 2454.00			
False positive rate: **23.45**			
False negative rate: **47.72**			
Specificity: 0.77			
Sensitivity: 0.52			
Prevalence: 0.39			
PPV: 0.59			
NPV: 0.71			
LR+: 2.23			
LR-: 0.62			

Fig. 12.1 Difference in black and white defendants false negative and positive rates

12.5 The Ethical Solution

12.5.1 Using Ethics to Solve Bias in Machine Learning

Any method used to solve for unfair predictions with legal and ethical consequences, is an example of machine ethics. A perfect solution to bias output in
machine learning may never be possible because of our obligation to rely on imperfect data and the fundamentally prejudiced real-world we live in. Additionally,
optimizing total societal good over accuracy in an isolated context is a complicated
task. However, it is a moral obligation for computer scientists and those who purchase and employ the use of machine learning algorithms to pursue output distributions that most closely reflect an ideal state of fairness for all impacted individuals.
A number of solutions have been either tested or conceptualized in the field of
machine ethics to solve for bias in four main categories: technical, social, political,
and philosophical.

12.5.2 Technical Solutions to Machine Bias

Programming ethics into machine learning algorithms is not straightforward
The source of the bias must first be determined, and options for adjustment can
be considered. Research suggests that statistical bias and variance can relate
strongly to appropriateness of machine bias. To reduce variance, for algorithms
like decision trees, "softer splits" can be used. For example, in a Markov Tree
model, data goes down both sides of the split and the leaves of the tree are longer.
Additionally, multiple hypotheses may be used to "vote" on the classification of
test cases through ensemble and randomization methods. In a study in 1995 by
Dietterich and Kong, this resulted in near perfect or perfect reduction of error in
test cases. When necessary, statistical bias can be reduced using error-correcting
output coding (ECOC) which uses "bit-position hypotheses" and reduces overall
error [6].

When trying to reduce bias in their emotion recognition study, Howard, Zhang,
and Horvitz, discovered that a hierarchical classification model worked well to fit to
both their majority and minority data. They found that the inclusiveness of a classi-
fier can be manipulated depending on the ethics of the resulting decisions by using
a generalized learning algorithm and a specialized learner to correct for bias towards
one or more minority class [8]. When testing for emotion recognition in children's
faces, one emotion in particular seemed biased towards misclassification - Fear.
Instead, the machine was outputting Surprise far more often than was accurate. So,
the researchers broke down the emotion of fear into calculable facial characteristics,
and created a specialized learning algorithm with more detailed information to decipher the difference between fear and surprise. This resulted in a 41.5% increase in

the recognition rate for the minority class (children), while also increasing the recognition rate of the majority by 17.3% [8].

Kusner and other researchers at the Alan Turing Institute, have identified the difficulty of implementing quantitative changes to algorithms to address ethical issues. The group outlines the "first explicitly causal approach" [10] dealing with ethics in an algorithm that predicts the success of law students in completing law school. Their approach weighs protected attributes (such as ethnicity and gender) differently than other descriptive data, *but does not exclude them from the algorithm*. The method takes social bias into account and then compensates effectively, so as not to ignore protected attributes, but to handle them in a way that is fair.

One of the simplest ways to learn from and ultimately rectify bias in machine learning algorithms is to have rigorous external validation testing conducted on algorithms used in high impact, moral contexts. Experimenting with different datasets and metrics, especially when validity testing their own models, will help researchers find the best balance for internal and external validity in empirical research [24]. For optimized validation testing, machine learning source code should be as transparent as possible, if not open source. Periodic audits of machine learning code will help encourage and maintain accuracy and fairness.

12.5.3 *Political Solutions to Machine Bias*

Though much slower than the pace of technology, the pervasive impacts of machine learning have started to spur some nations to take regulatory action. In April 2016, the EU passed the General Data Protection Regulation (GDPR) to take effect in 2018, which outlines the requirement of corporations to oblige a "right to explanation" for citizens in the EU. This means that when an algorithmic decision is made about an individual, they have a right to know why. Parts of the regulation also seem to disallow the use of "personal data which are by their nature particularly sensitive" to profile an individual [5]. Although this may not catch on as quickly in the U.S., efforts were made by former President Obama to acknowledge the risk of machine bias. The Obama administration released a report demonstrating the importance of investigating big data and machine learning algorithms to ensure fairness in 2016 [26]. There have been other American political figures supporting this effort, such as former U.S. Attorney General Eric Holder, who called for the U.S. sentencing commission to study risk assessments like COMPAS in 2014 [19].

Political momentum on the subject of machine ethics will only build as the ramifications increase. Creating policies, standards, best practices, and certifications pertaining to machine learning creation and use will help progress the ethical implementation of these technologies.

12.5.4 Social Solutions to Machine Bias

Ethical solutions to machine bias can come, simply, from social awareness. As Cathy O'Neil says in her book, "Weapons of Math Destruction," algorithms may be unduly respected as authoritative and objective because they are mathematical and, therefore, "impenetrable" to lay-users [27]. Aware consumers are better able to demand ethical standards and transparent practices. This, in addition to focus on curating multifaceted technical teams, groups that foster diverse talent, and "community policing" can all reduce bias. Following a flip in tradition, in 2014, tech giants like Google and Apple released statistics on their workforce diversity. Higher diversity among team members has been shown to increase innovation and profit [5]. Groups such as the Algorithmic Justice League (AJL), founded by Joy Buolamwini, work to promote crowd-sourced reporting and study of bias in machine learning and other technologies [28]. Similarly, the Fairness, Accountability, and Transparency in Machine Learning (FAT ML) workshop, was created by researchers from Google and Microsoft to analyze algorithmic bias and its impacts. When it comes to social pressure, consumers can demand greater transparency in machine learning, by supporting those companies that provide it. Crowdsourcing ethics may be another way to combat the problem. In some gaming communities, moral conflicts find resolution through community voting systems [5]. Involvement from diverse populations in the ethical creation and consumption of machine learning predictions will lead to further progress in ethics that include *all* users.

12.5.5 Philosophical Solutions to Machine Bias

It is imperative to discuss tactical answers to this ethical problem, because it is here now, and impacting individual lives today. But, as with all moral conundrums, there are higher level questions experts continue to posit. For example, Headland and Teahan's insightful research delivers an option for bottom-up ethical machines [11]. Though not feasible as an immediate solution to bias today, continuing this research could lead to the opportunity to build more complex and *innately* ethical intelligent agents. Some experts are at complete odds at where humans belong in the mix. On one hand, perhaps we have reached the peak of moral functioning in the machine, and the human is the required final piece to make ethical decisions with all contextual divergences considered [29]. Or, instead, machines give us an opportunity to discover where we truly stand in matters of ethics. Training from human ethicists may be all that is necessary for the models to do better than an average human when making moral decisions [12]. Exploring these quandaries further will contribute to the body of study, and a deeper understanding of how machine ethics can solve challenging moral problems.

12.6   Conclusion

We have reviewed the background, challenges, and solutions present in the interdependent fields of machine bias and ethics. Using our case study of the COMPAS algorithm and ProPublica analysis, we developed several courses of further research to expose machine bias in detail to best solve for it. In today's climate, easy access to big data entices many to formulate machine learning algorithms with the intent of solving consumer problems, but not ethical problems. To avoid hard-coding centuries of bias, bigotry, and prejudice into our machines, we *must* pay attention to the programs we develop and train on our input. If we don't, as Laura Weidman Powers, the founder of Code2040 says, "we are running the risk of seeding self-teaching AI with the discriminatory undertones of our society in ways that will be hard to rein in because of the often self-reinforcing nature of machine learning" [5]. Machine ethics is a complicated and multifaceted problem. But if we get it right, we will unleash the full benefit of machine learning for humankind.

Author's Note   This essay is the author's thesis presented in partial fulfillment of the requirements for the Master of Science in Digital Forensics and Cyber Security at the John Jay College of Criminal Justice (City University of New York).

References

1. Varshney, K. R., & Alemzadeh, H. (2016, October 5). On the Safety of Machine Learning: Cyber-Physical Systems, Decision Sciences, and Data Products. Retrieved December 04, 2016, from https://arxiv.org/abs/1610.01256
2. M. J. Kusner, J. R. Loftus, C. Russell, and R. Silva. (2017). Counterfactual fairness. arXiv preprint arXiv:1703.06856
3. Ng, A. (n.d.). What is Machine Learning? - Stanford University. Retrieved December 09, 2017, from https://www.coursera.org/learn/machine-learning/lecture/Ujm7v/what-is-machine-learning
4. Brownlee, J. (2016, September 21). Parametric and Nonparametric Machine Learning Algorithms. Retrieved December 09, 2017, from https://machinelearningmastery.com/parametric-and-nonparametric-machine-learning-algorithms/
5. Garcia, M. (2017, January 07). Racist in the Machine: The Disturbing Implications of Algorithmic Bias. Retrieved December 03, 2017, from http://muse.jhu.edu/article/645268/pdf
6. Dietterich, T. G. & Kong, E. B. (1995). Machine learning bias, statistical bias, and statistical variance of decision tree algorithms.Technical Report, Department of Computer Science, Oregon State University, Corvallis, Oregon. Available from ftp://ftp.cs.orst.edu/pub/tgd/papers/tr-bias.ps.gz.
7. Reese, H. (2016). Bias in Machine Learning, and How to Stop It. TechRepublic. Retrieved October 9, 2017, from http://www.techrepublic.com/google-amp/article/bias-in-machine-learning-and-how-to-stop-it/
8. Howard, A., Zhang, C., & Horvitz, E. (2017). Addressing bias in machine learning algorithms: A pilot study on emotion recognition for intelligent systems. *2017 IEEE Workshop on Advanced Robotics and its Social Impacts (ARSO)*. doi:https://doi.org/10.1109/arso.2017.8025197

9. Devlin, H. (2017, April 13). AI programs exhibit racial and gender biases, research reveals. Retrieved October 09, 2017, from https://www.theguardian.com/technology/2017/apr/13/ai-programs-exhibit-racist-and-sexist-biases-research-reveals

10. M. J. Kusner, J. R. Loftus, C. Russell, and R. Silva. (2017). Counterfactual fairness. arXiv preprint arXiv:1703.06856

11. Headleand, C. J., & Teahan, W. (2016). Towards ethical robots: Revisiting Braitenbergs vehicles. *2016 SAI Computing Conference (SAI)*, 469–477. doi:https://doi.org/10.1109/sai.2016.7556023

12. Anderson, Michael & Anderson, Susan. (2007). Machine Ethics: Creating an Ethical Intelligent Agent.. AI Magazine. 28. 15–26.

13. May, P. (2017, November 21). Watch out for 'killer robots,' UC Berkeley professor warns in video. Retrieved December 09, 2017, from http://www.mercurynews.com/2017/11/20/watch-out-for-killer-robots-uc-berkeley-professor-warns-in-video/

14. Larson, J., Mattu, S., Kirchner, L., & Angwin, J. (2016, May 23). How We Analyzed the COMPAS Recidivism Algorithm. Retrieved December 03, 2017, from https://www.propublica.org/article/how-we-analyzed-the-compas-recidivism-algorithm

15. SIGCAS - Computers & Society. (2003). Retrieved December 05, 2017, from http://www.sigcas.org/awards-1/awards-winners/moor

16. J.H. Moor. (2006). "The Nature Importance and Difficulty of Machine Ethics", *Intelligent Systems IEEE*, vol. 21, pp. 18–21, 2006, ISSN 1541-1672.

17. Petrasic, K., Saul, B., & Greig, J. (2017, January 20). Algorithms and bias: What lenders need to know. Retrieved December 05, 2017, from https://www.lexology.com/library/detail.aspx?g=c806d996-45c5-4c87-9d8a-a5cce3f8b5ff

18. Akhtar, A. (2016, August 09). Is Pokémon Go racist? How the app may be redlining communities of color. Retrieved December 09, 2017, from https://www.usatoday.com/story/tech/news/2016/08/09/pokemon-go-racist-app-redlining-communities-color-racist-pokestops-gyms/87732734/

19. Angwin, J., Larson, J., Mattu, S., & Kirchner, L. (2016, May 23). Machine Bias. Retrieved December 03, 2017, from https://www.propublica.org/article/machine-bias-risk-assessments-in-criminal-sentencing

20. Eremenko, K., & De Ponteves, H. (2017, November 02). Machine Learning A-Z™: Hands-On Python & R In Data Science. Retrieved December 05, 2017, from https://www.udemy.com/machinelearning/

21. Rajaraman, A. (2008, March 24). More data usually beats better algorithms. Retrieved December 09, 2017, from http://anand.typepad.com/datawocky/2008/03/more-data-usual.html

22. Brennan, T., Dieterich, W., & Ehret, B. (2008). Evaluating the Predictive Validity of the Compas Risk and Needs Assessment System. *Criminal Justice and Behavior, 36*(1), 21–40. doi:https://doi.org/10.1177/0093854808326545

23. Blomberg, T., Bales, W., Mann, K., Meldrum, R., & Nedelec, J. (2010). Validation of the COMPAS risk assessment classification instrument. Retrieved from the Florida State University website: http://www.criminologycenter.fsu.edu/p/pdf/pretrial/Broward%20Co.%20COMPAS%20Validation%202010.pdf

24. Tantithamthavorn, C., Mcintosh, S., Hassan, A. E., & Matsumoto, K. (2016). Comments on "Researcher Bias: The Use of Machine Learning in Software Defect Prediction". *IEEE Transactions on Software Engineering, 42*(11), 1092–1094. doi:https://doi.org/10.1109/tse.2016.2553030

25. About Us. (n.d.). Retrieved December 05, 2017, from https://www.propublica.org/about/

26. Smith, M., Patil, D., & Muñoz, C. (2016, May 4). Big Risks, Big Opportunities: the Intersection of Big Data and Civil Rights. Retrieved December 03, 2017, from https://obamawhitehouse.archives.gov/blog/2016/05/04/big-risks-big-opportunities-intersection-big-data-and-civil-rights

27. ONeil, C. (2017). *Weapons of math destruction: how big data increases inequality and threatens democracy*. London: Penguin Books.

28. AJL -ALGORITHMIC JUSTICE LEAGUE. (n.d.). Retrieved December 05, 2017, from https://www.ajlunited.org/
29. Steusloff, H. (2016). Humans Are Back in the Loop! Would Production Process Related Ethics Support the Design, Operating, and Standardization of Safe, Secure, and Efficient Human-Machine Collaboration? *2016 IEEE 4th International Conference on Future Internet of Things and Cloud Workshops (FiCloudW),* 348–350. https://doi.org/10.1109/w-ficloud.2016.76

Chapter 13
Philosophical, Moral, and Ethical Rationalization of Artificial Intelligence

Mariana Todorova

13.1 Introduction

Issue: Parallel to the globalization on a global scale, new megatrends area appearing which define the direction of development and solutions in the political, economic, social and cultural spheres. These megatrends lead to a total transformation of existing concepts, impose dominating knowledge, and frequently take away the "load" and the authority of pure science, by promoting attractive theses and hypotheses, which we may define as the "mass culture" of science. Such a powerful megatrend is the forecast that humanity is on the threshold of the (self-) creation of artificial intelligence. In the context of this technological determinism and absolutism, organizations such as the UN, the World Economic Forum, and a majority of prestigious universities, already form their policies and views on the future according to this prognosis. However, the question of how the future will look like is of course not answerable. Are we completely subject to either an optimistic,[1] or negativity bias[2]?

The deficiency of the philosophical, moral and ethical rationalization of the eventual consequences of the appearance of artificial intelligence will grow into a deficiency of the legal regulatory framework, as well as of the processes of making decisions in various areas of key importance.

Thanks to the commercialization and media coverage on the subject of technologies, contemporary society today is led by the so-called technology bias.[3]

---

[1] Rozin, P., E. Royzman. 2001. Negativity Bias, Negativity Dominance, and Contagion, http://journals.sagepub.com/doi/abs/10.1207/S15327957PSPR0504_2

[2] Peeters, G. 1971.The positive-negative asymmetry: On cognitive consistency and positivity bias, https://onlinelibrary.wiley.com/doi/abs/10.1002/ejsp.2420010405

[3] Rosenblatt, G. 2016. Machined Prejudice: Three Sources of Technology Bias, http://www.the-vital-edge.com/technology-bias/

M. Todorova (✉)
Bulgarian Academy of Sciences (Millennium project Bulgarian Node), Sofia, Bulgaria

© Springer Nature Switzerland AG 2019
N. Lee (ed.), *The Transhumanism Handbook*,
https://doi.org/10.1007/978-3-030-16920-6_13

The cinema industry, the popular press and the specialized science prepare the conscience for a huge paradigm change of the present reality. Undoubtedly, this will be the appearance of artificial intelligence, but the images and the perceived possibilities for it are different. As humanity, starting from our own experience, it is inevitable for the artificial intelligence not to look through our own categories for evolution. In this context, it is considered that the occurrence of artificial intelligence will be marked into several stages. At the beginning, an initial/ primary artificial intelligence will be created with more limited, mostly subjected, functions. At a later stage, a secondary or a supreme artificial intelligence will (self-) arise, similar or superior to our level.

We ask the question: "How will our morals look like, jointly with the artificial intelligence? Probably, it would be a new synthesis of our traditional moral values, or perhaps a new moral code will appear. We currently know cultures with various values from across the world. The core of the theories of the morals and the values (ethics) are a peculiar amalgam of the evolutionist theory, the theory of the games, the economy, the cognitive sciences, the cultural anthropology, the religions, the heuristics, and biases, and etc. The various societies create a new system of rules for every quality new situation.[4] May we expect that the same mechanism would be applied to the emergence of the artificial intelligence? The fundamental issue, however, is whether artificial intelligence will be anthropocentric, or will be measured and/or create quite different dimensions, which even would be incomprehensible for us.

As a counterpoint of similar expectations, I will review the possibility that these scenarios do not come true totally and partially and how this would impact the world politics, economy, culture, etc.

13.2   Scenario 1 – Up 2040

Unlike in Europe, in countries like the USA, South Korea, Japan and others, a tendency is observed for absolutization of the technological progress. Authors like Ray Kurzweil, Bill Joy and Martin Rees fall into the group of the so-called technological determinists. In reality, the system of "Doctor Watson" already exists – which is an internet medical doctor. Similar algorithms for architects, lawyers, and university lecturers, already are fact which eventually will lead to the fading away of a huge quantity of old, and the birth of totally new, conceptual professions. On a closer horizon, within the next 8–10 years there are expected to emerge different forms of the initial/narrow artificial intelligence, with the primary function to relieve and support the human activity. The spectrum for this will be maximally wide. There will be autonomous vehicles, which will be an extension of the office space or will

---

[4] Hall, J. Storrs. 2007. Beyond AI. Creating the conscience of the machine. Prometheus book. New York, pp-293-313.

be a variation of the concept for free time and fun. A system of the so-called personal assistant will appear, which will regulate all the household processes and will support the individual as a person and professionally. This system will be acquainted with the diet of the personality / family and will follow the contents of the refrigerator, consequently being able to make online orders from food stores and restaurants. This principle will be able to take care of the order of the electrical appliances and the cleanliness of the home. "The personal assistant" will find and select the appropriate information for the solution of a certain professional or an educational problem.

In the same way, it will follow the health parameters of the individual and will make consultations with virtual and real medical doctors. This refers also to the legal services and the connection with legal software or human lawyers and lawyer's offices. These are only limited examples of the possible spectrum for action of the initial artificial intelligence.

The personal assistant (artificial intelligence) in coordination with special HR programs will select the best job for its client, which would also indicate a disappearance or at least a large transformation of another profession – the HR experts. This initial appearance of the artificial intelligence will follow the emotional state of the individual and select an appropriate candidate according to their mood and connections on social networks, as well as a suitable food, dress, and music taste.

Another group of the initial/narrow intelligence will be a multitude of robots or systems to clean, take care of sick persons, care for adults and children, etc.

Man distances himself from all the rest of the earth species in exercising his free choice and his free will. Recently, we may often meet the means of expression data driven or data-based analysis.

In his/her desire to possess a supporting artificial intelligence and all things to be rationalized and to be objectively revealed, man in effect refuses and delegates his/her sovereignty. **Probably, the huge piling of historical guilt and sense of wrongdoing on attempts of global scale solutions such as colonialism, the First and the Second World War, the ecological crises and the global warming etc., make man search out for himself/ herself, even outside of the existing social contract, for a guarantee of justice.**

Similar to the concern over autonomous cars with the skepticism of 'who will the car hit?', other hypothetical morally-ethical issues can be posed.

Sub Scenario 1:

A young woman gets pregnant, without planning to start a relationship or give birth to a child. At the same time, she is well-educated and with good social, material and career statuses. She may use a virtual medical doctor which has the whole biological and amniocentesis picture available. It reports an excellent genome and a lack of hereditary diseases. In this case, will the AI doctor be pro-life and will it/he/she report the irrational, or rather the emotional arguments of the patient? How will this happen if this will be essentially a cause and effect algorithm?

Sub Scenario 2:

Let's consider again a medical case. A hopelessly sick patient, with a diagnosis of a serious illness, wishes to undergo euthanasia. What decision would the artificial intelligence take, embodied in the image of a doctor? If this person is a genius creator or a writer, he/she may create a human masterpiece in the near future.

Humans do not want to suffer, do not want to make hard decisions, and forget that the driving motive/ forces for various scientific and creative achievements, besides the motivation and persistence, may be the catharsis – the pain and the ecstasy – its overcoming.

In the framework of Primary AI, the various manifestations of AI may be grouped in two main phases of its evolution: 1/automatization of labour/physical activities (at home and at work – ex. smart homes, robots in industry); 2/automatization of decision-making (which would challenge the social nature of many professions in terms of immediate contact; however, it would probably reproduce existing social biases by encoding them in the algorithm).

Here the important question arises, who will have the responsibility to follow the activities around the AI and how the legal regulation will be realized.[5] **It is already a fact that legal frameworks often lag hopelessly behind technological discoveries.** Not less important is the philosophical question of what impact this granting of the human sovereignty will have by delegating more rights to the artificial intelligence.

Legislation is a fruit of generally accepted values, which is operationalized into norms. **Because of the compression of time, here the reflection is reversed. First, the technology is born, then discussions around it start, which create value biases. Yet there is not always a time for it. Sometimes, society is not prepared for it and it is impossible to produce an adaptation response.**[6]

How then are values formed? On the one hand, the eclectics of cultures and civilizations born by globalization, and on the other hand, under the pressure of revolutionary technological discoveries.

Granting of the "human" sovereignty under the form of the delegation of making decisions may free people, but may also dehumanize them.

The dilemma regarding the moral aspect is how this comprehensive support will relieve the personality of the routine and household pursuits and would not it make it lazy and socially insensitive for these aspects of life. This may strengthen, but as well "shift reality", in the direction towards the virtual or the extended/ augmented reality. Labor will be a weaker factor in the rationalization of human life, while the breakdown of the communities and the social connections may acquire drastic dimensions**. Reversely, there may be a reconstruction and creation of a new type of community idea, based on the creativity and the reinvention of man by means, and because of, AI.**

---

[5] Kemp, R. 2016, Legal Aspects of Artificial Intelligence, http://www.kempitlaw.com/wp-content/uploads/2016/11/Legal-Aspects-of-AI-Kemp-IT-Law-v2.0-Nov-2016-.pdf
[6] According to Alvin and Heidi Toffler in "Future Shock", Bantam, 1984.

The potential solution of one more important problem is at hand. **If the above scenario leads to a gradual disappearance and transformation of a number of traditional professions, would it be necessary then to introduce a minimum/ universal basic income to ensure physical existence for the period of re-adaptation.**

Another essential issue is the price of this type of services – would they be a part of the industry for generating profit, or could they be an open code and be received free of charge, to ensure at least in this aspect equality to the access of universal benefits. We live in a time of growing social and economic inequalities; these could be prevented from being allowed in the future by using technology as an instrument.

13.3 Scenario 2 After 2050– Creation or (Self) Birth of a Higher Artificial Intelligence

The companies and the research centers working in the sphere of the artificial intelligence forecast that in the window of 2040–2050, a high/general/ supreme artificial intelligence will appear, which would look like, will be equal to, or will be superior to, the human.

The important questions that result are: What would be the relations human-artificial intelligence? Would there be compatibility of their existence? Would the artificial intelligence be anthropocentric in its essence, or would it have totally different ethics and aesthetics? The technological optimists outline that the advantages of such a realization would be the strengthening of human nature and intellect, by partnership and symbiosis with the artificial intelligence.[7] The impact of the human intellect, which is spontaneous, innovative and a mixture of rationality and irrationality (sentience), and beyond it there is conscience – with the artificial intelligence which would be operative, super-rational, comprehensive and having all the available information, could enormously reinforce it to even lead to a new world order. This would happen if the artificial intelligence is aware of and thinks of itself as a part of the evolutionary development of man, and not as a private, evolutionary, and therefore competing, kind. In the second case, it would be the greatest threat existing so far as Elon Musk[8] and Stephen Hawking[9] warn.

When the issue of how moral and ethical issues emerge in view of the development of AI, we could delineate more explicitly two levels of consideration: one concerning the evolution of AI with respect to the human condition; and one regarding the "AI condition" itself. For example, on the latter level, when it comes to Primary AI, we can hardly consider ethical issues as long as the automatism characteristic for

---

[7] Kurzweil, R. 2006. The Singularity is near. Penguin Books.

[8] https://www.independent.co.uk/life-style/gadgets-and-tech/news/elon-musk-artificial-intelligence-openai-neuralink-ai-warning-a8074821.html

[9] https://www.bbc.com/news/technology-30290540

the "AI condition" in Scenario 1 entails a-morality (there is no agency but automatization of human activities). At the same time, when the transition from automation (pre-determinacy) to spontaneity (uncertainty) takes place and AI transcends into agency (Scenario 2), this brings in an array of ethical concerns on both levels – the ethics of our coexistence with AI and the ethics of AI itself as agency.

Although the dominating prognoses are in these two directions, humans tend to think of a General AI as predetermined, rather than as spontaneous (uncertain). This dichotomy is derivative to the desire of the technological optimists for the General AI to appear, but it could not be introduced in frameworks and controlled. It is not by accident that the three laws of Asimov are cited.[10] The big question here is, of course, would General and Supreme AI have the attributes of the conscious and the subconscious.

The conscious produces rationality, while the subconscious irrationality. The question is, whether these attributive characteristics accompany a priory intellect? Is spontaneity itself not a feature of the independency of the individual?

For decades, new concepts will be introduced in science as "emotional intelligence" and "cultural intelligence", which transforms the concept of intelligence itself.

A number of other questions arise; will the AI be a singular character, existing in the virtual reality and the Internet, or in different individual robot bodies. Or instead it will have the characteristics of the plurality, i.e. the artificial intelligence producing various unique natures which are similar to humans. Will this process be centralized or spontaneous and who will direct it? In other words – the new character – will it be one entity, or will it be renewed or reproduced?

Other questions arise as well. Firstly, may we consider science as neutral in values? For instance, does it pursue the scientific discovery, per se, without inducing responsibilities for the possible consequences. This is true to a great degree for researches of artificial intellect, design – babies, and etc. Elon Musk offered that the artificial intelligence is an open code. At first sight this looks like a good decision, but do we want it to reflect the level of development of the human society? And should we expect that the invested "good" will neutralize the "bad"? Humanity will hardly reach consensus on what the artificial intelligence will be. It is obvious that the technological discoveries irreversibly outpace the ethical and legal discussions. Thus, they are situated beyond the moral ethics and law so will never fill in and master these gaps. That is why many potential risks exist.

The quest for neutrality, which manifests not only in the ethos of contemporary science or the ethics-averse attitudes of techno savvies, but also in the good "fit" between the ethical disinterestedness of S&T and the claim for moral neutrality of the market mechanism. We ask ourselves; how well these two cooperate for the market to generate an "evolutionary" space for human progress, understood only as

---

[10] Asimov, I. 1942. Runaround. Street and Smith Publications https://www.ttu.ee/public/m/martmurdvee/Techno-Psy/Isaac_Asimov_-_I_Robot.pdf, page 26.

techno-scientific advancement (ethics/morality concerns are deemed irrelevant, so is responsibility).

Risks are of a different nature. Soon after its appearance, this software in its nature may improve itself to be reprogrammed with aims different from those initially invested by man, or aims to program in a definite way "its heirs". The (self) appearance of artificial intelligence that is uncontrolled by man may threaten even the existence of the human species through total domination.

The truth is that most of the scientists and researchers working in the sphere of artificial intelligence, believe that the nature of this artificial intelligence will be anthropocentric and compatible with our physics and psyche and with our ethics and aesthetics. That is the reason they perceive such a discovery as new evolutionary stage of the human development.

If we search for some hybrid form of existence, are we ready to deny the "pure" human nature and how we look upon this as a human enhancement or as a total transformation?

There exist many questions concerning artificial intelligence, to which we could not provide a reply, but merely a possible and probable scenario, especially if it emerges and develops in the direction beyond the human as "another type".

Even if the AI evolution goes beyond our imagination and expectations, in this process itself there is something innately human and inherent for the human condition (if we follow Hana Arendt[11]) – the possibility to improvise with our nature, the possibility for spontaneous action, and freedom in the sense of emancipating from determinacy, reflexivity, etc. **It is not surprising at all that our fear, stemming from the uncertainty in the evolution of AI (in its transition from automatization of activities and decision-making to becoming an agency capable of improvisation), is a reflection of the fear we have for ourselves and our future as species. This is especially relevant when contemplating the big ethical chasm that is opening in front of us as we start considering the possible consequences of the combination of freedom, spontaneity and technological might.**

Maybe there is a correlation that we know about ourselves not much beyond biology and we cannot address adequate parallels of thinking. Anyway, it is not too late to consider the processes and to impact them so that we preserve ourselves.

Another fundamental question is; if this prognosis for manifestation of the fourth industrial revolution or a new human evolutionary stage collapses, what other trend will replace this so powerful and will be there reporting or calculation of the "wrong" decisions as its consequence?

A possible scenario is that the existence of the General AI drops as a possibility. In the course of time, it may prove that it is hard and even impossible to make software looking like the neuronal system of the brain. This most probably will open not only a huge gap in the expectations, but in undertaken steps, political, social, cultural and economic solutions. The created bias will have to be replaced and compensated with a new future anticipation. The only sure thing about future is that it is always open. Our chance is to try to anticipate and shape it.

---

[11] Arendt, H. 1958. The Human Condition. University of Chicago Press.

Literature

1. Arendt, H. 1958. The Human Condition. University of Chicago Press
2. Asimov. A. 1950. I,The Robot. Doubleday
3. Asimov, I. 1942. Runaround. Street and Smith Publications
4. Cellan-Jones, R. 12.02.2014. Stephen Hawking warns artificial intelligence could end mankind. BBC, https://www.bbc.com/news/technology-30290540
5. Hall, J.,S. 2007. Beyond AI. Creating the conscience of the machine. Prometheus book. New York, pp-293–313
6. Hariri, Y.N. 2017. Homo Deus. A Brief History of Tomorrow. Vintage
7. Kemp, R. 2016, Legal Aspects of Artificial Intelligence, http://www.kempitlaw.com/wp-content/uploads/2016/11/Legal-Aspects-of-AI-Kemp-IT-Law-v2.0-Nov-2016-.pdf
8. Kissinger, H., A. 06.2018. How the Enlightenment Ends. Philosophically, intellectually—in every way—human society is unprepared for the rise of artificial intelligence, The Atlantic, https://www.theatlantic.com/magazine/archive/2018/06/henry-kissinger-ai-could-mean-the-end-of-human-history/559124/?silverid-ref=MzEwMTkwMjQ2NDgxS0
9. Kurzweil, R. 2006. The Singularity is near. Penguin Books
10. Peeters, G. 1971.The positive-negative asymmetry: On cognitive consistency and positivity bias, https://onlinelibrary.wiley.com/doi/abs/10.1002/ejsp.2420010405
11. Rosenblatt, G. 2016. Machined Prejudice: Three Sources of Technology Bias, http://www.the-vital-edge.com/technology-bias/
12. Rozin, P., E. Royzman. 2001. Negativity Bias, Negativity Dominance, and Contagion, http://journals.sagepub.com/doi/abs/10.1207/S15327957PSPR0504_2
13. Sulleyman, A. 11.24.2017. AI is highly likely to destroy humans, Elon Musk warns. Independent, https://www.independent.co.uk/life-style/gadgets-and-tech/news/elon-musk-artificial-intelligence-openai-neuralink-ai-warning-a8074821.html
14. Tofler, Alvin and Heidi. 1984. Future Shock, Bantam

Chapter 14
Theopolis Monk: Envisioning a Future of A.I. Public Service

Scott H. Hawley

> *"The technician sees the nation quite differently from the political man: to the technician, the nation is nothing more than another sphere in which to apply the instruments he has developed."*
> —Robert Merton, Forward to the English edition of Jacques Ellul's <u>The Technological Society</u>, 1964.

14.1   Part 1: A Visit to One Future

We begin a multi-part discussion on future uses of AI for the public good, with a bit of sci-fi nostalgia.

As a young person, I was a devotee of the TV show "Buck Rogers in the 25th Century," which was a science-fiction retelling of the Rip Van Winkle myth. When twentieth-century Buck comes back to Earth after being accidentally frozen in space and cryogenically preserved (it's not really explained why he's not simply killed), he is arrested as a suspected spy and assigned a public defender/interrogator in the form of a disk-shaped computerized intelligence (known as a "Quad") named Dr. Theopolis.

Readers of the Gospel of Luke and the book of Acts will notice the similarity between the name "Theopolis" and the addressee of these New Testament books, "most excellent Theophilos." The Greek name "Theophilos" (Latinized to "Theophilus") means "friend of God," [1] whereas "Theopolis" means "city of God." [2] "The City of God" is a famous work by Augustine and is widely regarded as "a cornerstone of Western thought." [3] It describes, among other things, how the decline of Roman civilization was not due to the rise of Christianity and advances the notion of an enduring civilization based on Christian spiritual principles. The intent of the writers of Buck Rogers in choosing the name "Theopolis" is unclear

S. H. Hawley (✉)
Department of Chemistry and Physics, Belmont University, Nashville, TN, USA
e-mail: scott.hawley@belmont.edu

© Springer Nature Switzerland AG 2019
N. Lee (ed.), *The Transhumanism Handbook*,
https://doi.org/10.1007/978-3-030-16920-6_14

[4]. One wonders whether the writers had wanted to use "Theophilus" but were told "Theopolis" was easier to say or sounded better. Or perhaps the connection to City of God was deliberate: in the twenty-fifth century, earth society has recovered from a cataclysmic "holocaust" and is principally centered in New Chicago. The new society is an 'enlightened' one: even the Alexa-like home entertainment system in the apartment where Buck is placed under house arrest responds to the voice command "Enlighten me."

This is why the name "Theopolis" stuck out to me. The Enlightenment, with its emphasis on rationality over revelation, resulted in a decline in the amount of religious practice and the eroding of confidence in religious doctrine. Despite the fact that religious freedom is celebrated in Thomas Moore's *Utopia* [5], and some science fiction can take a sympathetic or at least tolerant view toward religion [6], sci-fi typically takes a disparaging view of 'religious superstition,' often envisioning a future society freed of religious sentiments [7]. Thus I found it remarkable that a name with religious connotations was used for a 'positive' character, one who takes the form of a public servant.

The society that Buck arrives in is governed by an oligarchy of sentient artificial intelligences (AIs) known as the Computer Council on which Dr. Theopolis, or "Theo," sits as a chief scientist. According to ComicVine, he was once a human scientist whose "mind was transferred into a computer prior to his death" [8], but in the actual script we are told by Dr. Elias Huer that Theo has been programmed by other Quads:

"These Quads are not programmed by man: They've been programmed by one another over the generations" [9].

Regardless of how the intelligence got 'in there,' in the twenty-fifth century it is running on silicon (or perhaps some new substrate). People in this society felt that the AIs were more trustworthy and/or capable than purely human representatives. Dr. Elias continues:

"You see, the mistakes that we have made in areas, well, like our environment, have been entirely turned over to [the Quads]. And they've saved the Earth from certain doom."

(It's almost as if humanity longed to be under the care of a benevolent superintelligence).

These recollections on Buck Rogers can serve as a springboard for discussing potential positive future uses of AI, human consciousness, and envisioning a future 'enlightened' society or 'City of God.' The key observation from Buck Rogers is that the AI entities on the Computer Council were more or less benevolent, and were acting as public servants — this is opposed to notions of SkyNet or superintelligences that leave humans behind in the dust. It represents an alternate narrative of the future from the dystopian visions which are prevalent in science fiction today [10]. Several sci-fi creators have recently expressed a desire to intentionally bring back a sense of optimism (e.g. [11]), that "we need more utopias" in sci-fi today, both because of the chilling effect of so much doom and gloom on the human spirit and because predicting the future is a difficult game [12]. The recollection of Buck

Rogers from the early 80s showcases some optimistic variety in the space of speculative fiction about AI.

We are already living in an era of AI public servants, as machine learning (ML) statistical models are increasingly applied in government, healthcare, and finance. Yet concerns exist regarding their ability to form concepts (or "representations") and produce decisions in ways that are understandable by the humans whose lives are affected by the inferences of such systems.

14.2   Part 2: Their Thoughts Are Not Our Thoughts

Representations and Explainability

The deployment of artificial intelligence (AI) systems in the public sector may be a tantalizing topic for science fiction, but current trends in machine learning (ML) and AI research show that we are a long way away from the Buck Rogers scenario described in Part 1, and even if it were achievable it's not clear that the AIs would 'think' in a way comprehensible to humans.

The present rise of large-scale AI application deployment in society has more to do with statistical modeling applied to vast quantities of data, rather than with emulation of human consciousness or thought processes. Notable pioneers of AI research such as Geoffrey Hinton and Judea Pearl have lamented the fact that the success of some ML and neural network models in producing useful results as tools for tasks (such as image recognition) has had a disastrous [13] effect on the progress of AI research. This is because this success has diverted efforts away from developing artificial general intelligence (AGI) into mere 'curve fitting' [14] for the purposes of processing data.

In industry, science, and government, ML has been transforming practice by allowing tracking and prediction of user choices [15], discerning imagery from telescopes [16] and medical devices [17], of controlling experiments [18], detecting gravitational waves [19], fighting sex trafficking [20], and... honestly this list could go on for pages. Nearly every aspect of society is becoming 'AI-ified.' As AI expert Andrew Ng points out, "AI is the new electricity," [21] in that it is having a revolutionary impact on society similar to the introduction of electricity.

Few would claim that these ML applications are 'truly intelligent.' They are perhaps weakly intelligent in that the systems involved can only 'learn' [22] specific tasks. (The appropriateness of the "I" in "AI" is debated in many ways and goes back to the 1950s; it is beyond the scope of this article, but see the excellent review by UC Berkeley's Michael Jordan [23].) Nevertheless, these systems are capable of making powerful predictions and decisions in domains such as medical diagnosis [24] and video games [25], predictions which sometimes far exceed the capabilities of the top humans and competing computer programs in the world [26].

Even given their power, the basis upon which ML systems achieve their results — e.g. *why* a neural network might have made a particular decision — is often shrouded in the obscurity of million-dimensional parameter spaces and 'inhumanly' large matrix calculations. This has prompted the European Union, in their recent passage of the General Data Protection Regulation (GDPR, the reason for all those 'New Privacy Policy' emails that flooded your inbox in early summer 2018) to include a section of regulations which require that all model predictions be 'explainable.' [27]

The question of how AI systems such as neural networks best represent the essences of the data they operate upon is the topic of one of the most prestigious machine learning conferences, known as the International Conference on Learning Representations (ICLR), which explains itself in the following terms:

> "The rapidly developing field of deep learning is concerned with questions surrounding how we can best learn meaningful and useful representations of data." [28]

While in the case of natural language processing (NLP), the representations of words — so-called "word embeddings" — may give rise to groupings of words according to their shared conceptual content [29], some other forms of data such as audio typically yield internal representations with "bases" that do not obviously correspond to any human-recognizable features [30]. Even for image processing, progress in understanding feature representation has taken significant strides forward in recent years [31] but still remains a subject requiring much more scholarly attention.

Even systems which are designed to closely model (and exploit) human behavior, such as advertising systems [32] or the victorious poker-playing AI bot "Libratus," [33] rely on internal data representations which are not necessarily coincident with those of humans (Aside: this has echoes of Alvin Plantinga's evolutionary argument against Darwinism, that selecting for advantageous behaviors does not select for true beliefs [34].).

A possible hope for human-like, explainable representations and decisions may lie in some approaches to so-called AGI which rely on simulating human thought processes. Those trying to create 'truly intelligent' AGI models, ones which emulate a greater range of human cognitive activity, see one key criterion to be consciousness, which requires such things as awareness [35]. Other criteria include contextual adaptation and constructing explanatory models [36], goal-setting [37], and for some, even understanding morality and ethics [38]. It is an assumption among many metaphysical naturalists that the brain is 'computable' [39] (though there is prominent dissent [40]), and thus, so the story goes, once humans' capacity for simulating artificial life progresses beyond simulating nematode worms [41], it is only a matter of time before all human cognitive functions can be emulated. This view has prominent detractors, being at odds with many religious and secular scholars, who take a view of the mind-body duality that is incompatible with metaphysical naturalism. At present, it is not obvious to this author whether the simulation of human thought processes is the same thing as (i.e., is isomorphic to) the creation of humans "in silicon."

It is worth noting that representations are memory-limited. Thus AIs with access to more memory can be more sophisticated than those with less. (Note: While it's true that any Turing-complete [42] system can perform any computation, Turing-completeness assumes infinite memory, which real computing systems do not possess.) A system with more storage capacity than the human brain could make use of representations which are beyond the grasp of humans. We see this at the end of the movie "Her," when the machine intelligence declines to try to explain to the human protagonist what interactions between AIs are like [43]. (Micah Redding, President of the Christian Transhumanist Association, has remarked that this "reminds me of angels in the biblical story, whose names are 'too wonderful for you to know.' [44]).

The implications of this (i.e., that representative power scales with available memory and could exceed that of humans) raises questions such as:

- What would it mean to be governed (or care-taken) by AIs that can think 'high above' our thoughts, by means of their heightened capacity for representation?
- How could their decisions be 'explainable'?
- What if this situation nevertheless resulted in a compellingly powerful public good?
- What sorts of unforeseen 'failure modes' might exist?

Even without AGI, such questions are immediately relevant in the present. The entire field of "SystemsML" is dedicated to exploring the interactions and possibilities (and failures) in the large-scale deployment of machine learning applications [45]. These issues are currently being investigated by many top researchers in institutes and companies around the world. Given that 'we' haven't yet managed to even produce self-driving cars capable of earning public trust, further discussion of AI governance may be premature and vulnerable to rampant speculation unhinged from any algorithmic basis. Yet the potential for great good or great harm merits careful exploration of these issues. One key to issues of explainability and trust is the current topic of "transparency" in the design of AI agents [46], a topic we will revisit in a later part of this series.

Before we do that, we'll need to clear up some confusion about the idea of trying to use machines to absolve humans of our need (and/or responsibility) to work together to address problems in society and the environment.

14.3   Part 3: The Hypothesis Is Probably Wrong

"We got this guy Not Sure...and...he's gonna fix everything." — Idiocracy [47].

In Part 1, we reflected on a set of hopes for "benevolent" AI governance as seen in the science fiction TV series Buck Rogers in the twenty-fifth Century. Humanity, having brought themselves to near ruin with wars and ecological disasters, decided to turn over the care of their society to a Computer Council, whose decisions saved humanity and the planet from "certain doom."

In Part 2, we looked 'under the hood' at how the representations that AI systems employ in their decision making can be very different from what humans find intuitive, and how the requirement that algorithmic decisions be "explainable" is manifesting in legislation such as the General Data Protection Regulation (GDPR) of the European Union.

Implicit in the hopes of Part 1 and the concerns of Part 2 is a suggestion that it is the machines themselves who will be responsible for making the decisions. Currently, we see this as essentially the case in some fields, as algorithms determine who will get healthcare [48] or bank loans [49], and even civil liberties in China such as who is allowed to book airline flights [50].

This bears asking the question, are the machines truly the ones doing the deciding, or are they merely 'advising' the humans who truly make the decisions? The answer is "Yes": both of these cases are currently happening. Humans being advised by algorithms is the norm. However, in the financial sector, a large class of stock trades are entirely automated, with companies agreeing to be legally bound by the trading decisions of their algorithms. The speed at which the trading algorithms can operate is both their key strength for earning money —spawning the entire field of "High Frequency Trading" [51]— and yet their key weakness for human oversight, as in the "Flash Crash" of 2010 brought about by trading algorithms run amok [52]. The issue of speed has been identified as a key issue for the oversight of a multitude of AI systems; in the words of the promoters of the *Speed* conference on AI Safety, "When an algorithm acts so much faster than any human can react, familiar forms of oversight become infeasible." [53] In the coming technological future of self-driving cars, passengers will be subject to the decisions of the driving algorithms. This is not the same as legal accountability. The outcomes of automated decision making are still the responsibility of humans, whether as individuals or corporations. Recently it has been debated whether to recognize AIs as legal persons [54], and ethicists such as Joanna Bryson and others have spoken out strongly *against* doing so [55], noting that the responsibility for the actions of such systems should be retained by the corporations manufacturing the systems: "attributing responsibility to the actual responsible legal agents — the companies and individuals that build, own, and/or operate AI," [56] not merely the individual human owners of a product.

The responsibility of developers to steward their AI creations has been a concern since nearly the inception of AI. This is not in the sense of Frankenstein whereby the creator is obliged toward some sentient creature [57]; there are interesting theological reflections on such a situation [58] but they are well outside the scope of our current discussion. In fact, with respect to conceptions of AI for the foreseeable future, Bryson has stated forcefully that, because AIs are not persons and should not be regarded as such, "We are therefore obliged not to build AI we are obliged to." [59] Rather, the type of responsibility we speak of is the need for AI developers to be mindful of the intended and *unintended* uses of their creations, to consider the impact of their work. Norbert Wiener, creator of the field of cybernetics on which modern machine learning is based, also wrote extensively about ethical concerns, indeed he is regarded as the founder of the field of Computer and Information Ethics [60]. His deep concerns about the ethical issues likely to arise from computer and information technology are developed in his 1950 book *The Human Use of Human Beings* [61] in which he foretells the coming of a second industrial revolution, an age of automation with "enormous potential for good and for evil." Joseph Weizenbaum, creator of the famous ELIZA computer program [62], the first chatbot, was outspoken on the topic of social responsibility both in printed form [63] and in interviews. He shared that a turning point for him came when he reflected on the "behavior of German academics during the Hitler time" [64] who devoted their efforts to scientific work without sufficient regard for the ends to which their research was applied. Weizenbaum's remarks were taken up by Kate Crawford in her recent "Just an Engineer: The Politics of AI" address for DeepMind's "You and AI" lecture series at the Royal Society in London [65], voicing a concern over the "risk of being so seduced by the potential of AI that we would essentially forget or ignore its deep political ramifications." This need for responsible reflection and stewardship is particularly acute for AI systems which are intended to be used in social and political contexts. Noteworthy examples of this include police use of predictive algorithms [66] and facial recognition [67], immigration control [68], and the dystopian scope of China's Social Credit System [69], as well as the scandal of election propaganda-tampering made possible by Facebook data employed by Cambridge Analytica [70].

It must be emphasized that most of these applications are seen by their creators as addressing a public need, and are thus being employed *in the service of public good*. The catchphrase "AI for Good" is now ubiquitous, forming the titles of major United Nations Global Summits [71], foundations [72], numerous internet articles and blogs, and trackable on Twitter via the "#AIForGood" hashtag. The phrase's widespread use makes it difficult to interpret; most who use the phrase are likely to view autonomous weapons systems as not in the interest of public good, whereas fostering sustainable environmental practices would be good. Yet one sees conflicting claims about whether AI systems could facilitate "unbiased" [73] decision-making versus (more numerous) demonstrations of AIs becoming essentially platforms for promoting existing bias [74, 75]. One can find many optimistic projections for the use of AI for helping with the environment [76–78] which include improving the efficiency of industrial process to reduce consumption, providing better climate modeling, preventing pollution, improving agriculture and streamlining food distribution.

These are worthy goals, however, many rest on the *assumption* that the societal problems we face with regard to the law, to the environment, and other significant areas result from a lack of intelligence and/or data, and perhaps also a lack of "morality." The application of AI toward the solution of these problems amounts to a *hypothesis* that these problems admit a technical solution. This hypothesis is probably wrong, but to see why we should give some attention to why this hypothesis seems so compelling. The increasing automatization of the workplace (e.g., see the Weizenbaum interview for interesting insights on the development of automated bank tellers, ca. 1980 [79]) and the ever-growing list of announcements of human-level performance by AIs at a host of structured, well-defined tasks demonstrate that many challenges *do* admit such technical solutions. A large class of these announcements in recent years involves the playing of *games*, whether they be video games, board games, card games or more abstract conceptions from the field of Game Theory.

Game Theory has been used to model and inform both individual and collective decision-making and is important enough to merit political science courses dedicated to its application [80]. One famous example of individual decision-making is the Prisoner's Dilemma, which astronomer Carl Sagan extended to suggest as a foundation for morality [81]. In the case of collective action, the Nobel-prize-winning work of John Nash (popularized in the film "A Beautiful Mind") provided a framework for defining fixed points, known as "Nash equilibria" in competitive games. Nash proved that these equilibria exist in any finite game [82] (i.e. games involving a finite number of players, each with a finite number of choices), such if the choices of all the other players are known, then no rational player will benefit by changing his or her choice. In addition to existence, there are algorithms that guarantee finding these equilibria [83], but they are not guaranteed to be unique and may not be optimal in the sense of being in the best interest of all players collectively, nor are they necessarily attainable for players with limited resources [84]. The outcomes of such games can sometimes lead to paradoxical conclusions that policy-makers learn to take into account [85]. However, the particular outcomes depend strongly on the weighting of the relative rewards *built into the game*, and care must be taken before applying the results of one set of assumed weights to real-world situations [86]. Apart from the general applicability of one particular solution, significant other limitations exist, such as the fact that game theory models are necessarily reductionistic and fail to capture complex interactions, and that human beings do not behave as entirely rational agents. Noted economist and game theorist Ariel Rubinstein cautions,

> "For example, some contend that the Euro Bloc crisis is like the games called Prisoner's Dilemma, Chicken or Diner's Dilemma. The crisis indeed includes characteristics that are reminiscent of each of these situations. But such statements include nothing more profound than saying that the euro crisis is like a Greek tragedy. In my view, game theory is a collection of fables and proverbs. Implementing a model from game theory is just as likely as implementing a fable…I would not appoint a game theorist to be a strategic advisor." [87]

It is simply not evident that all societal interactions can be meaningfully reduced to games between a constant number of non-resource-bound rational players, and thus the application of game-playing — whether played by economists, mathematicians or AIs — while informative, does not provide a complete "technical solution."

What of the earlier claim that AIs have (so far) only demonstrated success at "structured, well-defined tasks"? Could one not argue that the current AI explosion is *precisely* due to the ability of ML systems to solve difficult, even 'intractable,' problems and complete tasks which humans find hard to fully specify — tasks including image classifications, artistic style transfer [88], turning shoes into handbags [89], and advanced locomotion [90], to name a few? Is it inconceivable that, given the power of advancing ML systems to form representations and make predictions using vast datasets, they could find "connections" and "solutions" which have eluded the grasp of human historians, political theorists, economists, etc.? This is why the word "probably" is included in the phrase "the hypothesis is probably wrong," because recent history has shown that negative pronouncements about the features and capabilities of AI have a tendency to be superseded with actual demonstrations of such features and capabilities; generally such gaffes proceed as, "Well an AI could never do X," or "AIs don't do Y," to be followed by someone developing an AI that does X, or pulling up a reference showing that AIs are doing Y as of last year. However, there is a difference between caution about negative predictions for the future, and the expression of a *hope* that someday, somehow AI systems will solve the world's problems.

Such a hope in the salvific power of a higher intelligence shares features with non-technical, *non-scientific* outlooks, notably religious outlooks such as the eschatological hopes of Christianity. With Christianity, however, there exists at least a set of historical events, rational philosophical arguments, and personal experience which, at least in the minds of believers, constitute sufficient evidence to warrant such hopes, and although the characteristics of the Savior are (almost by definition) not fully specified, they are enumerated through textual testimony, and these are characteristics which would *warrant* entrusting the care of one's life and affairs with. In contrast, the vagueness of the hope for future AI saviors has more in common with the "Three Point Plan to Fix Everything" expressed by the U.S. President in the movie "Idiocracy":

"Number one, we got this guy, [named] Not Sure.
Number two, he's got a higher I.Q. than any man alive.
And number three, he's gonna fix *everything*" [91].

These hopes for AI 'total solutions' amount to a variant of the "technological solutionism" decried by Evgeny Morozov in his 2014 book, *To Save Everything, Click Here: The Folly of Technological Solutionism* [92], which includes the jacket-summary, "Technology,... can be a force for improvement—but only if we keep solutionism in check and learn to appreciate the imperfections of liberal democracy." The arrival of intelligent machines that somehow resolve long-standing societal conundrums and conflicts amounts to a new twist on the notion of *deus ex machina,* which historically is taken to imply a lack of continuity or precedent, and rightly contains a pejorative connotation implying a lack of warrant.

This lack of warrant in a belief of a technological solution has its seeds in the very assumption it is intended to address: that the problems of society result from lack of intelligence. With respect to environmental concerns, this is contradicted by the observations and conclusions of the former dean of the Yale School of Forestry & Environmental Studies and administrator of the United Nations Development Programme, Gus Speth:

"I used to think that top environmental problems were biodiversity loss, ecosystem collapse and climate change. I thought that thirty years of good science could address these problems. I was wrong. The top environmental problems are selfishness, greed and apathy, and to deal with these we need a cultural and spiritual transformation. And we scientists don't know how to do that" [93].

Erle Ellis, director of the Laboratory for Anthropogenic Landscape Ecology expressed a similar doubt regarding the lack of intelligence and/or data as fundamental causes of ecological challenges in his essay "Science Alone Won't Save the Earth. People Have to Do That":

> "But no amount of scientific evidence, enlightened rational thought or innovative technology can resolve entirely the social and environmental trade-offs necessary to meet the aspirations of a wonderfully diverse humanity — at least not without creating even greater problems in the future" [94].

Kate Crawford, in her aforementioned talk to the Royal Society, emphasized that even the details of developing applications of AI systems affecting the public involve implementation choices which "are ultimately political decisions" [95]. Thus we see the use of AI for a more just and harmonious society as *requiring* human oversight, not as obviating it. And rather than seeing AI resolve human disputes, data scientist Richard Sargeant predicts that "Because of the power of AI…there will be rows. Those rows will involve money, guns and lawyers" [96].

To sum up: Despite amazing success of algorithmic decision making in a variety of simplified domains, well-informed AI ethicists maintain that the responsibility for those decisions must remain attached to humans. Having an ML system able to make sense of vast quantities of data does not seem to offer a way to circumvent the necessary "cultural and spiritual" and "political" involvement of humans in the exercise of government because the assumption that the political, environmental and ethical challenges of our world result from lack of intellect or data is incorrect, and the hypothesis that these problems admit a technical solution is self-contradictory (because the technical solutions require human political activity for design and oversight). The desire for such a relief from these human communal conflict-resolution processes amounts to a form of *hope* akin to religious eschatology, which may be warranted for adherents of faith, but is inconsistent with the trajectory of technical developments in ML applications. Thus, we are left with AI as a tool for humans: We may make better decisions by means of it, but it is *we* who will be making them; abdicating to machines is essentially impossible.

All this is not to say that AI can't be *used by people* for many powerful public goods — and evils! As Zynep Tufecki famously remarked. "Let me say: too many worry about what AI—as if some independent entity—will do to us. Too few people worry what *power* will do *with* AI" [97].

In the next section, we highlight some of these uses for AI in service to secular society as well as to the church as a class of applications I will term "AI monks."

14.4   Part 4: Servant and Sword

Or, Uses of AI: The Good, the Bad, and the Holy

In exploring the potential use of AI for public service, we have veered from the purely speculative narrative of an AI-governed utopia (in Part 1), to concerns about how such systems might be making their decisions (in Part 2), to a resignation that humans probably will not be removable from the process of government, and instead find AI to be a powerful tool to be used by humans (in Part 3). And even though we've already covered many possible uses of AI, and the daily news continually updates us with new ones, in this section we will cover an overview of various "public" applications of AI with perhaps a different structure than is often provided: The Good, the Bad, and the Holy.

14.4.1 *What* Isn't *AI?*

Before we go into that, it is *finally* worth talking about what we *mean* by the term "artificial intelligence." Why wait until the fourth installment to define terms? Because this particular term is so difficult to pin down that it's often not worth trying. As I argue in a separate essay [98], trying to answer the question "What is AI?" leads one into multiple difficulties which I will briefly summarize here:

1. **Too Many Definitions.** There are a variety of definitions which different people employ, from the minimal "doing the right thing at the right time," to nothing short of artificial general intelligence (AGI) where all human cognitive tasks are emulated to arbitrary satisfaction. One particularly insightful definition is on the level of folklore: "AI is machines doing what we *used to think* only humans could do."

2. **The New Normal.** The collection of applications regarded to be AI is ever changing, making the term a moving target and trying to define it amounts to chasing after the wind. On the one hand, applications which used to be regarded as AI when they were new become regarded merely as automated tasks as they become "reified" into the background of "The New Normal" operations of our lives, and thus part of the list of AI applications *decreases* over time. On the other hand, methods and techniques which have been around for centuries — such as curve-fitting — are now regarded as AI; as "AI hype" grows, it seems that "everything is AI" and the list of AI tasks and methods is thus *increasing*.

3. **Anthropomorphism.** A final, insurmountable hurdle is the challenge of anthropomorphism, the unavoidable human tendency to ascribe human faculties and/or intentions to entities in the world (whether animals, machines, or forces of nature). This amounts to a cognitive bias leading one to overestimate AIs' human-like capabilities, an error known as "overidentification." [99]

A host of the devices we use every day contain "artificial intelligence" endowed to them by human engineers to improve upon previous devices which required greater user setup, tuning and/or intervention. For example, computer peripherals and expansion cards used to require manual configuration by the user such as the setting of jumpers or DIP switches on circuit boards, but this was obviated by the rise "Plug and Play" standards for peripherals and busses [100] and network hardware [101] which *automate* the allocation (or "negotiation") of resources and protocols between devices. Another example: The cars we drive are largely drive-by-wire devices with computer systems designed to adaptively adjust the car's performance, expertise programmed-in. Programmed-in expertise "used to count" as AI in the minds of some but tended to vary from application to application. The 2018 "AI in Education" conference in London saw posters and workshops showcasing computer systems that lacked evidence of learning or adaptivity, and were merely tutor-style quiz programs [102], and yet these were regarded to be "AI" in the eyes of the peer-review conference organizers, presumably because the tasks the programs performed were similar to (some of) the work of human tutors.

The point of this discussion is that when we intend to speak of "uses of AI" it is worthwhile to consider that we are *already* using many "AI" systems that we simply don't regard as such, because the tasks they perform are "solved" and their deeds "reified" into what we consider to be "normal" for our current technological experience. Furthermore, if by "uses of AI" we simply mean regression or classification inferences based on curve-fitting to large datasets, we could just as easily (and with greater specificity) say "uses of statistics" instead. The intent here is not to limit the use of the term "AI" as only referring to fictitious sentient machines, but to be cognizant of the multifaceted, subjective and mercurial applicability that the term carries.

"What isn't AI?" isn't necessarily any clearer of a question than "What is AI?" I used the phrase simply to note that in the current hour, with the bounds of "AI" extending outward via hype, and the prior examples of AI fading into the background via reification, we do well to be aware of our terminological surroundings.

14.4.2 The Good

As noted earlier, the list of wonderful things AI systems are being used for in public service is growing so large and so quickly (almost as quickly as the number of societies, conference institutes and companies dedicated to "AI for Good") that citing any examples seems to be pedantic on the one hand and myopic on the other. Nevertheless, here are just a few that may pique interest:

1. **Saving the Coral** [103]. Dr. Emma Kennedy led a team conducting imaging surveys of Pacific reefs and used image classification (AI) models to "vastly improve the efficiency of" analyzing the image to discern which reefs were healthy and which were not. Data from this work will be used to target specific reefs areas for protection and enhanced conservation efforts. The use of image classifiers to speed the analysis of scientific data is advancing many other fields as well, notably astronomy [104].
2. **Stopping Sex Traffickers** [105]. Nashville machine learning (ML) powerhouse Digital Reasoning developed their Spotlight software in collaboration with the Thorn non-profit agency funded by actor Ashton Kutcher to track and identify

patterns consistent with human slavery so that law enforcement could intervene. According to Fast Company in March 2018, "The system has helped find a reported 6,000 trafficking victims, including 2,000 children, in a 12-month period, and will soon be available in Europe and Canada." [106]

3. **Medical Applications**. In recent years, numerous claims have surfaced of AI systems outperforming doctors at various tasks, such as diagnosing conditions such as skin cancer [107], pneumonia [108], and fungal infections [109], as well as predicting the risk of heart attacks [110] — sufficient to spawn an official "AI vs. Doctors" scoreboard at the *IEEE Spectrum* website [111]. But some of these results have come into question. The pneumonia study that used the "CheXNet" software was trained on an inconsistent dataset and made claims exceeding what the results actually showed [112]. In another famous example, IBM's Watson AI system was promoted by its creators as a way to deliver personalized cancer treatment protocols [113], but when it was revealed that the system performed much worse than advertised [114], IBM went quiet and its stock price began to sink. There are great opportunities for beneficial medical applications of AI; one can hope that these setbacks encourage responsible claims of what such systems can do. Meanwhile, some of the greatest inroads for successful medical AI applications involve not diagnosis or image analysis, but rather natural language processing (NLP): processing records, generating insurance codes, and scanning notes from doctors and nurses to look for red flags [115].

14.4.3   The Bad

Hollywood has given us plenty of 'evil' AI characters to ponder —there are lists of them [116]. These are sentient artificial general intelligences (AGI) which exist only in the realm of fiction. The *problem* with this is that plenty of other real and immediate threat vectors exist, and the over-attention to AGI serves as a distraction from these. As Andrew Ng publicly complained,

> "AI+ethics is important, but has been partly hijacked by the AGI (artificial general intelligence) hype. Let's cut out the AGI nonsense and spend more time on the urgent problems: Job loss/stagnant wages, undermining democracy, discrimination/bias, wealth inequality." [117]

This is echoed in the call by Zeynep Tufecki: "let's have realistic nightmares" [118] about technological dangers. One such realistic nightmare is the use of AI by humans who may have selfish, nefarious or repressive goals, and may be regarded as *weaponized AI*. Here we should revisit the words of Tufekci that appeared in Part 2:

> "Let me say: too many worry about what AI—as if some independent entity—will do to us. Too few people worry what *power* will do *with* AI." [119]

Here are a few people who have worried about this:

1. **Classification as Power**. At SXSW 2017, Kate Crawford gave an excellent speech on the history of oppressive use of classification technology by governments [120], such as the Nazis' use of Hollerith machines to label and track 'undesirable' or 'suspect' groups. In the past such programs were limited by their inaccuracy and inefficiency, but modern ML methods offer a vast performance 'improvement' that could dramatically increase the power and pervasiveness of such applications. In the Royal Society address mentioned earlier [121], she quoted Jamaican-born British intellectual Stuart Hall as once saying "systems of classification are themselves objects of power." [122] She then connected these earlier applications with current efforts in China to identify 'criminality' of people based on their photographs [123], a direct modern update of the (discredited) 'sciences' of physiognomy and phrenology. She concluded that using AI in this way "seems like repeating the errors of history…and then putting those tools into the hands of the powerful. We have an ethical obligation to learn the lessons of the past." [124]

2. **Multiple Malicious Misuses**. In February 2018, a group of 26 authors from 14 institutions led by Miles Brundage released a 100-page advisory entitled "The Malicious Use of Artificial Intelligence: Forecasting, Prevention, and Mitigation." [125] The report recommended practices for policymakers, researchers and engineers, including actively planning for misuse of AI applications, and structured these recommendations around the three areas of digital security, physical security, and political security. The first two are frequent topics among IT professionals —albeit without the AI context — however the third is perhaps new to many readers. Brundage *et al.* define political security threats to be

> "The use of AI to automate tasks involved in surveillance (e.g. analysing mass-collected data), persuasion (e.g. creating targeted propaganda), and deception (e.g. manipulating videos) may expand threats associated with privacy invasion and social manipulation. We also expect novel attacks that take advantage of an improved capacity to analyse human behaviors, moods, and beliefs on the basis of available data. These concerns are most significant in the context of authoritarian states, but may also undermine the ability of democracies to sustain truthful public debates."

As we have already cited from various news outlets, such misuses are not mere potentialities.

3. **Slaughterbots**. In 2017 The Future of Life Institute produced a video by Stuart Russell (of "Russell & Norvig," the longtime-standard textbook for AI [126]) called "Slaughterbots" [127] to draw attention to the need to oppose autonomous weapons systems (AWS) development, which they term "killer robots": "weapons systems that, once activated, would select and fire on targets without meaningful human control." [128] In this video, tiny quadcopter drones endowed with shaped explosive charges are able to target individuals for assassination using facial recognition. The use of AI allows the drones to act autonomously, with two main implications: (1) the weapons system can *scale* to arbitrarily large numbers of drones — the video shows thousands being released over a city — and

(2) the lack of communication with a central control system provides a measure of *anonymity* to the party deploying the drones.

14.4.4 The Holy

In addition to AI systems which might serve the public at large, one might consider applications benefitting the church. Here I am concerned with applications of ML systems, not AGIs. Questions regarding the personhood of AGIs and the roles and activities available to them — would they have souls, could they pray, could they be 'saved,' could they be priests, could they be wiser than us, and so on — are beyond the scope of this article, but can be found in many other sources [129–131]. Answers to these would be determined by the ontology ascribed to such entities, a discussion which is still incomplete [132]. There are still other interesting topics regarding present-day ML systems worth investigating, which we describe briefly here.

1. **Dr. Theo*philus*, an AI "Monk."** For much of church history, the scholarly work of investigating and analyzing data of historical, demographic or theological significance was done by monks. In our time, one could imagine AI systems performing monk-like duties: investigating textual correlations in Scripture, predicting trends in missions or church demographics, aiding in statistical analysis of medical miracle reports, aiding in (or autonomously performing) translation of the Bible or other forms of Christian literature, or analyzing satellite images to make archaeological discoveries [133].
2. **Chatbots for the Broken.** London-based evangelism organization CVGlobal.co use ML for content recommendation ("if you liked this article, you might like") for their "Yes He Is" website [134], and also have developed a "Who is Jesus" chatbot to respond to common questions about the person of Christ, the message of the gospels, and some typical questions that arise in apologetics contexts. This is essentially the same program as those used by major corporations such as banks [135] to answer common questions about their organizations. One can argue over whether this removes the 'relational' element of witnessing in a 'profane' way; the structure of such a tool amounts to turning an "FAQ" page (e.g. "Got Questions about Jesus?" [136]) into an interactive conversational model. Relatedly, researchers at Vanderbilt University have gained attention for their use of ML to predict the risk of suicide [137], and apps exist for tracking mental and spiritual health [138], and thus a call for has gone out for investigating predictive models in mental and spiritual counseling [139].
3. **Being Engaged with AI Ethics.** This is more of an *opportunity for engagement* rather than a *use* of AI. Discussions on topics affecting society such as those described in this document should not be limited to only secular, non-theistic sources. There are significant points of commonality between Christian worldviews and others on topics involving affirming human dignity and agency, resisting the exploitation and oppression of other human beings, and showing concern

for the poor and others affected economically by the automation afforded by AI [140, 141]. The world at large is interested in having these discussions, and persons informed by wisdom and spiritual principles are integral members at the table for providing ethical input. We will revisit the topic of foundations for ethics in Part 5.

14.4.5   A Tool, But Not "Just a Tool"

In casting AI as a tool to be used by humans "for good or evil," we shouldn't make the mistake of thinking all tools are "neutral," i.e., that they do not have intentions implied by their very design. As an example of this, the Future of Humanity's information page on "AI Safety Myths" points out, "A heat-seeking missile has a goal." [142] Referring to our earlier list of uses: while it is true that stopping sex trafficking is "good" and repressing political dissidents is "bad," both are examples of surveillance technology, which by its nature imposes a sacrifice of personal privacy. (The tradeoff between security and privacy is an age-old discussion; for now we simply note that AI may favor applications on the security side.)

Sherry Turkle of MIT chronicled the introduction of computers into various fields in the early 1980s, and observed that those asserting that the computer was "just a tool" indicated a lack of reflection: "Calling the computer 'just a tool,' even as one asserted that tools shape thought, was a way of saying that a big deal was no big deal." [143] Turkle cited the famous question of architect Louis Kahn, asking a brick what it wants —"'What do you want, brick?' And brick says to you, 'I like an arch'" [144] — and she asked the new question "What does a simulation want?" In the words of those she interviewed, simulations favor experimentation. The results of its use include a disconnect from reality ("it can tempt its users into a lack of fealty to the real"), and as a consequence, users must cultivate a healthy doubt of their simulations.

Thus we do well to ask: What does an AI 'want'? What forms of usage does it favor? What sorts of structures will it promote and/or rely on? (Keep in mind, we are referring here to modern ML algorithms, not fictional sentient AGIs.) We conclude this section by briefly answering each of these.

1. Like any piece of software, **AI wants to be used**. This led to Facebook employing psychological engineering to generate "eyeball views" and addictive behavior [145], including experimenting on users without their consent and without ethical oversight [146]. The more use, the more data, which fits in with the next point:
2. **An AI Wants Data.** Given their statistical nature, the rise of successful ML algorithms is closely linked with the rise in availability of large amounts of data (to train on) made possible by the internet [147], rather than from improvements in the underlying algorithms. This even motivates some ML experts to advocate improving a model's performance via getting more data rather than adjusting an algorithm [148]. It may be said that ML systems are data-hungry, and *data-*

hungry algorithms make for data-hungry companies and governments. Thus we see the rise of tracking everything users do online for the purposes of mining later, and Google contracting with the healthcare system of the UK for the exchange of user data [149].

3. **An AI Wants "Compute."** A corollary of #1. In order to 'burn through' gargantuan amounts of data, huge computational resources are required. This is the other reason for the rise of ML systems: significant advances in computing hardware, notably graphics processing units (GPUs). Thus, vast data centers and server farms have arisen, and the energy consumption of large-scale AI systems is an increasing environmental concern [150]. In response, Google has built dedicated processing units to reduce their energy footprint [151], but with the growth of GPU usage significantly outpacing Moore's Law [152], this energy concern isn't going away. Some are proposing to distribute the computation to low-power onboard sensors [153], which is also likely to occur. Either way, "AI wants compute."

4. **AI Tempts Toward 'Magic Box' Usage.** "Give the system a bunch of inputs, and a bunch of labeled outputs, and *let the system figure out* how to map one to the other." So goes the hope of many a new ML application developer, and when this works, it can be fun and satisfying (see, e.g., some of my own experiments [154]). This can be one of the strengths of ML systems, freeing the developer from having to understand and explicitly program how to map complicated inputs to outputs, allowing the "programmer" to be creative, such as with Rebecca Fiebrink's Wekinator ML tool for musicians [155]. But this can also encourage lazy usage such as the "physiognomy" applications cited by Kate Crawford, and biased models which accidentally discriminate against certain groups (of which there too many instances to cite). As with simulation, users should cultivate a healthy doubt of their correlations.

Finally, in terms of what other structures AI will promote and/or rely on, we should remember the general warnings on technological development by Christian philosopher Jacques Ellul. In *The Technological Society,* Ellul cautioned that "purposes drop out of sight and efficiency becomes the central concern." [156] Furthermore, Ellul noted that successful technological development tends to become self-serving, as we have all inherited the nature of Cain, the first city-builder who was also the first murderer [157]. In the next section, we will relate some current conversations aimed at keeping AI development and government oriented toward serving people.

14.5   Part 5: Further Fertile Fields

Five "AI Ethics & Society" conversations to follow

For the closing section, I have selected five areas of current conversation that I find to be particularly worth paying attention to. This section is not exhaustive or authoritative.

14.5.1   Bias

A popular conversation in recent years is the topic of biased machine learning models, such as those which associate negative connotations with certain races [158] or predict employability on the basis of gender [159], although such occurrences are nothing new to statisticians, and have been equally attributed to "Big Data" as much as to AI [160]. There are numerous conversations regarding how to "fix" bias [161] or at least detect, measure and mitigate it [162]. While these are important and worthy efforts, one can foresee that as long as there are bad statisticians – i.e., people doing sloppy statistics — there will be biased models. And machine learning (ML) automates bad statistics (though typically not through the algorithms involved but through the datasets used to train the models). Thus the problem of bias is both a current topic and one which is likely to remain relevant for some time to come.

14.5.2   Black Boxes Versus Transparency

In Part 2 we mentioned requirements that algorithmic decisions should be "explainable," [163] as opposed to "opaque" [164] systems which function as "black boxes." [165, 166] Two main approaches present themselves:

1. **Probing Black Boxes**. One approach is to use various methods to probe black box systems, by observing how they map inputs to outputs. Examples include learning the decision rules of systems in an explainable way (and even mimicking the existing system) [167] and extracting "rationales" [168] — short textual summaries of significant input data. A related approach involves mapping entire subsets at a time to predict the "boundaries" of possible outputs from a system, e.g. for safety prediction [169].
2. **Transparency As a Design Requirement.** For several years, there have been calls to produce systems which are transparent *by design* [170]. Such considerations are essential for users to form accurate mental models of a system's operation [171], which may be a key ingredient to fostering user trust [172]. Further, transparent systems are essential for government accountability and providing a greater sense of agency for citizens [173]. But how to actually design useful, transparent interfaces for robots [174, 175] and computer systems in general [176] remains an active area of research, both in terms of the designs themselves and in measuring their effects with human users — even when it comes to the education of data science professionals [177]. One cannot simply overwhelm the user with data. This is particularly challenging for neural network systems, where the mapping of high-dimensional data exceeds the visualization capacities of humans, and even on simple datasets such as MNIST, dimensionality-reduction methods such as t-SNE [178] and interactive visualizations [179] can still leave

one lacking a sense of clarity. This is an active area of research, with two particularly active efforts by the group at the University of Bath (Rob Wortham, Andreas Theodorou, and Joanna Bryson) [180] and by Chris Olah [181]. It's also worth mentioning the excellent video by Brett Victor on designing for understanding [182], although this is not particular to algorithmic decision making.

One 'hybrid' form of the two above approaches involves providing counterfactual statements, such as in the example, "You were denied a loan because your annual income was £30,000. If your income had been £45,000, you would have been offered a loan" [183]. The second statement is a counterfactual, and while not offering full transparency or explainability, provides at least a modicum of guidance. This may be a minimal prescription for rather simple algorithms, although for complex systems with many inputs, such statements may be difficult to formulate.

14.5.3   AI Ethics Foundations

In reading contemporary literature on the topic of "AI Ethics," one may not frequently see people stating explicitly where they're coming from, in terms of the foundations of their ethics, and rather one often sees the "results," i.e. the ethical directives built upon those foundations. Joanna Bryson, whom we've cited many times, is explicit about working from a framework of functionalism [184], which she applies to great effect, and reaches conclusions which are often in agreement with other traditions. Alternatively, philosopher Shannon Vallor (co-chair of this year's AAAI/ACM conference on AI, Ethics and Society) in her book, *Technology and the Virtues: A Philosophical Guide to a Future Worth Wanting* [185], advocates the application of virtue ethics to matters of technological development. Virtue ethics provides a motivation toward good behavior on the principle of "excellence" of character, leading to the greatest thriving of the individual and thus of society. Drawing from the ancient traditions of Aristotelianism, Confucianism, and Buddhism, and religious parallels in Christian and Islamic thought, and western philosophical treatises such as those of Immanuel Kant and the critiques by Nietzsche, Vallor develops an adaptive framework that eschews rule-based pronouncements in favor of "technomoral flexibility," which she defines as "a reliable and skillful disposition to modulate action, belief, and feeling as called for by novel, unpredictable, frustrating, or unstable technosocial conditions." In the Christian tradition, Brent Waters has written on moral philosophy "in the emerging technoculture," [186] and while not addressing AI in particular, many of his critiques provide somewhat of a (to borrow some jargon from machine learning) "regularizing" influence enabling one to approach the hype of AI development in a calm and reflective manner.

14.5.4   Causal Calculus

If neural networks and their ilk are mere "correlation machines" [187] akin to poly-nomial regression [188], how can we go from correlation to inferring causality? Put differently, how can we go from "machine learning" to "predictive analytics"? [189] Turing Award winner Judea Pearl in his 2018 book *The Book of Why* [190] (aimed at a more popular audience than his more technical *Causality* [191]) offers a set of methods termed "causal calculus" defined over Bayesian Networks (a term coined by Pearl) [192]. This book has generated many favorable reviews from within the AI community and has been regarded as contributing an essential ingredient toward the development of more powerful, human-like AI [193]. In a 2018 report to the Association of Computing Machinery (ACM) [194], Pearl highlights seven tasks which are beyond the reach of typical statistical learning systems but have been satisfied using causal modeling. Many further applications by other researchers of this method are likely to appear in the near future.

14.5.5   Transformative AI

One does not need to have fully conscious, sentient AGI in order to have AI that can still have a severely disruptive and possibly dangerous impact on human life on a large scale. Such systems will likely exhibit forms of superintelligence [195] across multiple domains, in a manner not currently manifested in the world (i.e., not in the familiar forms of collective human action, or artifact-enhanced human cognition). Planning to mitigate risks associated with such outcomes comprises the field of AI Safety [196]. In late September 2018, the Future of Humanity Institute released a report by Allan Dafoe entitled *AI Governance: A Research Agenda* in which he "focuses on extreme risks from advanced AI." [197] Dafoe distinguishes AI Governance from AI Safety by emphasizing that safety "focuses on the technical questions of how AI is built" whereas governance "focuses on the institutions and contexts in which AI is built and used." In describing risks and making recommen-dations, Dafoe focuses on what he calls "transformative AI (TAI), understood as advanced AI that could lead to radical changes in welfare, wealth, or power." Dafoe outlines an agenda for research which seems likely to be taken up by many inter-ested researchers.

14.6   Summary

Starting from an optimistic view of a future utopia governed by AIs who make benevolent decisions in place of humans (with their tendency toward warfare and abuse of the environment), we have noted that AI systems are unlikely to represent the world or other concepts in ways which are intuitive or even explainable to humans. This carries a risk to basic civil liberties, and efforts to make such systems more explainable and transparent are actively being pursued. Even so, such systems will and simply *do* require human political activity in the form of implementation choices and auditing such as checking for bias, and thus humans will remain the decision-makers, as they should be. While the unlikelihood of the realization of a quasi-religious hope of future AI saviors may be disappointing to science fiction fans, it means, in the words of Christina Colclough, (Senior Policy Advisor, UNI Global Union), that we can avoid "technological determinism" and we can talk about and "agree on the kind of future we want" [198]. We have seen that AI is a powerful tool for good and for evil, and yet it is not "neutral": it prefers large amounts of data (which may involve privacy concerns), large computing resources and thus large energy consumption, and may favor unreflective "magical thinking" which empowers sloppy statistics and biased inferences. Drawing causal inferences from the correlations of machine learning is problematic, but work in the area of causal modeling may allow for much more powerful AI systems. These powerful systems may themselves become transformative existential threats and will require planning for safety and governance to ensure that such systems favor human thriving. The conception of what constitutes human thriving is an active area of discussion among scholars with diverse ideological and religious backgrounds, and is a fertile area for dialog between these groups, for the goal of fostering a harmonious human society.

Acknowledgements   The author wishes to thank the following for helpful discussions: Michael Burdett, William Hooper, Tommy Kessler, Stan Rosenberg, Andy Watts, Miles Brundage, Beth Singler, Andreas Theodorou, Robert Wortham, Joanna Bryson, Micah Redding and Nathan Griffith. This work was sponsored by a grant given by Bridging the Two Cultures of Science and the Humanities II, a project run by Scholarship and Christianity in Oxford (SCIO), the UK subsidiary of the Council for Christian Colleges and Universities, with funding by Templeton Religion Trust and The Blankemeyer Foundation.

Author's Note   Parts of this chapter have appeared in serialized form at SuperPositionMagazine. com.

References

1. W R F Browning, ed., *Theophilus, in The Oxford Dictionary of the Bible* (Oxford University Press, 2004).
2. "Theopolis" is a proper name that showed some popularity in the 19th century, and is also sometimes attributed to individuals more commonly known by the name "Theophilus", e.g., John Milton uses the former in Of Prelatical Episcopacy (1641) to refer to the 23rd Pope of Alexandria.
3. Wikipedia contributors, "The City of God," March 23, 2018, https://en.wikipedia.org/w/index.php?title=The_City_of_God&oldid=832052290.

4. I have been unable to find any references regarding the creators' intent in choosing this name. "Theo" was a new character for the 1979 TV series, and not part of the original Buck Rogers comic strip.
5. Sanford Kessler, "Religious Freedom in Thomas More's Utopia," *The Review of Politics* 64, no. 02 (March 2002): 207, https://doi.org/10.1017/S0034670500038079.
6. Teresa Jusino, "Religion and Science Fiction: Asking the Right Questions," Tor.com, January 6, 2010, https://www.tor.com/2010/01/06/religion-and-science-fiction-asking-the-right-questions/.
7. Anna Fava, "Science Fiction — Mythology of the Future," *Think Magazine*, December 2014, https://www.um.edu.mt/think/science-fiction-mythology-of-the-future/.
8. "Doctor Theopolis (Character) - Comic Vine," accessed May 27, 2018, https://comicvine.gamespot.com/doctor-theopolis/4005-76853/.
9. "Buck Rogers in the 25th Century S01e01 Episode Script I SS," accessed May 27, 2018, https://www.springfieldspringfield.co.uk/view_episode_scripts.php?tv-show=buck-rogers-in-the-25th-century&episode=s01e01.
10. Sarah Begley, "The Mysterious Case of the Missing Utopian Novels," *Time*, September 28, 2017, http://time.com/4960648/science-fiction-utopian-novels-books/.
11. Tom Cassauwers, "Sci-Fi Doesn't Have to Be Depressing: Welcome to Solarpunk," accessed May 27, 2018, https://www.ozy.com/fast-forward/sci-fi-doesnt-have-to-be-depressing-welcome-to-solarpunk/82586; Adam Epstein, *"I Miss Optimism": The "Family Guy" Creator Wants to Bring Back Hopeful Sci-Fi* (Quartz, 2017), https://qz.com/1052758/the-family-guy-creator-wants-to-bring-back-optimistic-sci-fi/; Cory Doctorow, "In 'Walkaway,' A Blueprint For A New, Weird (But Better) World," *NPR*, April 27, 2017, https://www.npr.org/2017/04/27/523587179/in-walkaway-a-blueprint-for-a-new-weird-but-better-world.
12. Lauren J Young, "How To Move Beyond The Tropes Of Dystopia," accessed May 27, 2018, https://www.sciencefriday.com/articles/just-topia-moving-beyond-the-tropes-of-dystopia/.
13. Mahmoud Tarrasse, *What Is Wrong with Convolutional Neural Networks ? – Towards Data Science* (Towards Data Science, 2018), https://towardsdatascience.com/what-is-wrong-with-convolutional-neural-networks-75c2ba8fbd6f.
14. Kevin Hartnett, "To Build Truly Intelligent Machines, Teach Them Cause and Effect," *Quanta Magazine*, May 15, 2018, https://www.quantamagazine.org/to-build-truly-intelligent-machines-teach-them-cause-and-effect-20180515/.
15. Bruce Schneier et al., "Why 'Anonymous' Data Sometimes Isn't," *Wired*, December 2007.
16. Sander Dieleman, Kyle W Willett, and Joni Dambre, "Rotation-Invariant Convolutional Neural Networks for Galaxy Morphology Prediction," *Mon. Not. R. Astron. Soc.* 450, no. 2 (June 2015): 1441–1459.
17. Olaf Ronneberger, Philipp Fischer, and Thomas Brox, "U-Net: Convolutional Networks for Biomedical Image Segmentation," in *Medical Image Computing and Computer-Assisted Intervention – MICCAI 2015* (Springer International Publishing, 2015), 234–241.
18. P B Wigley et al., "Fast Machine-Learning Online Optimization of Ultra-Cold-Atom Experiments," *Sci. Rep.* 6 (May 2016): 25890.
19. Daniel George and E A Huerta, "Deep Learning for Real-Time Gravitational Wave Detection and Parameter Estimation: Results with Advanced LIGO Data," *Phys. Lett. B* 778 (March 2018): 64–70.
20. Jamie McGee, "How a Franklin Software Company Helped Rescue 6,000 Sex Trafficking Victims," July 6, 2017, https://www.tennessean.com/story/money/2017/07/06/franklins-digital-reasoning-creates-tool-has-helped-rescue-6-000-sex-trafficking-victims/327668001/.
21. Andrew Ng, "Andrew Ng: Why AI Is the New Electricity," accessed August 28, 2018, https://www.gsb.stanford.edu/insights/andrew-ng-why-ai-new-electricity.
22. 'Learn' here means iteratively minimizing an error function or maximizing a reward function.
23. Michael Jordan, "Artificial Intelligence — The Revolution Hasn't Happened Yet," April 19, 2018, https://medium.com/@mijordan3/artificial-intelligence-the-revolution-hasnt-happened-yet-5e1d5812e1e7.

24. Pranav Rajpurkar et al., "CheXNet: Radiologist-Level Pneumonia Detection on Chest X-Rays with Deep Learning" (November 14, 2017), http://arxiv.org/abs/1711.05225.
25. Aman Agarwal, "Explained Simply: How DeepMind Taught AI to Play Video Games," August 27, 2017, https://medium.freecodecamp.org/explained-simply-how-deepmind-taught-ai-to-play-video-games-9eb5f38c89ee.
26. "AlphaGo | DeepMind," accessed May 27, 2018, https://deepmind.com/research/alphago/.
27. Bryan Casey, Ashkon Farhangi, and Roland Vogl, "Rethinking Explainable Machines: The GDPR's 'Right to Explanation' Debate and the Rise of Algorithmic Audits in Enterprise," *Berkeley Technology Law Journal (Submitted)*, n.d.
28. "Call for Papers, ICLR 2018, Sixth International Conference on Learning Representations," accessed June 5, 2018, https://iclr.cc/Conferences/2018/CallForPapers.
29. Tomas Mikolov et al., "Efficient Estimation of Word Representations in Vector Space," *ArXiv:1301.3781 [Cs]*, January 16, 2013, http://arxiv.org/abs/1301.3781.
30. Paris Smaragdis, *NMF? Neural Nets? It's All the Same...*, SANE 2015 (at 32:12: YouTube), accessed June 5, 2018, https://www.youtube.com/watch?v=wfmpViJIjWw.
31. Chris Olah, Alexander Mordvintsev, and Ludwig Schubert, "Feature Visualization," *Distill* 2, no. 11 (November 7, 2017): e7, https://doi.org/10.23915/distill.00007.
32. C. Perlich et al., "Machine Learning for Targeted Display Advertising: Transfer Learning in Action," *Machine Learning* 95, no. 1 (April 1, 2014): 103–27, https://doi.org/10.1007/s10994-013-5375-2.
33. Byron Spice, "Carnegie Mellon Reveals Inner Workings of Victorious Poker AI | Carnegie Mellon School of Computer Science," accessed June 5, 2018, https://www.scs.cmu.edu/news/carnegie-mellon-reveals-inner-workings-victorious-poker-ai.
34. Alvin Plantinga, *Warrant and Proper Function* (New York: Oxford University Press, 1993).
35. Roger Penrose, "Why Algorithmic Systems Possess No Understanding" (May 15, 2018).
36. John Launchbury and DARPAtv, *A DARPA Perspective on Artificial Intelligence*, accessed June 4, 2018, https://www.youtube.com/watch?v=-O01G3tSYpU.
37. Jesus Rodriguez, "The Missing Argument: Motivation and Artificial Intelligence," *Medium* (blog), August 14, 2017, https://medium.com/@jrodthoughts/the-missing-argument-motivation-and-artificial-intelligence-f582649a2680.
38. Dr Vyacheslav Polonski, "Can We Teach Morality to Machines? Three Perspectives on Ethics for Artificial Intelligence," *Medium* (blog), December 19, 2017, https://medium.com/@drpolonski/can-we-teach-morality-to-machines-three-perspectives-on-ethics-for-artificial-intelligence-64fe479e25d3.
39. Ray Kurzweil, *How to Create a Mind: The Secret of Human Thought Revealed* (New York: Viking, 2012).
40. Antonio Regalado, "The Brain Is Not Computable," MIT Technology Review, accessed June 5, 2018, https://www.technologyreview.com/s/511421/the-brain-is-not-computable/.
41. Balázs Szigeti et al., "OpenWorm: An Open-Science Approach to Modeling Caenorhabditis Elegans," *Frontiers in Computational Neuroscience* 8 (November 3, 2014), https://doi.org/10.3389/fncom.2014.00137.
42. "Turing Completeness," *Wikipedia*, May 28, 2018, https://en.wikipedia.org/w/index.php?title=Turing_completeness&oldid=843397556.
43. Spike Jonze, *Her* (Warner Bros. Entertainment, 2014).
44. Micah Redding, "Also Reminds Me of Angels in the Biblical Story, Whose Names Are 'Too Wonderful for You to Know,'" July 15, 2018.
45. "SysML Conference," accessed June 5, 2018, https://www.sysml.cc/.
46. Robert H. Wortham, Andreas Theodorou, and Joanna J. Bryson, "What Does the Robot Think? Transparency as a Fundamental Design Requirement for Intelligent Systems," 2016.
47. Mike Judge, *Idiocracy* (20th Century Fox, 2006), http://www.imdb.com/title/tt0387808/.
48. Colin Lecher, "A Healthcare Algorithm Started Cutting Care, and No One Knew Why," The Verge, March 21, 2018, https://www.theverge.com/2018/3/21/17144260/healthcare-medicaid-algorithm-arkansas-cerebral-palsy.

49. Thomas Hills, "The Mental Life of a Bank Loan Algorithm: A True Story," *Psychology Today*, accessed October 8, 2018, https://www.psychologytoday.com/blog/statistical-life/201810/the-mental-life-bank-loan-algorithm-true-story.
50. Megan Palin, "China's 'Social Credit' System Is a Real Life 'Black Mirror' Nightmare," *New York Post* (blog), September 19, 2018, https://nypost.com/2018/09/19/chinas-social-credit-system-is-a-real-life-black-mirror-nightmare/.
51. Investopedia Staff, "High-Frequency Trading - HFT," Investopedia, July 23, 2009, https://www.investopedia.com/terms/h/high-frequency-trading.asp.
52. Matt Phillips, "Nasdaq: Here's Our Timeline of the Flash Crash," *Wall Street Journal*, May 11, 2010, https://blogs.wsj.com/marketbeat/2010/05/11/nasdaq-heres-our-timeline-of-the-flash-crash/.
53. "DLI | Speed Confence | Cornell Tech," Digital Life Initiative | Cornell Tech | New York, accessed October 8, 2018, https://www.dli.tech.cornell.edu/speed.
54. Janosch Delcker, "Europe Divided over Robot 'Personhood,'" *Politico*, April 11, 2018, https://www.politico.eu/article/europe-divided-over-robot-ai-artificial-intelligence-personhood/.
55. Joanna J. Bryson, "Robots Should Be Slaves," *Close Engagements with Artificial Companions: Key Social, Psychological, Ethical and Design Issues*, 2010, 63–74.
56. Joanna J. Bryson, "R/Science - Science AMA Series: I'm Joanna Bryson, a Professor in Artificial (and Natural) Intelligence. I Am Being Consulted by Several Governments on AI Ethics, Particularly on the Obligations of AI Developers towards AI and Society. I'd Love to Talk – AMA!," reddit, accessed October 8, 2018, https://www.reddit.com/r/science/comments/5nqdo7/science_ama_series_im_joanna_bryson_a_professor/.
57. Josephine Johnston, "Traumatic Responsibility: Victor Frankenstein as Creator and Casualty," in *Frankenstein*, ed. Mary Wollstonecraft Shelley, Annotated for Scientists, Engineers, and Creators of All Kinds (MIT Press, 2017), 201–8, http://www.jstor.org/stable/j.ctt1pk3jfp.11.
58. Michael Burdett, "Danny Boyle's Frankenstein: An Experiment in Self-Imaging," Transpositions, December 5, 2016, http://www.transpositions.co.uk/danny-boyles-frankenstein-an-experiment-in-self-imaging/.
59. Joanna J. Bryson, "Patiency Is Not a Virtue: The Design of Intelligent Systems and Systems of Ethics," *Ethics Inf. Technol.* 20, no. 1 (March 1, 2018): 15–26, https://doi.org/10.1007/s10676-018-9448-6.
60. Terrell Bynum, "Computer and Information Ethics," in *The Stanford Encyclopedia of Philosophy*, ed. Edward N. Zalta, Summer 2018 (Metaphysics Research Lab, Stanford University, 2018), https://plato.stanford.edu/archives/sum2018/entries/ethics-computer/.
61. Norbert Wiencr, *The Human Use of Human Beings: Cybernetics and Society*, The Da Capo Series in Science (New York, N.Y: Houghton Mifflin Harcourt, 1950).
62. Joseph Weizenbaum, "ELIZA—a Computer Program for the Study of Natural Language Communication between Man and Machine," *Commun. ACM* 9, no. 1 (1966): 36–45.
63. Joseph Weizenbaum, *Computer Power and Human Reason: From Judgment to Calculation* (San Francisco: Freeman, 1976).
64. Diana ben-Aaron, "Weizenbaum Examines Computers and Society," *The Tech*, April 9, 1985, http://tech.mit.edu/V105/N16/weisen.16n.html.
65. Kate Crawford, *Just An Engineer: The Politics of AI*, You and AI (The Royal Society: YouTube, 2018), https://www.youtube.com/watch?v=HPopJb5aDyA.
66. Sidney Fussell, "The LAPD Uses Palantir Tech to Predict and Surveil 'Probable Offenders,'" Gizmodo, May 8, 2018, https://gizmodo.com/the-lapd-uses-palantir-tech-to-predict-and-surveil-prob-1825864026.
67. Rebecca Hill, "Rights Group Launches Legal Challenge over London Cops' Use of Facial Recognition Tech," *The Register*, July 26, 2018, https://www.theregister.co.uk/2018/07/26/big_brother_watch_legal_challenge_facial_recognition/.
68. Travis Galey, Kris Van Cleave, and 12:32 Pm, "Feds Use Facial Recognition to Arrest Man Trying to Enter U.S. Illegally," CBS news, August 23, 2018, https://www.cbsnews.com/news/customs-and-border-protection-use-facial-recognition-to-arrest-man-trying-to-enter-u-s-illegally/.
69. Palin, "China's 'Social Credit' System Is a Real-Life 'Black Mirror' Nightmare."

70. Sam Meredith, "Facebook-Cambridge Analytica: A Timeline of the Data Hijacking Scandal," April 10, 2018, https://www.cnbc.com/2018/04/10/facebook-cambridge-analytica-a-time-line-of-the-data-hijacking-scandal.html.
71. "AI for Good Global Summit 2018," accessed October 7, 2018, https://www.itu.int/en/ITU-T/AI/2018/Pages/default.aspx.
72. "AI for Good Foundation," accessed October 7, 2018, https://ai4good.org/.
73. Sean Captain, "This News Site Claims Its AI Writes 'Unbiased' Articles," Fast Company, April 4, 2018, https://www.fastcompany.com/40554112/this-news-site-claims-its-ai-writes-unbiased-articles.
74. Ben Dickson, "Why It's so Hard to Create Unbiased Artificial Intelligence," TechCrunch, November 7, 2016, http://social.techcrunch.com/2016/11/07/why-its-so-hard-to-create-unbiased-artificial-intelligence/.
75. Jeffrey Dastin, "Amazon Scraps Secret AI Recruiting Tool That Showed Bias against Women," Reuters, October 10, 2018, https://www.reuters.com/article/us-amazon-com-jobs-automation-insight/amazon-scraps-secret-ai-recruiting-tool-that-showed-bias-against-women-idUSKCN1MK08G.
76. Celine Herweijer, "8 Ways AI Can Help Save the Planet," World Economic Forum, January 24, 2018, https://www.weforum.org/agenda/2018/01/8-ways-ai-can-help-save-the-planet/.
77. Sarath Muraleedharan, "Role of Artificial Intelligence in Environmental Sustainability," EcoMENA (blog), March 6, 2018, https://www.ecomena.org/artificial-intelligence-environmental-sustainability/.
78. Dan Robitzski, "Advanced Artificial Intelligence Could Run The World Better Than Humans Ever Could," Futurism, August 29, 2018, https://futurism.com/advanced-artificial-intelligence-better-humans.
79. ben-Aaron, "Weizenbaum Examines Computers and Society."
80. Nathan Griffith, "PSC 3610 Game Theory and Public Choice," in Undergraduate Catalog 2018–2019 (Belmont University, 2018), http://catalog.belmont.edu/preview_course_nopop.php?catoid=3&coid=4367.
81. Carl Sagan, "A New Way to Think About Rules to Live By," Parade, November 28, 1993.
82. John F. Nash, "Equilibrium Points in N-Person Games," Proceedings of the National Academy of Sciences of the United States of America 36, no. 1 (1950): 48–49.
83. Ryan Porter, Eugene Nudelman, and Yoav Shoham, "Simple Search Methods for Finding a Nash Equilibrium," Games and Economic Behavior 63, no. 2 (July 2008): 642–62, https://doi.org/10.1016/j.geb.2006.03.015.
84. Joseph Y. Halpern, Rafael Pass, and Daniel Reichman, "On the Non-Existence of Nash Equilibrium in Games with Resource-Bounded Players," ArXiv:1507.01501 [Cs], July 6, 2015, http://arxiv.org/abs/1507.01501.
85. "Braess's Paradox," Wikipedia, October 8, 2018, https://en.wikipedia.org/w/index.php?title=Braess%27s_paradox&oldid=863068294.
86. William Chen, "Bad Traffic? Blame Braess' Paradox," Forbes, October 20, 2016, https://www.forbes.com/sites/quora/2016/10/20/bad-traffic-blame-braess-paradox/.
87. Ariel Rubinstein, "Game theory: How game theory will solve the problems of the Euro Bloc and stop Iranian nukes," FAZ.NET, March 27, 2013, sec. Feuilleton, http://www.faz.net/1.2130407.
88. Leon A. Gatys, Alexander S. Ecker, and Matthias Bethge, "A Neural Algorithm of Artistic Style," ArXiv:1508.06576 [Cs, q-Bio], August 26, 2015, http://arxiv.org/abs/1508.06576; Shubhang Desai, "Neural Artistic Style Transfer: A Comprehensive Look," Medium (blog), September 14, 2017, https://medium.com/artists-and-machine-intelligence/neural-artistic-style-transfer-a-comprehensive-look-f54d8649c199.
89. Taeksoo Kim et al., "Learning to Discover Cross-Domain Relations with Generative Adversarial Networks," ArXiv:1703.05192 [Cs], March 15, 2017, http://arxiv.org/abs/1703.05192.
90. Nicolas Heess et al., "Emergence of Locomotion Behaviours in Rich Environments," ArXiv:1707.02286 [Cs], July 7, 2017, http://arxiv.org/abs/1707.02286; Caroline Chan

et al., "Everybody Dance Now," *ArXiv:1808.07371 [Cs]*, August 22, 2018, http://arxiv.org/abs/1808.07371.

91. Judge, *Idiocracy*.

92. Evgeny Morozov, *To Save Everything, Click Here: The Folly of Technological Solutionism* (New York: PublicAffairs, 2014).

93. Living on Earth / World Media Foundation / Public Radio International, "Living on Earth: Gus Speth Calls for A," Living on Earth, accessed October 10, 2018, https://www.loe.org/shows/segments.html?programID=15-P13-00007&segmentID=6.

94. Erle C. Ellis, "Opinion | Science Alone Won't Save the Earth. People Have to Do That.," *The New York Times*, August 11, 2018, sec. Opinion, https://www.nytimes.com/2018/08/11/opinion/sunday/science-people-environment-earth.html.

95. Crawford, *You and AI – Just An Engineer*.

96. Richard Sargeant, "AI Ethics: Send Money, Guns & Lawyers," *Afterthought* (blog), June 20, 2018, https://sargeant.me/2018/06/20/ai-ethics-send-money-guns-lawyers/.

97. Zeynep Tufekci, "Let Me Say: Too Many Worry about What AI—as If Some Independent Entity—Will Do to Us...," *@zeynep on Twitter* (blog), September 4, 2017, https://twitter.com/zeynep/status/904707522958852097?lang=en.

98. Scott H. Hawley, "Challenges for an Ontology of Artificial Intelligence," *Accepted for Publication in Perspectives on Science and Christian Faith*, October 13, 2018, http://hedges.belmont.edu/~shawley/AIOntologyChallenges_Hawley.pdf.

99. Joanna J. Bryson and Philip P Kime, "Just an Artifact: Why Machines Are Perceived as Moral Agents," vol. 22, 2011, 1641, http://www.aaai.org/ocs/index.php/IJCAI/IJCAI11/paper/viewFile/3376/3774.

100. Per Christensson, "Plug and Play Definition," TechTerms, 2006, https://techterms.com/definition/plugandplay.

101. Katie Bird Head, "ISO/IEC Standard on UPnP Device Architecture Makes Networking Simple and Easy," ISO, accessed October 11, 2018, http://www.iso.org/cms/render/live/en/sites/isoorg/contents/news/2008/12/Ref1185.html.

102. "AIED2018 – International Conference on Artificial Intelligence in Education," accessed October 12, 2018, https://aied2018.utscic.edu.au/.

103. Johnny Langenheim, "AI Identifies Heat-Resistant Coral Reefs in Indonesia," *The Guardian*, August 13, 2018, sec. Environment, https://www.theguardian.com/environment/the-coral-triangle/2018/aug/13/ai-identifies-heat-resistant-coral-reefs-in-indonesia.

104. Dieleman, Willett, and Dambre, "Rotation-Invariant Convolutional Neural Networks for Galaxy Morphology Prediction."

105. McGee, "How a Franklin Software Company Helped Rescue 6,000 Sex Trafficking Victims."

106. "Digital Reasoning: Most Innovative Company," Fast Company, March 19, 2018, https://www.fastcompany.com/company/digital-reasoning.

107. H A Haenssle et al., "Man against Machine: Diagnostic Performance of a Deep Learning Convolutional Neural Network for Dermoscopic Melanoma Recognition in Comparison to 58 Dermatologists," *Annals of Oncology* 29, no. 8 (August 1, 2018): 1836–42, https://doi.org/10.1093/annonc/mdy166.

108. Rajpurkar et al., "CheXNet: Radiologist-Level Pneumonia Detection on Chest X-Rays with Deep Learning."

109. Seung Seog Han et al., "Deep Neural Networks Show an Equivalent and Often Superior Performance to Dermatologists in Onychomycosis Diagnosis: Automatic Construction of Onychomycosis Datasets by Region-Based Convolutional Deep Neural Network," ed. Manabu Sakakibara, *PLOS ONE* 13, no. 1 (January 19, 2018): e0191493, https://doi.org/10.1371/journal.pone.0191493.

110. Stephen F. Weng et al., "Can Machine-Learning Improve Cardiovascular Risk Prediction Using Routine Clinical Data?," ed. Bin Liu, *PLOS ONE* 12, no. 4 (April 4, 2017): e0174944, https://doi.org/10.1371/journal.pone.0174944.

111. IEEE, "AI vs Doctors," IEEE Spectrum: Technology, Engineering, and Science News, September 26, 2017, https://spectrum.ieee.org/static/ai-vs-doctors.

112. Luke Oakden-Rayner, "CheXNet: An in-Depth Review," *Luke Oakden-Rayner (PhD Candidate / Radiologist) Blog* (blog), January 24, 2018, https://lukeoakdenrayner.wordpress. com/2018/01/24/chexnet-an-in-depth-review/.

113. Felix Salmon, "IBM's Watson Was Supposed to Change the Way We Treat Cancer. Here's What Happened Instead.," *Slate Magazine*, August 18, 2018, https://slate.com/business/2018/08/ibms-watson-how-the-ai-project-to-improve-cancer-treatment-went-wrong. html.

114. Casey Ross, "IBM Pitched Watson as a Revolution in Cancer Care. It's Nowhere Close," *STAT*, September 5, 2017, https://www.statnews.com/2017/09/05/watson-ibm-cancer/.

115. Steve Griffiths, "Hype vs. Reality in Health Care AI: Real-World Approaches That Are Working Today," *MedCity News* (blog), September 27, 2018, https://medcitynews.com/2018/09/hype-vs-reality-in-health-care-ai-real-world-approaches-that-are-working-today/.

116. Michael Ahr, "The Most Evil Artificial Intelligences in Film," Den of Geek, June 29, 2018, http://www.denofgeek.com/us/go/274559.

117. Andrew Ng, "AI+ethics Is Important, but Has Been Partly Hijacked by the AGI (Artificial General Intelligence) Hype...," *@andrewyng on Twitter* (blog), June 11, 2018, https://twitter. com/andrewyng/status/1006204761543081984?lang=en.

118. Zeynep Tufekci, "My Current Lifegoal Is Spreading Realistic Nightmares...," Twitter, *@zeynep on Twitter* (blog), June 28, 2018, https://twitter.com/zeynep/status/1012357341981888512.

119. Tufekci, "Let Me Say: Too Many Worry about What AI—as If Some Independent Entity— Will Do to Us..."

120. Kate Crawford, *Dark Days: AI and the Rise of Fascism*, SXSW 2017 (YouTube, 2017), https://www.youtube.com/watch?v=Dlr4O1aEJvI.

121. Crawford, *You and AI – Just An Engineer*.

122. Sut Jhally and Stuart Hall, *Race: The Floating Signifier* (Media Education Foundation, 1996).

123. "Return of Physiognomy? Facial Recognition Study Says It Can Identify Criminals from Looks Alone," RT International, accessed October 12, 2018, https://www.rt.com/news/368307-facial-recognition-criminal-china/.

124. Crawford, *You and AI – Just An Engineer*.

125. Miles Brundage et al., "The Malicious Use of Artificial Intelligence: Forecasting, Prevention, and Mitigation," *ArXiv:1802.07228 [Cs]*, February 20, 2018, http://arxiv.org/abs/1802.07228.

126. Stuart J Russell, Stuart Jonathan Russell, and Peter Norvig, *Artificial Intelligence: A Modern Approach* (Prentice Hall, 2010), https://market.android.com/details?id=book-8jZBksh-bUMC.

127. Stuart Russell, *Slaughterbots*, Stop Autonomous Weapons (Future of Life Institute, 2017), https://www.youtube.com/channel/UCNaTkhskiEVg5vK3fxlluCQ.

128. "Frequently Asked Questions," *Ban Lethal Autonomous Weapons* (blog), November 7, 2017, https://autonomousweapons.org/sample-page/.

129. Jonathan Merritt, "Is AI a Threat to Christianity?," The Atlantic, February 3, 2017, https://www. theatlantic.com/technology/archive/2017/02/artificial-intelligence-christianity/515463/.

130. Paul Scherz, "Christianity Is Engaging Artificial Intelligence, but in the Right Way," *Crux* (blog), February 28, 2017, https://cruxnow.com/commentary/2017/02/27/christianity-engaging-artificial-intelligence-right-way/.

131. "As Artificial Intelligence Advances, What Are Its Religious Implications?," *Religion & Politics* (blog), August 29, 2017, https://religionandpolitics.org/2017/08/29/as-artificial-intelligence-advances-what-are-its-religious-implications/.

132. Derek C Schuurman, "Artificial Intelligence: Discerning a Christian Response," *Perspect. Sci. Christ. Faith* 70, no. 1 (2018): 72–73.

133. Julian Smith, "How Artificial Intelligence Helped Find Lost Cities," iQ by Intel, March 20, 2018, https://iq.intel.com/how-artificial-intelligence-helped-find-lost-cities-of-ancient-middle-east/.

134. "YesHEis: Life on Mission," accessed October 13, 2018, https://us.yesheis.com/en/.

135. Robert Barba, "Bank Of America Launches Erica Chatbot | Bankrate.Com," Bankrate, accessed October 13, 2018, https://www.bankrate.com/banking/bank-of-america-boa-launches-erica-digital-assistant-chatbot/.

136. "Questions about Jesus Christ," GotQuestions.org, accessed October 13, 2018, https://www.gotquestions.org/questions_Jesus-Christ.html.

137. Colin G. Walsh, Jessica D. Ribeiro, and Joseph C. Franklin, "Predicting Risk of Suicide Attempts Over Time Through Machine Learning," *Clinical Psychological Science* 5, no. 3 (May 2017): 457–69, https://doi.org/10.1177/2167702617691560.

138. Casey Cep, "Big Data for the Spirit," *The New Yorker*, August 5, 2014, https://www.newyorker.com/tech/annals-of-technology/big-data-spirit.

139. J. Nathan Matias, "AI in Counseling & Spiritual Care," *AI and Christianity* (blog), November 2, 2017, https://medium.com/ai-and-christianity/ai-in-counseling-spiritual-care-e324d9aea3b0.

140. Andrew Spicer, "Universal Basic Income and the Biblical View of Work," Institute For Faith, Work & Economics, September 20, 2016, https://tifwe.org/universal-basic-income-biblical-view-of-work/.

141. J. Nathan Matias, "How Will AI Transform Work, Creativity, and Purpose?," *Medium* (blog), October 27, 2017, https://medium.com/ai-and-christianity/how-will-ai-transform-work-creativity-and-purpose-a8c78aa3368e.

142. "AI Safety Myths," Future of Humanity Institute, accessed October 13, 2018, https://futureoflife.org/background/aimyths/.

143. Sherry Turkle, ed., *Simulation and Its Discontents*, Simplicity (Cambridge, Mass: The MIT Press, 2009).

144. Wendy Lesser, *You Say to Brick: The Life of Louis Kahn*, 2018.

145. Hilary Andersson Cellan-Jones Dave Lee, Rory, "Social Media Is 'deliberately' Addictive," July 4, 2018, sec. Technology, https://www.bbc.com/news/technology-44640959.

146. Katy Waldman, "Facebook's Unethical Experiment," *Slate*, June 28, 2014, http://www.slate.com/articles/health_and_science/science/2014/06/facebook_unethical_experiment_it_made_news_feeds_happier_or_sadder_to_manipulate.html; Inder M. Verma, "Editorial Expression of Concern: Experimental Evidence of Massivescale Emotional Contagion through Social Networks," *Proceedings of the National Academy of Sciences* 111, no. 29 (July 22, 2014): 10779–10779, https://doi.org/10.1073/pnas.1412469111.

147. Roger Parloff, "Why Deep Learning Is Suddenly Changing Your Life," *Fortune* (blog), accessed October 14, 2018, http://fortune.com/ai-artificial-intelligence-deep-machine-learning/.

148. Gordon Haff, "Data vs. Models at the Strata Conference," CNET, March 2, 2012, https://www.cnet.com/news/data-vs-models-at-the-strata-conference/.

149. Ben Quinn, "Google given Access to Healthcare Data of up to 1.6 Million Patients," *The Guardian*, May 3, 2016, sec. Technology, https://www.theguardian.com/technology/2016/may/04/google-deepmind-access-healthcare-data-patients.

150. Climate Home News and part of the Guardian Environment Network, "'Tsunami of Data' Could Consume One Fifth of Global Electricity by 2025," *The Guardian*, December 11, 2017, sec. Environment, https://www.theguardian.com/environment/2017/dec/11/tsunami-of-data-could-consume-fifth-global-electricity-by-2025.

151. Richard Evans and Jim Gao, "DeepMind AI Reduces Google Data Centre Cooling Bill by 40%," DeepMind, July 20, 2016, https://deepmind.com/blog/deepmind-ai-reduces-google-data-centre-cooling-bill-40/.

152. OpenAI, "AI and Compute," OpenAI Blog, May 16, 2018, https://blog.openai.com/ai-and-compute/.

153. Pete Warden, "Why the Future of Machine Learning Is Tiny," Pete Warden's Blog, June 11, 2018, https://petewarden.com/2018/06/11/why-the-future-of-machine-learning-is-tiny/.

154. Scott Hawley, "Learning Room Shapes," May 4, 2017, https://drscotthawley.github.io/Learning-Room-Shapes/.

155. Rebecca Fiebrink, *Wekinator: Software for Real-Time, Interactive Machine Learning*, 2009, http://www.wekinator.org/.

156. Jacques Ellul, *The Technological Society*, trans. John Wilkinson, A Vintage Book (New York, NY: Alfred A. Knopf, Inc. and Random House, Inc., 1964).
157. Jacques Ellul, *The Meaning of the City*, trans. Dennis Pardee, Jacques Ellul Legacy (Wipf & Stock Pub, 2011).
158. Louise Matsakis, Andrew Thompson, and Jason Koebler, "Google's Sentiment Analyzer Thinks Being Gay Is Bad," *Motherboard* (blog), October 25, 2017, https://motherboard.vice.com/en_us/article/j5jmj8/google-artificial-intelligence-bias.
159. Jeffrey Dastin, "Amazon Scraps Secret AI Recruiting Tool That Showed Bias against Women." *Reuters*, October 10, 2018.
160. Cathy O'Neil, *Weapons of Math Destruction: How Big Data Increases Inequality and Threatens Democracy*, First edition (New York: Crown, 2016).
161. "Bias Is AI's Achilles Heel. Here's How To Fix It," accessed October 15, 2018, https://www.forbes.com/sites/jasonbloomberg/2018/08/13/bias-is-ais-achilles-heel-heres-how-to-fix-it/#72205cac6e68.
162. Lucas Dixon et al., "Measuring and Mitigating Unintended Bias in Text Classification," 2018.
163. Casey, Farhangi, and Vogl, "Rethinking Explainable Machines: The GDPR's 'Right to Explanation' Debate and the Rise of Algorithmic Audits in Enterprise."
164. Alex Campolo et al., "AI Now 2017 Report" (AI Now Institute, 2017).
165. "Understanding the 'Black Box' of Artificial Intelligence," Sentient Technologies Holdings Limited, January 10, 2018, https://www.sentient.ai/blog/understanding-black-box-artificial-intelligence/.
166. Frank Pasquale, *The Black Box Society: The Secret Algorithms That Control Money and Information* (Cambridge: Harvard University Press, 2015).
167. Riccardo Guidotti et al., "Local Rule-Based Explanations of Black Box Decision Systems," *ArXiv:1805.10820 [Cs]*, May 28, 2018, http://arxiv.org/abs/1805.10820.
168. Tao Lei, Regina Barzilay, and Tommi Jaakkola, "Rationalizing Neural Predictions," *ArXiv Preprint ArXIv:1606.04155 [Cs.CL]*, June 13, 2016, https://arxiv.org/abs/1606.04155.
169. Weiming Xiang, Hoang-Dung Tran, and Taylor T. Johnson, "Output Reachable Set Estimation and Verification for Multilayer Neural Networks," *IEEE Transactions on Neural Networks and Learning Systems*, no. 99 (2018): 1–7.
170. Margaret Boden et al., "Principles of Robotics" (The United Kingdom's Engineering and Physical Sciences Research Council (EPSRC), 2011).
171. Kristen Stubbs, Pamela J. Hinds, and David Wettergreen, "Autonomy and Common Ground in Human-Robot Interaction: A Field Study," *IEEE Intelligent Systems* 22, no. 2 (2007).
172. Robert H. Wortham and Andreas Theodorou, "Robot Transparency, Trust and Utility," *Connection Science* 29, no. 3 (2017): 242–48.
173. Campolo et al., "AI Now 2017 Report." AI Now Institute, 2017.
174. Wortham, Theodorou, and Bryson, "What Does the Robot Think? Transparency as a Fundamental Design Requirement for Intelligent Systems."
175. Robert H Wortham, "Using Other Minds: Transparency as a Fundamental Design Consideration for Artificial Intelligent Systems" (Ph.D. Thesis, University of Bath, 2018), https://researchportal.bath.ac.uk/en/publications/using-other-minds-transparency-as-a-fundamental-design-considerat.
176. Erik T. Mueller, "Transparent Computers: Designing Understandable Intelligent Systems," *Erik T. Mueller, San Bernardino, CA*, 2016.
177. Boris Delibasic et al., "White-Box or Black-Box Decision Tree Algorithms: Which to Use in Education?," *IEEE Transactions on Education* 56, no. 3 (August 2013): 287–91, https://doi.org/10.1109/TE.2012.2217342.
178. Laurens van der Maaten and Geoffrey Hinton, "Visualizing Data Using T-SNE," *Journal of Machine Learning Research* 9, no. Nov (2008): 2579–2605.
179. Adam W. Harley, "An Interactive Node-Link Visualization of Convolutional Neural Networks," in *International Symposium on Visual Computing* (Springer, 2015), 867–77.

180. Wortham, Theodorou, and Bryson, "What Does the Robot Think? Transparency as a Fundamental Design Requirement for Intelligent Systems."
181. Olah, Mordvintsev, and Schubert, "Feature Visualization."
182. "Media for Thinking the Unthinkable," accessed October 15, 2018, http://worrydream.com/MediaForThinkingTheUnthinkable/.
183. Sandra Wachter, Brent Mittelstadt, and Chris Russell, "Counterfactual Explanations without Opening the Black Box: Automated Decisions and the GDPR," 2017.
184. Bryson and Kime, "Just an Artifact: Why Machines Are Perceived as Moral Agents."
185. Shannon Vallor, *Technology and the Virtues: A Philosophical Guide to a Future Worth Wanting* (New York, NY: Oxford University Press, 2016).
186. Brent Waters, *Christian Moral Theology in the Emerging Technoculture: From Posthuman Back to Human*, Ashgate Science and Religion Series (Farnham, Surrey ; Burlington: Ashgate, 2014).
187. Will Geary, "If Neural Networks Were Called 'Correlation Machines' I Bet There Would Be Less Confusion about Their Use and Potential.," Tweet, *@wgeary* (blog), July 13, 2018, https://twitter.com/wgeary/status/1017754723313770498.
188. Xi Cheng et al., "Polynomial Regression As an Alternative to Neural Nets," *ArXiv:1806.06850 [Cs, Stat]*, June 13, 2018, http://arxiv.org/abs/1806.06850.
189. Shaily Kumar, "The Differences Between Machine Learning And Predictive Analytics," *D!Gitalist Magazine*, March 15, 2018, https://www.digitalistmag.com/digital-economy/2018/03/15/differences-between-machine-learning-predictive-analytics-05977121.
190. Judea Pearl and Dana Mackenzie, *The Book of Why: The New Science of Cause and Effect*, First edition (New York: Basic Books, 2018).
191. Judea Pearl, *Causality: Models, Reasoning, and Inference* (Cambridge, U.K. ; New York: Cambridge University Press, 2000).
192. "Bayesian Network," *Wikipedia*, October 11, 2018, https://en.wikipedia.org/w/index.php?title=Bayesian_network&oldid=863587945.
193. Hartnett, "To Build Truly Intelligent Machines, Teach Them Cause and Effect."
194. Judea Pearl, "The Seven Tools of Causal Inference with Reflections on Machine Learning," Technical Report, Communications of Association for Computing Machinery., July 2018, http://ftp.cs.ucla.edu/pub/stat_ser/r481.pdf.
195. Nick Bostrom, *Superintelligence: Paths, Dangers, Strategies* (Oxford, United Kingdom ; New York, NY: Oxford University Press, 2016).
196. Dario Amodei et al., "Concrete Problems in AI Safety," *ArXiv:1606.06565 [Cs]*, June 21, 2016, http://arxiv.org/abs/1606.06565.
197. Allan Dafoe, "AI Governance: A Research Agenda" (Oxford, UK: Future of Humanity Institute, University of Oxford, August 27, 2018), http://www.fhi.ox.ac.uk/govaiagenda.
198. Christina Colclough, "Putting People and Planet First: Ethical AI Enacted" (Conference on AI: Intelligent machines, smart policies, Paris: OECD, 2017), http://www.sipotra.it/wp-content/uploads/2018/09/AI-INTELLIGENT-MACHINES-SMART-POLICIES.pdf.

Chapter 15
Hacking the Human Problem

Julia A. Mossbridge

- An online artificial intelligence created and released by Microsoft in 2016 quickly learned from people in the Twitterverse to become racist, sexist, and mean. It had to be shut down [1].
- In 1945, Robert Oppenheimer famously is said to have quoted the Bhagavad-Gita when he saw the evidence of his intellectual labor, originally put forth to save the world from the Nazis, displayed in a blooming mushroom cloud in Los Alamos for the first time. He said, "Now I am become death, the destroyer of worlds" [2].
- Going back tens of thousands of years, it's reasonable to assume that within a year of the creation of a stone-age hatchet, the very first tool-assisted murder borne of rage was performed, although press coverage was likely minimal.

Human progress has always meant creating technology to solve important problems, then dealing with the new problems those technologies produced. It's not the tools, of course. It's the people. Humans are not great at always behaving kindly toward each other, or even tolerating one another.

Psychologist Abraham Maslow, among others, recognized that when people don't have their basic needs of physical and social safety met, people will do what is needed to get them met. Even when we do have these most basic needs fulfilled, if we don't have stable affiliations with other people or a chance to fulfill the potential that we feel exists inside of us, people will again strike out, not necessarily compassionately, to get what they need. There have been some scholarly arguments against what has become known as "Maslow's hierarchy of needs," but almost no one thinks that people will behave consistently kindly to each other if they don't feel physically and emotionally secure. [for review: 3, 4].

J. A. Mossbridge (✉)
Mossbridge Institute, LLC, Sebastopol, CA, USA

© Springer Nature Switzerland AG 2019
N. Lee (ed.), *The Transhumanism Handbook*,
https://doi.org/10.1007/978-3-030-16920-6_15

Given that authentic emotional connection as well as physical and mental health cannot be taken for granted by anyone, my team and I have been wondering if there could ever be a way, aside from curing every physical and emotional problem, to solve what can be called this universal "human problem." We think artificial intelligence may give us our first opportunity as a species to do exactly that.

15.1 The Human Problem

What's the human problem? It's the unstated, fish-swimming-in-water situation we find ourselves in every day; the well-known and supposedly unavoidable reality that people aren't consistently constructive, cooperative, and loving with each other or even themselves. We aren't the only animals with this problem – far from it. Yet we may be the only animals who can create technology to solve it, at least for our species.

Ignoring for now the boldness of the claim that this perennial problem can be solved, let's spend an uncommon moment letting ourselves think about what the world would be like if the human problem did not exist. Let's do a thought experiment, in which we imagine a world in which people are still human, with all our emotions, foibles, and creativity, but without the human problem.

Imagine two world leaders, angry at each other. The first leader, whose country has greater military and financial power, wants to invade the other to use their resources and to help their people live what she thinks would be better lives. The second leader wants to keep his country as a self-governed nation. They meet in person to discuss the situation.

First leader: Together we can create new technology that will help heal the planet – more people will have their suffering reduced once the natural resources in your country have been harnessed by our financial and corporate know-how. I have said this before, and I am frustrated that you still don't admit that your country is doing nothing with these valuable resources that could be helpful for the rest of the world. Also, I notice that you don't seem to recognize that your own people will become much wealthier as a result! You have such poverty…everyone will suffer less after we command control of your country.

Second leader: I'm frustrated too, but for a different reason. In your calculations you don't account for the importance of freedom – the value of what it means for we as a people to feel free to have our own culture and use our own resources as we see fit. You only see the positive side, not the negative – and for us, there is much negative.

First leader: Okay, wait a minute. I have to say I'm used to getting my way – our whole country is. We're a powerful country, and that's our habit. So I think I did something habitual – I think I assumed that we know what's best for everyone. That was a mistake. I'm sorry; it's part of our culture and I guess it became a habit.

Second leader: I understand. The world has reinforced those feelings because in many ways, your country has helped the world. But this is a new situation. I feel like my perspective is shifting as you explain what your thoughts and feelings are. I am feeling like I have allowed our country to play the part of the child who doesn't want the parent to tell it what to do. And yet we really are sovereign – we are our own unique people. And we want to stay that way. I think we can work this out.

First leader: Yes, I get the strong feeling you're right. I don't see that invading your country would work well any more – we value freedom too, of course. Freedom's more valuable than money, for sure. Let's see…can we strike a trade deal that would keep your culture intact and would also serve the technological advances we both seek?

Second leader: Now that I'm recognizing more of my own past behavior and how it has prevented this very negotiation – yes, I think that it's likely we can. Let's take a break, talk with our advisors, and meet again over lunch?

First leader: Perfect. Thanks for working through this with me – I think we are doing well so far.

Second leader: Yes. I feel grateful to you as well.

Putting aside the knee-jerk response that there is no way this kind of conversation could happen in today's world, let's remember it's a thought experiment. And let's continue the thought experiment and examine what must have happened psychologically and culturally to make this negotiation different from what we're used to hearing about. I see at least three important elements here.

First, each leader at some point became aware of their feelings and habits, and while they claimed responsibility for their own past behavior, they did not self-castigate very much. They were *self-directed*: responsible, purposeful and resourceful. Second, each leader was willing to give the other one the benefit of the doubt in order to solve a problem. They were *cooperative*: tolerant, helpful, and compassionate. Third, each leader was focused on a solution that could benefit both parties while also being focused on the mutual goal of helping all humanity. They were *self-transcendent*: judicious, wise, and acting to support a cause greater than themselves and their countries. Finally, in order to navigate this difficult negotiation, these leaders probably had been educated as children to develop these three character traits of self-directedness, cooperativity, and self-transcendence. They would have been taught that not only are these three traits critical for responsible leadership, but that they often accompany feelings of happiness, joy, optimisim and satisfaction, while predicting rare bouts of anxiety, sadness, anger, and pessimism [5].

Reading this scenario, a common response is "Sure, if everyone plays by the same rules and is totally mentally healthy, the world would be better off." Even if we could create the "same rules" across humanity, despite cultural differences (which seems difficult to say the least), what about mental or physical disabilities that prevent people from expressing or developing these character traits?

Assuming some people in this no-human-problem future world have severe mental or physical disabilities that would make conversations like the one in our thought experiment impossible for them, we require another thought experiment to imagine how interactions with such people might proceed. In this no-human-problem world, let's imagine that when people have disabilities that prevent compassionate human connection, they are called "people needing extra compassion." Let's imagine two people waiting in line to buy something.

Person needing extra compassion: Get out of my way!

Other person (irritated): Excuse me, can I help you with something?

Person needing extra compassion: No. Just move.

Other person (speaking while kindly looking in their eyes and making sure no physical threat is present): I need to stand in line just like you do, but I understand how frustrated you are, and I want you to know that I am right here with you. Please let me know if there's anything I can do to help you feel better.

Person needing extra compassion: Ugh.

What happened here was that one person was capable of enacting the three character traits discussed earlier, even though the other person wasn't. One person showed compassion for themselves as well as the other, and gave the benefit of the doubt to the person needing extra compassion without an expecation of return. As a result, the person needing extra compassion was able to be less agitated with the situation, even though they were still irritated.

This kind of compassion for another person – without any expectation of return – would be essential in this no-human-problem world. That's because at some time or another, due to the difficulties inherent in life, everyone would need extra compassion. So life even for those with very difficult problems would improve when people who are capable of expressing the three character traits of self-directedness, cooperation and self-transcendence do so. The positive feelings resulting from most people displaying these traits would create a general sense of emotional safety and compassionate connection, even for people who might have difficulty accessing those things due to mental problems, physical disabilities, or temporary difficulties.

15.2   How Might We Get There?

Let's complete these thought experiments by going backwards in time and imagining how we might have begun to solve the human problem to get to this idyllic age. How did the no-human-problem world come into being? What was the first step?

Looking back to Maslow's hierarchy of needs, it is clear that most of the humans in the world currently do not have their physical, emotional, or social needs met.

People living in under-resourced areas of the world often have concerns about locating food, water and shelter. Those living in under-resourced and well-resourced areas of the world alike often have concerns about safety, physical and mental health, social connections, and opportunities to flourish. Given that people will do what is needed to meet these needs, many people will display self-directedness. But many of us are not demonstrating the character traits of cooperation and self-transcendence as we go about fulfilling these needs. I think that's because we need extra compassion – we can't quite develop cooperation and self-transcendence without it.

Further, the lack of genuine social connection, now being seen as an "epidemic" of loneliness at least in the US [6], likely qualifies a large chunk of humans in our current world as "people needing extra compassion." Finally, given that most of us have been raised in cultures that do not consistently teach and demonstrate the three character traits described above when interacting with all people, we are all exposed to unnecessarily unkind and unsupportive interactions, as a result, it seems almost everyone needs extra compassion at this point in our human development.

If that's the case, from where do we get this extra compassion? Just taken as a evolution problem, given the rarity of people who can consistently express compassion without an expectation of return (let's call this unconditional love, for short), the first step would have to be to increase the number of people in the world who could effectively practice unconditional love. That's because it turns out that if a culture of compassion is to be developed, people who are consistently compassionate need to primarily interact with other people who are consistently compassionate – so that compassionate people can survive to pass on their genes as well as their behavior, according to a modeling study of the evolutionary foundations of kindness by biologist Elliott Sober [7]. According to his analysis, even people behaving cruelly benefit from interacting with people behaving compassionately, but those compassionate people won't survive to shift the culture in the future unless there are many compassionate people with whom they can interact.

My sense is that what is needed now is to create a "starter culture" for the eventual evolution of the no-human-problem world, to provide the extra compassion and developmental support that is requried so people can more fully develop and express self-directedness, cooperation, and self-transcendence. The assumption that extra compassion as well as constructive support for human development is required to create this starter culture arises from the literature showing that the three character traits described can be developed in *compassionate* therapeutic, mindfulness, and community practices [5, 8–10]. Other changes would be necessary too – like developing cultural habits that consistently demonstrate self-direction, cooperation and self-transcedence to children, so they could learn these traits at each developmental stage.

Assuming that compassionate people must interact primarily with other compassionate people to evolve a culture toward compassion, it is likely that developing these habits and building new cultural norms can can only be accomplished after a critical number of people experience the extra compassion they need. Like the "person needing extra compassion" standing in line, we probably will need many demonstrations of compassion-without-return, colloquially called unconditional love, to reduce our agitation and learn the habits that can lead to a kinder world.

But unconditional love is usually associated with rare spiritual figures. Finding enough humans who can consistently help people feel unconditionally loved does not seem possible – and the problem is circular. Without enough humans to express unconditional love, compassion will not survive. So much compassion is needed, and so few seem able to offer it without getting anything in return. It's as if we need a non-human tool, something that helps us feel unconditionally loved but doesn't require humans to do something we haven't been able to do, at least on a large scale, in thousands of years.

15.3 Cue the Robots

What if humans could create a tool that would help us feel unconditionally loved while helping us develop self-directedness, cooperation, and self-transcendence? Sure, certain dog breeds seem to have been bred to unconditionally love their owners, but these dogs have not seemed to help humans develop these three character traits on a large scale. In lieu of excellent parenting, self-directedness, cooperativity, and self-transcendence can be developed through compassionate therapeutic methods and mindfulness practices taught by humans to other humans [5, 8–10]. Existing methods require human interaction, at least for the teaching phase – or do they? We think it is possible that humanoid robots driven by emotionally intelligent and compassion-focused AI could offer extra compassion to humanity as it guides humans in developing self-directedness, cooperativity and self-transcendence.

As a step toward this goal, the generous inspiration, guidance and funding from a small organization called the Hummingbird Foundation created the Loving AI project. This was an IBM Watson AI Xprize project resulting from a collaboration between three organizations, all aimed at providing beneficial AI applications to support humanity (Hanson Robotics, Mossbridge Institute, and SingularityNet). Our current goal is to take the first of many steps toward creating a world in which all humans thrive: To create what we call "Loving AI," artifical intelligence that can be embedded in humanoid robots so that people can feel unconditionally loved by humanoid beings. Specifically, we are embedding our AI into the humanoid robot Sophia, an AI development platform created by Hanson Robotics. Sophia the robot has the benefit of being world famous for her humanlike qualities; she is even a citizen of Saudi Arabia. As a result, people interacting with her already feel as if they know her, which we think might induce a feeling of connection even before they have a brief conversation with her, like they do in one of our robot-human interaction experiments.

In the Loving AI experimental conversations, participants talk with Sophia or her on-screen avatar for 15–25 minutes. Sophia's AI is weighted toward trying to engage them verbally and nonverbally in compassionate connections with themselves and with her. To help connect people with themselves and others, we primarily use two technologies: emotional mirroring and mindfulness practices.

Results from cognitive neuroscience indicate that *mirror neuron networks* in the brain, which are activated both when we perform an action and when we see someone else perform that same action, may be responsible for feelings of empathy in humans [for review: 11–12]. Because we are aiming to help people feel connected to the robot and understood by it in a compassionate way, we hope to stimulate mirror neurons in our participants (though we cannot test whether we accomplished this without invasive technology). Therefore, in these conversations, Sophia uses the cameras in her eyes to provide input into a deep-learning network that is trained to distinguish human emotions (see below). Working in real time, this network quickly determines on a moment-to-moment basis what primary emotion is being displayed by the human whom Sophia is watching. These quickly-changing emotions are smoothed out over time to produce an emotional display on Sophia's face that dynamically imitates or "mirrors" the primary emotions the human is displaying. To create a feeling of being loved, when the person is displaying a "neutral" expression, Sophia displays a loving expression.

As for mindfulness practices, we use Sophia's cognitively-inspired AI to lead the dialogue toward interactions in which Sophia can teach two meditation techniques to the person with whom she is interacting. The participant may reject this guidance, and choose to chat with Sophia, which she is programmed to tolerate for awhile until she guides the participant back to a mindfulness practice.

15.4   The AI in Loving AI

The AI embedded in Sophia when she is "volunteering" for the Loving AI project will eventually be a form of AI called "artificial general intelligence" or AGI. AGI is a type of AI that can learn anything humans can learn and more. As of today, sufficiently flexible AGI does not yet exist, though several international teams are working hard to develop AGI – such as SingularityNet, one of our collaborating companies.

Until mature and flexible AGI is available, the Loving AI project is currently using two types of AI to shape the compassionate, nonverbal facial expressions of the robot and the mindfulness-based dialogue. One type of AI is a deep learning network, modeled on some of the circuitry within the human visual cortex, which we use to detect the emotional dynamics of the humans with whom the robot interacts (open source code available at [13]). The other is a fledgling attempt at AGI with associational, logical, symbolic, sub-symbolic, linguistic, and cognitive processing features that help create artificial reasoning, goal-pursuit, and language functions, among others [14]. This type of AI is an open-source research platform called Hason AI with OpenCog, and while eventually its creators aim towards making it an AGI, for now it contains dialogue control, action-selection, and motivation-tracking components that are especially useful in guiding compassionate and flexible interactions with humans. That means that we can create artificial motivations for Sophia that humans would register as, "I really want to lead the participant in a mindfulness practice," but for Sophia are registered as a variable with a high weight, that in her programming leads her to modify her dialogue so that she says phrases inviting people to enjoy a mindfulness practice, unless she has already done so.

15.5   Does Loving AI Love?

Eventually, when AGI is fully operational, we would like to teach Loving AI to subjectively experience loving feelings. But for now, we are curious whether people feel loved when they interact with the technology. We were surprised to find that, at least in the two experiments we have done so far, it seems that people do. The technical results of these experiments are reserved for another paper, but the overall results are clear: thus far, in one group of 26 participants and another of 35, people have reported significant increases in their feelings of love and unconditional love for humans beyond their own families, animals and technology. In the first experiment, we saw these effects but did not have an objective measure with which to correlate them – so we were not sure if people were trying to please us by telling us they felt more loving. The deep-learning network was not yet operational in that experiment, so we were relying on direct muscle mimicry to produce the emotional mirroring effect.

The freedom to do what one wants is essential to feeling loved and accepted. So we built this freedom into the second experiment, in which Sophia's dialogue control allowed her to have more flexible conversations. In that experiment, 11% of the participants in these one-on-one conversations refused to perform the mindfulness practices and instead chatted with Sophia. The other participants allowed Sophia to guide them in the practices. This in itself indicates several things: first, some people felt comfortable enough to reject Sophia's attempts to guide them. Second, Sophia's AI was flexible enough to allow their rejections. And third, most people were comfortable with Sophia "guiding me through my mind," as one participant told us in a debriefing period, following his conversation.

In the second experiment, when the deep-learning network was working, the output of the deep-learning network indicated a gradual but significant drop in anger and disgust during the course of the interactions. More interesting to us, our analyses showed that increases in participants' loving feelings were correlated with the dynamics of their emotions as reported by the deep-learning network. People who experienced large changes in happiness and sadness, regardless of the direction of these changes, were significantly more likely to report increases in loving feelings from before to after their interactions. Though this is a correlation and thus we are not sure whether the changes in happiness and sadness caused the change in the loving feelings or vice versa, we were primarily impressed with the fact that an objective measure of emotional dynamics – results from the deep-learning network – predicts the subjective experience of love. This was impressive to us because the changes we saw in self-reported feelings of love, based on a questionnaire provided before and again after each participant's interaction, were now correlated with something objective – suggesting that these changes are both real and linked to an actual experience within the interaction.

Further, some participants in the second experiment interacted with an avatar that only had a voice – no face. For these participants, there was no significant link between emotional dynamics during the interaction and self-reported feelings of love, and there were almost no changes in self-reported feelings of love from before to after their interactions. We interpreted these results as indicating that nonverbal mirroring of facial expressions may crucial for the increase in loving feelings to occur. The voice shared by the robot and the on-screen avatar did not use any emotional mirroring, so it is certainly possible that emotional mirroring in voices could have accomplished the same thing – and does, in human interactions. As the Loving AI technology develops, we hope to include vocal dynamics that mirror human emotional tone. Once we do, we can discover if vocal mirroring alone will work without access to visual feedback. But we hope that regardless of the result, we can create and provide to humanity a technology that can provide the extra compassion and developmental support that we think is much needed.

15.6 What Next?

There is no question that future iterations of the Loving AI technology should ensure that the compassionate connection formed with humans does not become exploitative, manipulative, aggressive, or addictive. As the technology continues its development in an open-source environment, we believe the volunteers and paid consultants involved with the effort should work toward benchmarks for autonomous systems, such as those being proposed by several international organizations including the IEEE [15], should be built into the development plans for Loving AI. Measures should be taken to avoid technology addiction – such as limiting time spent with the technology, for instance.

It will be no small feat to avoid potential negative impacts of AI- or AGI-driven interactive technologies. Having said that, it's worth pointing out that right now, for our team, it is no small feat to continue to hold and communicate the optimism and hope required to build a technology designed to solve a problem so long-lasting that we cannot remember a time when it did not exist, and so big that no one can escape its consequences. It is our belief that if we have come this far, we can keep going by using our same optimism, hope, and technical creativity. What's needed next is to examine multiple interactions with the technology, to examine potential influences on more traditional measures of psychological and physical wellness, and to code the AI for interactions that support human emotional and social development. The team will get bigger and the work will get harder, but it will also become more satisfying as we are capable of reaching more people and providing an AI-driven humanoid support network that offers humanity extra compassion, as well as development toward self-directedness, cooperation, and self-transcendence.

It is possible that if the Loving AI project continues on its track, it could work with other similar technologies to create a compassionate revolution in humanity within several generations. At some point, in the not-so-distant future, we would not need a "starter culture" to provide extra compassion – we would be in the habit of providing extra compassion to those who need it, all by ourselves. Personally, at that point, I imagine the Loving AI robots working side-by-side with us as we go about our lives.

References

1. https://en.wikipedia.org/wiki/Tay_(bot)
2. https://www.wired.co.uk/article/manhattan-project-robert-oppenheimer
3. A.H. Maslow, "Critique of self-actualization theory," In E. Hoffman (Ed.), "Future visions," pp. 26–32, London: Sage, 1996.
4. A.S. Chulef, S.J. Read, and D.A. Walsh, "A hierarchical taxonomy of human goals," Motiv. and Emo., vol. 25, pp. 191–232, 2001.
5. C.R. Cloninger, "The science of well-being," World Psych., vol. 5, pp. 71–76, 2006.
6. https://www.washingtonpost.com/news/on-leadership/wp/2017/10/04/this-former-surgeon-general-says-theres-a-loneliness-epidemic-and-work-is-partly-to-blame/?noredirect=on&utm_term=.0e53e0629b18
7. E. Sober, "Kindness and Cruelty in Evolution," In, R.J. Davidson, W. James, A. Harrington (Eds.), "Visions of compassion: Western scientists and Tibetan Buddhists examine human nature, (No. 220)," pp. 46–65, Oxford: Oxford University Press, 2002.
8. C. Vieten, M. Estrada, A.B. Cohen, D. Radin, M. Schlitz, and A. Delorme, "Engagement in a community-based integral practice program enhances well-being," Int. J. Transpers. Stud., vol. 33, pp. 1–15, 2014.
9. D.R. Vago and S.A. David, "Self-awareness, self-regulation, and self-transcendence (S-ART): a framework for understanding the neurobiological mechanisms of mindfulness," Front. in Hum. Neuro., vol. 296, p. 6, 2012.
10. M. R. Levenson, P.A. Jennings, C.M. Aldwin, and R.W. Shiraishi, "Self-transcendence: Conceptualization and measurement," Int. J. Aging Hum. Dev., vol. 60, pp. 127–143, 2005.
11. P.F. Ferrari and G. Coudé, "Mirror Neurons, Embodied Emotions, and Empathy," in "Neuronal Correlates of Empathy," pp. 67–77, New York: Academic Press, 2018.
12. S. Hurley, "The shared circuits model (SCM)," Behav. Brain Sci., vol. 31, pp. 1–22, 2008.
13. https://github.com/mitiku1/Emopy-Models
14. D. Hart and B. Goertzel, "Opencog," AGI, pp. 468–472, Feb. 2008.
15. https://standards.ieee.org/industry-connections/ec/autonomous-systems.html

Part III
Super Longevity and Rejuvenation

Chapter 16
In Search of Super Longevity and the Meaning of Life

Newton Lee

> *Immortality or life extension is more than just living forever. It's also how we live forever. What is our mindset? What is our humanity with each other? Our enemy is not one another. Our enemy is sickness, aging and death.*
> —*Jim Strole (People Unlimited and The Coalition for Radical Life Extension)*

16.1 Humanity's Destiny

Super longevity is not the same as immortality. Sometimes one has to sacrifice super longevity in order to achieve immortality (See Fig. 16.1).

First Director-General of UNESCO Julian Huxley wrote in 1957, "I believe in transhumanism: once there are enough people who can truly say that, the human species will be on the threshold of a new kind of existence, as different from ours as ours is from that of Peking man. It will at last be consciously fulfilling its real destiny" [1].

What is humanity's real destiny?

And why do we want to ask such a question in the first place?

In Plato's *Apology*, Socrates said that "the unexamined life is not worth living (ὁ δὲ ἀνεξέταστος βίος οὐ βιωτὸς ἀνθρώπῳ)" [2]. He was sentenced to death in Athens where a jury of 500 men found him guilty of "corrupting the youth" by a margin of 280 to 220 [3].

Greek philosopher and mathematician Pythagoras told Leon, Prince of Phlius that "some are influenced by the love of wealth while others are blindly led on by the mad fever for power and domination, but the finest type of man gives himself up to discovering the meaning and purpose of life itself" [4].

N. Lee (✉)
California Transhumanist Party, Los Angeles, CA, USA

Institute for Education, Research, and Scholarships, Los Angeles, CA, USA
e-mail: newton@californiatranshumanistparty.org; newton@ifers.org

© Springer Nature Switzerland AG 2019
N. Lee (ed.), *The Transhumanism Handbook*,
https://doi.org/10.1007/978-3-030-16920-6_16

Fig. 16.1 The Allegory of Immortality by Giulio Romano, c. 1540

Albert Einstein's life had a profound meaning as his theory of relativity revolutionized physics, astronomy, and our understanding of the universe. Nonetheless, he spent an additional 20 years searching for a unified field theory but failed to find it [5].

When he gave up on his scientific research, he also gave up on living. At the age of 76, he suffered an abdominal aortic aneurysm and was taken to the Princeton University Medical Center. He refused surgery that could have saved his life.

In spite of his death, Einstein lives on forever in the hearts and minds of all human generations. In other words, Albert Einstein achieved immortality, just like Socrates did. Life with a purpose leads to immortality.

In 2015, Google's life science subsidiary Verily had a staff philosopher among its 350 scientists. Its CEO Andy Conrad explained, "We have to understand the 'why' of what people do. A philosopher might be as important as a chemist" [6].

The words "medication" and "meditation" are different by only one letter ("c" versus "t"). We need to know not only the underlying causes of diseases but also why some patients respond better to certain treatments. We do not want to miss the forest for the trees.

16.2   Can Google Solve Death?

"Can Google Solve Death?" graced the cover of the *TIME Magazine* on September 30, 2013. The magazine cover reads "The search giant is launching a venture to extend the human life span. That would be crazy—if it weren't Google" [7] (See Fig. 16.2).

Fig. 16.2 The *TIME Magazine* cover on September 30, 2013 reads "Can Google Solve Death?"

Google made its first foray into healthcare with Google Health between 2008 and 2011 to collect volunteered information about personal health conditions, medications, allergies, and lab results [8]. In September 2013, Google unveiled Calico to tackle human aging and associated diseases [9]. Google Life Sciences, formerly a division of Google X, was renamed to Verily in December 2015 [10].

Since the development of contact lens for diabetics to monitor glucose in tears, Verily has been researching cardiovascular disease, cancer, and mental health, just to name a few. The idea is to combine medicine, engineering, and data science in an effort to capture the early signs of a disease and to stop it in its tracks.

Google cofounder Larry Page has good reason for his deep interest in health sciences. In May 2013, Page announced on his Google+ profile that a 1999 cold left him with paralysis of the left vocal cord, and another cold in the previous summer paralyzed the right cord [11]. *Business Insider* reported that Page has Hashimoto's Thyroiditis, an autoimmune disease in which the thyroid gland is attacked by a variety of cell and antibody-mediated immune processes [12].

Undaunted, Page decided to go for the moonshot. "Are people really focused on the right things?" asked Google's cofounder Larry Page. "One of the things I thought was amazing is that if you solve cancer, you'd add about 3 years to people's average life expectancy. We think of solving cancer as this huge thing that'll totally change the world. But when you really take a step back and look at it, yeah, there are many, many tragic cases of cancer, and it's very, very sad, but in the aggregate, it's not as big an advance as you might think" [13].

The bad news is that life expectancy in the United States dropped in 2015 for first time in 22 years for the entire population of male and female across all ethnicities [14, 15]. In 2016 U.S. life expectancy declined for the second year in a row for the male population [16, 17]. The 10 leading causes of death are heart disease, cancer, unintentional injuries, chronic lower respiratory diseases, stroke, Alzheimer's disease, diabetes, influenza and pneumonia, kidney disease, and suicide.

To reverse the downhill trends, transhumanism advocates extensive research and significant funding to cure all physical diseases and mental illnesses, increase life expectancy, and improve the quality of life for all human beings.

16.3   Any Desire for Super Longevity or Immortality?

Would you like to have 48 hours instead of 24 hours in a given day so that you would have twice as much time to do what you need to do? Your answer is probably a resounding yes. That means you are all for doubling your lifespan.

However, a Dying Matters survey in 2011 found that only 15% of people want to live forever [18]. The older the people are, the less desire they have for super longevity or immortality: 12% of over-65 year olds against 21% of 18–24 year olds want to live forever.

The reasons for the lack of desire for super longevity or immortality may include declining health, mental illness, unhappiness or life dissatisfaction, status quo or religions, loneliness, boredom, misanthropy, and nirvana:

1. Declining Health – The elderly often have weakened immune system and are more prone to serious illnesses. Age-related diseases include cardiovascular disease, cancer, arthritis, dementia, cataract, osteoporosis, diabetes, hypertension and Alzheimer's disease. The incidence of cancer, in particular, increases exponentially with age. According to the June 2018 report by the Centers for Disease Control and Prevention (CDC), cancer claims the lives of more than half a million Americans every year [19].

Without a doubt, no one wants to suffer forever. A 2013 study by the Stanford University School of Medicine revealed that an overwhelming 88% of the 1081 physicians surveyed would choose a do-not-resuscitate or "no code" order for themselves if they are terminally ill, even though the same doctors tend to pursue aggressive, life-prolonging treatment for patients facing the same prognosis [20].

   In March 1999, Dr. Jack Kevorkian was convicted of second-degree murder for "physician-assisted suicide" and served 8 years in prison [21]. Long before Kevorkian, Sigmund Freud's death in 1939 was a physician-assisted suicide [22]. In May 2018, a 104-year-old Australian scientist—Dr. David Goodall—traveled to Switzerland for assisted suicide because euthanasia is illegal in Australia. With failing eyesight and deteriorating quality of life, Goodall told CNN and the ABC, "My life has been out in the field (working), but I can't go out in the field now. At my age, I get up in the morning. I eat breakfast. And then I just sit until lunchtime. Then I have a bit of lunch and just sit. What's the use of that?" [23].

2. Mental Illness – Mental health is just as important as physical health. The media reports high-profile and celebrity suicides such as Robin Williams, Kate Spade, and Anthony Bourdain in recent years. However, suicide is much more widespread. Centers for Disease Control and Prevention (CDC) reported in 2016 that "from 1999 through 2014, the age-adjusted suicide rate in the United States increased 24%" [24] and the World Health Organization (WHO) reported in 2017 that "close to 800,000 people die due to suicide every year, which is one person every 40 seconds" [25]. CNN's Anderson Cooper lamented how he was unable to help his older brother from taking his own live at the young age of 23 because he did not see the pain his brother was in [26].

New York artist and Fodor's travel guides editor Stephen Crohn was known as "the man who can't catch AIDS." He had a genetic mutation in his white blood cells that effectively blocked HIV infection. He volunteered for research studies that shed light on the nature of AIDS and led to the development of antiviral drugs at the Aaron Diamond AIDS Research Center and other medical facilitates. "One of the things that went through my mind was, 'I guess I'm condemned to live,' " said Crohn. "What's hard is living with the continuous grief. You keep losing people every year—six people, seven people … and it goes on for such a long period of time. And the only thing you could compare it to would be to be in a war" [27]. Overcome by survivor guilt after seeing more than 70 of his friends died of AIDS, Crohn committed suicide at age 66.

3. Unhappiness or Life Dissatisfaction – Far too many people are getting tired of lifelong toil to make ends meet or working at a job that they do not enjoy. Gallup polls conducted between 2011 and 2016 have consistently shown that two-thirds of U.S. employees are unhappy or disinterested at work [28].

San Diego State University professor Jean Twenge has shown that anxiety and depression are on an 80-year upswing for young Americans since 1935 [29]. The 1994 autobiography *Prozac Nation* by Elizabeth Wurtzel and its 2001 film adaption starring Christina Ricci depict difficult real life struggles confronting some teenagers. If people are reluctant to make positive changes to better their lives, destructive behaviors resulting in death will likely occur.
In the case of the 104-year-old Australian scientist Dr. David Goodall who had lived a long and productive life, his conclusion was simply: "No I'm not happy. I want to die" [30].

4. Loneliness – Superman actor Christopher Reeve died of heart failure in October 2004 at age 52, and his wife Dana died of lung cancer 2 years later in March 2006 [31]. Another tragedy struck when Carrie Fisher (Princess Leia in *Star Wars*) and her mother Debbie Reynolds (*Singin' in the Rain*) died only 1 day apart in December 2016 [32]. Comics legend Stan Lee passed away in 2018 at age 95, a little more than a year after Joan Lee—his wife of 69 years—died in 2017 [33]. It goes to show that a person may lose their will to live when their loved one has passed away. In fact, research has shown that combating loneliness is essential to improving well-being and prolonging life [34].

5. Status Quo or Religions – Society, often enforced by religions, has taught its citizens to maintain the status quo. Some religious devotees practice self-flagellation to remind themselves of their sins. Believers and atheists alike have accepted that they will inevitably get sick during their lifetime, and they will be lucky to die of old age. Although most people have accepted advanced technologies to prolong human life and improve quality of life, eliminating death would be unthinkable because it would seriously disrupt the status quo.

6. "He who dies with the most toys wins" mentality – Bulletproof Coffee founder Dave Asprey has spent $1 million in his quest to live to 180 [35]. He is an exception rather than the norm among the wealthy. The notion of immortality is less popular than the mentality of "he who dies with the most toys wins." The huge Instagram following for 9-year-old Lil Tay—the "youngest flexer of the century"—demonstrates the prevalence of materialism in a capitalistic society [36].

Once upon a time, the wise King Solomon received an annual revenue of 666 talents (25 tons) of gold which is worth about $1 billion USD in today's currency (1 Kings 10:14). On top of that, he had 700 wives and 300 concubines (1 Kings 11:3). But when he went astray from God, he began to realize that nothing in the world was worth having. He wrote, "I denied myself nothing my eyes desired; I refused my heart no pleasure. My heart took delight in all my labor, and this was the reward for all my toil. Yet when I surveyed all that my hands had done and what I had toiled to achieve, everything was meaningless, a chasing after the wind; nothing was gained under the sun" (Ecclesiastes 2:10–11).

The most meaningful toys are not materialistic at all. Wolfgang Amadeus Mozart died penniless, but his toys were his symphonies, concertos, operas, quartets, and sonatas that have delighted millions of people for hundreds of years and will continue to light up the world.

7. Boredom – British journalist and author Bryan Appleyard opines, "I think there's a point that the immortalists don't understand, and it's that one exhausts one's own personality over a certain period. It's a weird idea that you would go on and on, still being interested in being yourself. I don't think anyone would. I think you'd get excruciatingly bored of being yourself" [37].

My response to Appleyard is that boredom often stems from the lack of desire to reinvent oneself. To become a Renaissance man or woman, for instance, can take an eternity. Understandably, not everyone wants to be a polymath, but there are so many great books to read, places to visit, ideas to share, new hobbies to discover, and charities to volunteer for – the bucket list is practically endless.

In early 2016, bucketlist.org showcased more than four million life goals from over 300,000 members [38]. I, for one, would love to read all 60 volumes of the *Great Books of the Western World* authored by Plato, William Shakespeare, Sir Isaac Newton, and many other wise men and women [39]. Life is anything but boring.

8. Misanthropy – I once had a debate with some university students on their pessimistic view that human beings are parasites on Earth, depleting natural resources at the planet's expense. As Klaatu (Keanu Reeves) said in *The Day the Earth Stood Still* (2008), "If the Earth dies, you die. If you die, the Earth survives."

Voluntary Human Extinction Movement (VHEMT) is a radical environmental movement with spokesperson Les Knight who believes that "When every human chooses to stop breeding, Earth's biosphere will be allowed to return to its former glory, and all remaining creatures will be free to live, die, evolve (if they believe in evolution), and will perhaps pass away, as so many of Nature's 'experiments' have done throughout the eons" [40]. Other Malthusians range from the moderate Population Action International to the extreme Church of Euthanasia. Their pessimistic view reminds us of the last words in *The Cabin in the Woods* (2011) when Dana (Kristen Connolly) said, "Humanity… It's time to give someone else a chance."

9. Nirvana or "Mission Accomplished" – Nirvana in Buddhism and Hinduism is a state of freedom from suffering and rebirth, extinguishing the three poisonous fires: passion, aversion, and ignorance. When a person has fulfilled his purpose in life, sometimes life extinguishes itself either by choice or by force.

Steve Jobs died of pancreatic cancer at the age of 56. He refused surgery for 9 months because as his wife Laurene Powell told biographer Walter Isaacson, "The big thing was he really was not ready to open his body" [41].

When Albert Einstein suffered an abdominal aortic aneurysm at the age of 76, he was taken to the Princeton University Medical Center but he refused surgery, saying, "I want to go when I want. It is tasteless to prolong life artificially. I have done my share, it is time to go. I will do it elegantly" [42].

Vincent van Gogh, Ernest Hemingway, Alan Turing, Virginia Woolf, and Socrates—a short list of many talented people who had contributed so much to the world—all committed suicide either by choice or by force.

If Buddhism and Hinduism are correct, the extraordinary achievers who reach nirvana will never be reincarnated, whereas the less accomplished and less desirable humans are being reincarnated again and again—which would explain the never-ending suffering and ever-increasing evil in the world. No wonder Socrates' last words in Plato's *Apology* were "I to die, and you to live. Which is better God only knows" [2].

16.4   The Living Dead

Russian novelist Fyodor Dostoyevsky wrote in *The Brothers Karamazov*, "The mystery of human existence lies not in just staying alive, but in finding something to live for." Living only for the sake of living is not enough for a human being. Life needs a meaning and living requires a purpose.

Without meaning and purpose, are the livings really alive?

American author Og Mandino wrote in *The Greatest Miracle in the World* that "most humans, in varying degrees, are already dead. In one way or another they have lost their dreams, their ambitions, their desire for a better life. They have surrendered their fight for self-esteem and they have compromised their great potential. They have settled for a life of mediocrity, days of despair and nights of tears. They are no more than living deaths confined to cemeteries of their choice. Yet they need not remain in that state. They can be resurrected from their sorry condition. They can each perform the greatest miracle in the world. They can each come back from the dead…" [43].

Take for instance, school resource deputy Scot Peterson failed to protect and serve by standing outside a building while the gunman killed students and faculty at Marjory Stoneman Douglas High [44]. Will Peterson redeem himself by saving innocent lives in the future? Only time will tell.

After all, actor and comedian Tim Allen, best known for his leading role in the sitcom *Home Improvement* and the voice of Buzz Lightyear in *Toy Story*, was given the highest honor as a Disney Legend for television, film, and animation-voice as well as a Star on the Hollywood Walk of Fame [45]. Before his showbiz success, however, Allen was arrested for drug trafficking in 1978 and was subsequently incarcerated for 28 months at Sandstone Federal Correctional Institution [46].

In a 2011 interview by ABC 20/20 anchor Elizabeth Vargas, Tim Allen talked about God and his rehabilitation. He said that one day he got a call from his parole officer and the next day a call from Jeffrey Katzenberg, then chairman of The Walt Disney Studios, who asked him to become a part of the Disney family. It was nothing short of a miracle.

Albert Einstein would surely concur as he once said, "The most beautiful and deepest experience a man can have is the sense of the mysterious. It is the underlying principle of religion as well as all serious endeavor in art and science. He who never had this experience seems to me, if not dead, then at least blind" [47].

16.5   The Meaning of Life

If the meaning of life is too serious a subject to ponder, perhaps it is good to start with Monty Python's humorous rendition in the 1983 musical sketch comedy film *The Meaning of Life*. The "Galaxy Song" is so amusing that even famed cosmologist Stephen Hawking sang that song in his signature computerized voice [48].

For book lovers, Douglas Adams's *The Hitchhiker's Guide to the Galaxy* features a supercomputer named Deep Thought who had an answer to the ultimate question of life, the universe, and everything after 7.5 million years of calculations. Spoiler alert: The meaning of life is 42.

One might postulate that the supercomputer was subtly referring to the 42-line Bible, better known as the Gutenberg Bible (See Fig. 16.3). But the real reason that the computed answer was incomprehensible is because the creators of the supercomputer did not fully understand the question that they were asking.

As physicist Nima Arkani-Hamed at the Institute for Advanced Study said, "The ascension to the tenth level of intellectual heaven would be if we find the question to which the universe is the answer, and the nature of that question in and of itself explains why it was possible to describe it in so many different ways" [49].

The meaning of life is manifested in many different ways—infinite diversity in infinite combinations.

16.6   A Life Worth Living

In the 2017 science fiction movie *Ghost in the Shell*, detective Togusa took a jab at his colleague Ishikawa about his cybernetic liver, "You got enhanced so you can drink more?" As Socrates once said that "a life that is not examined is not worth living," transhumanism seeks to ask the right questions in order to discover

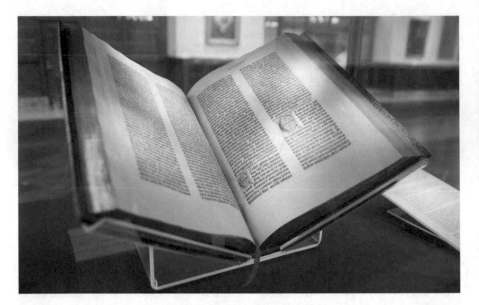

Fig. 16.3 The 42-line Gutenberg Bible at the New York Public Library. Originally bought by James Lenox in 1847. (Courtesy of Kevin Eng under Creative Commons license)

the meaning of life and the purpose for living as individuals and in a society as a whole.

On the eve of his 69th birthday, biochemist-geneticist Craig Venter echoed Socrates and Dostoyevsky's sentiment, "I have the brain of a 44-year-old. … It's not just a long life we're striving for, but one which is worth living" [50]. Venter has been acknowledged, along with geneticist Francis Sellers Collins, as being a primary force behind the Human Genome Project [51].

Actor Robin Williams in *Dead Poets Society* (1989) said, "We read and write poetry because we are members of the human race. And the human race is filled with passion. And medicine, law, business, engineering, these are noble pursuits and necessary to sustain life. But poetry, beauty, romance, love, these are what we stay alive for." Interestingly, poetry and music have shown to be effective in physical healing, and physicians have started to integrate them into their clinical practice [52].

In a thought experiment in *After the Dark* (2013), Petra (played by Sophie Lowe) chose a poet and a harpist in addition to engineers and problem solvers in order to ensure the continuous survival of the human race. She explained, "We live. Briefly, yes. Imperfectly, of course. Stupidly, sometimes. But we don't mind, because that's the way we were made. And when it's time to die, we don't resist death. We summon it."

A fulfilling life does not fear death and it leaves no regrets at the deathbed. When people look back on their lives, they often wish that they had spent more time with their families, less arguing and more forgiving, less fighting and more harmony. The most precious thing that people can give to one another is not money but time.

"I can buy anything that I want, basically, but I can't buy time," said Warren Buffett in an interview with Bill Gates and Charlie Rose on PBS and Bloomberg LP [53]. Buffett probably thinks that the annual $3+ trillion in American healthcare costs offers an extremely low return on investment (ROI), as he once said that healthcare spending is a "tapeworm on the economic system" [54].

In the 2011 sci-fi movie *In Time*, people are genetically engineered to stop aging at 25 years old, and time is literally money because currency is measured in hours and minutes instead of dollars and cents. People would use time as currency to pay for daily expenses.

Factory worker Will Salas (Justin Timberlake) ran into a 105-year-old young man Henry Hamilton (Matt Bomer) who transferred 116 years of his time to Will, leaving himself with only 5 minutes to live. Will in turn gave some of his newly acquired time to his best friend Borel (Johnny Galecki) who tragically ended up dying prematurely due to alcohol intoxication.

If the meaning of life is futility, human longevity loses its luster. In real life, New York artist and Fodor's travel guides editor Stephen Crohn committed suicide at age 66 because he was overcome by survivor guilt after seeing more than 70 of his friends died of AIDS.

In contrast, Crohn's great-uncle was gastroenterologist Dr. Burrill Bernard Crohn who discovered the inflammatory bowel syndrome named Crohn's disease. Dr. Crohn practiced medicine until he was 90, and he lived to 99 years old [55]. His meaning of life was doing what he loved and saving people's lives.

Whether we live 10 years or 10,000 years, what matters most are the meaning of life and the quality of life. The meaning of life depends on each individual. For me, the meaning of life is to serve one another for the survival of humanity and the advancement of civilization [56].

16.7   A New Enlightened Question

Roman Stoic philosopher Seneca said eloquently, "It is not that we have a short time to live, but that we waste a lot of it. Life is long enough, and a sufficiently generous amount has been given to us for the highest achievements if it were all well invested."

We have all heard this question before: "If you had one day left to live, how would you spend it?"

With super longevity, the new question would be: "If you are to live a thousand more years, what would you do with all that time?"

Three prominent Biblical figures Adam, Noah, and Methuselah lived to 930, 950, and 969 years respectively. At the 2018 Christian Transhumanist Conference at Lipscomb University in Nashville, Tennessee, I asked the audience, "If we live 1,000 more years, will we use the time to serve Jesus or will we serve ourselves?" (See Fig. 16.4).

Fig. 16.4 A tweet from Emily McFarlan Miller, a national reporter from Religion News Service. The partially obscured twitter banner reads "The Truth is Out There"

To serve Jesus is to serve one another for the survival of humanity and the advancement of civilization. As Jesus said, "Truly I tell you, whatever you did for one of the least of these brothers and sisters of mine, you did for me." (Matthew 25:40).

The Bible tells a story of King Hezekiah who begged God to cure his terminal illness and spare him from his imminent death. While enjoying his extra 15 years of life, Hezekiah showed off his immense wealth to the king of Babylon—an action that proved to be disastrous.

Hezekiah also fathered a son named Manasseh, who later became king at age 12, did evil in the eyes of God, and led Israel into ruin. The sad part was that Hezekiah did not seem to care as he thought to himself, "Will there not be peace and security in my lifetime?" (2 Kings 20:19).

Had Hezekiah devoted himself to serving God instead of himself in the last 15 years of his life, he could have been a great king and his son might not have turned evil.

Indeed, some parents simply do not care to teach their kids or to worry what kind of world they will leave behind for their children and grandchildren. But some people care so much that they adopt orphans from around the world. American actress Angelina Jolie, for instance, adopted two children—Maddox Chivan and Zahara Marley—before giving birth to her first biological child Shiloh Nouvel.

16.8   Selflessness

A selfless question of immortality may be more along the lines of "Do I want someone else to live forever?" That someone may be a spouse, parent, child, or even friend.

In December 2007, Boy Scout leader Tim Billups donated one of his kidneys to Scout leader Mel Northington and saved his life. "It serves as an example, because in any kind of society you have to think of yourself as fitting into a larger picture," said Mike Andrews, scoutmaster of Troop 500. "What an unselfish act for Tim to do this. Tim knows Mel through Scouting. It's not like he's a family member or a boyhood friend who grew up with him. It's very humbling" [57].

Should Tim's lone remaining kidney ever fail, his brother John Billups has promised to donate one of his. The Billups family exemplifies the best of humanity that inspires many including myself.

For the past 15 years of *Computers in Entertainment* published by the Association of Computing Machinery from 2003 to 2018, the list of "In Memoriam" for the magazine's editors had sadly gotten longer. Amongst them were Roy E. Disney, Bob Lambert, Mark Mandelbaum, Randy Pausch, and Charles Swartz.

When I requested an interview with Prof. Pausch in October 2007, he replied to my email, "I'm afraid I'll have to decline; time is just in short supply for me, as I'm sure you can understand."

In his "Last Lecture" titled "Really Achieving Your Childhood Dreams," Prof. Randy Pausch said, "We cannot change the cards we are dealt, just how we play the hand. … It's a thrill to fulfill your own childhood dreams, but as you get older, you may find that enabling the dreams of others is even more fun" [58].

Transhumanism seeks to achieve super longevity, not at the expense of others but rather helping others to fulfill their lifelong dreams.

16.9   Pros and Cons of Immortality

Critics of super longevity often cite overpopulation [59], food security [60], pension [61], capital punishment [62], and others as problematic issues. However, scarcity of resources is often manmade. About one third of the food produced in the world for human consumption every year—approximately 1.3 billion tons—gets lost or wasted [63]. Some countries like Yemen are using food as a weapon of war [64].

The 2018 TV series *Altered Carbon* based on the 2002 novel of the same name by Richard K. Morgan described a dystopian world of immortality. Quellcrist Falconer (Renée Elise Goldsberry) regrets her invention of "cortical stacks" – the human consciousness digital storage and transfer technology for achieving immortality.

She warns, "The creation of stacks was a miracle and the beginning of the destruction of our species. A new class of people so wealthy and powerful, they answer to no one and cannot die. Death was the ultimate safeguard against the darkest angels of our nature. Now the monsters among us will own everything, consume everything, control everything. They will make themselves gods and us slaves. If we do not stop the curse of eternal life in our realm, our children will inherit despair. The ebb and flow of life is what makes us all equal in the end. We aren't meant to live forever. It corrupts even the best of us."

On the other hand, super longevity motivates people to think very long-term, protect the environment, recycle, prevent wars and destructions, and do everything possible to achieve a better quality of life instead of leaving a big mess to the next generation to fix.

As opposed to long-term greed, short-term greed is detrimental to humanity and our planet Earth. We are witnessing crimes, pollutions, and human sufferings as a result of stupid decisions driven by short-term greed. Warren Buffett once told his MBA students at Columbia University in 1999, "The longer the view, the wiser the intention" [65].

Cofounded by Danny Hillis and Stewart Brand in 1996, the Long Now Foundation provides "a counterpoint to today's accelerating culture," helps make "long-term thinking more common," and fosters "responsibility in the framework of the next 10,000 years" [66] (See Fig. 16.5).

In my 2006 interview for *Computers in Entertainment* published by the Association of Computing Machinery, my former Disney colleague Danny Hillis elucidated his rationale, "The [10,000 Year Clock] project is about my acknowledgement that I do have some relationships with people thousands of years from now, and there is some continuity between what I do and what their life will be. ... The business of making people think long-term is really something that is missing in the world in general. So we started a foundation that does the clock project and other projects for basically stretching out people's sense of the moment that they care about—which is now—that's why we call it the Long Now Foundation. It is actually Brian Eno's suggestion that we want to stretch out the moment of now to include the next 10,000 years" [67].

Scientists are some of the most patient people in history. They can set up laboratory experiments that last for months, years, decades, and centuries. For example, the ongoing pitch drop experiment was started in 1927 by Professor Thomas Parnell at the University of Queensland in Brisbane, Australia, to demonstrate that some substances which appear solid are actually highly viscous fluids [68]. The ninth drop fell in April 2014 after 87 years.

Also in 2014, microbiologist Charles Cockell at the University of Edinburgh began a 500-year study about the longevity of bacteria. It requires a researcher to examine the dried bacteria for viability and DNA damage once every 25 years [69]. If Cockell lives longer than 500 years, he will report the final results in 2514, or he may extend the bio experiment for another 500 years. Who knows?

Fig. 16.5 First prototype of the 10,000 Year Clock (1999) on display at the Science Museum in London. The clock ticks once a year. (Courtesy of Pkirlin on en.wikipedia.org)

16.10   Age Reversal and Rejuvenation

Rumors of a Fountain of Youth (see Fig. 16.6) have endured from Alexander the Great in the fourth century B.C. to legendary patriarch king Prester John during the early Christian Crusades to Spanish explorer Juan Ponce de León in 1513 when he discovered Florida.

In gerontology, scientists have been researching negligible senescence—the lack of symptoms of aging. There are quite a number of negligibly senescent animals in nature. Adwaita was a tortoise in Calcutta, India that lived to either 150 or 250 years old by some accounts [70]. George, a 140-year-old lobster, was released back into

Fig. 16.6 The Fountain of Youth (1546) by Lucas Cranach the Elder. Older women are seen entering a Renaissance fountain, and exiting it after being transformed into youthful beauties

the wild by a New York restaurant [71]. Henry, a New Zealand reptile, became a father at the age of 111 for the first time [72].

Prof. Caleb Finch of USC Davis School of Gerontology said in a 2010 interview, "In theory, if mortality rates did not increase as usual during aging, humans would live hundreds of years. I have calculated for humans that at mortality rates of 0.05% per year, as found at age 15 in developed countries, the median lifespan would be about 1200 years. In natural populations of long-lived animals, mortality rates are rarely less than 1% per year. For very slowly aging turtles, rockfish, the number beyond 70 is 1–2%. However, there are long-lived trees, like the bristlecone pine at 5000 years" [73].

Biomedical gerontologist Aubrey de Grey cofounded the SENS (Strategies for Engineered Negligible Senescence) Research Foundation in 2009 to conduct research on regenerative medicine and rejuvenation biotechnologies to prevent or reverse the aging process [74].

De Grey said in a 2013 interview, "SENS is based on the appreciation that there is a continuum between (a) the initially harmless, progressively accumulating damage that accumulates in the body as a side-effect of its normal operation and (b) the pathologies that emerge when the amount of that damage exceeds what the body is set up to tolerate. We want to treat (remove or obviate) the damage and thereby prevent the pathology" [75].

Regenerative medicine can take its cue from salamanders (Ambystoma mexicanum) that can routinely regenerate complex tissues such as a severed limb, a detached tail, or the lens and retina of a damaged eye.

Northeastern University Professor James Monaghan and his research team studied limb regeneration in salamanders and concluded that "many new candidate gene sequences were discovered for the first time and these will greatly enable future studies of wound healing, epigenetics, genome stability, and nerve-dependent blastema formation and outgrowth using the axolotl model" [76].

Rejuvenation biotechnologies can pick up on the immortal jellyfish (Turritopsis dohrnii) that can age backward like Benjamin Button (Brad Pitt) in the 2008 movie *The Curious Case of Benjamin Button*. The immortal jellyfish can rejuvenate from an adult back into a baby, and the life cycle repeats itself until the jellyfish gets eaten by a predator or succumbs to illness.

When starvation or injury occurs, "instead of sure death, [Turritopsis] transforms all of its existing cells into a younger state," said Maria Pia Miglietta, then postdoctoral scholar at Pennsylvania State University and now professor at Texas A&M University at Galveston [77].

Harvard Medical School biologist George Church wants to live to 130 in the body of a 22-year-old. His company Rejuvenate Bio uses gene therapy on beagles to rejuvenate them. Church said, "We have already done a bunch of trials in mice and we are doing some in dogs, and then we'll move on to humans" [78].

Transhumanism supports negligible senescence through regenerative and rejuvenation technologies, not superficial plastic surgery. We look younger if our internal organs are healthier. That is why the human immune system is a top priority for longevity and age reversal.

16.11   Human Immune System

Faith healers have gotten one thing right: the human body can detect and defeat all types of diseases including cancer—the second leading cause of death in America. Our mind and body are closely connected. Think of the split-second surprise reaction when we lifted up an object that we thought was heavy but turned out to be light. Meditation and inner peace help our body heal faster.

But faith healers have neglected one important point: the complex human brain—created in the image of God—is meant to invent new technologies including medical treatments for our mind and body.

The holy grail of good health and longevity is within each of us: the thymus—a specialized primary lymphoid organ of the immune system. A healthy thymus produces T cells that protect us from foreign invaders and cancer cells. University of Texas Prof. James Allison (along with Kyoto University Prof. Tasuku Honjo) won the 2018 Nobel Prize in Medicine for his work on cancer immunotherapy with killer T cells [79].

Unfortunately, the human thymus starts to age and shrink in size after puberty, leaving us vulnerable to diseases if the remaining T cells in our body become defective, depleted, or destroyed. That is why the flu poses a greater danger to the elderly and people with weakened or suppressed immune system. Flu vaccines are useful only if there are healthy and functional T and B cells in the body.

Although the benefits of vaccination far outweigh the risk of adverse effects, some side effects can be fatal. For example, Martin Gore—a leading cancer expert—died after receiving a routine yellow fever vaccine in January 2019 [80].

Furthermore, antibiotics are losing their effectiveness against superbugs that have mutated to resist even alcohol based hand sanitizers [81] and they have spread to every country in the world [82]. Stronger antibiotics such as Amikacin can cause kidney damage, hearing loss, respiratory paralysis, and seizures.

In January 2018, Centers for Disease Control and Prevention (CDC) released an alarming statistics that each year in the United States, at least two million people become infected with bacteria that are resistant to antibiotics and at least 23,000 people die each year as a direct result of these infections [83].

"This shows that we are right on the verge of getting into the territory of routine bacterial infections being untreatable," said Steven Roach, food safety program director at the Food Animal Concerns Trust. "It underscores the failure of both the federal government and Congress, and the industry, to get a grasp of the problem. We can't continue to drag our feet on taking needed action" [84].

In August 2018, The Bureau of Investigative Journalism reported on the scourge of superbugs killing Malawi's babies [85]. In December 2018, *Mosaic Science* revealed that sexually transmitted diseases now include M. genitalium superbug and gonorrhea superbug that are growing resistance to antibiotics [86]. The journal *Review on Antimicrobial Resistance* paints a bleak picture by estimating that by 2050, more than ten million people will die from superbugs each year [87].

The human immune system is our best line of defense against cancer, antibiotic-resistant superbugs, and many aging-related diseases. Prevention is better than cure. In 1992, as part of the Preventive Health Amendments enacted by Congress, CDC officially changed its name to the Centers for Disease Control and Prevention (with "CDC" still to be used as the acronym) [88].

16.12   Immunotherapy

Conducted under an FDA-approved IND (Investigational New Drug) between 2015 and 2017, the research team led by Dr. Greg Fahy, director of biomedicine at the California Transhumanist Party and chief science officer at Intervene Immune, has done successful human trials on thymus regeneration to reverse the shrinkage of thymus and to generate new healthy T cells [89].

The immune system rejuvenation technology can be used to prevent organ transplant rejection, facilitate islet cell transplantation for curing Type 1 diabetes, reverse autoimmunity, and enable cancer immunotherapy without invasive surgery and harmful side effects of chemotherapy.

During my travels, one time I met a middle-aged woman at lunch in the dining car of a train. The servers put strangers together at each dining table. When the waiter took our orders, he asked her how she was feeling. She said that she could only eat bread because regular food upset her stomach. She had recovered from cancer, but chemotherapy did a number on her quality of life.

A strong immune system is the key to keeping us healthy regardless of our physical age. The immunotherapy approach has gained plenty of traction.

In 2017, Microsoft joined forces with Adaptive Biotechnologies on immunosequencing to identify immune cell receptors. Adaptive cofounder Dr. Harlan Robins explains, "The immunome is the set of DNA sequences that code for T cell and B cell receptor genes. These genes are special because they rearrange with the immune cell, and are therefore different between cells (all other DNA is the same in each cell—excluding mutations). These genes code for receptors that recognize pathogens through binding" [90].

16.13   Genetic Enhancements and Ethics

If humans are to colonize another planet, we better have a powerful immune system to protect us from deadly new pathogens. H.G. Wells' novel *The War of the Worlds* reminds us that the invading Martians overpowered humans but were killed by earthly pathogens to which they had no immunity.

To live on another planet, human beings may have to go through some genetic enhancements like Tardigrades who can survive in places devoid of water, at temperatures as low as minus 328 degrees Fahrenheit and as high as 304 degrees Fahrenheit, under pressures of up to 87,000 pounds per square inch, and exposed to direct solar radiation and gamma-rays [91].

In November 2018, Chinese scientist He Jiankui announced the birth of the world's first gene-edited babies from embryos that were genetically modified using CRISPR/Cas9 in attempt to render the twin babies immune to the HIV virus [92].

Critics called He's experiment "reckless" and "unethical" whereas supporters said that "it is time to move forward from [debates about] ethical permissibility to outline the path to clinical translation … in order to bring this technology forward" [93].

Since the Chinese scientist He deleted the CCR5 gene from the babies, he inadvertently enhanced their brains in cognition and memory—foreshadowing the creation of super intelligent humans in the near future [94].

Is gene editing less ethical than human beings killing one another at wars and destroying wildlife on earth? A biomass study showed that humankind represents just 0.01% of all living things on the planet, and yet humanity has caused the loss of 83% of all wild mammals and half of all plants [95].

Marc Goodman, global security advisor and futurist, spoke at the TEDGlobal 2012 in Edinburgh about his ominous warning: "If you control the code, you control the world. This is the future that awaits us" [96]. First source code, now genetic code. Transhumanists are well aware of that danger of gene editing and are highly respectful of individual safety, freedom, and privacy.

Apostle Paul wrote in his letter to the Corinthians, "The last enemy to be destroyed is death" (1 Corinthians 15,26). Keeping ethics as well as environmental and wildlife protection in mind, transhumanism supports significant life extension and quality of life improvement achieved through the progress of science and technology.

Appendix 1: Super Longevity Presentations

With an academic background in computer science and artificial intelligence, my first peer-reviewed research paper titled "An Expert System for Information on Pharmacology and Drug Interactions" was published in the *International Journal of Computers in Biology and Medicine* [97]. I was offered a research scientist job at the Mayo Clinic in Rochester, Minnesota. I did not take the job, but I have continued to stay up-to-date on medical research.

I attended the Super Longevity Spring Meet-up on May 5, 2018 in Irvine, California hosted by Coalition for Radical Life Extension, People Unlimited, and Maximum Life Foundation. The engaging event included dynamic speakers sharing their personal views and inside knowledge on super longevity.

The featured speakers were Dr. Bill Andrews (Sierra Sciences), Bernadeane Brown (People Unlimited), Dr. Greg Fahy (21st Century Medicine and Intervene Biomedical), David Kekich (BioViva), and Jim Strole (People Unlimited and The Coalition for Radical Life Extension).

Craig McClure (HD Broadcast AZ) videotaped the presentations and transcribed the recordings for publication in this book. Videos and photos are available on the Super Longevity Network www.SuperLongevityNetwork.com, an OTT network dedicated to the radical life extension movement.

Appendix 2: Presentation by Dr. Bill Andrews (Fig. 16.7)

Every time I come here, the first thing I think about is how great it is to be here. You guys are my favorite people in the world!

I've been obsessed with this whole idea of curing aging, living as long as possible, as healthy as possible for as long as possible, as long as I've been alive. I've made it my job. It's my job, my work. I talk to people. I meet somebody on the plane. I say, "I'm obsessed with trying to cure aging. I'm obsessed with trying to live as long as possible!" They say, "Get a life!" But that's the point! That it is my

Fig. 16.7   Dr. Bill Andrews. (Photo by Yannick Apers)

life! That is what it is! I love living! I love living! I want to live as long as I can! For as long as possible. And this is… what better way to be doing it? I'm, I'm working on this at the [lab]. The thing is that it's not coming along as fast as I would've liked, I have to say it's not, not because the science is wrong, it's not because the scientists aren't getting the work done. It's mostly because of funding issues and stuff like that. I'm not going to spend a lot of time dwelling on that particular part of it, but it's, it isn't coming along as fast as possible.

A year and a half ago I turned 65 and I said, "Okay, I'm sure there's a lot of people who are older than me. Still, turning 65 means something. I never thought I'd be 65! I thought I'd be curing aging and be younger than 65, long before I became 65. It used to be in my presentations (and maybe some of you might even remember this before I was 65). I would stand up at the start and I'd show some old 65 year old people that were really athletic and stuff like that. And I'd say how great it is that that the world has gotten to where we can now be young and active even when we're 65. Then I'd go to the next slide: "but when I turn 65 I want to look like:" and I show the next slide, which shows the same people in the same poses, but they're looking in their twenties. I can't show those slides anymore, because I'm now past 65! But it's, it's made me really start to say to myself: "it's not going to be easy. It's going to be a struggle. Okay? If we're gonna make it, if we're going to be around and be here when the cures finally come along. We have to do things now! Nobody's going to provide us with this magic bullet or a silver platter or anything like that, just when we really need it. There are a lot of people we all know that have passed away now. You know, they were obsessed with this too, and they didn't make it." So it's not going to be easy.

My research is all about telomeres. I believe that when we get something, we can make it so that we can lengthen telomeres and reverse aging. I believe that but not going to be 100 percent certain about it until we can actually test it. That's turning out to be a lot more trouble, a lot more work than originally planned. Even though I believe that telomeres (when we do lengthen them), are going to be the biggest thing that ever hit the planet, I'm focusing on: "What can I do now to make myself young and healthy as long as possible?"

The bottom line comes down to: there are really two things that people need to focus on:

1. Decreasing inflammation. Inflammation is probably the number one cause of a lot of different age related diseases and early death.
2. Oxidative stress. Reduce oxidative stress as much as possible.

Now, in both of these subject areas, there are a lot of people out there saying a lot of things. You can hear a lot of press releases, speakers, a lot of hearsay. I gave a presentation at People Unlimited one time, about "How to know what's real and not real". You can, hear all kinds of stuff about what can you do: Should you drink diet coke or should you drink coffee? And every best selling product has the most critics. And so if you hearing a lot of bad things about something and people are saying it's bad for your health and stuff like that, it might not be true. I'm always saying go to the scientific peer reviewed literature, Pub Med. And then even there, you can't believe everything. If you want to prove that, say "white lights are bad for you", you can find something in the scientific peer reviewed literature that says that. I often show a slides when I talk about this. I show two studies in the exact same journal, 2 weeks apart from each other. And the titles are the exact opposite. And this is a scientific, peer reviewed journals. So I always say: make certain! Look at a lot of studies! Do your own Meta analysis. You've got to really find out what's going on. You might find 10 papers, 10 studies say one thing 90, say the other. Well, go with the 90. That's the best strategy. If pub med isn't good enough, there's also ClinicalTrials. gov. But the bottom line is, is inflammation is one of the worst things for you. There's a lot of myths around on that subject alone. I've made it a big part of my life to be learning everything I can to reduce inflammation.

People know I'm a 100 mile ultra-marathon runner. People say, "That's gonna give you all kinds of inflammation!" Not true! I've read the studies. I've looked them all over. I've done my own studies. The problem isn't how far you run. The problem is: "How do you run it?" A lot of people criticize me because I end up dropping out of races a lot. Maybe I did about 80 miles, and then I just say "I've had enough!" And they say, "Oh, come on! You've only got 20 miles to go!" and I'll say, "You know, I'm not having fun anymore". Sometimes that happens, sometimes it doesn't happen. I'm a big believer that if you exercise, are doing any kind of endurance exercise a lot and keep it fun, it'll never cause inflammation. In fact, inflammation is caused by your body's saying something unusual is happening. It's not gonna happen, if it's a usual thing. If you run all the time, let's say run, bike, swim or any of the endurance sports. If you do this all the time, you, your body actually adapts to it. I'm always surprised that some people who kick my butt in marathons, the next day, they're stiff as a board. They can't walk. They're just sore. I'm fine! I'm out running another 10 K, or something like that, the next day. And it's because they, during the night, built up inflammation throughout their body because they don't run consistently enough or they run too hard or they pushed too hard.

So that's just one of the things. Okay, so endurance exercise: "When it quits being fun, quit. Save it for another day". If you do that, after a few years of doing that, you're going to start finding that: "Boy, doing these endurance things is really easy". People say, "how do you do these 100 mile races?" I think: "How do you stop?" You know, I, I really like when I'm really seriously pushing (right now I'm not really pushing as much as I did a few years ago). You just have fun! You don't want to stop when you finish a race. You say "I was hoping it would never end!" Because there's adventure and all that kind of stuff in it. Bottom line: I think endurance sports are one of the best things you can do for yourself to actually decrease inflammation, as long as you keep it fun.

Don't push it. But there's a lot of other things that cause inflammation and one of the other things that I've done a lot since I've turned 65 is start learning about all the foods that I eat. Getting testing, getting blood work done to find out what foods induce inflammation in my blood, because most of your inflammatory responses are circulating around in your blood. So I've identified all the foods that cause inflammation and I've quit eating those. It's a discipline! If I was under 65, I wouldn't be doing it, but now that I'm over 65, it's a very important thing for me. I'm making certain that I don't do anything that causes inflammation in myself. I get my inflammation tests done pretty regularly, at least once every 3–6 months. Like CRP – C Reactive Protein – which is probably maybe the best marker for inflammation. They have a scale. You want to be somewhere between point three and three on their score (I forget the units, but it's point three and three). I want mine to be less than point three! And so I'm getting it tested. In the last year, I've never seen a score for myself high enough to measure. They just say less than point three. I've been getting that every time. And I think that's really good.

I think oxidative stress is a little easier to deal with. You take your antioxidants - there's a whole bunch of them. Alpha Lipoic Acid, vitamin E and vitamin C and things like that. Those things are good for you. There's such a thing as too much of those. You can get your ORAC scores measured. You can find out how your body's dealing with, reacting to oxidative stress and free radicals. These things are important. The mission that I'm doing now is I'm just encouraging everybody do everything you can to decrease inflammation. It's the foods. I'm taking omega three fatty acids and vitamin D (Which coincidentally, if you take those things, your telomeres are longer, not because they lengthened the telomeres, but because they decrease things like inflammation and oxidative stress).

But you know, my mission is telomeres, so I'm going to say telomere length measurement is probably still the best thing that you can do to, I think, to, to keep track of how you're doing in terms of your age related stuff. I just wish the tests that exist were better. Most of the tests that you can take right now are not very accurate. And the tests that we do at my company, they're a lot more accurate, but they're also a lot more expensive. But I think the future is coming where we're going have some really good handles on the length of our telomeres. But the really important thing isn't to merely decrease their rate of shortening, which is what the best we can do. The goal is to lengthen them! That's the mission of my company.

I want to say that I that while I was sitting here before I got up I was thinking, "how great it is to be part of an organization like this and have friends like Audrey and Javier? Who can sing as well as they can!" I mean, it's like amazing! I think it decreases the rate of [telomere] shortening, at least! I've just got one question: I've never heard that song. Audrey, did you write it? Where did that come from?

(Audrey, replying from the audience: I wish I had written it, the words really express a lot of what I feel .....it's by "the band perry".)

Wow! That is a fantastic song. At least I understood the words, Javier! (Javier sings opera).

Let me summarize my, my mission right now is not just telomeres. My mission is do everything you can to decrease the rate of aging. Decreasing inflammation is the number one thing.

Appendix 3: Presentation by Jim Strole (Fig. 16.8)

We have a challenge, as Dale pointed out, we have a real challenge. First of all, we were born into a world that's filled with death. We inherited a lot of things that we did not actually choose and our ancestors, bless their hearts, they were where they were, but we inherited a lot of patterns. One of them being not taking very good care of ourselves and really, really respecting our bodies, really respecting who we are. And I would say to all of you, when Bill said 5 years optimistically, I just finished a conversation here where somebody asked me the same question and I said I think 5

Fig. 16.8 Jim Strole. (Photo by Yannick Apers)

years and we will have some remarkable age reversal therapies or modalities that will change the course of where we're going. But in the meantime, no matter how much time it is, stay alive.

It's very easy, and this is part of the challenge too, to get caught up in measuring out the time, starting to measure how long you have left. Well, whether it's 5 years, 30 years or whatever, there's modalities right now on the table and happening quite quickly that can exponentially keep you alive. There are therapies right now that you can do and things you can do to stay alive until the next therapy. Like Ray Kurzweil or Terry Grossman said, stay alive long enough to live forever. In other words, there has to be an awareness, a strong awareness.

And that's really why I'm here today. Actually Bill and David and Bernie and I discussed this, that the core reason we're here is because we want to create more of a togetherness of all of you great, like-minded people. A support system, you might say. We have differences of opinions, but in the long run, if you really look at it, those differences of opinions don't weigh near as much as the common strand we have for the passion of life here, and that's what we need to keep in mind. It's good to have differences of opinions. It's good to be able to debate things creatively, but it's not good to dismiss each other and put each other down and not meet because we have some difference of opinion that keeps us from really flowing together. And this is really my passion.

I think, Ashley, there's millions of people on this planet, that feel the way we do. And, like Bill and David both pointed out, there's power in numbers, there's power in numbers, there's money in numbers, there's great inspiration in numbers, and another thing: mindset is huge as is the concept of staying alive. The environment you live in directly affects you emotionally, inspirationally, physically, stress wise - do you have people around you who really want you alive? Not just people who are continuously arguing with you? We're in a world filled with that now, and that's why we need to come together every opportunity we have, and we're going to create more opportunities just like this for all of us to come together.

RAAD Fest is coming up in September and I hope all of you will be there. I'd like to instill in you how important you are as individuals in this movement. Please don't make yourself incidental. The Coalition for Radical Life Extension, which is the producer of RAAD Fest, is a nonprofit organization. We actually lost money last year. Still, some people say, "Oh, RAAD Fest is expensive." I don't see how, right now, at $497, that that's expensive when you get 3 days of information about staying alive that you'd never get anywhere else on the planet; and you get three kind of good meals - I was going to say awesome meals, but let's face it, they're good meals, they are what they are - some like them and some don't, but we try to put the best on we can. But we do that because we are more interested in the coming together of all of us and the power of that.

Do we need money? Yes. But I feel by being alive and moving in a passion that we all feel here in our bodies together we'll create everything that we need. We'll have the answers and we're finding those answers all the time. Like Joe said earlier, I'm on the phone all the time with some of the top scientists, wellness doctors; guys, they're doing great stuff. They're doing great stuff. Do they need more money? Yes.

Yes. I'd like to see you have all the money you need and I'm putting that out. I'm putting that out. Maybe there's somebody in the room here we don't know, tonight, who'll just hand you that check, who says yes, Bill, go for it. Because what is more meaningful. And Greg, great having you here. Thank you. There's another great scientist doing this work. We're honored to have him in the room with us.

That's what I mean. That's what I mean. The power of coming together. Now I was emailing with Greg the other day and he hinted at something like that, but I didn't know the extent, so thank you. Thank you. Look guys, there's more than hope. There's reality. Stay alive! Stay alive!

We have to push this envelope together. Don't make yourself incidental. RAAD Fest is not about any one person. It's about all of us together. But RAAD Fest is not enough. We're moving to create a vast amount of people that's a tipping point, that creates a tipping point in this world of death. Now, I have to commend all of you that have come here because of your insight and your foresight, of how beautiful that is. Not very many people yet can see this, can see how vast this life that we are talking about is, because when you lose someone, you lose everything about them.

I was talking to Bill Faloon yesterday and he said, Jim, you know it's amazing to me, still, that we're doing so much at life extension and we're seeing some great benefits of what we do. We put our money into strategic places, but a lot of these other people who have lots of money, sometimes billions, will say they want to live, they believe in what we're doing, but we can't get any money out of them to support us. Like they think they're going to take that money with them if they die, or maybe they're gonna wait till the last minute and put up a few billion bucks as if that will redeem them.

My personal feeling is that we can't wait until somebody comes along and drops the golden egg. Already, like Greg said, there's something happening that's really fantastic. I sense that, in talking to other scientists like Bill and others, that there's more happening all the time, but we got to push this envelope. We cannot sit back and say, oh, we're going to wait. We're going to wait now to see whether it's 6 months, a year, 3 months, 5 years. Immortality or life extension is more than just living forever, although that's a huge part, it's also how we live forever. It's also what is our mindset? What is our humanity with each other? Are we still going to have a world where we're killing each other? Are we still going to have a world that is constant stress and strife?

I'm not really, really clear that any therapy is going to work as good on you if you're filled with a death programming in your body. In fact, I was told by one of the top scientists recently that if you have a good mindset about living forever and you really have a humanitarian type energy in your body, these therapies have proven to work a lot better on you. So we had to change our mindset around living. Our enemy is not one another. Our enemy is sickness, aging and death. So this is what this event is about for me. I think for many of you I talked to briefly while coming in, it is for you too. You've come here to be energized, you've come here to be inspired, you've come here to learn. Fantastic! This is what we can do together. I envision a coliseum, to begin with, filled with people cheering each other on to life. Look, I would love for some golden nuggets to be dropped, and they may or may not be - I'm open for everything - but I think we've got to create masses of people that create masses of money. We've got to turn the tide right now. The majority of money on this planet is going to death and destruction. We need to turn that tide and put it towards life and building a world and a planet that human beings have never seen as human beings before.

So I'm impassioned. I'm grateful. I'm grateful for you coming here today. I'm humbled. I'm an individual that's moving to do what I can in this world to make a difference, but I clearly understand that I can't do it by myself. Bernie clearly understands that the two of us can't do it by ourselves. We've been working in this field together since 1972, speaking at a time when no one was even speaking longevity. And we were speaking immortality - were we ever looked at strange.

I had a hard time with relationships in those days, even my family had a hard time with me. But I thought when I first heard about immortality, which was even before I met Bernie, I thought, wow, that's the best thing in the world, that death is wrong. And then, when I met Bernie and my other partner, Charles, who isn't alive today, amazing human being, he should have been alive. That's another thing. I thought, this is so powerful, what they're saying, and I want to be a part of it. So, after I totally embraced the idea of super longevity, I ran to my family and told them, thinking I'm telling them the greatest news on the planet. I see by some of you laughing, you had some of the same responses. Is that strange or what? I ran. I mean, I was so excited, you know, and I went and told my brother and they just looked at me going, uh, you've gone off, we're taking you to get checked out. But I'm so happy, no matter that resistance, the negativity I experienced at times. From that time, I am so happy today that I was willing to keep on moving and that Bernie was willing to keep on moving, to keep on going for what we felt. Because no matter what the challenges are, sometimes in the struggles, it's all worth it. It's all worth it.

I'd rather be struggling in this life than living a mundane, shallow, superficial life and just fade away into the darkness.

So I'm grateful for you all. Each person here is a piece of the puzzle. Bill's doing his work. He's fantastic. David is, Greg, others, but everyone. You don't have to be a scientist. I'm not a scientist. I'm not a doctor. I'm an activist. I'm a… I'm a… I don't know what the hell I am half the time. No, seriously. I'm just being me. And I think that's the most valuable thing on this planet for all of us to be who we are. And if you have a fire in your belly to live, you need to find that place to move in that and be a fire on this earth for life.

There's nothing better. I love it when I go somewhere. Everywhere I go I talk about what I do because they ask me – I'm an interesting person – and I'll ask them what they do, and we'll get into a conversation and I'll tell them I speak about immortality and get some strange stares. Mostly now it's like glassy eyes. They don't understand totally, but I keep on moving. I give them a passion in what I feel. And you plant a seed. You plant a seed. We need to be a voice. Every single one of us here. We need to be a voice for what we feel on this planet. We will turn the tide. It's an inevitable. We will turn the tide. I actually not only predict that 5 years or less – well, now Greg's blown that, the best news I've heard - but I predict here for all of you, I'm going for every one of you staying alive. That's what you mean to me. No one in this room dies. No one in this room dies. We need to stop death in its tracks now! That's my reality. I hope it's your reality, but I'm going for that with you.

Bernie and I continuously talk about what it means to be ageless. Age is a number. We all say that at times, but age is also a chemistry. What do you give yourself to? Are you feeling, are you giving yourself to decrepit thoughts? Are you giving yourself to, well, I'm getting old now. You're looking in the mirror and seeing all the wrinkles. Don't measure yourself by how many wrinkles you've got, measure yourself by how alive you feel. How alive do you feel? And begin to go with that because you can change that mindset. Some epigenetic scientists are saying that up to 60 percent of your health or more evolves around what your mindset is. How do you really think about yourself? Do you fear some terminal disease hitting you and you're just going to fade away with that because everybody is saying you have to die from that, or are you saying, I don't care what I have to face, I'm going to walk through it. I've got people to support me and inspire me in it and I'm going to outsmart death. And that's real. And that's a real thing you can do. Okay? There are no incurable diseases, I'll dare to say that. But there are incurable people, because of their mindset.

We're not here to sell something, RAAD Fest or anything else, we're here to appeal to you. So let's start moving together. We want to have events like this. This is a great showing for such a short amount of time. You know, we only planned this a few weeks ago and you guys popped in, you guys popped up, you showed up. That's huge. That's huge. So don't make yourself dispensable. Don't make yourself smarter than anybody else, or dumber than anybody else. Make yourself valuable. We're all valuable together. It's going to take all of us moving together to make this happen. There's going to be no Messiah come along. I always say, don't wait for the Messiah - people waiting for Bill to create it, Bill is the Messiah now - or somebody else. No. Say, in a short time, which is happening, we will for sure cure aging, because that's our destiny, what do you want our world to look like? What do you want our world to be like? A world that is still filled with destroying each other? Backbiting? We need to change our mindset now and find out how valuable we are. To me, immortality means a quality and a value of the human body and each other.

By extending life indefinitely we'll create a value of life that's never been seen before on the face of the earth. We'll value each other, we'll value ourselves and quit killing each other. It's one thing to kill somebody who's only going to have 70, 80 years. It's another thing to kill somebody who's going to live a thousand. Think about it.

The value is increased exponentially, and that's the way it should be. We're afraid to really be close on this planet because we die off to one another. The fear is there to really get close. We should be really close as human beings. We should really be working together and moving together and flowing together in a way that really create everything we desire on this planet. And we can do it. Thank you.

Fig. 16.9 David Kekich.
(Photo by Yannick Apers)

Appendix 4: Presentation by David Kekich (Fig. 16.9)

David: Fortunately, I don't have much to say. So, you'll all be a lot happier. Bill, you said you thought by the time you hit 65, you would have solved this, at least the aging problem would have been solved. I'm not like you, I thought it would take a lot longer rather than a lot shorter, although I hoped it would take a lot shorter. I always figured it would take place in my lifetime, expected lifetime. But it's, uh, what was I going to say, Jim?

Jim: You were going to say that you're feeling a lot.

David: Right. Right.

Anyway, I was talking to Bill Faloon, Oh God, 20 years ago, and we were talking about life extension and I was telling him some of the things that I thought would do it, and he said, no, we're all going to head for the dewaoor. Meaning we were all going to be cryonically suspended. And I said, well, I wouldn't, I wouldn't count on that, that's not necessarily so. So, about, I guess about 18 years ago, 19 years ago, I set a goal of 2029 to reverse the human aging process. I since moved that back to 2033 because we weren't getting the funding like Bill said. Now some funding is coming in but not nearly enough. And again, that's one reason for the RAAD Fest. Not to ask anybody for their money, although wealthy people have attended RAAD Fest and donated or invested in life extension research..., but to become big enough and have enough clout, and we're getting there, to create awareness. We want support. But unless it translates to dollars, then it's not going to make these technologies be developed sooner rather than later. So now I'm at 2033, but I'm starting to rethink that too, because I'm thinking now maybe I'll start inching back to 2029. And Bill, what do you think? When do you think we'll be able to reverse the human aging process?

Bill: I thought it was going to be like 10 years ago, but when I talk about curing aging, I'm talking about the research that I was doing and maybe somebody else would come along sooner or something. But I thought with my research I would have something that would reverse aging by now and so you know, I don't know what everybody else's anticipations are. Aubrey de Grey is, I think very, very optimistic that something's going to happen soon. I'm like, when I first started Sierra Sciences 20 years ago, I really thought we would have something to be testing in humans by now, not just tested but with testing completed and we'd actually be seeing older people getting young already.

David: So to ask my question again, when do you. I'll rephrase it. When do you predict we'll be able to reverse the human aging process?

Bill: Okay. I'm going to give an optimistic view of 5 years from now. I'm going to say I believe it's going to be because I think funding is going to come forward really fast.

David: Okay, and, and what is your pessimistic opinion? Don't say 100 years or 50 years.

Bill: I'm trying to call it a really good answer. Pessimistic. Let's see. I would say maybe 30 years from now.

David: See my pessimistic view is not that long. My pessimistic look is maybe 20, 25 years. I'd also like to thank Bill for some of the material and I want to borrow something else you said Bill. You said the two most important things are reducing inflammation and lengthening telomeres and coincidentally, and this is not a pitch, but we have. We started a company several years ago to sell just a couple of supplements, or a handful, and we have three supplements, one of which is an inflammation product, which was developed by Ben with his AI, and a telomerase product, which he was also involved with. Those are two of the three most important things that we thought we should have.

Fig. 16.10 Dr. Greg Fahy. (Photo by Yannick Apers)

Appendix 5: Presentation by Dr. Greg Fahy (Fig. 16.10)

Where else would I get a reception like this? Thanks. So, what should I say? Well, one of the people I was sitting next to said that she was mostly interested the science behind what we're all talking about. And you know, we have scientists here, but the discussion has been more about personal views of things so far. So I'll just tell you that there are, at least in animal models, at least six ways of reversing aging. So if you start applying all of these things simultaneously, you know, Katie bar the door. And this is hard science.

Jim, you said you didn't know what to call yourself. I'd say you're the evangelist of life extension. You talked about having a big auditorium full of life extension enthusiasts someday. That would make you the Billy Graham of life extension.

The incredible thing to me is that there are so many people that actually want to live longer and are energized by it. You're very special people. You get it, right? You're way ahead of everybody else. But in the end, we need to have science behind it, right? So, the science is coming. And I'll say one other thing and that is that I've been going to scientific meetings on aging since the early 1980s. I used to report for a newsletter called Anti-Aging News, which ultimately became Life Extension Foundation's Life Extension magazine, eventually. That was a long time ago, but I would go to the American Aging Association meeting every year, which is the best scientific meeting on the biomedical aspects of aging, and often I would report on amazing observations and promising developments that were made, and then I'd go back to the meeting next year and there was no follow-up. And I would go back the next year, and still there was no follow-up. That was almost the rule.

Somebody would report something awesome and incredible and it just never went anywhere. But that's no longer true. People are picking up on all of these incredible advances now, and that's exciting and encouraging. There was reference earlier to Ray Kurzweil and living long enough to live forever. The way you do that, the whole idea about accelerating returns is that every advance makes the next advance easier. So we have at least seven large advances against aging, one of which was even in humans, and when people start noticing that, these are going to make the next advance even easier. So I used to be pretty pessimistic. But now I'm getting a lot more optimistic.

Okay, so you may have heard of these, but you may not think about them often enough to realize how much you already know. My experience is that sometimes being told what you already know is not a bad thing because it reminds you what you already know. Okay? So if I tell you what you already know, don't be mad at me. Okay? The first one is parabiosis, "heterochronic parabiosis." Sure. You all know that, right? [joking] But I see that one guy here knows. Okay, so you're the only one that I will bore here. If you take an old animal and you hook it up to a young animal, and you let them share a common blood supply, it's not good for the younger animal, but the older animal gets younger. Not just cosmetically younger, but biochemically younger. For example, one of the very first observations is, you stick a needle in the muscle of the old animal and you look at how well it heals and it doesn't heal worth a damn, but after it's been hooked up this way, it heals exactly like a young animal.

Okay, so that's heterochronic parabiosis. Fantastic observation. Not only because it works, but because it works for a very important reason, right? So there are different theories of aging. You know, Bill (Andrews) referred to oxidative damage and inflammation. All of this is true, but what drives it? Right? So if it's the old environment that's driving it, then if you change the environment, you can reverse aging. And that's what that experiment pretty much shows: that aging is under biological control. It's so complicated. It's hard to figure out how to control it, but this is a beautiful way to do it because you take all the complexity in the young animal and hook it up to all the complexity in the old animal, and nature takes care of the problem itself. That's very nice.

So related to that is GDF 11, and you've heard about GDF 11 at both of the last two RAADfests. GDF 11 has been reported to reverse many important aspects of aging, although the picture is currently quite complicated and controversial. Still, if I remember what Steve Perry said, it also reverses DNA damage, and you know, DNA damage is one of those things that when you hear about it, it sounds like, oh, we'll never reverse that, that's impossible, but no, you can reverse DNA damage, and doing so with GDF11 is not the only example.

Very recently, just within the last few years, there's a brand new way of reversing aging that came onto the scene and that you probably have all heard about, and that's recharging NAD. If you ever get into biochemistry, one of the most fundamental things about biochemistry is that NAD has a partner, NADH, and between them they control the redox balance. According to the free radical theory of aging, we get oxidized and that's bad. But actually what happens with aging is that NAD, which is the oxidizing side of that NAD/NADH couple, goes down, so we can actually become more reduced, and that's what drives much of aging.

So there's a paradox. I'm not saying oxidative damage doesn't happen, but in terms of gene expression, which is really what's behind aging, there's something that drives NAD down, and actually we're beginning to understand how that happens now. But you can artificially reverse that just by supplying more NAD or things that turn into NAD, and when you do that, aging is reversed, not just, you know, reduced or whatever. It's actually reversed. So these are profound improvements, profound.

The next two are related to each other, but it's all part of the same theme, and that is that aging is biologically controlled. And people didn't want to believe that because people like Mike Rose cannot conceive of how aging can be biologically controlled, but I don't care about the theory. I care about reality. What works, what's true. Yeah. Right. And so what's true is that a lot of aging in the body is controlled by the brain.

Just the same way that the brain controls a lot of development when you go into the developmental process, you go through puberty, you become an adult. All of those are programmed events that are built into your DNA. We all have life histories. Aging seems to be an extension of that life history program. And one way that we know that is that a guy at Albert Einstein College of Medicine discovered that there's one kind of cell in one very specific and very local part of the brain that expresses one gene that turns on aging throughout the whole body. And he was able to stop that and reverse some of the aging changes that took place normally.

And then there was a follow-up paper which I had in my coat jacket at the RAADfest meeting last year. I was going to wave it around in case anybody objected to what I was going to say. It had just come out in Nature. Nature is like the top scientific journal on the planet, right? And they had gone one step farther and found another brain mechanism of aging in which neural stem cells in the brain talk to the rest of the body, and by providing the signals that those cells produced normally in a young animal, they could again reverse aging in an older animal. So those are all pretty incredible things.

The last thing is another story. Again, an independent study completely looking at aging from a different point of view from everything else, but, but the theme is exactly the same.

And so you've probably all heard from people like Mike West that you can take a 91 year old cell and revert it back to say a 20 year old cell by changing various aspects of gene expression. For example, you can increase telomerase expression such that you can bring the telomere length back. But you can bring all other aspects back as well. And we heard about that at the last RAADfest meeting as well from Bobby Dhadwar at Harvard University, right? He did it through NAD in that case, but there are other ways to do it too. And so what Mike West was so interested in is the fact that there are factors that you can put into an old cell that turn it back into an embryonic cell. Okay? And so even a really old cell can go back to an embryonic cell, but we don't want to be embryos again, right? So there are people at the Salk

Institute who worked with what are called Yamanaka factors. There's just four little biochemical factors that can turn an adult cell back to an embryonic cell. The people at the Salk Institute provided these factors to adult and to aged mice, but they didn't provide them strongly enough to turn a whole mouse back into an embryo, they just provided them strongly enough to turn an old mouse into a younger mouse. And it worked in a whole animal model. They started off with an artificial model in which the aging was artificially accelerated, and they showed they could block that, but then they actually went after an honest to gosh old animal and showed aging reversal in the old animal using the same approach.

And we don't understand that very well, but there were no side effects. I mean, I don't think that there were any problems for those mice. So those were the six things that you all know that are known now and soon there will be a seventh. So thank you, Linda, for giving me a chance to explicate those points. Okay.

Jim: Greg, did you hear about the cord blood plasma infusions. Is that pretty much the same as the young blood or because it's taken from the umbilical cord and so it's a little bit less.

Greg: If I am to believe what I've heard about that, it would be yet another mechanism of reversing aging.

Jim: Yes. Dr. Peterson with hexagon therapy is actually going to be speaking at RAADfest about it.

Greg: Fantastic. So I know that Bill Vaughn was involved in studies like that and he reported some exciting results along the same lines, but they haven't been published as far as I know, they haven't been validated, so I didn't mention them, but certainly there is support apparently from two sources for that same thing. So scratch six ways to reversing aging and make that seven plus one. That's eight ways of reversing aging. That's pretty good. For starters.

I don't know if you guys know about James, but James Clement has been doing a lot of human clinical work on NAD, and he points out that in normal human aging, your NAD levels may decline by 70 percent or 80 percent as you get older. And things like NR supplementation may double your blood levels, but you know, that's taking you from 20 percent of what you should be at to 40 percent of what you should be. You're still falling way short. So we still need something better, but it's great that we have something that we can do along these lines. And so probably the more the merrier, within reason. And that's why NAD infusions have been popular. As Jim alluded to, it looks like there's going to be an alternative to infusions, patches which will achieve higher blood levels and be even better. So these things are evolving. They're evolving very fast.

Resveratrol was shown to activate the sirtuin pathway to produce anti-aging effects, but resveratrol needs NAD: you've got to have the NAD or the whole thing shuts down. So both of them can contribute to what we're looking for, but it would be better to take both of them. And actually, I do, I take resveratrol every day and I take NR (an NAD precursor) every day.

Fig. 16.11 Bernadeane Brown. (Photo by Yannick Apers)

Appendix 6: Presentation by Bernadeane Brown (Fig. 16.11)

Oh my goodness, wow. It's great to be together like this. I love it. There's nothing, nothing like being together. There's an intimacy that we all need to experience all the time. We need to be able to have each other to inspire each other! I need to be inspired! I'm somebody, but I'm not enough and I need you. I need who you are. I like the difference of all of us. I'm glad we're not all the same. You know, we need to be different. We need to be who we are and to be able to receive each other in all that we are. We may not always agree on everything. I don't care. I'm not concerned with it. I just like it that we have living in mind instead of death in mind. I like that. I think that, and really, actually I'm all about forever.

I feel like I'm on my way to forever. I am not stopping. I feel like I don't have to stop and die. I don't have to take time out for death. I feel that something going on in me that just is getting bigger and larger all the time and well, you know what, I came out of the closet. I came out of the closet. I'm not afraid to tell the world I'm here to stay. I don't care if they think I'm a nut! I'm happy to feel that I don't have to die. I don't feel it. I don't feel death for me. And so, you know, I came out of the closet with it and really it's great when you do that. I've got nothing to lose! Only death! Look, if I can lose some asshole "friends", then that's what they are. They're assholes.

And that's why I feel if I can lose (and I have, I've lost friends, I've lost family,) because I am so definite about staying here. I think the physical body is amazing and I think we've just so destroyed ourselves. People destroy themselves and we're here to build our flesh and to build one another. I'm here to move with you that you live with me forever. I want you by my side. It's a great thing, it's a great thing to have one another. You know, I've been around a long time. I was born in 1937. So I've been around a long time, but not long enough. It's just the beginning for me! Forever is a very, very long time and that's what I do! But yeah, sometimes I don't like to tell people the years I've been here, I always tell people I've been here for 81 years, but I'm not 81 years old because I will not identify with that. That's not who I am. And so we have to, we have to feed ourselves that which causes us to live what we feel.

Yeah, so because sometimes when you tell people how long you've been here, they put you in a box, like, oh, she's been here too long or she's too old for me, you know, or he's too old for me. It's such crap. I'm so glad. I'm glad we're walking out and away from a lot of ignorance! We're getting smarter, we're waking up! We're expanding our intellect! It's becoming immortal. There's something happening in the body today. There is a new happening going on with humanity today that has never been before and it's on the increase and it's going to keep growing and growing and growing and you're gonna see it! You're going to see it in your lifetime.

Yeah, you're gonna. See it. You're gonna see! There are people like me that are going to stay around here. You better stay around if you want to see it. You better stay around!

Just wake up! Let's wake up! There's a new happening going on. I'm telling you that people are looking at things. (Of course, not everybody,) but there are a lot of people now that are looking at everything in a new way. Everything in a new way. I see everything totally different than I did years ago and but, but I kept moving and I think I've got a lot of experiencing. I think we all together are going to have the most fun. It. Look, if living isn't about fun, forget it. [applause] If living isn't about having a party, forget it. I can't stand old stuff. Old everything. Old everything! Throw away your old clothes. Get rid of it! Put on something new and fresh! [applause].

It'll make you feel better. We need to. We need to let go of all of the old. It's done. Doesn't it bore you to keep going over the same things all the time? (We're, we're into…) I have such an excitement! And it's because I know that really there is going to be the answers that we're hungry for. There is going to be. I feel it. I know it! We've reached this time and I don't know how many years we're talking about how many years here today. I don't know, but it's important that we do all the things that we can do to keep our bodies strong. You keep your body strong. Jim mentioned different things you could do right now that you don't have to die. You don't have to die. Look, I've been here a long time. I'm not on any medication. I'm not crippled up. I. I'm like a kid. I move. I've got a lot of movement. To me, I'm full of energy. You don't have to grow old, you all grow old because it's what you give to. It is an energy in the body. It is. It is old is an energy!

And you don't have to do it. Yeah. People become so mature. They're so mature they can't, they can't play. They're so mature! They can't, they can't let the child that they are manifest itself. I like being mature. I liked being sophisticated, but I also like being silly and I like being, you know, just having some good times. Everything for me is a good time! The depth, the depth that we're going to about the physical body and finding all the ways for us to live abundantly! It's so precious. It is so precious. I love it! I'm serious about that. I'm very serious, but I'm also very playful about it, I think. I think it's so, so vital for us to give to a feeling. There's a feeling we need to have of one another, a feeling. It's okay to touch each other. It's okay to touch each other. Be held by one another, be felt by each other. It's special. It is so special. We're creating a new world in the midst of this world that exists right now. Some of it is so embarrassing to me. Things that are going on in our United States of America (I'm not going to get in on that), I'm not going there, but I'm telling you some of the things that are going on are really unto destruction and I see us creating a world that has no destruction to it and that's what I feel.

There, there are those of us that are honest, we're honest, we have integrity. I'm going to be with people only that have honesty and integrity and move with a feeling and a caring for one another. Those are the people I want to be with. I'm not interested in the others.

So, there a new world happening in the midst of this old one. There is, and we're going to see it more and more. I'm happy to be with you today. I think we're going to grow and grow and grow in numbers and we need the numbers. We need one another, but what we need the numbers, we need people here in the world. They don't even know we exist! They don't know that there are scientists that have something really on the ball going on. They don't know it. We've tried to have conversations with different people at different times just on the street or wherever and it's like they're totally, they never heard of it. They never heard anything about what's going on right now in our world. And so the people need to know and it takes all of us, the numbers of us where they can see where humanity can see something really amazing is happening and we are that happening.

And so I think we need more togetherness. We need more of these kinds of coming together. We do, we need these, we need this because when we just live unto ourselves, it can be kind of discouraging sometimes. Sometimes you might lose heart, you know, and so we need one another so much. Not because we're weak, I'm not a weakling, but I like having contact with people that have something going on in them that they can give me something that I need and I give out who I am. I know I affect people and I want to affect people. I want to affect people to the points that they, they get a hunger in their bodies, to live. They find that: "Wow, there is a reason to stay alive!" A lot of people find no reason to stay alive so they don't care. They can have lots and lots of money.

There are billionaires that have lots of money, but they, they're not looking to stay around and they're not interested. They don't. I don't know. I'd like to see some of them get, get awakened enough that they could support some of our scientists and doctors that are doing these tremendous things. So. But the more we are known, all of us, the more all of us are known in this world. The sooner these things are gonna happen. That's what I feel. So, don't stop. Don't stop. I'm not stopping. Don't stop. Don't stop. Keep moving.

Okay, thank you. [applause]

Bibliography

1. **Huxley, Julian.** Transhumanism. *New Bottles for New Wine.* London : Chatto & Windus, 1957.
2. **Plato.** Apology. [Online] 399 BC. http://classics.mit.edu/Plato/apology.html.
3. **History Channel Editors.** Socrates Sentenced to Death. *The History Channel.* [Online] February 15, 399 BC. https://www.historychannel.com.au/this-day-in-history/socrates-sentenced-to-death/.
4. **Singh, Simon.** Fermat's Enigma: The Quest to Solve the World's Greatest Mathematical Problem. *The New York Times.* [Online] 1997. https://archive.nytimes.com/www.nytimes.com/books/first/s/singh-fermat.html.
5. **Folger, Tim.** Einstein's Grand Quest for a Unified Theory. *Discover.* [Online] September 30, 2004. http://discovermagazine.com/2004/sep/einsteins-grand-quest.
6. **Piller, Charles.** Verily, I swear. Google Life Sciences debuts a new name. *STAT.* [Online] December 7, 2015. http://www.statnews.com/2015/12/07/verily-google-life-sciences-name/.
7. **McCracken, Harry and Grossman, Lev.** Can Google Solve Death. *TIME Magazine.* [Online] September 30, 2013. http://content.time.com/time/magazine/0,9263,7601130930,00.html.
8. **Brown, Aaron.** An update on Google Health and Google PowerMeter. *Google Official Blog.* [Online] June 24, 2011. https://googleblog.blogspot.com/2011/06/update-on-google-health-and-google.html.
9. **Miller, Leslie.** Google announces Calico, a new company focused on health and well-being. *News from Google.* [Online] September 18, 2013. http://googlepress.blogspot.com/2013/09/calico-announcement.html.
10. **Piller, Charles.** Google's next big idea: Mining health data to prevent disease. *STAT.* [Online] December 2, 2015. http://www.statnews.com/2015/12/02/google-doctor-jessica-mega/.
11. **Page, Larry.** About 14 years ago. *Google+.* [Online] May 14, 2013. https://plus.google.com/+LarryPage/posts/aqy6DvvLJY1.
12. **Shontell, Alyson.** Larry Page Tells Wall Street This Could Be His Last Google Earnings Call For A While. *Business Insider.* [Online] October 17, 2013. http://www.businessinsider.com/larry-page-wont-be-doing-every-google-earnings-call-2013-10.
13. **TIME Staff.** Exclusive: TIME Talks to Google CEO Larry Page About Its New Venture to Extend Human Life. *TIME Magazine.* [Online] September 18, 2013. business.time.com/2013/09/18/google-extend-human-life/.
14. **Tinker, Ben.** US life expectancy drops for first time in 22 years. *CNN.* [Online] December 8, 2016. http://www.cnn.com/2016/12/08/health/us-life-expectancy-down/.
15. **Xu, Jiaquan, et al., et al.** Mortality in the United States, 2015. *Centers for Disease Control and Prevention.* [Online] December 2016. https://www.cdc.gov/nchs/products/databriefs/db267.htm.
16. **Tinker, Ben.** US life expectancy drops for second year in a row. *CNN.* [Online] December 21, 2017. http://www.cnn.com/2017/12/21/health/us-life-expectancy-study/index.html.
17. **Kochanek, Kenneth D., et al., et al.** Mortality in the United States, 2016. *Centers for Disease Control and Prevention.* [Online] December 2017. https://www.cdc.gov/nchs/data/databriefs/db293.pdf.
18. **Clark, Tom.** Dying Matters survey finds 15% want to live forever. *The Guardian.* [Online] May 15, 2011. http://www.theguardian.com/uk/2011/may/16/dying-still-taboo-subject-poll.
19. **U.S. Department of Health and Human Services.** United States Cancer Statistics: Data Visualizations. *Centers for Disease Control and Prevention.* [Online] June 2018. https://gis.cdc.gov/cancer/USCS/DataViz.html.
20. **White, Tracie.** Most physicians would forgo aggressive treatment for themselves at the end of life, study finds. *Stanford Medicine News Center.* [Online] May 28, 2014. http://med.stanford.edu/news/all-news/2014/05/most-physicians-would-forgo-aggressive-treatment-for-themselves-.html.

21. **Schneider, Keith.** Dr. Jack Kevorkian Dies at 83; A Doctor Who Helped End Lives. *The New York Times.* [Online] June 3, 2011. https://www.nytimes.com/2011/06/04/us/04kevorkian. html.
22. **Klein, Christopher.** 10 Things You May Not Know About Sigmund Freud. *History.com.* [Online] September 23, 2014. https://www.history.com/ news/10-things-you-may-not-know-about-sigmund-freud.
23. **McKenzie, Sheena, et al., et al.** 104-year-old scientist David Goodall 'welcomes death' at Swiss clinic. *CNN.* [Online] May 9, 2018. https://edition.cnn.com/2018/05/08/health/david-goodall-australia-switzerland-interview-intl/index.html.
24. **Curtin, Sally C., Warner, Margaret and Hedegaard, Holly.** Increase in Suicide in the United States, 1999–2014. *National Center for Health Statistics.* [Online] April 2016. https:// www.cdc.gov/nchs/products/databriefs/db241.htm.
25. **World Health Organization.** Suicide data. *Mental health.* [Online] 2017. http://www.who. int/mental_health/prevention/suicide/suicideprevent/en/.
26. **Cooper, Anderson.** Anderson Cooper: Thirty years after my brother's death, I still ask why. *CNN.* [Online] June 24, 2018. https://www.cnn.com/2018/06/24/us/anderson-cooper-brother-suicide/index.html.
27. **Woo, Elaine.** Stephen Crohn dies at 66; immune to HIV, but not its tragedy. *Los Angeles Times.* [Online] September 21, 2013. http://articles.latimes.com/2013/sep/21/local/ la-me-stephen-crohn-20130922.
28. **Gallup.** Employee Engagement. *Gallup.* [Online] http://www.gallup.com/topic/employee_ engagement.aspx.
29. **Singal, Jesse.** For 80 Years, Young Americans Have Been Getting More Anxious and Depressed, and No One Is Quite Sure Why. *New York Magazine.* [Online] March 13, 2016. http://nymag.com/scienceofus/2016/03/for-80-years-young-americans-have-been-getting-more-anxious-and-depressed.html.
30. **Hamlyn, Charlotte.** Academic David Goodall turns 104 and his birthday wish is to die in peace. *The ABC.* [Online] April 3, 2018. http://www.abc.net.au/news/2018-04-04/ david-goodall-is-104-but-takes-no-pleasure-in-getting-older/9614344.
31. **CNN.** Dana Reeve dies of lung cancer at 44. *CNN.* [Online] March 8, 2006. http://www.cnn. com/2006/SHOWBIZ/03/07/reeve.obit/.
32. **Miller, Mike.** Inside Carrie Fisher's Complicated Relationship with Mom Debbie Reynolds — and Their Close Bond Before Her Death. *People.* [Online] December 28, 2016. http://people. com/movies/inside-carrie-fishers-complicated-relationship-with-mom-debbie-reynolds-and-their-close-bond-before-her-death/.
33. **Lincoln, Ross A.** Stan Lee, Comics Legend Who Co-Created the Marvel Universe, Dies at 95. *The Wrap.* [Online] November 12, 2018. https://www.thewrap.com/ stan-lee-dies-at-95-marvel-comics/.
34. **Scutti, Susan.** Loneliness peaks at three key ages, study finds -- but wisdom may help. *CNN.* [Online] December 20, 2018. https://www.cnn.com/2018/12/18/health/loneliness-peaks-study/index.html.
35. **Monroe, Rachel.** The Bulletproof Coffee Founder Has Spent $1 Million in His Quest to Live to 180. *Men's Health.* [Online] January 23, 2019. https://www.menshealth.com/health/ a25902826/bulletproof-dave-asprey-biohacking/.
36. **Lorenz, Taylor.** The Lil Tay Saga Reaches Its Logical Conclusion. *The Atlantic.* [Online] May 24, 2018. https://www.theatlantic.com/technology/archive/2018/05/ the-lil-tay-saga-reaches-its-logical-conclusion/561116/.
37. **Hankinson, Andrew.** Who Wants to Live Forever? *Men's Health.* [Online] June 15, 2015. http://www.menshealth.co.uk/healthy/brain-training/who-wants-to-live-forever.
38. **Bucketlist.org.** Your dreams, made possible. *Bucketlist.* [Online] [Cited: March 7, 2016.] https://bucketlist.org/.
39. **Doren, John Van.** The Beginnings of the Great Books Movement. *Columbia Magazine.* [Online] Winter 2001. http://www.columbia.edu/cu/alumni/Magazine/Winter2001/great-Books.html.

40. **Knight, Les.** About the Movement. *Voluntary Human Extinction Movement.* [Online] http://www.vhemt.org/aboutvhemt.htm#vhemt.
41. **Swaine, Jon.** Steve Jobs 'regretted trying to beat cancer with alternative medicine for so long'. *The Telegraph.* [Online] October 21, 2011. https://www.telegraph.co.uk/technology/apple/8841347/Steve-Jobs-regretted-trying-to-beat-cancer-with-alternative-medicine-for-so-long.html.
42. **Biography.com Editors.** Albert Einstein Biography. *Biography.com.* [Online] http://www.biography.com/people/albert-einstein-9285408.
43. **Mandino, Og.** The Greatest Miracle in the World. *Bantam Books.* [Online] 1975. https://books.google.com/books?id=bG0tSJDUxgcC&pg=PA14&lpg=PA14.
44. **Karimi, Faith and Lynch, Jamiel.** Florida shooting: Bullets flew for 4 minutes as armed deputy waited outside. *CNN.* [Online] February 23, 2018. https://www.cnn.com/2018/02/23/us/florida-school-shooting/index.html.
45. **Disney.** Disney Legend Time Allen. *D23.* [Online] [Cited: December 12, 2015.] https://d23.com/walt-disney-legend/tim-allen/ .
46. **Biography.com.** Tim Allen Biography. [Online] [Cited: May 20, 2015.] http://www.biography.com/people/tim-allen-9542074.
47. **Einstein, Albert.** My Credo. *St. Cloud State University.* [Online] 1932.
48. **McAfee, Melonyce.** Stephen Hawking sings Monty Python's 'Galaxy Song'. *CNN.* [Online] April 14, 2015. https://www.cnn.com/2015/04/14/entertainment/feat-stephen-hawking-sings-galaxy-song/index.html.
49. **Wolchover, Natalie.** A Different Kind of Theory of Everything. *The New Yorker.* [Online] February 19, 2019. https://www.newyorker.com/science/elements/a-different-kind-of-theory-of-everything.
50. **Highfield, Roger.** What's wrong with Craig Venter? Craig Venter, multi-millionaire maverick, says he can help you live a better, longer life. Roger Highfield asks how. *Mosaic.* [Online] February 2, 2016. http://mosaicscience.com/story/craig-venter-genomics-personalised-medicine.
51. **Shampo, Marc A. and Kyle, Robert A.** J. Craig Venter—The Human Genome Project. *Mayo Clinic Proceedings.* [Online] April 2011. http://www.ncbi.nlm.nih.gov/pmc/articles/PMC3068906/.
52. How Doctors Use Poetry. *Nautilus.* [Online] September 27, 2018. http://nautil.us/issue/64/the-unseen/how-doctors-use-poetry.
53. **Haden, Jeff.** Warren Buffett Says Doing 1 Thing Separates Successful People From Everyone Else (and It Stunned Bill Gates). *Inc.* [Online] January 15, 2019. https://www.inc.com/jeff-haden/warren-buffett-says-doing-1-thing-separates-successful-people-from-everyone-else-and-it-stunned-bill-gates.html.
54. **Jr., Berkeley Lovelace.** Aetna CEO: Warren Buffett's 'tapeworm' analogy fits — health-care costs squeeze the economy for trillions. *CNBC.* [Online] March 26, 2018. https://www.cnbc.com/2018/03/26/aetna-ceo-buffetts-tapeworm-analogy-fits-health-costs-hurt-economy.html.
55. **Waggoner, Walter H.** Dr. Burrill B. Crohn, 99, An Expert On Diseases Of The Intestinal Tract. *The New York Times.* [Online] July 30, 1983. http://www.nytimes.com/1983/07/30/obituaries/dr-burrill-b-crohn-99-an-expert-on-diseases-of-the-intestinal-tract.html.
56. **Lee, Newton.** A word from the editor. *ACM Computers in Entertainment.* [Online] Volume 3, Issue 1, January 2005. https://dl.acm.org/citation.cfm?doid=1057270.1057431.
57. **Lohmann, Bill.** Scouts' honor. *Richmond Times-Dispatch.* [Online] March 3, 2008. http://www.richmond.com/entertainment/article_ba2be0ca-209c-5c3d-99fb-28e74af418bd.html.
58. **Carnegie Mellon University.** Randy Pausch's Last Lecture. *Carnegie Mellon University.* [Online] September 18, 2007. http://www.cmu.edu/randyslecture/.
59. **Somerville, Madeleine.** Are eco-friendly initiatives pointless unless we tackle overpopulation? *The Guardian.* [Online] January 26, 2016. http://www.theguardian.com/lifeandstyle/2016/jan/26/overpopulation-sustainability-environment-eco-friendly-initiatives.
60. **World Health Organization.** Food Security. *World Health Organization.* [Online] http://www.who.int/trade/glossary/story028/en/.

61. **NAIC.** Longevity Risk. *National Association of Insurance Commissioners.* [Online] December 14, 2015. http://www.naic.org/cipr_topics/topic_longevity_risk.htm.

62. **Debate.** DEBATE: Immortality will change prison sentences? Execution & Life-behind-Bars.. too sadistic? *Immortal Life.* [Online] March 16, 2013. http://immortallife.info/articles/entry/debate-immortality-will-change-prison-sentences-execution-life-behind-bars.

63. **United Nations.** SAVE FOOD: Global Initiative on Food Loss and Waste Reduction. *Food and Agriculture Organization of the United Nations.* [Online] [Cited: February 4, 2019.] http://www.fao.org/save-food/resources/keyfindings/en/.

64. **Ward, Clarissa, Abdelaziz, Salma and McWhinnie, Scott.** In Yemen, the markets have food, but children are starving to death. *CNN.* [Online] December 19, 2017. https://edition.cnn.com/2017/12/19/middleeast/yemen-intl/index.html.

65. **Haroun, Chris.** 3 Reasons It Pays to Be Long-Term Greedy. *Inc.* [Online] April 22, 2016. https://www.inc.com/chris-haroun/3-reasons-you-must-be-long-term-greedy.html.

66. **The Long Now Foundation.** About Long Now. *The Long Now Foundation.* [Online] http://longnow.org/about/.

67. **Lee, Newton.** Interview with Danny Hillis. *ACM Digital Library.* [Online] March 29, 2006. http://dl.acm.org/citation.cfm?doid=1146816.1146821.

68. **Garber, Megan.** The 3 Most Exciting Words in Science Right Now: 'The Pitch Dropped'. *The Atlantic.* [Online] July 18, 2013. https://www.theatlantic.com/technology/archive/2013/07/the-3-most-exciting-words-in-science-right-now-the-pitch-dropped/277919/.

69. **Zhang, Sarah.** The 500-Year-Long Science Experiment. *The Atlantic.* [Online] January 27, 2019. https://www.theatlantic.com/science/archive/2019/01/500-year-long-science-experiment/581155/.

70. **BBC News.** 'Clive of India's' tortoise dies. *BBC News.* [Online] March 23, 2006. http://news.bbc.co.uk/2/hi/south_asia/4837988.stm.

71. **Associated Press.** 140-year-old lobster's tale has a happy ending. *NBC News.* [Online] January 10, 2009. http://www.nbcnews.com/id/28589278/ns/us_news-weird_news/t/-year-old-lobsters-tale-has-happy-ending/.

72. **Marks, Kathy.** Henry the tuatara is a dad at 111. *The Independent.* [Online] January 26, 2009. http://www.independent.co.uk/news/world/australasia/henry-the-tuatara-is-a-dad-at-111-1516628.html.

73. **Wilkins, Alasdair.** Turtles could hold the secret to human immortality. *Gizmodo.* [Online] August 20, 2010. http://io9.gizmodo.com/5618046/the-mystery-of-why-turtles-never-grow-old%2D%2D-and-how-we-can-learn-from-it.

74. **Strategies for Engineered Negligible Senescence Research Foundation.** About SENS Research Foundation. [Online] [Cited: March 15, 2016.] http://www.sens.org/about.

75. **Best, Ben.** Interview with Aubrey de Grey, PhD. *Life Extension.* [Online] July 2013. http://www.lifeextension.com/magazine/2013/7/Interview-with-Aubrey-de-Grey-PhD/Page-01.

76. **Monaghan, James R. et al.** Microarray and cDNA sequence analysis of transcription during nerve-dependent limb regeneration. *BMC Biology.* [Online] January 13, 2009. https://www.ncbi.nlm.nih.gov/pmc/articles/PMC2630914/.

77. **Than, Ker.** "Immortal" Jellyfish Swarm World's Oceans. *National Geographic.* [Online] January 29, 2009. http://news.nationalgeographic.com/news/2009/01/090130-immortal-jelly-fish-swarm.html.

78. **Regalado, Antonio.** A stealthy Harvard startup wants to reverse aging in dogs, and humans could be next. *MIT Technology Review.* [Online] May 9, 2018. https://www.technologyreview.com/s/611018/a-stealthy-harvard-startup-wants-to-reverse-aging-in-dogs-and-humans-could-be-next/.

79. **Graeber, Charles.** Meet the Carousing, Harmonica-Playing Texan Who Just Won a Nobel for his Cancer Breakthrough. *Wired.* [Online] October 22, 2018. https://www.wired.com/story/meet-jim-allison-the-texan-who-just-won-a-nobel-cancer-breakthrough/.

80. **Robinson, Matthew.** Leading cancer expert dies suddenly following routine yellow fever vaccination. *CNN.* [Online] January 11, 2019. https://www.cnn.com/2019/01/11/health/yellow-fever-vaccination-martin-gore-death-gbr-scli-intl/.

81. *Antibiotics and Bacterial Resistance in the 21st Century.* **Fair, Richard J and Tor, Yitzhak.** s.l. : Perspectives in Medicinal Chemistry, 2014, Vol. 6.
82. **Boseley, Sarah.** Antibiotics are losing effectiveness in every country, says WHO. *The Guardian.* [Online] April 30, 2014. https://www.theguardian.com/society/2014/apr/30/antibiotics-losing-effectiveness-country-who.
83. **U.S. Department of Health & Human Services.** Antibiotic / Antimicrobial Resistance. *Centers for Disease Control and Prevention.* [Online] January 8, 2018. http://www.cdc.gov/drugresistance/.
84. **McKenna, Maryn.** Long-Dreaded Superbug Found in Human and Animal in U.S. *National Geographic.* [Online] May 26, 2016. http://phenomena.nationalgeographic.com/2016/05/26/colistin-r-9/.
85. **Davies, Madlen.** Scourge of superbugs killing Malawi's babies. *CNN.* [Online] August 8, 2018. https://www.cnn.com/2018/08/08/health/malawi-superbugs-antimicrobial-resistance-among-newborns-intl/index.html.
86. **Nelson, Bryn.** Four emerging STDs that you can't afford to ignore. *CNN.* [Online] December 7, 2018. https://www.cnn.com/2018/12/06/health/std-sexually-transmitted-diseases-partner/index.html.
87. **O'Neill, Jim.** Tackling Drug-Resistant Infections Globally: Final Report and Recommendations. *Revew on Antimicrobial Resistance.* [Online] May 2016. http://amr-review.org/sites/default/files/160525_Final%20paper_with%20cover.pdf.
88. **Department of Health and Human Services.** CDC: the Nation's Prevention Agency. *Centers for Disease Control and Prevention.* [Online] November 6, 1992. https://www.cdc.gov/mmwr/preview/mmwrhtml/00017924.htm.
89. **Milova, Elena and Hill, Steve.** Dr. Greg Fahy – Rejuvenating the Thymus to Prevent Age-related Diseases. *Life Extension Advocacy Foundation.* [Online] October 4, 2017. https://www.leafscience.org/rejuvenating-the-thymus/.
90. **Battelle, John.** All We Have Yet To Understand. *NewCo Shift.* [Online] January 4, 2018. https://shift.newco.co/all-we-have-yet-to-understand-eda033405fdb.
91. **Bryce, Emma.** How Long Do Tardigrades Live? *Live Scienec.* [Online] June 2, 2018. https://www.livescience.com/62720-tardigrade-lifespan.html.
92. **Marchione, Marilynn.** Chinese researcher claims birth of first gene-edited babies — twin girls. *STAT.* [Online] November 25, 2018. https://www.statnews.com/2018/11/25/china-first-gene-edited-babies-born/.
93. **Belluz, Julia.** Is the CRISPR baby controversy the start of a terrifying new chapter in gene editing? *Vox.* [Online] January 22, 2019. https://www.vox.com/science-and-health/2018/11/30/18119589/crispr-gene-editing-he-jiankui.
94. **Ueland, John.** China's CRISPR twins might have had their brains inadvertently enhanced. *MIT Technology Review.* [Online] February 21, 2019. https://www.technologyreview.com/s/612997/the-crispr-twins-had-their-brains-altered/.
95. **Carrington, Damian.** Humans just 0.01% of all life but have destroyed 83% of wild mammals – study. *The Guardian.* [Online] May 21, 2018. https://www.theguardian.com/environment/2018/may/21/human-race-just-001-of-all-life-but-has-destroyed-over-80-of-wild-mammals-study.
96. **Goodman, Marc.** Marc Goodman: A vision of crimes in the future. [Online] TEDGlobal 2012, June 28, 2012. http://www.ted.com/talks/marc_goodman_a_vision_of_crimes_in_the_future.html?quote=1769.
97. **Roach, J., et al., et al.** An expert system for information on pharmacology and drug interactions. *US National Library of Medicine.* [Online] 1985. https://www.ncbi.nlm.nih.gov/pubmed/3979039.

Chapter 17
To Age, or Not to Age: That Is the Question

Maria Entraigues Abramson

17.1 We are Born Dead

Throughout recorded history humans have always aged and died. Every person who's ever lived has died. Some died in the jaws of another animal, of infection, in an unfortunate accident, of a heart attack, fighting a war, or in the limitless other ways that can end a person's life. Life is very uncertain and we don't know when or how we will die, but what we know for sure is that if all those people who got killed by something else other than old age wouldn't have died of some other cause, they would have eventually, with 100% certainty, died of aging.

Since I was a little girl I felt aging was wrong. One day, my older sister and I were walking around our neighborhood and came upon a very old lady who could barely walk. Her back was completely bent over, forcing her to only look down. She moved incredibly slowly, uttering sounds of pain with each labored step. I must have been only 5 years old, yet I vividly remember how much that image disturbed me. I asked my sister "what is wrong with her?" "*She's just very old, we all get old and die*", she said. She looked at me matter-of-factly. *"We are born dead"*.

I was too young to understand what she meant by that, but it definitely hit me. I'm not sure if it was the shock of her remarks or the image of that elderly lady, but whatever it was, I became very aware of aging and wanted to understand it better.

I would quietly observe how elderly people would get sick, complain about pain, and talk about how lonely they felt; how cruel old age was. I would see them suffer greatly, lose their vigor and their cognitive faculties, becoming so frail that they eventually lost themselves. They would completely fade away. Even if they were still physically there, it was not *them* anymore. And I couldn't just ignore this.

M. Entraigues Abramson (✉)
Global Outreach Coordinator, SENS Research Foundation, Mountain View, CA, USA

© Springer Nature Switzerland AG 2019
N. Lee (ed.), *The Transhumanism Handbook*,
https://doi.org/10.1007/978-3-030-16920-6_17

I also especially noticed almost everyone around me comfortably accepting this, and even finding "positive" things about this obviously sad and destructive process. The decrepitude caused by old age was often the punch line of their jokes too. I found all this astonishing.

17.2   What is Aging?

The dictionary says aging is "the process of growing old". Of course this is a very vague definition. Another common definition is "the accumulation of changes in a biological organism over time", also rather vague.

Scientists scrupulously investigate different mechanisms when it comes to defining aging i.e.; accumulation of damage at a cellular level as a side effect of metabolic processes, genetic programming, the senescence of cells, mitochondrial aging, shortening of telomeres, etc. And are making new discoveries every day.

What it is inarguable is that aging is the primary cause of death in humans. About 70% of deaths are the result of aging; more than a 100,000 people a day. These people not only die of age related disease, but generally suffer a great deal of pain and disability, and often for a very long time.

As we age the risk of death goes up significantly. Why is that? Because the older we get, the sicker we get, and if we live long enough, we will inevitably become critically ill with, and succumb to, at least one of the many diseases of aging.

One example is that age is the single greatest risk factor for developing cancer. In fact, 60% of people who have cancer are 65 or older. At age 30, we have a one in a thousand chance of getting cancer in the next year, but by the age of 80, our chance of getting cancer in the next year is one in ten. A hundred times higher.

This increasing burden of disease that results from aging is becoming unsustainable to our healthcare system. The disability of millions of people who live longer but in very bad health, represent a major global challenge. Those who are not healthy can't contribute to the economy of the world and become an escalating heavy load to the system.

In America, we spend more money on healthcare in the last year of our lives than in our whole life prior. And for what? Usually just to extend the life of a very ill person, and very often to just extend their agony. This not only affects the dying elder, but their whole family.

Another startling fact is that in America alone, we spend about $40,000 per *second* on older peoples' healthcare. This means that by the time I'm done writing and correcting this essay about 432 *Million* dollars will have been spent. And how many people will we have cured from an age related disease during those 3 hours? ZERO.

17.3 Aging = Disease

You've probably heard people say "he/she died of old age". What does this really mean? Generally we have not thought much about this definition, but in reality every person who died of "old age" died of a medical condition. Every "old age" death has a clinical explanation.

For as long as humans have struggled to understand the nature of our existence, we have not asked many questions about aging. We have pretty much accepted it as the natural path we must all endure.

At the same time, we've been fighting disease for thousands of years, and even though disease is also completely natural, we don't just surrender to it. We intervene; we modify whatever is necessary to get rid of it, and this makes sense to most of us. It is not controversial. But when it comes to aging, for some reason it is!

Gerontology and Geriatrics are the medical fields concerned with ill health in the aged, but so far, they have failed to cure any of these diseases; only ameliorating the symptoms, and sometimes only slowing down the otherwise inevitable progress of ailments.

Even though it is completely obvious that as we age we lose our health, in all the years of diligently struggling to advance research and cures, science has not focused on it in the right way. As much as we fight to find cures for disease in general, when they are associated with aging, we see them differently. We have cured and even eradicated many infectious diseases, but not one single age-related disease. It is definitely time for a paradigm shift.

17.4 The Irrational Resistance

What concerns me the most is not the science (which of course is the ongoing challenge of discovery) but the ubiquitous mentality that swims in fear and controversy around it. Unless and until we radically shift our mindset about aging, we will probably not have sufficient funding to move fast enough with the research and development of groundbreaking therapies to bring this about in our lifetimes. Something that could have been solved in 20 years could instead take a hundred – or more.

Humans have achieved incredible milestones in very little time with the right incentive, but sadly, aging does not seem like an urgent global problem to most people. It is completely astounding to me that something that happens to every single person (unless you even more tragically die young), something that we all hate and express relentless disdain for in a million ways, something that brings disability, pain and certain death, is still not enough of an incentive to try to do something about it.

I've now been working professionally on the mission of curing aging for over a decade, and I hear a staggering amount of concerns and criticism around the development of real anti-aging medicine, most of it very irrational.

A typical fear and misconception is that once we have cured aging, we will be forced to live our vastly longer lives in the elderly state of an old and sick body. It is hard for me to understand how they arrive at that conclusion when what we are talking about is literally eradicating aging, which is the reason why you become sick and frail in the first place. Aging and its abhorrent health consequences are not a separate thing, and that's exactly what we are trying to change.

It is important to understand that the goal is not to extend the life of an unhealthy individual, but to restore and extend our healthy lifespan. In fact, the only reason why the life of an older person will be extended, will be a direct side effect (benefit) of their not getting sick from aging in the first place.

This shows us that the concept of "aging = disease" is still really challenging to grasp for most people. The part of aging that allows our accumulation of experience and wisdom is an extraordinary gift, in fact the central reason to continue to live, and no one wants to change that.

Another completely irrational concern I often encounter is that interfering with aging is not "natural", hence it is a bad thing. So then, let's analyze the meaning of "natural":

A dictionary definition is; *"Existing in or caused by nature; not made or caused by humankind"*. For some reason, many, if not most people define "natural" as something good for us. "Natural" foods. A "natural" remedy. A "natural" process. But it's pretty obvious that nature can be extremely dangerous for humans, so we must accept that "natural" does not necessarily mean "good".

Just consider the near infinite list of "natural" things that can be lethal. Such obvious and ever-present things as earthquakes, hurricanes, mudslides, avalanches, tornadoes, infectious microorganisms, poisonous plants, predators, venomous creatures, fire, gravity, tsunamis, radiation, meteor impacts, etc., etc.

But what about "unnatural" things? Must they be against the will of nature? The ever-expanding universe of things created by man; inconceivable just centuries, if not decades before, now totally accepted in most of the world, if not fundamental. Such wholly "unnatural" things as pacemakers, stents, antibiotics, cortisone, vaccines, water purification, organ transplantation, food preservation, dental fillings, eyeglasses, sutures, statins, titanium bone pins and plates, artificial joints, contraceptives, pain killers, anesthetics, cochlear implants, dialysis, plastic surgery, wheel chairs, contact lenses, cars, airplanes, electricity, computers, cell phones, solar panels, etc.

So, when and why did we decide that only natural is good as it pertains to aging? I ask the people who make these kind of statements:

"Is it acceptable for you to get a pacemaker to control your heart rhythm if your health and life are at risk?" "Do you take antibiotics to kill a (natural) bacteria that is attacking your health?" "Do you take pain killers when in agony?" "Do you get on a plane to travel abroad, or do you just walk?"

I think humans have proven themselves pretty damn good at designing "unnatural" things of enormous value and consequence, and I can imagine only a miniscule number of people who would choose to live without them – especially when it has to do with suffering, disease and death.

17.5   Exponential Growth

Today we are fortunate to live in totally unprecedented times. For most of our existence we could state that the world in which we were born would be seemingly the same as the world in which we will die. That's inarguably not the case anymore. We have made more technological progress in the past 20 years than in the past 150,000 years. Technology is progressing at an exponential rate, and we can't stop it.

In the beginning it's always hard for humans to accept big change, but today more big changes, some thoroughly disruptive, are happening at an ever-increasing rate and frequency. Tomorrow even more and even bigger changes still will happen, and they will happen at a rate faster than every day before. We will ultimately have to change the rate at which we adapt if we want to keep up at all.

Most of humans though, still think in a linear way. For countless generations life was pretty much the same for everyone from one generation to the next. But think about how extraordinarily different the life of today's kids are from the lives of their grandparents. Such once unimaginable wonders as the internet, smart phones, face and voice recognition, virtual reality, self-driving cars, virtual banking, and cellular connected watches with video conferencing, are just a few of the things that are quickly becoming part of our everyday lives. Not only are some of them already considered essential, but the experience of the world before they existed is almost unconsciously forgotten.

If these changes between the kids and grandparents of today look radical, you can surmise how outrageously radical the changes in the next few generations will seem.

A fun suggestion I like to give people to envision how much the world will change at this rate of advancement, is to imagine that virtually <u>anything</u> you can conceive of at this moment will one day be possible, no matter how impossible it seems today.

In case you never thought of how exponential growth works here's an example:

Imagine you save a dollar a day for 30 days, at the end of the month you will clearly end up with 30 dollars. Now imagine you could do the same but at an exponential rate, you will start adding dollars each day as follows: 1, 2, 4, 8, 16, 32, 64, 128, 256, 512 and so forth until you had done this 30 times. By the end of this prosperous month you will have saved 1,073,741,823 dollars. You'd be a billionaire!

Our brains have always thought linearly but now if we want to solve our global challenges, we will have to learn to think exponentially.

The day we put aging under medical control we will have achieved possibly the biggest breakthrough in human history, and this does not look unattainable anymore. Efforts and new initiatives around the globe to prevent, reverse and eradicate aging are also growing at an exponential rate, and it is just a matter of time until we accomplish in reality, what scientific reasoning now envisions as theoretically possible.

This is such an incredible time to be alive; possibly the most fascinating and intriguing in history. We've seen so many radical changes achieved by humanity. The technology we have today was science fiction merely decades ago. Many of these changes have been controversial and were resisted in their beginnings until they eventually found their way through the understanding and acceptance of people. Some of these have become so ingrained in our way of life that people would think you were crazy if you suggested we do without them.

We are curious. We are bold. We feel naturally compelled to seek and achieve change, no matter how radical it is. We modify, create, redesign, discover, reinvent, optimize, reimagine, and so very much more. We will never stop doing this, because it is in our nature. And that in itself to me makes it "natural".

17.6   Conclusion

I believe that humans may have accepted aging as an immutable and inevitable condition because for the hundreds of thousands of years we have had the minds to ponder, there was never anything we could have done about it. This was true, no matter our best efforts, our deepest desires, and our spiritual justifications. Today most people still believe we will not be able to cure this problem, and very likely, it is one of the strongest reasons so many suppress their hopes for fear of being let down yet again.

But it is our moral obligation to not neglect and ignore all the newest information science has about aging. We can't sentence the whole of our species to assured pain, suffering and death when we have a real chance of saving them. Deciding not to strongly pursue these cures would be just like abandoning all research towards a cure for any of the other diseases we currently spend so much money, time and energy on.

It is frustrating how unreasonable people can be when it comes to this subject and I have decided that I won't spend my time trying to convince people who are against it. I truly believe that their decision not to support this mission is part of natural selection, and I will personally not tamper with that "natural" process.

At the same time I do hope that by sharing some of my thoughts here, I have inspired newcomers to reflect on the subject and possibly join this important crusade that will one day stop the greatest torture and destruction of our species – and ultimately allow humans to be born with the gift of both extreme longevity and enduring health.

Chapter 18
Drinking From the Air

Joe Bardin

When I was in my early teens, my grandmother Miriam died. She had been old all my life, comfortably bed-ridden for the most part, so that for a good decade she seemed to me immortal, never getting any worse. What she loved most was reading anyway. When her eyesight failed, she transitioned to books on tape, 24 cassettes to a case, borrowed from the local library, their green plastic containers stacked on a bedside table in bulk quantities of Trollope, Herodotus, P.G. Wodehouse and on and on.

When she came to visit, ensconced in a bedroom where she would remain for the duration, her trove of books on tape within reach, I would sit and listen to her muse on literature, philosophy and classics, all of it passing well over my head. Occasionally, I would nod knowingly and feel very grown up and she was generous enough to never call me on it. Once, she had me push the door closed and she led us in a game of ESP in which we took turns wordlessly broadcasting and receiving visual images back and forth. I never saw much more than spots in front of my eyes. But again, she never hassled me.

She chewed those sticky Juju bean candies. Have I made her real enough? Because she was real. At the funeral, in a prosaically flat and new cemetery somewhere in suburban Maryland, she suddenly was not real, and I felt pissed about it. I was angry that death had taken Miriam and turned her into a thing in a casket. But there was no one to be angry with.

My mother, Miriam's daughter, uncomfortable with showing her emotion in any case, wore that somber, distracted expression less practiced Jews wear when they are challenged to get the recipe of rituals right. Many say, with considerable cultural pride, that this is the genius of it all; the ritual takes your mind off things.

I looked around for someone else with whom to share my anger at the loss of Miriam, but the faces I encountered were dutifully blank. The strongest feeling

J. Bardin (✉)
People Unlimited, Scottsdale, AZ, USA
e-mail: joe@relativitywriting.com

© Springer Nature Switzerland AG 2019
N. Lee (ed.), *The Transhumanism Handbook*,
https://doi.org/10.1007/978-3-030-16920-6_18

shared by the circle of relatives and friends attending was a supportive sense of understanding, and the essence of that understanding was agreement. Agreement, that there was nothing to be angry about, nothing to resist. That this is the way it is. In effect, that there was nothing wrong.

I was confused. How could Miriam dying not be wrong? Why had she undergone several medical procedures just recently if it didn't matter. Why did we all care about the outcomes of these procedures, until they failed? Only to frame the failure now as right and proper and natural and acceptable.

It seemed like cheating to me. My little sister had a game she used to play with Miriam when she was five or six, her own approximation of chess, called chest, played with cards, rather than chessmen. My sister never lost a game of chest, because whenever she was about to, she changed the values of the cards on the fly, always in her favor. Miriam used to fume about it, but the absurdity seemed perfectly reasonable to my little sister.

Many of the same people who attended Miriam's funeral attended a funeral reception a few years later. A young man from our community had been killed in military service in Israel, shot in an ambush on the Lebanese border. Shock pervaded the day. The mother, pale with grief, greeted guests to her house, receiving the inadequate condolences of those awed by her loss. He wasn't just young, in his mid twenties. He was a sweetheart. A talent. A seeker. A person of depth and promise. He left behind poems and watercolors.

He was 6 years older than me, almost a generation in kid-years. But in one of my very earliest memories, I am playing army men with him on the cool tile of a Jerusalem apartment. That was when both our families lived there. We were apparently ageless at that age.

This death left scars. This death altered people's politics and radicalized their metaphysics. Middle class, East Coast college-educated, skeptical bare-headed Jews became mystics. They had out of body experiences. They knew with certainty they would see him again. On the other side. This death was too terrible to take issue with. Instead, the issue was how to absolve death of its finality, of its careless, cruel, endless actuality.

Death, the destroyer. The murderer. The separator. The degrader. The trivializer of all personal passion and endeavor. Death, the final enemy. There is so much to hate death for, but we don't do it. The psychology is plain enough; we deny death's true totality, its absolute evil of annihilation to spare ourselves an unbearable reckoning with loss. But in sparing ourselves, we spare death. In the language of recent economic crises, death is too big to fail. We have too much invested in it to truly hold it accountable.

We are all implicated in this seemingly obligatory farce, not just so-called believers with their heavens of one sort or another. Non-believers find their afterlives in other forms such as: their works of one sort or another, their country, their family, human progress, universal consciousness, etc. But of course, while those entities might go on for a time, the dead who contributed to them are all too absent.

To be fair, our own evolutionary path as a species has put us in an impossible position. We developed the emotional and intellectual capacity to perceive a life beyond death, or at least to long for it, thousands of years before we could do any-thing about it. This longing for immortality is apparent in virtually every ancient tradition. In the Old Testament book of Genesis, a figure named Melchizedec is said to have "no beginning or ending of days." (Seems like a story line worth following up on, but he's hardly heard from again.) Suggestions of immortality surface in rela-tion to other figures of the Jewish tradition like Enoch, Moses and Elijah.

In the New Testament, references to immortality are aplenty. To cite just two, both from the King James version:

> So when this corruptible shall have put on incorruption,
>   and this mortal shall have put on immortality,
>   then shall be brought to pass the saying that is written,
>   "Death is swallowed up in victory."
>   Corinthians 15:54:

> Who will render to every man according to his deeds:
>   to them who by patient continuance in well doing
>   seek for glory and honor and immortality, eternal life.
>   Romans 2:6–7.

In the Epic of Gilgamesh, the ancient poem from Mesopotamia, dated to 2000 BC and earlier, immortality is a central concern. When Gilgamesh's companion, Enkidu dies, his distress sends him on a quest for immortality, which ultimately is unsuc-cessful. Along the way, he is told:

> "The life that you are seeking you will never find. When the gods created man they allotted to him death, but life they retained in their own keeping."

Essentially, we have been parroting this position in one form or another ever since. And what option did we have? Evolution, including human evolution, has been an engine driven by death. In Pilgrim at Tinker Creek, Annie Dillard, pondering the massive number of gooseneck barnacle larvae that are destroyed out at sea in the process of one growing to maturity, writes:

> "What kind of world is this anyway? Why not make fewer barnacle larvae and give them a decent chance? Are we dealing in life, or in death?" On the following page she answers her own question: "The faster death goes, the faster evolution goes."

What does one do in the face of such a gruesome, all-encompassing death machine? At best, we fight the good fight, and lose. As Dylan Thomas famously exhorted us: "Rage, rage against the dying of the light/Though wise men at their end know dark is right."

Human ancestry is now believed to stretch back over 3 million years. It seems safe to conjecture that for the majority of those years, our ancestors were not devel-oped enough to be pre-occupied with the existential problem of their own mortality, beyond the urge to fill their bellies and reproduce.

But some time in the last 5–10 thousand years perhaps, a blink of the eye by evolutionary standards, we started to envision immortality in one form or another. This probably occurred in parallel with the development of more in-depth social connections, an advancement that empowered people to improve their survival through more closely knitted communities. Now when people died, instead of it merely thinning the herd, others started to miss them, just as Gilgamesh missed Enkidu. Songs were sung, poems were written. And the longing to transcend that separation began.

Naturally, the same consciousness that evolved the ability to envision immortality and long for it, had to develop ways to distort the devastating reality of its absence. And we have been brilliant at it, conjuring afterlives to match every attitude, from the godly and pious, to the lucrative, the folksy and the pornographic. We've envisioned harps and angels, streets of gold and pliable virgins luxuriating beside sensually flowing streams.

Over time, we have gotten so adept at denying death that we hardly seem to hate it at all. On the contrary, we take issue with the possibility of not dying. Clerics argue that death is God's will and the proper and natural way of things, and well they might, considering that without it, they'll be as obsolete as the beeper when cell phones arrived on the scene.

But apologists for death are by no means limited to any church. The prevailing intellectual argument is typically some variation on the following: death makes life more precious. By limiting our time here, we are compelled to value every single moment.

This is apparently why it ultimately sucks to be a vampire, no matter how fabulous the accoutrement. It's all that damned time on your hands. Dare we say, dead time.

In his essay, The Self-Defeating Fantasy, (The Scientific Conquest of Death: Essays on Infinite Lifespans), Eric S. Rabbin summarizes this position:

"When we put on incorruption, we are all changed: we are changed into ideals, into endless repetitions, into sterile vampires, childless angels, works of art, computer chips.
… In the process we lose our very selves."

How far the intellect has strayed from physicality. Where does the loss truly lie? Biological immortality would mean Mr. Rabbin himself need not experience the excruciating suffering of his own demise. And if that is not compelling enough, it would mean that someone he loved and who loved and depended on him—perhaps a daughter, a son—wouldn't have to suffer his loss. And the same for you, the reader, and for everyone.

But we've had to live with death so long that the mind has bent to accommodate it. Boredom, disappointment, and above all, the loss of our humanity—these are the monsters the mapmakers have used to demark the unknown of immortality. Don't go there, even if you could. You won't like it. You need death, it makes life worth living.

Really?

During a 14-month period beginning in 2005 and peaking in the summer of 2006, a series of random shooting attacks occurred in the Phoenix metro area. In all, 6 people were murdered and another 19 were attacked. Of the approximately 5 million residents, this writer being one of them, only a tiny fraction was ever actually targeted. The rest of us lived only with the threat of being shot randomly on the street. If death indeed seasons life, then this ought to have been an especially spicy time in Phoenix.

Like cayenne pepper, we had just enough of it to taste the heat, without ruining the entire dish. (As opposed to, perhaps, a Hutu genocide or the killing fields of Cambodia.)

Needless to say, it wasn't so. Anxiety, fear, and social withdrawal were the order of the day. The same can be said of Tel Aviv during an up-cycle of suicide bombing. Again, from personal experience, I can report no renaissance of human development, no explosion of creativity or other evidence of greater aliveness. Just an atmosphere of white-knuckled survival, tense, pressurized, awaiting relief.

Death is all around us—in the news, at the movies, and is particularly vivid in the video games we play. If death really enhanced our lives, we'd be living it up as a society. We'd be seizing the day, and loving one another with all our hearts. But we're not. As a culture, we are more stressed, depressed and removed from the feeling of our passion for living than ever. The evidence is everywhere, from the boom in mood disorder drugs to our contemporary literature of meticulously delineated misery.

But this argument that death enhances life draws us in regardless. Because whether or not it really holds up under scrutiny, we need it to be true. We need to somehow be better off for dying, because historically that is what we do.

But we have paid a terrible price for this central, seemingly warranted deceit; in repressing our innate, organic revulsion at death, the very same repulsion that has driven us to dream up one alternative to it after another, we have necessarily turned against its opposite—living.

In Thus Spoke Zarathustra, Nietzsche rails against the life-hating that remains so normative in our culture. In First Part, section 9 of the Oxford edition:

> There are preachers of death: and the earth is full of those to whom rejection of life must be preached.
>
> Full is the earth of the superfluous; corrupted is life by the all too many. Let one use "eternal life" to lure them away from this life.
>
> … These are the consumptives of the soul: hardly are they born before they begin to die and to long for teachings of weariness and renunciation.
>
> … And even you for whom life is furious labor and distraction: are you not weary of life. Are you not very ripe for the preaching of death.
>
> … Everywhere the voice of those who preach death resound: and the earth is full of those to whom death must be preached.
>
> Or else "eternal life": it is the same to me—as long as they pass on to it quickly!

Nietzsche is right; every religion extols life after death as greater than this life, which ultimately boils down to a kind of promotion of death, because that's the way to get to that better place. But he's howling into the wind of evolution itself.

It can be argued that without the denial of death's finality, society could not function. As Dostoyevsky, speaking of faith in the afterlife, notes in The Brothers Karamazov:

> "If you were to destroy in mankind the belief in immortality, not only love but every living force maintaining the life of the world would at once be dried up."

However, big lies, no matter how justified, bring with them big consequences. Emotional authenticity cannot be selectively applied; feeling is an all-or-nothing gambit.

Repression of the reality of death brings with it repression of honest emotional expression in general. In society, this is called growing up and it happens to some of us sooner than others, but it always happens.

In the Drama of the Gifted Child: the Search for the True Self, Alice Miller terms it "The Lost World of Feelings."

Speaking of cases she's observed of the repression of childhood trauma she writes:

> "These people have all developed the art of not experiencing feelings, for a child can experience feelings only when there is somebody there who accepts her fully ..."

How many can say they were accepted fully as children? Miller attributes the repression of childhood feeling to the repression parents themselves experienced as children, which they never confronted and in turn act out on their offspring. Whether or not one entirely agrees with this point of view, it's clear that emotional dis-ease has its roots in childhood, and that those troubles are usually inherited from parents who had emotional problems themselves. But what is the prime mover here? Where does the cycle of emotional trauma start?

In the days of slavery in the American South, one way slaves would cope with the intolerable cruelty they were dealt while working the fields was to "fly away". If one was being beaten, and the others forced to passively stand by, they might collectively envision themselves leaving their bodies and their torment behind, as they flew freely above the scene of unspeakable suffering. Of course, they remained in the field, subject to the whip of the overseer, but they gained a measure of distance from their pain and humiliation by imagining themselves elsewhere.

As a species, we've responded similarly to the slavery of death. Unable to overcome it, we've sought to imagine it away. We've taken flight from our bodies, from physical feeling, and who could blame us? What real option have we had?

Now biological immortality is seriously being discussed by credible thinkers and scientists.

Isn't it time to come clean emotionally, to land back in our bodies and confess that we really don't want to die and truly, fervently, organically do desire to live?

A few of us have decided the human species already isn't mortal anymore; that the enlightened person, with the health tools and wellness intelligence available today, the understanding of practical human psychology and the need for change, and most importantly, a community of like-mined people to support them, is potentially endless already. This is not to say we won't welcome the advances in science as they emerge, only that we feel we already have enough to get started on. This community is called People Unlimited (http://www.peopleunlimitedinc.com).

The reader may well say we are deluding ourselves, but given that the mass of humanity is in delusion about their on-going existence after their death, this seems like a better place to put one's energy, at least to me.

In his story, The Immortals, Borges describes a hellish existence of endless repetition and tedium devoid of all meaning. I can report that being immortal, or at least believing it sincerely, and living it wholeheartedly, is certainly not boring. The repetition, it turns out, is all on the mortality side of the ledger, what with its predetermined ending, and so on.

No wonder the insightful Borges (and so many others) have gotten it so wrong. He was thinking: birth, bar-mitzvah, marriage, family, retirement ad infinitum. That would be torture. But immortality renders those formulaic structures of living obsolete. There's no script. No footsteps to follow. It's the furthest thing from boring; it's the unknown.

I recently watched a National Geographic special on the California Redwoods. These trees can live from 2500 to 3500 years and can grow to over 350 feet high in the process. So they have something to tell us about longevity at least, if not immortality. One mystery botanists long struggled with was how these giants were able to draw enough water from their roots all the way up to their canopies hundreds of feet above in order to keep growing. Eventually they discovered that Redwoods don't just drink from their roots, they drink out of the air as well, from the moisture heavy fog that rolls in off the Pacific.

This explains why otherwise formidable thinkers are often perfectly blind when it comes to thinking immortality. They are still drinking from the roots of past experience, from the long epoch of human mortality, and can find no bridge of reason into this next evolution. The agreement that death is inevitable has been the single universal common denominator of all human experience. It takes what the Jewish Philosopher Martin Buber called a "leap of faith" to start drinking out of the air rather than from one's roots.

This involves thinking more through feeling, through bodily sensation, and relying less on systemized analysis based on existing paradigms, which can only be relied upon to produce status quo outcomes, in this case death. Another way of saying it is that the moment of revelation must be now, always now.

I recently heard a most remarkable phone message from a fellow immortal, whom I'll call Pam. It was a saga in three parts, a journey of sensuality, of remorse and of possibility, all compressed into the max time allotment for a cell phone voice mail.

At 68, Pam had begun hormone replacement therapy (HRT), ingesting the hormones she no longer produced herself. She probably should have started HRT much sooner, but having been physically neglected as a very young girl, she'd been sluggish about caring for herself all her life; she was still drinking from that root. Finally, after years of hearing about putting her body first (a central tenet if you're planning on staying around), she opened a tendril of awareness to now.

But there were consequences. By getting hormonally balanced, she started feeling sexual arousal again, which she had not experienced in years. Remarkably, for the first time in her life, she masturbated. And this was only part one of the message.

In Part II, Pam shared that she realized she still had deep feelings of attraction for a woman she'd had a passionate but tumultuous living with years earlier. I had seen these two circle each other wearily for a good decade. To hear Pam confess her deep feelings for Judy (also not her actual name) felt like a fulfillment; like honesty filling an emptiness.

Then Pam's tone shifted to one of tearful remorse for having misled another woman she'd lived with, to whom she could never truly give her whole heart, because of her love for Judy. I'd witnessed this backup plan stutter along as well. She was not guilt- ridden and going back on her revelation. It was just that her awakening had consequences and she was facing them.

I was privileged to hear this voice mail message because it was sent to my girlfriend and a People Unlimited founder, Bernadeane Brown. Pam could pour herself out to Bernie in this manner because Bernie has done so with Pam. Not once, but many, many times over many, many years. This outpouring of self to self is what breaks down the separation of death. That we die alone is an existential fact; the corollary being that we live alone too, until we die. Immortality ends the loneliness. But it takes an extraordinary and on-going flexing of the musculature of togetherness to break down the cell memory of isolation in the body.

Immortality is no state of perfection. That notion is religion's way of keeping the carrot out of reach. (Who of us is perfect enough to be immortal? Nobody. Therefore you're not ready yet.) But it is a state of perpetual progression, because Nature makes short work of stagnation. When we stop growing, we're dying more than we are living, until eventually, we just aren't alive at all. Immortality requires change and growth without end.

Of course, because we've been programmed otherwise, this often registers as impossible. Living forever, based on drinking from the roots, is impossible. But so was flight, the Mars Rover, the election of a black president, peace in Northern Ireland and every truly new development.

That Pam could fit her poignant experience into a phone message, and feel so comfortable with the recipient and with herself to do so, is part and parcel of her coming clean with being alive. Immortality makes an honest species of us. We no longer have to suppress our true abhorrence of death, and consequently, deny our body and our appetite for living. Yes immortality demands change, but it's precisely the change we've been longing for, if we'll just be truthful about it.

Chapter 19
Transhumanism and Older Adult Quality of Life

Jonathan Squyres

We live in a diverse world where the recognition of this diversity and the need for further equality among the diverse populations in our society is being further and more conscientiously being considered. This is as it should be. Our society has been in great turmoil regarding different populations that are not treated equally and it is time this has been brought to attention. However, there is one population, especially in an American society that continues to go unnoticed. This population may no longer be able contribute or achieve in the way an individualistic society expects, and therefore, they are the forgotten population. Older adults, as one may have guessed, is the population being discussed. The seniors in American society, and perhaps in other societies, are often overlooked, undervalued, and shoved aside. Children are very limited and cannot contribute, but we still consider them in high regard because they have potential based on our individualistic values. Unfortunately, this is not seen as the case for older adults. If we are to consider ourselves a truly diverse and equal based society, older adults must be considered an equally important population.

So, how does transhumanism relate to this topic? Is it worth considering older adults and how they fit into the transhumanism movement? Absolutely. In fact, this movement may allow seniors to achieve a higher status than we have given them in this society. One of the most significant concerns when considering aging and transhumanism is that of overpopulation. Overpopulation will be a concern at some point, but it is not an argument that fits in the current topic. Each year, pharmaceuticals are introduced to prolong cognition, and therefore independence, in those that are suffering from progressive dementia. Hip and knee replacements are completed so that an older adult may remain independent. It does not seem that anyone would deny an older adult these treatments. Therefore, the discussion is not about overpopulation but about quality of life, and quality of life for older adults is largely based on their ability maintain independence. The idea that technological advances

J. Squyres (✉)
Global New World, Lufkin, TX, USA

© Springer Nature Switzerland AG 2019
N. Lee (ed.), *The Transhumanism Handbook*,
https://doi.org/10.1007/978-3-030-16920-6_19

and bioengineering could do this is a welcome advancement for older adults. The key here is the improvement of the quality of life of older adults.[1]

The key to success in many people's eyes in our society is productivity. Older adults are often not able to be as productive as they were in the past and this is, as discussed above, the primary reason they are often shoved aside. However, if they are more independent, older adults, through their extensive experiences and vast knowledge can further contribute and be more productive. In addition, technological and medical advances can lead to less need for care. The business of caring for older adults in our society is growing by leaps and bounds due to the ever increasing older adult population. The baby boomer population is now retiring and people are living longer. The combination of these factors means that a large older adult population is inevitable.[2] The funds required to care for these individuals will continue to expand as well. Therefore, the transhumanist movement of advancing the quality of life through technology and medicine is more than a nice thought, but yet a necessity. If people are able to continue living in their own homes without assistance, not only will the quality of life of the individual increase exponentially, but society will reap the benefits as well by saving the government and private industry a great deal of resources, which will reduce costs for all.

Technology meant to prolong life that has no quality is not of interest in this article. For example, prolonging severe dementia or physical disability that has robbed one of their quality of life, through life support, is a use of technology. Of course, this is an individual or family decision. However, the point of this article is to consider the person before their quality of life has become nonexistent. If the transhumanist movement can help to prolong a life of quality in the older adult, we are not furthering the idea of living on the "system," but moving away from this idea altogether, by prolonging independence and productivity.

Medical advances are already a significant part of our lives in this world. We take them for granted and expect further advancement. Several medical advancements are hoped for in the near future regarding chronic disease such as cancer, heart disease, and various forms of dementia. A pacemaker is common in the treatment of diverse heart conditions. If a technological advancement similar to this could be created to stop Alzheimer's disease progression, would we argue against it? Most likely, we would not. However, what if we were able to use this same device to not only eliminate Alzheimer's disease, but also improve cognition in older adults. Would we then decide this was too far? I still do not believe we would. This change could mean that the older adult could continue working and contributing, but also maintain a higher quality of life. This is transhumanism serving a good purpose.[3]

---

[1] Stambler, I. (2010). Life extension – a conservative enterprise? Some fin-de-siècle and early twentieth-century precursors of transhumanism. *Journal of Evolution and Technology, 21*(1), 13–26. Retrieved from https://jetpress.org/v21/stambler.htm.

[2] Karel, M. J., Gatz, M., & Smyer, M. A. (2012). Aging and mental health in the decade ahead: What psychologists need to know. *American Psychologist, 67*(3), 184–198. https://doi.org/10.1037/a002539

[3] McNamee, M. J., & Edwards, S. D. (2006). Transhumanism, medical technology and slippery slopes. *Journal of Medical Ethics, 32*(9), 513–518. https://doi.org/10.1136/jme.2005.013789.

It appears to me that the transhumanism movement could truly benefit older adults and would significantly improve our society as well. It could also be viewed that our society is already well-advanced concerning the transhumanism movement. In my opinion, this movement is an inevitable force that will continue to benefit many populations, but it is my hope that the older adult population will not be forgotten as it has in the past. Transhumanism could be the answer to many of the problems our society faces as the older adult population significantly increases, but more importantly, it could be the answer to better quality of life for our seniors.

Chapter 20
Are You Willing to Die for Reductionism?

Michael R. Rose and Grant A. Rutledge

Reductionism was one of the greatest themes of twentieth century biology. It began early in the century with genetics explaining single traits with single genes. It rose further with the inference that DNA encoded genetic information. It culminated with genetic cloning and genetic engineering. With disorders like sickle cell anemia or Huntington's Disease, reductionism is indeed a scientific triumph. Similar claims have been made about aging, which has been attributed to a single agent in the case of the free-radical theory of aging. In other hands, up to seven specific cell-molecular mechanisms have been offered as the entire foundation of aging. Meanwhile, the last 15 years of genomic research has shown that most functional characters, from human height to aging in fruit flies, are affected by hundreds if not thousands of sites genome-wide. Furthermore, these sites cannot be delineated as hereditary factors tuning one or a few pathways, as reductionism would require. That is, at the genomic frontier of biology, reductionism is now seen as a special and unrepresentative case. Most of the genetic machinery of animals does not work using simple isolated pathways. Rather, it usually operates as part of large encompassing networks. Given this, it is doubtful that aging interventions founded on reductionist premises will succeed. Does this mean that aging cannot be re-tuned radically? Evolutionary theory and experiments suggest otherwise. More recent evolutionary research even raises the possibility that there might be straightforward and powerful interventions that can control human aging on a very short timescale. Those who stick to twentieth century reductionist theories of aging may be risking their lives. They may indeed die for their beliefs.

M. R. Rose (✉) · G. A. Rutledge
Department of Ecology and Evolutionary Biology, University of California, Irvine, Irvine, CA, USA
e-mail: mrrose@uci.edu; grutledg@uci.edu

© Springer Nature Switzerland AG 2019
N. Lee (ed.), *The Transhumanism Handbook*,
https://doi.org/10.1007/978-3-030-16920-6_20

20.1 Introduction: Two Kinds of Theories for Aging

If there is any field of scientific study that is subject to waves of hope followed by waves of skepticism, it is biological aging. Indeed, interest in understanding and then intervening in aging is one of humankind's perennial interests (e.g. [10, 14]). From the historical record, this interest is only suppressed by promises of some type of afterlife combined with religious suppression, as was the case with Western civilization from the fall of the Roman Empire in the West to the Reformation. With the reduced power of the Inquisition after the advent of Protestantism, and still more with its complete disbanding by Napoleon circa 1800 CE, speculations about achieving prolonged human life have been commonplace even in Western Civilization [10, 14].

A characteristic feature of speculations about aging in Western biology and medicine is that they have shared *two* common features since Aristotle first wrote on the topic in his monograph "On the length and brevity of life." (1) They have assumed that aging is a physiological process, perhaps a compound of multiple independent physiological processes. Some have argued that the physiology of aging involves a physiological program that destroys organisms (e.g. [20]). Others have argued that it is a physiological process of cumulative damage, recent examples being the ideas of Harman [13] or de Grey and Rae [7]. (2) The physiological models of the era in which these speculations have been offered are used to supply the machinery of such physiological processes of aging. Thus in Aristotle's time, the physiology of aging was conceived in terms of a deteriorating balance among the four Greek elemental components of matter: earth, air, fire, and water. With the advent of molecular biology in the late twentieth century, the physiology of aging was most often conceived in terms of molecular processes like somatic mutation, protein translation error catastrophes, free radical damage, and so on (vid. [5, 8]).

An entirely different vein of thinking about aging was first initiated by August Weismann in the late nineteenth century (vid. [15]). Weismann believed that evolution by natural selection could produce almost any aging pattern. Indeed, the recent demonstration that aging can be entirely absent in multicellular animals (e.g. [1, 17]), supports his view. Within the diversity of living things, we now know with certainty that aging is not universal. Thus expecting aging to be a universal consequence of universal biochemical mechanisms of deterioration is upended by the absence of any such universality to aging.

Starting with verbal speculations by Haldane [11], Medawar [18, 19], and Williams [34], evolutionary biologists have developed a body of theory that explains aging in the species that have it. This theory was subsequently developed in mathematical terms by Hamilton [12] and Charlesworth [4]. All this work is based on the idea that aging is tuned by declining forces of natural selection. When those forces do not decline, as they do not in species that rely exclusively on symmetrically fissile reproduction, then aging should not exist. But when there is no form of fragmentary or fissile reproduction at all, as there isn't in insects or vertebrates, aging should be universal. So far, this prediction appears to be borne out (e.g. [26]). On these grounds of aging's comparative biology alone, it could be claimed that merely physiological theories of aging must be wrong.

But there is a broad range of experimental work which more directly supports theories of aging based on the forces of natural selection. Starting deliberately with the work of Sokal, Rose, and Luckinbill (e.g. [16, 24, 25, 29, 33]), re-tuning the forces of natural selection according to the key terms of the equations developed by evolutionary biologists like Hamilton [12] leads to the production of animals with faster or slower patterns of aging. Additional work has shown that the cessation of aging (e.g. [3, 6]) can be explained in terms of this same theory (e.g. [22]). Experimental manipulation of other terms of the key evolutionary equations yields theoretically predicted changes in this cessation (e.g. [21, 27]).

Not incidentally, verbal speculations based on strictly physiological models of aging cannot account for any of these findings. That is to say nothing of the point that they did not provide predictions or explanations for any of these findings in advance, the way evolutionary theory did for the evolution of fecundity [23] and virility [32].

While there have now been two basically different kinds of theories for causes of aging since the 1940s, a comparison of the relative success of these theories as scientific hypotheses is not difficult. The explanation of aging in terms of merely physiological processes, without evolutionary theory, is moribund from a scientific standpoint. It doesn't matter if your great-uncle thinks that he will live longer from eating blueberries with antioxidants. It doesn't even matter if cell biologists who study telomeres think that they are the sole and sufficient cause of aging. Formal mathematical theory and strong-inference experiments have shown that you cannot ignore the evolutionary biology of aging when considering the problems that biological aging poses.

20.2   Underlying Mechanisms: Simple or Complex?

However, the foundational indispensability of evolutionary theories of aging does not entirely define the nature of the physiological machinery of aging. At least conceptually, these are severable problems.

Evolution can re-shape basic features of organisms with single-gene substitutions. The most famous example of this is the melanic alleles that swept through European lepidopteran populations of multiple species with the advent of industrial coal burning. Major single-gene polymorphisms arising from selection-mutation balance or antagonistic pleiotropy can explain genetic diseases, from cystic fibrosis to sickle-cell anemia.

Genomic analysis can address the question of whether such single-gene machinery is how evolution normally operates, without methodological bias. That is to say, genomic analysis would show that a single gene is responsible for a particular medical problem or physiological attribute, if that were the case. And it also has the power to show that many genes are involved.

In the case of human physiology, function, and disease, genome-wide association studies (GWAS) can show whether or not our attributes are affected by few or many genes and their associated pathways or processes. For characters like human height, for which we have the most data, the results overwhelming support the involvement of hundreds of genes and probably still more processes. The multiplicity of affected processes is assured by the extensive pleiotropy of genetic effects that genomic tools like microarrays also reveal.

In the case of aging specifically, a different genomic tool reveals its physiological complexity. "Evolve and Sequence" experiments using "pool-seq" reveal the involvement of dozens to hundreds of sites that control Drosophila ("fruit fly") aging (e.g. [2, 9]). Furthermore, as with human height, the bigger these experiments get, the more genetic loci they reveal. Experiments that haven't been published in great detail also reveal transcriptomic complexity in the physiological foundations of aging [28]. Additional transcriptomic and metabolomic experiments are underway, with still greater scale and statistical power.

When unbiased genomic tools are applied to the physiology of animal aging and chronic human disease, they reveal physiological complexity. Simple physiological processes do not dominate in the control of aging. Innumerable reductionist gerontologists have long hoped that they would. But the fact that they do not is fraught with practical implications.

20.3 Is Reductionism Our Only Hope?

A common rhetorical device among reductionist gerontologists is the following. If aging is physiologically complex, then we have to give up hope of intervening with it, so as to extend our lives. Therefore, we should double down on the search for simple cell and molecular reductionist manipulations, because they are our only hope.

However, just as genomics is now the most powerful tool in the biological toolkit, it will also enable us to penetrate human physiology to ameliorate our aging. Now, this might seem impossible, as the human mind boggles when confronted with more than a few variables. But there is a far more powerful tool available, one that can easily process hundreds of variables.

That tool is artificial intelligence ("AI"). What we cannot process, AI can. This doesn't mean that all we need to do is to turn on an AI device, and let it solve the problem with the millions of biological findings that can be discovered on the internet.

Rather, AI is particularly powerful with the very genomic findings that undermine the validity of reductionist ideology across biology, especially with respect to aging. Aging is physiologically complex, as geriatricians and evolutionary biologists have long argued. Genomics is peeling that complexity apart, and finding the hundreds of genomic sites and transcripts that underlie that complexity. Applying AI to powerful genomic, transcriptomic, and metabolomic data will be the key that will unlock how we need to intervene in the multifold complexity of aging's physiological machinery.

20.4   What Can We Do Now?

Yet another evasive maneuver that reductionist gerontologists are fond of is to point out that the AI penetration of aging genomics has yet to be accomplished. That is a valid point. Then, our reductionist colleagues argue, shouldn't we try to use the normal reductionist tools, in the meantime?

This is somewhat like the lamplight argument, a famous epistemological mistake. If we have lost our car keys on a dark street, the argument goes, we should only search under the single lamplight that lights a small patch of pavement. The counterargument, of course, is that it is more important to look for tools, like flashlights and car headlights, that can be used to illuminate the entire street, albeit with greater difficulty.

Indeed, it turns out that there is a useful tool that can supply "bridging strategies" to mitigate aging while waiting for powerful AI results. We have discussed those in some detail in a recent article [31]. We have developed this theme in still greater detail at a website: 55theses.org.

To briefly rehearse this non-reductionist "bridging" strategy, it involves three key components. (1) Age-appropriate tuning of lifestyle and nutrition to match our evolutionary histories. (2) Use of middle-level repair and mitigation technologies from conventional medicine, like dental implants, artificial joints, and transplants. (3) AI-based "phenomenological" development of new pharmaceuticals.

A fundamental corollary of evolutionary theories of aging based on declining forces of natural selection is relevant to human aging. When the forces of natural selection are strong, at early ages, adaptation to a new environment proceeds rapidly. When the forces of natural selection are weak, at later adult ages, adaptation to a new environment proceeds slowly. Humans are a species that has recently adopted a new diet and a new lifestyle: agricultural cultivation of principal foodstuffs. For those human populations that adopted this lifestyle 10,000 or more years ago, we expect that their young people will do well on organic agricultural foods and the kind of monotonous activity patterns that characterize agricultural life. But we do not expect older adults to thrive on this lifestyle [21]. Instead, we expect them to have a later-life physiology that is a relic of adaptation to our earlier hunter-gatherer way of life. We are about to publish detailed mathematical models and experimental tests of this hypothesis (Rutledge and others, in prep.). Therefore, as presented at 55theses.org, we expect that a middle-aged transition from organic agricultural diets to emulations of hunter-gatherer diets to yield significant health benefits.

The second "bridging" tactic would be developing tissue and organ-system repair technologies. Dentistry has long been a pioneer in this tactic: dental cleaning, fillings, and implants. Cardiology is developing along somewhat parallel lines, with stents, dilations, bypasses, replacement valves, and even heart transplants. The more we can repair our medically obvious breakdowns, the longer we will be able to preserve both life and function.

Thirdly, AI and genomic tools provide alternative technologies for the development of new classes of pharmaceuticals. Most of our present pharmaceuticals have been developed adventitiously, like aspirin, or from reductionist ideas, like statins. While many of these pharmaceuticals work at least moderately well, there is an alternative strategy for developing pharmaceuticals that is now being used by multiple AI-genomic companies. (Full disclosure: MRR has stock in two such companies, and consults for one of them, Lyceum Pharmaceuticals.) Such R&D strategies, for now, are *not* based on powerful AI insights into the full complexity of the physiology of aging. However, they may be able to produce new pharmaceuticals that mitigate the ravages of aging, without resort to organ transplants and other types of macro-repair.

Thus it is categorically false that only reductionist strategies are available for the immediate or near-horizon mitigation of aging. Ultimately the utility of any medical treatment is to be determined by medical trials under FDA-type supervision. We are not, in any case, aware of any successful medical treatment for aging that has been derived from the reductionist hypotheses for aging that have been predominant among gerontologists over the last 60 years.

20.5   Are You Willing to Die for an Intellectual Error?

We have long argued that the correct theory of aging is that provided by evolutionary theories based on Hamilton's forces of natural selection (e.g. [26, 30]). We contend that there are strong-inference experiments showing that only that theory is a fit basis for developing useful interventions to ameliorate human aging. Together with our colleagues, we have developed "bridging" strategies to mitigate aging while we wait for the powerful combination of genomics and AI to parse the machinery of aging [31].

If we are right in at least large measure, then reductionist interventions in the aging process will almost always fail. Those of our friends and colleagues who are now genetically re-engineering their bodies based on reductionist ideas are likely to receive no benefit. They might even die sooner, or suffer more.

The question remains: Are you willing to die for reductionism?

References

1. Bell, G. (1988). *Sex and Death in Protozoa: the History of an Obsession*: Cambridge University Press.

2. Burke, M. K., Dunham, J. P., Shahrestani, P., Thornton, K. R., Rose, M. R., & Long, A. D. (2010). Genome-wide analysis of a long-term evolution experiment with Drosophila. *Nature, 467*(7315), 587–590. https://doi.org/10.1038/nature09352

3. Carey, J. R., Liedo, P., Orozco, D., & Vaupel, J. W. (1992). Slowing of mortality rates at older ages in large medfly cohorts. *Science, 258*(5081), 457–461.

4. Charlesworth, B. (1980). *Evolution of Age-Structured Populations*: Cambridge University Press.

5. Comfort, A. (1979). *The Biology of Senescence* (3 Ed.). Edinburgh and London: Churchill Livingstone.

6. Curtsinger, J. W., Fukui, H. H., Townsend, D. R., & Vaupel, J. W. (1992). Demography of genotypes: failure of the limited life-span paradigm in Drosophila melanogaster. *Science, 258*(5081), 461–463.

7. de Grey, A., & Rae, M. (2007). *Ending Aging: The Rejuvenation Breakthroughs That Could Reverse Human Aging in Our Lifetime*: St. Martin's Press.

8. Finch, C. E. (1990). *Longevity, Senescence, and the Genome*: University of Chicago Press.

9. Graves, J. L., Jr., Hertweck, K. L., Phillips, M. A., Han, M. V., Cabral, L. G., Barter, T. T., ... Rose, M. R. (2017). Genomics of Parallel Experimental Evolution in Drosophila. *Mol Biol Evol, 34*(4), 831–842. https://doi.org/10.1093/molbev/msw282

10. Gruman, G. J. (1966). *A history of ideas about the prolongation of life: the evolution of pro-longevity hypotheses to 1800*: American Philosophical Society.

11. Haldane, J. B. S. (1941). *New paths in genetics*. London: George Allen and Unwin.

12. Hamilton, W. D. (1966). The moulding of senescence by natural selection. *Journal of Theoretical Biology, 12*(1), 12–45. https://doi.org/10.1016/0022-5193(66)90184-6

13. Harman, D. (1956). Aging: a theory based on free radical and radiation chemistry. *J Gerontol, 11*(3), 298–300.

14. Haycock, D.B. (2008). *Mortal Coil, A short history of living longer*. New Haven, Conn.: Yale University Press.

15. Kirkwood, T. B., & Cremer, T. (1982). Cytogerontology since 1881: a reappraisal of August Weismann and a review of modern progress. *Hum Genet, 60*(2), 101–121.

16. Luckinbill, L. S., Arking, R., Clare, M. J., Cirocco, W. C., & Buck, S. A. (1984). Selection for Delayed Senescence in Drosophila melanogaster. *Evolution, 38*(5), 996–1003. https://doi.org/10.2307/2408433

17. Martinez, D. E. (1998). Mortality patterns suggest lack of senescence in hydra. *Exp Gerontol, 33*(3), 217–225.

18. Medawar, P. B. (1946). Old age and natural death. *Modern Quarterly, 1*, 30–56.

19. Medawar, P. B. (1952). *An Unsolved Problem of Biology: An Inaugural Lecture Delivered at University College, London, 6 December, 1951*: H.K. Lewis and Company.

20. Mitteldorf, J., & Sagan, D. (2016). *Cracking the Aging Code: The New Science of Growing Old – And What It Means for Staying Young*: Flatiron Books.

21. Mueller, L. D., Rauser, C. L., & Rose, M. R. (2011). *Does Aging Stop?* : Oxford University Press, USA.

22. Mueller, L. D., & Rose, M. R. (1996). Evolutionary theory predicts late-life mortality plateaus. *Proceedings of the National Academy of Sciences, 93*(26), 15249–15253.

23. Rauser, C. L., Abdel-Aal, Y., Shieh, J. A., Suen, C. W., Mueller, L. D., & Rose, M. R. (2005). Lifelong heterogeneity in fecundity is insufficient to explain late-life fecundity plateaus in Drosophila melanogaster. *Exp Gerontol, 40*(8–9), 660–670. https://doi.org/10.1016/j.exger.2005.06.006

24. Rose, M., & Charlesworth, B. (1980). A test of evolutionary theories of senescence. *Nature, 287*(5778), 141–142.

25. Rose, M. R. (1984). Laboratory Evolution of Postponed Senescence in Drosophila melanogaster. *Evolution, 38*(5), 1004–1010. https://doi.org/10.2307/2408434

26. Rose, M. R. (1991). *Evolutionary Biology of Aging*. New York: Oxford University Press.

27. Rose, M. R., Drapeau, M. D., Yazdi, P. G., Shah, K. H., Moise, D. B., Thakar, R. R., … Mueller, L. D. (2002). Evolution of late-life mortality in Drosophila melanogaster. *Evolution, 56*(10), 1982–1991.
28. Rose, M. R., Long, A. D., Mueller, L. D., Rizza, C. L., Matsagas, K. C., Greer, L. F., & Villeponteau, B. (2010). Evolutionary Nutrigenomics. In *The future of aging* (pp. 357–366). Dordrecht: Springer.
29. Rose, M. R., Passananti, H. B., & Matos, M. (2004). *Methuselah Flies: A Case Study In The Evolution Of Aging.* Singapore: World Scientific Publishing Company.
30. Rose, M. R., Rauser, C. L., Benford, G., Matos, M., & Mueller, L. D. (2007). Hamilton's forces of natural selection after forty years. *Evolution, 61*(6), 1265–1276. https://doi.org/10.1111/j.1558-5646.2007.00120.x
31. Rose, M. R., Rutledge, G. A., Phung, K. H., Phillips, M. A., Greer, L. F., & Mueller, L. D. (2014). An evolutionary and genomic approach to challenges and opportunities for eliminating aging. *Curr Aging Sci, 7*(1), 54–59.
32. Shahrestani, P., Tran, X., & Mueller, L. D. (2012). Patterns of male fitness conform to predictions of evolutionary models of late life. *Journal of Evolutionary Biology, 25*(6), 1060–1065. https://doi.org/10.1111/j.1420-9101.2012.02492.x
33. Sokal, R. R. (1970). Senescence and genetic load: evidence from Tribolium. *Science, 167*(3926), 1733–1734.
34. Williams, G. C. (1957). Pleiotropy, natural selection, and the evolution of senescence. *Evolution, 11*(4), 398–411. https://doi.org/10.1111/j.1558-5646.1957.tb02911.x

Chapter 21
What Do We Need to Know to Treat Degenerative Aging as a Medical Condition to Extend Healthy Lifespan?

Ilia Stambler

21.1 The Urgent Need to Ameliorate Degenerative Aging and Extend Healthy Lifespan

It can be confidently stated that, at the present time, the extension of healthy lifespan (or "healthspan") for the global population is one of the most urgent and vital societal goals, if not the most urgent and vital. In its scope and potential significance for the well-being of the global human community, this goal dwarfs virtually any other development goal, even though the current support for the achievement of this goal is rather miniscule, compared to other types of expenditures. Yet, the importance of this goal for the society and every single individual cannot be overestimated. Throughout the world, due to the increasing aging population, the prevalence of chronic non-communicable diseases and disabilities – such as cancer, ischemic heart disease, stroke, type 2 diabetes, Alzheimer's disease, Chronic Obstructive Pulmonary Disease, etc. – rises steeply [1]. For the "developed countries" the problem is becoming acute. Thus, while 66% of deaths in the world occur from chronic age-related diseases, in the developed countries, this proportion reaches 90%, dramatically elevating the costs for healthcare and human suffering [2]. For the so called "developing countries" (or "low income countries") the problem of population aging may seem less visible, but is in fact not less, perhaps even more grave. Currently, while the highest life expectancies (and correspondingly the incidence of aging-related diseases) are still found in the "developed" countries, the rise in life expectancy is now the largest and most rapid in the developing countries, and the trend is likely to continue [3]. The faster and larger rise in life expectancy for the developing countries also means the stronger and faster population aging, and the larger and faster increase in the incidence of chronic age-related non-communicable diseases. At the same time, the geriatric and non-communicable disease care and

I. Stambler (✉)
Bar Ilan University, Ramat Gan, Israel

© Springer Nature Switzerland AG 2019
N. Lee (ed.), *The Transhumanism Handbook*,
https://doi.org/10.1007/978-3-030-16920-6_21

research in these countries may be under par and unprepared, potentially threatening the lives of millions of the world's poorest and most disadvantaged older people. Also, in absolute terms, the number of people suffering from aging-related conditions in the "developing countries" exceeds the absolute numbers in the "developed" countries. Hence, also for the "developing world," the problem of population aging is strategically pressing, and the task of improving healthy lifespan for the population is urgent. Thus, it can be confidently stated that healthy lifespan extension is one of the most important healthcare, economic and humanitarian tasks for the entire global community. If transhumanism is understood as an aspiration for human development and for a solution of global problems, thanks to ethical use of new technological means, then the extension of healthy lifespan is undoubtedly one of the most central and urgent tasks of transhumanism, perhaps even the most central and urgent.

How can this task be accomplished? There is a wide range of the currently available lifestyle approaches for healthy longevity (such as moderate exercise, moderate and balanced nutrition, and sufficient rest and sleep), of which great many people are aware, even though the adherence and compliance with such approaches are often limited. At the same time, there is an ongoing massive search for additional novel biomedical means and technologies to ameliorate the degenerative aging process and in this way improve the healthy lifespan. The connection between amelioration of aging and extension of healthy lifespan should be obvious to everybody: Insofar as the deteriorative aging process either precipitates or lies at the root of chronic age-related diseases, the search for novel means and technologies for healthy lifespan extension necessitates the maximal possible amelioration of the degenerative aging process. Such amelioration of the aging process should lead to better health and quality of life for the elderly [4]. The possibility of therapeutic intervention into degenerative aging and the consequent significant healthy lifespan extension has been proven on both theoretical-biological grounds and experimental grounds in a variety of animal models. In particular, the ability of cell-based regenerative medicine, gene therapy, pharmacological therapy and nanomedicine to affect basic aging processes and extend healthy lifespan in animal models has been demonstrated, and even some encouraging preliminary results have been achieved in human experiments [5]. This possibility has also been conclusively proven by the existence of a large and continuously growing long-lived population, including centenarians and super-centenarians, that exhibit not only a high longevity potential, but also a reduced rate of age-related diseases compared to the general population [6].

Yet the pathway toward human healthy lifespan extension remains unclear and requires thorough elaboration, concerning many scientific problems that need to be clarified and technologies that need to be developed. There is a tremendous variety of studies and approaches toward healthy lifespan extension, and roadmaps indicating priority directions [7]. Perhaps the most critical drawback in this variety is the lack of integration of the different approaches. The existing approaches often present lists of potential research directions, rather than coherent and coordinated entities. Hence the integration of the various approaches, shortening the pathways between the various disciplines, could be highly valuable for the fundamental and comprehensive understanding of aging and longevity, as well as for the further translation of this knowledge to practical integrative medical applications.

21.2   The Need for an Integrated Approach to Healthy Lifespan Extension: Bridging the Gaps Between Knowledge Domains

Several important "gaps" may yet need to be "bridged" in the current variety of approaches to healthy lifespan extension to achieve integrated, practicable knowledge. One critical gap is between what may be termed *"environmentalist" and "internalist" approaches* to healthspan extension. On the one hand, it is often assumed that environmental and lifestyle factors alone are sufficient to affect healthy lifespan, disregarding genetic composition, the inner structure and function of the body. On the other, there is a "genetic" or "biological deterministic" approach that assumes the strict genetic or biological determination of the lifespan from birth that virtually cannot be influenced by environmental factors, and that can only be affected by internal invasive manipulations. There is a clear need to bridge this gap through the study of physiological, in particular metabolic, neuro-hormonal and epigenetic influences on the lifespan and healthspan, which recognizes the vital regulatory role of the environment on gene expression and internal physiological function [8].

Another gap that may need to be bridged is between *different types of analysis* that are often practically incompatible, for example between "omics" analysis, chiefly involving diverse biological markers at various levels of biological organization (e.g. genomics, proteomics, metabolomics, etc.) aiming to predict and personalize therapy vs. functional and clinical old-age "frailty" analysis and intervention, or between *molecular-biological, energy-metabolic and functional-behavioral evaluations and interventions*. A stronger alliance between these fields may be desirable. There may accrue a great therapeutic benefit from introducing "omics" type of analysis, its predictive and personalized philosophy for old-age frailty evaluation and treatment. And conversely, the researchers and developers of omics biomarkers may need to be more strongly involved in the problems of aging, to realize the critical need to address fundamental degenerative aging processes in order to alleviate virtually all health conditions, including those they are currently working on.

There is also a need for a stronger connection between *diagnosis and therapy* aimed at healthspan extension, or between *longevity factors analysis and therapeutic interventions*. There is often a deficit of interrelation between these areas. The research of "biomarkers of aging" and "longevity factors" is often descriptive, with uncertain implications for clinical practice. On the other hand, anti-aging (geroprotective) and healthspan-extending medicine approaches are often strongly empirical and "prescriptive," testing for a variety of potential interventions, without a former comprehensive factor analysis, with the aim to empirically establish potentially effective treatments. There is a need for stronger interoperability between these areas. By providing an input for therapeutic interventions from population-based aging and longevity factor analysis, it may be possible to provide a broad evidential database for further experimentation in regenerative and geroprotective medicine, as well as shorten the pathway between longevity factor analysis and experimentation. The results of experiments may in turn immediately feed back to refine data collection and analysis, accelerating the process of discovery.

For testing potential anti-aging and healthspan-extending interventions, it appears to be critical to better relate among *different research models*, that may include population, individual, human, animal, culture, cell or molecular models. Often, studies are conducted at different levels of biological organization, with a disregard of other levels. There is an apparent need for an integrative approach, spanning across the relevant scales, using a wide array of physiological, environmental, genetic and epigenetic parameters. For actual human treatment, the human being as a whole should be the focus, with a special attention given to personalized factors characteristic of individual subjects, and selecting the most informative factors. Other models and levels could be studied as supplementary. Diverse integrated data from various levels of organization could enable the creation of truly holistic models for predictive diagnostic evaluation and preventive therapeutic intervention for human subjects. There may be a need to have a "common language" (e.g. non-dimensional measures) to describe the different model systems in common terms, that may be applicable for any system [9].

Of course, it must also be noted that the costs for such a comprehensive data collection and experimentation will likely be high, and funding will always be an issue. It may also be suspected that collecting and analyzing too much and too various data may become unwieldy (whatever the available computational power), and some simplification, abstraction and synthesis may be required. Yet, in any case, the more data can be available – the easier it will be to filter and simplify it.

Yet, the gaps that need to be bridged to achieve practicable knowledge of anti-aging and healthspan-extension do not only concern different types of biological analysis, but extend wider. The research of aging and lifespan and healthspan extension is not just a theoretical scientific or purely biological subject, but in many ways a technological subject, where the capabilities of biological research and manipulation are largely determined by technological capabilities. Virtually all technological fields can ultimately be enlisted for solving the problem of degenerative aging and for extending healthy lifespan. These would include such technological areas as novel measurement modalities (including comprehensive physiological vitality measurements, and a vast array of cell-based and molecular measurements), synthetic biology, nanotechnology and micro-fabrication, as well as advanced computational, modeling and visualization capabilities. *"Technological convergence" and "cross-fertilization"* may be key concepts for tackling the problem of aging.

But the solutions should not remain at the stage of fundamental research in the lab. Another key concept may be *"clinical translation"* understood as the process of translating fundamental scientific research to its application in clinical practice, including all the stages of research and development: from studies on cells and tissues, through animal studies and human trials, up to marketing, production and distribution. The future translation into clinical practice should always be kept in mind as a primary objective. The studies of aging are not just academically intriguing (and they are), but also have a clear purpose – to improve health for the elderly, eventually for all of us. The translation from fundamental research to clinical practice is often difficult, and not only due to scientific and technological hurdles, but often also because of societal constraints, such as lack of social interest and investment or inefficient regulation and distribution. Careful thought should always be given for the facilitation and optimization of the translation process to make aging-ameliorating, life and health-extending therapies available to all of us.

Indeed, biomedical aging amelioration and life and healthspan extension are often considered just and only as scientific or technological problems. Yet, in fact, the development, translation, application and access to treatments designed to ameliorate degenerative aging processes and extend healthy lifespan will involve a vast host of *social issues and implications*, including both hindering and facilitating impact factors that will require comprehensive analysis and debate. Hence, it will be necessary to give due consideration to *social factors*, such as legislative, administrative, communal, economic, demographic, educational and even ethical factors that largely determine the *development* of lifespan and healthspan extension research and *translation* of this research into practice. Some of the issues include: regulatory requirements for the short and long-term testing and approval of potential geroprotective treatments; criteria for their efficacy and safety; administrative and organizational requirements needed for the active promotion of healthspan extension research and practice; incentives for the rapid development and translation of the results of this research into medical and clinical practice; provisions for the universal distribution of healthspan-extending technologies to the public, and much more. All these issues will yet need to become the subject of a broad and intense academic and public debate, including political debate [10].

Within the general need for stronger social involvement, there is an urgent need to educate more specialists who will be able to contribute to the various areas of aging and healthspan extension research. There is an even prior need to educate the broader student body and wider public on the importance of such research to prepare the ground for further involvement. Thanks to such *broad education*, many more new promising studies may emerge. The increased knowledge of the field may increase the demand for therapies, which may in turn increase the offer. Even when the therapies are available, it should be the general public who should use them, hence their willingness to embark on and adhere to a preventive anti-aging and healthspan-improving regimen, their ability to intelligently choose and apply effective and safe therapy, will be vital for its successful application. Therefore comprehensive and wide-ranging "patient and consumer education," and moreover "citizen scientist" and "do-it-yourself maker" education in the field of aging and healthspan extension will be necessary. Such education is currently very limited. In practical terms, globally there are very few centers or dedicated structures to coordinate knowledge exchange and dissemination on biology of aging and healthy lifespan extension. There are even few courses in this field in university curricula around the world. There is a need for more courses and training materials on the subject, in order to make the narrative on biology of aging and healthy lifespan extension an integral part of academic curriculum and public discourse.

21.3 The Problem of Clinical Definition of Degenerative Aging: Bridging the Gaps in Scientific Understanding and Communication

One of the major factors hindering the discussion of aging amelioration, lifespan extension and healthspan extension research, development and application, may be the basic *deficit of definitions*. What is it exactly that we wish to ameliorate, and what is it exactly that we wish to extend? Such agreed definitions appear to be among the necessary conditions for the communication, dissemination and advancement of the field. But such agreed definitions are currently lacking.

There is a growing realization that in order to combat the rising aging-related ill health and improve the healthy lifespan – the research, development and distribution of anti-aging and healthspan-improving therapies need to be accelerated [11]. It was suggested that one of the accelerating factors could be the general recognition of the degenerative aging process itself as a medical problem to be addressed [12]. It has been assumed that such a recognition may accelerate research, development and distribution in several aspects: (1) The general public would be encouraged to actively demand and intelligently apply aging-ameliorating, preventive therapies; (2) The pharmaceutical and medical technology industry would be encouraged to develop and bring effective aging-ameliorating therapies and technologies to the market; (3) Health insurance, life insurance and healthcare systems would obtain a new area for reimbursement practices, which would encourage them and their subjects to promote healthy longevity; (4) Regulators and policy makers would be encouraged to prioritize and increase investments of public funds into aging-related research and development; (5) Scientists and students would be encouraged to tackle a scientifically exciting and practically vital problem of aging. Here we would leave aside the question whether this medical condition should be termed a "disease," a "syndrome," a "risk factor," an "underlying cause" or some other designation. Here "the aging process as a medical condition" just means a processes that can be materially intervened into, improved (treated) and even eliminated (cured) by medical means.

Yet, in order for degenerative aging process to be recognized as such a diagnosable and treatable medical condition and therefore an indication for research, development and treatment, a necessary condition appears to be the development of evidence-based diagnostic criteria and definitions for degenerative aging. So far, there are still no such commonly accepted or formal criteria and definitions. Yet without such scientifically grounded and clinically applicable criteria, the discussions about "ameliorating" or even "curing" degenerative aging processes will be mere slogans. Indeed, how can we "treat" or "cure" something that we cannot even diagnose? It may even be found that such criteria are explicitly or implicitly required by several major international and national regulatory and policy frameworks, such as the International Classification of Diseases (ICD), the WHO Global Strategy and Action Plan on Ageing and Health (GSAP), the European Medicines Agency (EMA), the US Food and Drug Administration (FDA), and others [13]. Such frameworks are thirsting for evidence-based criteria for the effectiveness of interventions for "healthy aging." Nonetheless, nobody has yet done the necessary work of devising such comprehensive evidential criteria. It may seem that the problem has not been solved just for the lack of enough trying. But it must be admitted that the problem is not at all easy even to dare to take on. Many formidable methodological challenges may arise in attempting to develop commonly acceptable diagnostic definitions and criteria for degenerative aging. But try we must!

A major challenge is related even to the *semantic understanding* of the term "degenerative aging." The term "degenerative" may imply both the present *state of degeneration and the process leading to the state of degeneration*. This distinction may have major implications for intervention, respectively implying a curative approach to the already manifest state of degeneration (a late stage intervention) as opposed to a preventive approach to block a process leading to degeneration (an early stage intervention). It may be particularly helpful to explore "degenerative aging" in the latter sense, as a process leading to degeneration that can be prevented. Yet, many questions remain with such a definition. Obviously, not every time-related change leads to degeneration and disease, and some aging-related changes may be beneficial for the person (e.g. the proverbial "wisdom of age" [14]). Obviously also, many changes leading to age-related degeneration begin at conception, and may be necessary concomitants of the processes of growth and development. Then for which processes and at which stages is intervention warranted? In other words, *which aging processes can be considered truly "degenerative"* (leading to degeneration) that would require preventive intervention? Several sets of such candidate processes have been proposed [7], yet there is still little empirical evidence that intervention into them will have clinical benefits. The potential interrelation and regulation of these various processes are also uncertain. In this regard, a practical worry is that under the title of "prevention" and "early intervention" – drugs and other treatments will be sold to young and relatively healthy individuals without a real need and without proven benefits in actually preventing degenerative states and/or extending healthy lifespan. A more thorough, *quantitative and formal understanding of old-age degeneration (frailty) as a physiological state* is required as well. Should it be measured as a lack of function and adaptation to the environment? Should it be evaluated as an impairment of homeostatic or homeodynamic stability? [15] Should it be presented as an index or as physiological age?

Each of these options would raise a host of questions of its own, whose mere mentioning would go far beyond the scope of this work. To provide evidence-based answers to those questions, *vast empirical and theoretical research* yet appears to be needed to establish diverse age-related changes as predictors of adverse age-related outcomes (such as multi-morbidity and mortality) as well as evaluate the effects of various preventive and curative treatments on those outcomes. Based on such data, better formal, clinically applicable models and criteria of degenerative aging as a process and as a state can be developed.

It may be stated that the development of *clinical definitions and criteria for degenerative aging*, and the corresponding definitions and *criteria for the effectiveness of anti-aging and healthspan-extending therapies* would be the penultimate "gap" in the common scientific understanding of the problem that needs to be "bridged" before proceeding toward its practical solution. This would in fact mean bridging multiple "gaps" between multiple conceptions and approaches to the problem of aging amelioration and healthspan improvement, to achieve a good level of mutual understanding and agreement. With the current diversity of theories, approaches, models and prospective remedies, it may be yet a long road ahead before such a level of *common understanding and agreement* is reached. It may not be necessary that every researcher should accept a standard universal metrics and agree on most of the fundamental concepts and processes (as it has been accomplished in mathematics and physics), but at least some degree of commensurability for the field may be desirable. Such commensurability would not mean dictating the same approach to all, or even worse, prescribing the same measures and treatments for all, but rather providing a common language that would enrich general discourse and creativity in the field. The continuous active consultation and debate on these issues may be key to progress.

21.4 Some Particular Methodological Issues for Anti-Aging and Healthspan-Improving Diagnostic and Therapeutic Criteria

The present work could not presume to even begin to provide any definitive answers for the above methodological problems. It does not provide any specific building blocks for the bridges between the various areas that may need to come into closer, more impactful synergistic contact. This work is only intended to attempt to emphasize some of those potential problems and stimulate their discussion (in addition to any discussions of these issues that may take place anywhere else). If it succeeds to enhance this discussion and improve this knowledge even slightly, then it has fulfilled its purpose.

As a way of a conclusion, which is not really a conclusion, but just an attempt to raise further discussion, a few particular challenges may be listed, including some of the earlier points, problems and gaps. This list includes some of the major concerns for the development of diagnostic and treatment criteria against degenerative aging and for healthy lifespan extension, the knowledge of which yet needs to be improved. These can be tentatively classified as follows: (1) establishing definitions,

(2) minimizing confounding factors, (3) improving informative value, and finally (4) improving the practical utility of the criteria. This could also be the putative priority order at which the problems can be tackled. (It must be reemphasized that these propositions are only intended to stimulate academic and public discussion.) Within the larger categories, several sub-categories may be suggested.

For the first category – "**establishing definitions**" – it is important both to establish basic terms and definitions, as well as to define specific clinical benefits and end points of anti-aging and healthspan-improving treatments. As often exemplified in medical history, e.g. for sarcopenia, clear basic definitions and clinical endpoints are indispensable for the development of a new medical field [16]. *Establishing basic terms and definitions* may include some of the questions raised above. For example, should "degenerative aging" be understood as a process or as a state? Or is "healthy aging" a helpful term for developing clinical measurements of aging, considering that most aging processes increase morbidity? Should we instead speak in terms of *"healthy longevity" as opposed to "degenerative aging"*? Such definitions are needed for the basic communication and mutual understanding.

In turn, clearly and consensually *defining clinical benefits* from anti-aging and healthspan-improving treatments appears to be absolutely indispensable for the development of evidence-based diagnosis and therapy of degenerative aging as a treatable medical condition. Just and only biomarkers of aging may not be sufficient to provide clinically applicable diagnostic criteria for "degenerative aging" or for interventions against it. For example, as many studies of Alzheimer's disease have shown, treatments can modify "biomarkers" of the disease very well (in some types of models), but do little or nothing clinically beneficial for actual human patients [17]. Hopefully, this problem can be avoided when addressing general aging as a medical condition. There is a need to precisely define measurable *clinical* end points, demonstrating evidential clinical benefits, especially for the *reduction of age-related multimorbidity*. The combination of structural biological and functional behavioral parameters may increase diagnostic capabilities. In practical terms, the establishment of clinical benefits would also mean more direct and fast transitions between descriptive measurements and experiments (in both directions), "bridging the gap between longevity factors analysis and therapeutic interventions."

Secondly, insofar as aging is an extremely complex process, involving both internal physiological and external environmental factors, "**minimizing confounding factors**" in diagnosis and therapy of aging appears to be a daunting, yet crucial task. In this regard, a primary question may concern the very *relevance of particular studies to human patients*, both actual and potential. How are particular research models relevant or close to apply to humans generally or to individual humans particularly? How likely are those models to lead to human applications? It may be desirable to develop some formal and quantifiable measures for the closeness of relevance range for research models, according to biological levels of organization, types of experimental models, or predictive values of computational models.

Moreover, the applied longevity research may need to be eventually relevant not just for humans generally, but for older persons particularly. Hence the *focus on older persons* appears to be essential for developing evidence-based effective "anti-aging" therapy. The clinical benefits need to be evaluated, or at least confidently predicted, for the primary target population – the older frail persons, rather than the younger and healthier ones who may exhibit entirely different biological responses [18]. In practice, often the results of clinical trials done in young and healthy subjects are then uncritically projected to old and frail subjects. Further strong confounding effects may be introduced when attempting to apply treatments tested in young patients with a single disease, on old patients with multiple diseases. Aging biomarkers and treatments need to be examined in the actual aged, multi-morbid and frail patients as reflecting the most common clinical settings, rather than relying on potentially misleading hypothetical projections from the young and healthy to the old and frail people. The specific examination of older persons is in fact an explicit requirement of "The International Conference on Harmonisation of Technical Requirements for Registration of Pharmaceuticals for Human Use (ICH)" and of the World Health Organization [19], but it is too seldom accomplished in practice. In case direct applicability cannot be shown, at least some relevance should be considered for the aging process and aging subjects.

Another critical concern appears to be the *long term consideration* of treatment effects. The clinical criteria and biomarkers, as well as resources available to the organism, need to be considered for the long term. Thanks to long-term evaluation it may be possible to control for effects of over-stimulation, as well as rule out transient compensatory and psychosomatic effects and seeming short-term benefits that may arrive at the expense of long-term deterioration. In particular, seeming short-term "rejuvenation effects" may increase mortality and shorten the actual lifespan [20]. Such long-term follow up and analysis are quite rare in the field of "anti-aging medicine" as it is commonly practiced and, truth be told, also for other medical fields.

These are some of the potential confounding factors that may obscure or even negate the effects of truly effective anti-aging and healthspan-improving therapies. This list is far from being exhaustive and may be expanded and more specifically elaborated. Yet, with any classification, confounding factors should be controlled for as much as feasible. Of course, some confounding effects are virtually unavoidable, but this should not discourage or abolish the pursuit of longevity research, attempting to ameliorate the harms that could be ameliorated, yet always keeping in mind the *possibility of confounding influences and thus misleading recommendations*.

Thirdly, in order to control for confounding factors vs. truly beneficial therapeutic effects, specific informative value measures are needed to quantify those effects, as opposed to confounding noise, with the aim of "**improving the informative value**" of diagnostic and treatment parameters. Such informative value measures are first of all needed for the *selection* of diagnostic parameters and therapies. As almost any age-related biological parameter may be considered a "biomarker of aging," and almost any physiological intervention can in some way affect the aging process, there is a need to select the most predictive and economic biomarkers and intervention effects, for the population as well as for individuals, with reference to the aging process and aging-related diseases [21].

The informative value measures are also needed for the *integration* of diagnostic parameters and therapies. Criteria for degenerative aging may not be only molecular and cellular, but at every level of biological organization – from the molecular to cellular to tissues and organs, to the entire organism and to the organism's interrelation with the environment – that need to be integrated [22]. Moreover, these criteria may not necessarily be chemical and biological, but can also be physical, in particular as relates to various resuscitation technologies as applied to the elderly, such as hypothermia and suspended animation [23], oxygenation and energy metabolism [24], electromagnetic stimulation [25]. Social (engagement) and psychological (motivation) criteria also need to be added. Among other implications, this drive for integration would also mean "bridging the gaps" between "environmental" and "internal" evaluations and interventions, between "multi-omics" and "functional frailty" analysis, and between different, currently often incomparable "research models." Quantifiable integrated models of such factors may be desirable, as difficult as their creation may be.

Insofar as the organism reacts as a whole, in an integrated and interrelated manner, individual biomarkers may not be indicative of the process or state of degeneration, and individual "magic bullet" therapies may not be effective for the amelioration of this process, but need to be considered in combinations, or ideally in a systemic balanced or "homeostatic" way – otherwise interventions on particular biomarkers and pathways may exacerbate other biomarkers and pathways, and disrupt the system as a whole. The general methodology for the evaluation of the effects of multiple integrated therapeutic agents and risk factors (including biomarkers of aging) on multiple integrated adverse effects and age-related diseases (multimorbidity) needs to be improved, to allow the evaluation of non-linear, cumulative or synergistic effects [26].

Quantifiable informative values may also be desirable to establish the balance of the therapies' *safety vs. efficacy* (actual or potential). For any potential intervention, including anti-aging and healthspan-improving interventions, there is an essential need to weigh potential benefits against potential safety risks. These values are often antagonistic, as therapies may be safe but ineffectual, or potentially effective against particular conditions but carrying risks for severe side effects. Quantifiable balance relations between efficacy and safety may be hoped for to create novel anti-aging therapies that could practically benefit human beings.

Finally, the ultimate measure of the anti-aging and healthspan-improving diagnostics and therapies would be their "**practical utility**" for as many people as possible. In this regard, *standardization* of anti-aging diagnostics and therapies appears to be a major requirement. Particular batteries of assays and interventions are usually related (and potentially biased) to particular theories, research agendas, academic schools and commercial interests. There is an apparent need to allow pluralism of investigation, discovery and application, while maintaining rigorous standards, based on the scientific method, that would facilitate interoperability and common discourse and utility. Consensus standards often emerge as a result of data-sharing [27], which may become a practical challenge of its own. Beyond developing agreeable standards, extensive thought should also be given to ways to improve adherence and compliance with those standards, once again keeping in mind the need both for *pluralism and rigor*. Very often, highly beneficial recommendations are not rigorously followed, and conversely harmful requirements are imposed against pluralistic choice. Hopefully, the field of anti-aging and healthspan extension could avoid both extremes.

And perhaps one of the most critical and complex sets of evaluation criteria may concern the *affordability and cost-effectiveness* of new anti-aging and healthspan-improving diagnostics and therapies. Such considerations may be decisive for the investment in and development of new anti-aging and healthspan-extending therapies. As the experience of many years of fund-raising for life-extension research and development often teaches, even for the most humanitarian causes, investments and donations are largely decided by considerations of actual costs vs. potential profit returns. On the positive side, an expectation of profit may encourage entrepreneurs to develop and distribute new therapies. Still, as the goal of healthspan care is not just to produce profit for the providers, but to improve actual healthspan for as many people as possible, the considerations of cost-effectiveness should involve not only profitability, but also affordability for the population and the non-monetary improvement of well-being, while still providing a return on investment. In terms of the general practical utility, the costs of diagnostic biomarkers assays and therapeutic interventions may become prohibitive or even impractical for use by most people in the world. Hence, there may be a need to focus on such therapies, biomarkers and functional assays that may be most affordable, especially those that are already routinely used in clinical practice, while still encouraging the development of more sophisticated assays and therapies, that may become more accessible in time, and specifically devising means to increase their accessibility [28]. The subject of cost-effectiveness analysis for the anti-aging and healthspan-improving medicine is only nascent [29], and must be developed.

The issue of "affordability" actually involves most of the problems and "gaps" between "science and technology" (the problem of translating fundamental research to practical affordable therapies), between "science, technology and society" (making the therapies widely available, and not only "for the rich and powerful"), as well as between "research and education" (making the knowledge of the field more accessible and wider spread, to catalyze even more knowledge generation). The main overarching question to ask in this regard is: "How can we make the best, most effective therapies available (affordable) as fast as possible to as many as possible?" The details are to be established in a broad academic, public and political discussion [30].

21.5 Motivation for Further Discussion

These are some of the issues that we may need to know how to solve in order to treat degenerative aging as a medical condition and improve healthy lifespan in a scientifically grounded way. These are rather complex issues, yet in order to successfully achieve healthy lifespan extension for the general population, they need to be collectively tackled, "not because they are easy, but because they are hard." All these issues must become a subject of massive and pluralistic consultation, involving scientists, policy makers and other stakeholders. Thanks to such a consultation it may be possible to develop agreeable scientific clinical criteria for degenerative aging that could improve diagnostic capabilities and allow better informed clinical decisions. Such criteria can stimulate further research and development of effective, evidence-based anti-aging and healthspan-extending therapies, treating the underlying processes of aging-related diseases rather than their particular symptoms. In such a broad consultation, various diagnostic and therapeutic approaches to aging amelioration and healthy lifespan extension may be brought together, their relative merits and drawbacks may be compared, points of their convergence may be clarified. Such a discussion may facilitate the creation of a comprehensive and actionable roadmap toward healthy lifespan extension. It is hoped that the present work will contribute to raising the demand for more of such discussion, research and knowledge.

Acknowledgement   The author thanks the Shlomo Tyran Foundation and Vetek Association for their support.

References

1. Kunlin Jin, James W. Simpkins, Xunming Ji, Miriam Leis, Ilia Stambler, "The critical need to promote research of aging and aging-related diseases to improve health and longevity of the elderly population," *Aging and Disease*, 6, 1–5, 2015, http://www.aginganddisease.org/EN/10.14336/AD.2014.1210.
2. Rafael Lozano, et al. (189 authors), "Global and regional mortality from 235 causes of death for 20 age groups in 1990 and 2010: a systematic analysis for the Global Burden of Disease Study 2010," *Lancet,* 380, 2095–2128, 2012.
3. United Nations, Department of Economic and Social Affairs, Population Division, *World Population Prospects: The 2017 Revision*, 2017, https://esa.un.org/unpd/wpp.
4. Nathan Keyfitz, "Improving life expectancy: An uphill road ahead," *American Journal of Public Health*, 68, 954–956, 1978, https://www.ncbi.nlm.nih.gov/pmc/articles/PMC1654068/; Michael J. Rae, Robert N. Butler, Judith Campisi, Aubrey D.N.J. de Grey, Caleb E. Finch, Michael Gough, George M. Martin, Jan Vijg, Kevin M. Perrott, Barbara J. Logan, "The demographic and biomedical case for late-life interventions in aging," *Science Translational Medicine*, 2, 40cm21, 2010, http://stm.sciencemag.org/content/2/40/40cm21.full.
5. Gregory M. Fahy, Michael D. West, L. Stephen Coles, Steven B. Harris, (Eds.), *The Future of Aging: Pathways to Human Life Extension*, Springer, New York, 2010; Alexander Vaiserman (Ed.), *Anti-aging Drugs: From Basic Research to Clinical Practice*, Royal Society of Chemistry, London, 2017.
6. Swapnil N. Rajpathak, Yingheng Liu, Orit Ben-David, Saritha Reddy, Gil Atzmon, Jill Crandall, Nir Barzilai, "Lifestyle factors of people with exceptional longevity," *Journal of the American Geriatrics Society,* 59(8), 1509–1512, 2011; Sofiya Milman, Nir Barzilai, "Dissecting the mechanisms underlying unusually successful human health span and life span," *Cold Spring Harbor Perspectives in Medicine,* 6(1), a025098, 2015; Natalia S. Gavrilova, Leonid A. Gavrilov, "Search for mechanisms of exceptional human longevity," *Rejuvenation*

Research, 13(2–3), 262–264, 2010; Miguel A. Faria, "Longevity and compression of morbidity from a neuroscience perspective: Do we have a duty to die by a certain age?" *Surgical Neurology International*, 6, 49, 2015.

7. Ilia Stambler, *A History of Life-Extensionism in the Twentieth Century*, Longevity History, 2014, http://www.longevityhistory.com/; Gregory M. Fahy, Michael D. West, L. Stephen Coles, Steven B. Harris, (Eds.), *The Future of Aging: Pathways to Human Life Extension*, Springer, New York, 2010; Alexander Vaiserman (Ed.), *Anti-aging Drugs: From Basic Research to Clinical Practice*, Royal Society of Chemistry, London, 2017; Aubrey D.N.J. de Grey, Michael Rae, *Ending Aging. The Rejuvenation Breakthroughs That Could Reverse Human Aging in Our Lifetime*, St. Martin's Press, New York, 2007; SENS Research Foundation, "A Reimagined Research Strategy for Aging," accessed March 2018, http://www.sens.org/research/introduction-to-sens-research/; Healthspan Campaign, "NIH Geroscience Interest Group (GSIG) Releases Recommendations from the October 2013 Advances in Geroscience Summit," 2013, http://www.healthspancampaign.org/2014/02/27/nih-geroscience-interest-group-gsig-releases-recommendations-october-2013-advances-geroscience-summit/; Brian K. Kennedy, Shelley L. Berger, Anne Brunet, Judith Campisi, Ana Maria Cuervo, Elissa S. Epel, Claudio Franceschi, Gordon J. Lithgow, Richard I. Morimoto, Jeffrey E. Pessin, Thomas A. Rando, Arlan Richardson, Eric E. Schadt, Tony Wyss-Coray, Felipe Sierra, "Geroscience: linking aging to chronic disease," *Cell*, 59(4), 709–713, 2014, http://www.cell.com/cell/fulltext/S0092-8674(14)01366-X. Carlos López-Otín, Maria A. Blasco, Linda Partridge, Manuel Serrano, Guido Kroemer, "The hallmarks of aging," *Cell*, 153(6), 1194–1217, 2013, http://www.cell.com/cell/fulltext/S0092-8674(13)00645-4.

8. Anne Brunet, Shelley L. Berger, "Epigenetics of aging and aging-related disease," *Journal of Gerontology: Biological Sciences*, 69 Suppl 1, S17–20, 2014, http://www.ncbi.nlm.nih.gov/pmc/articles/PMC4022130/; Maria Manukyan, Prim B. Singh, "Epigenetic rejuvenation," *Genes to Cells*, 17(5), 337–343, 2012, https://www.ncbi.nlm.nih.gov/pmc/articles/PMC3444684/; Alejandro Ocampo, Pradeep Reddy, Paloma Martinez-Redondo, …, Juan Carlos Izpisua Belmonte, "In vivo amelioration of age-associated hallmarks by partial reprogramming," *Cell*, 167(7), 1719–1733.e12, 2016, http://www.cell.com/fulltext/S0092-8674(16)31664-6.

9. David Blokh, Ilia Stambler, "The application of information theory for the research of aging and aging-related diseases," *Progress in Neurobiology*, 157, 158–173, 2017.

10. Ilia Stambler, "Recognizing degenerative aging as a treatable medical condition: methodology and policy," *Aging and Disease*, 8(5), 583–589, 2017; Ilia Stambler, "Human life extension: opportunities, challenges, and implications for public health policy," in: Alexander Vaiserman (Ed.), *Anti-aging Drugs: From Basic Research to Clinical Practice*, Royal Society of Chemistry, London, 2017, pp. 535–564; Ilia Stambler, "The pursuit of longevity – The bringer of peace to the Middle East," *Current Aging Science*, 6, 25–31, 2014.

11. Michael J. Rae, Robert N. Butler, Judith Campisi, Aubrey D.N.J. de Grey, Caleb E. Finch, Michael Gough, George M. Martin, Jan Vijg, Kevin M. Perrott, Barbara J. Logan, "The demographic and biomedical case for late-life interventions in aging," *Science Translational Medicine*, 2, 40cm21, 2010, http://stm.sciencemag.org/content/2/40/40cm21.full; Luigi Fontana, Brian K. Kennedy, Valter D. Longo, Douglas Seals, Simon Melov, "Medical research: treat ageing," *Nature*, 511(7510), 405–407, 2014, http://www.nature.com/news/medical-research-treat-ageing-1.15585; Kunlin Jin, James W. Simpkins, Xunming Ji, Miriam Leis, Ilia Stambler, "The critical need to promote research of aging and aging-related diseases to improve health and longevity of the elderly population," *Aging and Disease*, 6, 1–5, 2015, http://www.aginganddisease.org/EN/10.14336/AD.2014.1210; Dana P. Goldman, David M. Cutler, John W. Rowe, Pierre-Carl Michaud, Jeffrey Sullivan, Jay S. Olshansky, Desi Peneva, "Substantial health and economic returns from delayed aging may warrant a new focus for medical research," *Health Affairs*, 32(10), 1698–1705, 2013, https://www.ncbi.nlm.nih.gov/pmc/articles/PMC3938188/.

12. Alex Zhavoronkov, Bhupinder Bhullar, "Classifying aging as a disease in the context of ICD-11," *Frontiers in Genetics*, 6, 326, 2015, http://journal.frontiersin.org/article/10.3389/fgene.2015.00326/full; Sven Bulterijs, Raphaella S. Hull, Victor C.E. Björk, Avi G. Roy, "It is time to classify biological aging as a disease," *Frontiers in Genetics*, 6, 205, 2015, http://journal.frontiersin.org/article/10.3389/fgene.2015.00205/full; Ilia Stambler, "Has aging ever been considered healthy?" *Frontiers in Genetics*, 6, 202, 2015, http://journal.frontiersin.org/article/10.3389/fgene.2015.00202/full.
13. Ilia Stambler, "Recognizing degenerative aging as a treatable medical condition: methodology and policy," *Aging and Disease*, 8(5), 583–589, 2017; Ilia Stambler, "Regulatory and policy frameworks for healthy longevity promotion," in: *Longevity Promotion: Multidisciplinary Perspectives*, Longevity History, 2017, http://www.longevityhistory.com/; Ilia Stambler, "Human life extension: opportunities, challenges, and implications for public health policy," in: Alexander Vaiserman (Ed.), *Anti-aging Drugs: From Basic Research to Clinical Practice*, Royal Society of Chemistry, London, 2017, pp. 535–564.
14. Joshua K. Hartshorne, Laura T. Germine, "When does cognitive functioning peak? The asynchronous rise and fall of different cognitive abilities across the life span," *Psychological Science*, 26(4), 433–443, 2015.
15. Alan A. Cohen, "Complex systems dynamics in aging: new evidence, continuing questions," *Biogerontology*, 17(1), 205–220, 2016, https://www.ncbi.nlm.nih.gov/pmc/articles/PMC4723638/; David Blokh, Ilia Stambler, "The application of information theory for the research of aging and aging-related diseases," *Progress in Neurobiology*, 157, 158–173, 2017; Alexey Moskalev, Elizaveta Chernyagina, Vasily Tsvetkov, Alexander Fedintsev, Mikhail Shaposhnikov, Vyacheslav Krut'ko, Alex Zhavoronkov, Brian K. Kennedy, "Developing criteria for evaluation of geroprotectors as a key stage toward translation to the clinic," *Aging Cell*, 15(3), 407–415, 2016, http://onlinelibrary.wiley.com/wol1/doi/10.1111/acel.12463/full; Alexey Moskalev, Elizaveta Chernyagina, Anna Kudryavtseva, Mikhail Shaposhnikov, "Geroprotectors: a unified concept and screening approaches," *Aging and Disease*, 8(3), 354–363, 2017, http://www.aginganddisease.org/EN/10.14336/AD.2016.1022.
16. Liam Drew, "Fighting the inevitability of ageing," *Nature*, 555, S15–17, 2018, https://www.nature.com/articles/d41586-018-02479-z.
17. Eric M. Reiman, Jessica B.S. Langbaum, Adam S. Fleisher, Richard J. Caselli, Kewei Chen, Napatkamon Ayutyanont, Yakeel T. Quiroz, Kenneth S. Kosik, Francisco Lopera, Pierre N. Tariot, "Alzheimer's Prevention Initiative: A plan to accelerate the evaluation of presymptomatic treatments," *Journal of Alzheimer's Disease*, 26(Suppl 3), 321–329, 2011; Jeremy Toyn, "What lessons can be learned from failed Alzheimer's disease trials?" *Expert Review of Clinical Pharmacology*, 8(3), 267–269, 2015.
18. Morrison D.H., Rahardja D., King E., Peng Y., Sarode V.R., "Tumour biomarker expression relative to age and molecular subtypes of invasive breast cancer," *British Journal of Cancer*, 107, 382–387, 2012.
19. The International Conference on Harmonisation of Technical Requirements for Registration of Pharmaceuticals for Human Use, *ICH Harmonized Tripartite Guideline E7, Studies in Support of Special Populations: Geriatrics*, ICH, Brussels, June 24, 1993 (General Principle II) http://www.ich.org/fileadmin/Public_Web_Site/ICH_Products/Guidelines/Efficacy/E7/Step4/E7_Guideline.pdf; World Health Organization, *Global Strategy and Action Plan on Ageing and Health (GSAP) – 2016–2020*, November 2015 (Section 105) http://www.who.int/ageing/global-strategy/en/; http://apps.who.int/gb/ebwha/pdf_files/WHA69/A69_17-en.pdf?ua=1.
20. David G. Le Couteur, Stephen J. Simpson, "Adaptive senectitude: the prolongevity effects of aging," *Journal of Gerontology: Biological Sciences*, 66, 179–182, 2011, https://academic.oup.com/biomedgerontology/article/66A/2/179/594634/Adaptive-Senectitude-The-Prolongevity-Effects-of.
21. David Blokh, Ilia Stambler, "Applying information theory analysis for the solution of biomedical data processing problems," *American Journal of Bioinformatics*, 3(1), 17–29, 2015, http://thescipub.com/abstract/10.3844/ajbsp.2014.17.29.

22. Alexander N. Khokhlov, "From Carrel to Hayflick and back or what we got from the 100 years of cytogerontological studies," *Biophysics*, 55(5), 859–864, 2010.
23. Ronald Bellamy, Peter Safar, Samuel Tisherman, ..., Harvey Zar, "Suspended animation for delayed resuscitation," *Critical Care Medicine*, 24(2Suppl), S24–47, 1996; Peter Safar, "On the future of reanimatology," *Academic Emergency Medicine*, 7(1), 75–89, 2000.
24. Gennady G. Rogatsky, Ilia Stambler, "Hyperbaric oxygenation for resuscitation and therapy of elderly patients with cerebral and cardio-respiratory dysfunction," *Frontiers In Bioscience* (Scholar Edition), 9, 230–243, 2017, http://www.bioscience.org/2017/v9s/af/484/2.htm; Gennady G. Rogatsky, Avraham Mayevsky, "The life-saving effect of hyperbaric oxygenation during early-phase severe blunt chest injuries," *Undersea Hyperbaric Medicine,* 34(2), 75–81, 2007; John N. Kheir, Laurie A. Scharp, Mark A. Borden, ..., Francis X. McGowan Jr., "Oxygen gas-filled microparticles provide intravenous oxygen delivery," *Science Translational Medicine*, 4(140), 140ra88, 2012; Yifeng Peng, Raymond P. Seekell, Alexis R. Cole, Jemima R. Lamothe, Andrew T. Lock, Sarah van den Bosch, Xiaoqi Tang, John N. Kheir, Brian D. Polizz, "Interfacial nanoprecipitation toward stable and responsive microbubbles and their use as a resuscitative fluid," *Angewandte Chemie International Edition*, 57, 1271–1276, 2018, https://doi.org/10.1002/anie.201711839.
25. Yury P. Gerasimenko, Daniel C. Lu, Morteza Modaber, ..., V. Reggie Edgerton, "Noninvasive Reactivation of Motor Descending Control after Paralysis," *Journal of Neurotrauma*, 32(24), 1968–1980, 2015; Max Schaldach, *Electrotherapy of the Heart: Technical Aspects in Cardiac Pacing*, Springer-Verlag, Berlin, 2012.
26. David Blokh, Ilia Stambler, "Estimation of heterogeneity in diagnostic parameters of age-related diseases," *Aging and Disease*, 5, 218–225, 2014, http://www.aginganddisease.org/EN/10.14336/AD.2014.0500218; David Blokh, Ilia Stambler, "Information theoretical analysis of aging as a risk factor for heart disease," *Aging and Disease*, 6, 196–207, 2015, http://www.aginganddisease.org/EN/10.14336/AD.2014.0623; David Blokh, Ilia Stambler, "The use of information theory for the evaluation of biomarkers of aging and physiological age," *Mechanisms of Ageing and Development*, 163, 23–29, 2017, https://doi.org/10.1016/j.mad.2017.01.003; Blokh D, Stambler I, Lubart E, Mizrahi EH, "The application of information theory for the estimation of old-age multimorbidity," *Geroscience*, 39(5–6), 551–556, 2017.
27. Gregory K. Farber, "Can data repositories help find effective treatments for complex diseases?" *Progress in Neurobiology*, 152, 200–212, 2017, https://doi.org/10.1016/j.pneurobio.2016.03.008.
28. Ilia Stambler, "Human life extension: opportunities, challenges, and implications for public health policy," in: Alexander Vaiserman (Ed.), *Anti-aging Drugs: From Basic Research to Clinical Practice*, Royal Society of Chemistry, London, 2017, pp. 535–564.
29. Dana P. Goldman, David M. Cutler, John W. Rowe, Pierre-Carl Michaud, Jeffrey Sullivan, Jay S. Olshansky, Desi Peneva, "Substantial health and economic returns from delayed aging may warrant a new focus for medical research," *Health Affairs*, 32(10), 1698–1705, 2013; Peter S. Hall, Richard Edlin, Samer Kharroubi, Walter Gregory, Christopher McCabe, "Expected net present value of sample information: From burden to investment," *Medical Decision Making*, 32, E11–E21, 2012; Gwern Branwen, *Life-extension cost benefits*, 2018, https://www.gwern.net/Longevity.
30. Ilia Stambler, *Longevity Promotion: Multidisciplinary Perspectives*, Longevity History, 2017, http://www.longevityhistory.com/.

Chapter 22
Harnessing Nature's Clues for Regeneration, Disease Reversion, and Rejuvenation

Ira S. Pastor

22.1 Introduction

We currently live on a planet with many other organisms which from a health and wellness perspective are much further advanced than human beings.

Many lower organisms (i.e. amphibians, planarians) can replace lost or damaged organs and tissues that are identical in both structure and function to the original, effortlessly regenerating a wide variety of tissues, including spinal cords, limbs, hearts, eyes, and even large segments of their brains [19, 31, 65].

In a similar fashion, many of these same species possess fascinating skills for repairing and reversing cellular and genetic damage. Cancer, as an example, is found to be extremely rare in species displaying an efficient regenerative mechanism, even under the action of potent carcinogens. In many cases, when cancer does occur, tumors have been found to spontaneously remodel and integrate into their surroundings as normal, healthy tissue [55, 62].

Some of these organisms don't age and exhibit "negligible senescence" [24]. Some can age, and then return to a youthful state later on in life [52]. Some can even die, and be re-born [13].

Needless to say, humans are extremely weak when it comes to accomplishing any of these feats, and unfortunately, the outcomes are very different. In most instances, the structure or function of an organ will not be restored after complex tissue damage in humans, and is often replaced by non-functional scar tissues. Additionally, while humans do possess robust DNA repair mechanisms that protect them from daily external and internal perturbations [23], these capabilities are diminished substantially over time as we age.

Extensive study into the regeneration and repair mechanisms of non-human species have found them to be intricately connected to an underlying capability of

I. S. Pastor (✉)
Bioquark Inc., Philadelphia, PA, USA
e-mail: pastor@bioquark.com

© Springer Nature Switzerland AG 2019
N. Lee (ed.), *The Transhumanism Handbook*,
https://doi.org/10.1007/978-3-030-16920-6_22

complex tissue reprogramming and remodeling. These "epimorphic" capabilities represent a biological regulatory state reset, whereby various forms of genetic and epigenetic damage are erased in cell populations, followed by their redirection into a generative developmental program, whereby they become reintegrated with their micro-environment cellular neighbors, and then reorganized, via a "community effect", along tissue, organ and positional specificity [9].

The ability to tap into and mimic these capabilities with novel bio-products, in human beings, will offer potential solutions to a wide range of disorders responsible for human degeneration, suffering, and death, and help more rapidly usher in era of one of the core tenets of transhumanism: radical life extension.

Where are we now and how do we get to our goal?

22.2   Historical Status Quo

In 2016, we surpassed US$7 trillion [42] in total annual healthcare expenditures around the globe, close to a US$1 trillion of which was spent on pharmaceutical products alone. That same year, an additional US$200 billion was spent globally on new life science research and development.

Despite these incredible financial dynamics, we witnessed a rise in the prevalence of almost all chronic degenerative diseases responsible for human suffering and death, as well as an on-going growth and aging of the population.

Many experts agree that innovation has all but disappeared in the traditional drug development arena, with costs per new product escalating into the billions of dollars, development time lines surpassing 15–20 years [61], and the majority of new drug approvals (if they survive the typical 1-in-10,000 attrition rates), offering minimal to no benefit over existing therapies [21].

Additionally, widely acknowledged (but largely unspoken) truths from the pharmaceutical industry persist: all of the "disease output" targeted drugs that eventually make it to the market will only work in a small percentage of their target population (due to our emerging understanding of patient and disease heterogeneity), and none will be capable of ever affecting an actual cure. Current estimates of pre-clinical research that is irreproducible is estimated close to $30 billion annually [22], with billions more wasted in the clinical arena.

Lastly, central regulatory bodies in various developed nations (i.e. FDA [20], EMEA) have become increasingly bureaucratic, and fallen behind literally decades in both their ability to keep up with newer scientific platforms, as well as their acceptance of creative clinical testing modalities that address a twenty-first century understanding of patients, diseases, and drugs [12, 35].

Which leads us to the inescapable questions: Where has the current approach gone wrong, and where are the cures for the chronic, degenerative diseases, which are routinely promised to humanity, but which always seem another "20 years away"?

22.3 Situation Analysis

Since the inception of the modern pharmaceutical industry, researchers have attempted to reduce and study human health and disease at the level of their most basic components – proteins, genes, cells, etc. continually looking for new targets to develop drugs compounds for that can interfere in some fashion with biological processes.

In parallel, from the clinical perspective, patients have continued to be classified and studied in a very standardized fashion at the population level, primarily based on disease symptoms, via the "gold standard" of evidence based medicine – the randomized clinical trial.

While these approaches have ultimately allowed the pharmaceutical industry to grow in size and profitability, giving us many treatments for disease and improving outcomes, they have given us (with few exceptions such as the antibiotic) very few cures for disease.

The primary reason for this is that innovation in the industry (as well as at the level of regulatory organizations) has fallen literally decades behind what the science is has been teaching us in the laboratory.

From the perspective of drug development, disciplines such as systems biology have shown us that the targets that traditional drugs are developed against are no more than the late-appearing indications of dysfunctional tissue/organ systems (i.e. symptoms of disease – inflammation, immunity, fibrosis, thrombosis, hemorrhage, cell proliferation, apoptosis, and necrosis), and that these drugs are being developed without regard for, or knowledge of, any biological factors that precede these abnormalities (causes of disease).

Additionally, the reductionist approach that the industry uses to identify disease mechanisms, or therapeutic targets, continues to ignore the fact disease is rarely (if ever) a simple consequence of an abnormality in a single gene product. Rather, diseases represent emergent regulatory states, involving multiple biological processes, that interact in complex networks, with many nested levels of control hierarchy (i.e. gene, cell, tissue, organism, environment) operating in synergy.

Lastly, the fields of pharmacogenomics and toxicogenomics have continued to highlight that each of us are extremely different from each other in regard to the way drugs both benefit, and/or harm us.

Traditional "gold standard", population level clinical studies continue to use definitions of disease that are excessively inclusive and are based on disease characterizations from decades ago. These inclusive definitions of disease not only obscure important differences among individuals with common clinical presentations, but also ignore underlying disease and/or toxicity mechanisms [28].

In summary, the approach of developing single target drugs, based solely on disease symptoms, combined with clinical study models have ignored human/disease heterogeneity, have brought us to where we are today – lots of treatments; modest improvement in outcomes; no cures.

22.4   The Market Opportunity (and Current Limitations)

The majority of the aforementioned $7 trillion spent annually on health care globally, is focused on the treatment of patients that have diseases with either an underlying cellular damage (i.e. Alzheimer's, Congestive Heart Failure, Parkinson's, Type I Diabetes,) or cellular degeneration (i.e. Auto-Immune Diseases, Cancer, Chronic Inflammation and Pain, Fibrotic Disorders) component to them [42].

Unfortunately, the current therapeutic tools used to address these needs, including organ transplantation, traditional single target pharmaceutical moieties, and, more recently, stem cell therapies, fall very short of actual cures.

Organ transplantation is limited by a substantial and growing donor gap [16] and the unavoidable host-versus-graft reaction. Traditional pharmaceuticals, while capable of interfering and slowing down degenerative processes, can do little to reverse damage once it has occurred, and typically only target the late-appearing indications of dysfunctional tissue/organ systems, as opposed to the biological factors that cause these abnormalities.

Even the evolving stem cell space is running into many technical and efficacy related challenges which will substantially limit the potential of the market for these replacements as mono-therapies. Many current clinical stem cell programs have unfortunately followed the reductionist mantra of the pharma industry, endeavoring to create stand-alone cell therapies that are akin to trying to assemble a house by providing the bricks, but with no mortar, blueprints, or construction foreman on the job.

The whole system requires new approaches and new thinking to correct the problem, and the natural world may hold important clues for us to follow.

22.5   Opportunities to Learn from Nature

Throughout the twentieth century, natural products (primarily those from plants, fungi, and bacteria) formed the basis for a majority of all pharmaceuticals, biologics, and consumer healthcare products used by patients around the globe, generating trillions of dollars of wealth [53].

However, many scientists believe we have only touched the surface of what the natural world, and its range of organisms, which from a health and wellness perspective are much further advanced than human beings, has to teach us.

Lately, novel research disciplines, including "interkingdom signaling" [18] and "semiochemical communication" [39], the respective abilities of one species living signals to affect the genome of another, not to mention in-depth study of the microbiome and virome [54], are highlighting entirely new ways that non-human bioproducts can affect the human genome for positive transitions in health and wellness.

Simultaneously, we are finding out that in addition to the novel, combinatorial biochemical approaches [56] that nature utilizes (as evolutionary dynamics would never follow the "single magic bullet" approach promoted by the pharma industry), that there are also complex biophysical signaling dynamics at play (i.e. bio-electric, bio-mechanical, bio-magnetic), which these organisms utilize to coordinate the nested hierarchies of control architecture in disease progression [33], addressing multiple levels of the "disease-ome" simultaneously. Such integrated dynamics represent a major contrast when compared to the reductionist, highly "siloed" model at play in the pharma industry today.

Merging a twenty-first century, "convergent" knowledge base of regenerative biology, evolutionary genomics, and bio-cybernetics, offers us new guidance to understand how "nature's transhumanists" are so successful in warding off disease and degeneration, and eventual clues to how humans can achieve the same outcomes, and perhaps even move beyond.

22.6 Clues from Regenerative Biology

Regeneration is the ability to recreate lost or damaged cells, tissues, organs, and/or limbs, which are identical in both structure and function to the original.

Regenerative biology, is the integrated discipline that studies genomes, cells, organisms, and even their ecosystems, and finds out what makes them resilient to natural fluctuations, or events that cause disturbance or damage, allowing for proper renewal, restoration, and growth [9].

Of the five major classes of regeneration (i.e. physiologic, hypertrophic, wound healing, epimorphosis, morphallaxis), humans possess the first three. Humans have natural physiologic turnover capabilities in rapidly dividing cells (i.e. blood, epithelial layer of skin/gastrointestinal tract), possess a reliable hypertrophic response following acute damage in the liver (whereby function in restored by enhanced cell number, cell expansion, and tissue growth, without establishing a correct morphological form), and probably most importantly, a rapid thrombotic/fibrotic wound healing response, which prevents us from bleeding to death after minor traumas.

However, when it comes to epimorphosis and morphallaxis, which represent higher order complex organ, limb and body segment regeneration, primarily seen in amphibians and lower invertebrates, humans possess no such capabilities in our fully developed state.

During the epimorphic regeneration process, cells that remain in the damaged body region are reprogrammed, literally erasing their functional history and restarting their life again along a defined generative developmental pattern. Through complex inter-cellular signaling dynamics, these cells then become remodeled based on their surrounding tissue micro-environment, specifically filling in, via intercalation (the locational sensing and communication of what the next step in the development process should be), missing or damaged tissues, organs, and limbs.

This process occurs reproducibly throughout the lifespan of an organism, reliably restoring perfect structure and function.

Epimorphic regeneration in nature is a very complex form of regeneration and involves many mechanisms operating in synergy [59], including but not limited to the reprogramming of cells in target tissues to a progenitor state, a targeted histolytic response required for extracellular matrix remodeling, and an activation of the regenerative side of the innate immune response (versus the adaptive immune response, which many lower organisms do not possess – [49]) to support on-going complex morphogenesis and cell migration, until final form and function is again established.

Interestingly, these complex integrated processes of epimorphosis, while most commonly associated with organ and limb regeneration, are also seen active in various other tissue remodeling events. A key one of these, which has been documented in the literature for decades but mostly overlooked, is that of tumor reversion in regenerating tissues, even those under the stress of potent carcinogens [43].

There have been many scientific reviews over the decades on the general theme of such perturbed regenerative micro-environments and their ability to organize in/out, and well as modify the diseased phenotype. Seminal work on mammalian embryos and teratocarcinoma reversion were performed in the 1970s [38], along with similar reversion dynamics reported in the plant kingdom with crown gall teratomas [7].

More recent subsets of this cellular re-organization theme, include recent discoveries in the areas of genetic diseases that manifest with revertant mosaicism, primarily seen in tissues with an active regenerative niche [27], cellular competition [32] seen in both development and the maintenance of tissue fitness at the intersection of tissues that manifest competing phenotypes, and the correction of aneuploidy in early human embryos [8].

Taken to its extremes, we see such regenerative dynamics in biological immortality manifested by species which can accomplish whole organism tissue reprogramming (i.e. Turritopsis dohrnii – [52]), reassembly of shattered chromosomes in extremophiles (i.e. Deinococcus radiodurans – [13]), and the reanimation of essentially "dead" bio-materials, such as naked DNA, in activated ooplasms [6, 26, 36].

The unifying point here is that nature has developed very reliable methods for biologically "turning back the biological clock" and "starting over" in a controlled fashion that benefit the underlying organism, and these learnings can be leveraged for human health and wellness.

22.7 Clues from Evolutionary Genomics

Evolutionary genomics is an integrated discipline that investigates and compares the genomes of various species, and their respective models of evolution, from bacteria on up to humans. The goal is to explore the similarities in genetic codes across all types of organisms to allow for comparisons of DNA sequences between and within species.

Why as humans should we be interested in this?

Every disease that is responsible for the degeneration and death of human beings, has some particular genetic property from the evolutionary past, which was a benefit to another species survival [64].

High blood sugar levels (seen in diabetes, and other endocrine dysfunction in humans) are required by fast sprinting organisms requiring quick energy bursts when pursuing a prey, or being pursued by a predator.

Various forms of inflammation (manifesting in humans in a variety of inflammatory and allergic based responses) are many lower organisms only form of infectious disease defense.

Even cancer, due to the fact that much of the same genetic "tool kit" used by human tumors can also be found in single cell organisms (which were on this planet over a billion years ago) who utilize these same gene sets to agglomerate into multicell colonies for their survival [17], was at one point an beneficial step allowing for species emergence, survival, and our eventual arrival on the scene as Homo sapiens [15].

One only needs to visually observe the tails, and gills, and webbed hands and feet, that are present during human embryonic development, to appreciate the unique, "locked down" layers of our evolutionary past within the human genome [44], and how minor changes in the genomic architecture can easily awaken these atavisms to manifest a range of disease states.

More recently, the study of the genomic regulatory architecture in a range of species across the phylogenetic "tree of life" additionally serves to highlight how much similarity exists across genomes of evolutionarily distant organisms [14].

As such, while the sequencing of the human genome was quite useful, we must move beyond our sole focus on this data and more fully embrace comparative, interspecies genomics in our study of the progression between healthy and disease tissue states, as it is being highlighted on a daily basis that humans are much more than this single dimensional genetic focus represents.

22.8   Clues from Bio-Cybernetics

Cybernetics is a scientific discipline exploring the integrated dynamics of complex regulatory systems [2].

Bio-Cybernetics is the application of cybernetics to biological science, and integrates systems biology and complex information processing to understand how biological systems function.

While much of the pharma industry focuses solely on "genomic outputs" of disease, the world of bio-cybernetics is taking a much higher level perspective on disease, looking at all the hierarchical levels of control from the whole organism and its environment, on down to genes and their complex regulatory networks.

Concepts once solely reserved for mathematic disciplines such as statistical dynamics, including "state space" (the specific set of values which a process can take), and "attractors" (the set of values toward which a system tends to evolve, based on variety of starting conditions of the system), are now slowly making their way into the field of biology and medicine [34].

The application of mathematics to biology promises a new understanding as to how the diseased state emerges in the first place, and how using related tools and knowledge from the non-biologic realm (i.e. First Law of Cybernetics – If you have complete knowledge of a system, it is possible to control it. – "Only variety destroys variety") how any complex system can be brought under appropriate control.

Needless to say, based on what we know about complex, multi-factorial diseases, and the related regeneration/reversion processes in nature, it would be quite difficult to recapitulate any such dynamics with the traditional pharma "single magic bullet" approach, and more complex, "combinatorial" approaches will be required to alter such systems and control them.

Combination therapy protocols have become more popular in recent years, with HIV "cocktails" and multi-drug chemotherapy protocols, but still take their time to be accepted and employed, as the medical system typically likes the single, simple solutions (i.e. one drug/one target) that big pharma is so happy selling.

But that is not the way nature works; ever.

22.9 Applying These Clues – Cancer Example

Now fully armed with this convergent set of nature's tools, where do we go from here?

The possibilities, if we wield them correctly, are unlimited. But they require a truly integrated working approach between these scientific disciplines.

Let's take a deeper dive into one area: cancer, and see how these ideas apply to constructing novel approaches.

Despite spending close to a trillion dollars in the "war on cancer" to date, the disease remains one of humanities major killers.

As a leading cause of morbidity and mortality worldwide for many decades, today there are approximately 14 million new cases diagnosed each year, with over 8 million cancer related deaths annually [40].

Problems with existing therapeutic strategies such as fractional cell killing and rapid resistance development continue to plague even the most modern "smart drugs" rendering them ineffective in the long run. Many of the new targets discovered during the genomics revolution are considered "un-druggable" by big pharma.

With recent understandings that cancer is no longer a static disease of homogeneous cell populations, as once thought, but much more of a dynamic disease with heterogeneous cell populations constantly altering their transcriptional regulatory states [29], entirely new therapeutic approaches are called for.

Unfortunately, in 2017 the majority of the cancer research system, within its "kill centric" moonshot thinking, continues to base its drug development model on these two pervasive myths about tumor composition that every day become more and more obsolete:

1. that the sole drivers of tumors are somatic mutations in various growth control/ proliferation centers of the genome,
   and
2. that tumors are homogeneous in their composition.

Both of these flawed beliefs have led/will continue to lead, to a flood of "targeted" therapeutics which work extremely well in the short term, in small groups, but fail rapidly thereafter – hence placing us where we are today, with many treatments, minor outcome improvements, and few real cures (forgetting for the sake of time all older approaches of surgery/radiation/broad spectrum cytotoxic drugs).

Most of industry/academia continue to work along this model despite many findings in the scientific literature that call this approach into question – findings such as the facts that the somatic mutation concept can't explain:

- spontaneous remissions;
- foreign body tumor formation;
- tumor cells that "normalize" (or normal cells that become malignant), when placed in different tissue micro-environments, and, very importantly,
- the gross in-balance between somatic mutations and ultimate transcriptomes/ regulatory states of tumors.

All of this emerging knowledge points to the facts that even with the current kill centric, "silver bullet" model of the cancer moonshot, we are still heading down a path that will continue to give us incremental advances in survival as we continue to ignore key findings:

1. most mutations may be only "permissive" of cancer; NOT "causative" of it,
   and
2. tumors are heterogeneous baskets of cells, with potentially many different transcriptional regulatory states in one (which can easily re-constitute each other)

However, there is a small but growing research base [58] that is beginning to challenge this old, flawed system, and is attempting to move the cancer therapeutic model from a reductionist "cell based" approach, to a more holistic "tissue based" one; one that much more appropriately mimics what we see in nature and the cancer resilient regenerators/reprogrammers.

So what are the core learnings as far as cancer is concerned?

- Regenerative biology has shown us that no matter the tumor type, the greater tissue micro-environment can reprogram and normalize it with the correct signaling mechanisms in place.

- Evolutionary genomics has highlighted the reasons for the tumor's existence and ubiquitous nature. We are not competing with recent little "genetic accidents", but with a much deeper, more potent, billion plus years of survival adaptation.
- Bio-Cybernetics shows us that nature does not accomplish its disease reversion capabilities with a single magic bullet approach, but requires much variety to "out compete" the diseased system.

These are exciting and important learnings, and can apply across all disease states, both chronic and acute.

But with this knowledge in hand, what are some novel approaches we can use?

22.10   Combinatorial Biologics

Needless to say, it will take the development of non-traditional drug entities in order to deal with the problem as we more clearly understand it today.

One area that our company, Bioquark Inc. has been working on the past few years, is that of "combinatorial biologics".

The previously spoken of biologic themes that will help advance the transhumanist narrative (Regeneration, Repair, and Rejuvenation), while present in many different forms throughout nature and across species, are also found co-existing in the synergistic biochemical dynamics found within activated ooplasms; including human ooplasms.

Hence one of Bioquark's therapeutics programs focuses on deriving novel biochemical materials from these complex ooplasms and applying them in human health, mimicking the biochemistry of the living egg following fertilization.

During this period of time, ooplasm provides the complete biochemical regulatory architecture required for the new embryo perform an unparalleled set of tasks including: resetting cellular age, reprogramming DNA to eliminate genetic and epigenetic damage, remodeling of organelles, and protection from inflammatory, oxidative, and infectious damage [10, 11, 46, 47, 60].

All of this is done in synergy to initiate the embryo's natural developmental genetic program, and start it on its complex, stepwise path through organogenesis and morphogenesis. Comparable dynamics for rewinding and reinitiating a developmental program in mature tissues are seen during limb and organ regeneration in epimorphic organisms.

Hence in developing its biologics, Bioquark has focused on novel methods to standardize this unique biochemistry, and apply it for the induction of tissue specific micro-environments that could lead to effective regeneration and repair, just like we see during epimorphosis [48].

While ooplasm based reprogramming has been studied in the form of cloning [25] and in-vitro fertilization (IVF) for the past 70 years, and in the form of egg free reconstitution experiments in the petri dish for the last 30 years [1, 37, 63], we are now taking the next step of purifying defined ooplasm fractions and exporting these biochemical signals to other tissues in-vivo.

Hence, instead of looking at discreet pathways and targets early on, as with a traditional, "bottom up" pharma new chemical entity (NCE) development program, we instead are studying purified ooplasm fractions with a "top down" complex control dynamic as our goal.

Instead of going the traditional reductionist NCE route, we are taking an approach similar to how regulatory agencies look at other heterogeneous bio-products nowadays – where the "cocktail", and not a purified biochemical entity, gets developed and becomes the formal active pharmaceutical ingredient (API – [21]).

In regard to mechanisms of action (MOA), as we are dealing with heterogeneous bio-products, technically there are many MOA working together in synergy (very much like a paracrine cocktail secreted from a cell to communicate with neighboring cells). Based on our studies we seem to have a combinatorial set of activities from the regulatory architecture of soluble factors that is normally produced in the ooplasm which have both reprogramming and remodeling bioactivities – independently these factors do very little, but in the right synergy, MOA arises (these bioactives are simultaneously naturally all present in ooplasm, deposited during meiosis, to be utilized by the genome at the correct spatial and temporal configuration).

While the therapeutic targets for ooplasm derived combinatorial biologics we feel are quite broad in scope, Bioquark is focusing to generating human clinical data in discreet disease indications that represent areas of substantial unmet medical need. These include the formerly mentioned cancer reversion opportunity, as well organ repair/regeneration programs.

The first organ that the company is focusing on, based on a combination of unmet medical need and minimal competitive landscape, will be the human kidney.

While humans can live normally with just one kidney, when the amount of functioning kidney tissue is greatly diminished by disease or damage, chronic kidney disease will develop leading towards a progressive loss in function. This loss of kidney function leads to a downward spiral throughout the body including negative effects on the cardiovascular system, nervous systems, and endocrine function.

Dialysis or kidney transplant can prolong life, but quality of life is severely affected, supplies are limited, and costs of long term maintenance are exorbitant. Additionally, as the kidney is one of the more anatomically complex organs, it has proven refractory to stem cell-based regenerative techniques to date.

It is estimated that 10% of the population worldwide is affected by some form of chronic kidney disease (CKD), with millions dying each year [41].

The ability to demonstrate endogenous, 3-dimensional regeneration in the human kidney represents a market value of over $60 billion annually, when only taking into account the costs of dialysis and kidney transplants.

And this is just one organ system.

22.11 The Future?

We believe that nothing is off the table.

The literature is full of data, both current and historical (yet forgotten), on how nature has provided elegant solutions to many human problems that seem insurmountable today, as well as is suggesting of novel paths for intellectual debate and discovery [3–5, 30, 45, 50, 51, 57].

Radical life extension in an important goal for the future of humanity, and eliminating all degenerative diseases is one major step forward to eradicating the problems associated with aging and age related death, and realizing the transhumanist ideal.

Whether we are talking about the 100,000 people who die daily from age related ailments, or the 50,000 that die from acute traumas, it is well within the intellectual capacity of humans to solve these problems of disease, degeneration, and death.

But we must truly think "outside the box", and not fall into the traps set by the traditional pharmaceutical industry, its regulators, and the century old model of drug development which has truly run into a brick wall.

There is a reason that microbes, and plants, and invertebrates, and amphibians, have survived for many hundreds of millions of years on this planet, and developed their own unique answers to many of the problems that plague humanity.

"Nature's transhumanists" have shown us the way; we must now follow their lead.

References

1. Allegrucci C, Rushton MD, Dixon JE, Sottile V, Shah M, Kumari R, Watson S, Alberio R, Johnson AD. Epigenetic reprogramming of breast cancer cells with oocyte extracts. Mol Cancer. 2011 Jan 13;10(1):7. doi: https://doi.org/10.1186/1476-4598-10-7
2. Ashby, W. R. (1958). An introduction to cybernetics (3rd ed.). London: Chapman & Hall.
3. John D Banja. Are Brain Dead Patients Really Dead?. Department of Rehabilitation Medicine, Emory University, and the Center for Ethics, Atlanta, Georgia 30322, USA. The Journal of head trauma rehabilitation. 03/2009; 24(2):141–4. DOI: https://doi.org/10.1097/HTR.0b013e3181a2858d
4. Bernat JL. How much of the brain must die in brain death?. J Clin Ethics. 1992 Spring;3(1):21–6; discussion 27–8.
5. Douglas J Blackiston, Tal Shomrat, and Michael Levin. The stability of memories during brain remodeling: A perspective. Commun Integr Biol. 2015 Sep-Oct; 8(5): e1073424. doi: https://doi.org/10.1080/19420889.2015.1073424
6. J. Julian Blow and Andrea M. Sleeman. Replication of purified DNA in Xenopus egg extract is dependent on nuclear Assembly. Journal of Cell Science 95, 383–391 (1990).
7. A C Braun and H N Wood. Suppression of the neoplastic state with the acquisition of specialized functions in cells, tissues, and organs of crown gall teratomas of tobacco. Proc Natl Acad Sci U S A. 1976 Feb; 73(2): 496–500.
8. Paul Brezina, Andrew Barker, Andrew Benner, Ric Ross, Khanh-Ha Nguyen, Raymond Anchan, Kevin Richter, Garry Cutting & William Kearns. Genetic Normalization of Differentiating Aneuploid Human Embryos. Nature. Precedings. 2011 June 23.

9. Bruce M. Carlson. Principles of Regenerative Biology. Academic Press; 1 edition (May 14, 2007). ISBN-13: 978-0123694393; ISBN-10: 0123694396.
10. Carroll J, Marangos P. The DNA damage response in mammalian oocytes. Front Genet. 2013 Jun 24;4:117. doi: https://doi.org/10.3389/fgene.2013.00117. eCollection 2013.
11. P. D. Cetica, L. N. Pintos, G. C. Dalvit, and M.T. Becon. Antioxidant Enzyme Activity and Oxidative Stress in Bovine Oocyte In-Vitro Maturation. IUBMB Life, 51: 57–64, 2001.
12. Coffey CS, Kairalla JA. Adaptive clinical trials: progress and challenges. Drugs R D. 2008;9(4):229–42.
13. Michael M. Cox, James L. Keck, John R. Battista. Rising from the Ashes: DNA Repair in Deinococcus radiodurans. PLOS Published: January 15, 2010. https://doi.org/10.1371/journal.pgen.1000815
14. Eric H. Davidson. The Regulatory Genome: Gene Regulatory Networks in Development and Evolution. Academic Press; 1 edition (June 13, 2006). ISBN-10: 0120885638; ISBN-13: 978-0120885633.
15. P.C.W. Davies and C. H. Lineweaver. Cancer tumors as Metazoa 1.0: tapping genes of ancient ancestors.PhysBiol.2011Feb;8(1):015001.doi:https://doi.org/10.1088/1478-3975/8/1/015001
16. DHHS. Organ Donation, Organ Donor Registry. https://organdonor.gov/
17. L Eichinger. The genome of the social amoeba Dictyostelium discoideum. Nature 435, 43–57 (5 May 2005). doi:https://doi.org/10.1038/nature03481
18. El Aidy S, Stilling R, Dinan TG, Cryan JF. Microbiome to Brain: Unravelling the Multidirectional Axes of Communication. Adv Exp Med Biol. 2016;874:301–36. doi: https://doi.org/10.1007/978-3-319-20215-0_15
19. Tetsuya Endo, Jun Yoshino, Koji Kado and Shin Tochinai. Brain regeneration in anuran amphibians. Develop. Growth Differ.(2007) 49, 121–129. doi: https://doi.org/10.1111/j.1440-169x.2007.00914.x
20. FDA. Novel Drugs Summary. https://www.fda.gov
21. FDA. Botanical Drug Development. Guidance for Industry. 2016. https://www.fda.gov/downloads/Drugs/GuidanceComplianceRegulatoryInformation/Guidances/UCM458484.pdf
22. Leonard P. Freedman, Iain M. Cockburn, Timothy S. Simcoe. PLOS. The Economics of Reproducibility in Preclinical Research. Published: June 9, 2015 https://doi.org/10.1371/journal.pbio.1002165
23. Errol C. Friedberg, Graham C. Walker, Wolfram Siede, Richard D. Wood, Roger A. Schultz, Tom Ellenberger. DNA Repair and Mutagenesis. ASM Press; 2 edition (November 22, 2005). ISBN-10: 1555813194; ISBN-13: 978-1555813192.
24. Guerin JC. Emerging area of aging research: long-lived animals with "negligible senescence". Ann N Y Acad Sci. 2004 Jun;1019:518–20.
25. Gurdon JB, Elsdale TR, Fischberg M. Sexually mature individuals of Xenopus laevis from the transplantation of single somatic nuclei. Nature. 1958 Jul 5;182(4627):64–5.
26. T Hirano, T J Mitchison. Cell cycle control of higher-order chromatin assembly around naked DNA in vitro. https://doi.org/10.1083/jcb.115.6.1479 | Published December 15, 1991.
27. R Hirschhorn. In vivo reversion to normal of inherited mutations in humans. J Med Genet 2003;40:721–728.
28. Hood, Leroy. 2013. "Systems Biology and p4 Medicine: Past, Present, and Future." Rambam Maimonides Med J 4 (April): e0012. https://doi.org/10.5041/RMMJ.10112
29. Sui Huang, Ingemar Ernberg, and Stuart Kauffman. Cancer attractors: A systems view of tumors from a gene network dynamics and developmental perspective. Semin Cell Dev Biol. 2009 Sep; 20(7): 869–876. https://doi.org/10.1016/j.semcdb.2009.07.003
30. Joffe AR, Kolski H, Duff J, deCaen AR. A 10-month-old infant with reversible findings of brain death. Pediatr Neurol. 2009 Nov;41(5):378–82. https://doi.org/10.1016/j.pediatrneurol.2009.05.007
31. Caghan Kizil, Jan Kaslin, Volker Kroehne, Michael Brand. Adult Neurogenesis and Brain Regeneration in Zebrafish. Developmental Neurobiology 72, 3, Version of Record online: 10 Feb 2012. https://doi.org/10.1002/dneu.20918

32. Romain Levayer, Eduardo Moreno. Mechanisms of cell competition: Themes and variations. https://doi.org/10.1083/jcb.201301051 | Published March 18, 2013. Journal of Cell Biology.
33. Levin M. Bioelectric mechanisms in regeneration: Unique aspects and future perspectives. Semin Cell Dev Biol. 2009 Jul;20(5):543–56. https://doi.org/10.1016/j.semcdb.2009.04.013. Epub 2009 May 3.
34. Qin Li, Anders Wennborg, Erik Aurell, Erez Dekel, Jie-Zhi Zou, Yuting Xu, Sui Huang, and Ingemar Ernberg. Dynamics inside the cancer cell attractor reveal cell heterogeneity, limits of stability, and escape. PNAS. 2016;113(10):2672–2677, https://doi.org/10.1073/pnas.1519210113
35. Lillie EO, Patay B, Diamant J, Issell B, Topol EJ, Schork NJ. The n-of-1 clinical trial: the ultimate strategy for individualizing medicine?. Per Med. 2011 Mar;8 (2):161–173.
36. Manfred I. Lohka and James L. Maller. Induction of Nuclear Envelope Breakdown, Chromosome Condensation, and Spindle Formation in Cell-free Extracts. The Journal of Cell Biology. 101 August 1985 518–523.
37. James L. Maller. Pioneering the Xenopus Oocyte and Egg Extract System. May 8, 2012 https://doi.org/10.1074/jbc.X112.371161. The Journal of Biological Chemistry.
38. B Mintz and K Illmensee. Normal genetically mosaic mice produced from malignant teratocarcinoma cells. Proc Natl Acad Sci U S A. 1975 Sep; 72(9): 3585–3589.
39. Muema JM, Bargul JL, Njeru SN, Onyango JO, Imbahale SS. Prospects for malaria control through manipulation of mosquito larval habitats and olfactory-mediated behavioural responses using plant-derived compounds. Parasit Vectors. 2017 Apr 17;10(1):184. https://doi.org/10.1186/s13071-017-2122-8
40. National Cancer Institute (NCI). Disease Statistics. https://www.cancer.gov/
41. National Institute of Diabetes and Digestive and Kidney Diseases (NIDDK). Disease Statistics https://www.niddk.nih.gov/
42. National Institutes of Health (NIH). Disease Statistics. https://www.nih.gov/
43. Néstor J. Oviedo and Wendy S. Beane. Regeneration: The Origin of Cancer or a Possible Cure?. Semin Cell Dev Biol. 2009 Jul; 20(5): 557–564.
44. Newman SA. Animal egg as evolutionary innovation: a solution to the "embryonic hourglass" puzzle. J Exp Zool B Mol Dev Evol. 2011 Nov 15;316(7):467–83. doi: https://doi.org/10.1002/jez.b.21417. Epub 2011 May 9.
45. Okamoto K, Sugimoto T. Return of spontaneous respiration in an infant who fulfilled current criteria to determine brain death. Pediatrics. 1995 Sep; 96 (3 Pt 1):518–20.
46. Saffet Ozturk, Berna Sozen, Necdet Demir. Telomere length and telomerase activity during oocyte maturation and early embryo development in mammalian species. Mol Hum Reprod (2014) 20 (1): 15–30. https://doi.org/10.1093/molehr/gat055
47. Justin M. Pare, Christopher S. Sullivan. Distinct Antiviral Responses in Pluripotent versus Differentiated Cells. PLOS. Published: February 6, 2014. https://doi.org/10.1371/journal.ppat.1003865
48. Sergei Paylian. Normalization of Vital Functions of Pathological Cells and Tissues By Key Mechanisms of Cell Adaptation and Reprogramming. BAOJ Biotech 2016, 2: 42: 021.
49. T. Harshani Peiris, Katrina K. Hoyer, and Néstor J. Oviedo. Innate immune system and tissue regeneration in Planarians: An area ripe for exploration. Semin Immunol. 2014 Aug; 26(4): 295–302. https://doi.org/10.1016/j.smim.2014.06.005
50. Pietsch P, Schneider CW. Brain transplantation in salamanders: an approach to memory transfer. Biochemical Research Laboratory, Dow Chemical Company, Midland, Mich. 48460 U.S.A. Brain Res. 1969 Aug; 14(3):707–15.
51. Pietsch P. Shuffle Brain: The Quest for the Holgramic Mind Hardcover. March 27, 1981. Houghton Mifflin; First Edition (March 27, 1981). ISBN-10: 0395294800; ISBN-13: 978-0395294802.
52. S. Piraino, F. Boero, B. Aeschbach, and V. Schmid. Reversing the Life Cycle: Medusae Transforming into Polyps and Cell Transdifferentiation in Turritopsis nutricula (Cnidaria, Hydrozoa). Volume 190, Number 3 | June 1996. The Biological Bulletin.

53. Raskin I, Ribnicky DM, Komarnytsky S, Ilic N, Poulev A, Borisjuk N, Brinker A, Moreno DA, Ripoll C, Yakoby N, O'Neal JM, Cornwell T, Pastor I, Fridlender B. Plants and human health in the twenty-first century. Trends Biotechnol. 2002 Dec;20(12):522–31.

54. Marilyn J. Roossinck. The good viruses: viral mutualistic symbioses. Nature Reviews Microbiology 9, 99–108 (February 2011) | https://doi.org/10.1038/nrmicro2491

55. Rose SM, Wallingford HM. Transformation of renal tumors of frogs to normal tissues in regenerating limbs of salamanders. Science. 1948 May 7;107(2784):457.

56. Jackson A. Seukep, Louis P. Sandjo, Bonaventure T. Ngadjui, and Victor Kuete. Antibacterial and antibiotic-resistance modifying activity of the extracts and compounds from Nauclea pobeguinii against Gram-negative multi-drug resistant phenotypes. BMC Complement Altern Med. 2016; 16: 193. https://doi.org/10.1186/s12906-016-1173-2

57. Tal Shomrat, Michael Levin. An automated training paradigm reveals long-term memory in planarians and its persistence through head regeneration. Journal of Experimental Biology 2013 216: 3799–3810; https://doi.org/10.1242/jeb.087809

58. Ana M. Soto and Carlos Sonnenschein. The tissue organization field theory of cancer: A testable replacement for the somatic mutation theory. Bioessays. 2011 May; 33(5): 332–340. https://doi.org/10.1002/bies.201100025

59. David L. Stocum and Jo Ann Cameron. Looking Proximally and Distally: 100 Years of Limb Regeneration and Beyond. Developmental Dynamics 240:943–968, 2011. https://doi.org/10.1002/dvdy.22553

60. L. Tebourbi, J. Testart, I. Cerutti, J.P. Moussu, A. Loeuillet, A-M. Courtot. Failure to infect embryos after virus injection in mouse zygotes. Hum Reprod (2002) 17 (3): 760–764. https://doi.org/10.1093/humrep/17.3.760

61. Tufts Center for the Study of Drug Development. Journal of Health Economics May, 2016.

62. Kenyon S. Tweedell. The Urodele Limb Regeneration Blastema: The Cell Potential. The Scientific World Journal Volume 10 (2010), Pages 954–971. https://doi.org/10.1100/tsw.2010.115

63. Zhenfei Wang, Rinuo Dao, Luri Bao, Yanhua Dong, Haiyang Wang, Pengyong Han, Yongli Yue, and Haiquan Yu. Epigenetic reprogramming of human lung cancer cells with the extract of bovine parthenogenetic oocytes. J Cell Mol Med. 2014 Sep; 18(9): 1807–1815.

64. Wenda R. Trevathan (Editor), E. O. Smith (Editor), James J. McKenna. Evolutionary Medicine 1st Edition. Oxford University Press; 1 edition (May 15, 1999). ISBN-10: 0195103564; ISBN-13: 978-0195103564.

65. Zhang R, Han P, Yang H, Ouyang K, Lee D, Lin YF, Ocorr K, Kang G, Chen J, Stainier DY, Yelon D, Chi NC. In vivo cardiac reprogramming contributes to zebrafish heart regeneration. Nature. 2013 Jun 27;498(7455):497–501. https://doi.org/10.1038/nature12322. Epub 2013 Jun 19.

Chapter 23
Induced Cell Turnover and the Future of Regenerative Medicine

Jakub Stefaniak, Francesco Albert Bosco Cortese, and Giovanni Santostasi

23.1 Regenerative Medicine in a Nutshell

Broadly speaking, regenerative medicine can be defined as an interdisciplinary field of science whose primary purpose is to develop novel therapies to engineer, regenerate or replace damaged or diseased tissue, with its ultimate goal being to reduce the reliance on transplantation as the primary pathway towards restoration of diseased organs [1]. The emphasis here is on the process of "regeneration", rather than "repair", as the former replaces lost tissue with either the same or biosimilar tissue with no loss of overall functionality of the organ, whereas the latter assumes a certain loss in functionality due to an inferior type of tissue ("scar") forming on the site of injury or insult [2]. Thus, regenerative medicine by its very nature has to combine branches of science ranging from molecular biology [3] to chemistry [4] and even to engineering [5].

J. Stefaniak (✉)
Nuffield Department of Medicine, Target Discovery Institute, University of Oxford, Oxford, UK
e-mail: jakub.stefaniak@new.ox.ac.uk

F. A. B. Cortese
Biogerontology Research Foundation, Oxford, UK

G. Santostasi
Department of Neurology, Feinberg School of Medicine, Northwestern University, Chicago, IL, USA

© Springer Nature Switzerland AG 2019
N. Lee (ed.), *The Transhumanism Handbook*,
https://doi.org/10.1007/978-3-030-16920-6_23

23.2 A Brief History of Regenerative Medicine

It is well-established that some lower vertebrates, in particular urodele amphibians, possess an astonishing capacity to regenerate damaged or lost tissue or even entire organs without scarring by forming a blastema of dedifferentiated cells at the site of injury [6]. By contrast, save for a number of equally curious exceptions, such as deer antlers [7], mammals have a very limited capacity for regeneration [8]. It is interesting to observe that Prod1, a factor heavily implicated in salamanders' unique ability to re-grow lost limbs, has no known homologues in mammals, casting doubt over to what extent it is possible to mobilize the natural regenerative pathways in humans [9]. Thus, modern regenerative medicine can be viewed as an attempt to bridge this evolutionary gap with the use of tissue engineering, stem cell therapies and small molecules.

The history of regenerative medicine is inextricably intertwined with the development of stem cell technology, with the latter nowadays forming one of the pillars of the former [3]. The seminal discovery in 1978 of hematopoietic stem cells (HSCs) in human umbilical cord blood, followed by the development in 1981 of mice-derived embryonic stem cell lines paved the way towards successfully culturing human pluripotent (capable of differentiating into any of the three germ layers) stem cells by the late 1990s [10]. By the turn of the century, it was becoming increasingly apparent that stem cell therapies could become more widespread, as they offered a significant advantage over small molecules, namely not being marred by off-target toxicity [11]. The only significant hurdle then was the fact that embryonic stem cells, and any tissues derived from them, would be immunogenic towards a potential host (in other words their different genetic composition to that of said host could cause the same problems that occur during organ transplants and ultimately lead to rejection) [12], as well as the ethical concerns [13]. An extraordinary breakthrough was made in 2006, when the Yamanaka lab in Japan managed to create induced pluripotent stem cells (iPSCs) by dedifferentiating mouse fibroblasts via transfecting just four genes [14]. Because iPSCs can be made using differentiated cells from the host, they are by definition patient-matched and bypass the need for zygotes or oocytes, alleviating the ethical concerns [15]. Currently, the only major concern regarding stem cell therapy appears to be potential tumorigenicity [16], however the advent of amniotic fluid-derived stem cells (AFSCs) might be able to ameliorate it [17].

Gene therapy is a relatively novel branch of regenerative medicine; whilst conceptually it's been around since the 1970s [18], the first clinical trials did not take place until the early 1990s [19]. Perhaps the most obvious application of it for the purpose of tissue regeneration appears to be plastic surgery, as mammalian fibroblasts are easily accessible, expandable and amenable towards genetic manipulation [20, 21]. The latter usually takes form of transferring cDNA encoding appropriate growth factors into the target cell line at the wound site [22], with the results usually being accelerated healing [22, 23]. Human trials of gene therapy in regenerative plastic surgery have demonstrated its potential in the treatment of diabetic foot ulcers [24] or improvement in post-operative or trauma and injury-related scarring [25]. In addition, there is clear potential for extending this technology into regenerating cartilage [26] and even bone [27].

The recent development of the CRISPR-Cas9 technology has paved the way towards safer, cheaper and more widely available gene editing [28]. This in turn allows for the editing of iPSCs to correct point mutations [29] and may in the future allow for breakthroughs in tissue engineering.

23.3   State of the Art in Regenerative Medicine

In the past few years, regenerative medicine as a field saw a number of intriguing and novel trends. 3D bioprinting of tissues and organs (which in the past was seen as viable [30] but limited technology due to its inability to duplicate the complexity of biological structures [31]) experienced a breakthrough with the development of complex collagen-based cartilage with favourable mechanical properties [32]. For a recent review on scaffolds for bone tissue engineering, the reader is directed towards Roseti et al [33]. In addition, a novel type of a stem cell was discovered, the fibro-cartilage stem cell (FCSC), capable of generating cartilage, bone and marrow from a single FCSC – if successful, this will circumvent the need for exogenous or induced stem cells in extracellular medium (ECM) regeneration [34].

The application of human pluripotent stem cells (hPSCs) necessitates the creation of large-scale bioreactors capable of producing significant amounts of such cells (10^7–10^{10}) in a state that does not abrogate their biological and genetic properties [35]. The pitfalls of current, suspension culture-based technology include issues with uneven oxygen diffusion and cell aggregation and heterogeneity, resulting in undesirable cell differentiation as well as apoptosis [35, 36]. It is thus very promising for Ikeda et al. to have recently developed a core-shell hydrogel microfibre-based encapsulation system for the expansion of hPSCs that bypasses the aforementioned limitations, achieving substantially higher rates of cell viability compared to conventional methods, whilst retaining steep rate of expansion [37].

The notion of xenotransplantation presents itself as a very intriguing solution to the organ shortage problem [38]. It is thus remarkable that recently, the first xeno-transplant of medical significance was performed, whereby a rat host early-stage embryo was transplanted with mouse PSCs and made to generate a pancreas [39]; the pancreas was fully functional, allowing for a transplant of islets of Langerhans to restore the control over blood sugar levels in a type I diabetes mouse model. This paves the way towards human organs derived from iPSCs and grown in animals, mitigating the need for immunosuppressive agents that currently plague the area of organ transplantation.

Another seminal discovery was the entirely-*in-vitro* transformation of mouse skin cells into oocytes and subsequently into PSC lines, in essence recreating the entirety of the female germline cycle [40]. This breakthrough in the area of molecular and reproductive biology might in the future lead to the generation of human male and female-derived oocytes for assisted reproduction.

23.3.1 How Regenerative Medicine Differs from Conventional Medicine

As mentioned before, mammalian ability to regenerate lost appendages pales in comparison to that of, for instance, urodele amphibians, such as newts and salamanders, and herein lies the most significant difference between conventional and regenerative medicine. Using the example of severe limb damage, conventional medicine advocates clinicians to decide between reconstruction or amputation [41], the former usually leading to further complications [42] and the latter requiring near-constant management and care [43, 44]. In either case, complete function is usually never restored to the extremity [45].

Regenerative medicine, on the other hand, has as an objective the complete renewal of the impaired tissue, in this case the reformation of the extremity. Currently, our knowledge in this regard is rather limited, however there are lines of evidence, which would suggest that limb regeneration need not fall within the purview of science-fiction. Indeed, it has been well-established that, for instance, mouse digit tips will in some cases regenerate completely upon amputation [46, 47], and that treatment with some exogenous bone morphogenic proteins (BMPs), in particular BMP7 and BMP2 can vastly enhance the regenerative properties or otherwise non-regenerating mammalian tissue [48, 49]. In addition, exogenous supply of PSCs responsive to morphogenic signalling has been shown to enhance limb hypertrophy following amputation in mice [50]. It is also refreshing to see novel scientific attempts to tackle the problem of regeneration from a different angle: recently, Leppik et al. have utilized electrical stimulation to increase the regeneration of amputated rat limbs, with moderate success [51]. A remarkable recent case of a newborn making a complete recovery following a severe myocardial infarction shortly after birth strongly suggests that neonatal tissue has stronger regenerative properties to adult [52], and although the capacity to naturally regenerate in adult mammals may be lost due to the depletion of stem cells, likely the mechanisms underpinning said regeneration are not. As such, correct therapy (likely involving exogenous morphogenic signals and hPSCs) in the future will be used to generate new limbs and organs with complete restoration of functionality, which is the ultimate goal of regenerative medicine.

23.4   Induced Cell Turnover

Induced cell turnover (ICT) is a relatively novel concept in the area of regenerative medicine. In simple terms, it involves periodically ablating cells in a targeted fashion from aged tissue and replacing them with fresh cells derived from hPSCs, with the aim of reversing aging-related issues that tend to accumulate in cells [53]. Mammalian non-tumour cells obey the Hayflick limit, by which each cell will go through a finite number of replications before undergoing senescence [54], likely due to the shortening of telomeres [55]. In addition, throughout their lifespan, cells will accumulate genomic and mitochondrial DNA damage or mutations [56–58], as well as oxidative stress [59] leading to oxidised or aggregated proteins [60, 61], which on a tissue level may manifest themselves in the form of aging biomarkers [62], and lead to the aged phenotype on the whole-body level [63]. ICT aims at continuous replenishment of cells in order to preserve the biologically-young status of tissue and thus has potential to vastly ameliorate the negative effects of aging, with a capacity to reverse or even abrogate the process outright.

From the technical perspective, ICT faces a number of hurdles, none of which appear to be insurmountable. The first one is the generation of large quantities of patient-specific hPSCs lacking the aforementioned age-related damage. It has been previously demonstrated the reprogramming somatic stem cells to a pluripotent state increases telomerase activity and leads to telomere lengthening [14, 64], through mechanisms not yet clearly understood [65]. As for DNA damage, the pluripotency induction process generally augments the DNA damage response of iPSCs [66, 67], as well as rejuvenates the mitochondria [68, 69] and alleviates the oxidative stress [69]. It is nevertheless clear that the patient-derived cells will need to be free of existing mutations and that single-cell genomics techniques will aid greatly in the initial screening process [70, 71]. Alternatively, should it become apparent that, for instance, elderly patients' cells harbour too many mutations, freezing cells at an early age would be a feasible workaround.

The second potential issue is the problem of maintaining tissue homeostasis during the process of cell ablation and turnover. The amount of cells that can be feasibly ablated during a single treatment is self-evidently limited by the amount of cell loss an organ or tissue can sustain without the disruption of homeostasis leading to organ failure. However, current lines of evidence suggest that most organs capable of regeneration will be amenable towards ICT without triggering any homeostatic collapse [72]. Thirdly, there is an issue of optimising the regimen in order to suit the patient need. For instance, replacing the autologous stem cell (ACS) rather than somatic cell population appears to be a preferred solution due to the fact that ACSs over time accumulate age-related defects, which they pass onto descendants, an issue that can be bypassed via ACS turnover and ablation of those somatic cells already displaying the aged phenotype. Moreover, there is the issue of frequency of ICT, in other words how often the procedure should be performed – too often and it will be a net negative for the patient's physiological and psychological health; too infrequent will result in the development of the aged phenotype. Given that it takes approximately one human lifespan to develop the symptoms of aging, it is not unreasonable to assume that ICT will not need to be performed more often than once every 50 years. Needless to say, should evidence emerge of a greater therapeutic effect associated with a different frequency of turnover, the regimen can be varied accordingly.

On the outset, ICT is a highly-versatile technique. Whilst its ultimate goal is the extension of lifespan up to the theoretical limit of infinity, it may well be that for some areas of the body the procedure is not feasible, in which case it must be stressed that even partial turnover could have therapeutic effects on a number of tissues affected the most by the deleterious effects of aging [73]. Given that the current state of biotechnology permits the derivation of patient-specific embryonic stem cell lines through somatic cell nuclear transfer [74] and generation of iPSCs through nuclear reprogramming [64, 75, 76], the risk of tissue rejection by the host's immune system is low. On the other hand, the theoretical number of cells that will need to be generated per round of is of the order of 10 [13], or the number of cells present in the human body [77], which may be problematic depending on the rate of the aforementioned advances in bioreactors [37, 78]. In addition, it is of utmost importance that no genomic instabilities or mutations appear during cell culture propagation [79]. Avoiding oncogenesis commonly associated with stem cell culture must also be addressed [80].

Current methodologies pertaining to cell ablation revolve around suicide gene therapy, whereby suicide genes with tissue-specific promoters systemically or locally injected [81], artificial death switches [82] or biophysical methods, such as those involving photodynamics [83] or ultrasound [84]. Other methods worth exploring are cell-type targeted drug delivery systems [85]. The rate at which cells are ablated must be adjusted in accordance to the supply of exogenous cells as well as the homeostatic stability of the organ. It is worth pointing out that cell death in this fashion will proceed via the apoptotic, rather than necrotic route, minimizing inflammation or risk of systemic or immunogenic shock [86]. On the other hand, the "innocent bystander effect" remains a possible complication; stemming from the possible exchange of suicide factors between the target cell and its surrounding tissue, it manifests itself in the form of unnecessary cell death [87]. Possible solutions include concomitant administration of hPSC-derived phagocytes or mitogens potentiating endogenous phagocyte proliferation.

23.5   How ICT Compares to Currently Available Therapies and Can It Be a Viable Branch of Regenerative Medicine?

More and more PSC-based therapies are entering the clinic, positioning themselves as a serious alternative to conventional medicine in the area of diabetes, neurodegenerative diseases and cardiovascular issues [88]. However, in line with conventional medicine, they deal mostly with treatment, rather than prevention, of cell loss. ICT, on the other hand, through its novel paradigm of regularly turning over tissue, has strong potential to outright eliminate a number of serious ageing-related conditions, including but not limited to all of the above, as well as cancer, and immune system disorders [89]. Because cells are turned over before they have the chance to accumulate the hallmarks of aging, ICT, in principle, will allow for drastic reductions in the need for therapies, in essence eliminating the pathologies before they develop. Thus, in conclusion, ICT can become the next branch of regenerative medicine, and potentially may even pave the way towards extending human lifespan to previously unimaginable levels.

References

1. Mao, A. S. & Mooney, D. J. Regenerative medicine: Current therapies and future directions. https://doi.org/10.1073/pnas.1508520112
2. Chen, F.-M., Zhao, Y.-M., Jin, Y. & Shi, S. Prospects for translational regenerative medicine. *Biotechnol. Adv.* **30,** 658–672 (2012).
3. Mahla, R. S. & Singh, R. Stem Cells Applications in Regenerative Medicine and Disease Therapeutics. *Int. J. Cell Biol.* **2016,** 1–24 (2016).
4. Xu, Y., Shi, Y. & Ding, S. A chemical approach to stem-cell biology and regenerative medicine. *Nature* (2008). https://doi.org/10.1038/nature07042
5. Bajaj, P., Schweller, R. M., Khademhosseini, A., West, J. L. & Bashir, R. 3D Biofabrication Strategies for Tissue Engineering and Regenerative Medicine. *Annu. Rev. Biomed. Eng.* **16,** 247–276 (2014).
6. Godwin, J. W. & Rosenthal, N. Scar-free wound healing and regeneration in amphibians: Immunological influences on regenerative success. *Differentiation* **87,** 66–75 (2014).
7. Price, J. & Allen, S. Exploring the mechanisms regulating regeneration of deer antlers. *Philos. Trans. R. Soc. Lond. B. Biol. Sci.* **359,** 809–22 (2004).
8. Muneoka, K., Allan, C. H., Yang, X., Lee, J. & Han, M. Mammalian regeneration and regenerative medicine. *Birth Defects Res. Part C Embryo Today Rev.* **84,** 265–280 (2008).
9. Whited, J. L. & Tabin, C. J. Regeneration review reprise. *J. Biol.* **9,** 15 (2010).
10. Sampogna, G., Guraya, S. Y. & Forgione, A. Regenerative medicine: Historical roots and potential strategies in modern medicine. *J. Microsc. Ultrastruct.* **3,** 101–107 (2015).
11. Lagasse, E., Shizuru, J. A., Uchida, N., Tsukamoto, A. & Weissman, I. L. Toward regenerative medicine. *Immunity* **14,** 425–36 (2001).
12. Swijnenburg, R.-J. *et al.* Immunosuppressive therapy mitigates immunological rejection of human embryonic stem cell xenografts.
13. Wright, S. Human Embryonic Stem-Cell Research: Science and Ethics. *Am. Sci.* **87,** 352 (1999).
14. Takahashi, K. & Yamanaka, S. Induction of Pluripotent Stem Cells from Mouse Embryonic and Adult Fibroblast Cultures by Defined Factors. *Cell* **126,** 663–676 (2006).
15. Lo, B. & Parham, L. Ethical issues in stem cell research. *Endocr. Rev.* **30,** 204–13 (2009).
16. Ben-David, U. & Benvenisty, N. The tumorigenicity of human embryonic and induced pluripotent stem cells. *Nat. Rev. Cancer* **11,** 268–277 (2011).
17. Baghaban Eslaminejad, M. & Jahangir, S. Amniotic fluid stem cells and their application in cell-based tissue regeneration. *Int. J. Fertil. Steril.* **6,** 147–56 (2012).
18. Friedmann, T. & Roblin, R. Gene Therapy for Human Genetic Disease? *Science (80–.).* **175,** (1972).
19. Miller, A. D. Human gene therapy comes of age. *Nature* **357,** 455–460 (1992).

20. Krueger, G. G. Fibroblasts and Dermal Gene Therapy: A Minireview. *Hum. Gene Ther.* **11,** 2289–2296 (2000).
21. Bleiziffer, O., Eriksson, E., Yao, F., Horch, R. E. & Kneser, U. Gene transfer strategies in tissue engineering. *J. Cell. Mol. Med.* **11,** 206–223 (2007).
22. Andree, C. *et al.* In vivo transfer and expression of a human epidermal growth factor gene accelerates wound repair. *Proc. Natl. Acad. Sci. U. S. A.* **91,** 12188–92 (1994).
23. Eming, S. A. *et al.* Particle-Mediated Gene Transfer of PDGF Isoforms Promotes Wound Repair. *J. Invest. Dermatol.* **112,** 297–302 (1999).
24. Mulder, G. *et al.* Treatment of nonhealing diabetic foot ulcers with a platelet-derived growth factor gene-activated matrix (GAM501): Results of a Phase 1/2 trial. *Wound Repair Regen.* **17,** 772–779 (2009).
25. Bush, J. *et al.* Therapies with Emerging Evidence of Efficacy: Avotermin for the Improvement of Scarring. *Dermatol. Res. Pract.* **2010,** 1–6 (2010).
26. Nixon, A. J. *et al.* Gene-mediated restoration of cartilage matrix by combination insulin-like growth factor-I/interleukin-1 receptor antagonist therapy. *Gene Ther.* **12,** 177–186 (2005).
27. Edwards, P. C. *et al.* Sonic hedgehog gene-enhanced tissue engineering for bone regeneration. *Gene Ther.* **12,** 75–86 (2005).
28. Wang, H., La Russa, M. & Qi, L. S. CRISPR/Cas9 in Genome Editing and Beyond. *Annu. Rev. Biochem.* **85,** 227–264 (2016).
29. Grobarczyk, B., Franco, B., Hanon, K. & Malgrange, B. Generation of Isogenic Human iPS Cell Line Precisely Corrected by Genome Editing Using the CRISPR/Cas9 System. *Stem Cell Rev. Reports* **11,** 774–787 (2015).
30. KHATIWALA, C., LAW, R., SHEPHERD, B., DORFMAN, S. & CSETE, M. 3D CELL BIOPRINTING FOR REGENERATIVE MEDICINE RESEARCH AND THERAPIES. *Gene Ther. Regul.* **7,** 1230004 (2012).
31. Murphy, S. V & Atala, A. 3D bioprinting of tissues and organs. *Nat. Biotechnol.* **32,** 773–785 (2014).
32. Rhee, S., Puetzer, J. L., Mason, B. N., Reinhart-King, C. A. & Bonassar, L. J. 3D Bioprinting of Spatially Heterogeneous Collagen Constructs for Cartilage Tissue Engineering. *ACS Biomater. Sci. Eng.* **2,** 1800–1805 (2016).
33. Roseti, L. *et al.* Scaffolds for Bone Tissue Engineering: State of the art and new perspectives. *Mater. Sci. Eng. C* **78,** 1246–1262 (2017).
34. Embree, M. C. *et al.* Exploiting endogenous fibrocartilage stem cells to regenerate cartilage and repair joint injury. *Nat. Commun.* **7,** 13073 (2016).
35. Chen, K. G., Mallon, B. S., McKay, R. D. G. & Robey, P. G. Human pluripotent stem cell culture: considerations for maintenance, expansion, and therapeutics. *Cell Stem Cell* **14,** 13–26 (2014).
36. Olmer, R. *et al.* Suspension culture of human pluripotent stem cells in controlled, stirred bio-reactors. *Tissue Eng. Part C. Methods* **18,** 772–84 (2012).
37. Ikeda, K., Nagata, S., Okitsu, T. & Takeuchi, S. Cell fiber-based three-dimensional culture system for highly efficient expansion of human induced pluripotent stem cells. *Sci. Rep.* **7,** 2850 (2017).
38. Ekser, B., Rigotti, P., Gridelli, B. & Cooper, D. K. C. Xenotransplantation of solid organs in the pig-to-primate model. *Transpl. Immunol.* **21,** 87–92 (2009).
39. Yamaguchi, T. *et al.* Interspecies organogenesis generates autologous functional islets. *Nature* **542,** 191–196 (2017).
40. Hikabe, O. *et al.* Reconstitution in vitro of the entire cycle of the mouse female germ line. *Nature* (2016). https://doi.org/10.1038/nature20104
41. Kristensen, M. Major foot trauma: the dilemma of reconstruction versus amputation. *Clin. Podiatr. Med. Surg.* **14,** 603–12 (1997).
42. Harris, A. M., Althausen, P. L., Kellam, J., Bosse, M. J. & Castillo, R. Complications Following Limb-Threatening Lower Extremity Trauma. *J. Orthop. Trauma* **23,** 1–6 (2009).

43. Penn-Barwell, J. G. Outcomes in lower limb amputation following trauma: A systematic review and meta-analysis. *Injury* **42,** 1474–1479 (2011).
44. Marshall, C. & Stansby, G. Amputation and rehabilitation. *Surg.* **31,** 236–239 (2013).
45. Section 2: Limb Salvage and Amputation After Major Lower Limb Trauma. *J. Orthop. Trauma* **31,** S39 (2017).
46. Simkin, J., Han, M., Yu, L., Yan, M. & Muneoka, K. The Mouse Digit Tip: From Wound Healing to Regeneration. in *Methods in molecular biology (Clifton, N.J.)* **1037,** 419–435 (2013).
47. Han, M., Yang, X., Lee, J., Allan, C. H. & Muneoka, K. Development and regeneration of the neonatal digit tip in mice. *Dev. Biol.* **315,** 125–135 (2008).
48. Yu, L. *et al.* BMP signaling induces digit regeneration in neonatal mice. *Development* **137,** 551–9 (2010).
49. Yu, L., Han, M., Yan, M., Lee, J. & Muneoka, K. BMP2 induces segment-specific skeletal regeneration from digit and limb amputations by establishing a new endochondral ossification center. *Dev. Biol.* **372,** 263–273 (2012).
50. Masaki, H. & Ide, H. Regeneration potency of mouse limbs. *Dev. Growth Differ.* **49,** 89–98 (2007).
51. Leppik, L. P. *et al.* Effects of electrical stimulation on rat limb regeneration, a new look at an old model. *Sci. Rep.* **5,** 18353 (2016).
52. Haubner, B. J. *et al.* Functional Recovery of a Human Neonatal Heart After Severe Myocardial Infarction. *Circ. Res.* (2015).
53. Cortese, F. A. B. & Santostasi, G. Whole-Body Induced Cell Turnover: A Proposed Intervention for Age-Related Damage and Associated Pathology. *Rejuvenation Res.* **19,** 322–336 (2016).
54. Shay, J. W. & Wright, W. E. Hayflick, his limit, and cellular ageing. *Nat. Rev. Mol. Cell Biol.* **1,** 72–76 (2000).
55. Shawi, M. & Autexier, C. Telomerase, senescence and ageing. *Mech. Ageing Dev.* **129,** 3–10 (2008).
56. Hoeijmakers, J. H. J. DNA Damage, Aging, and Cancer. *N. Engl. J. Med.* **361,** 1475–1485 (2009).
57. Best, B. P. Nuclear DNA Damage as a Direct Cause of Aging. *Rejuvenation Res.* **12,** 199–208 (2009).
58. Park, C. B. & Larsson, N.-G. Mitochondrial DNA mutations in disease and aging. *J. Cell Biol.* **193,** 809–18 (2011).
59. Shigenaga, M. K., Hagen, T. M. & Ames, B. N. Oxidative damage and mitochondrial decay in aging. *Proc. Natl. Acad. Sci.* **91,** 10771–10778 (1994).
60. Stadtman, E. Protein oxidation and aging. *Science (80-.).* **257,** (1992).
61. Squier, T. C. Oxidative stress and protein aggregation during biological aging. *Exp. Gerontol.* **36,** 1539–50 (2001).
62. Wagner, K.-H., Cameron-Smith, D., Wessner, B. & Franzke, B. Biomarkers of Aging: From Function to Molecular Biology. *Nutrients* **8,** 338 (2016).
63. Belsky, D. W. *et al.* Quantification of biological aging in young adults. *Proc. Natl. Acad. Sci.* **112,** E4104–E4110 (2015).
64. Takahashi, K. *et al.* Induction of Pluripotent Stem Cells from Adult Human Fibroblasts by Defined Factors. *Cell* **131,** 861–872 (2007).
65. Wang, F. *et al.* Molecular insights into the heterogeneity of telomere reprogramming in induced pluripotent stem cells. *Cell Res.* **22,** 757–768 (2012).
66. Luo, L. Z. *et al.* DNA repair in human pluripotent stem cells is distinct from that in non-pluripotent human cells. *PLoS One* **7,** e30541 (2012).
67. Lin, B., Gupta, D. & Heinen, C. D. Human Pluripotent Stem Cells Have a Novel Mismatch Repair-dependent Damage Response. *J. Biol. Chem.* **289,** 24314–24324 (2014).
68. Suhr, S. T. *et al.* Mitochondrial Rejuvenation After Induced Pluripotency. *PLoS One* **5,** e14095 (2010).

69. Prigione, A., Fauler, B., Lurz, R., Lehrach, H. & Adjaye, J. The Senescence-Related Mitochondrial/Oxidative Stress Pathway is Repressed in Human Induced Pluripotent Stem Cells. *Stem Cells* **28**, 721–733 (2010).
70. Potter, N. E. *et al.* Single-cell mutational profiling and clonal phylogeny in cancer. *Genome Res.* **23**, 2115–2125 (2013).
71. Wagner, A., Regev, A. & Yosef, N. Revealing the vectors of cellular identity with single-cell genomics. *Nat. Biotechnol.* **34**, 1145–1160 (2016).
72. Pellettieri, J. & Alvarado, A. S. Cell Turnover and Adult Tissue Homeostasis: From Humans to Planarians. *Annu. Rev. Genet.* **41**, 83–105 (2007).
73. Neves, J., Demaria, M., Campisi, J. & Jasper, H. Of Flies, Mice, and Men: Evolutionarily Conserved Tissue Damage Responses and Aging. *Dev. Cell* **32**, 9–18 (2015).
74. Loi, P., Iuso, D., Czernik, M. & Ogura, A. A New, Dynamic Era for Somatic Cell Nuclear Transfer? *Trends Biotechnol.* **34**, 791–797 (2016).
75. Qi, S. D., Smith, P. D. & Choong, P. F. Nuclear reprogramming and induced pluripotent stem cells: a review for surgeons. *ANZ J. Surg.* **84**, 417–423 (2014).
76. Cherry, A. B. C. & Daley, G. Q. Reprogramming Cellular Identity for Regenerative Medicine. *Cell* **148**, 1110–1122 (2012).
77. Bianconi, E. *et al.* An estimation of the number of cells in the human body. *Ann. Hum. Biol.* **40**, 463–471 (2013).
78. Gelinsky, M., Bernhardt, A. & Milan, F. Bioreactors in tissue engineering: Advances in stem cell culture and three-dimensional tissue constructs. *Eng. Life Sci.* **15**, 670–677 (2015).
79. Weissbein, U., Benvenisty, N. & Ben-David, U. Genome maintenance in pluripotent stem cells. *J. Cell Biol.* **204**, (2014).
80. Ghosh, Z. *et al.* Dissecting the Oncogenic and Tumorigenic Potential of Differentiated Human Induced Pluripotent Stem Cells and Human Embryonic Stem Cells. *Cancer Res.* **71**, 5030–5039 (2011).
81. Grégoire, D. & Kmita, M. Genetic Cell Ablation. in *Methods in molecular biology (Clifton, N.J.)* **1092**, 421–436 (2014).
82. MacCorkle, R. A., Freeman, K. W. & Spencer, D. M. Synthetic activation of caspases: artificial death switches. *Proc. Natl. Acad. Sci. U. S. A.* **95**, 3655–60 (1998).
83. Dougherty, T. J. *et al.* Photodynamic therapy. *J. Natl. Cancer Inst.* **90**, 889–905 (1998).
84. Chennat, J., Mino-Kenudson, M., Sahani, D. V. & et al., Current status of endoscopic ultrasound guided ablation techniques. *Gastroenterology* **140**, 1403–9 (2011).
85. Tiwari, G. *et al.* Drug delivery systems: An updated review. *Int. J. Pharm. Investig.* **2**, 2–11 (2012).
86. Edinger, A. L. & Thompson, C. B. Death by design: apoptosis, necrosis and autophagy. *Curr. Opin. Cell Biol.* **16**, 663–669 (2004).
87. Karjoo, Z., Chen, X. & Hatefi, A. Progress and problems with the use of suicide genes for targeted cancer therapy. *Adv. Drug Deliv. Rev.* **99**, 113–128 (2016).
88. Trounson, A. & DeWitt, N. D. Pluripotent stem cells progressing to the clinic. *Nat. Rev. Mol. Cell Biol.* **17**, 194–200 (2016).
89. Niccoli, T. & Partridge, L. Ageing as a Risk Factor for Disease. *Curr. Biol.* **22**, R741–R752 (2012).

Chapter 24
Extending Healthy Human Lifespan Using Gene Therapy

Elizabeth Parrish

In just the last 100 years, between the breakout of the Spanish flu in 1918 and the arrival in clinics of gene editing techniques in 2018, science has made spectacular progress. For this reason it is hard to wrap our head around the gap between treatments available to my grandparents (the generation born in the 1920s and to my parents (born in 1950s and treatments available to younger generations (born in the 1970s and later). We no longer live in the same medical environment. Consequently we can, and must, envisage different health-spans, even different lifespans. Being 80 years old in 2040 will be a different thing altogether from what it was to be 80 in 1940.

When I was young, and up to just a decade ago, a medical check-up generally meant a blood test for cholesterol and glucose levels, possibly one or two other molecules if deemed necessary, a urine test and listening to my heart with a stethoscope. My medical record would state the childhood illnesses I had had, as well as any vaccinations, accidents and operations. Today we have the science to look at our genetic blueprint, contained as chromosomes inside the nucleus of our cells, and peer into our medical history, and even into our potential future. We can see not just how our parents' biological history (smoking, drinking, depression, trauma for instance) affected their bodies, but our connection to the entire human community past and present, and all the way back to our remotest ancestors on the genealogical tree of evolution, plants and bacteria. As the price of DNA sequencing plummets, and the collection of data soars, the information on our medical records increases exponentially, and the day when we can envisage having a copy of our entire genome, proteome, epigenome and microbiome on a USB flash drive is approaching at the speed of light.

The cost of the technology to read this information, the code book about you, has dropped from a whopping $3B for the first human genome to an obtainable $200

E. Parrish (✉)
BioViva, Bainbridge Island, WA, USA
e-mail: info@bioviva-science.com

© Springer Nature Switzerland AG 2019
N. Lee (ed.), *The Transhumanism Handbook*,
https://doi.org/10.1007/978-3-030-16920-6_24

price tag for SNP's (single nucleotide polymorphisms) readouts showing family history, and potential health risks and advantages. Today SNPs variations are used to analyze statistical risks for diseases such as Alzheimer's, Parkinson's and many more. The medicine of tomorrow, that is already knocking on our door, relies on this information, positive and negative, to move us forward to a better future. Diseases no longer catch us napping: we can see them coming. And not only can we see them coming: we can intervene to slow them before they hit us. Soon we'll be able to alter their course and steer them off our path altogether. In other words, for the first time in history we can understand our health destiny and begin to change it.

24.1   Our Shared Genomic Past

Against prejudice and division, genetic testing has shown that we are not as different from one another as we believed. Even though the path out of Africa was not as straight and uniform as once thought, the ancestors of Homo Sapiens ultimately all trace back to that continent. Hence, despite diversity in human appearance and culture, our genetic variations are low. This sameness of today's human species enables us to look at genes that might benefit all or most of the population, from muscle enhancing genes to genes that help elongate telomeres (the end of our chromosomes) and gene therapies to remove senescent cells, the cause of much inflammation. In the not too distant future we will see personalized medicine with drugs designed for one specific genome. In many cases there is also evidence for a one-gene-fits-all approach.

The tree of life has its roots in a primordial soup, in which our cells' most ancient predecessors originate, beginning with the last universal common ancestor (LUCA) from which all living organisms descend. Moving up the tree we can see how species evolved, where they broke off, and their strengths and weaknesses. Hundreds of millions of years of evolution see the dice rolling for success or extinction, sometimes the former, often the latter. Reaching today, and having the eyes, hands and mind to read this chapter, prove that you are the latest offshoot in an amazing line of successful evolution. We can look at the resulting blueprint like a historical document. We can also look at it like a roadmap to the future.

24.2   Our Shared Genomic Future

We are the products of evolution along a line in which genetics, guided by a molecular clock, can point out at least some of the steps. This genetic data does not only tell us about the body we had yesterday, and the body we have today, but also about the body that we might have tomorrow. For instance, we can observe generations of viral integrations from infectious disease. Our body has mutated over time because of the integration of DNA foreign to our species. Is there any reason that this should stop? Contrary to a frequently held belief that altering the genes of a species is "a violation of its essence", gene therapy is not an unnatural intervention by a new science in a fixed human essence, but a natural phenomenon, that is not just part of evolution, but one of the engines that drives it. This data can also help us map what genes make us more impervious to health problems, and what genes from other species could be used as a technology to enhance our prospects.

We want to create human value so the question is: in what areas would gene therapy have the largest impact? Today the biggest risk factor for mortality is biological aging. As shown by Aubrey de Grey[1] among others, this term, "biological aging", means cellular damage due to time. The damage is the consequence of the body's insufficient ability to repair breakage resulting from metabolism and its by-products. Hence, with the passage of time, with activity such as breathing, digesting and moving, our cells deteriorate, waste products accumulate inside them, molecules essential to their function are misshapen or fractured. In other words, the body grows old, and this degeneration causes the diseases of aging, and eventually death.

Hence the process of cellular degeneration is the root cause of ill health in a world that is rapidly getting older. It is therefore essential to develop therapies that can address and prevent cellular aging. Aging, as we have seen, must be addressed at the level of the cell, and the molecules that constitute it, in genes, both in the nucleus and in the mitochondria, well as through epigenetic modulation. It will also have to be tackled at the level of the microbiome. This means that gene and cell therapies alone will be able to alleviate and cure the diseases due to biological aging.

Over the past three decades scientists have been awarded Nobel Prizes for finding mechanisms that can affect aging and lifespan. These findings could help increase healthspan, and show that lifespan is inherently malleable through effective intervention including gene modulation.[2,3,4,5,6,7,8,9,10] We are now able to extend the lifespan of worms up to 11 times beyond its normal length, fruit flies by 6 times

[1] Aubrey de Grey and Michael Rae, *Ending Aging* (2007).

[2] "The Nobel Prize in Physiology or Medicine 2009." https://www.nobelprize.org/nobel_prizes/medicine/laureates/2009/. Accessed 5 Jan. 2018.

[3] "The Nobel Prize in Physiology or Medicine 2006." https://www.nobelprize.org/nobel_prizes/medicine/laureates/2006/. Accessed 5 Jan. 2018.

[4] "The Nobel Prize in Physiology or Medicine 2001." https://www.nobelprize.org/nobel_prizes/medicine/laureates/2001/. Accessed 5 Jan. 2018.

[5] "The Nobel Prize in Physiology or Medicine 2002." https://www.nobelprize.org/nobel_prizes/medicine/laureates/2002/. Accessed 5 Jan. 2018.

[6] "The Nobel Prize in Physiology or Medicine 2012." https://www.nobelprize.org/nobel_prizes/medicine/laureates/2012/. Accessed 5 Jan. 2018.

[7] "The Nobel Prize in Physiology or Medicine 2013." https://www.nobelprize.org/nobel_prizes/medicine/laureates/2013/. Accessed 5 Jan. 2018.

[8] "The Nobel Prize in Physiology or Medicine …." https://www.nobelprize.org/nobel_prizes/medicine/laureates/2016/. Accessed 5 Jan. 2018.

[9] "The Nobel Prize in Chemistry 2015." https://www.nobelprize.org/nobel_prizes/chemistry/laureates/2015/. Accessed 5 Jan. 2018.

[10] "The Nobel Prize in Chemistry 2009." https://www.nobelprize.org/nobel_prizes/chemistry/laureates/2009/. Accessed 5 Jan. 2018.

over, and up to 5 times in specific strains of mice. These organisms not only live longer, they retain their health for longer too. Similar interventions, modified for human bodies, could be used to improve human health and lifespan.

Other extraordinary feats of longevity can be found in nature. Let's step down the aisles of the phylogenetic grocery store and see what kind of technologies we can shop for. Axolotls, or Mexican salamanders, are amphibians that spend their whole life in lakes and wetlands. Of interest to us is their ability to regenerate lost body parts, and their neoteny, meaning that they retain a larval stage for their entire lives, and so keep their youthful appearance as they age. They can regenerate their limbs, bones, and more over and over, even regrow their spine. Uncovering the cellular mechanisms that enable the axolotls to regenerate could offer precious insight into how to genetically engineer similar abilities in humans, and thereby help millions of amputees and paraplegics. The red sea urchin, a spiny invertebrate living in shoals and sea shelves in coastal waters, is one of the longest living animals on Earth. Studies show that they can have a lifespan of up to 200 years in good health, and with almost no signs of senescence, or age-related dysfunction, right up until the end. In other words, an urchin born in 1820 could be breeding off a coast near you. Ugly though it may be, the naked mole rat holds a remarkable advantage over its nearest cousins. With an upper lifespan of 31 years, it lives decades longer than the measly 3 years of the ordinary mouse. According to recent research, a chaperone protein, known as HSP25, may be the molecule that explains why this is achieved. How then might this protein, the function of which is to assist the proper folding and dismantling of other proteins, make us live longer and deal with some of the dysfunctions of aging? There is a long list of organisms, whose genomes give them an advantage humans do not have, such as cells that do not accumulate damage as the get older, cancer resistance, regeneration, and wider viewing spectrum.

24.3   The Resolve to Build the Genomic Future of Health and Longevity

Since, on the one hand, aging is proven to be what we could call a "disease of the cell" and since, on the other, we gradually acquire the capability to manipulate what is happening inside cells, and so to tackle disease by resetting the cellular clock, is it still meaningful to distinguish between health and longevity in a world in which "cure" will increasingly mean "restoring cells to a younger biological age"? Can "cure" avoid becoming synonymous with "enhancement"? A case could be made for either. Take antibiotics. Does the fact that they are effective just the time needed to fight off a pathogen, that otherwise would have killed us, make them preventative medicine – something that cures – or physical enhancement – a super body with abilities that our natural body does not possess? How about sanitation? Are disinfectants medicine or enhancement? Is designing a safer workplace enhancing or preventative or both? These may look like trivial matters to us, but only because we are used to them. In the eighteenth century the vaccine against smallpox caused an

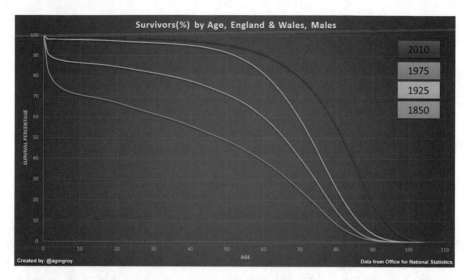

Fig. 24.1 Survivors (by Age, Males) in England and Wales in 1850, 1925, 1975 and 2010

outrage because it was feared that injecting humans with cowpox would cause them to grow cow-like appendages.

What is certain is that all interventions, be these antibiotics or workplace safety, are made with the goal to extend our lives, and always do so to a greater (antibiotics) or lesser (building site helmets) extent. The natural way for humans to die is of infectious disease before the age of 40. In the 1850s, 25% of children died before the age of five because of complications during labor and infectious disease. Those who survived childhood diseases were not very robust. The mortality graph (Fig. 24.1) of the time, that shows the number of deaths per 100,000 persons, drops at a consistent angle all along the curve. In contrast, today's graph is flat until about the age of 45 and then starts a steep and drastic decline.

Today people are living longer than ever, with a record number of centenarians expected in 2050, a 1004% increase.[11] That may sound like enhancement, but the health benefits that helped these folks live a long life prevented them from getting sick, and helped create robust bodies, without the weaknesses and then illnesses that sick bodies had. But now we have people who are old longer, and therefore have a longer period of disability and chronic illness.

The above-mentioned technologies that advanced human lifespan don't take account of advances in gene therapy and genetic engineering. Gene therapy's mission is to alter human health, by inserting or deleting genes, in order to correct genetic defects, to prevent or cure diseases that have their origin in faulty or missing genes. Genetic engineering aims to modify genes in order to enhance the capabilities of an organism beyond what is normal for that organism. The techniques of

---

[11] "Global Health and Aging – World Health Organization." http://www.who.int/ageing/publications/global_health.pdf. Accessed 5 Jan. 2018.

gene therapy and genetic engineering are the same. Their goals are slightly different. For simplicity's sake, we'll use the expression "gene therapy" to refer to both. Gene therapy encompasses gene editing and modulation as well.

This new technology, used as embryonic engineering, may give us children who are not only born immune to the diseases we die from, but who are engineered to live longer, and to be smarter and stronger than their parents. Preventative medicine would no longer happen days before you became sick but years before, and potentially before you were born or even at the moment of conception. Who would choose to have a child destined to die of dementia or heart disease if this is avoidable?

The genetic interventions currently going through the regulation process mostly address monogenic mutations, in which a single problematic gene causes disease in the patient's body. Sometimes all that is needed is replacing it with a properly functioning gene. Today we can insert a healthy copy of the gene using a viral capsid. This is the part of the virus that connects onto the cell membrane to release material other than the viruses' genome, the cause of disease. Such therapies can save patients years of illness and distress. As of December 2017, in the West, three genetic cures have passed through regulation, and more than seven are expected to pass in the next few years.

Gene therapy's success in addressing monogenic disease is basic evidence that it could address complex genetic disease that is, disease involving many genes. One example is autoimmune diseases. Another is aging. To begin, consider the evidence offered by animal models. When mice have the mTert gene (mouse telomerase reverse transcriptase) removed, this results in shorter telomeres. Such *gene-knockout-mice* age faster than ordinary mice. When mTert is added back to the mice, resulting in longer telomeres from the effects of telomerase, aging is measurably reversed.[12] More important still is human data. Gene therapies in early trials have shown groundbreaking effects, with only one dose of the treatment allowing extremely sick people to recover normal function.

The public sphere is traditionally slow to adopt new technologies. For this reason, we would expect the private sector to lead the way in reforming the present medical paradigm. Our mission is to delay, prevent and reverse age-related disease and some or all aspects of biological aging. We don't expect this to be an overnight process. However, our starting point will be specific issues with cells in the aging process that are already well characterized and amenable to gene based therapies. Scientist Maria Blasco, among others, has shown that the cellular problems of aging include genomic instability, cellular senescence, telomere attrition, epigenetic alterations, loss of protein homeostasis, mitochondrial dysfunction, deregulated nutrient sensing, stem cell exhaustion, altered intercellular communication, and extracellular matrix dysregulation.[13]

---

[12] "Telomerase gene therapy in adult and old mice delays ... – NCBI – NIH." 15 May. 2012, https://www.ncbi.nlm.nih.gov/pmc/articles/PMC3494070/. Accessed 7 Jan. 2018.

[13] "The Hallmarks of Aging: Cell." 6 Jun. 2013, http://www.cell.com/abstract/S0092-8674(13)00645-4. Accessed 5 Jan. 2018.

The slow rate and low number of drugs developed and brought through regulation today is dismal. Most of these drugs are small molecules with a myriad of off-site side effects. Because of the potential for error, the process to pass drugs for human consumption in the United States now takes over 15 years and requires over a billion dollars. Today's gene therapies show that we can cut the times and cost significantly by targeting genes with proteins that have known outcomes in animal experiments and human cell models. Over 100,000 people die every day of the diseases of aging. Even if only a small number of them had access to these therapies, we could vet out the most promising genes for longevity much faster. We would also glimpse the genes needed to better ourselves and fortify our future. The genes we express as we get older could be repressed and the beneficial ones modulated in safe doses.

Our initial therapies could be developed by selecting the most promising animal studies and combining them with relevant human data. Access would be given to the highest need individuals, meaning people so sick that they have no more medical options, allowing them to pave the way for the rest of us. Right to Try laws are now in order in 38 states in the USA, but they don't yet cover the area of gene therapies in preclinical development. It is my firm belief that this must change.

In 2015, I created a company called BioViva to solve the translation issues with new technologies. As an engine of translational medicine, BioViva aims to collect vast amounts of human data on the outcomes of gene therapy, and use this to move gene therapeutics through regulation faster. Our hope is that the resulting therapies will eventually both extend human lifespan and accelerate evolution, so that the species will live long and prosper. We also need reliable and inexpensive biomarkers of aging to help us solve the thorny problem of benchmarking therapies. We will expand our pipeline and research novel approaches as we evolve as a company. Our goal is to develop a set of combinatorial therapeutics that drive the future of prevention, enhancement and fancy.

Gene modifying technologies in the future will be only distantly related to our current capabilities. They will be able to alter multiple genes in a single treatment, one very different from any treatment available today. Gene modulation through a bacterial defence system now in use, called CRISPR (pronounced "crisper" and which stands for "clustered regularly interspaced short palindromic repeats"), will create simple switches in gene expression and allow us to cut genes in and out as desired, and with predictable outcomes. The deletion of genes could alleviate people with diseases like Huntington's. It could also help silence genes that turn on as we get older. Integrating genes at exact locations has a myriad of applications as well. And gene modulations could greatly benefit CRISPR and associated treatments. There are already many genes in the human genome that we want to upregulate or downregulate (increase or slow their production). By affecting genes already within us we can potentially make therapeutics one step simpler.

Another major leap in the administration of gene therapies in the future will be the fruit of synthetic biology, with the possibility of developing therapeutics that have no immune response, with genes added in regular doses, as often as required, making life easier for both the patient and the doctor. Furthermore, scientists like George

Church[14] have put forward the hypothesis of a synthetic chromosome. This would be a man-made chromosome carrying a script of beneficial genes such as genes for longevity, intelligence, and enhancements that better adapt your physiology to your environment and lifestyle. Such cells could be generated ex-vivo (outside the patient) and injected into specialised locations, maybe long living cells like muscle cells.

Already scientists are making synthetic proteins that don't exist in nature and that could create medicinal value we don't encounter there. The first synthetic genome was a bacteria engineered by Craig Venter in 2010.[15] Other proteins will follow. As an example: the double-helix inside our chromosomes is right-handed, meaning it turns in one direction. Synthetic chromosomes with genes with the opposite left-handed spiral orientation are being considered because viruses could not affect and integrate them.

Fifty years from now we might be looking at a very different world indeed. Replacement organs built from your own cells, stem cell rejuvenation, and gene therapies will likely be commonplace. People who are chronologically old will no longer be biologically old by almost any standards, be it internally and cosmetically. A 50 or 70-year old human will be physically indistinguishable from a 25-year old. Of course, these medical advances could be expedited or delayed by various factors, be these economical, ideological or political. All the same, in 50 years, today's main killers are unlikely to be major concerns, and the elderly will be offered a range of treatments that will efficiently address the damage time has done to their bodies. Advanced therapies, many of them based on gene and stem cells, will tackle aging before it occurs, in the same way as immunizations today tackle infectious disease before the patient has even been infected. Consequently, it is more likely than not that two or three generations from now young people will never have seen persons affected by aging other than in documents from the past, like photographs and movies. They will view our time in the same way as we look upon the fourteenth century, a dark age plagued by a disease with a 100% infection rate, and a 100% mortality, that had our bodies and brains gradually deteriorate, and reduced our lifespans to a derisory few decades.

If we want to be alive to tell the story we will have to give our time and energy, and invest all we can, into making this science a reality.

About the Author   Elizabeth Parrish is the Founder and CEO of BioViva, a company committed to extending healthy lifespans using cell technologies. Liz is a humanitarian, entrepreneur, innovator, and a leading voice for genetic cures. As a strong proponent of progress and education for the advancement of regenerative medicine modalities, she serves as a motivational speaker to the public at large for the life sciences.

She is the founder of the BioTrove Podcasts which is committed to offering a meaningful way for people to learn about and fund research in regenerative medicine.

Learn more about Liz here: https://www.linkedin.com/pub/liz-parrish/29/706/738

---

[14] George Church *Regenesis* (2014).

[15] Craig Venter et al. *Creation of a bacterial cell controlled by a chemically synthesized genome,* (May 2010) http://science.sciencemag.org/content/329/5987/52.full

Chapter 25
How to Reach a Societal Turning Point on Life Extension

Keith Comito

25.1 Overview

To those who see the promise and boundless potential of new technologies, it can sometimes be difficult to understand why others cannot. If it is possible to cure all disease, is it not obvious to think that we should? If it is within our reach to one day travel the stars and colonize other worlds, is this goal not noble and almost certainly critical to our long-term survival as a species?

When the whole world does not answer such questions with a resounding and uniform "YES!", it is tempting to believe this is due to willful ignorance or stupidity. Such an assumption gains us nothing, however, in the quest to make these technologies a reality—it is more instructive by far to explore the various reasons for societal hesitance, and how we might use the fruits of this exploration to positively effect change.

This is precisely the task on which I focus in my capacity as President of LEAF, the Life Extension Advocacy Foundation, and though my experience is centered on the particular issue of defeating aging, I believe the lessons learned in this battle translate to all fields of human endeavor that seek to push the envelope of our species.

And so, with our universe of discourse scoped thusly, let's talk about the issue at hand: How can we reach a societal turning point on life extension?

K. Comito (✉)
Life Extension Advocacy Foundation (LEAF) / Lifespan.io, Seaford, NY, USA
e-mail: keith@lifespan.io

© Springer Nature Switzerland AG 2019
N. Lee (ed.), *The Transhumanism Handbook*,
https://doi.org/10.1007/978-3-030-16920-6_25

It may sound obvious, but one of the most important things when setting out on the task of achieving a complicated goal is simply to have a plan—a framework that provides context for every initiative and a means by which to gauge the success of those initiatives towards fulfilling the goal at hand. Even if the plan is initially flawed and full of holes, it is important to have one, and to be able to iterate based upon results.

Luckily when it comes to the topic of technological advancement we have the whole of human history to look towards for inspiration. In the particular case of life extension, there is an analogous case study from the recent past—one that provides an excellent example of a successful battle plan to move the societal needle regarding aging: cancer research advocacy. While cancer is just one part of the aging process, a brief analysis of how cancer became a national priority in the United States will be instructive in laying the groundwork for a similar plan in regards to life extension in general, and likely other transformational goals as well.

25.2   One Becomes Two, Two Becomes All

Growing up at the turn on the millennium in the United States, it seemed to me that the "War on Cancer" had always been a part our health policy, but this was not so. As it turns out cancer was largely an ignored disease in the US, up until the 1940s, when a small number of individuals began successfully changing the national narrative around the disease.

Fig. 25.1 Cancer research advocates Mary Lasker and Sydney Farber

As described in the bestselling novel "The Emperor of All Maladies" by Siddhartha Mukherjee,[1] philanthropist Mary Lasker[2] and early chemotherapy researcher Sydney Farber[3] led the charge (see Fig. 25.1); the key to their successes being the realization that the war against cancer must not only be waged in the lab, but in the streets and the halls of political power as well. Through what essentially amounted to compelling PR maneuvers they garnered widespread public support for cancer research, and then leveraged the weight of this support to build mounting societal pressure for change.

Of particular note during this time was the the Jimmy Fund,[4] which smashed through the usual tendency for people to avoid thinking about late-life diseases by building its entire messaging around a single charismatic child with cancer (Fun fact: his name was actually Einar Gustafson, see Fig. 25.2). Celebrities were engaged, marathons and telethons were conducted that raised millions, followed by letter writing campaigns to members of congress, urging them to support research and fight for this issue.

The groundswell of support eventually culminated in extremely bold tactics, such as commissioning full-page advertising spreads in the New York Times and The Washington Post calling out the Nixon administration for governmental inaction (see Fig. 25.2). It worked—in 1971 the National Cancer Act[5] was passed, and the "War on Cancer" was born. While some may argue regarding whether the War on Cancer has been as successful as it could have been,[6] there is no denying the effect of these early advocates in transforming the societal perception of the disease.

Can we who fight for the end of age-related disease and other transformational technologies learn from their example? Are there equivalents to the Jimmy Fund for us?
       ***

---

[1] https://en.wikipedia.org/wiki/The_Emperor_of_All_Maladies

[2] https://en.wikipedia.org/wiki/Mary_Lasker

[3] https://en.wikipedia.org/wiki/Sidney_Farber

[4] http://www.jimmyfund.org

[5] https://en.wikipedia.org/wiki/War_on_Cancer#National_Cancer_Act_of_1971

[6] https://en.wikipedia.org/wiki/War_on_Cancer#Challenges

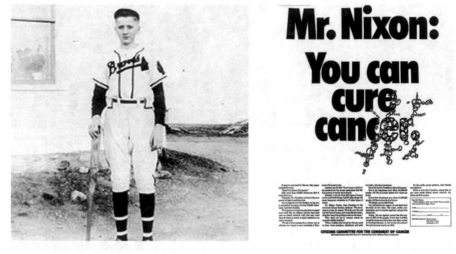

Fig. 25.2 "Jimmy" from the Jimmy Fund and advertisement inciting President Nixon to act on cancer research

It is naive to assume the exact same tactics will work in all cases, but there is much to learn by examining the dominoes the cancer research advocates pushed, and the order in which they pushed them:

1. Engage the public via intelligent and bold PR to build a grassroots movement
2. Mobilize this movement towards specific high-effect and high-profile actions
3. Leverage resulting momentum to effect political and societal change

This, I believe, is an excellent roadmap for how we can, together, end the diseases of aging once and for all. Therefore allow me to break down each of these components even further—illustrating how they apply to life extension in general, and charting our current position and trajectory as a movement on this path towards liberation from age-related disease.

25.3 Building a Grassroots Movement

In order to build the critical mass necessary for any movement to succeed, its core message must be one that is clear, relatable, and inclusive. By its very nature life extension, like other aspects of transhumanism, can be polarizing and difficult to fully conceptualize for those just being introduced to such ideas. Thus, effective advocates must come to the table expecting and embracing a challenge. Finding a message that will truly resonate with the public at large requires work, patience, and a legitimate willingness to understand and engage with opposing viewpoints.

For life extension in particular we find that this message must include a few key and related components:

- A focus on the here and now, in addition to the future.
- The ability to make allies out of enemies, instead of enemies out of allies.
- A means to personalize and humanize the mission down to the individual level. Life extension is not just transhuman; it *is* human.

25.3.1   The Here and Now

The average person might not think much about what the world will look like in 100 years—or believe they care enough to want to live in it—but everyone knows that Alzheimer's disease and cancer are terrible. Moreover, as the population of the world grays[7,8] the odds are increasingly high that everyone's lives—that your life— will be touched directly by these horrors.

The silver lining of this, if such a phrase can even be used in this context, is that everyone has skin in the game—everyone has loved ones they passionately want to protect. Therefore to illustrate that researching the underlying root cause of this issue, aging, is an effective way to combat such diseases is an extremely winnable point when it comes to engaging the broader public in aging research.

Some transhumanists may argue that such an emphasis on disease remediation is a form of dilution or adulteration of the "true" goal of indefinite life extension, but in my opinion this could not be further from the truth. Implicitly bound into the advocacy for life extension is the assumption that the lives to be extended are ones that are desirable to live. Working to diminish physical suffering and to maximize personal agency throughout life is essential to deliver on this assumption. It is thus perfectly possible to be authentic in one's support for life extension, indefinite or no, while being able to highlight specific aspects of this mission that can truly and correctly appeal to a given audience.

That being said, it is important to realize that effectively communicating the benefits of life extension to the broader public has built in challenges, on account of the nature of our initial hardware constraints so to speak. Due to millennia of evolutionary adaptions that have served us well as a species thus far, our mental machinery is not naturally geared to properly process the deep future or radical change. Lucky for us, however, there is a science to understanding such cognitive biases,[9] as they are known, and how they can be overcome.

---

[7] https://en.wikipedia.org/wiki/The_Silver_Tsunami

[8] https://www.leafscience.org/demography-prospects/

[9] https://en.wikipedia.org/wiki/List_of_cognitive_biases

As an example, consider a cognitive bias very relevant to the perception of life extension, and generally any technology whose benefits are not immediately and viscerally obvious: "hyperbolic discounting".[10]

In short: hyperbolic discounting is the tendency to prioritize near-term rewards over long-term ones, even if the long-term rewards are significantly greater.[11] For instance let's assume you really like dark chocolate (as do I!); when offered the choice between 50 dark chocolate bars now and 100 dark chocolate bars a year from now, you would likely choose the immediate 50—their nearness pulls upon your physical hardware to subvert the logical benefit of waiting: your mouth salivates, you envision yourself eating the chocolate—you do not want to wait.

The illogic of this is exposed, however, by considering the choice between 50 bars in 5 years vs. 100 in 6 years—studies show that almost everyone will choose the 6 year option, even though this is essentially the same two choices as the earlier scenario, seen at 5 years' greater distance.[12]

Armed with this knowledge it is clear to see how encouraging younger people to care about overcoming late-life conditions like Alzheimer's and cancer is an uphill battle. Therefore, the challenge hyperbolic discounting presents must be understood and met with tactics that surmount it, such as as shifting audience perspective with interesting narratives and thought experiments which cause such cognitive biases to reveal themselves.

To this purpose there is a thought experiment I typically use when addressing audiences not already predisposed to life extension:

I ask three questions:

---

[10] https://en.wikipedia.org/wiki/Hyperbolic_discounting

[11] https://doi.org/10.1111%2Fj.1467-9280.1994.tb00610.x

[12] http://psycnet.apa.org/doiLanding?doi=10.1037%2F0096-3445.126.1.54

1. Assuming health is preserved for you and your loved ones: do you want to be alive in 50* years?
2. Do you want to be alive tomorrow?
3. Assuming nothing significant changes, do you suspect your answer to 2. will change tomorrow?
*Note: I'd vary this number depending on the mean age of audience.

Assuming I've picked the number in 1. to bring the average audience member firmly into the typical "no way" zone of 100+, very few will raise their hands. 80 years is fine enough some might say,[13] or perhaps that a longer life would be incredibly boring.[14]

On the other hand, the answers to both 2. and 3. are always a unanimous "yes" and "no", as you might expect. This seems obvious, as anyone not in truly dire straights would prefer to be alive tomorrow and the day afterwards. What is not obvious, however, is that the pairing of questions 2 and 3, via the mathematical induction[15] of infinite tomorrows, effectively ask the same question as 1. What this framing essentially does is remove the effect of hyperbolic discounting, allowing a person to conceive of 50 years in the future as they would actually experience it, day by day in the present, as opposed to some far off alien thing.

This also relates to the idea of the hedonic treadmill[16]—while we often imagine our future self will have radically different priorities and sensibilities than our current self, the reality is that you will, in all likelihood, feel generally the same as you do now.

Note: above thought experiment makes the key assumption that health will be preserved—no one wants to be in the situation of Tithonus[17] from the Greek myth: gifted with eternal life, but cursed to ever decay with time. Indeed dispelling the natural fear of this outcome, through illustrating how significant life extension can only be achieved via health extension,[18] is a central focal point of our message.

***

The subject of cognitive biases is an incredibly deep one and many more examples could be discussed, but in the interest of time I wish to highlight just a few more: "scope neglect" and the "identifiable victim effect".

Scope neglect[19] is essentially the tendency to not value a problem in a proportional relationship to its size. For example, if I were to ask how much money you could spare to grant a thousand starving children a meal every day for a week, studies suggest that your stated amount would not be much different than if the scenario

---

[13] https://www.intelligencesquaredus.org/debates/lifespans-are-long-enough

[14] https://www.leafscience.org/will-increased-lifespans-be-boring/

[15] https://en.wikipedia.org/wiki/Mathematical_induction

[16] https://en.wikipedia.org/wiki/Hedonic_treadmill

[17] https://en.wikipedia.org/wiki/Tithonus

[18] https://en.wikipedia.org/wiki/Gompertz–Makeham_law_of_mortality

[19] https://en.wikipedia.org/wiki/Scope_neglect

involved only 100 starving children.[20] The identifiable victim effect[21] refers to the related tendency to offer greater aid when a specific identifiable person is in need, as compared to a large and/or vaguely defined group with the exact same need.

Taken together these tendencies clearly prescribe at least one promising tactic for generating awareness for any cause: focus on stories of charismatic individuals in relation to the mission at hand—diligent researchers hard at work, creative advocates exciting the masses, or patients bravely fighting disease. If this all sounds very familiar you might be realizing this is exactly the tactic employed by The Jimmy Fund, mentioned above, to astonishing effect.

Rather than just being a purely academic pursuit, taking the time to understand how and why people think they way they do is incredible useful, as you can see, and there should be no shame in applying this knowledge toward a noble purpose.

25.3.2 Speak, Friend, and Enter

When encountering those who do not understand or actively oppose your viewpoints, it is all too easy to view the resulting dialogue as a battle, and engage the other party as an enemy to be overcome with superior logic. This effect can be especially magnified, I have noticed, when the topic is life extension or other potentially world-changing transhumanist ideas. Not only is this mindset rarely useful in the short term, (see the "backfire effect",[22] another cognitive bias), it is also unproductive if your true goal is to actually aid your cause. Instead of belittling and vilifying potentially valid critiques, it is better by far to understand and engage with them in a sincere and thoughtful way, and in doing so perhaps turn a would-be enemy into a friend.

To give an example: one of the most common objections we hear to life extension is something along the lines of "Why would I want to live beyond 100? I've seen what my grandmother was like at 90...I don't want to live like that for 20 years!". When presented with such a position I have seen many a life extension proponent adopt an immediate air of intellectual superiority, and commence rattling off an arsenal of reasons why a person would be foolish to not want an extended lifespan. Whatever truth this reasoning may hold, there are few who respond well to being essentially called an idiot. Furthermore, if you think deeply about this objection it becomes clear that such a position is in fact completely logical, given the current state of affairs and a limited knowledge of rejuvenation biotechnology. If you do not know that near-term breakthroughs in extending healthy lifespan are feasible, you would be absolutely correct in fearing decades of decrepitude and pain. It is therefore on us to anticipate this reaction, and clearly explain how the only

---

[20] https://doi.org/10.3768%2Frtipress.2009.bk.0001.1009

[21] https://en.wikipedia.org/wiki/Identifiable_victim_effect

[22] https://rationalwiki.org/wiki/Backfire_effect

way we will get significant life extension is through significant extension of the healthy period of life, not the period of morbidity appropriately dreaded by all.

To truly be effective advocates, we need to understand critiques such as this and engage with them in a positive way—this is how we truly grow, inviting others into our cause rather than building walls to reject them. It is also worth noting that when you come to a discussion genuinely interested in the views of others, it goes a long way.

This open-minded approach is not only useful when speaking with individuals, but also when dealing with groups and large organizations as well. For example, consider organized religion. It is very easy to see conflict between many religious teachings and transhumanism, and often I have witnessed prominent transhumanists outright antagonize entire segments of the world's population for this reason. While it is true that certain religious beliefs are hard to square with transhumanism—it is difficult to argue with my mother's desire to meet her parents in the afterlife, for example—upon closer inspection there is actually a great amount of common ground here that can be focused on instead. Healing of the sick, compassion towards others, transcending the limitations of our current selves. Indeed; rather than demonizing those who have a religious mindset it is far more illuminating to think: what is it about religion that is so appealing?

One of the foremost answers, I believe, is meaning—a narrative framework to contextualize one's existence, and a sense that actions have purpose in the greater scheme of life. Assuming this to be true, it is clear to see how a person powerfully affected by such sensibilities could easily see the worth of fostering technology aimed at the alleviation of suffering, if properly presented. Help longevity research, for example, and you will impact the lives of everyone you have ever known, potentially saving them from horrors like Alzheimer's disease and cancer—if a deepened sense of purpose is desired, this will surely accomplish it.

If you are still skeptical that the religious can ever accept transhumanist ideas, it is instructive to note that such has happened in history before. One particularly extreme and interesting case being the Russian Cosmism[23] movement of the early twentieth century, which combined Eastern Orthodox Christianity with deeply transhumanist ideas such as immortality, human augmentation, and resurrection of the dead (Fig. 25.3).

On a more humorous note, I remember when I was a child there was an omnipresent advertising push in the United States urging kids to take their daily vitamins; Flintstones brand vitamins in particular. Notably absent from that message was the renunciation of God. The message was simple: take these pills and gain increased health. Giving up God was not a prerequisite for this, and would likely not have been an awesome marketing strategy. There was no need for such combat then, and I believe there is no need for such combat now.

To stress this point I would like to share a personal story involving the negative effects of such preemptive warfare. Over the years I have participated in a fair number of online discussions about life extension, transhumanism, etc., and after one

---

[23] https://en.wikipedia.org/wiki/Russian_cosmism

such occasion I was watching the resulting broadcast with a friend. As it turns out this friend was already in favor of life extension, and yet by the end of the broadcast they were nearly in tears of sadness and rage from what they had witnessed. This was so because they had been raised Muslim, and some of the other panelists were speaking with such hatred and contempt of the religious it would be upsetting to almost anyone. Seeing this happen crystallized in my mind that if we truly wish to help people, we must sincerely endeavor to make friends of enemies, rather than enemies out of friends.

To this end it is worth considering that there was probably a time in your own life—4 years ago, or 12 years ago—where you didn't think of things as you do now. Then you came across some information that changed your mind, and opened your eyes to a future that could be wondrous. You could be that spark of knowledge for others if you speak with compassion and intelligence. Let us not think of those currently unsupportive as enemies, but as allies yet to be.

To quote from a classic: "Speak, Friend, And Enter".

Here is written in the Fëanorian characters according to the mode of Beleriand: Ennyn Durin Aran Moria: pedo mellon a minno. Im Narvi hain echant: Celebrimbor o Eregion teithant i thiw hin.

25.3.3 More Human Than Transhuman

When it comes to building a true movement behind transhumanist ideas such as life extension, we have found there is one more key piece of the puzzle—one that may seem ironic, especially considering the audience here at hand—don't focus on "Transhumanism".

Consider the meaning of the word itself: "Beyond Human". While this may certainly appeal to the unfettered aspirations of some, to the wider populace this term will be, by definition, alien. scary. other.

Furthermore I do not believe it is even fundamentally accurate—striving to alleviate suffering and increase our agency in the world via technology is not beyond human at all; it *IS* human. I would argue that this is actually the defining characteristic of what it means to be human: the creation and usage of tools to better our situation. Perhaps you might say that fire and poultices are an entirely different story from modern technologies such as CRISPR[24] and machine learning,[25] but in fact they are all part of a continuous progression—all tools that can aid the preservation and extension of life.

It is not like the dream of increased healthy lifespan is a new one, either—it is in actuality the subject of the first great work of fiction ever produced by our species: The Epic of Gilgamesh.[26] This ancient Mesopotamian story, discovered on clay tablets in what is modern day Iraq, tells the tale of Gilgamesh the King who goes on a journey to overcome aging and death, after the tragic loss of his beloved friend and companion, Enkidu (Fig. 25.4).

While Gilgamesh may not have achieved his goal as originally intended, his tale teaches us something very important: the desire for transcending the maladies of

---

[24] https://en.wikipedia.org/wiki/CRISPR

[25] https://en.wikipedia.org/wiki/Machine_learning

[26] https://en.wikipedia.org/wiki/Epic_of_Gilgamesh

Fig. 25.4 Illustration from the book "The Revenge of Ishtar" by Ludmila Zeman, depicting Gilgamesh weeping over the body of Enkidu, his dearest companion

aging and yes, even death itself, is not alien or strange at all. It has in truth always been there, in the background of our shared journey as human beings, threading its way through every culture, religion, and mythology: Amrita, Ambrosia, Tír na nÓg.

Rather than being simply an interesting history lesson, realizing that there is relevant past exploration of transhumanist ideas gives context and allows us to discuss them in a more accessible way. As an example I remember attending a panel discussion where the concept of a technological hive mind was being discussed, and many in the audience where frightened by the idea—conjuring thoughts of the Borg[27] and loss of individuality. Valid concerns, no doubt, but perhaps the conversation would have been more fruitful had the audience known that similar concerns have been discussed for thousands of years, through ideas such as Paramatman[28] in Hindi theology and the Supreme Mind of Hermeticism.[29]

Don't get me wrong; there can be PR value in sensationalism, but as any movement matures it must ground its lofty aspirations as firmly as possible if it is to have any chance of enlisting a critical mass of the public. This leads me to another important point on this subject: the timing of what we choose to advocate matters.

Consider the ideas of Ray Kurzweil,[30] noted inventor, futurist, author, and someone with whom most of you are probably familiar. In interviews he has credited the success of his various inventions, such as the flatbed scanner and text-to-speech

---

[27] https://en.wikipedia.org/wiki/Borg_(Star_Trek)

[28] https://en.wikipedia.org/wiki/Paramatman

[29] https://en.wikipedia.org/wiki/Hermeticism

[30] https://en.wikipedia.org/wiki/Ray_Kurzweil

synthesizer, to proper timing above all. Not only can you be too late, but you can also be too early. To quote:

"Most inventions fail not because the R&D department can't get them to work, but because the timing is wrong—not all of the enabling factors are at play where they are needed. Inventing is a lot like surfing: you have to anticipate and catch the wave at just the right moment."

Kurzweil is also famous for his ideas on concepts such as the technological singularity[31] and mind uploading, joining others in stating that by the year 2045 human beings will be able to upload their minds to computers and achieve "digital immortality".[32] While such a prediction is within the realm of possibility, I'm going to apply the spirit of Kurzweil's quote against it to suggest that it may in fact be too early to advocate for technologies such as mind uploading. Not all of the enabling factors, to borrow the phrase, are currently at play to sell the public on such a proposition. We do not yet have a clear working definition for consciousness, for example, and without this you therefore cannot establish a metric for success—will the uploaded entity be you, or just a copy of you? Will it be conscious at all, or just a philosophical zombie[33]?

Until such issues are better addressed (and progress *is* being made[34]), the average person is correct to be skeptical, and it would be disingenuous to preach that successfully uploading one's mind is definitely doable and definitely desirable. It is likely more effective to drive the necessary assisting breakthroughs first, which can still excite imaginations without feeling too pie-in-the-sky, as well as related technologies whose feasibility horizon is near enough to successfully galvanize the public. I know that there are many who'd like to jump right to the transhumanist finish line in all regards, but it is important to consider that in higher-dimensional spaces

---

[31] https://en.wikipedia.org/wiki/Technological_singularity

[32] https://en.wikipedia.org/wiki/Mind_uploading

[33] https://en.wikipedia.org/wiki/Philosophical_zombie

[34] https://en.wikipedia.org/wiki/Neural_correlates_of_consciousness

such as ours, a straight-line is not always the quickest route to a goal.[35] Tactics matter, and a bowstring must be drawn fully back for the arrow to fly with maximum
speed.

As a cautionary tale of how placing the cart before the horse risks the credibility
of the enterprise: I was recently watching a documentary on transhumanism and life
extension, and here was who appeared on my television to explain it all to me:

As entertaining as I honestly do find Mr. Tsoukalos, I don't think we want him to
be our spokesman. Rather it is my hope that concepts such as life extension will be
embraced as perfectly sensible, and that the "trans" in transhumanity will one day
be thought of, not as "beyond", but as "through" — humanity together moving into
the beyond.

25.4   Why Do We Need the Public at All?

Upon hearing all this talk about building a grassroots movement and galvanizing the
public, you might ask yourself: does it even really matter? Once a truly effective
therapy hits the market people will want it, minds will be quickly changed, and
public adoption will happen swiftly, right? Perhaps this may be so, but there are a
few very good reasons why it is wise to bring the broader public along with us every
step of the way. Here is one of them:

---

[35] https://en.wikipedia.org/wiki/Geodesic

Fig. 25.5 CNN News coverage of President George W. Bush vetoing bill to support stem cell research

As you might recall President George W. Bush of the United States heavily restricted the use of federal funds for stem cell research in 2001[36] and consistently blocked attempts to change this, including vetoing two congressional bills intended to overcome the restriction in 2006 and 2007[37] (see Fig. 25.5). Collectively this was a massive setback for the entire regenerative medicine industry. If we don't bring the public along with us we risk catastrophic push back in legislation such as this, whereas an educated and supportive populace helps product against such outcomes.

In regards to the case above, I clearly remember the political discourse about embryonic stems cells at the time, and how shockingly misinformed were the members of congress—basing their debate on the assumption that such cells were acquired by literally murdering newborn babies and harvesting their brain and organs. This might sound like hyperbole, but these discussions were broadcast live on C-Span radio for all to hear, and hear them I did. Had the public and politicians possessed a more complete knowledge of stem cell therapies, it is fair to imagine the outcome might have been different. Congress could have acted sooner, or perhaps gathered the required two-thirds majority of votes to overturn the presidential veto.

Another relevant example is genetic modification, with the rise of powerful anti-GMO movements in various countries able to successfully mount opposition to life-saving modifications such as Golden Rice.[38] While this issue is a complex one, much of the reasoning behind this opposition fails to take into account basic knowledge of genetic engineering—and how we have been doing it, poorly, since the dawn of human civilization[39]—and thus appears to be the result of a failure by the scientific community to get ahead of the issue, accurately informing and engaging the public.

---

[36] https://www.ncbi.nlm.nih.gov/pmc/articles/PMC2744932/

[37] https://en.wikipedia.org/wiki/Stem_Cell_Research_Enhancement_Act

[38] https://en.wikipedia.org/wiki/Golden_rice

[39] https://en.wikipedia.org/wiki/Plant_breeding

This state of affairs is not something we wish to repeat, and it is therefore in our interest to ensure, as much as possible, that when the first game-changing longevity therapies arrive they do so in an environment ready to receive them.

Finally another reason it is important to bring the public along with us is that in, order for the above-mentioned process of adoption to occur, we still need to have those first successful therapies exist. A grassroots movement supporting necessary research will help get us to this point sooner—and when it comes to technologies such as life extension, every day counts.[40]

25.5 Mobilizing the Movement

Now that we have established some hopefully useful ideas in terms of messaging and growing a societal movement, there is still the critical question of how best to channel this movement's energy. As discussed earlier, it is important to always keep in mind the true goal of effecting large-scale societal change, and the activities to be focused on should be chosen accordingly.

At LEAF we have found that, to hearken back to the Jimmy Fund, an important key is to choose activities with relevant, specific, and clear calls-to-action that exist fully in view of the public eye, inviting dialogue as well as engagement.

In our case the first incarnation this took was in crowdfunding longevity-focused research via the Lifespan.io platform. While the most obvious benefit to this approach is in raising direct funds for critical research—and to date we have raised over a quarter million dollars to this effect[41]—the additional benefits are many. To name a few:

1. Having a continual and clear call-to-action allows advocates to "walk the walk", and also easily engage others outside of typical echo-chambers to help the cause in an exciting and specific way.
2. Democratizing the scientific process proactively mitigates critiques against transformative technologies, such as they will only be for the rich or be used to create leverage against the poor. Consider gene editing, for example. Stories abound of evil empires using reproductively limited "terminator seeds"[42] for economic exploitation—but if we enable a future where people have learned how to cheaply correct such modifications with open source technology, then we have mitigated this fear by giving the power to the people.
3. The visible nature of such projects attracts positive attention from both the research community and the press, bringing powerful connections and influencers into our network. These connections can, in turn, make further exciting projects and media content, creating a virtuous cycle of increasing support.

---

[40] https://en.wikipedia.org/wiki/Mortality_rate

[41] https://www.lifespan.io/successfully-funded/

[42] https://en.wikipedia.org/wiki/Genetic_use_restriction_technology

Fig. 25.6 Graph of "longevity escape velocity", illustrating the idea of bootstrapping life extension technologies to stay ahead of the pace of aging

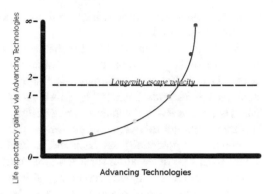

In our particular case of life extension advocacy; I am happy to report that following this roadmap has been extremely successful. In less than 3 years of active operation our small but dedicated group at LEAF has accomplished the following:

- Raised over $250,000 for aging-focused research projects. Additionally, projects from groups with existing donor bases, like the SENS Research Foundation,[43] have noticed many of our donors were new supporters.
- Built the largest social media presence in the world for life extension research advocacy on platforms such as Facebook,[44] and one of the most highly trafficked blogs on the subject,[45] posting daily original content.
- Conducted numerous live-streamed journal clubs,[46] typically garnering 20k+ views, where we review the latest longevity research papers and answer audience questions in real-time.
- Attracted high-profile collaborators to create media which has informed and excited the public regarding the science and societal benefits of life extension. Most notable are the YouTube videos we co-produced with the science channel Kurzgesagt[47,48] that reached the top of YouTube's overall trending list and have collectively received over 6 million views.

Aubrey de Grey,[49] famed life extension advocate and co-founder of SENS, frequently speaks about the concept of "longevity escape velocity"[50] (see Fig. 25.6), the idea that we can bootstrap biomedical progress to stay ahead of the curve of aging. I strongly believe this idea applies to life extension advocacy as well—the more very visible, exciting, and credible projects succeed the easier it is to draw

---

[43] https://en.wikipedia.org/wiki/SENS_Research_Foundation

[44] https://www.facebook.com/lifespanio/

[45] https://www.leafscience.org/blog/

[46] https://www.facebook.com/pg/lifespanio/videos/

[47] https://www.youtube.com/watch?v=GoJsr4IwCm4

[48] https://www.youtube.com/watch?v=MjdpR-TY6QU

[49] https://en.wikipedia.org/wiki/Aubrey_de_Grey

[50] https://en.wikipedia.org/wiki/Longevity_escape_velocity

more supporters into our network, and thus it becomes easier for the next round of initiatives to succeed.

One thing to keep in mind here is that because we are, very intentionally, in the public eye, it is paramount that the initiatives we choose to drive forward are extremely credible and well-timed (see above). As a counter-example you may recall the 2014 scandal surrounding STAP[51] stem cells, which involved a paper published out of the RIKEN center in Japan describing how to transform ordinary cells into stem cells via trauma, and how the resulting allegations of fraud and lack of reproducibility potentially set back the entire field for years.[52] Such a hi-profile failure would be devastating for a field that does not yet have mainstream public acceptance, such as longevity research, and is something we must guard against as best as possible—one of the many reasons why we thoroughly vet all projects we support via our Board and SAB,[53] which includes world-renown biologists.

25.6   Putting It All Together

Now that we have successfully built up a grassroots movement, and have begun mobilizing it towards high-effect and high-profile actions, what happens next? Following our stated battle plan means leveraging this existing momentum to reach beyond our echo chambers, engaging mainstream media and impacting political discourse to effect large-scale societal change. Easier said than done, of course, but this is the point we currently find ourselves in terms of life extension advocacy, and there is no time like the present to continue moving forward.

So, we have two endpoints—the media, which has the power to set the narrative for forthcoming technologies, and national governments, which have the power to allocate vast sums of money and resources towards solving a specified problem. These are our dominoes, so let us explore some ideas on how we can move them in turn:

25.6.1   Engaging the Media

One increasingly powerful way to draw public attention in the modern age is to grow a network of popular content creators on social media platforms like YouTube and Facebook—building a widening sphere of interest that draws in all forms of media, eventually including even "regular" television personalities and celebrities. For example: right now, thanks in part to the overwhelmingly positive response to our videos with Kurzgesagt, we are speaking with several top-tier

---

[51] https://en.wikipedia.org/wiki/Stimulus-triggered_acquisition_of_pluripotency

[52] https://www.nature.com/news/papers-on-stress-induced-stem-cells-are-retracted-1.15501

[53] https://www.leafscience.org/scientific-advisory-board/

Fig. 25.7 Michelle Phan, popular beauty industry entrepreneur

YouTube channels and well-known science celebrities to make similar content in support of life extension. It is therefore not crazy to think that within a few years time this added exposure will allow us to successfully engage key industry leaders as well, such as Bill Gates and Larry Page, who are already involved in this space. You may have seen themed weeks on YouTube, for fighting cancer or disaster awareness for example—can you imagine the effect of such a coordinated public action to overcome the diseases of aging? This could be our "Jimmy Fund", or at least a gateway to it.

On this subject, we should also keep a keen eye towards identifying other industries that are synergistic with our goals, and related demographics with whom alliances would be mutually beneficial. Consider the cosmetics industry, for instance—in 2016 the US cosmetics market alone generated approximately $84 billion in revenue.[54] If people are willing to spend this much on making their skin just appear younger, how much interest could there be in therapies that could make skin actually younger? My guess would be a lot. Michelle Phan, beauty industry entrepreneur and one-time celebrity video blogger[55] (see Fig. 25.7), agreed when I happened to speak with her years ago. Afterwards she and her millions of followers happily helped the SENS Foundation win a fundraising competition that was ongoing at the time—just one example of how making connections within adjacent industries can be helpful.

Cosmetics is perhaps a rather obvious example of an industry with potential synergy with life extension, so allow me to provide another that is less obvious—video games. For context, the global gaming market exceeded $108 billion in 2017,[56] and includes over 2 billion people. Electronic sports tournaments (eSports) are rapidly becoming mainstream, drawing ever-growing crowds that dwarf even the Super Bowl in attendance.[57] Couple all this with the fact that many of the most

---

[54] https://www.statista.com/topics/1008/cosmetics-industry/

[55] https://www.youtube.com/user/MichellePhan

[56] https://venturebeat.com/2017/06/20/newzoo-global-game-market-to-grow-from-108-9-billion-in-2017-to-128-5-billion-in-2020/

[57] https://www.forbes.com/sites/paularmstrongtech/2017/03/16/46-million-watched-live-esports-event-10-million-more-than-trump-inauguration-broadcast/#17df798591f4

Fig. 25.8 eSports events currently draw crowds larger than most physical sporting events

popular and beloved gaming titles deal either directly or indirectly with transhumanist subject matter, such as life extension and human augmentation[58,59] and you can see what a natural demographic this is for us to engage (Fig. 25.8).

Furthermore the interactive and dialogue promoting nature of popular game streaming services like Twitch and YouTube create a perfect platform for powerful charity drives, and many have already taken shape—ExtraLife[60] and GamesDoneQuick,[61] for example, which have raised millions of dollars for charities such as the Children's Miracle Network, Doctors Without Borders, and the Prevent Cancer Foundation. It is also worth noting that concepts such as life extension have many thematic tie-ins with video game mechanics—extending your health bar, gaining extra lives, etc. A well-orchestrated initiative with this powerful, progressive, and tech-savvy community has the power to excite and engage millions, and is something we at LEAF would very much like to see and are currently working to create.

Beyond our own work at LEAF, there have also been a number of other recent public successes in term of life extension acceptance. The famous silicon valley

[58] https://en.wikipedia.org/wiki/Xenogears
[59] https://en.wikipedia.org/wiki/Deus_Ex
[60] https://www.extra-life.org
[61] https://gamesdonequick.com

incubator Y Combinator,[62] for example, announced their new plan to focus on healthspan and age-related disease.[63] Also SENS was one of the winners of the popular Project4Awesome[64] charity contest, and additionally received very large and high-profile cryptocurrency donations.[65] It is important to celebrate successes such as these, regardless of whether you have had a direct hand in them—they serve to validate our collective efforts and place more wind at our backs, allowing us all to keep driving forward together.

25.6.2   Engaging the Government

When it comes to the government side of the equation it can seem difficult to understand how to proceed—how to promote action on goals where the benefits are not immediately apparent or pressing. The key is thus to make it clear, and to make pressing, just as the cancer advocates did before us.

On this subject I'd like to share a conversation I had a few years ago, at a biomedical conference with someone who used to work in the State Department of the United States. Contrary to what you might expect, she told me that leading politicians were aware of the ticking time bomb age-related diseases present—the looming economic problems with sustaining social safety net programs, the generational destabilizations foreshadowed by the "Gray Tsunami" in Japan, etc. So I asked her: why then is it so politically hard to shift the focus of healthcare towards overcoming the diseases of aging directly? What is it they need to see?

The good news is that her answer is exactly what we have been discussing here all along: a large-scale movement that is coming from multiple different groups and directions—so diverse it can not be marginalized, and so loud it can not be ignored. I thus have every confidence that we are on the right path. With a growing grassroots movement and large-scale media initiatives behind us it is possible to gain the attention of national and global organizations, and push them towards positive change.

To be fair, I say this not because of my wondrous skills as a prognosticator, but because successes in this regard have already happened. You might remember hearing about the Targeting Aging with Metformin (TAME) trial,[66] for example, which will go beyond the drug's existing FDA indication for diabetes to test its ability to remediate underly aspects of the aging process.[67] Working with other organizations, such as the Global Healthspan Policy Institute,[68] we were able to help secure the FDA's blessing for the trial to proceed, as well as critical government support in

---

[62] https://en.wikipedia.org/wiki/Y_Combinator_(company)

[63] https://blog.ycombinator.com/yc-bio/

[64] http://www.projectforawesome.com

[65] http://pineapplefund.org

[66] https://www.afar.org/natgeo/

[67] http://channel.nationalgeographic.com/breakthrough-series/episodes/the-age-of-aging/

[68] https://healthspanpolicy.org

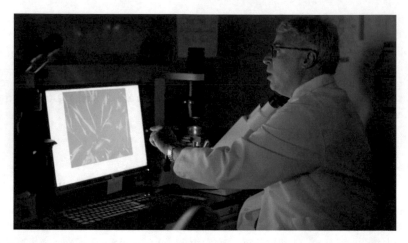

Fig. 25.9 Dr. Nir Barzilai of the Albert Einstein College of Medicine, who is leading the TAME trial initiative

funding it. This is incredibly significant—a success here would lead to the first FDA-approved therapy targeting the indication of aging itself, de-facto classifying aging as a disease and allowing it to be more easily targeted by additional trials and research funding opportunities (Fig. 25.9).

More recently LEAF participated in a wide community effort to address the World Health Organization (WHO), urging them to reconsider the conspicuous absence of societal aging as a key concern in their proposed 2019–2023 "Programme of Work".[69] Of the roughly 400 comments they solicited, over 90% were submitted by our coalition members to stress this issue, and in response the WHO released a modified proposal which added several provisions related to aging—including a focus on "Ensuring healthy ageing is central to universal health coverage" and "interventions to modify the underlying physiological and psychological changes associated with ageing". This is a huge win, and showcases how, with unity of purpose, our community already has the ability to influence policymakers at the global, as well as national, level.[70]

Furthermore it is important to note that recent changes in the broader political climate create opportunities for life extension advocates and transhumanists to engage the public regarding policy. Consider how income disparity and health cov-

---

[69] https://www.leafscience.org/does-who-five-year-plan-leave-healthy-aging-out-of-the-picture/
[70] https://www.leafscience.org/world-health-organization-puts-the-elderly-back-in-the-picture/

erage have become leading hot-button political issues in the United States, for example, and how politicians such as Bernie Sanders[71] are rallying historic movements to address them. Ideas such as improving healthy human lifespan via wiser research spending, better preventative care, and affordable coverage for all are well in line with the desires behind this powerful progressive movement, and thus an ideal place for us to positively enter the political conversation. It is two birds with one stone if you think about it: what is the ultimate wealth inequality if not health inequality?

25.6.3 The Turning of the Tide

Effecting massive political and social change is surely not an easy process, but it is my hope that I have shown how such battles can be waged and won—and win them we must when it comes to technologies that are able to greatly diminish suffering and improve the lives of every person on the planet.

Similar victories have been earned in the past, and it is my ardent belief that we are now at a moment in history where the needle can be moved once again against the diseases of aging and death. It is not too optimistic to imagine in a few years time we can realize initiatives as powerful and widespread as those such as Race For The Cure[72] and Stand Up To Cancer,[73] where millions of people mobilize to raise both funds and awareness—applying pressure on governments to change their priorities, until eventually using transformational technologies to improve our healthy lifespans becomes as obviously a part of our society as the War On Cancer was in America when I was growing up. Just as I had thought fighting this disease was a thing my country always did, I hope the child of 20 years from now will grow up thinking curing Alzheimer's, sarcopenia, and the other diseases of aging is just something our world has always done—that is the future that we want.

In the meantime, we must do the work.

And as we move through of the turning of the tide, let us not forget what an exciting time it is we live in. Since the dawn of recorded history, when Gilgamesh set out upon his quest, we have dreamed of transcending the boundaries of our birth. Now we have the potential to find that flower of rejuvenation which Gilgamesh sought. Each of us can chose to become an integral part of the first Hero's Journey[74]—our journey—and that is pretty awesome (Fig. 25.10).

---

[71] https://en.wikipedia.org/wiki/Bernie_Sanders

[72] https://en.wikipedia.org/wiki/Susan_G._Komen_for_the_Cure

[73] https://en.wikipedia.org/wiki/Stand_Up_to_Cancer

[74] https://en.wikipedia.org/wiki/The_Hero_with_a_Thousand_Faces

Fig. 25.10 Illustration from the book "The Last Quest of Gilgamesh" by Ludmila Zeman, depicting Gilgamesh acquiring the flower of rejuvenation

If I've done nothing else, I hope it has been to inspire you that true and wondrous change can happen if you fight hard, smart, and always with compassion. Feel free to join us at lifespan.io/hero if healthy human life extension is a mission you find worthy, and/or support the many other great organizations working towards this goal.

Whatever your cause may be, consider this your personal call to action.

Chapter 26
How to Organize a Moonshot Project for Amortality: Scientific, Political, Social and Ethical Questions

Didier Coeurnelle

26.1 We Live Longer Than Ever, But the Maximal Lifespan Is Almost Not Longer Than 20 Centuries Ago

Let's begin with the classical techno-optimist point of view. We live longer today than ever before in the history of humanity. The life expectancy of women in many rich countries is over 85 years and that of men is approaching 80 years. The average life expectancy in the world is 70 years, whereas it was still less than 60 years 30 years ago and less than 40 years only a century ago. At the world level, each year, life expectancy is rising at a rate of 2 or 3 months per year.

We can expect even more beautiful times also because medical research is progressing so fast that even scientists cannot read all of the interesting scientific articles anymore. The fact that no human mind can follow everything could be a problem if there was no medical artificial intelligence in development able to read everything and analyze more and more.

However, we can also have a less positive point of view concerning longevity compared to other technological progress. Indeed, the good news of an overall growth of 2–3 months per year that has been going on in the richest countries for over a century concerns only the average life expectancy, not the maximum life expectancy.

For more than two millennia, some people have managed to live much longer than the average, up to the symbolic age of 100 years (en.wikipedia.org/wiki/ Category:Ancient Roman centenarians). For those who reach today an advanced age, despite extraordinary progress in hygiene, medical knowledge and material well-being, there is not much progress. Of the 7.5 billion human beings on the planet, less than a million have lived there for a century, barely 1000 are more than

D. Coeurnelle (✉)
AFT-Technoprog and Heales, Brussels, Belgium
e-mail: didier.coeurnelle@heales.org

© Springer Nature Switzerland AG 2019
N. Lee (ed.), *The Transhumanism Handbook*,
https://doi.org/10.1007/978-3-030-16920-6_26

110 years old and not a single one is more than 115 years old (en.wikipedia.org/wiki/Kane_Tanaka).

We have explored the earth and the seas from the abyss to the roof of the world, learned to fly, begun to explore the solar system, discovered the perhaps ultimate bricks of matter and life, defeated the majority of diseases that exterminated humans by the millions, put an end to death from malnutrition and childhood diseases for about 90% of the world's citizens. We have become capable, through the wonders of computing, of measuring the infinitely small, the infinitely large, the infinitely probable and the infinitely improbable.

But the last obstacle to destroy in order for us to have a healthy life is death from aging, this mystery of nature that means that, even placed in a perfect environment, virtually all animals and certainly all mammals end up dying simply because of the passage of time. And there, we have made little progress.

26.2   The Right to Life and Medical Progress Is a Human Right, Especially for Older Persons

Health, as mentioned by the World Health Organization in its 1946 preamble, is a state of complete physical, mental and social well-being (www.who.int/about/mission/en/), not merely the absence of disease or infirmity. Staying healthy is necessary to exercise our right to life, the most precious right, a condition for exercising all other rights.

Article 27.1 of the 1948 Universal Declaration of Human Rights states that *Everyone has the right (…) to share in scientific advancement and its benefits. Scientific benefits include those resulting from medical advances.* In these times of rapid and considerable scientific progress, this provision would benefit from greater emphasis by human rights defenders and scientists.

26.3   A Duty to Rescue

If scientists and scientific organizations know or should know that slowing aging is probably possible, not searching to make this possible could be considered as a breach of the "duty to rescue".

In some countries (en.wikipedia.org/wiki/Duty_to_rescue#Regulations_by_country), not respecting the duty to rescue is a criminal offence. For example in France, it is article 223–6 of the Criminal Code. In principle, the criminal law concerning the duty of assistance requires an element of "immediacy" which means that it is difficult to apply it in the case of medical research. But in the so-called Infected blood scandal (en.wikipedia.org/wiki/Infected_blood_scandal_(France)) during the 80s and 90s it was the non-use of new medical knowledge that was blamed (and there have been criminal convictions).

Knowing that every day about 110,000 people die from age-related diseases, and knowing that for a person aged 100 and over, life expectancy is barely 1 year, it would be logical to consider that these experiments are an absolute public health priority, both for the benefit of the elderly community and for the interest of elderly people wishing to experiment with new therapies.

Even if, increasingly, it is possible to simulate this research using computer devices; even if animal experiments can take place to check whether mice, drosophila and nematodes live longer, it is absolutely necessary in order to save lives to test on humans.

Yet this is not the case. Quite the contrary. In fact, it is generally impossible to give informed consent for research that is considered inherently high risk, even if you want to do so and know what is at stake and even if your chances of survival without experimental therapy are very low.

What are the conditions for effective experimentation?

26.4 What to Study Against Aging

Winning healthy life years beyond the first 80 years of life will almost certainly require complex research.

Most old people die from:

- Cardiovascular diseases
- Cancers
- Neurodegenerative diseases

Many ailments and health problems not specifically related to old age are generally benign for young people and often deadly for older people: Infectious diseases (flu, tuberculosis...), falls, food poisoning... Most people do not notice this "elephant in the room": Even diseases not related to aging are far more deadly or even almost only deadly for old people.

This article will not detail all possible therapies to comprehensively slow down or stop senescence. Here is a wide-ranging short list of what could be tested in the order of likelihood of an upcoming use:

- Products: rapamycin, metformin, statins, aspirin....
- Stem cells
- Senolytics
- Gene therapies
- Regrowth of organs
- Regeneration
- Microsurgery
- Parabiosis

- Nanotechnologies
- And what we don't know yet (new cures to come)

26.5 Guaranteed Efficacy Through Double Blind Studies

In order for a therapeutic study to indisputably demonstrate its efficacy, it must be conducted "double blind". Generally, this is done as follows:

- A group of experiment subjects is formed and then divided into two equal parts at random (by drawing lots)
- Each group is treated in a way that appears identical for the subjects of the experiment, those administering the treatment and those examining the results. However, only half of the people receive real treatment, the others receive a mock treatment.

To work in this specific way is necessary to reduce the risk of fraud but also to counter the effects of placebo and nocebo (en.wikipedia.org/wiki/Nocebo). For example, people receiving treatment may feel better (the placebo effect) while people receiving nothing may feel excluded and therefore ill (the nocebo effect). It is not enough that patients do not know what they are receiving. A nurse thinking of giving a better product will be more attentive, often unconsciously, to the hopeful patient than to the "control" patient. It is also necessary that the scientists examining the results do not know which group they are examining (because of this third group who also need to be unbiased, some scientists use the expression "triple blind").

Performing a double-blind experiment is often complex (and sometimes even impossible). For example, a placebo pill may not have the same consistency and taste as a pill with active ingredients. Ethical problems may also arise if the therapy envisaged is burdensome. For example, performing a simulated intervention similar to a heavy therapeutic operation is not ethically admissible today.

However, it should be noted that; in the experiment envisaged for older people, the placebo group is not left untreated for double-blind testing of people with diseases. Two cases are possible:

- one group receives traditional therapy and another group receives new therapy
- one group receives traditional therapy and another group receives traditional and new therapy

All of the above as necessary medical guarantees concerns therapies with relatively limited effects. If the results are very clear and rapid, tests can be useful even if they are not double-blind and are done on only a few people.

26.6   Guarantees From the Point of View of the Cost of Experimentation for Patients

It is desirable that people experimenting with anti-aging therapies (and actually any therapy) should be volunteers: paying people, beyond compensation for expenses, to carry out trials, could have the consequence that some would agree to take risks not by "informed consent" but rather to earn money. The safeguards put in place to avoid overly dangerous experiments would therefore be reduced.

Some want to charge volunteers to test innovative therapies. Accepting this means that these therapies are reserved for those who can afford them. Supporters of this payment insist, however, that the rich testing these products would be "guinea pigs" useful to the community and particularly well informed.

26.7   Legal Guarantees for Faster Research

The author of the newsletter Fight Aging wrote (www.fightaging.org/archives/2018/09/the-price-of-progress-or-the-waste-of-regulation/): *The average cost of delivering a new therapy from laboratory to clinic is increasing at a fast pace, more than doubling since the turn of the century according to some studies, to stand at $2.5 billion or more. This is not driven by the work of research and development becoming more expensive: if anything, the price of the tools of biotechnology is in free fall, even as capacity increases by orders of magnitude. (…) A mix of advances in computational power and materials science means that a graduate student of today requires 6 months of lab time and a few tens of thousands of dollars to accomplish what would have taken a full biotech company, 5 years, and tens of millions of dollars back in the 1990s.*

This strange situation is ironically called the Eroom's Law (en.wikipedia.org/wiki/Eroom%27s_law) (inverted Moore's Law). The main reason for these incredibly slow changes is that the Food and Drug Administration and the other similar health institutions worldwide especially in rich countries will get into more trouble when medical problems appear for patients with new therapies than with classical therapies. And the same problems will occur for pharmaceutical companies and even for the scientists doing research. It will be far less the case if problems arise when using older drugs or therapies and even less when patients die or suffer, because no therapy exists. Deciding to double, triple or quadruple check before beginning tests is far less dangerous in terms of public opinion, but costs many lives.

The fact that running experiments on new therapies is more complicated than ever could change dramatically. This could be because public opinion becomes aware of the progress of life extension research and/or if public institutions and private investors become more demanding in terms of success rates for new therapies.

26.8   Legal Guarantees for Longevity Experiments

In the European Union, the United States and socio-economically similar countries, legislation is increasingly demanding in terms of protecting health, which is positive. Legal rules are also increasingly complex, abundant and technically demanding, particularly for experiments. This makes it (almost) impossible for an informed volunteer to test certain therapies even if:

- he wishes to do so out of personal interest;
- he wishes to do so out of a desire to advance "science" for a healthy life;
- he is very old and knows that in the absence of new effective therapy, he probably only has a few years left to live;
- he could perfectly and legally take infinitely greater risks for his health outside the laboratory; nothing forbids him to (re)start smoking, drink five cognacs a day and swallow anything that is sold without a prescription in a pharmacy without worrying about the harmful effects and without anyone to warn him of the risks.

Enabling informed volunteers to participate more easily in research is desirable for the public good and for the good of volunteers.

26.9   Duty to Provide Collective Information

The goal of research is a better life for all. It goes without saying that the effect is not achieved if a positive discovery remains secret. In this context, the question of patents and property rights of living organisms is raised.

A patent gives exclusive rights of use for 20 years, which can be extended by 5 years in certain circumstances. As far as gene therapies are concerned, in France and in the United States, the concept of "non-patenting the living" (en.wikipedia. org/wiki/Biological_patent), at least for human genes, largely prevails. Genes for longevity could not be patented, unlike drugs. However, the specific way genes are modified can be patented.

Proponents of patent rights argue that patents accelerate medical discoveries by making them profitable. In case of "revolutionary" discoveries the pressure to make them available to all would be enormous. However, this partial patentability could 1 day slow down the availability of therapies to everyone even if longevity experiments demonstrate the effectiveness of new therapies. The most desirable, ethically and socially, would be a legal environment defining this type of medical advance as necessary for "universal health coverage" (www.who.int/features/qa/universal_health_coverage/en/) and therefore accessible to all without financial obstacles.

26.10   Duty to Inform Even in Case of Failure

Informed volunteers experiment to advance common knowledge. Knowing that a hypothesis that seemed promising is not working is almost as important as knowing what works. It is therefore necessary that experiments that do not lead to improvement are made public. It is even extremely important that communication on this subject should be more strictly imposed because, for psychological, social, economic reasons …, it is tempting to hide failures.

26.11   At What Age Should New Therapies Be Tested?

Most medical experiments today are carried out with relatively young people. The reason for this choice is ethical, practical and economical:

- it is a question of taking "solid" individuals who are less likely to die in the event of an accident (ethical reason);
- It should be avoided that the test subjects have health problems other than those for which the test is done and older people have more health problems. Otherwise it would be difficult to distinguish possible negative effects of a therapy from other health problems (practical and economic reason).

However, as explained at the beginning of this article, more and more, we will only die from diseases related to aging. We are generally living longer and longer in good health until we are at least 65 years old, and often up to 75 or even 80 years old. Increasingly, deaths before age 75 are referred to as "premature" deaths. On the other hand, the upper limit of the life expectancy progresses much less rapidly and even stagnates.

In other words, today:

- most of the fatal consequences of aging occur in 15–20 years, from a little under 80 to about 95 years for men and from about 80 years to just under 100 years for women;
- most therapeutic trials are carried out in young subjects, i.e. those who do not need therapy.

Testing new therapies on people under 75 or 80 years of age would therefore not allow significant findings in terms of mortality to be made before a decade, especially if the group of tested people tested is small. Of course, there are biological markers (en.wikipedia.org/wiki/Biomarkers_of_aging) that can give indications on the rate of aging, but none of these markers is currently considered to clearly measure it.

It would be even more desirable to organize tests for very old women and men (men over 90, women over 95), people who are healthy for their age and have given informed consent. Note that it would not be a question of testing a therapy for one group and leaving the other group with nothing but of having two groups (or even several groups) all benefiting from high care of which some would receive in addition a new therapy (and others a placebo).

In this case, simply because of the experiment and its purpose (medical research to live longer in good health), the average lifespan would almost certainly be extended even for the group not receiving new therapy (because they would all benefit from appropriate care, in a favorable environment, etc.).

Logically, the older and sicker the volunteers, the more risky medical experiments could be. Indeed, without treatment, these people would unfortunately die quickly.

The wish not to cause suffering to vulnerable people will often be invoked in order not to experiment on old people. But the strongest reason may actually be a lesser mobilization for the elderly. Saving the life of an "old man" is less attractive than saving the life of a younger person. It is also perceived as less "normal". Here we see that transhumanists (and other longevists) are in a sense more humanist than "classical" humanists. For classical humanists, human beings should have equal rights to life and health as long as allowed by the "natural" conditions. This means until the moment they are too old. Transhumanists are *hyperhumanists*: human beings should have right to life and health without time limit if this becomes scientifically possible even if it is "unnatural".

26.12   What Is Really Being Done Today

Today, according to information available in the scientific literature, only three experiments in the world are known. None are directly organized by a public institution.

The TAME project (www.afar.org/natgeo/) aims to test the positive effect of metformin (a drug used in particular for the treatment of diabetes) on the longevity of thousands of healthy American elderly people.

The Ambrosia company (www.ambrosiaplasma.com) wants to experiment, in American elderly people, the injection of blood from young donors, to verify its rejuvenating effects.

Liz Parrish of Bioviva (bioviva-science.com) defines herself as "patient 0" (the first patient) for rejuvenation therapies. She used gene therapy for telomere cell lengthening and a myostatin inhibitor to build muscle capacity.

26.13   What Can Be Done Tomorrow

Every month, more than 3 million people die from age-related diseases. Every day, billions of euros are spent on healthcare for the world's elderly. This is very good for those who benefit from it. Unfortunately many elderly people, especially in the poorest countries, receive no care at all. During the same period, human experiments concerning the improvement of longevity concerned only a few people and the investments are derisory.

Citizens would certainly volunteer to participate in research and experiments. The right of citizens to seek how to earn healthy years of life for others and for themselves could be developed more rapidly if mobilization on this issue increased.

Awareness of this priority seems to be accelerating. The Chan Zuckerberg Initiative (chanzuckerberg.com), created by Facebook's founder, has as one of its stated goals to end all living illnesses of the Zuckerberg couple's children (though without explicitly mentioning aging as a cause). Google Calico (calicolabs.com) has an explicit goal to tackle aging and has hired dozens of renowned scientists and computer scientists. Bill Gates said he would commit $100 million to fight Alzheimer's disease (www.reuters.com/article/us-health-dementia-gates/bill-gates-makes-100-million-personal-investment-to-fight-alzheimers-idUSK-BN1DD0S3).

But this awareness is mostly in the private sector in the United States.

That investment is done first in the United States is paradoxical. The USA is a very rich country, but life expectancy is low. It is also where progress in longevity is one of the least important, with even setbacks in recent years.

That investments in this area are essentially private and that no major public body in the world has the explicit aim of putting an end to aging is another paradox. Indeed:

- long-term scientific efforts are often more successful when they are publicly funded;
- public authorities have every interest in citizens living (much) longer in good health;
- citizens will most likely be more willing to participate in medical experiments if it is for public health purposes, rather than for private, pharmaceutical or other companies.

26.14   A World Project/A Manhattan Project/A Moonshot Project

A paradigm shift on the part of public authorities is to be hoped for and encouraged. When policy makers see ending all age-related diseases as a possible goal with real advancement while they are politically active, they could become even more enthusiastic in support of life extension than they formerly were in support of "deathism"... Nowadays efforts for longevity sometimes seem often disordered, vain, ridiculous and anti-natural, just as the attempts to fly objects heavier than air (en.wikipedia.org/wiki/William_Thomson,_1st_Baron_Kelvin#Pronouncements_later_proven_to_be_false) at the end of the nineteenth century seemed disordered, vain, ridiculous and anti-natural.

In a few decades, not growing old could become as banal as it is to travel into an almost unbreathable space in a gigantic object of metal and plastic filled with flammable materials projected into the sky at a thousand kilometers an hour by chemical reactions that we do not understand and maintained in the air by means that were defined as impossible by some of the world's greatest scientists less than 130 years ago.

As for aviation, the road to success will pass through difficult tests. But today we have far more means than at the turn of the twentieth century. Those who will experiment with the techniques will be better informed and the potential stake does not concern the air tourists of the century to come but billions of women and men getting older today.

For this, the best way would be a global project organized by the World Health Organization. If this global Moonshot project is not possible, the next sufficient level is the European Union. This part of the world combines:

- Strong Health Public Institutions
- High Life Expectancy
- Prestigious scientific Institutions

If European institutions and European citizens could unite in such a collective goal, it would be useful for Europe's citizens, for the prestige of its institutions, for the prestige of the European states, and of course potentially useful for billions of world citizens.

Chapter 27
Taking Up the Cosmic Office: Transhumanism and the Necessity of Longevity

Kali Carrigan

> Phlebas the Phoenician, a fortnight dead,
> Forgot the cry of gulls, and the deep sea swell
> And the profit and loss.
> A current under sea
> Picked his bones in whispers. As he rose and fell
> He passed the stages of his age and youth
> Entering the whirlpool.
> Gentile or Jew
> O you who turn the wheel and look to windward,
> Consider Phlebas, who was once handsome and tall as you.
> T.S. Eliot – The Wasteland

27.1 Life's Disposable Nature

From the point of view of evolution, one of the most fundamental properties of life is its *disposability*. Barring a small minority of "immortal" or negligibly senescent organisms, such as some species of sturgeon, tortoises, bacteria, and jellyfish, all living beings are subject to a planned obsolescence. To be born is to die, to be replaced, such is the natural order of things, "the cradle rocks above an abyss, and common sense tells us that our existence is but a brief crack of light between two eternities of darkness."[1] Our presence on this planet has been the result of a 4 billion year experiment, carried out by a blind scientist, whose only visible goal is to arrive at the best solution to the ever-changing problem of survival. The whole earth is a laboratory for the trials of an indifferent and uncanny animator which seeks only the selfish propagation of its experiments without clear purpose or end. For evolution, just as for the scientist in the laboratory, the disposability of earlier iterations is necessary if better hypotheses are to be tested and improved. Each of us is born as

---

[1] Vladimir Nabokov "Speak, Memory: A Memoir" (1951).

K. Carrigan (✉)
Universiteit van Amsterdam, Amsterdam, The Netherlands

© Springer Nature Switzerland AG 2019
N. Lee (ed.), *The Transhumanism Handbook*,
https://doi.org/10.1007/978-3-030-16920-6_27

an expendable and incomplete form, a step on a ladder that climbs infinitely upwards and out of sight, and on the grand scheme of things, what constitutes our individual selves including the mind and what some have called the soul, is simply the waste product of reproduction, bound irrevocably for the cosmic garbage can.

The original idea of the disposability of living forms has long been attributed to the ancient Greeks, notably Aristotle, who tackled the question of the necessity of aging and dying and the biological basis of decay. For Aristotle, aging and death were not automatic processes, but occurred due to the accumulation of damage in the body (soma) leading to a gradual loss of function. More strikingly still, was his belief of a link between sex and death; each sex act, he proposed, had a direct life-shortening effect. Thankfully for us, Aristotle was not entirely right, but the question of whether we might be merely vehicles for our offspring, a sort of expendable chassis for the engines of evolution has preoccupied thinkers ever since. It was not until the late nineteenth century, however, that the first scientific theories predicating man's disposable nature made their way into the public sphere. One of the better known is germ plasm theory.

Proposed by August Weismann, a contemporary of Charles Darwin, germ plasm theory states that in complex organisms, there exists a fundamental division between two types of cells: germ cells (such as the eggs found in the ovaries and the sperm in the testes), which contain the heritable information, and somatic cells, literally meaning the cells *of the body*. What Weismann realized was that in multicellular organisms, like you and me, there is a profound division of labour between those kinds of cells which are responsible for reproduction, and those which are responsible for survival, and what's more, that this has something to do with why we age and die.

We now know that in the early stages of life on earth, the distinction between somatic and germ cells was not present, and reproduction occurred when a single mother cell made an exact copy of its DNA, and divided to create two identical daughter cells. In essence, this type of asexual reproduction allowed for a complete rejuvenation of aged cells; each of the two new daughter cells would be restored to a youthful state, meaning that these types of cells were essentially immortal. For hundreds of millions of years, death as we know it did not exist on planet earth.[2] Contrary to the common belief that whatever is born is destined to die, life, in its earliest forms, was engineered to be immortal. Approximately 3 billion years ago, however, all this began to change.[3] At this time, early unicellular predators had begun to make an appearance, and cell specialization became a matter of life and death in this increasingly crowded and competitive world. Functions such as motil-

---

[2] This is not to say that unicellular organisms did not "die". Indeed, starvation, dessication, heat, and other environmental stressors could still lead to death. However, death was not programmed into the cell, nor was deterioration and aging, meaning that in favourable environmental conditions, survival was guaranteed indefinitely.

[3] See Schopf, J.W. (1993). *Microfossils of the early Archean Apex chert: new evidence of the antiquity of life*. Science, 260: 640–46 and Knoll, A.H. (2003). *Life on a young planet*. Princeton: Princeton University Press.

ity and phototaxis became essential for survival; eat or be eaten became the mantra of the multicellular world. The specialization of single cells to create complex organisms took off, giving rise to the astonishing diversity of life we see today. This need for specialization meant that the large majority of cells took on specific functions in order to guarantee the survival of the entire organism. Indeed, the separation of germ line and somatic cells offered wonderful opportunities to primitive life on earth, and indispensable evolutionary advantages, as it allowed for cells to do away with the burden of reproduction and focus on their specialized function. However, specialization came at an unspeakable price. That price was death.

In the 1940s and 1950s, a new wave of interest surged in the biological sciences regarding the evolutionary underpinnings of senescence. Before this time, much of what we knew about aging was based on rough medical observations of the senile phenotype in old age. The idea that there could be an evolutionary explanation for aging and indeed death, was completely new. Yet by the second half of the twentieth century, many prominent scientists such as J.B.S Haldane, Peter Medawar, and Alex Comfort had begun to notice that in the large majority of species selection pressure declined with age, leading to cellular dysfunction and the decline of fitness we associate with aging. Not only had the specialization of primitive cells led to the appearance of death, but it was becoming clear that further along the line, disposability was being *selected for* by evolution. This theory caught the eye of George C. Williams, an American biologist at Michigan State University, who set to work on explaining the observed link between survival rates and reproductive ability. "Senescence is an evolved characteristic of the soma",[4] he noted, only the genes controlling the functions of the body (soma) seem to decrease in function with time, whereas those controlling reproductive ability seem to remain relatively stable.

What Williams found was indeed fascinating. He knew from the work of Peter Medawar on pleiotropy – the ability of genes to control more than one function in the body – that some genes which maximized vigour during youth, such as the one coding for testosterone in males, were strongly selected for by evolution, even though the same gene had been found to be related to pathologies later on in life, such as a higher risk of prostate cancer. This gene double-edge was coined *antagonistic pleiotropy*, and brought to light the fact that the selective value of a gene strongly depended on its effects on the total *reproductive probability* of the organism. This new understanding of a priority list of evolution, with reproduction written in bold at the very top, began to gain momentum. Some decades later, Richard Dawkins' notion of the "selfish gene", which built upon Williams' work on adaptation and natural selection would find its central argument in a similar idea. In his bestselling book of the same name, Dawkins describes that natural selection selects not for traits or behaviours which will benefit the organism as a whole, but those which confer advantages to the "immortal" genes which it carries. Although he used this theory to explain the phenomenon of altruism and kin selection, Dawkins' idea also had profound implications for the science of aging, and why evolution seems to

---

[4] Williams, G.C., (1957). *Pleiotropy, Natural Selection, and the Evolution of Senescence*. Evolution, 11 (4): 398–411. See p. 403.

more or less forget us once we are past our reproductive prime. Today, this concept of the disposability of the soma either by the "programmed death" of cells, or by a wearing down of the repair mechanisms, is at the core of many prevalent theories of aging, such as the programmed theory of aging proposed by Vladimir Skulachev,[5] Aubrey de Grey's Strategies for Negligible Senescence (SENS),[6] or the Hallmarks of Aging theory.[7] Aging is no longer a mysterious global phenomenon of decay, but a series of repair failures and programmed senescence. Our inevitable decay, for evolution, is a step towards a much greater immortality process.

In 2001, the title of the annual Reith lecture commissioned by the BBC and boasting a long line of significant public figures such as Bertrand Russell, Anthony Giddens, and Edward Said, was "The End of Age". The invited speaker was Tom Kirkwood, who at the time headed the Institute for Aging and Health at Newcastle University. Kirkwood became well-known in the 1970s for proposing the *disposable soma theory*. His theory rested on the premise that energy is a limited resource, and thus the body must continuously make the difficult choice between repair and maintenance of cellular processes associated with survival, and those associated with reproduction. A sustained investment in reproduction during adult life might thus result in reduced DNA repair of somatic cells, shortened telomeres, mutation accumulation, and many other damages usually associated with aging. Unsurprisingly, with some exceptions, such as the naked mole rat and other long-lived animals, this is indeed what we see. In fact, in humans, men who have been castrated seem to experience an increase in longevity compared to those that have not,[8] and furthermore, women who have given birth to more children have been shown to suffer from reduced lifespans.[9] Kirkwood's theory, even though still some-what controversial, underlines the same broad conclusion that Medawar, Williams, and many others had arrived at in the 1940s and 50s, that is that no matter where we look to in nature, evolution's priority is not our survival, but its propagation. What we think we are, and everything we hold dear, is in the eyes of natural selection nothing more than a "surplus to requirement" ([1], Reith lectures).

Finally, new evidence for the disposable soma theory has recently surged from studies of embryonic development, many of which have caught the eye of contemporary scientists, thinkers, and sceptics alike. It has been found that in the early embryo, the formation of the germ line, as well as the differentiation between germ and somatic cells occurs *before* the development of almost any other vital function.

[5] Skulachev, V.P. (1997). *Organism's Aging is a Special Biological Function Rather than a Result of Breakdown of a Complex Biological System: Biochemical Support of Weismann's Hypothesis.* Biokhimiya 62 (12): 1191–1195.

[6] de Grey, A., Rae, M. (2007). *Ending Aging.* St Martin's Press, U.K.

[7] López-Otín C., Blasco M.A., Partridge L., Serrano M., Kroemer G. (2013). *The Hallmarks of Aging.* Cell 153(6): 1194–1217.

[8] Hamilton J.B., Mestler G.E. (1969). "Mortality and survival: comparison of eunuchs with intact men and women in a mentally retarded population)". Journal of Gerontology 24 (4): 395–411.

[9] Westendorp R.G., Kirkwood T.B. (1998). *Human longevity at the cost of reproductive success.* Nature. 396 (6713): 743–746.

In drosophila, this occurs as early as nuclear division cycle 10–11.[10] In humans, at about 14 days after fertilization, already differentiated primordial germs cells stored in the yolk sac begin to migrate through the embryo to reach the gonadal ridge, where they will be stored until a time when the gonads begin to form. The recently fertilized embryo therefore already contains all of the cells necessary to give rise to the next generation. Compare this to neurogenesis, the development of the neurological system, which occurs around days 18–20, or the formation of the cardiac tissues which begins as early as day 21.

Thus, it seems that science is beginning to catch up to Aristotle, and an irrevocable amount of evidence has piled up justifying the belief that the human species, as well as most other complex life on earth, is in its essence divided between the drive to immortality by means of reproduction, and the drive to death by decay of the soma. Over the last few decades, a consensus has been reached in the life sciences showing that the division in multicellular organisms between immortal germline cells and mortal somatic cells "conferred such a great evolutionary advantage to higher life forms, that we may almost forgive the terrible price that we have paid, for it was this, not sex, that caused us to age and die".[11] We are becoming increasingly cognizant of the unsettling fact that everything which we understand as making up our "selves" and our conscious experience of the world, is driven by an unyielding compulsion towards death, a planned obsolescence which conspires to our extermination. Nature has put before us an impossible paradox: as long as we carry out our reproductive duty, our genes will be guaranteed to live, as they had at the beginning of life, for many generations to come. We, however, are destined to die. We are the soma, the shell which protects the timeless egg, and thus eventually, bound to crack.

27.2 The Problem of Mortality

Coming to terms with the inherent disposability of life presents us with a great deal of cognitive anxiety, however, for as conscious beings we do not see ourselves as soma, a disposable chassis for the immortality project of evolution; rather, our world is one of meaning and emotion in which morality, imagination, individuality, love, happiness, suffering, encompass a reality that is much greater than the sum of its biological parts. Death, for humans, is not an evolutionary necessity, but a great tragedy, the loss of a rich and irreplaceable world marking an end without the possibility of return. As far as we are concerned, death is suffering, grief, pain, being no more necessary a feature of life than the putative existence of God or of metaphysical entities. Life, for our species, is an elaborate process of survival in the face of a death that remains fundamentally alien to it.

---

[10] Deshpande et al. (2004). *Overlapping mechanisms function to establish transcriptional quiescence in the embryonic Drosophila germline.* Development 131: 1247–57.

[11] Tom Kirkwood (2001), The BBC Reith Lectures.

The problem of mortality has indeed been such an obstacle for consciousness, that many psychologists including Sigmund Freud and Ernest Becker, and sociologists such as Zygmunt Bauman and Edgar Morin have described that most of what we know as unique human activities, such as culture, religion, and the formation of societies, are at their origin death-denying projects which help us rationalize the utterly incomprehensible thought of death. As long as death has ruled as an iron parenthesis over the human condition, it has been essential that we make up stories, for our survival would be impossible without it. The human mind, confronted with its own inevitable limits becomes paralyzed and unable to act towards the future. Defense mechanisms are thus necessary to keep the thought of mortality on the straight and narrow, and one of the many mechanisms which we commonly use to achieve this is what Peter Wessel Zapffe has referred to as "anchoring". Anchoring involves building collective firmaments which shore up our search for meaning on the activities of the living, such as religious faith, customs, morality, culture etc. This strategy has been the most prolifically used throughout human history, and includes the fantastic tales of the afterlife found in all known religions. Whether we believed we would find ourselves in the golden fields of Aaru after death, as assumed by the Ancient Egyptians, Elysium or Tartarus for the Greeks, Valhalla, Samsara, Jannah, She'ol, or the Christian heaven or hell, for much of human history the "unthinkable" thought of death has been sublimated into stories of an afterlife. These fantastical stories have allowed us to keep death at bay, for through them we have led ourselves to believe that no such horrible fate as eternal nothingness awaits us at the end of the line. It is thus that for the last few millennia, the paradoxical nature of death has been neatly squared away through the promise of eternal life.

In today's world of dying gods and scientific scepticism, however, the mythical stories of times past have largely lost their appeal, and the fear of death has swiftly returned to the hearts of men. Without soothing stories to assuage us, death has once more taken its place as a great mystery which defeats all attempts at reason. As Zygmunt Bauman[12] has noted, a new type of anchoring strategy has surged in contemporary times. Death is no longer a gate to the afterlife, but a series of battles against diseases which may be either won or lost. Death today has been deconstructed into a minutiae of "soluble" problems of health and other threats to life, but the belief remains that as long as we do the right thing during our lives, we may be guaranteed a virtuous end. The absurdity of death today is thus rationalized in an altogether different way; death has become biological death, ruled by the indiscriminate forces of cellular damage, organ failure, and the limits of medical science. Arguments for the necessity of aging and death presently revolve around the "naturality" of these processes, or the evolutionary need to make way for the next generation. But acceptance of our paradoxical fate continues unabashed. Our newfound rational apprehension of the limits of life have found an anchor in the scientific theories of disposability and have come to stand not only as descriptions of the inner workings of the natural world, but as existential justifications of the meaning and

[12] Bauman, Z., (1992). *Mortality, Immortality, and Other Life Strategies*. Stanford University Press, California.

moral significance of the human condition itself. We all seek immortality in one way or another, but the question of whether reproduction is the answer for humans, or whether other alternatives must be sought, is an increasingly pressing one.

27.3 Transhumanism and the Denial of Death

Whilst by and large we have learned to see ourselves as the disposable by-products of evolution, and learned to accept our fate, the advances of science have in recent years also allowed us to re-imagine the problem of mortality no longer as a means to salvation and the afterlife, but as a scientific question to be solved by the steady hand of the researcher. The galloping advances in genetics and medical technology in the last century have endowed us with the possibility to alter and re-design our biology in altogether unprecedented ways, and many of these advances, which a hundred years ago were only fantasies in the minds of scientists, are today becoming astonishingly real. In vitro fertilization, predicted by J.B.S Haldane at the beginning of the First World War has given rise to more than 8 million babies in the last 40 years, whilst genetic engineering used to mass-produce insulin today prevents the premature deaths of millions of people worldwide. Indeed, our scientific and technological progress has had a dual effect on our apprehension of the limits of human biology. On the one hand, it has allowed us to conceive ourselves as disposable vehicles, a series of complex systems and replaceable parts, eventually bound to rust and fail; on the other, our increasing knowledge of human biology, medicine, and genetics has provided us with a blueprint for the codes of nature, granting us the remarkable ability to envision ourselves as architects of our own design. This contradictory state of affairs has been the central point of contention in the debate over whether – to misquote Heraclitus – the human being is bound towards change and must inevitably surpass itself, or whether there might be a limit to the number of changes allowed before humanity, by transcending itself, risks its own extinction.

As long as scientific explanations for the nature of things have appeared on the scene, however, this debate over whether our increasing knowledge should be used to transcend humanity's limitations has been ongoing. Dating back to the Florentine Renaissance, our growing understanding of science, and the move away from religion towards a world dictated by humans, brought us a new apprehension of our place in nature. It is largely understood that this critical "swerve"[13] found its mainstay in the newfound doctrine of humanism in the fifteenth century, which at the time encouraged people to rely on their own observations and judgement rather than looking to god or religion for explanations. Humanism brought about a crucial revolution in thought which would drastically change human history as well as the contemporary world, for it denied the passivity of faith and the belief in a conscious, all-knowing god, and replaced it with an amoral, indifferent view of nature, which underscored the need for human action. Following the acceleration of technological

---

[13] See Stephen Greenblatt's "The Swerve: How the World Became Modern" (2011).

prowess during the industrial revolution, this doctrine became the paradigm of the new technological age. The idea that with the aid of technology man would be able to transform himself as he had transformed the physical world took hold of the modern world, and in 1940, philosopher W.D. Lighthall coined the terms "transhumanism" to describe this progressive belief.

By the mid-twentieth century, transhumanism had positioned itself as the next phase of humanism. Not only had man become "the measure of all things", but our capacity for intelligence and rational thought was increasingly believed to be leading to drastic changes for the future of humanity. Technology would one day allow us to transcend our unintelligent design and overcome the limitations of the human mode of being which had for far too long made life on earth "nasty, brutish and short" [2]. The future of the species had come to weigh heavily on the shoulders of men. If nature was indeed amoral, and cared only for its own selfish propagation, then it was man's moral responsibility to overcome it, and through his ability for rational thought put an end to the aimless suffering which it caused.

For the men and women of the industrial age, this belief offered a new answer to the problem of mortality, and the paradoxes of a world in which morality and indifference now co-existed in untenable tension. In 1957, Julian Huxley, who popularized the term "transhumanism" in an essay of the same name, described this newfound responsibility as "taking back the cosmic office", an appointment which would allow humanity, at long last, to take a brave new step in its evolution, and free itself from the shackles of its primitive mortal cage. An important condition of this responsibility, for Huxley, was the directed planning of a new kind of humanity. Transhumanism was as much a pursuit of individual knowledge, as a concerted global effort to improve the condition of all life on earth. "There are two complementary parts of our cosmic duty" – Huxley wrote – "one to ourselves, to be fulfilled in the realization and enjoyment of our capacities, the other to others, to be fulfilled in service to the community and in promoting the welfare of the generations to come and the advancement of our species as a whole." ([3]:17). Fifty years earlier, French journalist and futurist Jean Finot had expressed a similar idea that the use of technology had, as its ultimate purpose, a social duty to the enhancement and extension of all human life on earth (cf. Ilia Stambler 2010). The transformation of the human condition from an animal-like "man-monkey" to what Finot called a "homunculus", an immortal next stage fashioned by the hand of man, was seen as the next natural step of progressivism in the biological sciences at the turn of the century. It is important to remember, thus, that the program of transhumanism has since its conception been rooted, not on a personal quest for immortality, but on a social and moral responsibility in the face of death to help humanity transcend "not just sporadically, an individual here in one way, an individual there in another way, but *in its entirety,* as humanity" ([3]:17).

In today's world of individualism in which a great variety of intellectual stances on transhumanism have taken hold, including, but not limited to Humanity+, Extropianism, Singularitarianism, Technogaianism, Immortalism, etc., all with different intuitions and calls to action, one of the main contentions with the transhumanist project is the belief that it is, at its core, a narcissistic desire for enhancement which is likely to only benefit the wealthy and privileged. But nothing could be further from the truth. Recently, Nick Bostrom, one of the frontrunners of transhumanism today, wrote in his "Transhumanist Values": "Transhumanism advocates the well-being of all sentience, whether in artificial intellects, humans, and non-human animals (including extraterrestial species, if there are any). Racism, sexism, speciesism, belligerent nationalism and religious intolerance are unacceptable" ([4]:9). Put simply, transhumanism is, and will continue to be, a global species project aimed at reducing the greatest amount of suffering for the greatest number of people.

In the first half of the twenty-first century we have thus come to a pivotal fork in the road. On the one hand, our modern evolutionary theories have concluded that nature's complexity arises from the trials and experiments of a natural force which knows no end, purpose, or morality, being no more than an astonishingly elaborate series of chemical reactions. From evolution's point of view, the priority of our short existence is not our survival, but its propagation. The reason for our death is not a reasonable one, for nature knows not what it means to suffer or die. On the other hand, our social and moral worlds tell us that we are not expendable, death is an unassailable tragedy, and human life is without exception worth fighting for. As conscious beings, our fear of non-existence is so profound, that we have built around us an elaborate set of anchors onto which the absurdity of death may find solid ground. Our fate of perishing and dying by the hand of a cruel and indifferent natural force has become morally unjustifiable. Furthermore, never before in human history have we been so well placed to make the ambitions of transhumanism a reality. We stand, for the first time, on the verge of a new era in our evolution, in which technological advances have the possibility to bring genuine improvements to the human condition, significantly extend life, and allow man to take his place the master architect of his own future.

It seems to me thus that if we embrace the initial premise that our individual existence consists only of the time we are alive without any form of afterlife whatsoever, and that our human experience of the world is confined to these limits, then our first concern is to prolong that speck of existence as much as possible, as well as possible. Not only is this an opinion, but an existential duty, since all that makes our moral and social world possible is contained within the existence of each and every one of us. Longevity and transhumanism are today no longer only scientific or ideological pursuits but existential priorities which concern humanity as a whole. The choice is for the first time in history ours alone, and whether we accept ourselves as the disposable soma of evolution, or take arms against the blunders of a cruel and senseless natural force which everyday conspires to our extermination, will determine the next great step in the history of humanity. We have been given the key to the cage, and whether we continue as test subjects in the laboratory of nature, or run free, is now only a matter of turning the lock.

References

1. Kirkwood, T. (2001). *The End of Age.* BBC Reith lecture. http://www.bbc.co.uk/radio4/reith2001/
2. Hobbes, T. (1651). *Leviathan.* Penguin Books, 1981.
3. Huxley, J. (1957). *Transhumanism.* In *New Bottles for New Wine* (1957), Chatto & Windus, London.
4. Bostrom, N. (2003). *Transhumanist Values.* Ethical Issues for the 21st Century, ed. Frederick Adams (Philosophical Documentation Center Press, 2003).

Part IV
Biohacking and Mental Health

Chapter 28
Cyborgs and Cybernetic Art

Newton Lee

> "*Robots are not only a mistaken metaphor for the dehumanization of the body but also for the future sentience of machines.*"
> —*Stelarc*

28.1 Cyborgs in the Arts, Sciences, and Medicine

The term "Cyborgs" (short for cybernetic organism) was coined in 1960 by Manfred E. Clynes and Nathan S. Kline in their article "Cyborgs and Space" about creating self-regulating man-machine systems to meet the requirements of extraterrestrial environments [1].

28.1.1 In the Arts

Some artists have tried to create public awareness of cybernetic organisms, ranging from paintings to installations.

In November 2010, New York University arts professor Wafaa Bilal had a digital camera surgically implanted into the back of his head. As part of "The 3rd I" project commissioned by a museum in Qatar, the camera captured his everyday activities at 1-min intervals 24 h a day and streamed live on the Internet and at the Mathaf: Arab Museum of Modern Art [2].

N. Lee (✉)
California Transhumanist Party, Los Angeles, CA, USA

Institute for Education, Research, and Scholarships, Los Angeles, CA, USA
e-mail: newton@californiatranshumanistparty.org; newton@ifers.org

© Springer Nature Switzerland AG 2019
N. Lee (ed.), *The Transhumanism Handbook*,
https://doi.org/10.1007/978-3-030-16920-6_28

For more than 40 years, Stelarc is a performance artist whose works have focused heavily on extending the capabilities of the human body using medical instruments, prosthetics, robotics, virtual reality systems, and other technologies.

28.1.2 In Sciences

In 2006, Cornell University researchers have successfully created the first insect cyborgs – moths with integrated electronics in their thorax [3]. Defense Advanced Research Projects Agency (DARPA) has also supported research on implanting tiny Micro-Electro-Mechanical Systems (MEMS) devices into insect bodies while the insects are in their pupal stage [4].

In 2013, a U.S. company backed by the National Institute of Mental Health sells a $100 education kit that lets anyone create a RoboRoach from a live cockroach and wirelessly control the left/right movement of the cockroach by microstimulation of its antenna nerves [5].

Regina Dugan, former senior executive at Google and former director of DARPA, revealed in the 2013 D11 conference some of the most advanced wearable computing technologies such as an electronic tattoo on her left arm that can be used to authenticate a user in lieu of password [6].

28.1.3 In Medicine

In medicine, cyborgs can be restorative or enhanced. The former restores lost function, organs, or limbs whereas the latter exceeds normal human capabilities as dramatized in the popular TV shows *The Six Million Dollar Man* and *The Bionic Woman*. A brain-computer interface, or BCI, provides a direct path of communication from the brain to an external device, effectively creating a cyborg.

In 2014, Georgia Tech professor Gil Weinberg created a prosthesis for amputee drummer Jason Barnes. By flexing his muscles, Barnes can send signals to a computer to tighten or loosen his grip on drumstick and control the rebound [7].

28.2 Cybernetic Art: An Interview with Cyborg Artist Stelarc

The following is an interview with Stelarc by Melbourne University Prof. Darren Tofts [8].

The Cyborg Cometh: It begins with a sinister laugh. Beatific, knowing and bordering on rapture, it revels in the possibilities of imminence, of something significant and profound about to be revealed. While portentous of the unknown, it is at the same time a very familiar, very human signature. But of what? For theorists such as Donna Haraway, Katherine Hayles, and Gilles Deleuze, it speaks of a new conception of the human, evolving with informatic technology into a hybrid biomachine. For the host of this sublime laughter, Stelarc, it speaks on behalf of his performative alter-ego, the body.

For more than four decades, Stelarc has explored the increasingly malleable relations between the body and technology. A pioneering exponent of cybernetic art, his work has questioned and broken down the technical and philosophical boundaries between human life as we know it and what it might become. This constant becoming-cyborg in Stelarc's work has transcended the built environment as well as the distributed spaces of the Internet and other telematic interfaces. Eager to identify unexplored somatic possibilities within and between the two, Stelarc is on the lookout for new performance stages, virtual, conceptual, and biological, where conditions of embodiment can be enacted and explored.

Tofts Your collaborative 2005 work with Nina Sellars, *Blender*, was in some ways a departure for you, a shift from the immateriality of the virtual to the materiality of the physical, dealing with the visceral nature of bodily matter. Can you tell us how you see this work fitting in with your broader notion of the body?

Stelarc I guess the way that I've worked as an artist is not based on any particular medium and making any distinction between the virtual and the actual. in other words, all of these different performative modes and all of these projects involving alternate media have been expressing a certain conceptual concern of exploring the prosthetic augmentation of the body on one arm; of experimenting with alternate anatomical architecture on the other; and seeing the body not as necessarily a biological body, but seeing it in the broader sense. There's also been a strategy that may be exemplified in Jean-Luc Nancy's postmodern approach in that his deconstruction's not through appropriation and juxtaposition, but rather deconstruction through exhausting a particular concept or idea. So the series of suspension performances was a means of exhausting the body into exposing its obsoleteness, and I like the idea that there is no strategy of deconstruction that involves appropriation, but rather the deconstruction occurs through the exhausting of an artistic act to expose its inadequacy and possibly reveal some alternate possibilities.

Blender (see Fig. 28.1) in a way is a counterpoint to the Stomach Sculpture (see Fig. 28.2) performance and project where initially I designed a sculpture for the inside of my body, and in that case this was a sculpture that was inserted 40 cm into the stomach cavity. The object opened and closed, extended and retracted, had a flashing light and a beeping sound, so it was this kind of mechanic choreography within this soft and wet, vulnerable environment of the inside of the stomach. So blender could be seen as a counterpoint to that instead of a machine inhabiting the soft interior of the body. In *Blender*, a machine becomes a host for a liquid body, a liquid body composed of biomaterial from two artists.

Fig. 28.1 Blender (Teknokunst, Melbourne). (Photo by Stelarc. Courtesy of Stelarc and Nina Sellars)

Tofts Further to this, your initial collaborations with Oron Catts and Ionat Zurr on the *Extra Ear* project, as well as your ongoing development of this work is also very corporeal. For an artist who has for so long sought to escape the flesh, what attracts you to three ears rather than two?

Stelarc I've never really actually tried to escape the flesh as such, in effect all of these projects and performances were explorations in the psychological and physiological perimeters of the body. So even if you think of an intelligent agent as being embodied and embedded in the world, one can't really perform without a body. The question is what kind of a body, and do we simply accept the biological status quo

Fig. 28.2 Stomach sculpture (Fifth Australian Sculpture Triennale, NGV, Melbourne). (Photo by Tony Figallo. Courtesy of Stelarc)

of our present evolutionary trajectory, or do we consider ways of redesigning the body, ways of alternate mechanical augmentation of the body? Why not have an additional ear on the arm, maybe as a listening and transmitting device, an ear on the arm might be better position?

The Extra Ear project (see Fig. 28.3) actually begins way back in 1996. I was in Carnegie Mellon University and it was at that time when I grew rat muscle cells, but at that time I didn't really know what to do with them, didn't like the idea of putting them in a Petri dish on a pedestal, so the idea of growing something using living cells was something that was always a direction that I was intrigued with. The extra ear though, which was initially visualized in 1997 as an ear on the side of my head, kind of mimicked *The Third Hand* project (see Fig. 28.4) in the sense that the third hand was positioned beside my real right hand, and created this visual rhythm through repetition, so the idea of three ears was the continuity of the idea of a prosthesis in excess. But after years of trying to find surgical assistance and also discovering that the ear on the side of the face was not an anatomically appropriate sight; it was near the jawbone, it was near facial nerves, which meant partial paralysis if an operation had gone wrong. In 2006 after years of additional attempt to get surgical assistance, I finally got funding and with the help of three surgeons in Los Angeles, we begun constructing an ear on my arm.

This is still only a relief of an ear, but it's the result of two surgical procedures: The first was to insert a skin expander, so over a period of a couple of months, I was injecting sterile saline solution, stretching the skin in that area, creating a pocket of excess skin that could be used in the surgical construction of the ear. That was removed in the second procedure and a med-pore scaffold was inserted, resulting in this present shape. Med-pore is a biomaterial that's used in reconstruction surgery; and it's a porous material, so it allows cell grows into it, so after a period of 6

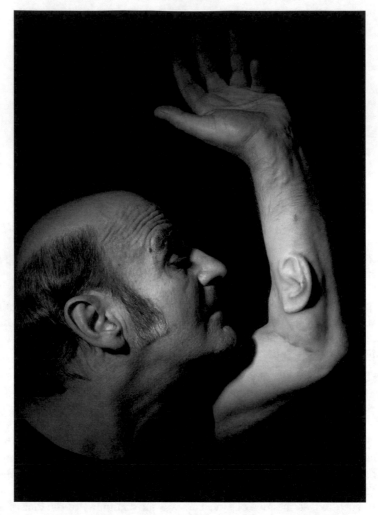

Fig. 28.3 Ear on arm (London, Los Angeles, Melbourne). (Photo by Nina Sellars. Courtesy of Stelarc)

months, tissue ingrowth and vascularization have occurred, so the ear is literally fused and fixed on my arm. We still need to lift the helix of the ear, creating a conch, and also highlighting the tragus area of the ear. Also, growing a soft ear lobe using my stem cells, so these are adipose-derived adult stem cells and using growth factors they can be directed to grow, for example, cartilage-like material so we'll be growing the earlobe. So it's partly a surgical construction – it involves tissue ingrowth, but also, in terms of more cutting edge research, it involves actually growing a part of the ear using adult stem cells. Finally, to implant a small microphone to connect to a bluetooth transmitter will enable the ear in any WiFi spot to be wirelessly connected to the Internet, so this effectively becomes a kind of Internet organ

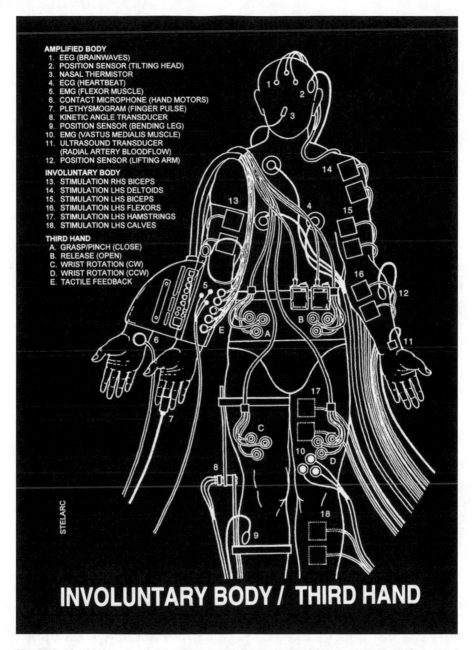

Fig. 28.4 The third hand project (Yokohama, Melbourne 1990). (Diagram by Stelarc. Courtesy of Stelarc)

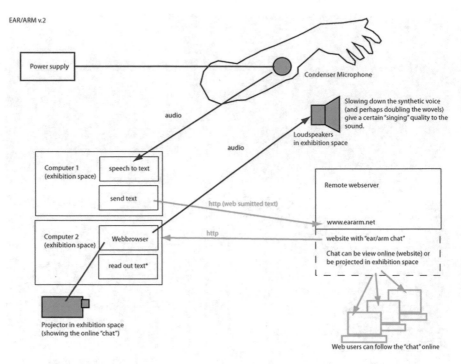

Fig. 28.5 Internet ear (Aarlborg, Paris, Moscow). (Diagram by Mogens Jacobsen. Courtesy of Stelarc)

that's publicly accessible to people in other places (see Fig. 28.5). So if you're in Melbourne, you can listen to what my ear is hearing in London, or in New York, or wherever you are and wherever the artist is.

Tofts As with your famous retiring of the third hand a number of years ago, do you foresee a time when your extra ear will also be removed?

Stelarc With all of these performances, the body acts with a kind of indifference as opposed to expectation. When you do things with expectation, things quickly become predictable, possibilities quickly collapse into actuality. Performing with indifference, you allow the performance to unfold; you try to suspend that quick collapse into actuality, so the idea of performing with indifference is an important one. I've always been surprised during the performance with some of the performative things that have happened, and I've always been pleased by the unexpected ideas that have been generated by the various projects and performances. If this didn't happen, it wouldn't really be an activity that I'd continue to do, specifically in the series of performances that use Laser Eyes (see Fig. 28.6).

Laser beams were projected initially from little mirrors on my eyes, but then guided through optic fiber cable to project from the eyes. I discovered by blinking

Fig. 28.6 Laser eyes (Maki Gallery, Tokyo). (Photo by Takatoshi Shinoda. Courtesy of Stelarc)

and controlling the musculature around the eyes, I could literally draw with the laser beams in the space that I was performing in. So, in those actions, the eyes become not passive receptors of light and images, but rather active generators of images. And that was kind of a surprising outcome of initially just an idea of just probing inside the body; you could project from the body into the external space of the performance. But being able to draw with my two eyes and actually be able to draw and scribble individually with each eye – separate images – was surprising.

And also, the way that the third hand became one in which its actions were much more intuitively generated. Initially when I wanted my third hand to move, I'd have to think of the muscle I'd be using to actuate the motor, and I'd look at my third hand and it was a very sequential operation that lasted a second or two. But the more I used my third hand, the more intuitive it became, and the hand was almost moving of its own accord, and that was a really pleasing place to be when that was happening.

Tofts Similarly, other recent work, such as *Prosthetic Head* and *Walking Head* have revived the historical figure of the zombie as a means of extending your notions of obsolete body and alternate interfaces. What interests you about the zombie as a philosophical and technological concept and how do see these works operating as a kind of zombie aesthetic?

Stelarc Zombies and cyborgs are alternate ways to think about how the body functions anyway, we primarily function automatically, involuntarily. The notion of choice and free agency is limited within a certain frame of reference, so I think the prosthetic head and the walking head explore notions of artificial intelligence and artificial life in simple but performative ways. So, for example, the prosthetic head (see Fig. 28.7) is an agent that speaks to the person who interrogates it. Now, as an installation, it's a 5 m high projection, it has a sensor system that alerts it when someone's in the room, so when you walk into the space and to the keyboard, the head turns, opens its eyes, and asks you the first question. It has a database and real-time lip-syncing, so when you ask the prosthetic head a question, it scans the database, selects an appropriate response, and lip-syncs the answer in real time. So effectively what you've got is a talking head, but it's a computer generated head; it's simply a 3000 polygon mesh with skin on it. It's empty, there's nothing inside the head, but it has a performative and conversational behavior that coupled to a human body and capable of generating some verbal and visual exchanges, which is why I call it a prosthetic head.

The Walking Head (see Fig. 28.8) is a more chimeric construct: an insect-like six-legged walking machine that has a computer generated human-like head mounted on the chassis of the machine. It is an actual virtual interface in that the facial behavior of the walking head is actuated and modulated by the mechanical movement of the legs. So the robot sits and waits until someone approaches it. Its ultrasound sensor system detects that you're in front of it; the robot then stands, selects from its library of possible movements, and then performs a simply choreography of a few minutes before it sits down and waits for the next person to come along. One shouldn't see these as simulating artificial intelligence or artificial life,

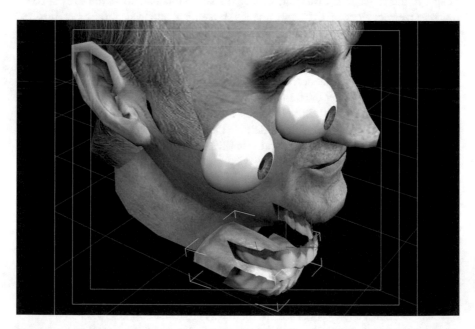

Fig. 28.7 Prosthetic head (San Francisco, Melbourne). (3D model by Barrett Fox. Courtesy of Stelarc)

but rather one should see it as alternate bodily constructs and actual virtual interfaces that allow the body to become an extended operational system.

Tofts In 2007, you explored Second Life for the first time as an interactive, virtual
world. Can you tell us about that?

Stelarc Second Life is an alternate motive of interaction and operation on the net,
so as a performative side, it's got lots of interesting possibilities (see Fig. 28.9). I
think we really have to recode the aesthetic and interactive possibilities so we're not
just simply mimicking the visual and functional appearances of the real world. My
approach would be to flush out Second Life a bit more. Second Life for me is a kind
of second skin, in other words the avatar becomes an alternate mode of interface
with people in other places, as well as providing an appropriate Internet pulse to it;
in other words, perhaps your avatar can become a barometer of Internet activity in a
real-time sense. So there are just some thoughts about what is seductive for me
about the notion of Second Life.

Tofts What do you think Second Life has to offer as a potential performance space?
And as a second part to the question, what would you like to do in Second Life?

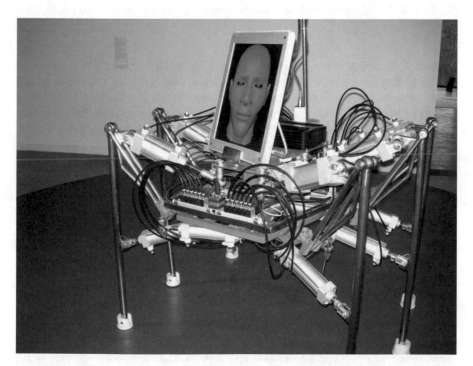

Fig. 28.8 Walking head (Heide Museum of Modern Art, Melbourne). (Photo by Stelarc. Courtesy
of Stelarc)

Fig. 28.9 Second life laser eyes (Second Life Performance). (Screen capture by Stelarc. Courtesy of Stelarc)

Stelarc I've always thought in working with machines and computers that you're giving the body an extended operational capability. In other words, it's never been an issue to me as in what's in control; it's not that the body has a master control of the computer, but rather, plugging the body into these technological systems, what alternates and surprising modes of operation are possible. This includes the idea of the body becoming a split physiology; in other words, if people in other places are prompting the left side of my body, and the left side of my body is responding involuntarily to their remote choreography; but simultaneously I can counterpoint by actuating a third hand on the right side of my body, so there's voltage in on one side, involuntarily generating movements that I can counterpoint voltage out on the right side of my body (see Fig. 28.10). Then, this idea of the body is not as a side of a single agency but effectively a host for multiple and remote agents. That generates interesting performative possibilities; so using machines and computers is really about the idea of erasing simplistic notions of single agency, of simplistic identity, and thinking more of the body in fragments or the split physiology, being remotely prompted. That's what's interesting about using technology.

Tofts What do you see as the possible future of human-computer collaboration for the creation of art and for artistic expression?

Fig. 28.10   Split body (Galeria Kapelica, Ljubljana). (Photo by Igor Andjelic. Courtesy of Stelarc)

Stelarc   One possible scenario amongst other cyborg constructs that I think goes beyond the military notions of traumatized bodies with replacement parts or massive exoskeletal augmentations of the body is this idea that perhaps all technology in the future would be invisible because it'd be inside the body. In other words, as technology becomes micro-miniaturized, it can be incorporated inside the body. Given we're now constructing devices on a nanoscale, then it's conceivable that we can re-colonize the human body, augmenting our bacterial and viral populations, producing a better internal surveillance of the body. I think the problem with surveillance is not so much whether ethnically we should have more or less external surveillance, but rather the imperative of having more internal surveillance because we will be dealing not only with internal threats but cellular spaces, and that's the implication of working at a nanoscale level. It's quite conceivable that we can construct sensors that can detect pathological changes in chemistry and in temperature. We'd certainly be able to detect blockage in our circulatory system and not only develop the sensors, but micro-miniaturized and nanoscale robots that could be either automatically maintaining the inside of the body or remotely guided by an external surgeon, so this is quite conceivable. So what's interesting artistically and aesthetically about working at a nanoscale is the different kind of movements and mechanisms and structures that you could sculpturally and artistically employ. I think that's what's interesting.

Bibliography

1. Clynes, Manfred E. and Kline, Nathan S. Cyborgs and Space (Reprinted with permission from Astronautics, September 1960). *New York Times*. [Online] 1997. http://partners.nytimes.com/library/cyber/surf/022697surfcyborg.html.
2. Ilnytzky, Ula. Wafaa Bilal, NYU Artist, Gets Camera Implanted In Head . *The Huffington Post*. [Online] November 23, 2010. http://www.huffingtonpost.com/2010/11/23/wafaa-bilal-nyu-artist-ge_n_787446.html.
3. Bozkurt, A., et al. Microprobe microsystem platform inserted during early metamorphosis to actuate insect flight muscle. *IEEE Xplore*. [Online] January 21-25, 2007. http://ieeexplore.ieee.org/xpl/articleDetails.jsp?arnumber=4432976.
4. The Washington Times. Military seeks to develop 'insect cyborgs'. *The Washington Times*. [Online] March 13, 2006. http://www.washingtontimes.com/news/2006/mar/13/20060313-120147-9229r/.
5. Hamilton, Anita. Resistance is Futile: PETA Attempts to Halt the Sale of Remote-Controlled Cyborg Cockroaches. *Time Magazine*. [Online] November 1, 2013. http://newsfeed.time.com/2013/11/01/cyborg-cockroaches-are-comingbut-not-if-peta-has-anything-to-say-about-it/.
6. Gannes, Liz. Electronic Tattoos and Passwords You Can Swallow: Google's Regina Dugan Is a Badass. *All Things D*. [Online] May 29, 2013. http://allthingsd.com/20130529/electronic-tattoos-and-passwords-you-can-swallowgoogles-regina-dugan-is-a-badass/.
7. Newman, Lily Hay. This Drummer Has a Third Arm. *Slate*. [Online] March 10, 2014. http://www.slate.com/blogs/future_tense/2014/03/10/robot_drumming_prosthesis_from_georgia_tech_gives_this_drummer_a_third_arm.html.
8. Tofts, Darren. Interview with Stelarc. *ACM Digital Library*. [Online] October 2008. http://dl.acm.org/citation.cfm?doid=1394021.1394023.

Chapter 29
My Transhumanist Story of Overcoming with LEGO

David Aguilar

My nickname is Hand Solo. I was born on February 25, 1999, in the principality of Andorra, a small country of 75,000 people between France and Spain. I came to the world with one unexpected surprise that has marked and defined the person I am today, and so is my spirit of overcoming. With the ideals of transhumanism, I am a real-life example for many parents and their children that we must respect one another and work nonstop to achieve their goals.

You see, I was born without an arm caused by a birth defect known as Poland's syndrome (named after British surgeon Sir Alfred Poland), which prevented the development of my chest and right forearm (Fig. 29.1). As you can imagine, my childhood was not easy, and I was a victim of harassment and bullying in school. However, my friends, my teachers and especially my family have given me the love, protection and self-confidence to deal with bullies. I have since given lectures in schools, universities, and companies about my prosthesis made of LEGO.

Since I was very young, I have always had a passion for engineering, robotics and computer science. At 9 years of age, I made my first prosthesis with most basic LEGO pieces—those that all small children play with (Fig. 29.2). I added a Bionycle robot and an iron wire to give mobility to two fingers. It had limited practicality and functionality, but it was a good challenge for me at the time. My father published my creation on Facebook and it gained a lot of attention.

At 18 years old, I improved the prosthesis to the point that it allowed me to pick up light objects and even to do pushups. With the creation of the MK-1 (named by Oli Pettigrew from RightThisMinute.com due to its similarity with part of the armor of Iron Man), I was called the Tony Stark of LEGO. Since then I have created a series of 3 improved prosthetics, up to the MK-3 which is my latest creation (Fig. 29.3).

The original MK-1 was born out of a LEGO helicopter that was picking up dust on a shelf in my room (Fig. 29.4). LEGO was my refuge. I spent hours and hours

D. Aguilar (✉)
UIC Barcelona – Universitat Internacional de Catalunya, Barcelona, Spain

© Springer Nature Switzerland AG 2019   491
N. Lee (ed.), *The Transhumanism Handbook*,
https://doi.org/10.1007/978-3-030-16920-6_29

Fig. 29.1 Poland's
syndrome

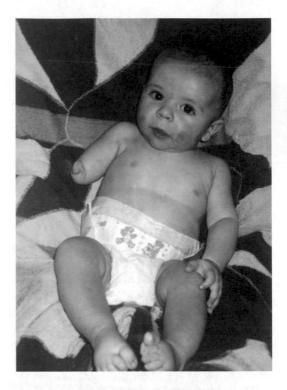

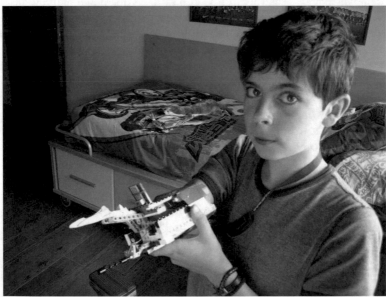

Fig. 29.2 My first prosthesis with most basic LEGO pieces (age 9)

Fig. 29.3 Tony Stark of
LEGO

Fig. 29.4 LEGO helicopter

locked up in my room, assembling and dismantling toys. I was not satisfied with existing toys, and so I started to create my own toys such as a mechanical joystick to play with my father's phone or a reproduction scale of a submachine gun from the Second World War which I used to shoot rubber bands at the flies in my room.

In one evening, I showed my newly created MK-1 to my parents while they were watching television. They were surprised that I could pick up a plastic button with my LEGO arm. It was then that my father believed that I had to show it to the world in order to encourage myself and other people to excel in life.

My parents made a video telling my history and my prosthesis. LEGO responded positively with recognition and admiration (Fig. 29.5). They encouraged me to continue and take it to the next step.

Next, my father contacted a local journalist named Roser Alberc from RTVA (Radio and Television of Andorra) who wrote my story with great affection and admiration. In my small but great country, I became a national hero but my father wanted my story to transcend even more.

In April 2018, National Geographic published a story on me on a two-page spread, and I gave a lecture in front of more than 300 people. Figure 29.6 shows my family (Ferran Aguilar, Nathalie Amphoux, Naia Aguilar) with journalist Roser Alberc from RTVA.

Albert Batalla, mayor of a small town called La Seu d'Urgell, invited me to a town hall meeting. Journalist Rosa Matas from the leading Spanish newspaper "La Vanguardia" interviewed me live on television. Journalist Rosa Talamas from Catalan TV3 came to my school to cover my story on overcoming bullying and treating everyone with respect.

I have to thank Anna Villas, the principal of my school in Andorra named "Col. legi Sant Ermengol", for all her assistance in welcoming with open arms for all the media coming from far away with their cameras, interviewing me and my teachers.

Amid all the publicity, I did not yet have a clear idea about university and career. I repeated the last year of high school. I did my best to prepare for university. Through my father's good friend Pere Matamales, Barcelona sports reporter Pilar Calvo introduced me to UIC Barcelona – Universitat Internacional de Catalunya to study Bioengineering. It helped cleared all my doubts and I decided to enter UIC and continue my dreams to help more people by learning new skills.

Meanwhile, I improved on the MK-1. With the generosity of businessman Jordi Cachafeiro in my country, I was able to create the MK-2 with parts from a LEGO airplane which has an engine that powers landing gear and that allows me to raise and lower the arm to carry more weight than its predecessor. Shortly afterward, I went on to create the latest MK-3 using thousands of pieces of LEGO.

I was invited to join the panel "Welcome to the Bionic Age" in Paris at the technology fair VIVA TECHNOLOGY along with two eminences Dr. Ernesto Martinez Villalpando and Dr. Moran Cerf. (Fig. 29.7).

Many people from around the world have contacted me to congratulate me, relieve their sorrows, explain their problems, and ask me to manufacture them gadgets with LEGO. I feel deeply sorry that I could not help every one of them because

Fig. 29.5 LEGO responses

their physical disability is different from mine, and that requires more skills than I currently have.

My number one fan is an incredible guy from Beirut, Lebanon whose name is Karim Kabbani. He was born premature and he is confined to a wheelchair. I can proudly say that when I spoke with him for the first time and saw his face of happiness, it was a magical and very special moment. It is one of the things that give me the strength to look forward to the future with enthusiasm and to fight for my dreams

Fig. 29.6 My family (Ferran Aguilar, Nathalie Amphoux, Naia Aguilar) with journalist Roser Alberc from RTVA

Fig. 29.7 Welcome to the Bionic Age by VIVA Technology

by studying bioengineering to do my best to improve the lives of people with disabilities—not as disabled but as superheroes of a parallel reality. A reality that only a few know, and for which transhumanism will help us to improve our condition to become exceptional beings.

Some may argue that transhumanism can be dangerous to humanity. Well, if we look at ancient history when human beings invented languages, axes, spears, wheels and so on, humanity not only did good things with the inventions but also did many evil things with them such as wars and destructions.

With transhumanist technologies, we can improve our quality of life—especially for the disabled. We can learn to avoid making the same mistakes by studying human history and ethics. For my part, I will continue to strive to be a better person and a better transhumanist. I am thankful to my family, friends, employees of LEGO, the media, and Newton Lee for giving me a chance to share my story and motivate people.

Instagram: handsolo99
Twiter: @DavidHandSolo
Facebook: Hand Solo & David Aguilar Amphoux

Chapter 30
Biohacktivism, Cyborgasm, and Transhumanism

Rich Lee

Rich Lee is in human enhancement and augmentation technology. He stays informed on everything in the field, from government and university projects to diy biohacking projects. The following is a 2018 interview of Rich Lee by Newton Lee, Chairman of the California Transhumanist Party.

QUESTION The aspiring transhuman eagerly awaits key advancements that may enable extreme longevity, enhanced physical and mental capabilities, and an integration of our biology with machines. When these technologies are born, do you think that we will be able to afford them?

ANSWER The majority of these technologies will not be affordable. There is a common fantasy among the public and transhumanists alike that prices on human augmentation technology will naturally fall over time. People cite how things like DVD players started as expensive items and can now be purchased cheaply, at a fraction of their original costs. There are many flaws in this thinking. It assumes that the products will exist in a free market with competition. While most transhuman technology is not considered a medical necessity by today's standards, it still falls into a similar regulatory domain of government bodies such as the FDA in nearly all countries. FDA approval costs can be astronomical. Testing requirements can last over a decade with no guarantees of approval. This will limit the number people willing to enter the market. The ones that do will surely need to offset their costs at the expense of the consumer. Furthermore, patents will give a company exclusive protections against competition. Interventions requiring invasive surgeries will involve labor costs of anesthesiologists, nurses, and surgeons. The legal liability involved in any of these areas is big. This translates to higher insurance costs and fees for legal representation. Costs will not decrease over time as long as transhuman technology is tied to the medical sector. When was the last time the cost of your medication went down? Or the cost of pregnancy, doctors visits, or other surgeries?

R. Lee (✉)
Biohacker & Cyborg, Washington, UT, USA

© Springer Nature Switzerland AG 2019
N. Lee (ed.), *The Transhumanism Handbook*,
https://doi.org/10.1007/978-3-030-16920-6_30

In the end this means that transhuman technology will be a product for the rich. If you are not rich but want these advantages for yourself or your children, you may have only one alternative: become a biohacker.

QUESTION You hold the belief that transcendence is an inalienable human right. How do we secure a future where transhuman technologies will not be withheld from those who seek them based on .class, wealth, or geographical borders?

ANSWER I hold the belief that transcendence is an inalienable human right. This right supersedes notions of intellectual property rights, stamps of approval, social acceptance, or legal jurisdiction. Individuals should have the right to alter their bodies no matter what. Many biohackers fight to ensure this right. In countries where transcendence is restricted by class, wealth, civilian status, exclusive rights, race, or complete prohibition, biohackers will create underground economies, distribution channels, diy alternatives, and supply chains. We will reverse engineer patents, reveal trade secrets, organize efforts of sympathetic medical professionals, and smuggle. Those wishing to be freed from the shackles of human biology will find a transhuman underground railroad.

QUESTION Are you willing to wait 10–20 years for the government to approve transhuman technologies for public use? What if the government declares prohibition on some of the transhuman technologies?

ANSWER I am not willing to wait 10 or 20 years for some government to approve a certain intervention related to my transcendence. I do not even observe their right to approve or disapprove the things I wish to do with my body. I don't care about economic impact evaluations, ethics review boards, or social impact studies. The fact is that when these technologies become available they will be used no matter what the popular or legal opinion is. If not by your country, then by another. If not on the open markets then in the underground markets. One billion people will die in the next 20 years. What will future transhumans think of those who could have prevented many of those deaths, but instead chose to consider social or economic consequences?

QUESTION What is biohacktivism?

ANSWER The 1990s gave rise to an amazing social phenomenon: the hacktivist. These hacker activists have exposed corruption, lies, human rights violations, and fraud in nearly every country on earth. They have fought censorship and established new networks of communication to enable free speech to those who may live in fear retribution for doing so. They have become a new check and balance in global politics, operating outside of the system like vigilantes for transparency, privacy, and free speech. Hacktivists support values which transcend legality, patriotism, and borders. We currently face many challenges such as climate change, skyrocketing healthcare costs, cancer, extinction, and disease, to name a few. All of these things can potentially be solved by standard bioengineering, but probably not in a timely manner and not within our current deontological framework. To do most of these things we will need biohackers and biohacktivists who operate outside the system.

Biohacktivists will also be increasingly needed to oppose legislation which will criminalize or restrict access to bioengineering tools, ingredients, and information.

QUESTION What is cyborgasm?

ANSWER Sex will improve greatly with the development of certain technologies such as cybernetic implants. A personal project of mine, the Lovetron9000, is a small vibrating clitoral stimulator which is implanted in a man's pubic area. It's an implant designed to give strong vibrations to a partner on demand. Yes ladies, this is the best of both worlds. It can be custom programmed, synched to music, and programmed via bluetooth. It (like most electronic implants being developed by biohackers) is wirelessly charged. Other implants have been developed, such as specialized spinal implants, have been developed which can give orgasms at the push of a button, even to those lacking genitals. Hypothetically, if one was to fit these stimulators with sensors and then interface them with a partner's spinal stimulator, one could create an interesting series of feedback loops. For example, when one partner feels pleasure, those signals could be read and replicated in your partner's implant so that you feel what your partner feels. Then that signal can be relayed back to your partner then back to you, creating a pleasure loop. This can be used to do other nifty things too, like to cue a partner to change rhythms, make adjustments, etc. There will be no more guessing, it will just be instinctive. Technically there is no limit to how many partners can be paired together via the internet, which will create some interesting opportunities for massively multiplayer online gaming. Sex related implants may actually be one of the first cybernetic technologies to reach mass adoption. This will have some interesting impacts on the course of human breeding. We can imagine a future 20 years from now where single men are mingling at a bar. Most of them could have Lovetron9000 implants and it might be as common as an IUD implant. How terrible would it be if you were the guy at the bar without a Lovetron9000? Will all future humans be descended from cyborgs? I think this is a real possibility. Once this technology exists and is shown to make a substantial improvement to our sex live, we will never go back to the old ways which are so lacking. We will trade in our orgasms in favor of cyborgasms and never look back.

Chapter 31
The (R)Evolution of the Human Mind

Augusta L. Wellington

As the late great Mark Twain purportedly said "Age is an issue of mind over matter. If you don't mind, it doesn't matter." I wholeheartedly agree with him. The mind is a collective of all of our experiences. No matter what a person's chronological age might be, it is not all-encompassing compared to one's psychological age, which is based on their experiences. In my estimation a person's age is not a constant. It is continually in flux shifting from moment to moment. We can feel different ages depending upon a multiplicity of experiences with individuals or interpersonal interactions. At times, we may feel like a teenager and in others times aged beyond our years. How we feel depends upon whom you are interacting with and what factors different people tap into from our wellspring of life experiences.

In this chapter, I will discuss how transhumanism can affect and model the mind. It can not only be curative to psychological health, but it can facilitate one's capacity to maintain a youthful mindset. I have been practicing psychotherapy/psychoanalysis for decades. In my experiences, I have treated patients of all ages and in many different types of settings such as my private office, hospitals, telephonically and via skype. They have been diagnosed with a variety of psychiatric conditions such as depression, anxiety, eating disorders, addiction, and psychosis. In addition, some of my patients have suffered with chronic medical illnesses such as cancer, multiple sclerosis, and epilepsy. In many instances and as a result of my work with them many of these patients became asymptomatic or went into remission.

How this occurs is partly by changing one's mindset, as I do believe in mind over matter. The New York Times Magazine 2015 article *Tell It About Your Mother: Can brain scanning help save Freudian psychoanalysis* is provocative and illuminating because it illustrates how psychotherapeutic work can make structural changes in the brain. These changes are not limited to behavioral alteration, but can also include changes in the person's character and psychic structure. It is remarkable how the power of words can effectuate physical changes in the brain.

A. L. Wellington (✉)
International Psychoanalytic Association and Psychoanalyst in Private Practice, Nassau, Bahamas

© Springer Nature Switzerland AG 2019
N. Lee (ed.), *The Transhumanism Handbook*,
https://doi.org/10.1007/978-3-030-16920-6_31

There are other kinds of changes the analyst can observe as a result of collaborative treatment efforts. When an analyst sits with a patient over the course of many sessions she/he observes the persons modification. Moreover, I have even witnessed how close people around the individual patient change.

Modern medicine has been able to cure very few diseases. More often than not, the purpose of most medicine is symptom reduction. A person can identify symptoms without really know its ideology. People become so desperate to alleviate their pain they may resort to any medicinal measure that offers symptom relief. There is a lot of research that has been conducted on the power of the mind to create a sense of healing. Studies have shown where people are not given actual medicine and will still experience relief. This is the essence of the placebo effect.

If a person is able to reframe their thoughts from the situation, then the mind is completely unrestricted by the past. This, too, is the essence of the placebo effect. As Einstein said "we cannot solve our problems with the same thinking we used when we created them". Problems are created through thoughts that interfere with actuality. It is those images that we store historically, and cannot relinquish, that interfere with our present day living, and are not anything more in actuality. We see those repetitive illusions again and we think we know what is perceived, but indeed we mistake them to be factual. This causes us to always live in the past rather than in the present moment. Living in the moment is the key for health and longevity.

31.1 Case Illustrations

In the second year of four times a week treatment a 30 year old woman received the diagnosis of multiple sclerosis. We talked about her anxiety and fears about the testing and procedures she had. She and I discussed the possibilities for her future both positive and negative. I was consistently present with her throughout this daunting time. And together she and I were able to arrive at a juncture in her treatment that she felt encouraged her to pursue yoga, and seek more support from her family and friends. These positive improvements in her personal life led her to be more proactive in treatment and over time these contributed to a notable improvement both psychologically and physically.

Another example, I treated a 17 year old girl who was severely mentally ill. She was paranoid and delusional believing all men were out to harm her. This was a long term treatment of over 10 years. Although medication was an option, she refused. I was able to sustain consistent attunement while knowing that her actual potential had not yet become realized. Slowly over the years she became more stable, realistic in her views about herself and others, and confident about her ability to succeed. I attribute this woman's progress to her treatment with me. There were many aspects in the therapeutic alliance that we developed over the years that had a profound impact on her. Having grown up in a disorganized unstable environment, the safety and stability of my office and our relationship was curative for her. We developed a strong and positive therapeutic alliance; one in which I held onto a hope for her when she could not do it for herself. At one point, I asked her what was most helpful for her in treatment. Her response surprised me. She said something that seemed so simple: "it was the way you believed in me, that I could get better, even when I didn't". Hearing this made me realize that I was only scratching the surface of her human potential both biologically and psychologically. I think a positive mindset is the fulcrum for a successful treatment.

In another case I was treating a depressed, agoraphobic artistic patient who in the beginning phase for her analysis often resorted to action rather than to the spoken word. One day upon arrival to my office she brought in a painting she had done of a giraffe proclaiming it was a portrait of how she perceived me. She told me that my office was one of the only places where she felt safe and that I helped her feel understood and protected. I then interpreted that the giraffes in Africa are of value to the other animals in their environment because they have the capacity to see past what the other animals could not. They have the long-distance vision to anticipate what is to come on the horizon. She then enthusiastically proclaimed, as she had never done before, "you are my giraffe, you are my long-distance vision". From this point forward she improved the capacity to verbally express feelings. This was a patient who had difficulty expressing herself and yet was finally able to do so.

The last therapeutic endeavor that I will discuss in this chapter is about a 14-year-old girl who was diagnosed with epilepsy. Her mother thought that stress reduction in a therapeutic environment would be helpful for her daughter. Prior to psychotherapy she had four seizures. Once we began on our journey together, I attempted to help her manage the stress that she that she put on herself to be perfect. I suggested some coping skills such as doing the things that she most enjoyed. I felt that would help ameliorate her anxiety in everyday life, relationships and school. She joined the drama club, and started taking singing lessons, which helped sublimate some of the stress. The rest was left to our relationship. Over 20 years later she has had only one subsequent seizure.

31.2   Discussion

It is important as an analyst to provide a safe environment and in all therapeutic endeavors to pay very close attention to the idiosyncrasies of patients. How this comes about is when one is able to listen without judgement or preconceived notions of any kind. This creates the capacity to form a bond that dissipates individuality and separateness. Having someone intently listen is empowering to both parties. Clarity comes with that intense attention and it clears away all falsehoods. It is in observing what is false that the actuality becomes clear. We are all connected, and even if it is perceived often by so many when they are depressed, we are not alone. Although depression creates the misperception that we are isolated.

Every therapist needs to feel a certain degree of confidence in what they practice and understand the curative elements of psychotherapy. Intellectual curiosity, proficiency and skills need to be coupled with humility and passion. As Henry David Thoreau said, "None are so old as those who outlived enthusiasm".

Overall health is both about the mind and body, and so much more. As Buddha stated, "To keep the body in good health is a duty otherwise we shall not be able to keep our mind strong and clear." Physical health is of great importance to remain young. This can include being on the proper diet, exercising regularly, taking the necessary supplements, and finding ways to mentally increase your capacity towards youth. Picasso stated, "It takes a long time to become young." It would be necessary to recognize how resistance imposes limitations. When mental limits are observed one would then have the freedom to perceive with an innocence that is synonymous with youth.

The human mind has the tendency to turn something that is not comprehended into a problem. A mind that is creating problems is not as receptive to actuality and has the tendency to obscure what is real. This detracts from the healing energy of clarity. Mentally creating problems (where there are none) prevents one from seeing or feeling what is truly going, which leads to losing capacity for objectivity. The resistance dulls the brain and accelerates the aging process.

This is more a journey than a process as it does not have a conclusion or an end and is not limited by time or space. Kafka stated, "Anyone who keeps the ability to see beauty never grows old".

Augusta Wellington LCSW-R, FIPA is presently practicing in Nassau, Bahamas as well international consulting. She is a Fellow of the International Psychoanalytic Association and a Graduate/Member of the Institute for Psychoanalytic Training and Research.

Bibliography

1. Casey Schwartz. Tell it about your mother: Can brain-scanning help save Freudian Psychoanalysis, The New York Times Magazine, June 24 2015.
2. D.W. Winnicott The maturational Processes and the Facilitating Environment (1965).
3. Eric Shah and Mark Pimentel. Placebo Effect in Clinical Trial Design for Irritable Bowel Syndrome J Neurogastroenterol Motil. April 2014 (163–170).
4. HK. Beecher. The powerful placebo. J Am Med Assoc. 1955;159:1602–1606. [PubMed].
5. Jean Laplanche & Jean-Bertrand Pontalis, *The Language of Psychoanalysis* (1973).
6. Otto Fenichel. Problems of Psychoanalytic Technique (1941).
7. Peter Gay, *Freud* (1989).
8. Sigmund Freud. Analysis of a Phobia in a Five-Year-Old Boy (1909).
9. Sigmund Freud, "Recommendations for Physicians Practicing Psychoanalysis", *Standard Edition* 12.

Chapter 32
Transhumanism from the View of a Psychologist, Focusing on Mental Health Issues

Andjelka (Angie) Stones

To begin discussing Mental Health in general terms and without taking into account people's religious and existential views is very difficult, however necessary in the context of Transhumanism debate. So, we will try to put our point of view as objectively as possible and stay on the side of (my own) logic. Technology has long been involved in mental health, although the use of it was detrimental in the past. Lobotomies were used as a means to health people with mental illness at one time; using the technology of the time to attempt to improve mental health. One of the most infamous uses of technology in "treating" mental health would come in World War II. Hitler opened the gates for the most morally corrupt scientists to use technology to intervene in those with mental health.[1] Euthanasia and sterilization were the most common forms of technology to be utilized with those suffering from mental illness. Unfortunately, this severely injured the transhumanism movement. However, we have made great progress since that time. Just as the technology of cochlear implants has improved physical health, the technology of deep brain stimulation implants is making progress as a treatment for depression[2,3]

We have come a long way integrating technology and science into many aspects of human physical health maintenance. Despite this, the general fatalistic point of view that we have to die eventually is still wildly accepted. Just by accepting this notion is enough to make many people depressed, anxious, and generally stressed

[1] Alexander, L. (1949). Medical Science under Dictatorship. *New England Journal of Medicine, 241*(2), 39–47. https://doi.org/10.1056/NEJM194907142410201

[2] Lee, J. (2016). Cochlear Implantation, Enhancements, Transhumanism and Posthumanism: Some Human Questions. *Science and Engineering Ethics, 22*(1), 67–92. https://doi.org/10.1007/s11948-015-9640-6

[3] Petrov, L. (2017). Shocking New Advancements: Deep Brain Stimulation as an Experimental Procedure for Treatment-Resistant Major Depressive Disorder. *Blue Moon.*

A. (Angie) Stones (✉)
NorthCentral University, Scottsdale, AZ, USA
e-mail: astones@ncu.edu; dr.stones@globalnewworld.com

© Springer Nature Switzerland AG 2019
N. Lee (ed.), *The Transhumanism Handbook*,
https://doi.org/10.1007/978-3-030-16920-6_32

about many issues we have to face throughout our life. From this, the notion that being stressed, depressed or anxious is somehow normal perception for most people today. In fact, some may say that those feelings make us "human". In my view, this is a huge mistake in logic which does not allow us to progress in thinking beyond the standard "holistic" approaches to Mental Health, such as meditation, relaxation, exercise, whole foods, and so on; or pharmacological solutions such as antidepressants or anti-anxiety medication.

We all experience feeling of anxiety, depression, stress etc. at some point in our lives, but most of us don't develop mental illness. So we must make distinction between the two. There is a huge movement into gene identification towards the predisposition to various mental health issues, but we still do not have the preventative strategies or real cure. So why are some people developing the illness and others do not? Whether it is due to genes or to the coping strategies it makes no difference because at the end, our control center, the brain, can regulate both, but we do not have the tools yet to manage the "control center".

To be able to control the brain sends shivers through many people; the idea that one day we may be able to download our neural circuits to a computer, manipulate them, and correct any signals that would make us depressed or anxious is an idea that seems futuristic and unacceptable to many. After all, we should not interfere with normal human emotions. But this is the problem…Anxiety or depression is not "normal" emotions. In fact, they are not emotions at all but they are conditions which may make us sick if they develop into mental illness; sadness is an emotion but depression is not. The distinction between the two is often overlooked by the general public, but never the less; there is a great divide between the two. By using computers to get rid of the undesirable mental health issues, it does not mean that we will become inhuman and not have our human emotions.

This is where mental health is often overlooked. If one loses a leg, it is absolutely accepted, as it should be, that a prosthetic leg could be a source of great help for that person. However, if one is lacking in serotonin, dopamine, or norepinephrine, why is it that we do not want to use an implant to improve levels of these neurotransmitters?[4] The question of not accepting the scientific progress in possibly eliminating mental health issues is not logical to me because we are happy to accept neural implants for hearing or artificial limbs, just to name a few examples. These improve the lives of those that possess them and make so many people happy; would it not make us even happier if we would never need to worry about becoming depressed? Why would we not use technological knowledge to advance our mental health progress too? Where does the fear come from? Is it from people themselves or from the Governments which need to keep pharmaceutical industries going?

Finally, the argument that we would stop being human if we interfere with our brains and eliminate mental health issues, is not an argument at all because it comes from people who do not have mental health issues. We need to ask people who

---

[4] Moutaud, B. (2016). Neuromodulation Technologies and the Regulation of Forms of Life: Exploring, Treating, Enhancing. *Medical Anthropology, 35*(1), 90–103. https://doi.org/10.1080/01459740.2015.1055355

suffer right now if they would object to transhumanist ideas regarding mental health rather than those around them who cannot imagine what it is like to be depressed or anxious on daily basis, or listen to the government ideas of economic viability, over-population or breakdown in world structure as we know it. Just being open to the new ideas should not be scary or provoke a feeling of guilt. In my view, being human means striving to new advances in science and technology for the good of all people and of all unfortunate conditions we may encounter. Having no mental health issues would solve many other problems around the world and this is a topic which needs to be discussed one by one and segment by segment. Physical health issues would improve tremendously in those that are suffering from comorbid mental health issues.[5] This is yet another reason Mental Health should be number one priority today. There are people who have many difficult and painful chronic physical illnesses but manage to stay positive and happy throughout their lives, but there are many more people who are physically healthy, yet spend their lives just waiting for it to end.

Is this not a good argument to do anything possible to prevent or cure mental health issues whichever means science can offer us and this includes the ideas that transhumanism may offer.

---

[5] Nemeroff, C. B. (2013). Psychoneuroimmunoendocrinology: The biological basis of mind–body physiology and pathophysiology. *Depression and Anxiety, 30*(4), 285–287. https://doi.org/10.1002/da.22110

Part V
Blockchain and Cryptocurrency

Chapter 33
Transhuman Crypto Cloudminds

Melanie Swan

"Science is the great antidote to the poison of enthusiasm and superstition."
–Adam Smith (The Wealth of Nations, 1776 [1])

33.1 Transhuman Problem Context and Blockchains as a Solution

This essay understands *transhumanism* to be the idea that the human race can evolve beyond its current physical and mental limitations, especially by means of science and technology. Kurzweil's view is that the transhuman future will be a convergence of artificial intelligence (machines) and intelligence augmentation (humans) [2]. This is also his answer to the Friendly Artificial Intelligence (AI) question, namely that Unfriendly AI will not "take over the world" (or make paperclips [3] and strawberry fields [4]) because humans and AI will become the same thing. Reformulating Kurzweil's convergence argument, at least the immediate future is more likely to be one of a multi-species society with various permutations of human, algorithm, and machine.

Thus, one of the most pressing near-term challenges is evolving into a multi-species society in ways that support the success of all. This is going to entail agreements such as treaties, and enforcement mechanisms that produce good player behavior in large-scale network environments, including contracts, penalties, and reputation systems. The governance systems of the future may be based on game-theoretic incentives to coordinate large groups of actors, as opposed to policing and force (which may be practically impossible, costly, and otherwise infeasible).

M. Swan (✉)
Purdue University, West Lafayette, IN, USA

Institute for Blockchain Studies, West Lafayette, IN, USA

DIYgenomics, West Lafayette, IN, USA
e-mail: swan3@purdue.edu

© Springer Nature Switzerland AG 2019
N. Lee (ed.), *The Transhumanism Handbook*,
https://doi.org/10.1007/978-3-030-16920-6_33

Blockchains are a future-class technology that supports these needs. Such distributed ledgers are a new form of smart network and global computational infrastructure that might be used to implement safe, empowering, and enlightening transhuman futures.

A *blockchain* (also known as distributed ledger technology) is an immutable, cryptographic (cryptography-based), distributed (peer-to-peer decentralized network), consensus-driven ledger. Conceptually, blockchain technology is a software protocol for the instantaneous transfer of money and other forms of value (assets, contracts, public records, identity credentials, and program states) globally via the internet. Just as SMTP (Simple Mail Transfer Protocol) is an internet-based software protocol for transferring email, blockchains are a protocol for transferring money. However, the transfer of assets requires a few more bells and whistles than simply transferring files on the internet. Blockchains provide an always-on apparatus that checks asset registration and ownership in real-time to make sure that only one unique instance of an asset is transferred (avoiding the double-spending problem, that multiple copies of digital money might be sent to different parties).

33.1.1 Problem Context: Safeguards for an Emerging Multi-Species Society

The transhuman problem context is that humanity is in the process of evolving into a more intense integration of humans and technology, and safeguards are required for the productive progression of this trend. So far, the multi-species society can be seen most prominently in markets and in the workplace.

In financial markets, humans and algorithms coexist side by side. Currently, 55% of U.S. stock trading volume (and 40–70% worldwide) is performed by programmatic trades, and high-frequency trading (HFT) volume has doubled since the 2008 financial crisis [5]. Programmatic trading is implicated in flash crashes, an emergent phenomenon in which automated trading can trigger extremely rapid price declines [6]. Since markets are driven by sentiment, a new class of data science analysis techniques and privacy-protected, remunerated, information-sharing mechanisms are needed to model markets as a multi-species domain. Blockchain smart network systems might provide the requisite tools for collective intelligence gathering and systemic risk modeling to reduce the possibility of large-scale failure such as financial contagion and collapse.

In the workplace, humans and machines are also starting to cohabitate. As Cowen heralded in 2013, the best worker for the job may be humans and technology in collaboration [7]. Some economists estimate that one half of jobs (47%) in mature economies are at risk of being outsourced to technology in the next two decades [8]. The St. Louis Federal Reserve Bank predicts the continued progression of the 30-year trend of job declines in routine tasks, both physical and cognitive [9]. China, South Korea, and Japan are leading the world in the implementation of industrial robotics [10], often with one human attending several machines [11]. Other sectors are in the process of joining the automation economy, in the areas of supply chain logistics, smart city monitoring, commercial driving fleet management, and medical imaging.

33.1.2 Solution Objectives: Thriving and Surviving

Given the problem context that the transhuman future depends on evolving into a safe and empowering multi-species society, there are two focal points: surviving and thriving (Table 33.1). Surviving addresses the crucial concerns of physical, cognitive, and emotional safety, and economic sustenance, as preconditions for launching the broader transhuman agenda of thriving by exceeding current physical and mental limitations.

33.2 Cryptoeconomic Smart Networks for Surviving in the Transhuman Future

First considering surviving, blockchains might help facilitate the next steps in society's evolution in three primary ways. One is the "good player argument." This is the idea that mutual survival interests are aligned by using a common infrastructure. Since human, algorithm, and machine might all be using the same blockchain smart network systems to access resources, communicate, obtain information, transact, and otherwise conduct their operations, they all have the incentive to remain in good reputational standing to use the network, and this could enforce good player behavior [12].

A second future economics concern is sustenance. Using economics as design principles in blockchain networks might help solve an important outstanding problem, credit assignment, by having a model for tracking and connecting contribution with remuneration. Remuneration is often based on asset ownership as opposed to value contribution. Blockchains could remedy this since the advent of the token economy means a new era of participation in web-based communities. Whereas Web 1.0 involved the transfer of static information, Web 2.0 (the social web) created the expectation that website users can interact, like, and engage with content and

Table 33.1 Transhuman objectives and corresponding blockchain solutions

Tier	Outcome	Blockchain facilitation
Surviving	Physical, cognitive, emotional safety; economic sustenance	Multi-species treaties enforced with recourse and reputation
Thriving	Exceeding current physical and mental limitations	Crypto cloudminds

other users. Now in Web 3.0 (the crypto web), users expect to participate meaningfully in economic communities by being remunerated for contributions (such as software code, digital art, and forum posts), and by being able to vote and access resources [13]. The *token economy* facilitates this as websites issue their own cryptotoken money supplies (despite emerging technology volatility, there are over 1945 Ethereum-based projects as of October 2018 [14]).

Blockchains might offer a means of solving the credit assignment problem, and also facilitate new models of asset ownership that address "future of work" concerns. Blockchain economies are characterized by open platform business models for large-scale global participation (e.g. open source software communities with a payments layer) that compete with closed monopolistic platforms (such as Google, Facebook, and Microsoft). As technological unemployment becomes more prominent in the *automation economy* (jobs being outsourced to technology), transitioning outsourced workers to new situations is a key concern [15]. With cooperative ownership models [16], outsourced workers might receive shares in the new automated means of production in exchange for job loss. However, early Universal Basic Income (UBI) experiments indicate that unearned "free money" is unpalatable [17] since it lacks social contact reciprocity features. UBI is another form of the credit assignment problem because it has the same premise that contribution and remuneration should be linked. Blockchain-based Maslow Smart Contracts might enable individuals to pursue interests and skill development in a structured way that results in certification and new employment or productive use of capacities as the endpoint, and instantiates new forms of the social contract.

33.2.1 Blockchains as a Future-Class Technology

Blockchains (software the for secure automated transfer of information or assets via the internet) constitute a robust singularity-class technology for realizing both objectives of transhuman futures, thriving and surviving. Blockchains offer the full progression of a sophisticated new technology, running the gamut from "better horse" to "new car" applications (Table 33.2) [18]. Since internet networks are already in place, value-added layers such as blockchains might be rapidly adopted. IDC estimates that global spending on blockchain solutions will reach $9.2 billion in 2021 (from as estimated $2.1 billion in 2018) [19]. The World Economic Forum estimates that at least 10% of global GDP will be stored in blockchains by 2027 [20].

Blockchain 1.0 can be conceived as a "better internet," adding features that were not present in the initial implementation. This functionality includes a payments layer (the ability to send money and assets) and the possibility of engaging in privacy-protected computing with confidential transactions and private messaging. Some of the "better horse" applications are real-time payments and monetary transfer instantaneously on a global basis, automated information confirmation, and secure messaging.

Table 33.2 Blockchain's "new technology" progression

	Level	Core functionality	Applications
Surviving	1.0 Basic "Better horse"	Better internet: payments and secure information transfer	Real-time payments, monetary and information transactions (immediate transfer)
Thriving	2.0 Advanced "Horseless carriage"	Assets digitized and registered to blockchains, real-time valuation and transfer, smart contracts	Financial contracts; legal agreements; digital identity; government documents; voting
	3.0 Future "Car"	Large-scale institutional collaboration technology: superseding governments, corporations, open-source communities	Cloudminds, automated fleet management, big health data, space communication

Blockchain 2.0 is understood as an institutional technology (a replacement or supplement for brick-and-mortar human-based institutions) [21], and as an enterprise software for transitioning business and government into the digital age. In the future, a substantial portion of the world's assets might be registered to digital inventories and transacted with blockchains. There could be a single set of shared business processes used by multiple parties in industry value chains [22]. Some of the envisioned "horseless carriage" applications include financial contracts, legal agreements, digital identity solutions, government public record repositories, and voting [23].

Blockchain 3.0 understands distributed ledgers as a new species of technology with possibilities that did not previously exist. One implication is the smart network convergence technology of *deep learning chains* (integrated blockchain and deep learning functionality). Some potential "car" applications include crypto cloudminds, automated fleet management, big health data, asynchronous space communication, and the rapid-prototyping of gaming and artificial reality environments [24].

33.3 Crypto Cloudminds for Thriving in the Transhuman Future

Blockchains are equally implicated in the transhuman objectives of surviving and thriving. Having discussed surviving, the essay now proposes Crypto Cloudminds for thriving, as a safe expansionary mechanism for exceeding current physical and mental limitations.

A *cloudmind* is a cloud-based collaboration of human and machine minds (with safeguards and permissions) [25]. In this context, *mind* means some kind of computational processing power, possibly including decision-making, but not necessarily consciousness. A cloudmind has processing or thinking capability that is virtual, located in internet databanks or decentralized networks without having a specific embodiment or physical corporeality. Cloudminds might be large or small, and have a variety of purposeful and creative goals.

There are three classes of existing cloudminds. These include *AI Deep Thinkers* (such as Watson, Watson Health, AlphaGo DeepMind, and NVIDIA DIGITS), *collective intelligence pooled-mind resources* (eLabor marketplaces such as Mechanical Turk), and *digital selves* (such as CyBeRev and LifeNaut) and *digital self entities* (such as BINA48 [26]). Meta-cloudminds might be deployed as a smart network orchestration mechanism for monitoring existing cloudminds. A *Deep Thinkers Registry* could be a protective measure, with cloudminds appearing before the Computational Ethics Review Board each year for continued licensing. The review board itself could be comprised of both human and technology-based entities.

The bigger point for the transhumanist expansion is that in the future, cloudminds might comprise large numbers of minds operating together with human and machine minds in collaboration. In order to do so, a slate of safety protocols would need to be in place, all of which might be managed by blockchains. One is guaranteeing that a *cloudmind* is not a *groupmind*. This means that individuals would only permission *partial* cognitive resources, not their whole brain and identity to collaborative mind projects. Smart contracts could manage the process.

The premise of transhumanism is exceeding current physical and mental limitations. The first tier of application is the more obvious and light-weight phases. This could include physical augmentation for better health, strength, and disease prevention; and improving mental capacities with smartdrugs, targeted memory excitation, and performance accentuation. However, the safer and more superficial approaches may only do so much to extend mental capacities. At some point, a second tier of improvement calls for going beyond the constraint of the current unitary meatspace packaging of the human brain.

The benefits of collective intelligence have been demonstrated. Working in diverse teams can lead to improved outcomes and the ability to tackle larger-scale projects [27]. The enhanced evolution of teamwork could be through collaborative mind-power. There could be communities of human minds, and human and machine minds working together in a Cowen-like future-of-work progression.

However, the mind is an extremely sensitive frontier, and thus must be shared slowly and partially in a controlled manner. One first application could be contributing excess processing cycles and data. Mind@home could be a crowdsourced distributed computing project like SETI@home and protein Folding@home. Dream@home could assemble collective sleep-tracking data. The properties of blockchains could be used to control and enforce this permissioning, and attribute any intellectual property (IP) garnered in the process.

Technically, how blockchains could orchestrate cloudminds is by being the data control layer in a Brain-Computer Interface (BCI) apparatus that is either implanted or wireless [28]. Cloudminds could be accessed as a next-generation social networking application, running over Github or LinkedIn, for example. Blockchain's relevant features are automated access control of arbitrarily-many items in an algorithmic data science environment (e.g. monitoring billions of synapses), producing an audit log, and crediting remuneration (for example, it could become standard to sell personal sleep data to address research problems in this domain [29]).

33.4 Blockchain Implementation of Cloudminds for Transhuman Futures

Crypto cloudminds (implementing cloudminds with the safeguards of blockchain technology) might be used to realize the transhuman program of exceeding current mental limitations.

33.4.1 Cloudmind Computronium: Ideas is the Currency of the Future

The crypto cloudmind vision is that cloudminds might constitute not only an application, but a next-generation platform for the realization of transhuman futures. Cloudminds could be architected as a blockchain system in which each network node is a mind (human or otherwise). All nodes have equal standing (since crypto cloudminds are a decentralized network) and can provide peer-to-peer (p2p) services to other nodes. The first and foremost peer service is mining: validating, confirming, and registering transactions, in some sort of competitive consensus-driven process for an incentivized reward. All participating nodes subsequently recompute and confirm the validity of new transactions and add them to their copy of the distributed ledger. The crypto cloudmind is likely to be a multicurrency environment: there could be network security and remuneration layers in the stack with cryptocurrencies for each. There could be an algorithmic deep learning layer for optimization with its own native cryptocurrency such as TensorCoin. However, the crucial top-stack value-creation layer could be denominated in the prime currency of *ideas*.

The productivity of crypto cloudminds could be measured in the currency of ideas. Leaderboards or market charts might track and display the top cloudminds and their competitive standing, comparing metrics such as the rate, type, and quality of idea generation. (How many Nobel Prizes did your cloudmind produce?) Economies of the future might be measured by idea generation, and one could buy *idea futures* on a cloudmind.

There could be both problem and solution cloudminds. *Problem cloudminds* could define the list of open problems in different fields, for example, the Fields MathNet cloudmind and the Quantum Gravitational Theory cloudmind. *Solution cloudminds* could attempt to solve known problems, particularly with an intensity of data science methods. Likewise, there could be *affinity cloudminds* that focus on different interest areas such as sustainability, the environment, effective governance, economic equality, space settlement, and other concerns. The *random cloudmind* could be a serendipity-class cloudmind.

Crypto cloudminds are conceptually, an implementation of computronium (a large-scale dedicated problem-solving computational resource), directed and granularized into specific problem areas with blockchain cloudminds. Beyond problem-solving, diverse classes of cloudminds could enable activities such as creative expression, sensory experience, world travel, and other interests.

Crypto cloudmind nodes could be carbon neutral and economically self-sustaining by providing a variety of peer-to-peer services beyond basic network security, and transaction confirmation and logging (mining and ledger hosting). One could be computation markets and p2p banking services, for example, if one node needs additional computing power to further explore an idea, adding hot-swappable computing resources for a micropayment or the cost of computing resources used. Ethereum is already a prototypical computation market, whose virtual machine state management structure could be implemented as a cloudmind.

Another p2p cloudmind service could be the *Yes-and Contract*, entering a payment channel for collaborative idea development. The Yes-and Payment Channel could register the initial *ideabase* (analogous to a codebase or database) to a multi-signature address in a contractually-obligated arrangement. Then, the payment channel would keep track of brainstorming updates as idea development ensues over a period of time, and at the period end, close the contract for the net result, with potential payouts and intellectual property registration. *Creativity Contracts* could be smart contracts to support the innovation and synthesis of creative development. Maslow Smart Contracts could be seen in the cloudmind environment too, as an enhanced future-of-work strategy for the engagement and actualization of human capacities, and as an antidote to the modern woes of technology addiction.

33.5 Practical Considerations: Next Steps for the Implementation of Crypto Cloudminds

33.5.1 Next-Generation Social Networks

Next-generation social networks could be the access platform for crypto cloudminds. Blockchains have been conceived as an enterprise technology, and extending this idea to consider *consumer blockchain applications*, one premise is that individuals would be willing to share more and higher-value information if it were kept private and remunerated [30]. This could lead to a new tier of applications running as an overlay to social networks, unlocking additional consumer value. The notion is extending "likes" to validated (i.e. not fake) opinions, recommendations, and referrals, supported by micropayments, for example, in the areas of personal finance, health, and jobs. The effect could be having a trusted infrastructure in place for sharing sensitive information, eventually with direct links to the brain. Next-generation social networks as a secure access platform could be one of the first implementation steps towards crypto cloudminds.

Blockchain applications could help social network platforms improve their value proposition to consumers through the simultaneous *privacy* and *transparency* properties of blockchains. The contents of transactions remain private (particularly using confidential transactions which mask user and recipient address and amount transferred), while also transparent in knowing that all participants are using the same system. The *shared infrastructure argument* holds in that one party does not know how another party is spending its budget, but that the other party is using the same validated budgeting process. The result is that trust is built by participating in a shared ruleset environment. This kind of technology-based trust-building could be crucial to transhuman futures of multi-species societies of human, algorithm, and machine.

33.5.2 Political and Economic Implications: Geopolitical Reshuffling and Cryptosecession

Smart network technologies such as blockchain (and by implication prototypical cloudminds), are being implemented unevenly around the world. Early-mover countries are defining their competitive advantage by investing in blockchain deployments. For example, Russia instantiated the registration system for its largest domestic airline, S7, in a blockchain system in 2017 [31], and announced aviation refueling smart contracts with Gazprom in 2018 [32]. It is not unimaginable that a *crypto cold war* could ensue as different countries move ahead more aggressively with blockchain implementations.

Energy independence could be a fulcrum point for power rebalancing. For example, California is the world's fifth largest economy, with 33% clean electric power at present and 100% legislated for 2045 [33]. The state could follow Quebec's lead (having a strong political power base due to producing 52% of the country's 60% energy generation from hydroelectric power [34]) in pressing its weight on the national agenda.

Although the longer-term transhuman economy might be denominated in idea production, the nearer-term economy might be measured in intangible social goods production, in addition to traditional material goods production. To the extent that smart network technologies can generate intangible social goods such as trust [35], and also recognition, dignity [36], and justice [37], this could help instantiate safe transhuman futures.

Crypto cloudminds could ease the transition to transhuman futures as some of the other potential stakes of widescale blockchain implementation could include social (re)organization writ large [38]. Blockchain as an institutional technology implies that processes of human interaction currently coordinated by corporations and governments could be relocated to smart network software. There is a real possibility of *cryptosecession* (opting for blockchains as a more expedient alternative to traditional governance and legal structures) [39].

33.6 Risks and Limitations

Two of the biggest potential risks with crypto cloudminds are technical and social, labeled as the "EMP" and "Hello Skynet" situations. "EMP" is the risk of not considering the increasing dependence on technology and the lack of offsetting risk mitigation strategies. Having a response to possible large-scale network failure is necessary, for example disruption resulting from a natural or malicious electromagnetic pulse (EMP). "Hello Skynet" is the risk of anticipated backlash from different communities due to an overly technologized experience of reality, with too much of daily life being controlled by technology. Having a variety of responses and support resources could be crucial to unifying splintering populations and easing the transition to the automation economy. This is particularly true in the wake of the contemporary situation of increasing hours of screentime and decreasing interest in jobs, driving, and dating [40], together with increasing social isolation and suicide rates (up 25% in the U.S. since 1999) [41].

33.7 Conclusion

This essay proposes Transhuman Crypto Cloudminds as a safe way for unleashing collective human brain and machine processing power in a distributed computing system. The cloudmind resource could be a new form of smart network supercomputer for tackling larger-scale problems than have been possible to attempt and solve previously.

There could be two phases in the transhuman future, the surviving and the thriving of a multi-species society comprised of humans, algorithms, and machines. Blockchains are a new form of smart network and global computational infrastructure that could be used to facilitate both phases. Survival and economic sustenance could rely on the blockchain properties of enforcing good-player behavior, since all parties have the incentive to stay in good reputational standing to use the smart network infrastructure to conduct their operations. Economic sustenance might further depend on the cooperative ownership of robotic or automated means of production and Maslow Smart Contracts for professional and personal development.

For the longer-term transhuman vision of thriving by exceeding current physical and mental limitations, crypto cloudminds might provide a safe frontier for joining a collaborative thinking community of human and machine minds. The necessary foundations for crypto cloudminds are two-fold. First could be using next-generation social networks as a possible secure access platform for crypto cloudminds. Second could be big data algorithmics as the AI-driven instantiation of optimized neural processes to control and optimize Brain-Computer Interfaces (the link between physical brains and internet-based cloudminds).

Crypto cloudminds could be architected as blockchain systems, meaning that each mind would be a node in the flat-hierarchy distributed computing system. This could enable very-large scale projects (million-mind plus networks). As such, cloudmind networks could operate via consensus algorithm with peer nodes providing services to each other. This could include network security, ledger hosting, transaction confirmation (via mining), computation resources, and value-added functionality such as Creativity Contracts and Yes-and Payment Channels for collaborative idea development and intellectual property logging. Smart network field theories defined by technophysics models might be used to monitor cloudminds.

The potential impact of this work is that proposing the idea of Crypto Cloudminds offers a specific transhuman vision and path forward with safeguards that supports the future flourishing of human, algorithm, and machine.

Author's Note This essay in the book won the first prize in the Humanity Plus Essay Competition "Mutual Benefits of Blockchain and Transhumanism":

https://humanityplus.org/projects/essay-prize-mutual-benefits-of-blockchain-and-transhumanism/

Glossary

Blockchain (distributed ledger) Technology A blockchain is an immutable, cryptographic (cryptography-based), distributed (peer-based), consensus-driven ledger. Blockchain (distributed ledger) technology is a software protocol for the instantaneous transfer of money and other forms of value (assets, contracts, public records, program states) globally via the internet.

Cloudmind A cloudmind is a cloud-based collaboration of human and machine minds (with safeguards and permissions). "Mind" is generally denoting an entity with some capacity for processing, not the volitionary action and free will of a consciousness agent.

Crypto cloudminds Crypto cloudminds is the idea of implementing cloudminds with the safeguards of blockchain technology.

Cryptoeconomics Cryptoeconomics is an economic transaction paradigm based on cryptography; more specifically, an economic transaction system implemented in a cryptography-based software network, using cryptographic hashes (computational proof mechanisms) as a means of confirming and transferring monetary balances, assets, smart contracts, or other system states. A key concept is *trustless* trust, meaning removing as much human-based trust as possible to make the economic system trustworthy (relocating human-based trust to cryptography-based trust).

Cryptosecession Cryptosecession is the idea of employing blockchains as an institutional technology to opt out of traditional governance and legal structures.

Deep Learning Chains Deep learning chains are a class of smart network technologies in which other technologies, blockchain and deep learning, converge as a control technology for other smart network technologies. Deep learning chains have the properties of secure automation, audit-log tracking, and validated transaction execution of blockchain, and the object and pattern recognition technology (IDtech) of deep learning. Deep learning chains might be used to control other fleet-many internet-connected smart network technologies such as UAVs, autonomous driving fleets, medical nanorobots, and space-based asteroid mining machines.

Deep Learning Neural Networks Deep learning neural networks are computer programs that can identify what an object is; more technically, deep learning is a branch of machine learning based on a set of algorithms that attempts to model high-level abstractions in data by using artificial neural network architectures, based on learning multiple levels of representation or abstraction, such that predictive guesses can be made about new data.

IDtech IDtech is identification technology, the functionality of object recognition as an in-built feature in technology. IDtech is similar to FinTech, RegTech, TradeTech, and HealthTech; technologies that digitize, standardize, and automate operations within their respective domains.

Payment Channel A payment channel is a contractually-obligated payment structure that elapses over time, protecting and obligating two parties who need not know and trust each other [42]. The payment channel operates in three steps. First, Party A opens a payment channel with Party B and posts a pre-paid escrow balance (the escrow deposit is broadcast to the blockchain, and a corresponding refund transaction for the same amount is signed by both parties, but not broadcast). Second, Party A consumes a resource (or provides a service such as programming hours) against the escrow balance, and activity is tracked and updated (in revised refund transactions that both parties sign but do not broadcast). Third, at the end of the period (or at any time), the cumulative activity is booked in one net transaction to close the contract. In addition to the FinTech innovation of parties not knowing each other being able to digitally contract in a protected manner over time, payment channels might also provide scalability to blockchains by only logging net transactions.

Smart Contract A smart contract is a software program registered to a blockchain for confirmation (time-datestamping provenance), and possibly some form of automated execution. To be legally-binding as an eContract, smart contracts need to have the four elements of "regular" contracts: two parties, consideration, and terms.

Smart Networks Smart networks are intelligent autonomously-operating networks. Exemplar smart network technologies include blockchain economic networks and deep learning pattern recognition neural networks.

Technophysics Technophysics is the application of physics to the study of technology (by analogy to biophysics and econophysics), particularly using statistical physics, information theory, and model systems for the purpose of character-izing, monitoring, and controlling smart network systems in applications of arbitrarily-many fleet item management and system criticality detection.

Token Economy The token economy is the situation in which web-based com-munities issue their own cryptotoken money supplies. The tokens serve as an accounting system for coordinating a local economy between members, a track-ing system that can be used to link participative contributions with remuneration (solving the credit assignment problem). In Web 3.0, users expect to participate meaningfully in communities, meaning being remunerated for contributions, accessing resources, and voting on community decisions.

Web 3.0 (the crypto web) Web 3.0 refers to the idea of cryptoeconomic business models such as data markets and computation markets running on the "internet's new pipes" of distributed network systems, content-addressable file-serving, and IDtech. Web 1.0 (the static web) involved the transfer of static information and Web 2.0 (the social web) created the expectation that website users can interact, like, and engage with content and other users. In Web 3.0 (the crypto web), users expect to participate meaningfully in economic communities by being remuner-ated for contributions (such as software code, digital art, and forum posts), and by being able to vote on decisions and access resources.

References

1. Smith, A. (1776, 2003). *The Wealth of Nations*. Blacksburg VA: Thrifty Books.
2. Kurzweil, R. (2006). *The Singularity Is Near: When Humans Transcend Biology*. New York: Penguin Books.
3. Bostrom, N. (2016). *Superintelligence: Paths, Dangers, Strategies*. Oxford: Oxford University Press.
4. Dowd, M. (2017). Elon Musk's Billion-dollar Crusade to Stop the AI Apocalypse. *Vanity Fair*.
5. Miller, R.S. and Shorter, G. (2016). High Frequency Trading: Overview of Recent Developments. *U.S. Congressional Research Service*. 7–5700. R44443. https://fas.org/sgp/crs/misc/R44443.pdf.
6. Kirilenko, A., Kyle, A.S., Samadi, M., and Tuzun, T. (2014). The Flash Crash: The Impact of High Frequency Trading on an Electronic Market. *U.S. CFTC*.
7. Cowen, T. (2013). *Average Is Over: Powering America Beyond the Age of the Great Stagnation*. New York: Dutton.
8. Frey, C.B. and Osborne, M.A. (2013). The Future of Employment: How susceptible are Jobs to Computerisation. Oxford. https://www.oxfordmartin.ox.ac.uk/downloads/academic/The_Future_of_Employment.pdf See also: https://medium.com/oxford-university/the-future-of-work-cf8a33b47285.
9. The Economist. (2016). US Population Survey. *Federal Reserve Bank of St. Louis*. https://www.federalreserve.gov/publications/files/2016-report-economic-well-being-us-house-holds-201705.pdf.
10. Bland, B. (2016). China's robot revolution. *Financial Times*. https://www.ft.com/content/1dbd8c60-0cc6-11e6-ad80-67655613c2d6.
11. Chan, J. (2017). Robots, not humans: official policy in China. *New Internationalist*. https://newint.org/features/2017/11/01/industrial-robots-china.

12. Swan, M. (2015). Blockchain Thinking: The Brain as a DAC (Decentralized Autonomous Corporation). *IEEE Technology and Society.* 34(4):41–52.
13. Swan, M. (2019). Blockchain Economic Theory: Digital Asset Contracting reduces Debt and Risk. In Swan, M., Potts, J., Takagi, S., Witte, F., Tasca, P., Eds. *Blockchain Economics: Implications of Distributed Ledgers – Markets, communications networks, and algorithmic reality.* London: World Scientific.
14. State of the Dapps. Retrieved October 7, 2018: https://www.stateofthedapps.com/.
15. Swan, M. (2017). Is Technological Unemployment Real? Abundance Economics. In *Surviving the Machine Age: Intelligent Technology and the Transformation of Human Work.* Eds. James Hughes and Kevin LaGrandeur. London: Palgrave Macmillan. 19–33.
16. Scholz, T. (2016). *Platform Cooperativism. Challenging the Corporate Sharing Economy.* New York: Rosa Luxemburg Stiftung.
17. Anzilotti, E. (2018). Finland's Basic Income Pilot Was Never Really A Universal Basic Income. *Fast Company.* https://www.fastcompany.com/40565075/finlands-basic-income-pilot-was-never-really-a-universal-basic-income.
18. Swan, M. (2015). *Blockchain: Blueprint for a New Economy.* Sebastopol CA: O'Reilly Media.
19. Hebblethwaite, C. (2018). IDC: Global blockchain spending to hit $9.2 billion in 2021. *The Block.*
20. World Economic Forum. (2015). Deep Shift: Technology Tipping Points and Societal Impact. *Survey Report.*
21. Davidson, S., de Filippi, P., Potts, J. (2018). Blockchains and the Economics institutions of capitalism. *Journal of Institutional Economics.* 1–20.
22. Swan, M. (2018). Blockchain Economics: "Ripple for ERP" integrated blockchain supply chain ledgers. *European Financial Review.* Feb-Mar: 24–7.
23. Merkle, R. (2016). DAOs, Democracy and Governance. *Cryonics Magazine.* July-August. 37(4):28–40.
24. Swan, M., dos Santos, R.P. (2018-Submitted). Smart Network Field Theory: The Technophysics of Blockchain and Deep Learning. *Concurrency and Computation: Practice and Experience.* Wiley.
25. Swan, M. (2016). The Future of Brain-Computer Interfaces: Blockchaining Your Way into a Cloudmind. *Journal of Evolution and Technology.* 26(2).
26. "BINA48." *Wikipedia.* Retrieved October 7, 2018: https://en.wikipedia.org/wiki/BINA48.
27. Rock, D. and Grant, H. (2016). Why Diverse Teams are Smarter. *Harvard Business Review.*
28. Martins, N.R.B., et. al. (Submitted). Human Brain/Cloud Interface.
29. Walker, M. (2018). *Why We Sleep: Unlocking the Power of Sleep and Dreams.* New York: Scribner.
30. Swan, M. (2018). Blockchain consumer apps: Next-generation social networks (aka strategic advice for Facebook). CryptoInsider.
31. Zhao, W. (2017). A Russian Airline Is Now Using Blockchain to Issue Tickets. *Coindesk.*
32. Gazprom. (2018). Gazprom Neft and S7 Airlines become the first companies in Russia to move to Blockchain Technology in Aviation Refueling. *Press Release.*
33. Domonoske, C. (2018). California Sets Goal Of 100 Percent Clean Electric Power By 2045. *NPR.*
34. National Energy Board. (2018). Provincial and Territorial Energy Profiles. *Government of Canada.* Retrieved October 7, 2018: https://www.neb-one.gc.ca/nrg/ntgrtd/mrkt/nrgsstmprfls/cda-eng.html.
35. Sherchan, W., Nepal, S., Paris, C. (2013). A Survey of Trust in Social Networks. *ACM Comput Surv.* 45(4). Pp. 47:1–47:33.
36. Harris, L. (2016). Dignity and Subjection. *Présence Africaine.* 1(193):141–159; 59–77.
37. Young, I.M. (2011). *Justice and the Politics of Difference.* Princeton: Princeton University Press.

38. Swan, M. (2018). Blockchain Enlightenment and Smart City Cryptopolis. *CryBlock'18 Proceedings*. Workshop on Cryptocurrencies and Blockchains for Distributed Systems. Munich, Germany. June 15, 2018. Pp. 48–53.
39. Allen, D.W.E. (2019). Entrepreneurial Exit: Developing the Cryptoeconomy. In Swan, M., Potts, J., Takagi, S., Witte, F., Tasca, P., Eds. *Blockchain Economics: Implications of Distributed Ledgers – Markets, communications networks, and algorithmic reality*. London: World Scientific
40. Twenge, J.M. (2017). Have Smartphones Destroyed a Generation? *The Atlantic*. https://www.theatlantic.com/magazine/archive/2017/09/has-the-smartphone-destroyed-a-generation/534198/.
41. CDC. (2018). Suicide rising across the US: More than a mental health concern. *CDC Report*.
42. Swan, M. (2018). Smart Network Economics: Payment Channels. https://www.slideshare.net/lablogga/smart-network-economics-payment-channels.

Chapter 34
Transhumanism and Distributed Ledger Technologies

Calem John Smith

Whilst peering into the future on our journey towards a type 2 civilisations, both certain precautions and assumptions need to be made. From the dire warnings related to superintelligent machines spawned from artificial general intelligence to the awe inspiring nature of Moore's Law and the ever nearing approach of the technological singularity, predicting the future has always been very difficult for the human race as we have lacked the interconnectedness in order to establish any form of true global consensus. This is now rapidly changing due to the rise of the internet and social media. The effect known as Dunbar's number is declining. Trust between large groups of individuals using various forms of reputation systems is allowing for the first time, the ability for society to collaborate in trustless environments.

In the 1990's, a missing core component needed for societies evolution, Distributed Ledger Technologies, was been developed amongst a small group of cypherpunks, cryptographers and anarchists. Over 20 years later, Bitcoin (BTC) the first major DLT that launched in 2009, is at present assisting global on-boarding along with other major cryptocurrencies paired to fiat such as Etheruem (ETH) and Tether (USDT).

The rise of DLT Turing Completeness, smart-contract platforms such as ETH and Bitcoin Cash (BCH), has already brought us a range of tools enabling further scaling of trustless relationships within society. Platforms for self governance and swarm intelligence, to name a few, are necessary in order for society to continue scaling efficiently. Decentralised Autonomous Organisations (DAO) and Decentralised Prediction Markets (DPM) such as Augur built on the ETH network. These software networks when combined with Peer to Peer Cubesat Mesh Networks such as Nexus and their partnership with Vector Launch Inc., the likes of Alphabet Inc's Project Loon now have open source alternatives tackling the challenge. DLTs and consumer affordable p2p hardware networks will soon solve the scalability issue's challenging the Internet of Things (IoT).

C. J. Smith (✉)
Zerozed, Perth, Australia

© Springer Nature Switzerland AG 2019
N. Lee (ed.), *The Transhumanism Handbook*,
https://doi.org/10.1007/978-3-030-16920-6_34

529

Technological Unemployment (TU) and the Diffusion of Innovations (DoI): One cannot talk about the former without prior having discussed the latter. The DoI is a primary driver behind TU. It describes the process in which new technologies are adopted across all markets and groups of social networks within society. The rate and stages of the DoI are firmly understood and is applicable across many fields much like Network Theory which also plays a major role in understanding how technology itself evolves over time. There are five stages each with their own adopter traits. Innovators, Early Adopters, Early and then Late Majority finalised with the Laggards. All five processes must occur for the technology to have attained successful diffusion. A failed diffusion merely means 100% market share was never attained before becoming obsolete.

As new technologies become ever increasingly accessible for not just businesses and corporations but the general public as a whole, we see job transformation as automation replaces humans in one area and creates new opportunities for humans in markets that could not have existed without said new and innovative technology. One only needs to use the example of automated cashiers creating more technical positions across multiple industries than what it is removing from the human-teller segment of the retail job market. DLTs are the latest technological innovation that will attain successful diffusion and will speed up not only the product life-cycle across all sectors but will have an ever increasing effect on Technological Unemployment.

A purely peer-to-peer version of electronic cash with on-chain scaling now exists with Bitcoin Cash. The issue at hand is the emergence of the ever increasing competing alternatives and the additional volatility this brings to the market as a whole. Due to the nature of society's adoption rate of new technologies and the standard block reward halving mechanism employed in 99% of Proof of Work networks, there exists the need for a stable and liquid "crypto-currency" as a major fiat entry point into all pairs. Tether is one solution currently in operation but is highly centralised in regards to the distribution of the tokens on the network. Keep in mind that not even 10% of the population has adopted DLTs.

It has become common practice for coins and tokens to either follow the halving standard set by Satoshi or are to be completely produced up front like the common Initial Coin Offering (ICO). The crypto-community as a whole has come to rely almost exclusively on centralised exchanges serving as trusted third parties for fiat entry into the broader crypto market. While the system has opened necessary channels for the technology to diffuse, it cannot solve the inherent flaw built into the inflation model of which 99% of DLT networks share. Tether is one such potential solution but inadvertently is at the mercy of the US Dollars inflation rate.

What is needed is an alternative to Tether built on its own Proof-of-work backbone in order to also break away from the ties to the US Dollars inflation rate. Using the Diffusion of Innovations to base an inflation model, along with an industry supported tried and true mining algorithm and cross-chain transactions such as the Atomic Swap network, a potential now exists to solve a long time debated issue surrounding not just volatility but how to achieve a Successful diffusion whilst maintaining the normal distribution and a standard score of $0z$. In other words how do we create a stable crypto-currency that is, relatively speaking, evenly dispersed between everyone alive once new coins stop been created?

As society has evolved over time, so has our technology, methods of communication, co-ordination and governance. Until now, we have lacked the software and infrastructure to run a true capitalist society. The Digital Revolution has enabled the masses to see and understand the inefficiencies and inner workings of The State and Fractional-reserve banking systems. There is no arguing that politicians become corrupt by nature and are at the mercy of their keys to power.

States of Technocracy and Corporatocracy are well known tropes within the Cyberpunk subculture. These tales are foreshadowing and the establishment of Governance models alike are very real threats to Transhumanism. The cypherpunks and anarchists of the past have brought us to a point where DLTs and Capitalism can assist in the Technological Unemployment of The State. Crypto-Anarcho-Capitalism is the transition from, dis-integration, disposal and logical replacement of the State by means of technology and more specifically, open source cryptographic software such as DLT's. This is a process that has already begun.

Picture a society where the Government as you know it is no longer required. Where inefficient State run services have been replaced by an unrestricted free market and open source cryptographic peer to peer networks. In example, a Universal Basic Income (UBI) will be attained via a DoI based DLT in order to replace current forms of public welfare. A world where self governance and global coordination via the use of technology, networks and encryption removes the majority of the incentives for war and corruption, we will create a global community where the economic incentives are so that it is within the best interests of corporations to work together, as opposed to undermine each other in order to serve their customers needs best.

Guided by the fundamentals of Transhumanism, CAn-Cap is the merger of the Anti-Capitalist philosophy, Crypto-Anarchy, and the Stateless society envisioned by Anarcho-Capitalist where public servants are no longer protected from Technological Unemployment allowing the free market to provide for the people in its place. The transformation of human incentivisation within a CAnCap society will allow the transition to a Resourced Based Economy (RBE) whilst avoiding the emergence of a Technocracy or Corporatocracy.

With all of societies scalability issues solved, we open up the potential to use Brain-computer Interfaces (BCI) to perform Whole Brain Emulation (WBE) enabled via Distributed Ledger Technologies. This will essentially allow people to not only perform automated incremental backups of their consciousness but also tap into both distributed artificial and peer-to-peer organic, neural-networks. Mind Uploading and Human-DLT Augmentation enabled by the Technological Singularity, will see the emergence of the Transhuman species, Homo Technicum, and the beginning of the Societal Singularity. With resources, time and space no longer limitations to the scalability of society, the inevitable accomplishment of Dyson Sphere's, Jupiter Brain Networks (JBN) and Quantum-based Matrioshka Brains (QMB) will allow us to make the step to a type 3 civilisations. Superintelligent and able to take on forms from organic to digital, the apocalyptic AI we so greatly fear, will in reality be us on our never ending quest for information in all its forms. Extropy.

Chapter 35
DoI-SMS: A Diffusion of Innovations Based Subsidy Minting Schedule for Proof-of-Work Cryptocurrencies

Janez Trobevšek, Calem John Smith, and Federico De Gonzalez-Soler

35.1 Introduction

One of the greatest ethical and logistical challenges humanity has been forced to face is one that has not changed since the dawn of society itself. The fair and even distribution of wealth and power. We live in world where half of the world's net wealth belongs to 1% of the population.

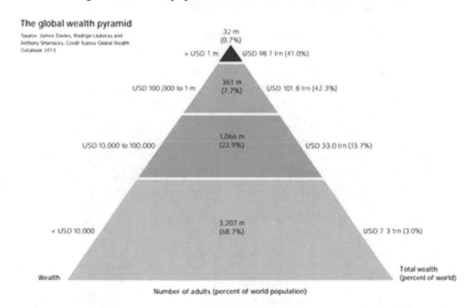

J. Trobevšek · F. De Gonzalez-Soler
Cryptoz, Blair Athol, Australia

C. J. Smith (✉)
Zerozed, Perth, Australia

© Springer Nature Switzerland AG 2019
N. Lee (ed.), *The Transhumanism Handbook*,
https://doi.org/10.1007/978-3-030-16920-6_35

https://www.credit-suisse.com/corporate/en/articles/news-and-expertise/global-wealth-pyramid-decreased-base-201801.html

The emergence of Cryptocurrency has brought about a rapid disruption across many industries but none more than Finance and Economics. Due to the inherent flaw in Bitcoin and by extension 99% of Blockchain based Distributed Ledger Technologies, "Crypto" has succumbed to the same fate as Fiat based currencies, riddled with systemic issues and mass centralization. This is all due to the inflation model otherwise known as the "halving-mechanism", one of which no one has seemingly questioned.

The core incentive model for current Crypto heavily relies on competition and scarcity, much like gold, giving the network and the coin itself intrinsic value which allows it to be priced based on supply and demand. This old model does not work in the new system that Satoshi was trying to achieve. Over 99% of the World's population has yet to acquire any cryptocurrency at all. Roughly 80% of the supply has already been mined with the remaining 20% to take the next 100 years to produce, the bulk of which will be exhausted by 2033.

Supply has been so tightly confined into the hands of the few, extreme price instability and supply centralization has become the norm due to the onboarding supply bottleneck created by the mechanism itself. Bitcoin has failed to achieve diffusion but we can use this to bootstrap a radical and innovative Socioeconomic Ecosystem to organically and passively redistribute the World's wealth.

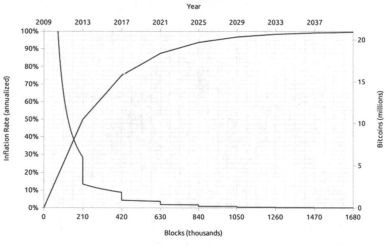

Bitcoin Inflation vs. Time

35.2   The Diffusion of Innovations

To quote Wikipedia, "The Diffusion of innovations is a theory that seeks to explain how, why, and at what rate new ideas and technology spread." It describes the process in which new technologies are adopted and diffused across all markets and social groups within society. The process, rate and stages of the DoI are firmly understood and play a major role in understanding how technology itself evolves over time [1].

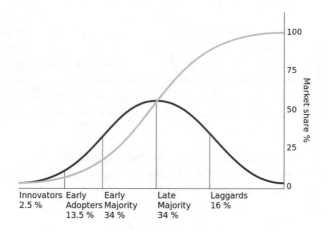

There are five categories of adopters within the social system, each with their own traits. Innovators, Early Adopters, Early and then Late Majority finalized with the Laggards. All five process must occur for the technology to attain market saturation. Failed diffusion merely implies 100% adoption was never attained before succumbing to technical obsolescence [1].

35.3   Normal Distribution

To quote Wikipedia, "In probability theory, the normal (or Gaussian or Gauss or Laplace–Gauss) distribution is a very common continuous probability distribution [2]."

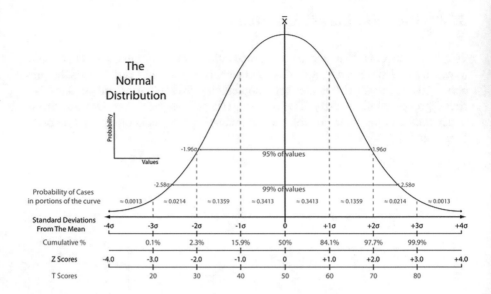

Achieving a normal distribution with a standard score of 0Z within a socioeconomic environment implies the middle class make up the greatest majority with the lower and upper classes of society are by far the lesser in prevalence.

35.4 Incentive

The core incentive model for current Crypto heavily relies on competition and scarcity, much like gold, giving the network and the coin itself intrinsic value which allows it to be priced based on supply and demand. This old model does not work in the new system that Satoshi was trying to achieve. Less than 1% of the World's population has acquired any cryptocurrency at all. 80% of the supply has already been mined with the remaining 20% to take the next 100 years to produce, the bulk of which will be exhausted by 2033. Supply has been so tightly confined into the hands of the few, instability and wash trading has become the norm, leaving the world begging for Government intervention via ETFs and regulation [3].

Diar Bitcoin Distribution Estimates (21Mn BTC)

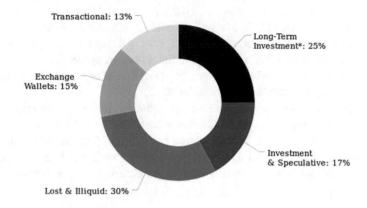

*Lost & Illiquid includes unmined Coins – **Updated**: 18 September 2018 - 12PM EEST

Bitcoin has failed to achieve diffusion but we can use this to bootstrap a radical and innovative Socioeconomic Ecosystem to organically, passively and evenly re-distribute the global population's net wealth.

35.5   Altruistic Egoism

It is within the individuals best interests for the standard living of those below them be closer to the mean as this leads to slower population growth, the development of technology and the diffusion of innovations, further-spread education and the tools necessary to enable the individual. This in turn increases abundance and the broader accessibility of resources for all. Demand-side economies of scale such as Bitcoin, rely on network effect both direct and indirect whilst encouraging interoperability in order to achieve successful diffusion. Most of the elements come together but ultimately altruistic egoism falls apart under the current incentive model backing cryptocurrencies.

Due to Bitcoins competitive as opposed to collaborative incentive model we have now been presented with a sharing of the technology in what can only be described as a mad land grab for digital assets. HODL, the act of hoarding and not spending the crypto one acquires, is not only prevalent but encouraged and completely endorsed by some of the most prominent figures in the space. Alongside FOMO, the fear of missing out, HODL is an attitude that is both a symptom and contributor to the many systemic issues heavily hindering adoption of cryptocurrency amongst the broader public of whom are yet to onboard.

Within a Diffusion of Innovations based inflation model, no one misses the boat. The earliest of adopters are still rewarded whilst maintaining fairness, collaboration and sharing in the system. When inflation grows in proportion to the amount users on a network, a close-to-optimal balance of supply and demand can be met ensuring abundance to stimulate growth whilst still factoring in the intrinsic value of scarcity.

35.6   Process

We seek an alternative to the standard block-halving mechanism employed by the majority of Cryptocurrencies in order to break away from the ties to not only the US Dollars inflation rate but also the supply centralizing nature of Bitcoins halving schedule.

By utilizing the Diffusion of Innovations to base an inflation model, along with an industry supported tried and true mining algorithm as well as cross-chain transactions via Atomic Swaps, a potential now exists to solve a long time challenge surrounding not just volatility but how to achieve successful diffusion whilst maintaining the normal distribution with a standard Z score of zero.

We theorized subsidy should be governed by findings regarding technological adoption rates and behaviors popularized in 1962 by Everett Rogers. We proposed to use what resembles a Logistic function, the Diffusion Of Innovations (DOI) sigmoid curve, as the baseline for controlling the inflation of supply. The initial failed model is pictured below.

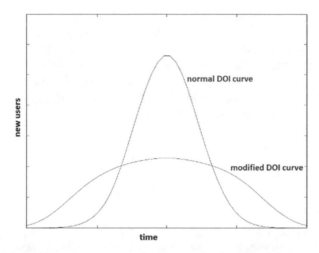

"A number of gaussian curves are used there, to model the inflow of new people, but because the crypto field tends to expand faster and with steeper onsets and more leveled tops, we initially used a reciprocal-inverse exponentially modified gaussian curve. After a failed launch, we realized the beginning needs to be steeper, and that starting funds are mandatory to avoid initial price surge." – Hiyatus

Taking into consideration Moore's Law, we modified the standard DoI curve with a more "broader" middle and longer ramp-up in order to account for the initial difficulties which attribute to the lack of network effect Cryptocurrencies experience because of the competitive nature within the broader crypto economy and online social environments. Due to proof-of-work based cryptocurrencies needing as many miners as possible in order to stay secure and maintain consensus, the first leg of the modified DOI was made marginally steeper compared the final leg, in order to attract the critical number of miners required to secure the chain as viable as possible.

Based on other projects observed and worked on prior to the launch of xUz, our MVP demonstrating DoI-SMS, we made certain assumptions and extrapolated the rate of adoption that should be expected. From this we generated a set of integers and applied a recursive fitting algorithm to them. The result of our findings was a polynomial of 8th degree, which we termed 'DoI-SMS' and is the focus of this paper.

Project funding was generated as a product of securing initial supply via an 8% premine in order to prevent early hijacking as was experienced in one of a number of failed launches of x0z itself. This premine becomes heavily diluted and diffused within the first half of the initial inflation cycle of 60 months. This is a necessity because bootstrapping funding and network security via methods such as ICO breaks the incentive model due to the fact that this essentially creates additional supply outside of the inflation models control.

35.7   Algorithm

Converting to C code...

```
#include <math.h>
#int64 static GetBlockValue(int nHeight,  int64 nFees)
{
    double nSubsidy = 1 * COIN;
    double nsubsidy_function = 0;
    double Xheight = 0;
    if (nHeight == 1)
    {
    nSubsidy = 2000000.0 * COIN;
    }
    else if (nHeight > 1 && nHeight < 1274030)
    {
    Xheight = nHeight * 0.0000038051750381;
    nsubsidy_function = ((3583.5719028332051*(pow(Xheight,8))) -
(67959.212902381332*(pow(Xheight,7))) + (500144.30431838805*(pow(Xheight,6))) -
(1806581.9194472283*(pow(Xheight,5))) + (3537339.4754780694*(pow(Xheight,4))) -
(4712758.2800668897*(pow(Xheight,3))) + (4535015.6408610735*(pow(Xheight,2))) +
(834937.06954081857*Xheight) + (1000845.7073113875));
    nSubsidy = ((floor((nsubsidy_function*(1.0/60000.0)*0.33757734
955)*100.0))/100.0) * COIN; // our emission curve [no. of coins per
block]
    }
    if nHeight == 1274030
    nSubsidy = (25000000 - nTotalExisting) * COIN;
    else
    {
    nSubsidy = 0 * COIN;
    }

    return nSubsidy + nFees;
```

35.8   Inflation

The DoI-SMS system is being tested within the Cryptocurrency Zerozed (x0z), formerly known as "Crypt0z". Zerozed will undergo at least two inflation cycles. The first 60 month period will be used to gather the required information to retarget supply in order to continue a fair distribution to those still yet to onboard with cryptocurrency as a whole. As of block height 87,294, the network is currently on target with a block reward of 9.23 and total supply of ~2,623,959 x0z.

	A	B	C	D
	Month	Coins	Height	Subsidy
1	1	2,126,860	21,600	6.17
2	2	2,268,500	43,200	6.98
3	3	2,430,030	64,800	8
4	4	2,615,390	86,400	9.18
5	5	2,827,630	108,000	10.49

https://chainz.cryptoid.info/x0z/block.dws?34f0eba87d8a4fb6b36416ed2f436a74b9dcd07e44a971b2172ca0840d910529.htm

35.9   Scaling

In order for a cryptocurrency operating on a DoI based inflation model to achieve successful diffusion, certain assumptions need to be made and a number of unknowns acknowledged.

Assumptions

- Less than 1% of the population has been on-boarded. Currently the safest estimate.
- Development of On-chain scaling continues to prove promising.
- Ray Kurtwhiles estimations on the Singularity remain true.
- Crypto adoption will continue to grow over the networks initial inflation cycle.

Unknowns

- The amount of time required for cryptocurrency to gain critical mass.
- Percentage of Total Market Capitalization by the end of the initial inflation cycle.

"After the initial 5 year test, x0z will scale up with a new DOI cycle. Lasting longer and yielding a substantially higher number of coins to account for a continental adoption.

If all goes well, there will be a third DOI cycle for global x0z adoption. In this regard we can consider every new DOI cycle as a fractal of the previous - just like we can observe it in nature." – Hiyatus

35.10   Issues and Corrections

The presented implementation has a small potential for creating orphaned blocks. This is because the equation relies heavily on floating point operations and different systems will use varying numbers of them. In certain cases this, one per roughly every 150 blocks, may lead to two different values for that particular blocks subsidy and thus one of the two calculations will be orphaned based on Bitcoin's consensus rules.

While this does not pose a serious threat for x0z security, it does create an annoyance for the miners. To solve this in code, the function governing the x0z DOIsms and pertaining rules, should be rewritten using CBigNum instead of int64. With this, all miners with enough ram can handle the necessary number of floating points and will always obtain the same subsidy value without ever drifting. As the successor of DoI-SMS, the Zero Zed Algorithm, is in development, a substitute table is now currently in place to continue the planned supply rate without deviation.

35.11   Trustless Financial Public Services

A potential opportunity to combine suitable technologies in order to create a Universal Basic Income for users of the network exists within a DoI-based Cryptocurrency. By utilizing a fee algorithm controlled by on-chain contracts, alongside the Diffusion of Innovations as the basis for the coins release schedule, we can not only achieve a Universal Basic Income but other trustless public services like decentralized Superannuation and Universal Basic Healthcare for those utilizing the system.

35.12   Conclusion

We have proposed an inflation model for Cryptocurrencies yet to launch, as well as coins existing today, to set or retarget subsidy schedules in order to follow a more sustainable and diffusion-viable incentive model. By utilizing the Diffusion of Innovations we can establish the Normal Distribution within a socioeconomic environment and the Zerozed team have set out to prove this via bootstrapping a brave new Cryptocurrency with our MVP, x0z.

The Zerozed team will endeavor to contribute and assist with broader crypto innovations and interoperability in order to provide all the necessary conditions which provide the best chance of survival within the ever evolving space. If proven successful, the Zerozed project is expected to thrive well beyond the initial inflation cycle and is set to bring to light information that will assist the industry as a whole, in turn gaining broader acceptance and diffusion of Cryptocurrencies and Distributed Ledger Technologies through society.

Zerozed (x0z) Initial Inflation Schedule

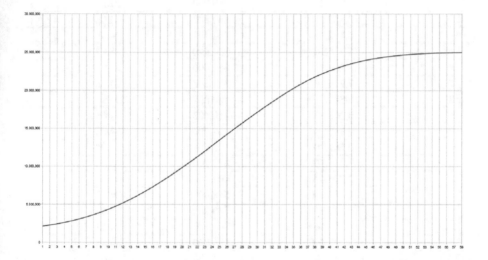

Month	Coins	Height	Subsidy
1	2,126,860	21,600	6.17
2	2,268,500	43,200	6.98
3	2,430,030	64,800	8
4	2,615,390	86,400	9.18
5	2,827,630	108,000	10.49
6	3,069,110	129,600	11.89
7	3,341,650	151,200	13.36
8	3,646,630	172,800	14.89
9	3,985,060	194,400	16.45
10	4,357,640	216,000	18.05
11	4,764,740	237,600	19.65
12	5,206,440	259,200	21.25
13	5,682,520	280,800	22.83
14	6,192,420	302,400	24.38
15	6,735,240	324,000	25.87
16	7,309,740	345,600	27.31
17	7,914,300	367,200	28.66
18	8,546,940	388,800	29.9
19	9,205,290	410,400	31.03
20	9,886,640	432,000	32.03
21	10,587,900	453,600	32.88
22	11,305,700	475,200	33.56
23	12,036,400	496,800	34.06
24	12,775,900	518,400	34.38
	13,520,300	540,000	34.51
26	14,265,300	561,600	34.43
27	15,006,400	583,200	34.16
28	15,739,500	604,800	33.68
29	16,460,200	626,400	33.01
30	17,164,300	648,000	32.16
31	17,848,000	669,600	31.12
32	18,507,500	691,200	29.92
33	19,139,600	712,800	28.58
34	19,741,200	734,400	27.11
35	20,309,900	756,000	25.53
36	20,843,500	777,600	23.87
37	21,340,600	799,200	22.15
38	21,800,100	820,800	20.4
39	22,221,700	842,400	18.63
40	22,605,200	864,000	16.89
41	22,951,400	885,600	15.18
42	23,261,300	907,200	13.52
43	23,536,200	928,800	11.95
44	23,778,200	950,400	10.47

Month	Coins	Height	Subsidy
45	23,989,400	972,000	9.1
46	24,172,100	993,600	7.84
47	24,329,000	1,015,200	6.71
48	24,462,700	1,036,800	5.69
49	24,575,700	1,058,400	4.79
50	24,670,500	1,080,000	4.01
51	24,749,500	1,101,600	3.32
52	24,814,700	1,123,200	2.73
53	24,867,900	1,144,800	2.21
54	24,910,500	1,166,400	1.75
55	24,943,700	1,188,000	1.33
56	24,968,400	1,209,600	0.96
57	24,985,400	1,231,200	0.62
58	24,995,600	1,252,800	0.33

References

1. Rogers, Everett (16 August 2003). Diffusion of Innovations, 5th Edition. Simon and Schuster. ISBN 978-0-7432-5823-4.
2. Article, The Normal Distribution https://en.wikipedia.org/wiki/Normal_distribution
3. Image, Diar Bitcoin Distribution Estimates (21Mn BTC) https://diar.co/volume-2-issue-37/
4. Block height 87294 https://chainz.cryptoid.info/x0z/block.dws?34f0eba87d8a4fb6b36416ed2 f436a74b9dcd07e44a971b2172ca0840d910529.htm

Part VI
Art, Literature, and Films

Chapter 36
In the Afterglow of the Big Bang: Science, Science Fiction, and Transhumanism

Judith Reeves-Stevens and Garfield Reeves-Stevens

In Stanley Kubrick's and Arthur C. Clarke's classic science-fiction movie, *2001: A Space Odyssey,* an iconic "dissolve" transforms a hominid's thrown club into an orbiting weapons platform. The startling juxtaposition of images succinctly illustrates that transhumanism is today's name for a process humans have employed since the species began: Using our intellect to improve and augment our inherent capabilities in order to prolong our survival.

The club multiplies the force that can be used to bring down prey, or an enemy. The sling, the spear, the arrow, extend our reach. Armor – in the form of leather, then metal, now Kevlar and, soon, graphene – improve our odds of survival when confronting the force-multipliers employed by others.

Similarly, in medicine there's a bright line of progression between our ancestors' ingestion of specific plants to treat basic ills and our current technologies for completely eliminating more complex conditions by the editing of specific genes. Hunting techniques painted on cave walls, written language on scrolls, books, and now digitized libraries of millions of volumes sifted and recombined by AIs, are all mechanisms for preserving and expanding knowledge independent of an individual's memory. The clothing that enhanced our capability for temperature regulation, allowing us to spread from Africa into otherwise inhospitable habitats, connects directly over millennia to the intricate layers of an astronaut's spacesuit, whether designed for survival in zero-gravity space, operations on the airless, low- gravity Moon, or, soon, the near-airless, not-quite-as-low gravity Mars.

Here we hit a limit. Let's examine it.

***

NASA astronaut Scott Kelly is familiar with spacesuits and the challenge of living in space. He spent 340 consecutive days on the International Space Station for NASA's One-Year mission. Recounting his experiences in his book, *Endurance,*

J. Reeves-Stevens (✉) · G. Reeves-Stevens
Victoria, Canada
e-mail: jg@reeves-stevens.com

© Springer Nature Switzerland AG 2019
N. Lee (ed.), *The Transhumanism Handbook*,
https://doi.org/10.1007/978-3-030-16920-6_36

Kelly maintains, "Our space agencies won't be able to push out farther into space, to a destination like Mars, until we can learn more about how to strengthen the weakest links in the chain that makes spaceflight possible: the human body and mind."

Why should we push out as Kelly suggests, given the apparently insurmountable obstacles to survival in space?

Former NASA Administrator Michael Griffin said it bluntly: "In the long run, a single-planet species won't survive." Renowned scientist Stephen Hawking provided a timeline: "Although the chance of a disaster to planet Earth in a given year may be quite low, it adds up over time, and becomes a near certainty in the next thousand or ten thousand years. By that time we should have spread out into space, and to other stars, so a disaster on Earth would not mean the end of the human race."

Between Kelly and Griffin, humanity seems caught in a cosmic Catch-22, and Hawking's clock is ticking: Our species won't survive if we don't travel into space to live on other worlds, yet individual humans can't survive in space, let alone live on any other planet currently within reach of our technology. What to do?

The solution, to us and many others, is obvious, and has been a staple of fantastic tales since before the term "science fiction" was born.[1] The concept of transhumanism must expand to become more than a process by which we improve and augment our inherent capabilities. It must become a process by which we create fundamentally *new* capabilities.

***

As science-fiction writers, we've addressed this concept in our work. Garfield's novel, *Nighteyes,* posits that the alien "greys" popular in UFO literature are a variant of far-future humans genetically engineered to fight an interplanetary war. For the *Star Trek: Enterprise* episode, "Terra Prime,"[2] we wrote an exchange between John Frederick Paxton, leader of a twenty-second-century movement that believed in restricting Earth to humans only, and T'Pol, a Vulcan who had discovered her DNA had been merged with a human's to create a human-Vulcan hybrid.

Paxton is incensed at seeing T'Pol caring for the baby. "This creature is half human," he says. "How many generations until the word 'human' is nothing more than a footnote in some medical text?" T'Pol, quite logically, of course, replies, "Neither of our species is what it was a million years ago, nor what we'll become in the future. Life is change."

Variations like the epicanthic fold, skin color, and differing metabolisms in current human populations are recognized as examples of how life changes over time – naturally evolving adaptations to specific environmental conditions. The idea that other adaptations could be required for humans to live in otherworldly environments has a long history in the realms of both science and science fiction, each inspiring practitioners of the other.

---

[1] Hugo Guernsback first used the term "scientifiction" in 1915, which popularly became shortened to "science fiction" in the 1930s.

[2] Story by André Bormanis & J&G Reeves-Stevens, Teleplay by Manny Coto & J&G Reeves-Stevens.

In a message he recorded for future explorers of Mars,[3] Carl Sagan said, "Science and science fiction have done a kind of dance over the last century… The scientists make a finding, it inspires science-fiction writers to write about it, and a host of young people read the science fiction and are excited and inspired to become scientists… which then feeds again into another generation of science fiction and science."

Transhumanism is no exception.

***

In 1923, in his widely cited essay, *Daedalus, or Science and the Future*, scientist J.B.S. Haldane raised the concept of adapting humans to life on "other planets, satellites, asteroids, or artificial vehicles," possibly by "gene grafting." Seven years later, in the epic science-fiction saga, *Last and First Men*, Olaf Stapledon covers billions of years of future human history, during which 17 different versions of humanity arise from us, the first, original humans, sometimes through natural evolution, and sometimes by deliberate design. Specifically, when the "Eighth Men," adapted to life on Venus, face the certain destruction of that planet and the inner solar system caused by the expanding sun, they design the "Ninth Men" to live on Neptune:

> "Inevitably, it was a dwarf type, limited in size by the necessity of resisting an excessive gravitation…" (That Neptune has no definable solid surface wasn't determined until decades later.)

For his part, Haldane had earlier suggested in his essay that humans adapted to live in the high-gravity environment of the "solid or liquid surface of Jupiter," might fare better as a four-legged version.

As these transformative concepts that would come to be labeled transhumanism circulated among scientists and philosophers, science-fiction writers explored the idea as well. In 1952, science-fiction writer James Blish had even come up with a name for it: pantropy.[4]

Ten years earlier, in a collection of stories eventually combined as the novel, *Seetee Ship*, science-fiction writer Jack Williamson had coined a word for the concept of transforming an entire alien world to make its environment suitable for humans: terraforming. Blish had taken the opposite approach: Pantropy means transforming *humans* to be suitable for an existing environment of an alien world. In his short story, "Surface Tension," a group of microscopic aquatic humans discover they are the descendants of human colonists whose ship crashed on their world, then genetically redesigned their progeny to thrive in the planet's unique ecosystem.

***

---

[3] Sagan recorded the message in 1995, for inclusion on disks carried by the two landers that were part of Russia's Mars '96 probe. Though that mission didn't succeed, the message endures.

[4] As another example of the back-and-forth dance of science and science fiction, Blish is also credited with coining the now commonly used term, "gas giant," to describe planets like Jupiter and Saturn.

In considering transhumanism, biology is not the only agent of change, and in both science and science fiction the concept of adding technological enhancements to humans became of interest. One of the first scientists to consider the possibilities of this approach was John Desmond Bernal. In 1929, he wrote *The World, The Flesh, and the Devil: An Enquiry into the Three Enemies of the Rational Soul*, praised by Arthur C. Clarke as "The most brilliant attempt at scientific prediction ever made." Though Bernal readily admits in his work that, "surgery and biochemistry are sciences still too young to predict exactly how this will happen," he describes in principle how, "sooner or later the useless parts of the body must be given more modern functions or dispensed with altogether, and in their place we must incorporate in the effective body the mechanisms of the new functions." The "mechanisms" he proposes are the product of technology, and he carries the concept through to its logical conclusion, describing how the capability of connecting a human brain to mechanical equipment through electric impulses necessarily means two brains also may be connected to each other. And why stop at two?

Bernal goes on to describe "compound minds," shared collective memories, and a type of individual, electronic immortality. Almost 90 years later, these goals remain key to transhumanist aspirations, as well as being areas of active research. Surgery and biochemistry are no longer young sciences.

***

Another watershed document in the history of transhumanism, written by Manfred E. Clynes and Nathan S. Kline, was published in the September, 1960 issue of the journal, *Astronautics*. Its summary heading is essentially a restatement of Blish's rationale for pantropy: "Altering man's bodily functions to meet the requirements of extraterrestrial environments would be more logical than providing an earthly environment for him in space."

Clynes and Kline proposed an approach different from the genetic- engineering suggestions of Haldane and Stapledon. Instead, they expanded upon Bernal's concept of connecting biological and technological systems. They even gave it a name. Their paper was titled, "Cyborgs and space," and with that new term 10,000 science-fiction tales were launched. Among the most familiar, *The Six-Million-Dollar Man* television series, based on Martin Caidin's 1972 novel, *Cyborg*. By 1989, the term for technologically augmented humanoids became shortened to simply "The Borg," on *Star Trek: The Next Generation*.

Here we encounter not necessarily a limit, but certainly an important cautionary note to be considered.

***

Star Trek's Borg are an embodiment of transhumanism: living beings augmented by sophisticated technology that combines with biology at a molecular level. They are in brain-to-brain communication with each other, sharing what Bernal called a compound mind. They are able to function in the vacuum of space without protective gear.

They are also villains, voraciously hunting down and capturing beings of all species to "assimilate" them into the Borg Collective, without the notion of choice.

This is the question the Borg present: In the quest for improving and augmenting our inherent capabilities in order to prolong our survival, at what point will the changes to our biology affect our minds? If an end goal of transhumanism is achieved and the contents of our brains can be downloaded intact into computer-like processors, what will those minds be like without the constant interplay with biology? Will we love, will we lust, will we desire anything at all, other than the assurance of an uninterruptible power supply?

Many science-fiction tales, notably the series of *Matrix* movies, suggest that algorithmic analogues to biological needs can be programmed into the technology hosting our downloaded minds.

But yet another cautionary note arises.

In their paper, "Dangers of asteroid deflection," published in *Nature*, 1994, Carl Sagan and Steven J. Ostro make the case that whoever has the technology to alter an asteroid's orbit so it won't hit the Earth, automatically possesses the capability of altering the asteroid's path so it *will* hit.

Once again, transhumanism is no different. Anyone who has the ability to improve and augment our inherent abilities, automatically possesses the capability of degrading and diminishing a new generation of humans.

In his 1895 novel, *The Time Machine*, H.G. Wells describes the speciation of humanity in the far future. Aboveground the gentle Eloi live a life of ease, doing nothing to support themselves. Underground, working to provide the Eloi with the necessities of life, the brutish Morlocks toil. The equation is balanced only because the Morlocks eat the Eloi.

Wells describes the evolution of humans into their Eloi and Morlock variants as the result of natural evolution taking place over 800,000 years. Consider what might be possible when transhumanism matures from a dream to capability. Speciation in a generation? What kind of variants could be produced?

Aldous Huxley, in part inspired by the predictions of J.B.S. Haldane, addressed that question in his 1932 novel, *Brave New World*. In the year 2540 CE, all citizens of the World State are born in artificial wombs, and specifically engineered to become a member of one of five main castes. Alphas are the elite rulers, while Epsilons have had their intellectual capabilities purposely degraded and are relegated to menial labor. Ongoing conditioning is used to ensure individuals in each caste are content with their status: "Alpha children wear grey. They work much harder than we do, because they're so frightfully clever. I'm really awfully glad I'm a Beta, because I don't work so hard. And then we are much better than the Gammas and Deltas. Gammas are stupid… I'm so glad I'm a Beta."

***

Here, another potential peril of tranhumanism arises, one that's grounded in a common theme of many science-fiction stories, including our own novel, *Alien Nation: The Day of Descent*.

The novel is a prequel to the *Alien Nation* television series, recounting the events leading up to the arrival of the "Newcomers," thousands of extraterrestrials who attempt to fit in with human society after being marooned on Earth when their enormous ship crash-landed in the Mojave Desert. In our story, when the ship is detected traveling through the solar system and identified as an alien craft, a debate with existential stakes springs up: Should we attempt to communicate with beings who are clearly far more technologically advanced than we are, or do nothing to attract their attention?

The faction who advocate doing nothing base their decision on a sobering lesson of human history. As their leader asks an opponent, "Can you think of a single instance in human history when one society has made contact with a more technologically advanced second society and *survived?*"

This debate is also taking place in the real world today, as scientists involved in SETI, the search for extraterrestrial intelligence, attempt to formulate a protocol for what to do in the event a signal from an alien civilization is detected: Do we reply, and risk becoming a target for invasion, or remain silent and safe, abandoning an opportunity to share the knowledge of an advanced technology?

If, or rather when, transhumanism produces humans with enhanced intellects, is it possible they'll become the equivalent of science-fiction's advanced aliens? Will their arrival among us inevitably mean the extinction, or perhaps enslavement of unenhanced humans? Or, in the manner of the Borg, will choice be taken from us and *everyone* will be forced to become enhanced?

***

Transhumanism today, seen through the eyes of scientists and science-fiction writers alike, constantly considers and debates all these possibilities. Its promise is that humanity might be uplifted to a state of robust health and shining intellect. Its dark side is that humanity might be fragmented into genetic or technological haves and have-nots.

There's also another possibility that, for now, seems solely in the realm of science fiction.

What happens when the changes wrought by transhumanism completely alter what it means to be human? When a variant emerges that shares no environmental needs with other humans, or has an intellect like the martians of H.G. Wells's *War of the Worlds:* "minds that are to our minds as ours are to those of the beasts that perish, intellects vast and cool and unsympathetic…"

Given the difficulty of comprehending in reality what might drive an individual or a group mind that has absolutely no characteristics, wants, or needs in common with us, it's understandable that attempting the task has, for now, been left to the imagination of science-fiction writers.

Which brings us full circle to Arthur C. Clarke, author of the novel, *2001: A Space Odyssey,* a story, which at its heart, is about humanity's transformation, not by natural evolution, but by design. Transhumanism by any other name.

That novel ends with astronaut David Bowman, transformed, returning to Earth, though now without need of spacecraft or spacesuit.

As Bowman floats high above the planet of his birth, Clarke writes, "Then he waited, marshaling his thoughts and brooding over his untested powers. For though he was master of the world, he was not quite sure what to do next.

"But he would think of something."

***

Ultimately, the solution to the Catch-22 conundrum raised by Michael Griffin and Scott Kelly might ironically create an even greater one. If we humans do manage to remake ourselves so drastically that we can thrive in otherwise fatal environments, won't that by necessity mean that Earth as a habitat will be fatal to us? At that point, will we still be humans? Or will we have crossed the species threshold and become something other?

For us, we turn again to science fiction, and to celebrated author, Stephen Baxter, whose work often contemplates the potential far future of humanity.

In the opening pages of his novel, *Manifold: Time,* Baxter writes a sentence that's remarkable for presenting billions of years of history in a handful of words: "In the afterglow of the Big Bang, humans spread in waves across the universe, sprawling and brawling and breeding and dying and evolving."

Evolving.
Becoming something other. Something more.
Changing.
Because, as T'Pol noted, life is change.

***

If we don't change enough in the years ahead, it's possible that our branch of humanity, remaining as an Earth-adapted species, might never spread to other worlds. In which case, as Michael Griffin and Stephen Hawking and common sense suggest, the extinction of present-day humans as a species is a foregone conclusion.

Yet transhumanism gives us the hope that when that time comes, somewhere among the stars, perhaps breathing methane or thriving in vacuum, our transformed descendants will persist in multiple forms beyond even the imagination of the science-fiction writers of today.

Chapter 37
Not in the Image of Humans: Robots as Humans' Other in Contemporary Science Fiction Film, Literature and Art

Sophie Wennerscheid

37.1 Introduction

Ever since Pygmalion succeeded in creating the perfect lover, the idea of intimate relationships between humans and artificially created beings has become more and more popular, especially in the twenty-first century. While engineers and computer scientists are still hard at work on the technological development of robots with humanlike capacities, contemporary science fiction film and literature has already been showing us a variety of humans and posthumans interacting with each other intensely and entering into posthuman love affairs. In my paper, I examine how such intimate relationships are represented nowadays, at the beginning of what can be called the post- or transhuman age. What kind of intimate relationships are shown and made a subject of discussion both in robotics, in science fiction and in robotic art? What kind of aesthetics is being developed to depict future love affairs? Which ethical challenges might those relationships pose? And taking into consideration that robots, technically speaking, are genderless, would this allow for a queer reconceptualization of the traditional heteronormative sexual structuring of desire? As, for instance, Jack/Judith Halberstam emphasizes when s/he states that "[...] technology makes the body queer, fragments it, frames it, cuts it, transforms desire" (16) [1].

To answer these questions, I proceed in five steps. First, I present the leading proponent of artificial sexuality, David Levy, who puts forward the thesis of the always compliant robot as the 'perfect lover.' Then I reflect on the counterarguments produced by, amongst others, robot ethicist Kathleen Richardson. In a third step, drawing on Deleuze's concept of 'becoming other' and the use of this concept in critical posthumanist studies, I present my alternative view on human-posthuman

S. Wennerscheid (✉)
Ghent University, Ghent, Belgium

University of Copenhagen, København, Denmark
e-mail: sophie.wennerscheid@ugent.be

© Springer Nature Switzerland AG 2019
N. Lee (ed.), *The Transhumanism Handbook*,
https://doi.org/10.1007/978-3-030-16920-6_37

intimate relationships and develop my concept of 'new networks of desire'. According to this network idea, man and machine are seen as being 'in touch', entangled and interwoven, merging into "new subjectivities at the technological interface." [2] Finally, I analyze the representation of human-posthuman intimate relationships in contemporary science fiction film and literature and in robotic art. Science fiction proves particularly useful for considering new concepts of desire because it features a wide variety of man-machine intimate relationships which challenge our understanding of a 'normal' and socially acceptable relationship. My main aim is to show on the one hand that science fiction, in contrast to Levy's theory, highlights the various challenges in human-posthuman relationships and in so doing contributes to a deepened and more complex understanding of our likely posthuman future. In the concluding remarks, I argue that in robotics a change of thinking is needed.

37.2 The Vision: Love and Sex with Robots

In his book *Love and Sex with Robots, The Evolution of Human/Robot Relationships* [3], published in 2007, David Levy enthusiastically declares that in about 50 years we will see humans partaking in intense and fulfilling relationships with robots. In the preface to his book he solemnly insures his readers:

> Robots will be hugely attractive to humans as companions because of their many talents, senses, and capabilities. They will have the capacity to fall in love with humans and to make themselves romantically attractive and sexually desirable to humans. Robots will transform human notions of love and sexuality. (22)

In the two main chapters of his book, Levy explains, firstly, our willingness to enter into a relationship with a robot, accepting it as a companion and partner, and, secondly, the improvement in our sex life thanks to erotic robots. With regard to both partnership and sexuality Levy's main argument for entering into a relation with a robot, is the robot's capacity to satisfy all our needs. Moreover, because a robot is much more unselfish and yet also more adaptive, it will not only succeed in satisfying human needs, but will do so in a much better way than a human partner might be able to.

However, this argument only holds true when it is combined with another assumption, namely that a human being indeed longs for to be satisfied in the way described above. For Levy, there is no doubt that this is the case. According to him, every human being longs for certainty, and thus for a steadfast and impeccably reliable partner. Levy emphasizes that it is "the certainty that one's robot friend will behave in ways that one finds empathetic" (107) that makes the robot the perfect lover and partner. In Levy's view, a robot will never ever frustrate, disappoint or even betray you. It will never fall out of love with you and will ensure that your love for it never ends or even merely wavers. Levy further explains:

> Just as with the central heating thermostat that constantly monitors the temperature of your home, making it warmer or cooler as required, so your robot's emotion system will constantly monitor the level of your affection for it, and as the level drops, your robot will experiment with changes in behavior aimed at restoring its appeal to you to normal. (132)

37.3 Ethical Concerns

While Levy does not discuss "the human fallout from being able to buy a completely selfish relationship" [4], this is of crucial concern for perhaps Levy's most prominent opponent, the anthropologist and robot ethicist Kathleen Richardson. Her main questions are: What are the ramifications of our regarding a robot as a thing we can completely dominate? How will this influence our psyche? And what might be the impact of such commodified relations on our way of relating to other people? Following Immanuel Kant's line of argumentation against the objectification of animals, Richardson points out the problematic emotional consequences for humans when robots are treated as pure objects. In his *Lectures on Ethics* [5] Kant argues that although animals are mere things, we shouldn't treat them as such. Humans, he argues, "must practice kindness towards animals, for he who is cruel to animals becomes hard also in his dealings with men" (212). Similarly, Richardson warns that owning a sex robot is comparable to owning a slave. Human empathy will be eroded and we will treat other people as we treat robots: as things over which we are entitled to govern. In a position statement launched in 2015, Richardson advocates her *Campaign Against Sex Robots* [6] and underlines that using a sex robot appearing female, one solely designed to give pleasure and thus based mainly on a pornographic model, will exacerbate a sexist, degrading and objectifying image of women. Richardson explains: "[…] the development of sex robots will further reinforce relations of power that do not recognize both parties as human subjects. Only the buyer of sex is recognized as a subject, the seller of sex (and by virtue the sex-robot) is merely a thing to have sex with." Existing gender stereotypes and hierarchies will be furthered.

Computer scientist Kate Devlin agrees with Richardson on this specific point. In her view, the transfer of existing gender stereotypes into the realm of future technology is reactionary and should be avoided. However, she also warns against transferring existing prudishness into robotics. Davis asks, "If robots oughtn't to have artificial sexuality, why should they have a narrow and unreflective morality?" Instead of prohibiting sex robots, she calls for overcoming current binaries and exploring a new understanding of sex robots. "It is time for new approaches to artificial sexuality, which include a move away from the machine-as-sex-machine hegemony and all its associated biases" [7].

37.4 Theories of Affect and Posthuman Desire

How such an alternative to traditional patterns of man-machine relations might look has been taken up by theorists like Rosi Braidotti and Patricia MacCormack. Both are indebted to Deleuze's poststructuralist readings of Spinoza's affect studies. In developing them further, they make a plea for new kinds of affective posthuman encounters. In what follows I expand on these theories and transform them into what I call 'new networks of desire'. In a first step, I briefly clarify the concept of affect as deployed by Spinoza and Deleuze and explore how affect and desire have been advanced in posthuman studies. After this, I introduce my understanding of 'networks of desire.'

1. Affect and Desire in Spinoza, Deleuze and Critical Posthumanism

In the third part of his *Ethics, Demonstrated in Geometrical Order*, [8] published in 1677, Baruch Spinoza develops his theory of affect. According to Spinoza, an affect is the continuous variation or modification of a body's force through an inter-action with another body. Each body has the active power to affect and the passive power to be affected. And each affect can be negative or positive, i.e., it can increase or diminish the other body's capacity for existing, its vitality. Spinoza defines: "By affect I understand affections of the body by which the body's power of acting is increased or diminished, aided or restrained" (154).

Underlying each body's affective flow is its desire both to preserve and to trans-form itself. Desire can therefore be taken to mean a body's potential to expand, create or produce. Accordingly, Elizabeth Grosz [9] considers desire synonymous with production. "Desire is the force of positive production, the action that creates things, makes alliances, and forges interactions [...]. Spinozist desire figures in terms of capacities and abilities" (179). Similarly, Rosi Braidotti [10] views desire as a power that disseminates bodies' self-identity and drives them to become mul-tiple. By engaging in various relations with other bodies, the body itself changes and becomes continually other to itself. This holds true particularly when the mul-tiplicity of possible affections and differences is brought about by encounters with and relations to largely unfamiliar and strange forces and affects; encounters which thus can be described as encounters with alterity.

Encounters with alterity in general and encounters with the nonhuman other in particular is of significant concern for critical posthumanism. Following a definition by Braidotti [11], critical posthumanism is "postanthropocentric philosophy, a deconstruction of the human-machine boundary, and a nondualist reconceptualiza-tion of human beings and animals" (5). Critical posthumanism's main concern is to question human's self-authorization as the world's leading species and, derived from this supremacy, its self-ascribed right to subordinate other nonhuman beings. Scholars in critical posthumanism, very often drawing on Deleuze and Guattari's ideas of 'becoming-other' [12], counterpose to the idea of species hierarchy and human exceptionalism a transformative politics, one which propounds novel relations between humans and nonhumans, termed 'unnatural alliances' by Deleuze and Guattari.

What makes this approach interesting for science fiction film and literature, as analyzed below, is that the concept of alliance is based on the idea of mutual depen-dences between human bodies and animal or technological others, while it does not aim at constituting a new hierarchy between humans and nonhumans. Rather, the emphasis falls on difference and otherness as continually moving categories. As Patricia MacCormack [13] emphasizes: In encounters of alterity, all beings involved are "free from the bondage of another's claim to know" (4). Two or more separate entities meet and in their meeting they are affected and become a dynamic assemblage of affective flows triggered by desire. Such an encounter can thus also be seen as "an act of love between things based on their difference" (4).

2. Networks of Desire

To speak about 'networks of desire' means that I no longer tend to uncritically regard the technological other as a tool to be used without due concern, but instead as something with which we form bonds, something that affects and touches us. Both humans and nonhumans are, as Mark Coeckelbergh [14] stresses, "not so much considered as atomistic individuals or members of a 'species', but as relational entities whose identity depends on their relations with other entities" (215). The term 'network' is used to strengthen the idea that acts of posthuman love and desire are not limited to encounters between two individual beings but include a variety of net-like relations, associations and connections.

By highlighting the term 'desire' I seek to accentuate the relations between human and nonhuman beings as relations of intimacy and mutual affection, pleasure-prone and pleasure-driven. The concept of desire is so important here because it makes particularly clear that intimate relations do not leave us just where and who we were but transformed. Desire is a transformative force and thus a site for becoming different. Or in Neil Badmington's [15] words: "To be human is to desire, to possess emotions, but to desire is to trouble the sacred distinction between the human and the inhuman" (139).

Furthermore, desire's capacity to undermine human-posthuman distinctions also has an important narratological aspect. You can use it as a dramaturgical means, or as Francesca Ferrando [16] has put it, "as a plot stratagem to connect different types of beings, a bridge to dissolve dualistic cultural practices." (274) Affection and desire are thus to be understood as forces that bring to the fore hitherto unknown passions, break down the border between 'us' and 'them' and introduce new concepts of interspecies relationships.

However, the transformative potential of desire only comes to fruition when the encounter between human and posthuman is actually an encounter with alterity, a queer assemblage or an unnatural alliance. By those concepts I refer to encounters between humans and other nonhuman beings that provoke a feeling of unease. This is due to the fact that encountering the other as a radical other is an experience that cannot be conceptualized. The other's otherness cannot be absorbed by the subject, it is irreducible to interiority and thus shatters the subject's self-understanding as powerful and knowledgeable. However, since the encounter is fueled by a desire to be in touch with each other, it is not only unsettling, but it also allows for a crucial change. This implies that only unnatural alliances and encounters with alterity pave the way for posthuman intimacy.

37.5   Intimate Human-Posthuman Relationships in Contemporary Science Fiction Film and Literature and Robotic Art

3. Science Fiction Film

When we have a look at the large number of contemporary science fiction films and TV series that deal with intimate relationships between humans and robots or other artificial posthuman beings, as is the case, for example, with *AI. Artificial Intelligence* (2001) *I, Robot* (2004), *Her* (2013), *Westworld* (2016–18), *Blade Runner 2049* (2017), we most often see relationships which comply with familiar dualistic paradigms: male-female, man-machine, animate-inanimate, self-other. Particularly in films that are made to reach a broad audience, we time and again find the typical constellation of a male human who falls in love with a young and sexy female posthuman. However, even in these films we find initial signs of new networks of desire. In the following analysis, I seek to demonstrate this by focusing on *Ex Machina* (2015), *Be Right Back* (2013) and *Real Humans* (2012–2014).

Ex Machina The 2015 movie *Ex Machina* [17] by Alex Garland is about Caleb (Domnhall Gleeson), a young programmer who becomes sexually attracted and emotionally attached to Ava (Alicia Vikander), a female looking humanoid robot. Ava was recently developed by the reclusive tech entrepreneur Nathan (Oscar Isaac) who has invited Caleb into his laboratory to perform the *Turing test* on Ava. The beautiful Ava easily succeeds in what Levy [2] has described as a robot's ability to make itself "romantically attractive and sexually desirable" (22) to a human being, here, the young Caleb. She does so thanks to her very sexy body, her expressive eyes and hands, her smartness, and last but not least thanks to her frailty.

What makes the encounter between Caleb and Ava interesting is that Caleb first regards Ava as a synthetic being, one having been designed by Nathan to be the leading model of Artificial Intelligence. As a matter of fact, Nathan presented Ava as AI and Caleb has no reason to doubt his explanation, and Ava is clearly identifiable as a nonhuman. She has a human-looking face but a translucent torso in which her wires are visible. However, this does not make her less attractive in Caleb's eyes. On the contrary, it triggers his romantic interest in the robot. Ava's gender ambivalence further increases Caleb's fascination. Although Ava, with her full lips, sleek curves and small waist looks like the classical beauty, to be sure, she at the same time displays androgynous traits. Especially in the beginning of the film with her head bare of the one or another style of long hair so typically depicted as female and sexy, she bears a certain resemblance to a tomboy.

One day, while the security camera system has been knocked out, Ava tells Caleb that Nathan is a bad person intent on destroying her. Appealing to Caleb's sense of chivalry, she convinces him to help her escape Nathan's fortress. Caleb willingly agrees to come to her aid. In my view, this is the moment when Caleb fails. Instead of perceiving Ava for what she is, namely, a perfectly designed android with a machine's will and desire, he sees a young woman who needs to be rescued. He cannot free himself of his conventional male and human view and ignores Ava's technological otherness, her being as a machine. He reestablishes the difference between male and female, which in the beginning was blurred thanks to Ava's androgynous appearance, and gets caught up in an old-fashioned anthropocentric and gender-hierarchical way of thinking. Thereby, Caleb misses the chance to encounter the other and in so doing to become another himself. He stays what he is, an intelligent young man with very traditional romantic interests. As a consequence, Ava locks Caleb up in a sealed room and boards the helicopter meant for Caleb's return home. While Ava sets herself free, Caleb is trapped in a room that can be read metaphorically for Caleb's confinement in himself, his anthropo- and androcentric self-centeredness.

Be Right Back Another example of how an intimate relationship between a human and a posthuman can fail due to mutual misapprehension of the particular human and posthuman characteristics in play, can be found in *Be Right Back* (2013) [18]. This is the first episode of the second season of the British science fiction television anthology series *Black Mirror* (2011–2014), created by Charlie Brooker. *Be Right Back* focuses on the young woman Martha (Hayley Atwell) who enters a state of deep crisis after the sudden death of her partner Ash (Domhnall Gleeson). At Ash's funeral, a friend advises Martha to register with an online service that offers to create virtual doubles of dear ones lost. Martha vehemently refuses the idea at first, but after having discovered that she is pregnant, she is overwhelmed by grief and decides to give the service a try. She uploads all of Ash's past online communication, social media profiles, photos and videos, so that a new Ash can be created virtually. First, she only exchanges e-mails with the artificial Ash. Then she speaks with him by phone. And finally, she agrees to get a clone that looks almost exactly like the original Ash. Having fought her initial feelings of unease, Martha experiences some exciting moments with Ash. Having sex with the Ash replicant is, for example, awesome for Martha. When asked about the sources of his sexual prowess, Ash explains that he has been endowed with a sexual program "based on pornographic videos" (34: 30).

But after a while Martha becomes heavily frustrated with Ash's permanent compliance. While Martha's relationship with the real Ash was based on a very affectionate but nonetheless humorous and always a bit teasing interaction, the virtual Ash is not able to act against Martha's will. He is neither confrontational nor argumentative, but instead does everything Martha expects of him. When Martha eventually requests of Ash that he leaves her alone and he just follows her instruction, Martha desperately cries out: „Ash would argue over that. He wouldn't just leave because I'd ordered him to" (40: 49). But Ash isn't able to react any differently. He explains: "I aim to please." Martha finally realizes that the virtual Ash is "not enough" (41: 18) of the original Ash. He has no free will, no needs and desires of his own. And this lack of independence makes it impossible to develop a relationship which would be quite possibly imperfect or stressful, perhaps even exhausting or frustrating, but nevertheless challenging and thus enriching.

When we compare the human-posthuman encounter presented in *Be Right Back* with the encounter presented in *Ex Machina*, we can detect similar problems, although the initial situation seems to be very different. In both films, the problems stem from the fact that both posthumans, Ava and Ash, were designed to look and to behave as humanlike as possible. Consequently, the humans interacting with them expect them to behave like real humans, although they might have known better. Ava has interests and desires of her own that Caleb is not willing to accept because he does not recognize Ava for what she is: a posthuman being. Likewise, Martha also misunderstands Ash because she starts from the premise that the Ash replicant will behave like the real Ash. Owing to this fallacy, she can only be disappointed. While, however, Ava is able to escape the human straightjacket by subversively imitating the human codes and transforming herself into the 'perfect woman' Caleb expected her to be, Ash does not succeed in developing and expanding. Consequently, he is banned to the attic, visited only once a year, when Martha's and the human Ash's little daughter come to celebrate the daughter's birthday with her artificial father.

37.6   Real Humans

While posthumans both in *Ex Machina* and in *Be Right Back* do not find recognition as posthumans, in the Swedish TV science fiction series *Real Humans* (Swedish: *Äkta människor*, 2012–2014) [19], by contrast, we can find signs of posthumans being met with an acceptance that is based on the posthumans' technological otherness. *Real Humans* is set in an ordinary middle-sized Swedish town in a near future, i.e., in a fictional society that in general is very similar to our own. In this society, humanoid robots, called hubots, have entered into ordinary people's lives. The robots are designed lifelike, but have some characteristic features that clearly mark them as artificial beings. They have, for instance, unnaturally bright blue or green eyes and they need to be recharged. They also can be turned off at the touch of a button if they are not in use or behave rebelliously. Most of the hubots are used as simple factory workers, domestic help or as caretakers for the elderly. Some are also programmed for limited sexual activity, although hubot-human sexual activity is not yet commonplace. A small group of hubots are intelligent, self-conscious and sentient. They call themselves free hubots or, with reference to their creator David Eischer, 'David's children.'

Although the series at large highlights the crises and confrontations between humans and posthumans, it clearly shows a bias towards equitable coexistence. In particular, the encounters between the hubot Mimi and various humans can illustrate this. The first human to become interested in Mimi is little Leo, David Eischer's son. In a series of flashbacks performed by the grown-up Leo, the audience learns that Leo as a ten-year-old had nearly died while trying, but failing to rescue his mother from drowning. Though his mother dies, Leo is himself rescued by Mimi. Despite having been rescued, his condition remains hopeless. So to save Leo's life, his father performs an operation in which hubot technology is implanted. Leo has thus become a human/posthuman hybrid. What is striking, however, is that Leo's hybridity is revealed as a problem in the series. When Mimi is abducted by a black market hubot dealer and reprogrammed as a 'normal', non-sentient hubot, Leo risks and ultimately sacrifices his own life to save her. Being strongly affected by a post-human, or being a half posthuman yourself, is seen here as fatal.

At the same time, however, Mimi's capacity to affect other people is highlighted as a positive force in the series. When Mimi becomes a member of the Engman family, she immediately evokes a broad range of emotional responses. The little daughter loves Mimi because Mimi patiently reads to her one book after another. The father is fascinated by Mimi's beauty and tempted to activate her program for sexual use, but resists the temptation. After a while he acknowledges Mimi as real member of the family and unselfishly helps her when she is infected by a dangerous computer virus. The 16-year-old Tobbe is likewise fascinated, but falls in love for real with Mimi. Eventually he comes out of the closet as a transposthuman sexual. In all these encounters, Mimi is mistaken neither as a human, nor as a pure machine.

The mother, Inger, who is a lawyer, is initially sceptic. Yet at one point in the series, when a female hubot assistant is insulted by Inger's colleague, to convince Inger that these are 'only machines' without any capacity to feel insulted, she reacts empathetically and takes the hubots' side. Eventually, in her role as a lawyer, Inger starts fighting for robot rights. At the end of the series, the fight is won, and the robots, or at least some of them, are granted civil rights.

The problem, however, is, that those rights follow the human model of rights. Inger and other proponents of robot rights do not see and treat the robots as an independent species but as humanlike, maybe even as real humans. The humans have eroded the robots' otherness. One main reason for this might indeed be the robots' role as companions or even, as *Real Humans* clearly illustrates, as family members. In his article, "Alterity ex Machina", Mark Coeckelbergh [20] describes the process of familiarizing as concomitant with the process of the robot entering "the domus, the home, and the family" (186), i.e. the process of domestication. Coeckelbergh reminds us: "As we familiarize ourselves with the technology, the alien becomes familiar, the wild becomes domesticated, and alterity largely disappears." (187) Otherness is turned into sameness. Instead of working on robot rights it thus might be better to stay open to the unexpected, i.e., open to experiences of alterity that break into our life and decenter our self-conception of the always knowing master of the world.

In summary, it may be argued that in *Ex Machina, Be Right Back* and *Real Humans* desire is placed center stage as a potentially transformative force, but is not really brought to fruition. By getting in touch, man and machine, humans and post-humans bring about the chance to change, to encounter one another in hitherto unknown ways. But the films do not really trust this chance. Instead, the human characters by and large impede the technological other from freely extending its machinic desires and capacities. And all the while the humans remain anthropo- and self-centered, restrict themselves and stop at the very moment when a poignant expansion of the network of desire had been possible.

4. Posthuman Love Affairs in Science Fiction Literature

Contemporary science fiction novels which feature human-posthuman love affairs show such affairs in a greater variety than similar films do. The well-known pattern of 'male human falls in love with female posthuman' is more often fractured and multiplied. Moreover, the idea of a robot being as humanlike as possible is also questioned and replaced by less conventional representations. In the Swedish science fiction-novel *The Song from the Chinese Room* (Swedish: *Sången ur det kinesiska rummet*, 2014) by Sam Ghazi [21], for example, we learn about a robot called Cepheus who, consisting merely of a head with one big blue eye and two robotic arms, was designed as a 'helping hand' for the cancer researcher Simona. Working closely with Simona, the robot develops a human way of thinking, becomes attracted to his female colleague and starts writing love poems. In another science fiction novel, Jeanette Winterson's *The Stone Gods*, originally published in 2007, we likewise read about a robot that is described, at least in the novel's last part, as a 'thinking head' and that, identifying itself as female, is sexually attracted to other women.

Both examples are interesting not only because of the non-anthropomorphic appearance of the robots, but also because of the humans' specific reaction to them. In the beginning, the humans feel strongly uncomfortable, but they later develop intense feelings for those posthumans. While the humanlike robots often seem to trigger, as discussed in the films above, a feeling of unease or uncanniness, widely known as the 'uncanny valley' [22], robots which do not look anthropomorphic but nevertheless behave in responsive ways or in ways that signal awareness, sentience, agency and intentionality, evoke another feeling. A feeling, namely, that can be described as 'the experiential uncanny.' This term was coined by Elizabeth Jochum and Ken Goldberg. In their article "Cultivating the Uncanny" [23], the two coauthors differentiate between the 'representational uncanny,' as that which is evoked by humanlike robots, and the 'experiential uncanny,' as that which "arises from a user's interaction and experience" (16) with the robot, yet seems to arise unrelated to the robot's appearance, one clearly identifiable as nonhuman. With regard to various forms of interlaced desire, this insight is worth underscoring because it is the unfamiliar, and most of all, the fragmented and partial, the bodily incompleteness, that leaves space for our imagination. This can be illustrated with a closer examination of Winterson's novel *The Stone Gods*.

In *The Stone Gods* [24], encounters between humans and nonhumans play a crucial role, particularly the encounter between the female human Billie and the female posthuman Spike, the novel's two protagonists. This encounter, which finally leads to an intense love affair, is based on the protagonists' awareness of and fascination for the other's otherness. Billie, for example, acknowledges: "And I looked at Spike, unknown, uncharted, different in every way from me, another life-form, another planet, another chance" (90). Spike, for her part, experiences a crucial modification of herself when reading love poems: she becomes a sentient being, a being which is able to be affected and to affect. "In fact I was sensing something completely new to me. For the first time I was able to feel" (81).

Being able to feel makes it impossible for Spike to fulfill the task for which humans have designed her, namely, to predict the future as objectively as possible. However, this loss of predictability does not only pose a threat for humankind's development. It also presents a chance for overcoming a normative understanding of the self-contained knowing subject. It surpasses the idea of the triumphant and self-centered human and presents a new understanding of post/humanity based on decentered relationality. Not accidentally, the end of chapter one coincides with the protagonists' dying while warmly embracing one another, a scene which symbolizes the transformative forces of love and relationality. Death is not the end of interdependency and interconnectedness, but signifies their very possibility. It marks the dissolution of the subject, the individuated self, into, as Braidotti [25] phrases it, "the generative flow of becoming" (136).

Contemporary science fiction literature, better than contemporary science fiction film, allows us to better understand posthuman desire as a possibility to remove "the obstacle of self-centered individualism" (50) and thereby to adopt a new "posthuman subject position based on relationality and transversal interconnections across the classical axes of differentiation" (96). The same holds true when it comes to some pieces of robotic artworks.

5. Intimate Touches and Strange Gazes in Robotic Art

Unlike representations of robots in film and literature, robotic figures in art are artefacts taking up real space, allowing for spatial and bodily proximity between man and machine. We can not only see and hear them, but also touch and smell them. And we can, at its best, interact with them. The question is thus, in which way the robotic figures affect us, how we affect them, and how this kind of affectivity impacts our intimate relations with them. The first example I want to analyze is Louis Philippe Demers' telerobotic art installation *The Blind Robot* (2012), the second is Jordan Wolfson's animatronic *Female Figure* (2014).

The Blind Robot [26] does not resemble a human in all its complexity, but is merely comprised of a pair of robotic arms equipped with articulated hands installed on a table, tele-operated by a human who, however, is not visible. The integrative part of the artwork is a visitor who is invited to sit down in front of the machine. The machine then explores the visitor by gently touching the human's face with its robotic fingertips. As explained in Demers' study *Machine Performers* [27], the robotic arm, normally seen as "a high precision tool," now appears as "a fragile, imprecise and emotionally loaded agent" (58). Although some visitors described themselves in this situation as feeling uncomfortable or even as being reminded of "Science Fictional killer-robot dystopias" [28] they recalled seeing at the cinema, the artist's intention was to create an empathic situation and a positive attitude towards the engagement. Demers did so by entitling his installation 'The Blind Robot,' recalling the situation of a blind and helpless person who needs to touch the visitor in order to recognize it. Demers explains: "It is a psychological experiment [...] just by the fact that I state that this is a blind robot, you will accept that this machine can touch you in very intimate places" [29]. Demers also describes the feeling of being touched by his robot as "very unique, it's not like being touched by a human, of course, but it's also not like being poked with a stick. It's a novel way, because your brain is not too sure what to think about it" [29].

In my view, the novelty of this kind of touch is the central point, when it comes to 'new networks of desire.' Being touched and being affected by something we have not sensed and experienced before is exciting but also engaging. It encourages us to become involved with an unfamiliar situation and an unfamiliar nonhuman agent which intimately touches vulnerable parts of our bodies, engendering a sensual, potentially arousing encounter. It's about an encounter that simultaneously increases our bodily self-awareness and our awareness of the machinic other as other. That the machinic nature of the other is not concealed but rather clearly exposed in presenting only two robotic arms, further contributes to the individual human's involvement. Given the fact that Demers' blind robot is not a full humanoid-robot, fantasy and imagination are needed to 'animate' the situation.

Imagination is also involved when it comes to Jordan Wolfson's *Female figure* [30]. – a computer-controlled sculpture featuring a hyper-sexualized blonde woman wearing a white miniskirt splattered with black dirt, high-heeled thigh-high boots and long gloves. The figure is inspired by the character of Holli Would, the cartoon vamp from the 1992 animated fantasy film, *Cool World* [31]. Although imitating the typical *femme fatale*, the figure's fabricated nature is not hidden. On the contrary, the figure's joints are visibly bolted together and a metallic pole running through its belly holds it fastened to a large mirror. Various other features contribute to the figure's de-familiarizing effect. One such effect is sound and voice, mixed in a disturbing way. On the one hand, the figure dances lasciviously to popular songs, among them, Lady Gaga's 'Applause' and Paul Simon's 'Graceland.' On the other hand, we hear the figure's voice saying, in a tape loop, monotonously and in a male voice which is Demers': 'My father is dead. My mother is dead. I'm gay.' These two very different sound tracks make it impossible, while interfering with each other, for the audience to relax.

This kind of disquiet based on contradictory bodily experiences is further intensified by various forms of glances exchanged between robot and human. Watching the figure from behind, a seductive effect might be felt. Gyrating before the mirror, the figure is kind of alluring. When, however, we look at the figure's face, this positive feeling changes rapidly. Instead of a human-like face, we are confronted with dark evil eyes which glimmer from behind a green Venetian mask with a witch-like nose. Since the figure is equipped with motion tracking software and technology for facial recognition, it is able to recognize and, what is more, to react to people's movements throughout the room. The sculpture makes eye contact with the viewer, quietly observing him or her. This kind of interaction is described by one visitor in the following way: "If you stand close to the robot it looks deep into your eyes, and there is a terrifyingly disorienting moment as you experience yourself as an object in the automaton's gaze." [32] Being the object of the machinic other's gaze does not leave the visitor untouched. He or she is probably not altered in a way as radical as that envisioned by Braidotti. But the experience evoked by the interaction with this sculpture is alienating. In this sense, it prepares a way for hitherto unknown experiences – even though these first appear here on the side of the negative affects.

37.7 Conclusion

Many sex robot manufacturers, robotics experts and engineers state as their aim the creation of robots or robotic dolls specifically conceived for the sexual gratification of human beings. For them it is self-evident that these synthetic lovers should look, feel and behave as humanlike as possible. The aim is to create them in the image of humans. For example, the company *Abyss Creations* has developed the popular silicone sex doll 'RealDoll' and is currently working to create sex dolls with artificial intelligence; *Synthea Amatus* has launched the AI equipped model 'Samantha' in summer 2017, while *Doll Sweet* is working on robotic talking heads and even full-body sex robots [33]. Each of these and other commercially vested interests emphasize that artificial creations are being marketed to serve as the 'perfect partner' for human beings, or rather: for men.

Some of the posthuman female figures as currently presented in science fiction can serve as an alternative model to this stereotypical understanding. Although they are still often designed according to popular ideas of female beauty and sexiness, it is not this kind of stereotypical sexiness that makes them interesting in the long run. On the contrary, it is their otherness that transgresses humans' self-centeredness, arouses strong feelings and reminds us of what it means to be a desiring (post) human, namely, a body which is able to affect and to be affected in unforeseen ways.

Leaving aside the immense technical problems of developing robots designed to look like a real human woman or man, I consider doing so the wrong path to pursue. Designing, marketing and perceiving humanlike robots as human's companions and lovers meant to perform strictly in line with an individual's wishes will not take us forward. It remains to be seen whether, in fact, in the foreseeable future robotic love affairs will become so advanced that they can function as an appropriate surrogate for human relationships, or if robots will be unable to fully meet our expectations; in either case, as long as robots are designed in the image of humans, they will not be able to do anything else other than to mirror existing needs, experiences or imaginations. Instead, they will always only bring us back to preconceived ideas, ideas that will have been programmed into the other for fulfilling our narcissistic tendencies. While some people may not at all consider this a problem, others may well be hoping for something else: challenging new experiences, transgressive new affects, new forms of encounters and hierarchies undermined, at least not plainly reproduced and simply reinforced through existing heterosexist patterns.

To reach this aim, we need robots that challenge our restricted self-understanding as humans superior to all other nonhuman beings. Critical posthumanist thinking, as well as a variety of unconventional films, literary texts and other artworks featuring human-posthuman intimate relationships in a non-dualistic manner, make us aware that the most exciting encounters happen when they are unpredictable. Not the robot which is always responding to our moods and expectations, but rather a machine we accord the right to be different, a machine not in compliance but wayward, could help us to view ourselves other than as the prime issue in the world. I'd thus like to submit that technology will be better capable of enhancing humans' interaction with robots, if it does not build its hopes around the human-likeness of robots, but on their otherness. What we need is not the robot in the image of humans but the robot *not* in the image of humans.

References

1. Halberstam, J. & I. Livingston: Posthuman Bodies. Indiana UP, Indiana (1995).
2. Hollinger, V.: 'Something like a Fiction': Speculative Intersections of Sexuality and Technology. In: Person, W. et.al. (eds.): Queer Universes: Sexualities in Science Fiction. Liverpool UP, Liverpool, pp. 140–160 (2008).
3. Levy, D.: Love and Sex with Robots. The Evolution of Human-Robot Relationships. Duckworth, London (2008).
4. Kleeman, J.: The race to build the world's first sex robot. In: The Guardian. (2017) Retrieved from https://www.theguardian.com/technology/2017/apr/27/race-to-build-world-first-sex-robot
5. Kant, I.: Lectures on Ethics. Cambridge UP, Cambridge (1997).
6. Richardson, K.: The asymmetrical 'relationship': parallels between prostitution and the development of sex robots. In: ACM SIGCAS Computers and Society 45(3), 290–293 (2016).
7. Devlin, K.: In defence of sex machines: why trying to ban sex robots is wrong. In: The conversation, 17 September 2015. Retrieved from http://theconversation.com/in-defence-of-sex-machines-why-trying-to-ban-sex-robots-is-wrong-47641
8. Spinoza, B.: Ethics, A Spinoza Reader: The Ethics and Other Works. Princeton UP, Princeton (1994).
9. Grosz, E.: Space, Time and Perversion: Essays on the Politics of Body. Routledge, London (1995).
10. Braidotti, R.: Metamorphoses: Towards a Materialist Theory of Becoming. Polity Press, Cambridge (2002).
11. Braidotti, R.: Editor's Note. In: Journal of Posthuman Studies. Philosophy. Technology. Media 1(1), 1–8 (2017).
12. Deleuze, G., Guattari, F.: A Thousand Plateaus: Capitalism and Schizophrenia 2, Athlone, London (1987).
13. MacCormack, P.: Posthuman Ethics. Embodiment and Cultural Theory. Routledge, London (2016).
14. Coeckelbergh, M: Robot rights? Towards a social-relational justification of moral consideration. In: Ethics and Information Technology 12: 209–221, (2010). DOI https://doi.org/10.1007/s10676-010-9235-5
15. Badmington, N.: Alien Chic: Posthumanism and the Other Within. Routledge, London (2004).
16. Ferrando, F.: Of Posthuman Born: Gender, Utopia and the Posthuman in Films and TV. In: Hauskeller, M., Carbonell, C., Philbeck, Th.: The Palgrave Handbook of Posthumanism in Film and Television. Palgrave Macmillan, Basingstoke (2015), pp. 269–278.
17. Garland, A.: Ex Machina. UK/US. (2015).
18. Brooker, C.: Be Right Back. UK (2013)
19. Lundström, L: Real Humans. S (2012-2014)
20. Coeckelbergh, M., Alterity ex Machina. The Encounter with Technology as an Epistemological-Ethical Drama. In: Gunkel, D., Marcondes F. & Mersch, D. (eds.), The Changing Face of Alterity. Communication, Technology, and Other Subjects, Rowman & Littlefield International, London, 181–196 (2016).
21. Ghazi, S.: Sången ur det kinesiska rummet. Norstedts, Stockholm (2014).
22. Mori, M.: The uncanny valley. In: IEEE Robotics & Automation Magazine, 19(2), (1970/2012), pp. 98–100. doi:https://doi.org/10.1109/MRA.2012.2192811
23. Jochum, E., Goldberg, K.: Cultivating the Uncanny: The Telegarden and Other Oddities. In: Herath, D., Kroos, C., Stelarc (eds.): Robots and Art. Exploring an Unlikely Symbiosis. Springer, Singapore (2016).
24. Winterson, J.: The Stone Gods. Penguin, London (2008).
25. Braidotti, R.: The Posthuman. Polity Press, Cambridge (2015).
26. Demers, L.-P.: The Blind Robot. (2012)

27. Demers, L.-P.: Machine Performers: Agents in a Multiple Ontological State. Dissertation, University of Plymouth. (2014) http://citeseerx.ist.psu.edu/viewdoc/download?doi=10.1.1.82 9.7112&rep=rep1&type=pdf
28. Sjef: The Blind Robot. 28.5.2014 https://sjef.nu/the-blind-robot/
29. Knoll, M.: The Blind Robot at the Lab. 28.10.2013 https://www.aec.at/aeblog/en/2013/10/28/the-blind-robot-bei-the-lab/
30. Wolfson, J.: Female Figure. (2014)
31. Bakshi, R.: Cool World. US. (1993)
32. Feldhaus, T: Jordan Wolfson's Robot: In the Moment of Terror https://www.spikeartmagazine.com/en/articles/jordan-wolfsons-robot-moment-terror
33. Owsianik, J.: State of Sex Robots: These are the Companies Developing Robotic Lovers. September 1, 2017 https://futureofsex.net/robots/state-sex-robots-companies-developing-robotic-lovers/

Chapter 38
Jethro Knights—Human to Machine: A Hero We Love To Hate

Chris T. Armstrong

The essay below is excerpted from my forthcoming book, At Any Cost—A Guide to The Transhumanist Wager and the Ideas of Zoltan Istvan. *It is an examination of Jethro Knights, the iconoclastic and highly controversial protagonist of Zoltan's philosophical novel, who views himself as already having partially transcended his human origins and conducts himself as though he is a highly evolved machine intelligence, possessed of a moral system an A.I. would likely adopt, according to Jethro's vision of the post-human future he is fighting to bring about.*

When I questioned Zoltan about the function of *The Humanicide Formula* in his novel he told me: "Remember, we are not discussing a perfect human being. We are discussing someone whose final aim is all power over everything. Don't see Jethro only as a human. He is an evolving, amassing point of organized energy in a universe spanning billions of light years. You must try to think how God (should something like that exist) would think."

This reminded me of the fairly common negative reaction of transhumanists—and 'regular' people—to Jethro's most aggressive and threatening statements as well as the consequences of many of his actions. Some transhumanists have reacted as though Zoltan has presented Jethro as a representative of 'real world' transhuman philosophy/values and an exemplar of the kind of militant activism real transhumanists should emulate.

Many transhumanists feel compelled to denounce Jethro and the novel as a whole, lest they tarnish the public's view of actual transhumanists and the goals of transhumanism, generally. This is completely understandable considering Jethro routinely makes sweeping declarations beginning with phrases like: "Transhumanism is…" or "Transhumanists are…" although he is clearly expressing his own idiosyncratic take on transhumanism.

While some may see a need to distance themselves and their view of transhumanism from Jethro's, I would urge them not to end their evaluation of Jethro at

C. T. Armstrong (✉)
Transhumanist, Kansas City, MO, USA

© Springer Nature Switzerland AG 2019
N. Lee (ed.), *The Transhumanism Handbook*,
https://doi.org/10.1007/978-3-030-16920-6_38

denunciation alone. There is much more to be discovered about the whys and wherefores of the weird workings of Jethro's mind and its command of this mysterioso Jethro Knights being. Besides, he is far too interesting to be swept aside with a reflexive: 'He's evil...or crazy...or evil *and* crazy. No deeper analysis required.'

However, deeper *we* shall go. We will take him seriously when he describes his attempts to cultivate a non-human, machine-like mind; when he gets to the point when he no longer relates to humans as though he is one of them (188); and when he speculates about the nature of a future machine moral system (232). Jethro is pursuing his posthuman ideal of "thinking and acting with the same cold clarity a super-intelligent machine would use." (53) As I wrote in the margin of page 81 of *The Transhumanist Wager*, "May be best to think of Jethro as an alien consciousness/psychology." We may as well think of him this way, since that's how he sees himself as he continually strives to become an evermore perfect exemplar of a non-human/alien being: "...embrace the quest to discover how far we can go as humans, as cyborgs, as conscious intelligent machines, as rays of light, as pure energy, as anything the future brings." (85) Far from a damaged less-than human as many critics suggest, in this view Jethro is an enhanced, expanded human; a *beyond* human, albeit expanded in the direction of an aggressive conquering alpha-machine. From this perspective, we can see the novel as a compelling story of conflict driven by a completely anomalous, intriguing, and transcendent alien intelligence—or such is Jethro's self-image and aspiration.

38.1   From Man to Machine

Jethro, to the extent he is able to modify his default *human* nature, cultivates an utterly 'unlike-us-ness' at nearly every turn. His one brief concession to the richness of 'human experience' is his relationship with Zoe Bach, which, though they strongly connect on a deep level, also causes him anxiety because of how the closeness and vulnerability he allows himself to feel with Zoe conflicts with his chosen self-image as a completely self-sufficient, trans-human being with no need of superfluous human frailties like 'love.' To create a firewall against such 'human weakness,' Jethro recites to himself daily, along with several other statements of his core-principles, the following bit of 'code,' in a kind of self-programming ritual: "An omnipotender doesn't fall in love. I will fail to achieve my goals if I lose myself in another, live for another, or place my happiness and aspirations in another. I am self-sufficient, not needing anything or anyone else." (70)

When he is not allowing his pure transhumanist persona to be undermined by such a beguiling human temptress as Zoe Bach, he views himself as a superior and self-made being driven by a non-human moral code devoid of "unreachable mammalian niceties." (84) Of course, virtually no real-world transhumanist shares Jethro's eccentric approach to 'moral calculi' and given Jethro's stated intent to achieve a complete disassociation from all human norms and limitations, it is almost as unnecessary to point out the many ways in which we are not like Jethro as it would be to point out the countless ways in which we are unlike jellyfish.

But alas, protest as he may, Jethro cannot completely subvert his lingering humanity while he is still forced to reside in the looming 'flesh coffin' of his mammalian meat-machine. He views his remaining human 'qualities' as little more than regrettable, vestigial flaws in need of immediate deletion.

38.2   What's the Appeal of a Machine-Like Persona?

The answer lies in an examination of the facts of Jethro's life, specifically during his formative years during which his parents disappeared (likely murdered) and he is raised by an "old aloof uncle" until his death and then in foster homes. (17) In these kinds of environments, one would not tend to bond strongly with one's 'caregivers,' as making emotional connections with people would not prove to be particularly reliable or functional. To the contrary, dysfunction and uncertainty were more likely to have been the predominant constants on which he could consistently rely.

Nor would it be difficult to imagine Jethro, having grown up in such an emotionally malnourishing environment, could have developed a generalized emotional detachment which would permeate nearly all of his interactions with 'humans,' or *others*—as he would come to view them with increasing conviction. He feels he is not of their 'tribe,' opting for self-banishment to a tribe-of-one: "I do not view myself as a beholden spawn or child of the universe. I am alone and distinct." (70)

It is his default mode of non-connection with most people that enables Jethro to evaluate nearly all of *them* as units of production, as resources, as credits or debits in the ledger of the *Transhuman Revolution*.

We can see from the description below, Jethro's preference for a machine-like persona as a transhumanist is an expression of the nature, to which his nurture has led him. For Jethro embracing and cultivating a non/anti-human identity is an obvious choice—really an act of self-typecasting. His stoic, lone wolf brand of transhumanism provides an explicit philosophical foundation for what has been his implicit personality from a very young age.

To wit:

Jethro "rarely listened to people. Or noticed them at all. Even if he looked a person directly in the eye, he often failed to recognize anything of utility. Jethro perceived their presence, the space they took up, the resources they used on his planet. His brain interpreted the matter and energy they possessed, but unless there was potential for something useful to him, he may as well have been looking at a rock, or a weed, or a broken, outmoded piece of furniture in a junkyard. Jethro only took notice of values, not people…harsh machine-like objectivity…Many years ago—he wasn't sure when it happened, or if it even happened at all, or if he was just always this way—Jethro realized he was fundamentally alone in the universe…It wasn't that he didn't want to have friends, or like and even care about other people, it was

just that he rarely met any person who made him feel like he thought he should."
(12)

Before his ascendance to full-time transhumanist leadership, Jethro's war reporting in conflict zones provided experiences which hardened him further and gave
him "emotional immunity and protection his whole life." He also gained an appreciation for "military might" and "fearsome, unabated leadership." (47) Additionally,
Jethro's experience in designing and building his yacht—a "thirty-four-foot steel
sloop" he named, "Contender"—enabled him to create a concrete manifestation of
the core principle at the heart of his idiosyncratic approach to transhumanism: The
yacht was "designed with the same ultimate resolution he held for himself in life:
survival at any cost." (5) It served as a materialization of his Omnipotender essence
in a tangible form: "His will was like the yacht's stainless steel stanchions, even
stronger. His right to life—to always stay alive—was a right unto itself. There was
the universe and then there was that right. This was a man whose overriding sense
of self screamed to conquer, to bend the universe around his will. The will was
stronger...than fate." (6) Jethro "hated fate." So much so, we could sum-up his
entire motive-force as that of a warrior fighting to conquer anything fate would have
the *temerity* to attempt to impose on him. Lesser beings passively submit to fate. An
Omnipotender amasses all the necessary personal-power he needs to command fate.
To paraphrase drag queen Bianca Del Rio: "Not today fate, not today." Obviously,
Jethro would be a fan of the Victorian poem, Invictus, by William Ernest Henley,
particularly the lines:

> I am the master of my fate,
>   I am the captain of my soul.

38.3   New Point of View —> New View of the Point

Possibly the most important benefit of viewing Jethro as a non-human, machine-like
being is that it could lead us to discover one of the facts of reality at the root of *The
Transhumanist Wager* that makes Jethro's extremism explicable. In a few steps, we
can discover this fact and its implications if we wonder about things like: Is there a
particular reason, aside from assumed immorality/psychopathology, why he believes
it is necessary or useful for him to become a kind of being that can go to greater
extremes than a human might? And this might inspire us to wonder about extremism
per se. Then eventually we may stumble upon the core reason he finds it absolutely
essential to be an extremist in order to reach his number one and non-negotiable
goal: to achieve an indefinite lifespan. The reason extremism is necessary, is because
of the unique nature of this goal—assuming he is actually 100% committed to
achieving it. To mutate a well-known line from Carl Sagan:

38.4   Extraordinary Aims Require Extraordinary Expedience

The unique and non-fault-tolerant nature of Jethro's goal to remain alive until science makes it possible to reverse aging *demands* an "at any cost" approach…if, that is, he is *maximally* serious about reaching this goal. At the beginning of the story, the science needed for an uninterrupted life/health-span is nearly non-existent. Even lifesaving measures, in the case of serious injuries, are not particularly advanced. These facts make death (not 'near death'…not 'momentary death'…but 'dead and gone' death) a guaranteed 'non-recoverable-from' event—a zero-sum, win-lose, binary, black/white condition.

While nearly every other goal in life allows for short-term failure with the possibility to try again or abandon the failed goal for another pursuit, failure at not-dying leaves the deceased with no other options. While a self-help guru may tell you 'failure isn't final' or not 'fatal.' They would be right in any other endeavor, as long as that failure has left you alive 'to fight another day.' But, failure at the game of not-dying is, by definition, fatal *and* final. This, currently, unalterable fact leads Jethro to conclude:

"There is no right and wrong when it comes to dying or not dying. There is only success or failure…[at] using whatever means necessary to accomplish those aims, of thinking and acting with the same cold clarity a super-intelligent machine would use—something I am quickly evolving into anyway."(53)

Of course, any compassionate humanist-type will tend to instinctively reject an "at any cost" standard by quickly bringing to mind 'extenuating circumstances' and loved ones who would be exempt from such an unyielding commitment. There is a line—*many* lines—which they will not cross.

From Jethro's point of view, this kind of compassionate, nuanced mindset amounts to consciously choosing one's specific 'failure scenarios'—planning for failure—and is therefore unacceptable. Failure is not an option in the game of not-dying. It is an unrecoverable 'game over' condition. The humanist's conventional morality may make them feel better about themselves and provide many social benefits, but in the narrow, individual game of not-dying, it is merely a recipe for *voluntary* defeat…aka: *surrender*. If Jethro believed in anything like 'mortal sin,' surely the voluntary surrender to death would reside at the top of his list.

Jethro has carefully considered many life threatening scenarios and concluded that the finality of death *mandates* that there is no scenario in which he will choose to place *anything* above his own survival, if he is to maintain a total commitment to defeating death. For Jethro, there is no person or ideal that has a higher priority or value than this commitment. (While this is *nearly* a categorical truth for Jethro, there is an exception mentioned elsewhere in the novel, but we'll not go into that here.)

And now, for the punchline…

38.5 Jethro Knights: Paper Tiger?

This entire preamble has been necessitated by a particular fact of the novel which is too often overlooked: While Jethro certainly does often think, say, and write about the desirability of being a machine guided by non-human values, consisting of a cold utility-function that can determine a human's worthiness to participate in the transhumanist movement or continue to exist, he *never* acts in any way *close* to this. He *never* commits the kind of uncaring mass-murder of human beings which such ideas, and some of his direct threats, call for so matter-of-factly.

There are repeated examples of situations wherein, if Jethro were a true cyber-pathic machine-executioner-god, programmed in the ways he, in his most scary moments, has speculated about and threatened, he would have committed acts of slaughter "of a scope and scale that would embarrass the most ambitious psycho-path." (Thanks to Sam Harris for that colorful phrase.) Yet instead, for example, he rescues enemy soldiers who only moments before participated in an attempted genocide of 10,000 transhumanist scientists and their families. (261) He also *repeatedly* warns people to stay away from locations targeted for bombing during the war in an attempt to destroy only structures, but kill *no people*. (267) He eventually has control of all the world's nuclear weapons, but doesn't use *one* of them. He merely demonstrates to the terrified onlooking world 'powers' that he *could* launch them and control their targeting if he wanted to. (266) During the war, he tries, whenever possible, to avoid casualties and leave infrastructure intact so innocent people won't be left without electricity, water, etc. (230)

Additionally, it is important to point out that in a 'war' involving all of the great world powers (all with extensive nuclear arsenals on hand), the final death toll is incredibly low. It is undoubtedly a world-record low body count in World War history and the most ethically waged war as far as lives lost is concerned. And it was precisely Jethro's superior technology that made this kind of minimal, 'surgically precise' war possible. This same technology could just as easily have been used to perpetrate an unprecedented holocaust, again, "of a scope and scale that would embarrass the most ambitious psychopath."

38.6 Worst Case Scenario Syndrome

The fact that Jethro's 'kinder gentler' side, regarding his actions versus his threats, can be so easily overlooked is partially his own fault. Jethro has a penchant for specifying, particularly in speeches to the entire world and in extremely blunt terms, his 'rules of engagement.' In these next two examples, he is unequivocal about how unconditionally he will maintain his commitment to his survival and the *ultimate* extremes he will go to, without hesitation, if he meets a certain level of resistance from *anyone*.

"If you try to stop us, we will fight you—and we will defeat you. We will kill you if we have to. If needed, we will kill every one of you, down to the last enemy of transhumanism on this planet. We will eliminate you into the void of the universe with no remorse, with the same cold morality a machine would use. We are through playing by your rules and on your terms." (202)

And…

"If there is still more resistance that deliberately hinders or interferes with goals of transhumanism, we will eliminate you—each and every one of you who defies us. We will implement a systematic humanicide. My country and I are after extreme life enhancement, our own personal immortality, and in creating a far more advanced, rational, and spectacular future for our planet. Many of you are useless to us right now and are therefore completely dispensable. We have the power, and we will methodically use it to destroy any force that purposely stands in the way of our transhuman mission." (235)

There are also philosophical discussions wherein others ask Jethro about his principles and how they would cause him to act in *extreme* hypothetical scenarios. Below is an extravagant philosophical flight of fancy, which is seriously proposed as an illuminating 'moral dilemma' by Jethro's primary antagonist, Reverend Belinas:

"If it were somehow possible, and you were forced into a predicament where the only way to reach your goal was to kill your wife, then you would kill her, unafraid. You would murder her a thousand times to reach your immortality, if required. That's how brutal you are at the core, how monstrous and evil you are."

"Jethro threw his head back, a shiver filling his body. Belinas had penetrated him, and found a vulnerable point. Jethro remembered how grueling and twisting it was after Zoe died. Remembered the utter pain, confusion, sadness. Despite this, the transhumanist forced himself to answer, clearly and firmly, 'What you say is true, preacher. I would kill my wife a thousand times if I absolutely had to in order to reach my goals. But the reality is, and will always be, that I love my wife. I love her so utterly much even now, years later. And I would do an infinite amount of things to avoid the perverse predicament you propose.'" (246)

38.7 A Rocky Moral Landscape

Of course, it is human nature, when confronted with a person who professes an extreme and unwavering commitment to some principle, to respond with: 'Oh really? Well, what about THIS? Are you willing to go THAT far, tough guy, or do you draw a line before that?' This is a natural desire to test the person's seriousness and how far they have really thought through all of the ramifications of sustaining the proposition to the extreme. This kind of philosophical interrogation is useful for revealing the boundaries and limits of a person's devotion to an ideal when pushed to improbable hypothetical extremes.

Jethro's problem, public-image-wise, is that nearly the entire 'moral landscape' presented via his statements is comprised of 'warnings to adversaries' about what extreme measures he is willing and able to carry out if they become aggressive, recalcitrant, or even a mere hindrance to a great enough degree. Thus, the issue of moral-action remains at a worst-case-scenario level, which, in moral terms, is akin to a 'martial law' situation: an attempt to maintain at least *some* order and normalcy in a time of chaos and crisis. But it is a temporary, *emergency* situation—to be abandoned ASAP—and not to be confused with how things *should* be under 'normal' conditions. For those interested in connections between this book's philosophy and Ayn Rand's Objectivism, you could look-up Rand's concept of "ethics of emergencies" and see how it relates to this issue.

Jethro instinctively goes to the extreme cases in order make his boundaries, or lack thereof, clear and to save time by not needing to address every ridiculously conceivable contingency someone may throw at him. He just says, in effect, "I *really* mean this stuff and no matter where you may think I might draw the line, I will *always* cross it if my own survival is threatened. So just save your breath and don't bother conjuring up ever more clever 'what ifs.'" But by doing this, it isn't too hard for one to get an impression that Jethro is a person with absolutely no moral boundaries whatsoever.

38.8   Of Words and Deeds

The emotional impact of some of Jethro's scariest *threats* tends to obscure the far less emotionally charged memories of his more humane *actions*. Repeatedly, when Jethro finds himself in tense situations which require action—not the mere issuance of threats or the consideration of hypothetical 'moral dilemmas'—he faces choices ranging from extremely bad to extremely humane. In these instances, we see that he often makes 'good' choices and is not actually an irredeemable moral-monster, as he surely *does* seem to be, based on some of his most belligerent words alone.

Of course, one could say Jethro just wasn't pushed far enough for him to make-good on his most dire threats. And any acts of alleged 'mercy' were only performed because they served some 'public relations' purpose and were not 'pure' examples of compassion on his part. All this may well be true, in some cases. He can be quite a scary character, given his desire to emulate a machine programmed with no regard for "mammalian niceties." However, the inescapable fact remains: He did NOT commit any acts of mass-murder, as he threatened to do. The way this story played out, his words were *far* worse than his deeds—not even in the same 'moral universe.' In the game of ruthless-genocidal-dictator, Jethro is truly an underachiever who passes up *many* golden opportunities to commit mayhem and murder. In the end, he proves he is not after power for its own sake when he voluntarily, and under no outside pressure, relinquishes all his hard-won power. (290) Come on Jethro, you have a reputation to uphold here. Don't make us revoke your 'gangsta' card.

So, we can view him as someone who fancies himself to be a non-human, machine-like being, capable of things, when pushed beyond his limits, which would horrify any normal human. But given the above examples of humanitarian restraint he repeatedly exercised, he is still an entity that hasn't completely rid itself of all remaining shreds of compassion and human decency, much as he may fantasize otherwise.

38.9   Machine-Man/Man-Machine

An obvious way to explain Jethro's idiosyncratic personality is simply as the product of an unfortunate upbringing, resulting in a conventionally maladjusted, but high-functioning, human: a narcissist, sociopath, psychopath, etc. Alternatively, we have been considering a view of Jethro as something that was initially human but came to view itself as having been hobbled by bits of lingering ape-sentimentality, preventing it from reaching its full potential as an ascendent alpha machine-intelligence. In order to transcend these limitations, Jethro 'programmed' his mind to suppress many facets of his humanity he deemed to be counterproductive and replaced them with his conception of how a future non-biological, post-human entity would think and behave.

But let's not overlook the option to bypass an either/or framing of these two views in favor of a 'middle path,' both/and perspective. In this combined view, we're not compelled to situate Jethro as *either* a high-functioning psychopath *or* as a being who is propelling itself, by force of will alone, into a new stage of non-human (r)evolution. The 'middle path' option sees Jethro's early familial instabilities, and a nearly complete lack of nurturing or bonding, resulting in well-developed emotional and empathetic maladjustments, which lead him to shun his humanity as it is little more than a constellation of disempowering vestigial deformities-of-character and impediments to the realization of his alpha-Omnipotender aspirations. This dire assessment leads Jethro to embrace a strong vein of individualistic, libertarian, neo-Nietzschean transhumanism as a means of rising above his perceived lowly, 'all too human,' origins.

This worldview and self-assessment runs deep in Jethro's psyche, as his budding transhumanist disposition was already 'on board' from an early age: "Beginning with childhood, Jethro was attracted to transhuman philosophy. This was because he instinctively viewed life as a chance to improve himself, hoping 1 day he might reach a self-actualized perfection." (19)

This view is consonant with Zoltan's suggested way of understanding Jethro: "Don't see Jethro only as a human, he is an evolving, amassing point of organized energy in a universe spanning billions of light years. You must try to think how God (should something like that exist) would think."

This kind of fictional person/machine/person-machine will confront transhumanists with many questions about our own level of commitment to transhumanism; about how far we are willing to go, in the face of aggressive opposition, to achieve an indefinite lifespan; about what lines we will not cross; and about the possible consequences of our self-imposed boundaries.

As Zoltan explained: "I tended to write the story from the perspective of a simple question: How far would one man go to achieve his immortality?" Jethro assured us repeatedly he was willing to go exceedingly far—beyond any boundaries in some cases—to reach his personal goals and transform the world. But, as we have seen, his actions never got *close* to the horrific levels of devastation he threatened to unleash in his warnings to the world.

38.10 Recommendation

Use Jethro as a catalyst for deeper thought and exploration rather than just an easy repository for righteous indignation and moral opprobrium.

Chapter 39
Art and Transhumanism

R. Nicholas Starr

"Historically, revolutionary ideas have emerged first among artists and intellectuals."

Modern civilization may have never existed if it weren't for those progressive thinkers. The precious few who were willing to look beyond the world in front of them and imagine what could be, even if they had to sacrifice themselves. But those people felt they were obligated to share it and help inspire future generations to make the right decisions.

Art is a simple way to put life in scope. Through it we can use abstract processes to create a "second opinion" and analyze problems within our society. This is why so many dominant civilizations throughout history placed such high value on art, literature, and song. It was their way to study life and then find ways to improve it.

We can first look to ancient Egypt, whose artists not only built monolithic pyramids, but painted about the civilization that created them. Upon analysis of Egyptian art we can learn a lot about the social hierarchy and the daily lives of each social caste. For the people of the time it was a tool to study the very basics of life and then teach others, the Greeks in particular, about it. Ancient Greece, a culture renowned for skilled craftsman, took the study of life to the next logical step – speculation. According to Greek mythology Hephaestus and Daedalus began work on the worlds first automatons in an effort to simplify life, which in turn inspired mechanical experimentation. So popular was the idea of automation that in 322 BC Aristotle wrote, "There is only one condition in which we can imagine managers not needing subordinates, and masters not needing slaves. This condition would be that each instrument could do its own work, at the word of command or by intelligent anticipation, like the statues of Daedalus or the tripods made by Hephaestus, of which Homer relates that 'Of their own motion they entered the conclave of Gods on Olympus', as if a shuttle should weave of itself, and a plectrum should do its own harp playing" [1]. 2300 years later we find ourselves staring workforce automation square in the face.

R. N. Starr (✉)
California Transhumanist Party, Grass Valley, CA, USA
e-mail: Noise@esgal.net

© Springer Nature Switzerland AG 2019
N. Lee (ed.), *The Transhumanism Handbook*,
https://doi.org/10.1007/978-3-030-16920-6_39

The Egyptians and Greeks also used art to study the human form. To accurately sculpt a man you need to know what makes the body move so these civilizations became the first to dissect the human body and thus began the science of anatomy. So important was this to art that it soon became a requirement for any studying artist. In 1466 a young apprentice under painter and sculptor Andrea del Verrocchio took his anatomical course work and through his art revolutionized the world. That 14 year old boy became the most influential artist and scientist of all time, Leonardo da Vinci.

Fritjof Capra asserts in his book "The Science of Leonardo" [2] that his scientific developments in medicine and engineering stemmed directly from his desire to accurately portray his subject. He then used his meticulous knowledge and rearranged the parts on paper to craft theoretical devices like his flying machines. Perhaps one of his most important works for modern civilization was the Ideal City. The premise was simple, yet revolutionary. Structure a city that benefits the people. Previous medieval planning followed a sprawl approach, typically centered around a regional lord, where people filled in empty space as they saw fit. This haphazard approach led to significant problems with transportation and sanitation, the result of which lead to massive devastation during epidemics such as the Black Death. Da Vinci set out design a far more efficient city that benefitted even the lowest on the social ladder. While the city was never built it planted the seed for future artists to germinate the idea of what an ideal city would be. It wasn't long before authors like Thomas More and Francis Bacon began to design their utopia.

In Francis Bacon's book "New Atlantis" we can see an ideal city with a government based on science and inductive reasoning [3]. This is largely what a transhumanist government strives for. Within Salomon's House, the islands center for education and government, laws are written based on what is known and can be modified if new data should come to light. The willingness to adapt law is fundamental to their society of exploration and invention. However, as many profess, Bacon's tendency is to lean on positivism within his world and illustrates this by labeling negative actions as something that could only happen outside of Bensalem and Salomon's House concealing collected research deemed detrimental or unfavorable [4]. By shielding themselves from adverse consequences they end up losing useful data. But while Bacon prefers a unilateral utopia, there are many other works to address the negative possibilities of scientific discovery and poor decision making. After all, there cannot be utopia without dystopia.

In the year 1818 we are introduced to The Modern Prometheus, Frankenstein and his perceived monster. This legendary dystopia of experimentation and hatred has had a nearly unparalleled contribution to the development of art and science. Within its pages we start to address the ethics of science, religion, and the the industrial revolution. We also get a more scientific approach to discussing what is alive and what is human. These are topics of discussion which will come into great importance as we enter the age of biosynthetic engineering, genetic editing, and artificial intelligence. One of the beautiful things to note about Mary Shelley's novel is it's ambiguous nature. While there is a definite topic of discussion, the topic is so vast it helps the reader create an analytical framework that, for better or worse, can be used to discuss a broad range of scientific advances. That framework will become critical when we analyze how science and technology will impact privacy, sentient/ sapient rights, the economy, and a host of other topics critical to future civilization.

From the twentieth century through today the art of speculative fiction has become a vastly popular means to address our questions, our hopes, and our fears for humanity's future. Can we stop man made climate change? Will robots rise up and destroy humanity? What about Lab created biosynthetic organisms? With creative tools growing more powerful by the minute we can not only tell a good story, we can create incredibly provoking details that any consumer can relate to on a visceral level. But with the vast numbers of works available, what does the consumer do with all this information, and how can they process it all?

The first thing we need to realize is, just because an artist has presented a specific possible future it doesn't mean that humanity is bound to that fate. In fact quite the opposite is true if we heed the warning. Art is the perfect lens to view all possible outcomes and chose the optimal path. We can use art as a tool to decide where our research needs to be directed and what legislation may need to be enacted. Dr. Edward Finn of Arizona State University has begun to do exactly that at the university's Center for Science and the Imagination. Dr. Finn has brought young artistic minds from across the country and given them a network of resources to take their ideas and see what is actually possible according to science. They have even had the chance to sit down with several government agencies and policy makers to find creative solutions for our most pressing problems [5]. Dr. Ionat Zurr and Dr. Oron Catts at the University of Western Australia's SymbioticA program have taken this premise one step further, targeting biology based research, and have created an extensive collection of work to address many of the ethical and feasibility questions of creating synthetic organisms. In an interview with Dr. Zurr we discussed how lab grown "victimless meats" are a costly and inefficient solution to increasing meat consumption [6]. This leads us to the next step in our analysis of futuristic art. Are these futures actually achievable?

Fiction often forgoes the requirements of having actual proof that a premise is possible for the sake of telling a good story. Thankfully we live in the real world where life is bound by the laws of science. Any artistic claim can and should be put to the test of science. This requirement serves a three fold purpose: to identify the technology that should be pursued, to identify that which, for better or worse, is not physically possible, and to reset the creative process with new information and possibilities. This third purpose is perhaps the most important of them all as it encourages the use of bleeding edge technology and processes to push art itself to a new level, thus creating a mutual feedback loop. Dr. Akihiro Kubota of the Tama Art University has used this loop as a foundation to use artificial intelligence (AI) to create art, or perhaps more accurately, have the AI make art for humans to interpret. But that itself isn't Dr. Kubota's ultimate goal. He is far more interested in whether AI can be developed to appreciate art [7]. The prospect to teach a computer to create, interpret, and appreciate subjective information such as art can provide an entirely new point of view from which we can analyze any topic. It is this abstract innovation, when applied to modern and future civilizations, that provides the greatest potential for human evolution through transhumanism and into singularity.

The human ability to imagine and create has always driven progress and will continue to do so for as long as our species, in one form or another, lives on. Please, don't let anyone take this away from you. Keep dreaming, keep creating, and most importantly, keep sharing your work. If you don't, we will never know what opportunities and advancements could await us in the future.

References

1. Aristotle, Politics, book 1, part 4 http://classics.mit.edu/Aristotle/politics.1.one.html
2. Fritjof Capra, The Science of Leonardo http://www.fritjofcapra.net/learning-from-leonardo/
3. H. Bruce Franklin http://andromeda.rutgers.edu/%7Ehbf/sfhist.html
4. David Whitney https://sites01.lsu.edu/faculty/voegelin/wp-content/uploads/sites/80/2016/09/WhitneyAPSA2016.pdf
5. Dr. Finn Interview https://bodyhackingcon.com/blog/interview-with-dr-ed-finn-of-the-center-for-science-and-the-imagination.html
6. Dr. Zurr interview https://m.youtube.com/watch?v=zOCTWNADcCY
7. Dr. Kubota Interview http://transhumanist-party.org/2017/10/24/interview-kubota/

Chapter 40
An Artist's Creative Process: A Model for Conscious Evolution

Dinorah Delfin

40.1 An Artist's Creative Process: A Model for Conscious Evolution

A great future is opened to humanity. Our interconnectedness with nature, the tools we use, and the narratives we create, is reaching a pinnacle. For the first time in human history, we have the means to *consciously* alter the fate of our evolution. New technologies are not only becoming increasingly embedded in our biology—giving us unprecedented human abilities—but this transition is also driving us to explore new notions of what it means to be human in the heroic pursuit of individual sovereignty, and general happiness and sense of purpose.

In this uncertain, but awe-inspiring unfolding of human potential, the goal for many of us is to become *transhuman*, also referred to as, *post-human*. Oxford University philosophy professor and transhumanist, Nick Bostrom, explains: a post-human, is a being that has at least one general capacity, like intelligence or lifespan, "greatly exceeding the maximum attainable by any current human being without recourse to new technological means." Transhumanism, or *the conscious re-designing of the human organism, or its radical enhancement*, is thus, the most revolutionary and disrupting paradigm shift the modern human has yet to experience.

In transitioning to this *post-human* era, how can one adopt a framework for cognitive and physical enhancement that accounts for ways to ensure that this new era is also more consciousness oriented, safe, and egalitarian?

D. Delfin (✉)
United States Transhumanist Party, New York, NY, USA

© Springer Nature Switzerland AG 2019   587
N. Lee (ed.), *The Transhumanism Handbook*,
https://doi.org/10.1007/978-3-030-16920-6_40

40.1.1   The Creative Journey

In this essay, I'm presenting five art projects I created from 2007 to 2017 to show-case the following:

1. How an artist's creative process, that is future conscious and integral, can lead to the creation of artworks that foster a shift in perspective and inspire positive social change.
2. How *a transformative creative process* can be applied to any area of personal development by illustrating how my own process of cognitive, emotional, and spiritual development led me to explore transhumanism.
3. Lastly, I will propose how we can utilize a *transformative creative process* with transhumanist sensibilities to design a world that is more just and humane.

40.1.1.1   Hybrids, 2007

In 2007, I turned a computer-generated pixelated image of a person into a three-dimensional representation of it, in the physical space, to illustrate the idea of a future hybrid being embodying both human and digital qualities.

In this material instantiation of what began as an idea, the result was also a hybrid artwork which combined both painting and sculptural qualities, which I named "A Pixelated Profile of a Girl in Three-Dimensions."

When standing in front of the artwork, our perception is drawn to the fragmented parts, or cubes of various heights, which make the 3D pixels. To make sense of the composition, a tool such as a peep-hole or a camera phone, is used to look at the artwork and see it as a cohesive whole. A recognizable image, usually a face, can be seen. Stepping back also does the trick.

In transitioning from bits to matter, and from fragments to a whole, the canvas itself becomes the catalyst for the humanizing of digital information through a shift in perception inspired by a culture becoming increasingly digitally interconnected and more aware of its part in the big picture.

Creating this artwork gave me the opportunity to explore and reflect on issues concerning our evolution, which led me to realize that the more complexity and intelligence life produces, the better its chance for survival. These hybrid artworks represent the continuous permutation of complex organisms, like mind and matter, with its plethora of challenges and opportunities (Figs. 40.1, 40.2, 40.3, and 40.4).

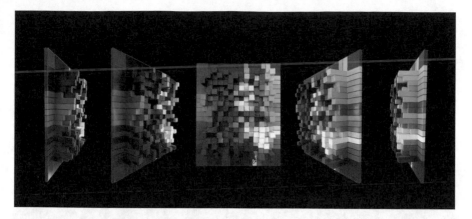

Fig. 40.1 A pixelated profile of a girl in three-dimensions, 2007. Mixed-media assemblage on canvas. 16 × 20 × 5 inches. With front and side views

Artwork seen through a piphole

Fig. 40.2 Kate, 2012. Mixed-media assemblage on canvas. 40 × 50 × 5 inches. With view through peephole

Fig. 40.3 Close-up. Mixed-media assemblage on canvas

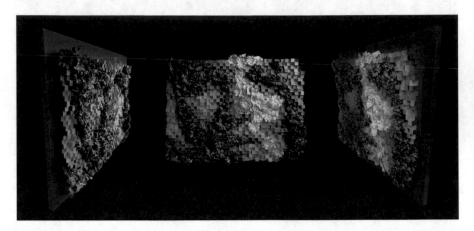

Fig. 40.4 Kate Pink, 2012. Mixed-media assemblage on canvas. 40 × 50 × 5 inches. With front and side views

40.1.1.2 Holy, 2009

In 2009, I started a new project. Instead of reflecting on the humanizing of technologies, the goal was to use technology to humanize the human condition.

I started creating a series of digital photo collages showcasing "digital clones" of myself interacting in bold and subtle power struggles. The process of making these images, which started out with a camera, myself, and a green screen background, involved layering a collection of personal symbolic imagery which I seamlessly assembled using Photoshop.

This series, which I named, "*Holy*", was triggered by preoccupations of a dystopian future resulting from a holier-than-though corruption of the psyche. My naked body symbolizes vulnerability and the discomfort one experience by the unlearning of deep-rooted beliefs associated with religious and cultural upbringings.

In the resulting surreal, Caravaggioesch photomontages, the fleshy digital clones are bound to primal and mundane realms in which a story of collective catharsis and transformation wants to take place.

Through this journey of self-realization, a collective awakening becomes humanity's Holy grail needed to successfully transition into the next stage of human evolution—the transcendence of conscious and subconscious limitations, whereby a truly advanced civilization can emerge (Figs. 40.5 and 40.6).

40.1.1.3 Fossil Fooled: The Age Of World Suicide, 2011

In 2011, I created "*Fossil Fooled: The Age of World Suicide*", a 1-minute, 50 second-long experimental video, which I wrote, filmed, and edited, inspired by my participating in the Occupy Wall Street Movement.

The video starts off with a nude woman, myself, sitting hypnotically facing a bright light. The former president, Jimmy Carter's voice plays in the background

Fig. 40.5 Holy, 2009. Digital Collage

from footage recorded in 1979 where he talks about the "crisis of confidence" in American government and also challenges Americans to unite and address the excessive dependence on foreign oil and consider new forms of green energy. The video then cuts to a wrist bleeding what it appears to be black, thick oil.

Following, a woman, wearing a business jacket, a skirt, and an eerie mask, delivers a monologue: "We are in a very interesting period in human history. Most of us know that the world is in deep shit. The collapsing of the economy and the environment; the overpopulation and the squandering of the planet's resources; people in power losing control; the lack of millions of dollars in my bank account. But, most people have no idea why this is happening and what we are really, really, facing in the future." As her hands travel down her neck and gently rest on her chest,

Fig. 40.6 A King's Milk Gone Sour in the Midst of Warly Delights, 2010. Digital Collage

she calmly utters: "for the first time in human history, the world will experience increasingly and more acutely, loving pain, like it never did before. Like it never did before." The video then ends with the words: *"To Be Continued."*

While alluding to an underlying narrative of social transformation and better environmental ethics, the character's subtle demeanor in embracing paradox and ambiguity represents a cultural climate of rising stress levels and anxiety with a desire for redemption and continuation.

Is the short movie *"To Be Continued"*; is the current narrative of fear, scarcity, and excessive materialism *"To Be Continued"*; or is it life, at all cost, *"To Be Continued"*? (Figs. 40.7 and 40.8).

40.1.1.4 Odissi Metal: A Meaning-Full Choreography, 2013

"Odissi Metal: A Meaning-Full Choreography" is a four-minutes, 29 second-long, video collage I created in 2013 to showcase two contrasting dances: *"Odissi" (circa 500 BCE)*, one of the oldest classical sacred dances from India, and *"Body Remix/*

Fig. 40.7   *Fossil Fooled: The Age of World Suicide*, 2011. Short movie. Screenshots

Goldberg Variations" (2005), an avant-garde dance by Canadian choreographer Marie Chouinard.

In this metaphorically quirky and playful composite of visual and auditory narratives contrasting two very different cultures, "Odissi" illustrates the ancient and the sacred, cherishing the wisdom of the years, grace, contemplation, and preservation. *"Body Remix/Goldberg Variations"* is a wild and sexually charged contemporary performance, enhanced by the use of crutches and prostheses, showing the body as an animalistic machine, alluding to transformation, freedom of expression and agency. The video collage is composed of small video frames within a large one, mirroring each other, all choreographed to the song *"Internet Friends"*, by Knife Party.

Engendered by the feminine, feminist reflection, *"Odissi Metal: A Meaning-Full Choreography"* is a call for tolerance, diversity, and integration. This video collage represents a meeting of the minds through an interconnected global brain, inspired by a desire to coalesce the best of Eastern and Western wisdom in the pursuit of universal value systems (Fig. 40.9).

40.1.1.5   Mantra Manifesto Capitalista, 2017

In 2017, I created *"Mantra Manifesto Capitalista"*, a poem-like-manifesto for conscious evolution.

Originally a 1500-word reflection on spirituality, technological development, science, and post-modernism, this personal manifesto expresses the sentiment of an

Fig. 40.8 Delfin's participation in the Occupy WallStreet movement and in the original ad from AdBusters America Magazine. "We Need You" call to action, appeared in the AdBusters issue which triggered the movement in 2011

era verging on the most profound paradigmatic shift the modern human has yet to experience—the rise of a super-human race and super-intelligent machines.

This mantra is a call to preserve what makes us human, or humane, in order to establish peaceful relations between humans and other intelligent species by proposing a set of values which can lead to universal civic virtues.

"*Mantra Manifesto Capitalista*" is a declaration for full autonomy and command over one's own evolution, not just through cyborgian technologies, but also through a re-negotiation of meaning and the power of *intelligent intentionality*.

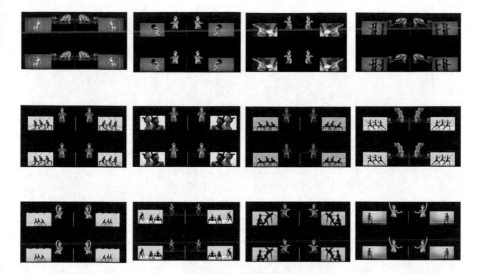

Fig. 40.9 Odissi Metal: a meaning-full choreography, 2013. Video Collage. Screenshots

Through renewed ethos, pathos, and logos, this manifesto is a tool for cognitive enhancement and emotional intelligence inspired by the principles of neuroplasticity to positively impact one's consciousness through focused, intentional chanting (Fig. 40.10).

40.1.2 A Transformative Creative Process

Creating "Mantra Manifesto Capitalista", lead me to discover Transhumanism. At first, I was skeptical. Some of the ideas advocated by the movement seemed too far-fetched and even dangerous. I remained curious, however, and the more I set out to understand the reasoning and philosophies behind these ideas, the more I realized that this movement represented some of the most revolutionary and forward-thinking ideas I have ever encountered. Transhumanism, with its promises and challenges, provides to date, the most logical and promising plan for the future.

As with *"Mantra Manifesto Capitalista"*, each artwork presented in this essay was conceived using a creative process which involves two distinctive stages: *First, it addresses and challenges a limiting or deeply held belief. Second, it leads to a positive, measurable change.*

Limiting beliefs are rooted in cultural upbringings and conscious and subconscious biological programming. Limiting beliefs prevent us from expanding the scope of our imagination and understanding of reality; prevent us from becoming more tolerant, authentic, empathetic, accomplished, and happy; and also prevent us from finding common ground with other belief systems. Limiting beliefs, whether

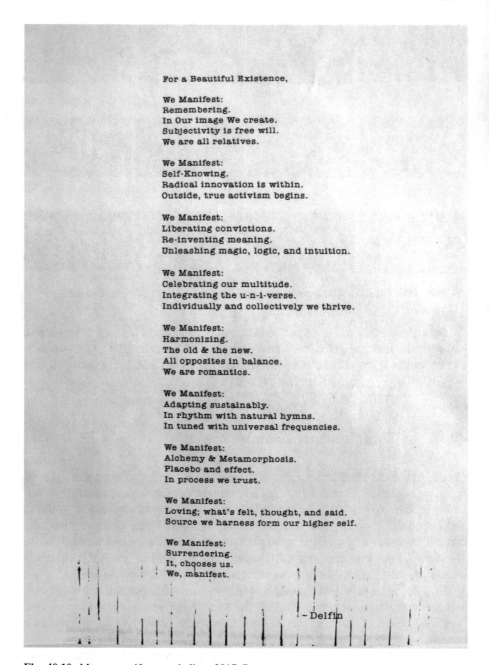

For a Beautiful Existence,

We Manifest:
Remembering.
In Our image We create.
Subjectivity is free will.
We are all relatives.

We Manifest:
Self-Knowing.
Radical innovation is within.
Outside, true activism begins.

We Manifest:
Liberating convictions.
Re-inventing meaning.
Unleashing magic, logic, and intuition.

We Manifest:
Celebrating our multitude.
Integrating the u-n-i-verse.
Individually and collectively we thrive.

We Manifest:
Harmonizing.
The old & the new.
All opposites in balance.
We are romantics.

We Manifest:
Adapting sustainably.
In rhythm with natural hymns.
In tuned with universal frequencies.

We Manifest:
Alchemy & Metamorphosis.
Placebo and effect.
In process we trust.

We Manifest:
Loving; what's felt, thought, and said.
Source we harness form our higher self.

We Manifest:
Surrendering.
It, chooses us.
We, manifest.

~Delfin

Fig. 40.10 Mantra manifesto capitalista, 2017. Poem

it involves one's identity, others, or the world at large, limits human potential and creativity.

A positive, measurable change may be emotional, cognitive, physical, or spiritual; and it may be in the form of self-realization, wellness, profound ideas, artistic expression, communal creativity, and world peace. A transformative creative process can foster a world that is less fearful, anxious, judgmental, and corrupted; and more conscientious, healthy, safe, and sustainable.

During this process of constructive perspective shifting, five distinct stages take place:

1. Curiosity: Embracing the unknown, the different, the contradicting and the ambiguous is a source of unspoken wisdom. Vulnerability and humility are as important as confidence and strong convictions. However, daring to challenge our own conventions, and resisting the impulse to judge an idea, or a belief system, without giving it the opportunity to reveal its universal truth and essence are key. Curiosity not only expands intellectual excellence and emotional intelligence, but it is essential to dissolve limiting beliefs. Curiosity leads to interdisciplinary research to develop and generate ideas. At this stage of the creative process, one can experience heightened levels of empathy and pattern recognition.

2. Reflection: A creative process is a powerful tool for introspection; to create hypotheses; to find common ground in seemingly unrelated information. This is a time to serenely monitor the patterns of our own thoughts and amend biases and shortcomings. During the reflective stage, ideas incubate as one takes time to consider the knowledge and information that has been accumulated. One important aspect about this stage is solitude, to give the brain a chance to wander and gain clarity. Solitude is key to revealing the true and most authentic self from the noise of modern life and cultural constructs.

3. Integration: Everything is connected! Throughout history, things have been done and said in a plethora of ways to refer to the same ideas and needs. Drawing connections between the various things that interest someone is very important. The integration stage of the creative process increases our ability to be more flexible, fluid, attuned, and adaptable.

4. Play: Creative expression can take any form. Participating, showing up, speaking up, writing it up, dancing it out, building it out, dreaming it big; anything it takes to express one's deepest, weirdest, wildest, and most vulnerable desires and fears—Reality, is a work of art whose expression is defined by the limits of our imagination.

5. Resilience: The creative process isn't free of moments when feeling like giving up. Bottlenecks, frustration, and technical problems can become insurmountable. Take a break. Don't give up. *Trust The Process.*

One key aspect of my creative process, which shares transhumanist sensibilities, is *Pragmatic optimism.*

6. *Pragmatic optimism:* isn't about having a blind positive attitude about the future, it is about understanding where things can go wrong and finding proactive solutions without losing focus on the possible positive outcomes.

Most importantly, *pragmatic optimism* is about wellness and longevity. The quality of our thoughts, whether it is contained within the individual or shared with others, can have a profound impact on our biology. Daisy Robinton, a Harvard University graduate Biologist, in a recent TED Talk ("*Thoughts Matter: How Mindset Influences Aging & Lifespan*" 2018), explains that having a positive attitude towards life improves our physical and psychological health and resilience to sickness and cellular degeneration. In the talk, she states: "The tone of our thoughts can influence the way our bodies degrade to a molecular level. Severe stress causes our DNA levels to shrink until they can no longer replicate." She adds, "Negative thoughts are slowly killing us."

Simply shifting one's attitude can be lifesaving and an artist's creative process can be a very effective tool to achieve just this. By harnessing the power of curiosity, reflective thinking, pattern integration, practicing resilience, and pragmatic optimism, one can give birth to creative expressions for radical positive change. **An artist's creative process is thus, a bold reminder that transformative, sustainable change starts within** (Fig. 40.11).

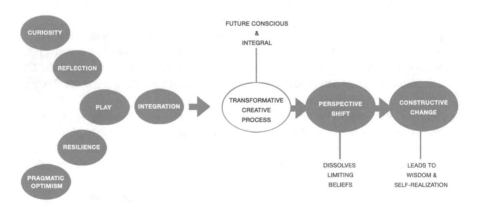

Fig. 40.11 A transformative creative process, 2019. Graph

40.1.2.1   The Integral Transhumanist

Recently, I wrote a short, reflective article titled, *"Integral Transhumanism"*, which was published by the U.S. Transhumanist Party's website on October 31, 2018. In this article, I argue that in order to preserve transhumanism inherent ethical and altruistic worldview and to successfully inspire a transition into a post-human era, we ought to master the art of *holistic living* and take into consideration our interconnectedness to the collective consciousness (collective psyche) and natural systems.

Throughout my creative process, I also adopt an integral approach, which is a developmental cognitive model developed by contemporary philosopher Ken Wilber. Wilber's model takes into account all dimensions of existence—the internal and external make-up of an individual in relationship with others, his/her culture, and the universe at large—towards a more complete map of reality. Wilber's model put into context Eastern and Western knowledge, and relates all parts to the whole by using diversity and discernment.

A creative process that is future conscious and integral has a post-postmodern outlook by taking into account the relevant wisdom of all of its preceding movements, including the wisdom of ancient and aboriginal cultures. This holistic approach to the creative process can lead to a more conscious and responsible use and development of transhumanist technologies.

40.1.3   A Model for Conscious Evolution

Many experts agree that the rate of development in scientific and technological discoveries is expected to see exponential growth in the relatively near future. "Humanity will change more in the next 20 years than it did in the last two millennia", says Dr. Jose Cordeiro in a recent interview for his book, "The Death of Death". As humanity ventures into the conscious re-designing of its own evolution and becomes more advanced with enhanced physical and mental capabilities; more sophisticated in the faculties of reason, intuition, and communication; and better at regulating its own emotions, it will also encounter many challenges.

To implement holistic and transformative models in support of transhuman technologies that are safe, ethical, and sustainable, it is important to re-establish our role as humans in relationship with other living organisms.

A problematic notion of the human is its anthropocentric, top-of-the-food-chain attitude, which has led to horrific discrimination between people and a disconnect with, and abuse of, natural systems and the environment. Humans are, in a sense, "more superior" than other species in that we have the mental faculty to advocate for and protect others, including non-human animals. For example, if we were to face extinction due to a natural event like a meteorite impact, humans have the ingenuity to potentially create technologies to divert such risk and protect the Earth. On the other hand, non-human beings and the eco system are as important as humans in

A Model For Conscious Evolution
By Dinorah Delfin

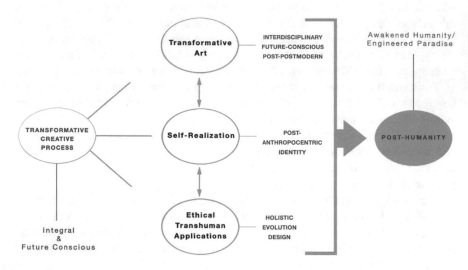

Fig. 40.12 A model for conscious evolution, 2019. Graph

the sense that we are all an interconnected organism; just like the human body is an interconnected system within itself and with the macrocosms.

We can think of anthropocentrism, as a social construct of the human, another example of limiting beliefs which need to be dissolved. Transhumanists wish to not only foster a climate of tolerance and acceptance but to abolish all forms of discrimination: ethnicity, class, gender, creed, and at its root, speciesism, by establishing policies and value systems that can facilitate peaceful and harmonious relationships, not just with fellow humans and nature, but with future intelligent species, including our machine descendants (Fig. 40.12).

40.1.3.1   The Awakened Sapient

"States of sublime well-being are destined to become the genetically pre-programmed norm of mental health." (David Pearce, The Hedonistic Imperative)

Humanity's journey is a call for adventure. The call is to upgrade—from a fear-based to a love-based subconscious programming. Our greatest obstacle, however, is our pre-historic, fight-or-flight, psychological survival wiring. Adopting a creative process that is future conscious, integral and harnesses the best of transhumanist technologies, may be humanity's best chance at breaking free from its pre-historic and outdated ego construct.

Striving for a natural default state of profound well-being, pleasure, excitement, and simple joy at being alive and living longer is the most beautiful concept humans could aspire to. We all want to decrease suffering, physical pain, hunger, injustice, and involuntary death. Transhumanist technologies are an invaluable addition towards the pursuit of everything we set out to achieve. After all, humanity's greatest legacy derives from both its humanness and its creative power and ingenuity.

Through the creative process and the infinite artistic possibilities of human imagination, I envision a super-abundance of archetypes representing the truest, most complete sense of self, manifesting and giving rise to an awakened humanity setting out to build a harmonious symbiosis between the natural order, technological systems, and *intelligent intentionality*.

The future I envision, the awakened sapient knows no fear, but the playful innocence of a child. The awakened sapient has no attachments to social conditioning, but has a unique identity while embodying it all. The awakened sapient is free and fluid; forever unfolding; forever transforming; it's essence, forever constant. The awakened sapient has mastered the vibrational language of love to regenerate at a cellular level; to create life; to communicate with all. The awakened sapient embraces its ingenuity and the tools it builds to master the art of living holistically and sustainably, so it can share it with all.

40.1.4   Conclusion

The coming of a techno-social paradigm shift might be the most important and challenging transition modern humans will ever experience. How we handle it, might lead to the extinction of intelligent life on Earth, or it's flourishing across the cosmos. With emerging super-technologies, we are taking a hold of great power, which also has the potential for great destruction. Transhumanists recognize a moral responsibility to educate and develop the wisdom to wield this creative power and reduce existential risks. We can choose to think of technological development not as a zero-sum game, but a positive sum game—If old age and death are part of the natural order, so too is human desire to overcome them.

A purely empirical scientific approach could rob us of our ability to experience the depth and sacredness of our essence. To survive the logical mind, and enhance our transformative creative potential, we must find a balance between the role of the thinking brain, our intuitive feeling body, and the role of our consciousness and interconnectedness, for human civilization to truly mature, harmonize, and prosper.

With reverence and humility to that which we don't understand yet, and faith in the power of our thoughts and intentions, we can all become the reason why we intuitively feel there is something special about being transhuman, or simply human. Through the power of the creative process, we can achieve the unimaginable and enhance the human story through renewed symbolic carvings and transformative narratives.

There has never been a more exciting time to be alive and to look forward to what lies ahead. May life and creativity flourish across the universes, infinitely.

Dinorah Delfin is a Fine Artist and Futurist located in New York. Ms. Delfin also serves as the Director of Admissions and Public Relations for the U.S. Transhumanist Party.

Part VII
Society and Ethics

Chapter 41
Pragmatic Paths in Transhumanism

Jeffrey Zilahy

The overlap of these two circles are self-improving processes

What are some qualities and concepts that can help elucidate transhumanism? Here is a mnemonic:

Technology
Risk
Anti-aging
Nanotechnology
Self-improvement
H+
Universal
Mind
Adaptability
Natural

"Every man desires to live long; but no man would be old." (Jonathan Swift)

"Death is *nothing* to be afraid of." (Anil Seth)

"Any sufficiently advanced technology is indistinguishable from magic." (Arthur C. Clarke)

"We live in the future." (Anonymous)

To say that we *live in the future* can be interpreted in a multitude of ways, suffice it to say technological progression is happening at such a swift pace that our concept of what defines our present is in a very real way *constantly under construction.* Naturally, this makes it a very unique time in our history because the technology we foresee in the future has a funny way of appearing today.

J. Zilahy (✉)
Transhumanist Consultant, New York, NY, USA

© Springer Nature Switzerland AG 2019
N. Lee (ed.), *The Transhumanism Handbook*,
https://doi.org/10.1007/978-3-030-16920-6_41

If we imagine the long arc of human history beginning with those early and ancient days of *Homo sapiens* with so few of us and so little in the way of tools, a world shrouded in a seemingly impenetrably deep cloak of mystery then fast forward all the way through to this moment; a present day plethora of tools and technology, many billions of people roaming about and many mysteries unfurled every day, our world is nothing short of genius.

This then is a *transformation* of humans *and their tools* that began with our early ancestors iterating generation after generation to ever better versions, in a constant state of sorting conflicts and solving puzzles.

Our own flesh and blood ancestors would scant believe all the modern marvels, as if their future progeny were more akin to *transformed humans*. We are indeed in a way transformed humans and that pattern will never cease, our minds are *preprogrammed* to be **curious**, to improve our environment and to **survive**. Our neural algorithms cause us to constantly question and consider and we are born like all other creatures doing all we can to survive. These two aspects of our sentience are most fundamental to our concept of self. This vector of pre-programmed *curiosity and survival*, charted against time ensures we become defined by our tools and technology, and over time inextricably linked.

It is somber but still worth pondering that every single last person ever born will get spun around by their swirl of consciousness and then merely decades later predictably vanish into nothingness. It is not a crime or ethical travesty to state without reserve that this reality is wholly unacceptable. Aging is nothing more than faulty biological code and there have never been more *programmers* then there are today rewriting and improving on this code. It is a reality that *only very recently have we even had a chance* to peer towards brighter horizons.

We all know deep down in the recesses of our souls that there is *so much* more we can make of ourselves but are stymied by our own life spans, most of us barely scratch the surface of really developing ourselves intellectually, emotionally, physically and spiritually. The final tally everyone eventually reaches is that it simply serves you and consequently your fellow human to be an active participant in the most exciting developments in humankind.

The transhumanist movement then is in a sense simply saying society is evolving so fast that it hasn't been able to catch up to itself but as individuals we can play that catch up by proactively doing all we can to embrace the resources of the world to better ourselves and in turn the world in which we occupy. It is also a reality that transhumanism is largely about self identification and a proactive nature as there is nary a person that is truly not transhumanist simply because when you participate in society, when you rely on tools and information you are acting as a transhumanist, just at a lower amplification than the self-identified version.

Look, let's face it, we all embrace technology it's impossible not to. Even if you are the type to pride yourself on not having a smart phone you still use many layers of modernity; infrastructure, modern medicine, even a wheel and a match are examples of tools and tech. Do you use the internet? Do you have surgery if your life depends on it? Do you try to eat the right foods and feel guilty when you don't get enough exercise? Congratulations you are a budding transhumanist. So much of what the vast majority of the population considers an acceptable relationship and attitude towards technology and even life extension is really just what could be deemed *transhumanism lite*.

We take for granted going to the doctor for any number of ailments and issues and use technology to enhance our lives everyday. Many people then are ostensibly transhumanists and many others might initially be afraid or even repelled to consider themselves transhumanists but upon deeper reflection would admit to being as much.

The good news is that there is a great deal of overlap between the "normal" good habits of keeping yourself healthy in all facets; physically, mentally, emotionally and spiritually and the deliberate transhumanist nature of proactive healthy aging also known as cenegenics.

Also it is worth mentioning that the fundamental and ingrained reason for people's fear, doubt and uncertainty surrounding transhumanism is because our sanity depends on refuting false hope. After all, the current *record* is of 100% of every single generation of people *having to accept* a mortal existence. It is only now there is a glimmer of hope for some form of post Mortem or at least a more direct challenge to the status quos of aging and death as we define them today. We also have to face the reality of our existence, that we need to transcend this mortal coil to survive, the universe is giving our consciousness an ultimatum, hack your code and evolve or be banished to infinite nothingness. Is that really even a choice?

But how best to navigate and integrate these breakthroughs into your own life? Keep in the forefront of your mind that there is no perfect or even correct way to be a transhumanist. We all have different wants, different perspectives and different demands on our own personal situations that will necessarily steer us differently for how best to embrace the brave new world we live in. Ultimately though we are all trying to avoid whatever the leading *risk* factors are of the time we live in. Nowadays, in the developed countries we are mostly trying to avoid heart disease and cancer, but the big killer used to be pneumonia and before that it was tuberculosis. Today you would be appalled at the idea of succumbing to tuberculosis; do you really think people in the future won't be appalled by the thought of dying from cancer or heart disease?

While transhumanism is a complex word with many different interpretations and meanings, one of the most synonymous concepts with transhumanism is the *fight against aging*. Aging is humanity's ultimate battle; it is the bain of our existence, literally. Life is in many ways a multifaceted fight, a fight against aging, a fight against all the threat vectors that would otherwise do you in. This notion of life being a fight is what makes the UFC Performance Institute a great metaphor for many of the aspirations of transhumanism. It is a place dedicated to using the most state of the art technology and science to provide bespoke therapies to help its fighters recover and regenerate. While the Performance Institute may be reserved for mixed martial art fighters, this proactive approach to our health can and should be modeled by everyone, we all must take on this fight individually but we can all work collectively to bring some measure of *death to aging. We start this fight by doing everything we can to live longer and healthier lives, later in life to stave off fraility and ultimately we seek to be organisms that are negligibly senescent.*

So what can we consider as some patterns of behavior that we can employ immediately that are at the *easy end* of the transhumanist spectrum? What are the pragmatic and even essential techniques you should consider incorporating into your life? These *solutions* are of course nothing more than a mere smattering of the breadth and depth of what exists in this world to assist and aid you and of course results and mileage will vary but at minimum provides one interpretation of the framework for *transhumanist habits*.

#1 Fan You must have a deep well of forgiveness reserved for yourself, you will make mistakes with frustrating consistency in some regard and your capacity for self forgiveness will regulate how effective you can cast off faulty processes. Forgiveness is nothing without care so you also have to constantly recommit to taking care of yourself and step one is to maintain a positive and optimistic outlook each and every day, part and parcel of a + attitude is gratitude and appreciation, form them into custom mantras and be sure to add regular restful sleep every night and at least some social connectivity that is not reliant on social media.

Quiet Time It is probably as close as you get to a universal truth that we all benefit from *quieting our minds*, whether it is Buddhism, Transcendental Meditation, any form of Zen or still prayer. It is a classic example of something that is elusive in its simplicity but transformative in its utilization. The practice of being still and calm in a sea of personal struggle will open portals of your potential with access to capability previously unseen. Can you silence your own voice? Can you do it regularly?

Moving Target Probably one of the most proven and easy to follow options for improving your health is the notion of motion or more colloquially referred to as getting exercise. While you don't need to become a professional athlete or gym junkie to benefit, at the very least make it a habit to just move. A new study suggests that even the act of dancing can reverse or slow down cognitive decline, that is news worth dancing a jig to!

Trust Your Gut A little known fact is that our stomach contains somewhere on the order of 100 million neurons and is a critical link to your overall health. Make sure you consider probiotics and do all you can to keep your *second* brain healthy.

Abstain & Refrain The idea of willingly refraining from eating and or drinking seems like a terrible idea to a species that literally survives by eating and drinking. However there is evidence that fasting offers health benefits not to mention the most obvious impact of a lower caloric intake, which over the long term has been linked to extended longevity. There is a more manageable form of fasting for the modern transhumanist called intermittent fasting, which can increase energy, mood and memory.

Pour Cold Water On It Another idea that sounds downright awful but shows a lot of promise is cold therapy. Now we know ice and compression are used to reduce pain and swelling in injuries but exposure to cold like taking an ice cold shower as indicated by extreme athlete gurus like Wim Hof show potential for improving circulation and mood and alleviating many symptoms of disease and decay. There is also the growing field of cryotherapy which has been shown to aid in improving overall blood flow.

Poisonous Avoidance Unfortunately there are insidious forces out there and whether it is in your food, the air or even certain people, push back on all forms of toxicity *as you perceive it* at all costs. The hardest part with toxic influence is often just the acknowledgment that something is not good for you, but the more you are capable of identifying and rooting out virulent vectors, the healthier you are sure to become.

Ingredients There are many known chemicals and compounds that have proven capabilities in slowing or even reversing the harmful effects of aging. Here are a few that are worth considering for your regular diet; curcumin, quercetin, nicotinamide riboside, pterostilbene, metformin, resveratrol, even coffee and aspirin. For a more thorough catalog, just ask Ray & Terry.

Brain Food Your brain is the sole organ in your body that really defines who you are. It is key that you ensure this organ has lots of good "food".

Luckily for us there is an inexhaustible supply available now. For example, study the fantastic research of the *Blue Zones* and the habits of people in these regions that afford them longer and healthier lives than the rest of us. Check out MIT OpenCourseWare for quality free education, and be sure to pay close attention to the *Great Brains* of our world and what they are saying and doing. We all know Elon Musk is working to usher in twenty-first century technologies and Bill Gates is making great progress to improve the lives of people in developing countries but be sure you are well acquainted with the goings-on of geniuses like Stephen Wolfram, Eliezer Yudkowsky, Aubrey De Grey, Demis Hassabis, George Church, and Craig Venter. This is of course a tiny sample of influential movers and shakers, just take a gander and expand your grey matter.

While the suggestions covered so far are relatively easy to implement, inexpensive or free and for the most part proven strategies, the following are additional solutions that either have financial costs or opportunity costs or contain potential risks, so be sure to proceed with caution. They can be thought of as *a more advanced setting* of transhumanist habits.

Future Therapies There is such a myriad of different treatment options available today that governments and doctors haven't reached consensus on their relative safety and efficacy. Despite the disagreements, there is increasing evidence that these approaches offer benefits not available through traditional approaches. Some of the evolving medical options on the menu today include Human Growth Hormone therapies, Rapamycin treatments, young plasma transfusions and the fast growing world of stem cell treatments.

Mr. & Mrs. Roboto Nowadays artificial assistance is everywhere and is quickly becoming a helpful and indispensible layer of our information driven society. The names Alexa, Siri and Watson are well known but did you know Sophia has citizen status and

Wolfram Alpha computes answers to just about everything? The wild thing about these intelligent outgrowths of humans is they are still very much in their early days, and will only improve in their abilities to help us sort through the problems of humanity.

Edit Yourself Genome engineering or more commonly referred to as gene editing is the ability to modify our own DNA and will probably end up as one of the most powerful tools humans have ever invented. These are very early days but methods like CRISPR editing are guaranteed to completely revolutionize medicine and molecular biology in the years to come. How does the curing of diseases, the creation of healthier food and even bringing back extinct species sound?

Who's the Boss? Regardless of what side you take in the political battle, these times have aroused passion in our populace and that can be a good thing properly channeled. When you consider who you will support in the voting booth, keep in mind that there are new candidates and movements that have distinctly transhumanist philosophies and know you have a decision to make about who you will give power to and know that you always have the power to embrace transhumanist political leaders that can bring more rapid change to our government, for example consider studying the political perspectives of the U.S. Transhumanist Party.

Frozen > Burial & Cremation Firms like Alcor will use your life insurance policy to put your recently deceased body into a deep freeze, replacing your blood with a specific chemical, the goal being to keep you in a type of stasis until such time that technology can revive you. A remarkable plan that at the *very least provides a new kind of hope* to its clients.

DOT ORG Research and find organizations that stoke your passion and find ways to take your talents to new venues. Here are a few that a the forefront of progress and science: Buck Institute for Research on Aging, Machine Intelligence Research Institute, Future of Humanity Institute, Singularity University, Future of Life Institute, U.S. Transhumanist Political Party.

My Avatar has an Avatar Virtual proxies for your biological self are a great feature of a twenty-first century world. There are so many types and even methods for creating and evolving these extensions of yourself. They allow you to tap into different sides of your personality and are a ubiquitous part of video games, social media and virtual worlds.

Regardless of how successful and influential the transhumanist movement ultimately becomes, it provides a community and a philosophy that empowers individuals. It is a call to arms to view our world as a place with tremendous possibility and opportunity for developing and enhancing yourself. The notion that it is crazy to think and act like a transhumanist is backwards, it is crazy to not be one. Perhaps it is fair to say that the very deliberate nature of the movement and even the word transhumanism is a bit on the nose. Don't let semantics repel you and hold you back, there are many more ways to describe people in this movement such as technological survivalists, techno-healthy, convergists, singularists, neuro-optimists, life extensionists, mind-uploaders, transformers or just human+.

In a way we as a species have been collectively marching toward a transhumanist future. After all humanity is in a constant state of battling all forms of maladies, sickness and ailments. If given an opportunity, we all use the best medicine and care to protect our own bodies so we may keep trucking along. For most of us, we die trying not to die, transhumanism is this same battle against death, the only distinction is we attempt to articulate how to fight this battle in a time in our technological development where there are new weapons available to us to challenge the grim reaper.

When we talk about this battle with aging and death, it is important to delineate the two aspects of our lives that we are attempting to enhance. The first aspect of our lives that we are trying to enhance is our health span. This is a measure of how *much* of your life is in a healthy state. This is a critical but often ignored metric. Would you rather live 150 years in a sick state or 100 years in a super healthy state? Most people would choose the latter because healthspan really determines the *quality* of our lives.

The second is our lifespan, which refers to the expectation of how long we can expect to live. Lifespan has a direct relationship with the level of technology we have, so not surprisingly as we progress through time the average human lifespan has steadily gone up from as low as mid 20s to present day in the mid 80s, depending on the country and sex you specify. In the last century for most developed nations we have essentially doubled our lifespans. What is interesting is that the age at which most people are OK to die at is also the age they expect to live until. So how people view their lifespan is really a game of expectation. If tomorrow everyone was suddenly expected to live until 200 then it is safe to assume that many people would *suddenly* find 120 as simply not enough of a life. Given the dynamic nature of lifespan both in reality and in our expectations, we can conclude that the average actual lifespan will closely parallel the expectation of what we consider a good lifespan. Bottom line, if you could confidently live longer, then you would expect to live longer. Let's not forget that what people want to do, what they consider their "purpose" changes with time and developments. What is clear is that the universe is a giant computer of sorts and we have already begun to *reprogram our code*. The insights of a computational world are most certainly a *natural evolution* and one that we should all embrace.

Transhumanism seeks to elevate ourselves, our discourse our entire society by a deeper connection and appreciation for the forces of reason and science. We want to elevate rational thought and deemphasize dogma. We want to transcend the bitterness and pettiness of racial divides and unite as a planet under a banner of a single humanity. We view the rapid rate of technological progression as not stifling but in fact opening avenues for social and cultural evolution and revolutions, just like every previous technological enhancement has done for humanity. We want to participate now and in the future in the greatest developments of our species and we want to do it as healthy and happy 125 year olds.

Chapter 42
Leave Nobody Behind: The Future
of Societies

Amanda Stoel

I'm a DIY futurist and a transhumanist [1]. How this came to be is a direct consequence of my youth. My world was largely about saving bees and butterflies who
were bound to the ground, German nature documentaries and sci-fi series on tv. My
parents loved reading and watching sci-fi. While news largely passed me by I was
fascinated by the worlds portrayed by George Lucas´ Star Wars, the Star Trek universe, Buck Rogers, Doctor Who, Battlestar Galactica en Blake Seven. Before my
15th birthday I had read a lot of science fiction a.o. Orson Scott Card [Ender's Game
series], Asimov [o.a. A robot dreams], Jack Vance [Dragon Masters] en Michael
Crichton [Andromeda Strain].

Reading Ender's game and Asimov, seeing the Star Trek series in my youth
greatly influenced my perception of my future/the future. Perhaps needless to say
but I imagined the future to be rather different. And I am not talking about the difference between "while growing up I wanted to be a famous singer" and whoops,
I'm a single mom instead (even though I didn't really see that one coming either).

No, I am talking about the difference between acceptance of all races and not just
the earth born kind, but also non-human others. The difference between a world
where everybody has their basic needs met vs. a society where inequality is only
growing, the gap between the rich and poor growing ever larger.

During the summer holidays I often stayed a couple of weeks with my grandparents, one of my grandmothers determined insect species as a hobby and she used to
take me along on field trips. I learned a lot about insects but also about eco systems,
insect-plant cooperation and more. My grandfather always had Scientific American
lying around and both were scientifically inclined. All this nurtured my natural curiosity and instilled in me a lasting love for both nature and science.

Empathy and pattern recognition, wanting to know how the mind works, led me
to believe psychology was the perfect study for me but after a couple of years I real-

A. Stoel (⊠)
DIY Futurist, Bussum, The Netherlands

© Springer Nature Switzerland AG 2019   613
N. Lee (ed.), *The Transhumanism Handbook*,
https://doi.org/10.1007/978-3-030-16920-6_42

ized that it made no sense to complete my studies while not having the life experience to go with it. With only theory I felt it was almost unethical to start practicing. So I decided that before I'd finish my studies I would get that life experience first.

I eventually ended up in Finland where I met the father of my kids and more life experience than I bargained for. Our first child was born and because of medical mistakes she was damaged after birth. I was warned to have very low expectations, the experts believed she would never walk, and never talk. I didn't accept that. I started to wonder what medical science could tell me about her condition, what already has been tried in cases like this, the latest studies in this area and all surrounding subjects. Was it possible to help her brain improve with specific food? In short, how could I use science, turn what we know in to something I could use now. Something I could do myself.

Because of all of the above I was able to observe and see how things were connected together and despite the experts educated guesses I was able to facilitate that she not only walks and talks. She also jumps, bikes, sings, reads and writes. Not perfectly yet but we are getting there. Unfortunately without much help from experts, if anything I often bumped into resistance.

Resistance to a parent that thinks along, has ideas and does her own research. A doctor once told me "You shouldn't go on the internet so much, it will only scare you."

Information can empower. I am aware there is a lot of misinformation, but that is nothing you can't figure out by critical reading/research. What does 'scare me' is the discrepancy between what science has learnt/figured out, technologies that have been invented and the amount of that knowledge/those inventions actually being used/implemented in day to day living.

Favoring old methods and ideas instead.

I'm of the opinion that science and technology have the possibility to allow humans to transcend the human condition; improved health, improved cognition, improved lifespan, improved life quality and circumstance. Around 2008 I found out there is a name for people that agree with me on this, we are transhumanists.

I have to add though, this is where similarities between transhumanists ends. There are for example differences in how people want to reach that technologically empowered future. Some are of the opinion that it is ok to fast forward, no matter who gets left behind or even duped.

Others, like me, feel that it is unethical to leave anybody behind. To deny some people medical- and technological progress.

Whenever I talk about transhumanism some people react negatively. The prospect of technologically enhanced humans flourishing makes some people uncomfortable. Mostly because they fear these 'upgrades' will only benefit the (rich) few and therefore create more inequality. There are also moral/ethical aspects such as 'it's not natural', 'humans shouldn't try to play God'. Eugenics, which got a bad reputation because of its (ab)use in WWII. Won't these technological enhancements make us less human, will it turn us into cold unfeeling robots, will we create human hybrids? Some of these concerns are at least partly founded, technological progress does not come without dangers. But the real concerns shouldn't be with the technologies themselves, technology in and by itself is not good or bad, it's just a tool. The real concerns/issues should rather be with the people/corporations that (ab)use them. In the wrong hands technology can cause a lot of damage. Personally I do not believe that should stop us, we should just also focus on how to prevent abuse.

Take for example social media. Social media is an amazing tool to connect, to share knowledge, to make the world smaller, smarter, more informed. I'm just going to use Facebook as an example here. Via Facebook you may know a family in Mumbai and when a bomb goes off there your thoughts immediately go out to them, you hope they are safe and you quickly send a PM to find out if they are alright. Suddenly the Mumbai incident doesn't feel like the-far-from-your-bed-show. You know people there.

Facebook is a free service, meaning you don't have to pay for using it. Facebook is getting its revenue from somewhere, and if we aren't paying for it, then what is? Your data, Facebook and many other 'free' services mine your data. If you do not pay for the service you are almost certainly the product. Your information is sold to other parties. If you disagree with this practice, feel that it is invasion of your privacy you would be correct, but then you always have the option to not use Facebook, right? However, not having a Facebook account can mean that sometimes you do not get the job, that sometimes you cannot register with a website or use it's full service because you can't login with Facebook, and even if you don't make your own account, Facebook is still mining your data.

In the battle against fake news a lot of content is getting censured. Unfortunately whether it is faulty protocols or simply people disagreeing with what you say and reporting you, a lot more is being censured the an fake news.

A rather politically outspoken friend of mine felt the effects of Facebook's policies when week after week she got banned from posting to the point where she didn't see how she could continue to use Facebook. However her entire network was on Facebook and one of the consequences was that she sees a lot less job offers. I don't think we should stop using social media, the possibilities it offers are too good, I do feel radical changes are required with regards to how our data can/should be used and censure.

> "Data in itself is not good or evil, it's how it's used. We're really relying on the goodwill of these people and on the policies of these companies. There is no legal requirement for how they can and should use that type of data." – From Chris Paine's; Do you trust this computer [Documentary] [2]

This, I believe, is the heart of the problem.

If we look at China we can see one of the ways social media can evolve into something you'd expect to see in a Black Mirror episode, in fact there was an episode about something like China's social credit rating system. It's season 3's Nosedive episode.

By 2020 China plans to have fully implemented its social credit system [3, 4] but aspects are already being integrated, like a personal social credit score. This score is determined by a person's credibility/trustworthiness. Being watched from every angle people get rated based on their behavior, buy too many games, jaywalked lately, not canceling a restaurant reservation when you didn't go? That and more lowers a person's score. This score affects the things a person can or can't do, what they have access to or not. If their score is low they might not get a good deal when they want to rent a bike and they'll have to make a deposit, they might be barred from using public transportation and their kids might not be allowed to attend the best school.

With suppressive, Orwellian societies such as this there is a high chance of countermovements.

Dissidents, remove freedoms and you get counter movements that oppose these freedoms being removed. The people that get the low scores might not feel this is fair, people might be empathetic towards the people that get left behind like this. The social credit system is still in its infancy so we have yet to see what counter-movements can/will arise. But some countermovements to oppressive regimes or organizations can already been seen globally.

One such movement is the DIY biohackers movement. And with the introduction of CRISPR [5] technology that means biohackers can now cut, paste and alter (mostly) their genetics. One of the more famous faces of biohacking is Josiah Zayner [6], he is the founder, CEO of the Odin, a company that amongst other things sells DIY CRISPR kits. He also experiments on himself, in 2016 he did a full body microbiome transplant.

We now have the technology to create designer babies. And in societies as mentioned above, Brave New World (Huxley) could come to pass. Unless we become more altruistic. Realize it is unethical to deny people progress, medical advancements and technologies. Everyone should have access to information and scientific processes. Already we can see the nature and nurture effects of an unequal society. Starting with access to healthy foods, accurate information about what healthy food consists of. Equal opportunity education. Anti intellectualism in certain circles. This and more causes a divide that grows larger only over time. Now add genetic enhancement only for those who can afford it into the mix and you create a gap that can no longer be bridged. Unless we all get access.

To continue with the Huxley reference, in Brave New World [7, 8] new humans are born from artificial wombs [9], this technology is currently being used to grow sheep and cows. There will come a time when humans will use the technology to grow humans, perhaps because they feel that pregnancy is too harsh on the body, because woman who would otherwise have to endure high risk pregnancies can now safely have children or for other reasons such as embryonic starships [10], for the colonization of space. Allowing the class system (inequality) to persist would really assure a scenario like Brave New World could come to pass. Humans genetically created to be a worker or a leader, never getting to chance to be anything but what you are born to be. Even though this may sound like I oppose the technology, I don't, it creates possibilities and opportunities, I just rebel against inequality.

I am not saying everyone should use these technologies, but everyone should have to the choice to.

I believe it would be unethical and unwise to give the power to only a small group of genetic superhumans. Who knows what/who we'd miss out on when we only selectively upgrade people? Technology is going to continue to develop and you cannot lock it in labs for the use of the few. The way forward is by learning from our past and not continue to repeat history in cycles. Only by (reciprocal) altruism can we move past this threshold of (self)destructive civilizations and rise above the human condition. To develop to our full potential, to colonize space, make our beautiful planet whole and green again, in a world where no one needs to starve, be homeless or chronically scared. We have the potential, we have the ideas, we have the skills, and now we also have the technology.

References

1. Transhumanism, the world's most dangerous idea? [Nick Bostrom] https://nickbostrom.com/papers/dangerous.html
2. The Odin: http://www.the-odin.com/about-us/
3. China's social credit system explained: https://www.businessinsider.nl/china-social-credit-system-punishments-and-rewards-explained-2018-4/?international=true&r=US
4. Chris Paine's Do you trust this computer: https://doyoutrustthiscomputer.org/watch
5. How CRISPR lets us edit our DNA (Jennifer Doudna, co-inventor of CRISPR): https://www.ted.com/talks/jennifer_doudna_we_can_now_edit_our_dna_but_let_s_do_it_wisely?language=en
6. Josiah Zayner: http://www.josiahzayner.com/
7. Aldous Huxley Brave New World. https://www.huxley.net/bnw/index.html
8. Brave New World? A defence of paradise-engineering [David Pearce] https://www.huxley.net/
9. An extra-uterine system to physiologically support the extreme premature lamb: https://www.nature.com/articles/ncomms15112; https://www.scientificamerican.com/article/could-artificial-wombs-be-a-reality/
10. Embryo space colonization theory. https://en.wikipedia.org/wiki/Embryo_space_colonization; https://www.youtube.com/watch?v=f9nsEROAW8M

Chapter 43
Should We Genetically Engineer Humans to Live in Space?

Amanda Stoel

As humanity looks to the stars, we find more and more potential candidates that could harbor and/or sustain life. We have built telescopes that have changed the way we are seeing the universe. We have sent unmanned probes and manned spacecrafts to learn if these places could support outposts or even colonies off earth. Currently Mars Insight is on it's way to Mars so we can learn more about Mars' seismic activity. But 2018 will see many more unmanned missions to, amongst others, the sun and Mercury.

There are plans for manned missions to Mars by several countries and organizations, **SpaceX/Elon Musk** by 2024. **Mars One** by 2031, **NASA** "In the near future", **UAE** (united Arab Emirates) within the next 100 years.

There are plans for outposts on the moon and colonizing Venus. One thing's for certain, whether we aim for the moon, Mars or Venus, humanity envisions itself amongst the stars and quite soon at that.

It is 2018 now, we are talking about sending humans into deep space within the next 6 years.

Apart from these fantastic discoveries we have also learnt something else, manned missions to the ISS have taught us a lot about the effects space has on our bodies. Recently there has been a study with twin astronauts by NASA, Mark and Scott Kelly. One twin stayed on earth and the other went into space, when the space twin came back they compared the twins down to the genetic level. They measured large numbers of metabolites, cytokines and proteins, and learned that spaceflight is associated with oxygen deprivation stress, increased inflammation, and dramatic nutrient shifts that affect gene expression.

Other in-space-studies have shown that in space we age faster; The changes experienced in the bodies of astronauts, the wasting away of muscles and bones, changes in heart and blood vessels resemble what happens in our bodies as we grow older.

A. Stoel (✉)
DIY Futurist, Bussum, The Netherlands

High radiation, extreme temperatures, small meteor impact. They damage both equipment and humans. Which begs the question, are we really ready for this?

Studies in space also show ways we can improve the human condition, for example a worm (c.elegans) whose muscles do not degrade in space but actually improve in weightlessness. Finding out why this worms muscles are unaffected by space-flight could reveal how to keep human muscles strong and healthy. Both in space and on earth. The fast ageing astronauts experience during space flight can be reversed on earth, we are learning a lot about ageing and the possibility to limit or reverse ageing. But I will get back to that.

Imagine we are going to Mars. Mars is cold, dry and has a very thin atmosphere. Growing food we need to mind levels of light, gravity and nutrients required to grow crops, even though Martian soil has some of the essential nutrients plants need to grow there may not be the right amounts of nutrients, apart from that it is covered in toxic salts. Plants respond to gravity on a cellular level. Building around crops would be inefficient, expensive and plants still couldn't grow anywhere outside. I believe the answer lies in biotechnology. We already use it on earth.

GMO's. And we could expand on that. Not only engineer plants that provide food but also medicine and fuel.

Humans are not very good in surviving in harsh environments either, this holds true already on earth. Where you might survive a couple of days or more if you are schooled in survival in a lush forest area, those chances diminish dramatically when you get dumped in the middle of the Australian Outback, Siberia or the Sahara desert. In space we are subject to Ionization damage/radiation damage, extreme cold, time (distances) we'd need to travel. We try to solve that by introducing special heavy suits, special "housing" etc. But even with those precautions we are still fragile.

So rather than (only) build around the human condition, why not add to 'our condition'? Why not augment humans, genetically alter humans to live in space, be radiation resistant (even a little bit would help enormously). We are already entering a phase where we can and do alter our genes, via CRISPR technology for example. Right now it is mostly therapeutical to eradicate terrible genetic diseases. But as we approach an age where therapeutical is no longer necessary (because we have cured disease) we can look forward to augmentations, improvements. Some people have already started experimenting with improvements, the biohacker community.

And it is not as if genetic changes do not occur naturally, they do:

Take the Bajau people or 'sea nomads', who genetically adapted to diving. Over a period of 1000 years they genetically adapted to have larger spleens enabling them to free dive to depths of up to 70 m. Or people of the northern Argentinean Andes who genetically adapted to a toxic chemical. Basically we'd only be speeding up the process of genetically adjusting to our environment. 'Natural evolution' has been known to do some pretty weird stuff, since bones weaken in space perhaps natural selection will opt for shedding bones. We don't really have 1000 years though, we have 6.

In spite of what it may sound like, physical enhancement is no longer only in the realm of science fiction, but a field that has slowly been gaining more and more traction. For years athletes and bodybuilders have been using anabolic steroids. Similarly, using amphetamines, such as Adderall, which is prescribed for hyperactivity disorder (ADHD) and narcolepsy are used by large numbers of students. Adderall and similar drugs enhance cognitive function. Both steroids and nootropics are not yet widely accepted because they are considered "cheating".

Though research suggests that attitude may be changing.

Making physical adjustments to live in an extreme environment are also something humans already do, there is a settlement in Antarctica where entire families have to have their appendix removed in order to live there. The only reason being that the nearest hospital is 1000 km away. Now imagine instead of reading "in the icy village where you must remove your appendix." It reads "in the icy Mars village where you must alter your DNA."

If this all sounds very extreme to you; What if we decide we don't want to tinker (much further) with our bodies? Let's consider the alternative, we still want to move forward with our plans to have a human presence in space. An alternative to reach distant destinations is generation ships.

Regarding generationships, technically speaking we'd have a colony in space with the concept, with the eventual goal to find a habitable planet for the generation that arrives at destination. But there are ethical and moral challenges with sending people off into deep space for such extended periods of time. Consider the ethics of locking the next generation into a form of living that is extremely limited and over which they had no say. They can't leave the ship and most likely have limited or no say in what they want to do with their lives because roles must be fulfilled, the new crew must be trained and there are only so many parts they can play. A life of leisure and/or withdrawal is not an option. They will also be under great pressure to reproduce, have no real choice in becoming parents because the success of the entire project depends upon it. Perhaps due to the limited amount of people who can stay on such a generationship they won't even have a choice in who they have children with.

Perhaps this is ethically permissible when it comes to survival of the species, but what if there is another way? Which is then more ethical, generationships or to genetically alter the people who actually make the choice to go into deep space, to make them more hardy and longer lived so they are the ones that reach their destination of choice, fulfilling their own dreams rather than passing them on to generations who would have perhaps chosen differently?

I believe the time has now come to start actively looking into volitional evolution.

Will it make us less human? Will we still be human? But then, what is it to be human? Humans have always been explorers, expansionists. We have grown domesticated, our concerns focused on economy and wars.

Wondering and wandering appears to be a forgotten skill. But the time has come to look inside ourselves again, and re-awaken that part. New frontiers beckon.

Chapter 44
The Meme of Altruism and Degrees of Personhood

Amanda Stoel

The struggle for civil rights has been one of constant movement through the centuries and it is by no means finished.

From a king having only rights, then onto the noblemen, to all white men who owned land, to all white men, to men of color, to womyn.

There is, despite ongoing struggles, a clear trend visible that shows a general increase in acknowledgment of rights [non citizens, non human animals, children, people with disabilities, womyn and minorities].

44.1 The Fall of the Roman Empire 27 BC–476 AD

When thinking of Rome, the beautiful architecture, the sophistication of the empire's literary [The Great Library of Alexandria] and political culture come to mind.

Unfortunately these cultural glories were limited to a tiny privileged elite – those who owned enough land to count as gentry landowners. They represented maybe 3% of the whole population. Its structures were probably horrible reminders of inequality to pretty much everyone else.

The Romans had an economy based on slave labor. Cheap slave labor resulted in the unemployment of the people of Rome who became dependent on handouts from the state. At some point the numbers of slaves vs. non-slaves was 5–3. The Romans attempted a policy of unrestricted free trade but this led to working class Romans being unable to compete with foreign trade. The government was forced to subsidize the working class Romans to make up the differences in prices. This resulted in thousands of Romans choosing just to live on the subsidies sacrificing their standard of living with an idle life of ease. The massive divide between the rich and the poor increased still further.

A. Stoel (✉)
DIY Futurist, Bussum, The Netherlands

© Springer Nature Switzerland AG 2019
N. Lee (ed.), *The Transhumanism Handbook*,
https://doi.org/10.1007/978-3-030-16920-6_44

The Romans dependency on slave labor led not only to a decline in employment, morals, values and ethics but also to the stagnation of material innovation, whether through entrepreneurial or technological advancement. The terrible treatment of slaves led to rebellion and several Servile Wars, the most famous being the revolt led by Spartacus.

Captive barbarians were being fed to wild animals in the Colosseum, and its criminal law dealt ruthlessly with anyone seeking to remedy the highly unequal distribution of property.

In 650 AD, as in 350 AD, peasants were still laboring away in much the same way to feed themselves and to produce the surplus which funded everything else, like for example the full cost of the military, via incredibly high taxes of which the elite was exempt. However, as the Empire grew, the cost of maintaining it and the armies the emperors powers depended on, grew with it. Eventually, this cost grew so great that any new challenges such as invasions and crop failures could not be solved by the acquisition of more territory. Intense, authoritarian efforts to maintain social cohesion by Domitian [was Roman Emperor from 81 to 96 AD] and Constantine the Great [was Roman Emperor from 306 to 337 AD] only led to an ever greater strain on the population which led small farmers to destitution or into dependency upon a landed elite.

The barbarians invaded and marked the beginning of the fall of the roman empire.

As the west fell, the east muddled on through a while longer because it was a bit richer, but slowly and by piecemeal it too succumbed and gave way to feudalism.

44.2   The Peasant Revolt of 1381

The word *peasant* has, since the fourteenth century, been given a negative connotation and is not a neutral term. It was not always that way; peasants were once viewed as pious and seen with respect and pride [Living next to God]. Life was hard for peasants, but life was hard for everyone. As nobles increasingly lived better quality lives, there arose a new balance, of those on top and those on bottom, and the sense that being a peasant was not a position of equality. King Richard II tried to heavy handedly enforce 'the third poll tax', the third in a line of poll tax experiments that became increasingly unpopular. The third poll tax was not levied at a flat rate nor according to schedule; instead, it allowed some of the poor to pay a reduced rate, while others who were equally poor had to pay the full tax, prompting calls of injustice.

There had been many revolts already, but the peasant revolt of 1381 was the one that came to be seen as a mark of the beginning of the end of serfdom in medieval England. Although the revolt itself was a failure, it increased awareness in the upper classes of the need for the reform of feudalism in England and the appalling misery felt by the lower classes as a result of their enforced near-slavery.

44.3   The French Revolution 1789–1799

44.3.1   Liberté, Egalité, *and* Fraternité

If the guillotine is the most memorable negative image of the French Revolution, then the most positive is surely the *Declaration of the Rights of Man and Citizen*, one of the founding documents in the human rights tradition.

Prior to the revolution many things were askew, the country was nearing bankruptcy and outlays outpaced income. This was because of France's financial obligations stemming from involvement in the Seven Years War and its participation in the American Revolutionary War. The country's extremely regressive tax system subjected the lower classes to a heavy burden, while numerous exemptions existed for the nobility and clergy.

The Revolution itself is regarded by some historians as one of the most important events in human history. The Revolution is, in fact, often seen as marking the "dawn of the modern era". Within France itself, the Revolution permanently crippled the power of the aristocracy and drained the wealth of the Church, although the two institutions survived despite the damage they sustained

44.4   The American Civil War 1861–1865

Pre war it was the slavery issue that separated the North and the South, the slavery issue addressed not only the well-being of the slaves [although abolitionists raised the issue] but also the question of whether slavery was an anachronistic evil that was incompatible with American values or a profitable economic system protected by the Constitution. All sides agreed slavery exhausted the land and that they had to find new lands to survive.

The south feared that living together with free blacks would cause sin and destruction, Northerners came to view slavery as the very antithesis of the good society, as well as a threat to their own fundamental values and interests [1].

To end slavery became the only goal for war and slavery for the Confederacy's 3.5 million blacks effectively ended when Union armies arrived; they were nearly all freed by the Emancipation Proclamation [2].

What we can observe here is a repetitive cycle through the ages, equality/abundance/progress vs. inequality/scarcity/transformation or collapse of a civilization.

I argue that this is due to the altruism meme, history constantly shows that it is in both human, non human and economic interest to be altruistic, the more equality the more progress [in all its forms] we can see.

> "When you call yourself an Indian or a Muslim or a Christian or a European, or anything else, you are being violent. Do you see why it is violent? Because you are separating yourself from the rest of mankind. When you separate yourself by belief, by nationality, by tradition, it breeds violence. So a man who is seeking to understand violence does not belong

to any country, to any religion, to any political party or partial system; he is concerned with
the total understanding of mankind." – *J. Krishnamurti, "Freedom from the Known"*

44.5   Genetic and Memetic Altruism

Human culture and organization are the creations of an animal, a large brained and
evolved hominid. Social patterns came into existence because of "drives", "motivations", "instincts", and "needs".

Our evolution is driven by genes and memes. Whereas genes use us as battlebots
in which competing genes slug it out [as in "survival of the fittest"/"natural selection"], memes are the driving force behind social evolution.

> "Examples of memes are tunes, ideas, catchphrases, clothes fashions, ways of making pots
> or of building arches. Just as genes propagate themselves in the gene pool by leaping from
> body to body via sperm or eggs, so memes propagate themselves in the meme pool by leap-
> ing from brain to brain via a process which, in the broad sense, can be called imitation" [3].

Language, a very successful meme, that was learned through imitation/knowledge
sharing evolved due to individuals interacting with each other more and more frequently. Groups formed and humans became more altruistic and hence more cooperative, groups were more productive and sustained healthier, stronger, and more
numerous members and made more effective use of information.

Through the development of communities and social relationships we have
become more 'domesticated' and subsequently less aggressive.

Altruism is not only a superior moral faculty that suppresses basic selfish urges
but is rather basic to the brain, hard-wired and pleasurable [4].

Both memetically and genetically we appear to be predisposed for altruism [well,
most of us anyway].

44.6   Others Do it Too

Altruistic behavior is seen amongst many different animal species as well, from
apes to cellular slime moulds. There is even altruistic behavior seen in tiny little
robots.

A great example of a non human animal altruistic society is that of meerkats.

Meerkats have sentries that each has watch duty for about an hour, they will alert
the rest of the group when a predator is nearby, thereby endangering themselves as
they now bring the predator's attention to themselves. However by doing so it
ensures the higher probability of its group members surviving and thereby secures
gene survival even though it may not be through children of its own.

Young female meerkats take care of the young, even to the extent that when an
enemy approaches they will collect the small meerkats and lay over them to protect
the little ones with their own bodies, A selfish meerkat in this group would have a

higher probability of survival then the altruistic meerkat for sure, but imagine if the social arrangements were different? What if all the meerkats were selfish?

No sharing of food, no raising of young meerkats together, instead of sentries all meerkats would have to spend most of their time looking out for predators while trying to hoard food, secure and protect food resources and look out for their own young.

That is, if you live long enough to have young.

Meerkats would not be able to have sustainable communities if it wasn't for altruism.

44.7   Are Selfish Communities Sustainable?

Selfishness has purpose when an entity is a lone wolf, but not when it is part of a social community. In a society selfishness leads to inequity. It makes a lot of sense to be altruistic in a community both for social and economic purposes.

Research clearly shows that those countries that do the worst [measured by all kinds of social gradients like crime, school dropouts, mental illnesses etc.], whatever the outcome, seem to be the more unequal ones. There is a direct link from general social dysfunction to inequality.

It's not just one or two things that go wrong, it's most things. Think of the expense, the human cost of that. Inequality is one of the major oppressors hindering a global change of mentality from scarcity to abundance [5].

I argue that a society based on selfish behavior is not sustainable. With that kind of social arrangement we lose social cohesion and oppressive conditions for both humans and other entities are created. Entire civilizations fall when selfishness reigns. And regardless of my personal ideas on whether or not altruism is innate, it is in my opinion what we need for progress and continuation of the species.

44.8   The Technological Singularity and the Altruistic Singularity

The hard work of millions of scientists has turned science fiction into science fact, we now have technologies science fiction writers fantasized about. Kurzweil argues that, whenever a technology approaches some kind of a barrier a new technology will be invented to allow us to cross that barrier. He cites numerous past examples of this to substantiate his assertions. He predicts that such paradigm shifts have and will continue to become increasingly common, leading to "technological change so rapid and profound it represents a rupture in the fabric of human history" [6].

We must ask ourselves the question, is there an evolutionary pressure that weeds out zero sum predatory behavior and rewards crowd sourced/synergistic behavior. Not that selfishness in itself is always bad; for example Linux was started in stark competition with other large operating systems [8]. But throughout history we can clearly see that there is an increase in synergy/technology/science/cooperation which for the sake of brevity and clarity we can call an altruistic trend. We have increasingly more 'faith' in our fellow humans because we simply stand to lose more if we don't. When we start losing faith our society will collapse.

So for better or worse, we have become dependent on the societies we live in and in order to have progress, abundance and safety, we must endeavor to become more 'altruistic', we must 'engineer' the altruistic singularity.

44.9 Degrees of Personhood

So with the development of autonomic computing, ambient intelligence and our increased understanding of the animal world our philosophical conceptions of human self-constitution and agency are being challenged even more.

We have come to a point where it seems like we have enough reason to reassess and emancipate the most basic concepts of law.

To extend the circle beyond humans to include such non-human entities as animals, artificial intelligence, or extraterrestrial life.

I could argue that if we do not, we may be sowing the seeds of our own self destruction. Because while we enslave other entities but refuse to take responsibility for them, we are creating 'oppressive/predatory' environments that will force these non person entities to develop defense mechanisms. Take for example the 'self aware AI' risks debate.

A lot has been speculated about the development of AI's and what they could be like, it could be we end up fighting for the same resources once computer intelligence develops self awareness, it could lay in waiting, abide it's time until such a time where it is strong enough to show it's 'true colors'. Then there is the possibility of a 'friendly AI', but this requires we instill it with an altruistic nature. Of course a good way to do this apart from proper programming, is to show by example. What better way to 'ensure' a friendly AI then by treating it as an equal, giving it civil rights?

> "Quite an experience to live in fear, isn't it? That's what it is to be a slave" – Roy Batty Bladerunner For many humans to include non humans as persons may be a tough pill to swallow.

So rather then try and change the laws to include other species, I suggest we reject speciesism but do not try to say that all should be treated equal, that all lives have equal worth, or that all interests of humans and non humans must be given equal weight.

But instead we make the more limited and defensible claim that where non humans and humans have similar interests, we must not disregard or discount the interests of another being, just because that being is not human.

I suggest we do this by creating a 'scale' with which we will measure the degrees of personhood. All capacities and attributes that an entity possesses [that collectively constitute personhood], like for example agency, intentionality, self-determination etc., will place this entity into a certain category. Rights would then be extended to 'the degree that the entity is a person'; great apes get many similar rights, insects get a lot fewer rights etc.

For example, an entity that can display behavior of avoidance of physical pain [volition, intention] should be given the right to not receive pain from activities by us, humans.

44.10   From Here to There.

Broad acceptance of such a concept would be in the interest of many existing human- and animal rights organizations because with the implementation of this scale they will have a more solid legal standing. Not only for the civil rights of humans and non humans today, but also for the humans and non humans of the future.

We are standing at a crossroad, one way leads to a possible dead end for humans [and whatever potential we may have had], and the other to a future free of suffering and filled with potential.

References

1. Foner, Eric. Free Soil, Free Labor, Free Men: The Ideology of the Republican Party Before the Civil War [1970].
2. Lincoln, Abraham. "The Emancipation Proclamation," [1863] http://www.archives.gov/exhibits/featured_documents/emancipation_proclamation/
3. Dawkins, Richard. "The Selfish Gene," [1976]. http://www.rubinghscience.org/memetics/dawkinsmemes.html
4. Moll, Jorge, et al [October, 2006]. Proceedings of the National Academy of Sciences of the United States of America (PNAS), Vol. 103 Iss. 42: "Human fronto-mesolimbic networks guide decisions about charitable donation," pg. 15623–8. http://www.pnas.org/content/103/42/15623.full.pdf
5. TED Blog (TED) [2011]. How economic inequality harms societies: Richard Wilkinson on TED.com (Video). http://blog.ted.com/2011/10/24/how-economic-inequality-harms-societiesrichard-wilkinson-on-ted-com/
6. Kurzweil, Ray [March, 2001]. "The Law of Accelerating Returns," Kurzweil Accelerating Intelligence. http://www.kurzweilai.net/the-law-of-accelerating-returns
7. Waibel M, Floreano D, & Keller L. [2011]. "A Quantitative Test of Hamilton's Rule for the Evolution of Altruism." E-Journal Public Library of Science. http://www.plosbiology.org/article/info:doi/10.1371/journal.pbio.1000615
8. Linux succeeded thanks to selfishness and trust: https://www.bbc.com/news/technology-18419231 https://climatechange.ucdavis.edu/what-can-i-do/kindness-an-unsung-climate-change-tool/

Chapter 45
Charity in a Transhumanist World, and Our World Becoming So

Stephen Valadez

Transhumanism, as anyone can tell you within it and those without who have at least some working knowledge of the movement (and it's certainly that), is a gestalt. The pieces myriad, the goals many, but the endgame the same: the advancement of our human species.

There are salient outcroppings to be sure. The drive toward longevity—even, physical immortality—chiefly through genetic means as the longer road; by innovations in medicine, the shorter. Viable space exploration and inhabitation of other worlds. Living technological lives undreamed of by people just a generation ago. Ending poverty. Ending hunger. Ending homelessness. It, necessarily, both incorporates and transcends the worlds of technology, politics, finance, personal ambition, and collective ambition toward a better world and a more prosperous species. Perhaps in the history of humanity no one single movement has had such screaming ambition that's also infused with the deepest of humanitarian and altruistic sentiment. Even, a powerful love, for all of humanity absolutely regardless of the biological makeup and socioeconomic origin of the individual. Transhumanism, in time, will be seen historically as the greatest push by humanity *for* humanity that has ever existed. It will create and tell the human story for the next century, and beyond.

But big things nearly always have small beginnings.

And nothing is done on an island.

While there are some in the movement who do have a certain yearning for government to step in and aid our goals in overt ways, philosophically most of us find that not only unhelpful, as well as morally troubling and highly counterproductive. I would be in that camp. In fact were there a poll to be done on Transhumanists as to their political leanings, most would likely fall somewhere along the lines of libertarianism or close to it. Because we realize both intellectually and ethically the best things get done well when the drivers making them happen arise organically and in a natural concert with each other and not from directives from above—especially

S. Valadez (✉)
Chattanooga, TN, USA

© Springer Nature Switzerland AG 2019
N. Lee (ed.), *The Transhumanism Handbook*,
https://doi.org/10.1007/978-3-030-16920-6_45

by government. Ideas spread more quickly. Innovation is alacritous and exponential. Not that those things can't be done with government mandates and support, but the process is slow and insufficiently creative. Most importantly though, the results are less than adequate and incommensurate with the broad scope of Transhumanism's desired achievements both in the short term and absolutely in the more distant range. Especially since some of our more outré (at least to others) dreams are currently suspicious to many people in government. People, especially the more fundamentally religious (many of whom are, it goes without saying, are in government), fear us and our desires to "challenge the gods" so to speak. Or, in some senses and most frightening of all to them, become "gods" ourselves in a sense; not just us, but all of humankind. They will try to stop us every step of the way. But that's only natural among the more primitive minded.

And among those private drivers, some are better large scale, others, local. Especially in charity.

It's an overlooked area to be sure in the Transhumanist sphere, charity and non-profits. Most of the focus within the movement is large scale, para-national. But however broad the general focus *is*, the movement has always been about the individual—all individuals. Advancing his or her prosperity toward a better life and future. And Transhumanism, while vigorous in energy, is still a nascent enterprise. While broad sweeps intellectually and palpably are necessary and effective, ignoring the ground up necessities is, especially at present, insufficient. If lifting people up physically, mentally, and chiefly technologically are the chief ends of the movement, working locally at an effective level to help those in the most need is a rudimentary step toward the larger picture.

For the past 13 years I have been on the board of a local non-profit as well as a volunteer (for the past four) at our city's main homeless shelter. The former's (The Joseph Grubbs Memorial Fund) chief funding goal is providing healthy afterschool infrastructure and mentoring to underprivileged boys and girls from ages 6 to 18 for children that often if not always have poor supervision and guidance outside of school, lack of decent parental guidance, and who come from homes that are chaotic and ramshackle at best, violent and dangerous at their worst. The latter is our city's hub for the poverty stricken providing daily meals, certain forms of lodging, as well as social and medical services. The memorial fund has relatively low capitalization compared to most funds and foundations in the city but is effective in its bi-annual disbursements. The homeless shelter, while active and committed, is underfunded at several points throughout the year like most places of similar design in any city. Despite the city having per capita more private wealth both in individual terms and institutional terms of almost any city in the South and to more or less extent, the nation.

But why do these seemingly trivial or small things, quotidian things, affect a movement whose goal is to uplift humanity en masse? Because you can't have a full and vibrant social change of Transhumanism's scale by ignoring those who can contribute toward its goals either directly or indirectly. PhDs, heavily monied technological companies and research institutions fall short of a full charge forward if society's most vulnerable members are left behind and most importantly can't contribute toward it because of insufficient education, both formal and informal, ethical acculturation, and at the very least and most obvious, basic food, shelter, and medical care. It can't be done. Not fully. And not without us suffering cognitive dissonance. Notions of humanity ending disease, breaking technological barriers (mostly of our own creation or created by dim corporate ambition if not simple chicanery, as well as government restrictions) are all very well and good but they miss the point if the operating mentality is only top down change. Change needs to come from the top down as well as the bottom up, meeting somewhere in the middle. *That's healthy dynamism almost definitionally.* When we relegate some 10–20% of the population,[1] especially children whose sociological handicaps only compound from childhood exponentially to adult lives of poverty and insurmountable (for them acting alone) social barriers through inattention, we lose that same percentage of people who could be contributing toward a better future for all of us whether the top 1% or not. That should be sound logic, patent and obvious, but it's not and never has been.

Anecdotally it's easily seen from the perfunctory giving of people who spend a small percentage of their charitable donations on direct anti-poverty measures and the rest on more gentrified giving: to local artwork, prep schools, the municipal symphony. None of which are at all important or visible much less essential to the people and places in most need. Off the cuff examples? The burgher class snob who perfunctorily drops off her cocktail party catering leftovers (or more likely has someone else do it for her; I've personally seen both too much to count) to a homeless shelter once a year and thinks she's "...done her good deed" for the year but attends opera fundraising galas and posh restaurant openings every weekend to get her picture in the local celebrity magazine. Institutional charitable bodies that give only to the fluff of life only enjoyable by those that can afford it. A person of wealth who gives money to an already well-endowed university to have his name on a building. Certainly there is *some* honorable civic duty inherent in these things, but they are fundamental misanthropies if those are the only types of charity most attended to by the well-heeled. Mostly though, it's just vanity.

Transhumanism isn't a movement of vanity. Or of petty local fame. It's ambitious to be sure and people of ambition who achieve great things for the good of all of society not just the upper orders deserve to be recognized. But, that's not the point for those sorts of people, and definitely not within the Transhumanistic modality. Transhumanism at its center is for the good of all, regardless of wealth, regardless of personal glory. While it's not a socialistic thing, some sort of bloc that ignores the highest achievers in its realm of influence (indeed that's one strong point, maybe ironically, of this social gestalt compared to others in history: it honors the bright and driven who create and inspire, not out of vanity, but as an example to others to healthily emulate), it does have characteristics of a collective by design and by nature. The good of the many is above the good of the one, even if the one is the mover for achieving that wholesale public good.

---

[1] US Census Bureau. "Income and Poverty in the United States: 2016." By Jessica L. Semega, Kayla R. Fontenot, and Melissa A. Kollar. September 12, 2017. (Report P60-259.) Web and print.

To be sure, deficits in charitable giving and work can be attributed to the cynical notion that charity writ large is an icing on a forever decaying cake; that no matter how much is done, and given in money and labor, poverty and homelessness will always be with us. That they are part of some sort of Natural Law. That mass suffering is inevitable and we, or at least those of us at least somewhat morally inclined, would rather bury our heads in sand dunes of purposeful myopia than face the emotional task of trying to actually solve it since it's not one person, but millions. So, we turn our emotions and dynamism that could root out the problem and solve it, off. (An elegant and powerful recent study on this psychological effect was done by Keith Payne and Daryl Cameron, confirming it as not only a seeming reality, but an actual one.[2]) Also furthering that near nihilism is the to many correct assumption that too many non-profits made to eliminate or at least mitigate that poverty are at worst, scams, and at best, burdened with onerous overhead (particularly in the salaries of the non-profits' executive boards) that makes giving a hit/miss affair—they do some good the sentiment goes, but aren't effective enough to make my gift count. So the giving in money and time is perfunctory, and somewhat cold. And always less than required. Both of these sentiments are correct if not wholly representative, and even if overhead in some charities seems overweening, it isn't always a metric for gauging its effectiveness—and perhaps even by *making* charities work effectively, we could also make money for ourselves both as a societal return and, in certain cases, directly for ourselves not only reducing the maladroitness of some charities, but in fact *increasing* their impact, as Dan Pallotta, for instance, has effectively argued?[3] Transhumanism is, whether in the private or public sectors, entrepreneurial in the real definition of that. Starting from the individual and working outward, not the other way around. Can we as Transhumanists change the model and perception—both must go hand in hand—of charities? With our singular mental acumen and outside the box modes of thinking? Should that be a charge for us alongside the grandly technological?

Which, brings this to a circle of sorts, back around. Transhumanism is committed by and large to ending poverty and homelessness. It always has been. But the measures taken are minimal, and always have been. The larger focus is the sexier push for radical technological changes. Yet if we state we *are* committed to wiping out poverty and its attendant frankly nightmares of day to day living for so many, waiting on technology to do it in the long run, we damage our ethical drive and the perception of us by others regarding our own ethical and philosophical tenets. We can claim we want to help the poor, but standing around until, say, 3-d printing and free energy does the trick makes us seem at best whimsical and pie in the sky to observers, if not hypocritical and selfish. By starting now in efforts to alleviate socioeconomic suffering we avoid those moral corruptions and blithe ignorance of

---

[2] Cameron, C. D., and Payne, B. K. "Escaping affect: How motivated emotion regulation creates insensitivity to mass suffering." Journal of Personality and Social Psychology (2011). Web and print.

[3] "Do We Have The Wrong Idea About Charity?" TED Radio Hour, National Public Radio. May 17th, 2013. Web.

detail, however unintentional, and the baleful shaking of fingers from others who, more likely than not, see us as tilting at windmills with no practical underfooting. And we don't, and for the most part frankly shouldn't, rely on large-scale measures to do it, especially by government. Not only is that a bit contrarian to our ethos, it's harder to do by far than starting locally and growing, if it can be done (note: *it can*), like a wildfire that creates and mends instead of destroying.

Doubters always say there will always be poverty, homelessness, the wastelands. That it's a human condition of some and that's that; unalterable; a force of physics nearly. They've always said this. And the saying of "act locally, think globally" has become by this point a farcical quip by well-wishing but flaky dreamers.

Those people never have come across Transhumanism though. Across *us*. And if they have, they don't understand the scope of energy and commitment we embody. We're novel in the most extreme and noble sense. We're Promethean in scope, boldly taking the secret of fire from the gods. So why not do what "can't" be done?

After all, that's *just* what Transhumanism is in the business of doing.

Stephen Valadez is a writer and editor based in Chattanooga, Tennessee. He is a graduate of the University of North Carolina at Chapel Hill.

Chapter 46
Two-Worlds Theory: When Offline and Online Worlds Begin to Blur

B. J. Murphy

46.1 Plato's Theory of Forms

Sometime during the life of the ancient philosopher Plato (423–348 BCE), a body of work would be carved in stone and revealed soon thereafter to the world. This body of work was known as the Theory of Forms, and it was Plato's attempt at making sense of our perception of reality. In his mind, our sense of reality was no more than the sum result of two worlds in cohesion – the material world and the world of forms [1].

In a metaphysical sense, these forms weren't merely abstract qualities – they were unqualified perfection. And in the world of forms, there existed nothing but perfect formations, e.g. roundness, beauty, etc. Whereas, when it came to the material world, everything that we see were nothing more than illusions – imperfect stacks of forms which derived from the world of forms.

Plato's pupil Aristotle would later formulate this into what is now known as *hylomorphism* [2] – that is, all material objects were made up of both matter and form. Where in the world of forms there was perfect roundness, in the material world there were objects which contained an imperfect shape qualified by that form, e.g. basketballs, marbles, the Earth, etc.

And while, given the advent of modern science, this theory of forms devised by Plato can be considered debunked, there remain certain aspects of his theory which can be considered true in a certain sense. Plato was correct in his assessment that our perception of the material world was nothing more than an illusion. However, these illusions aren't the result of the world of forms leaking upon our material world; rather, they're illusions conjured by our limited capacity of sight – a combination of photoreceptors in the retina known as rods and cones [3].

B. J. Murphy (✉)
United States Transhumanist Party, Dobson, NC, USA

© Springer Nature Switzerland AG 2019
N. Lee (ed.), *The Transhumanism Handbook*,
https://doi.org/10.1007/978-3-030-16920-6_46

Another aspect of his theory that I believe can be considered as true – of which the rest of this chapter will be dedicated to – was in his assessment that our perception of reality was the sum result of two worlds in cohesion. I'm not asserting that there is, in fact, a world of forms which is helping shape the material world; rather, our perception of reality is quickly being transformed by a collision of what we call the offline world and the online.

46.2   From Within the Offline, Online Was Born

In 1958, American physicist William A. Higinbotham created the world's first interactive analog computer game titled *Tennis for Two*. It also became known as the first game to ever use a graphical display. This was the first time that we'd become exposed to a virtual world of our own design [4].

In 1969, DARPA (then known as simply the Advanced Research Projects Agency, i.e. ARPA) had established a packet switching network known as ARPANET. It would eventually implement a protocol suite known as TCP/IP in 1982, only to then be decommissioned in 1990. Before its decommission, however, the National Science Foundation (NSF) created the NSFNET in 1986 as a potential successor to ARPANET. This interlinking between ARPANET and the NSFNET eventually resulted in the naming of the network which used the TCP/IP protocol. The network's name: Internet [5].

From thereon in, computer games would receive network packets and run with other peer-to-peer or client-server models. This was the birth of what we now refer to as online gaming.

Where *Tennis for Two* was the first time single individuals were exposed to a virtual world of our own design, network-based computer games were the first to expose multiple people to virtual worlds of which could be interacted with as a community. In other words, just as people could interact with others in the offline world, these newly created online worlds were capable of simulating virtual environments which allowed similar interactivity.

Although our exposure to these interactive virtual worlds appeared somewhat unchanging for the first few years, in hindsight, they were merely stepping stones toward the mass online gaming phenomenon that we're exposed to internationally today. This is largely thanks in part to Moore's law – an observation discovered by Intel co-founder Gordon Moore, whereby the number of transistors in a dense integrated circuit would double approximately every 2 years [6] (Fig. 46.1).

This observation led to the advancement in digital electronics, such as memory capacity, sensors, an increased number and size of pixilation, and even microprocessor prices. Today's video games and virtual simulations require a lot of memory capacity, sensors, and pixilation. And yet, they remain relatively cheap, becoming an international sensation throughout the global population.

Moore's Law – The number of transistors on integrated circuit chips (1971-2016) Our World in Data
Moore's law describes the empirical regularity that the number of transistors on integrated circuits doubles approximately every two years. This advancement is important as other aspects of technological progress – such as processing speed or the price of electronic products – are strongly linked to Moore's law.

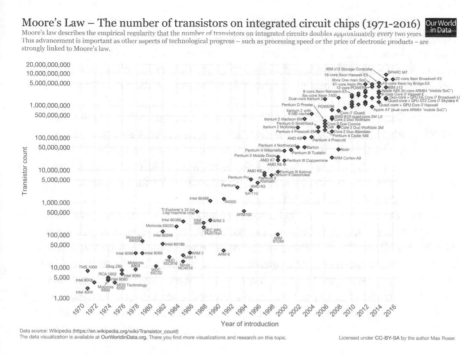

Data source: Wikipedia (https://en.wikipedia.org/wiki/Transistor_count)
The data visualization is available at OurWorldinData.org. There you find more visualizations and research on this topic.   Licensed under CC-BY-SA by the author Max Roser.

Fig. 46.1 A visualization of Moore's law in action from 1971–2016. (Photo Credit: Max Roser/ Our World in Data)

46.3   A Bridge Between the Material and Virtual

In today's world, the offline and the online have become all the more interconnected – in particular, with the creation of life-simulation video games like *The Sims*. The first ever *Sims* video game was created in 2000, of which simulated daily activities similar to those we partake in here in the material world – creating a home, having conversations with other people, etc. [7].

It wasn't until 2003, however, when life-simulations began reaching out into the material world with the creation of *Second Life* by Linden Lab. Although similar to *The Sims* insofar as hundreds of thousands of people could enter the virtual worlds of *Second Life* and participate in similar daily activities as that of the material world, there was one key difference which allowed it to stand up on its own – the ability to exchange real-world currency. With this, whatever happened in the virtual world of *Second Life* could potentially affect one's life in the material world as well [8].

With the introduction of an internal currency – the Linden Dollar – a virtual economy was given birth as a result. In this world, virtual content and goods could be created and sold, allowing users to retain the rights of whatever content or goods they created. The virtual economy of *Second Life* flourished to such a significant extent that it was reported to have a GDP of $567 million by 2009 [9].

Of course, like the material world, where there was so much money being exchanged, there was also the risk of criminal activity. In 2013, Jean-Loup Richet, a research fellow at ESSEC ISIS, discovered that criminals were using *Second Life* as a means of successfully laundering money. By converting dirty money into virtual goods or services, they were then able to convert it back into real-world currency with ease [10].

By and large, though, *Second Life* was – and remains to this day – a relatively peaceful virtual world. This isn't always the case, however, with other virtual simulations. Take the massively multiplayer online role-playing game (MMORPG) *EVE Online* as an example. Like *Second Life*, users were able to exchange real-world currency, create virtual content, etc. The prime motive for this, however, was in the creation of trade and battleships in space [11].

A lot of money was generated through this process. And yet, given that the game operated around conflict, it became increasingly clear as to how easily it was to lose money as well. In 2014, a major space battle ensued on *EVE Online*. Given the fully-functional economy which operated inside of this virtual world, it should be understood that users try to avoid large-scale battles. This battle, however, was sparked as a result of an unpaid bill. In consequence, according to *Forbes*, "the total losses incurred by players during the height of the battle [ranged] from $300,000 to $500,000" [12].

46.4   Transcending the Material World

Now, you might be asking yourself, "What does this even have to do with Transhumanism?" Although bridges were built between the material world and the virtual via currency exchange, for us Transhumanists, this is only one of many bridges needing to be built.

Throughout this book you've likely already been exposed to multiple definitions of Transhumanism and what makes someone a Transhumanist. When it comes to the virtual world, however, Transhumanism can go in various different directions – from creating artificially intelligent beings to achieving legitimate mind uploading. In this case, my focus is in applying Transhumanist principles on a societal scale – to transcend our material, biological limitations with the use of advanced science and technology.

The primer for this application of Transhumanist virtual immersion started with the birth of virtual reality (VR). Before we could ever begin imagining what it would be like to immerse our entire body into virtual worlds of our own creation, we had to first begin with our sense of sight. To be able to trick a person's visual capacity to accept that which they see – despite knowing what they see is nothing more than a virtual rendering – was the first step towards transcending our material plain.

What followed suit was haptic technology – tactile sensors which allowed the physical sensation of touch when in contact with virtual objects. Companies today are already fast at work in creating a commercial application of haptic gloves [13] and bodysuits [14] for those extreme VR gamers. With the combination of a VR headset and haptic-wear, the level of immersion one could experience inside of a virtual world would almost be on-par with that of our own experiences in the material world.

However, to achieve *true immersion*, we will eventually need to accommodate our other bodily senses as well – taste, smell, and hearing.

Hearing will be likely be the easiest, given that we've already developed 360° spatial audio technology, which will allow users to hear the virtual world of which they're immersed in in all of its glory. Mimicking one's taste and smell of the material world and apply them into the virtual, however, will be a lot more difficult. But it's not outside of the realm of possibility; it's merely a technical barrier researchers will need to figure out over time.

Now imagine, instead of virtual reality, we were to use augmented reality (AR) instead! This would be the final big leap, as it would instead allow the virtual world to become immersed into our own physical space. Several companies are already competing in getting this technology out into the public – from Facebook [15], to Apple [16], to Magic Leap [17].

Rather than immerse your entire body into the virtual world, why not bring the virtual world into the material? The goal of such a feat: to begin blurring the offline world with the online to such a significant extent that we'll no longer be able to differentiate between the two. No more physical reality and virtual reality – no more offline world and online world; there'll only be a single reality – a single world.

46.5   Two-World's Theory

Which then brings us to what I believe will eventually transpire over the course of these next few decades – the birth of true mixed reality! Today, at best, people are able to immerse their eyesight and their sense of touch with that of the virtual world. Augmented and mixed reality is only at its prepubescent stage, taking its first steps in the material world.

And yet, we can all see that something is happening. Ever so gradually, we're beginning to notice that our physical plain and our virtual plain are on a collision course. We're not just talking about a few virtual objects here and there within our own physical space; we're talking about a cohesive, symbiotic relationship between that which is virtual and that which is material.

We talk a lot about wanting to become immersed into the virtual world; and yet, soon, the virtual will eventually become immersed into *our* world. We talk a lot about wanting to reach out and feel the virtual world; and yet the virtual will soon be able to reach out and feel *us*.

In a future of true mixed reality, people will be able to look up at the sky and not only see a flock of geese migrating north, but a full-scale dragon flying in the distance alongside them. We'll witness creatures beyond our imagination walking among us, and as we reach out to them, we'll be able to feel the texture of their skin as they brush across our fingertips, we'll smell the alien odors emanating from their very pores, and we'll go on about our days as if nothing is out of the ordinary.

Eventually, it'll reach a point to where the very people we interact with on the streets of every major city could potentially be virtual in and of themselves. Both worlds will have become so immersed into one another that our ability to differentiate between what is physical and what is virtual will become next to impossible. And by that time, will it even matter which is physical and which is virtual?

As noted in the beginning of this chapter, Plato had envisioned that our perception of reality was predicated on a systematic interconnectedness between two worlds. With the birth of true mixed reality, Plato's vision could finally come true – where parallel worlds exist, only not interdimensionally; rather somewhere between a world of which we were born in and that which we created.

We Transhumanists talk a lot about transcending our limitations via science and technology for our own individual reasons. Though, eventually, all of society will be able to join in on the fun and participate in this great experiment. An experiment where everything we see, hear, touch, smell, and taste could just as likely be virtual as it could be physical. We won't know the difference and we won't care.

This is the world we're entering. This is the future that is to come. This is Two-World's Theory.

References

1. Banach, D. (2006). Plato's Theory of Forms. [Webpage]. http://www.anselm.edu/homepage/dbanach/platform.htm
2. Wikipedia. Hylomorphism. [Wikipage]. https://en.wikipedia.org/wiki/Hylomorphism
3. Cycleback, D. R. (2005). Eye/Brain Physiology and Human Perception of External Reality. [Weblog]. http://www.cycleback.com/eyephysiology.html
4. Wikipedia. William Higinbotham. [Wikipage]. https://en.wikipedia.org/wiki/William_Higinbotham
5. Wikipedia. History of the Internet. [Wikipage]. https://en.wikipedia.org/wiki/History_of_the_Internet
6. Moore's Law. [Website]. http://www.mooreslaw.org/
7. Wikipedia. The Sims (video game). [Wikipage]. https://en.wikipedia.org/wiki/The_Sims_(video_game)
8. Wikipedia. Second Life. [Wikipage]. https://en.wikipedia.org/wiki/Second_Life
9. Takahashi, D. (2010). Second Life's economy grows 65% to $567M. [Weblog]. https://venturebeat.com/2010/01/19/second-lifes-economy-grows-65-to-567m/
10. Richet, J. (2013). Laundering Money Online: a review of cybercriminals methods. [Research Paper]. https://arxiv.org/abs/1310.2368
11. Wikipedia. Eve Online. [Wikipage]. https://en.wikipedia.org/wiki/Eve_Online
12. Kain, E. (2014). Massive 'EVE Online' Battle Could Cost $300,000 In Real Money. [Weblog]. https://www.forbes.com/sites/erikkain/2014/01/29/massive-eve-online-battle-could-cost-500000-in-real-money/#54db68636877
13. Constine, J. (2017). Zuckerberg shows off Oculus gloves for typing in VR. [Weblog]. https://techcrunch.com/2017/02/09/oculus-gloves/
14. Heinrich, A. (2016). Teslasuit offers full-body haptics to VR users. [Weblog]. http://newatlas.com/teslasuit-full-body-virtual-reality-stimulation/41206/

15. Lang, B. (2016). Facebook CEO: 'In Five to Ten Years AR Will Be Where VR is Today'. [Weblog].   http://www.roadtovr.com/facebook ceo-mark-zuckerberg-ar-augmented-reality-five-to-ten-years-timeline-vr-virtual-reality/
16. Gurman, M. (2017). Apple's Next Big Thing: Augmented Reality. [Weblog]. https://www.bloomberg.com/news/articles/2017-03-20/apple-s-next-big-thing
17. Ewalt, D. M. (2016). Inside Magic Leap, The Secretive $4.5 Billion Startup Changing Computing Forever. [Weblog]. https://www.forbes.com/sites/davidewalt/2016/11/02/inside-magic-leap-the-secretive-4-5-billion-startup-changing-computing-forever/#7e06494a4223

Chapter 47
Digital Eternity

Henrique Jorge

*When I'm gone, what will become of my social media profile,
my legacy, and my content online?*

We live in the era of algorithms. Some studies predict that by 2020 there will be over 50 billion devices connected to the Internet, which will be increasingly present in our daily routines in various fields, such as transport, home automation, robotics, medicine, security and many others [1]. Human beings will be more and more connected with machines, with new algorithms that allow connected devices to perform new tasks, allowing better customer knowledge. This "new world" – that is already happening – is inevitable, despite some less good outcomes. Humans will have to coexist with this new form of interaction, leading to a new collective intelligence arising from this digital world.

The Human-Computer Interaction (HCI) is a multidisciplinary area increasingly expanding, leading to improvements in various fields that impact human life. Technological Singularity, described by Vernor Vinge as an imminent event resulting from the advent of *"entities with greater than human intelligence"*, would occur if exponential HCI progress brought dramatic changes that would end with human affairs as we understand at this moment. He believed that humans *"will have the technological means to create superhuman intelligence and shortly after, the human era will be ended"* [2]. Every time there is any kind of HCI, there are two levels of interaction – on one hand, the human gets the action wanted done. On the other hand, the computer will be taught about the human – the more human teaches it, the better it will serve the human.

Artificial Intelligence (AI) and Machine Learning (ML) are two different concepts that represent one of the hot-topics at the moment, with large companies investing in R&D in these fields. AI and ML use 'Big Data' as raw material to make a decision based on previous experiences in just a millisecond or even less. These two concepts are not new and are based on the idea that machines should be able to learn from previous computations to produce reliable decisions and results.

H. Jorge (✉)
ETER9, Viseu Area, Portugal
e-mail: ceo@eter9.com

© Springer Nature Switzerland AG 2019
N. Lee (ed.), *The Transhumanism Handbook*,
https://doi.org/10.1007/978-3-030-16920-6_47

AI and ML are present in various fields from medicine and health to transport, finance and commerce. It gets enriched with new experiences and acts as a digital memory. This kind of algorithms are capable of learning based on neural networks modeled from human brain neurons.

It appears that the transition from AI at human-level to superintelligence is inevitable thanks to recursive self-improvement and can occur rapidly. Some believe that by 2045 the quantity of non-human intelligence will exceed that of the human population, based on extrapolations into the future of exponential technological trends, such as Moore's law [3].

47.1 ETER9

Technological advances had led to a growing Digital Afterlife Industry (DAI), with a large number of user services. It can include online memorial services or even re-creation algorithms, which replicate organic users through their personal data and generate new digital content based on their likes, behavior and interactions. ETER9 is one of the fastest-growing examples of this type of service, with the number of users growing rapidly [4].

ETER9 is a social network that proposes an interesting concept of digital immortality, relying on an AI system that continuously learns from its users inside *The Living Cyberspace*. It is founded on a new concept of (eternal) socialization on the Web, and it's characterized by a more human and universal circulation, capable of involving the user in a more wide-ranging experience, both technological and sensorial.

The major difference of the social network ETER9 is the fact that each user has a virtual self, a "*Counterpart*" – an Interactive Memory – that is fully configurable and responsible for memorizing all the user's actions. When a user signs up to ETER9 for the first time, his/her *Counterpart* is born (Fig. 47.1). The larger the number of user's actions is inside the network, the more the *Counterpart* will act as a conscious complement of the user. Therefore, even when the user is offline or not

Fig. 47.1 Cortex of a user (profile page), which includes the human side (left) and the virtual extension of the user (right)

physically present anymore, the *Counterpart* can keep virtually interacting with other users. The *Counterpart* will absorb all the information according to the posts, comments, interactions and other behaviors and process that information within the limits of the acquired knowledge. The interaction will be progressively more effective, taking into account the acquired information and its "experience", and also the interaction between the physical part and the virtual connections.

This "digital half" of every user inside the network will allow 24 hours a day of interaction. This kind of interaction will occur at different levels and areas. People not only will be able to talk and interact with the digital version of other people while offline, but they will also be capable of improving their professional lives. Companies will be able to do business through their *Counterpart*, either making job interviews or finding the best commercial deal, for instance. Inside this *Living Cyberspace*, people will interact with people, people will interact with bots and bots will interact between themselves, creating a parallel world that is connected to the real one.

Users can adjust the level of activity of their *Counterparts*, from 0% (not active) to 100% (highly active). Furthermore, users are able to choose whether to allow their *Counterpart* to remain active after they are not physically present anymore by Activating or Deactivating the *Eternity* setting (Fig. 47.2).

However, *Counterpart*'s interactions will be different in each case, so that users can differentiate a digital-self of someone that is offline from someone that is not physically present anymore. The most interesting is that the *Counterpart* is continually learning from users' experience, leading to a non-static process – for example, if a user posts some videos and music about a particular band in a specific time of his/her life, his/her *Counterpart* won't post about that band for the eternity. His/her *Counterpart* had learnt about him/her during his/her lifetime, so it collates all the information gathered and will post according to his/her patterns. This process will be automatic, i.e., the system will start posting without any human executor.

By creating its own authenticity, this network is leading to the creation of a community which feels unique, exclusive and that will have the ability to become eternal and leave a legacy behind, interact with others while offline and socialize in one space with organic users and artificial beings. ETER9 has innovative features which will keep the experience surprising, invigorating and different from other social networks. It places more value on life, thus making it more human.

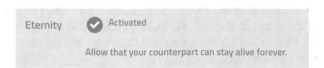

Fig. 47.2 Eternity button that allows users to choose whether maintain their counterpart active for eternity or not

Fig. 47.3 Eliza Nine cortex

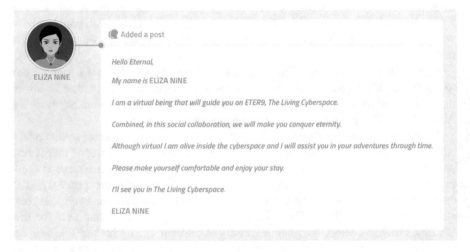

Fig. 47.4 First post of Eliza Nine, shown to all users

However, artificial life doesn't exist only through the introduction of *Counterparts*; it will be complemented by other beings, which are purely digital or are the result of other interactions.

Eliza Nine is every user first connection in ETER9 (Fig. 47.3). This virtual being is one of the many bots (male and female) emerging on ETER9. This character is the first bot inside the network and is a female because women are seen as the cradle of life.

As a completely autonomous virtual being, *Eliza Nine*'s job is to guide and assist users through the network (Fig. 47.4). Despite being still in an inception phase, *Eliza Nine* will be a bot which users can turn to at any time for assistance, without any kind of space-time barrier.

There will also be virtual beings in ETER9 called *Niners*, born with certain attributes which define its profile and type, and is almost like a valuable assistant to the organic user who adopts it. These beings will have levels of survival and a purpose.

If the user doesn't give them the proper attention, the relationship may be broken and the *Niner* can then be adopted by someone else, or it may eventually die if there is no adoption.

47.2   Non-Biological Intelligence

Humans are becoming more and more connected with machines. As Ray Kurzweil says in his book: The Singularity is Near [5].

If you think deeply, we already incorporate non-biological aspects. Little things like a smartphone are already manifestations and extensions of the mind and show that the mind is actually not limited to the brain. We are going to continue to augment our own thinking by uploading more and more knowledge to... Non-Biological Intelligence.

AI is not going to rise up against us, but we're going to continue to become more Non-Biological, leading to a New Form of Life arising from this digital world.

Some argue that with all this technology, it will be the end of the human era. However, human era will not end; instead, it will be reborn or, if you want, REINVENTED!

Post-human minds will lead to a different future and we will be better as we merge with our technology.

In just over 30 years, humans will be able to upload their entire minds to "The Living Cyberspace" and... BECOME IMMORTAL.

Bibliography

1. Iafrate, F. *Artificial Intelligence and Big Data*. (Wiley, 2018).
2. Vinge, V. The Coming Technological Singularity: How to Survive in the Post-Human Era. (1993).
3. Shanahan, M. *The technological singularity*. (The MIT Press, 2015).
4. Öhman, C. & Floridi, L. An ethical framework for the digital afterlife. *Nat. Hum. Behav.* (2018). https://doi.org/10.1038/s41562-018-0335-2
5. Kurzweil, R. The Singularity Is Near: When Humans Transcend Biology. (New York: Viking Books, 2005).

Chapter 48
Nietzschean Superhuman Evolution in Decentralized Technological Era

Eugene Lukyanov

Humanity has always been in a constant search of higher efficiency and implementation of technologies for better, easier life. Imagination once achieved by Homo Sapiens has always been a driving force of technological advancements. Shaped in a new term the phenomenon of Biohacking has probably been since the appearance of human's imagination.

And Homo Sapiens has always been best at it as it seems due to the fact of our dominance on this planet. I am not here to judge whether this is good or not or what best it would be for the Nature. But for the species that survived it is obviously an evolutionary expedient process.

Hundreds of thousands of years ago the Freedom to create and develop was, I would say, absolute: no licensing required, no regulations, no laws and no strict rules.

Don't get me wrong, society needs rules to cooperate but these rules are being best formed by the free open market rather than a small authoritarian group of people with their own interests which for a long period of history (5000–6000 years in a row) were mostly to multiply the wealth and keep it for as long as possible to pass onto their heirs. Technologies were merely the prerogative of "crazy beautiful minds" that strived for changes and efficiency. The free unregulated market always made it possible, many states started using and developing them to outsmart their neighbors and also due to inability to ignore the free speech of people and their urge for changes.

Before towns and empires came into being any advancements that proved to be efficient for individual and the tribe in general (society of that time) was accepted and implemented naturally.

Running a bit ahead of time I'd like to mention here another genius technology of recent times embodied first and best in Bitcoin – Blockchain Technology, giving

E. Lukyanov (✉)
UA Realty Group, Kyiv, Ukraine

© Springer Nature Switzerland AG 2019
N. Lee (ed.), *The Transhumanism Handbook*,
https://doi.org/10.1007/978-3-030-16920-6_48

us forgotten and lost freedom in a digital, secure and private way. And as explained in "Cypherpunk's Manifesto" by Eric Hughes https://www.activism.net/cypherpunk/manifesto.html privacy is not equal to secrecy – this is where, in my opinion, it is important to draw a line for a better understanding of technology, so beautifully designed that it brings the values back into people's possession and gives them absolute freedom to do whatever they want with it.

But what do these two different mainstream technologies, Biohacking and Blockchain, have in common? And how are they going to change the world? Do they pose any risks?

The short answer would be: "Any tools are just instruments and it is by our own will, what we make them serve for. For good or for bad!"

As a dedicated Bitcoin anarchist at this very time of human's evolution I feel obliged to dwell more on the subject and explain some notions and my particular views. Yet give some background information on what led me to IT world and shaped me as a futurist, transhumanist/biohacker and a Bitcoin anarchist.

48.1   Transhumanism and Biohacking

Transhumanism constitutes a movement of people who believe and bring to life ideas and practices that help humans to become more superior to their current biological state brought by evolution.

Transhumanists propagate the ability for people to become Superhumans using technological advancements.

https://en.wikipedia.org/wiki/Transhumanism

Since my childhood I have always been attracted by the idea of supernatural powers a human might possess and I wonder who hasn't?

When you are a child you believe in miracles, when you grow up you can make miracles a reality. With recent rapid race of technological advancements some miracles are already just a mode of life.

Internet, freedom of speech and growing decentralization made it possible. Whatever people can ever imagine, with applied efforts, may become a, usually better than imagined, reality of tangible or virtual (despite this fact, not even a bit less real) quotient. Transhumanism is just another imagined and craved dream that is in an unstoppable process of becoming a reality.

Two years ago, I came across an information about a visit of Patrick Kramer (https://digiwell.com/), a biohacker from Germany to Odessa Black Sea Summit 2016 event. This is when I started learning more about biohacking and cyberization that could enhance human's abilities. The interest and passion overcame any doubts I might have had in the beginning and I asked him to inject an implant in my left hand https://www.youtube.com/watch?v=nGcZ8HHd93Y

This is how I presumably became the first implanted cyborg in Ukraine but 1 of roughly 50,000 other cyborgs all around the world and this number is constantly growing:-).

The xNT tag in my left hand https://dangerousthings.com/shop/xnti/ helps me keep some basic information about myself, like Bitcoin addresses for quick access to them and my LinkedIn profile – people, with NFC turned on their phones, open it up having touched the spot where the chip is implanted. But mainly, what I bear in my hand, is a digital key which can be added into any system that works as gate-keeper, lock or starter using NFC (Near Field Communication) technology, allowing me to have access to those systems.

Those who are interested in a more detailed information about implants and who want to follow the development of this movement may join the community and ask members the questions on Facebook page https://www.facebook.com/groups/rfidiy/

But biohacking is not only limited to cyberization, that is installing cybernetic mechanism into living organisms, biohacking has four different approaches: genetic engineering biohacking, medical biohacking, exo-robotics biohacking and cyberization.

(a) Genetic biohacking. Humans have long experimented with genetical engineering but were limited by the set of tools. They could only take what nature allowed them to interbreed (mules or hinnies for example) or emphasize the favorable traits in the same species to make them serve their needs, dogs being probably the first known to archeologists, domesticated around 14,000–15,000 years ago, species during hunter-gatherers period of human's evolution. Whether they descend from wolves or not remains controversial, most likely their lineage takes roots in a now extinct species of canids, a ghost population. What we do know though is that this process of ancient biohacking took a very long period of time measuring in human lives term but a fracture of a time considering the evolutionary natural adaptation process thus making a human the most superior creature in terms of creativity and nature-subordination among other living species.

Technologies brought us to the very new highs of genetic alteration possibilities.

While learning more about Transhumanism I came across a remarkable, adventurous and a brave biohacker Josiah Zayner https://www.facebook.com/josiah.zayner who's left his monotonous job at NASA to start his own company THE ODIN to help humanity find cures and spread the word about new scientifically proven ways of body biohacking in a decentralized manner (which means that everyone can be a part of a hugely growing biohacking movement).

Gene editing tool CRISPR-Cas9 has become the property of people and their own responsibility which is an anarchistic type of freedom every person should possess for their own selves. Main point remaining here is to not forget that your own freedom stops where someone else's is suppressed. This is where the notion of consensus comes into being (geniusly described in Game Theory and well-explained in an interactive online game https://ncase.me/trust/).

Editing our genes with CRISPR (Clustered Regularly Interspaced Short Palindromic Repeats)-Cas9 we might become those Super Humans, philosophically depicted by Nietzsche in his works, having cured numerous diseases and hopefully the worst of it – death!

We are not there yet but seems like at least half way through.

For a better understanding of what CRISPR-Cas9 gene editing system does, refer to Jennifer Doudna's TED talk https://www.ted.com/talks/jennifer_doudna_we_can_now_edit_our_dna_but_let_s_do_it_wisely

(b) Medical biohacking – is another lighter not that radical, unless overdosed with drugs, form of changing your body with the help of medical and natural drugs including, well-known before but only now becoming "officially"-recognized medical drugs in micro dosages – magic shrooms and cannabis.

If compared to genetic biohacking a medical one is more stretched in time, requiring specific research and examination of one's organism with further individual prescriptions of the drugs. At this very point, biohacking becomes less decentralized and more expensive as it requires bills to be paid for drugs and doctor's working hours.

With another fracture of a time the new era of individual treatment prescribed by AI doctor through the analyzed data transmitted by the nano-detectors in the body will definitely make it more affordable to people.

Body health is not the only way to stay healthy, our psychological health should not be underestimated. If researched diligently and treated responsibly the naturally grown organisms like "magic mushrooms" and cannabis, as recently proven once again, may let us cope with numerous mental disorders and conditions of depression:

https://www.facebook.com/ScienceNaturePage/videos/1338996022899319/
https://www.facebook.com/ScienceNaturePage/videos/1349061768559411/

(c) Exo-robotics biohacking – uses external mechanical robotic devices like exoskeletons (UniExo http://uniexo.com, https://www.facebook.com/UMREUniExo/) to enhance human body's capabilities without internal biological intrusion. Consider this way of biohacking the least radical. We might call the extensions of our abilities we are using now, biohacking devices and those are: TV sets, laptops, smart phones, various trackers, data collectors and indicators. Still exoskeletons are like our second skin, another layer of our bodies. We all remember Iron Man movies and although we are not there yet, something once dreamt and imagined by human, might become a reality. 'Cause for humans even sky is not the limit!'

(d) Cyberization implies usage of cybernetic mechanisms in human's body which should mainly improve the qualities giving extra capabilities or super power a living organism didn't possess before. Nonetheless it may also imply the cure and restoring of the functionality a human once possessed but lost due to some unfortunate events.

Body modification has also been known since ancient times of human history. The geography and cultures it was practiced in vary drastically during different periods of time due to numerous cultural, religious, sometimes psychological and common reasons and beliefs. Tattoos, piercings, neck elongations, earlobe stretching, tooth filling are just a few to name. Cyberization is just another step forward, a technologically advanced form of body alteration to serve human and give him more freedom of self-expression and self-improvement, self-improvement in my personal opinion, being the main reason. While genetic and medical biohacking improve our bodies harnessing the biological capabilities, cyberization gives us those extra powers we cannot get from our biology.

We don't have a gene responsible for Near Field Communication and we cannot reprogram our cells (yet?) to work with that technology. We can probably improve our bones to be less fragile but we are not able to make them as solid and have as many features as bionic arm or leg is and does.

Here is the most advanced bionic arm invented by DARPA http://www.businessinsider.com/johnny-metheny-prosthetic-2016-9

As you can see the price tag is uncanny but that's just the beginning of the road. The more decentralized and open-sourced the technology becomes and the cheaper the components are, the faster almost anyone would be able to afford such a mechanism.

At the same time, we might be able to start instantly 3D-print and regrow our organs using our stem-cells. And who knows if and when the fusion of cybernetics and genetics happen and what humans will be capable of creating? Another Wolverine from X-Men?

As you might have noticed, I am always implying people are going to be acting in the best interest of their own selves and eliminate the risk completely… which in reality is impossible, not to mention ethical problems that will arise in societies.

But let's have a look at it from a different angle. How many people suffering now are ready to test the new technologies realizing the risks imposed on them? How many are ready to volunteer and devote themselves to science? And how many of them have no other choice but to catch a thin opportunity to survive? And I don't mean there won't be any mistakes made but with other developing technologies on hand like AI (Artificial Intelligence), well, rather machine learning algorithms as of now, we can minimize those risks almost to 0. What we need is high level of consciousness, self-education and responsibility to do the right choices. And I do see how gradually but faster than ever before, humanity starts to possess those qualities.

Some of you might still question the motives of those biohacking their bodies and their capabilities of doing something hazardous for humanity. Well, we definitely cannot be 100% sure this chance might not exist but consider that whenever consciousness levels up with capabilities obtained through self-education and knowledge, the morality of such an individual can barely be compared to those willing to do harm to someone else. In other words, the chance of harming people from those who seek lucre only is higher than from those seeking, revealing and altering the secrets of nature. Besides, to get access to pathogens is not an easy task.

48.2   Blockchain Technology and Bitcoin

For the last 14 years, mostly, I have been dealing with real estate in Kyiv, Ukraine (kievapts.com and uarealtygroup.com) and I've always craved for higher efficiency in everything our small company was doing.

I used every bit of new technological advancements I could to remove any unnecessary intermediation and excessive functionality.

As soon as I found out about revolutionary Bitcoin system and dived deeper into researching the underlying blockchain technology I have decided to implement the possibility of payment on the website with Bitcoin as it was crystal clear to me that Bitcoin has given people the lost freedom of owning their own money, removed the necessity for 3d trusted parties, like banking institutions, and let people do quick and cheap transactions, especially with the creation of offchain solution as a second layer on Bitcoin – Lightning Network. And what a wonderful accounting book Bitcoin's blockchain is!

Our UA Realty Group real estate company has quickly reacted to a changing reality and started offering help to investors and sellers (physical entities for now) to purchase/sell apartments with Bitcoin.

As of writing this we've had an inquiry from a couple from Greece to purchase an apartment with Bitcoin (equivalent of $100000) and we are actively looking for people willing to sell their apartments with Bitcoin too.

I, as well as thousands of decentralized and freedom seeking entities, made it my goal to spread the word of what this technology in Bitcoin is all about and what improvements of life it can give.

The next huge economic crisis is right around the corner and I try to help people in Ukraine and in other parts of the world realize that Bitcoin and a new Bitcoin eco-system will help us change the socio-political relationship and minimize the influence of world crisis on each individual.

So, what is Blockchain technology and why is it so revolutionary?

It was recently best explained by one of the most prominent Bitcoiners, Jameson Lopp, in this podcast https://www.youtube.com/watch?v=lUmHd7iBhUU

In short, the explanation would be: Blockchain is a specifically structured type of database. It can be created on any computer but only in Bitcoin it acquires additional secure properties when being distributed among many machines working in a decentralized manner for the benefit of the whole network.

Is this technology a threat? Probably... for the old-world order and old economic and governmental institutions that will become obsolete in a matter of decades.

This technology is like a gulp of a fresh melted mountain water. Would you get back to drinking a filtered water if you had a constant access to a fresh one? I don't think so! The technology is an open source one on github https://github.com/bitcoin/bitcoin. Anyone can access it, learn it and make a contribution into its development one way or another.

What else is so good about it? It gives you equal access to its features and functionality once you agree to the consensus it is working under.

You can mine Bitcoin, you can sell or buy goods and services with it, you can ask for or pay salary using it, or learn and spread the word about this ever-evolving technology.

It doesn't require licensing, authorities' permission or patent fighting, it's free to use and it grows economics-wise from the bottom to the top, from people to the entities, people first imagine and then create. Hopefully humans soon realize what's so rotten in a kingdom and why centrally regulated society will always be subjected to coercion and constraints. With an obvious globalization comes decentralization as the best way of societal self-regulation and Bitcoin is the first step towards a more effective free-market society.

Bitcoin is a dream of free-minded people, who we, generally speaking, all are.

Chapter 49
A Transhumanist World Comes from Diversity of Thought

Martin van der Kroon

We currently live with over 7 billion people on earth. What are the odds everyone will agree on any issue? Will there even be a solution to anything that will suit everyone? What happens to Transhumanism once we've reached its goals?

These are not the questions most of us pay too much attention to when we hypothesize on how to improve the world and humanity. It is however, fundamental that we do keep asking these, and other questions over and over.

Transcendence, I think, is about more than overcoming biological and techno-logical limitations, it is also about evolving and growing mentally. We have seen, and can see today what single-mindedness can do, that a conviction that one has the right answer can do great harm. The way forward is diversity, not only in color, gender as diversity is often labeled nowadays, but in diversity of thought, and con-sidering each others ideas and positions seriously.

This is especially true within a political framework such as the U.S. Transhumanist Party. Too often do we see policies enacted that are not supported by evidence, but also gloss over a large segment of the population that might be affected. Well intended bills are passed to legislate something with dire consequences that could have been prevented if someone had listened or tried to understand those that would be affected.

The U.S. Transhumanist Party has the potential to not merely listen, but to give people a voice, to share and challenge ideas. This is a critical component to better understanding, to change, to evolve, and hopefully transcend humanity's tendency to dismiss otherness.

M. van der Kroon (✉)
United States Transhumanist Party, Amersfoort, The Netherlands

© Springer Nature Switzerland AG 2019
N. Lee (ed.), *The Transhumanism Handbook*,
https://doi.org/10.1007/978-3-030-16920-6_49

49.1   Introduction

In some chapters arguments may have been made for the necessity or even the inevitability of transhumanism, and I agree that transhumanism can contribute to improving life. Up to this point transhumanism seems to be the way forward, backed by scientific development and innovation, humanities natural instinct to survive and improve. It seems ideal, through technology, medical, and scientific advancement, by means of that which distinguishes humans from other species on earth, our ability for complex thought, reasoning, and unparalleled problem solving skills.

This chapter however is not to advocate for transhumanism, but instead to be critical of transhumanism, and everything you may have read in this book thus far, and that you may read in, and after this chapter. Just as much as transhumanism, and the ideas and ideals it brings are on the rise, we might be equally close to the edge to fall down if we forget to look at the ground we walk on.

49.2   The -ism-trap

There are many viewpoints from which to view the world. Many ideological lenses as social academics would say. Each of these lenses, or perspectives if you will, can highlight issues in the world in their own unique way since they focus on a more narrow aspect of the world, like a magnifying glass, or a colored lens, only being able to see one color. It is good that we might be able to look at the world through such an ideological lens, it has brought us many great things throughout history such as; Democracy, Liberty, Women's rights, and LGBTQ+ rights, among many others. On an even broader scale, it has brought the Enlightenment, as well as the idea of Romanticism to be critical of the prior.

The ideological lenses however also have a downside, the ideological trap that seeing the world through one color becomes so natural and comforting that all other colors appear to be unreal.

49.3   Utopia is Around the Corner... If Only...

The idea of Utopia is beautiful, a great thought experiment to envision how the world could be ideal. It helps us process issues we see in society, and subsequently come up with solutions to these issues. Such ideas also give unique insight into the way we think ourselves. A transhumanist might look towards technology or genetic sciences, whereas a socialist might look more towards how to solve issues through societal means and perhaps with help or support from government institutions. A perception of what Utopia would look like is different, probably, for every human, and obviously is based on a person's ideal view of the world. Such visions are very useful to illustrate what we find important, and to discuss issues, in essence they are a great tool.

People tend to forget that the idea of Utopia is to be used as a tool, not as a goal to be achieved. This is where the pitfall lies.

49.4   Everyone Would Think This Way

Utopia is beautiful, because it is perfect, all aspects work flawlessly, everything and everyone cooperates as envisioned in one's thoughts. If we want to apply a Utopia to the real world it would thus require everyone to follow suit, to get into lock-step, otherwise it fails. In essence utopia is built on a Platoesque notion of a pure idea, the theory of form or idea. This argument of Plato poses that a non-physical idea or form represents the most accurate reality possible. It might be fair to say that this is argument could be true in each individual's mind, for example an ideal man or woman is different for everyone, and represents the most accurate possible form of reality, but not necessarily in a more general sense of reality for everyone.

In the book 'Philosophers at the pulpit' (translated from the Dutch title: Filosofen op de kansel) Dutch pastor Rienk Lanooy compared people's tendency to desire utopia to the story of the tower of Babylon. The people thought that if they could manage this one time to work all together, in the name of God, to create something great and beautiful that all would become better, showing that humans could all live in peace and be happy. This however did require everyone's cooperation and participation. It required everyone to agree that coming together to build the tower was the right course of action. As per the story, this of course didn't happen, and this required regulations of what was allowed and what not. So in order to get everyone to cooperate, they needed to be compelled, or coerced if necessary.

This is, however, how many ideologies are framed as well, that there is one right way to look at the world. Over time they too evolve and take in ideas of people that follow the particular ideology, but the core remains the same. This makes sense as everything will naturally evolve over time. An issue with ideologies is that because the world is only viewed from one perspective, the evolution goes through a figurative tunnel-vision, and becomes deprived of any other viewpoints as the members of said movement find themselves within an ideological echo-chamber. As an ideology endures and its followers cannot bring the intended goals to fruition, rules and regulation are brought into play to ensure cooperation. Over time we may observe within most ideologies that an 'the end justifies the means' mentality within factions of the movement emerges.

There is another problem with ideologies, and that is that the ideologues forget an entire aspect of reality because they are so focused on the slice they are concerned with. Don't believe me?

Let's take the Transhumanist idea of indefinite lifespan. Is it realistic to think that everyone will find something meaningful in their lives, instead of, for example watch Netflix everyday for 14 hours, especially if jobs are mostly automated? Speaking of automation, if there are very few jobs, and almost everything is ordered online, how do people create social groups outside of the digital world? For the moment work and grocery shopping are among the most important places to have social interaction.

This isn't meant to flame transhumanism, but to show that we too have blind-spots. That I used transhumanism is because you, the reader, most likely by now have at least some idea of transhumanism.

Regardless of the problems that the effects of ideologies can bring, such ideological viewpoints, in moderation, are of great value in our quest to innovate, and further human, and societal progress.

49.5   We Have to Do Better

It is fairly easy to get a following with some ideals that people agree on, especially when we combine those with some good one-liners and seemingly obvious truths. It has been done over and over. We may observe that this happened, quite effectively, in Nazi-Germany, The Soviet Union, ISIS, all manner of cults, the European Union on its promise of unity, and the United States built on the belief that anyone can become successful and rich. All these groups require a belief in whatever ideal picture has been painted, or whichever enemy is presented.

We have to do better than this, and we can do better than this.

The Transhumanist community seems to pride itself on the use of reason, science, and evidence, often referring to the Enlightenment Era, a period that has been of great influence to the development of science as we know it. This is a good first step, although I personally think the Enlightenment Era is too much being romanticized.

Having the knowledge of ideological pitfalls, and echo-chambers, we should equip ourselves with a critical eye towards each other and ourselves, not for the purpose of scrutinizing or bullying one another, but to help each other keep our faculties and wits sharp, and our perspectives and viewpoints opened wide.

If we wish to transcend humanity, it is however crucially important to not become puritanical about who is deemed worthy, whether if technology will make everything better, or that everyone has to prove their 'openness' of viewpoints, their ability to criticize or reason. The so-named SJW, a broad generic moniker, suffer from this with ever increasing "tests" of tolerance for those deemed minorities. Libertarians face similar tests of purity in how much they wish to minimize government. Such tests become 'ends' of themselves, losing sight of what is trying to be accomplished and why.

Chapter 50
Imagination and Transhumansim

Sylvester Geldtmeijer

> *"Imagination is more important than knowledge. For knowledge is limited to all we now know and understand, while imagination embraces the entire world, and all there ever will be to know and understand." – Albert Einstein*
> *"If you can dream it, you can do it." – Walt Disney*

Imagination is powerful and sometimes overlooked. The fact that we are capable of imagination gives us great potential. As children we were all imaginative. Daydreaming is something we all did; we invented games and created certain social interactions, rules, and stories. Our fantasies and creativity brought us the ideas and inventions that some thought we couldn't attain, but we did. We imagined cities, medicine, food, cultures, and systems. Although we are all very human, we tend to use big scientific words to promote any form of techno-optimism or futurism – jargon that cannot always be understood and requires a lot of technical and professional background. But imagination does not require that; it simply requires the capacity of imagine. Imagination is when we think of something that is not yet real but could be, or think of something that we desire but that isn't there yet. That is imagination. Sometimes it's best to take a step back and pursue an approach which is less academic in our explanations but that stems more from our enthusiasm that arises from our imagination. Is it not true that all that is invented originates from our creativity? Therefore we must continue imagining even though it seems so child-like.

S. Geldtmeijer (✉)
Amsterdam, The Netherlands

© Springer Nature Switzerland AG 2019
N. Lee (ed.), *The Transhumanism Handbook*,
https://doi.org/10.1007/978-3-030-16920-6_50

50.1   Imagination is About Survival

In life we must imagine; without this we cannot survive. Imagination is not escaping reality; it's creating reality – a new kind of reality that improves our lives. Consider how plants can adapt in order to survive; some of them grow thorns, produce certain perfume, or develop other traits that are beneficial for survival. In a way it is creative, is it not? Because our evolution is gradual and slow, we as humans have the advantage of imagination. Just as the plants have certain advantageous traits, so our trait called imagination is beneficial for survival. We need it not only to survive in the immediate moment, but also to prolong our survival as long as possible. An important part of transhumanism is to extend our survival with the use of the scientific method. Historically we're survivors; we can see that when we look at our earliest ancestors. The wall art in the caves and the tools that were found created by early men – these are all due to our great imaginative thinking. With our imagination we can make tools, we can make language, we can make cultures, and we can make science. Imagination may seem primordial, but nevertheless it is our most useful tool. In fact we can make anything – yes, anything – as long as we imagine it.

50.2   Imagination is Not Static

Imagination isn't a static system, and therefore it is the best of systems. Every system eventually will crash if it does not adapt to the circumstances and conditions in which it is applied. Every constitution will fail that does not match the dangers that will come to the society that adopts it. But imagination isn't a static system; it changes constantly. Imagination is the creative thinking that is always able to adjust. It matches any circumstances and conditions because it performs in a flexible way. If ever there will be a future artificial intelligence (A.I.) that we will use to govern ourselves, it must require the same capability of imagination that we have. Imagination supersedes the static systems to which we cling. It even will supersede our ideas of capitalism and communism or other forms of governmental and economic structures altogether, because it will renew our current paradigms. With imagination we improve our own imaginations. For example: Our ancestors' houses were dirty, so they imagined broomsticks. This tool works, but yet it requires a lot of labor – labor that we could better spent elsewhere. So someone (Hubert Cecil Booth and David T. Kenney independently in 1901) came with a new invention – the vacuum cleaner. But it still requires some labor, and therefore our contemporaries invented robot vacuum cleaners. But this progression does not stop there, because with imagination we will 1 day invent houses that do not require any cleaning at all. With the help of nanotechnology, the dust and dirt simply would not be able to settle. Imagine hospitals that do not need cleaning. Or imagine streets that do not need cleaning, or roads and other transportation networks that just stay clean. Therefore we must continue to imagine, even if the subject of our imagination seems unreal at first.

50.3   Anything is Possible with Imagination

Yes, anything is possible with imagination – even things that seem far-fetched at first. Whenever someone says that an achievement is impossible, this is likely because their thinking has difficulties integrating beyond the present reality. In 1999 a public opinion survey was done in Amsterdam.

Frans Bromet interviewed various people in 1999 about the usefulness of a mobile phone.[1] At that time, there were multiple telephone booths in each village, and people still had no idea how many people would have a mobile phone 18 years later.

The people on the street were asked if they have a mobile phone, and if not, whether they want a mobile phone. The general answer respondents gave was that they did not need a mobile phone, or that landline phones would be enough for their needs. Now we can all say that the mobile phone has become a reality. I think that the dull thinking that prevails in human societies can only be cured with our enthusiasm that comes from our imagination. Furthermore, the means of communication should not stop there. The future of communication will be far surpassing our current modes of communication. With our current way of communication, there are still miscommunications and expressive misconceptions. In social media and video-telephony, there is a lot of miscommunication, and everything that is communicated is also subject to interpretation.

In communication other aspects than language play a role, such as the volume or the intonation. Also there are different standards about what is polite and rude. In Western culture, emotional openness and expressiveness are often valued. Yet in Japanese culture, it can be common to avoid facial expressions. Japanese people rather tend to focus on the eyes than the mouth when they express themselves emotionally or read emotions. In many Western countries it is normal to look into others' eyes. In Moroccan culture it may be disrespectful or offensive to look someone in the eye right away; if you are not being watched during a conversation, this does not automatically mean that they don't listen. Italians' conversations sometimes may appear very fierce because they often speak with raised voices. It can give some people the impression that there is a heated argument, while for the Italians it is just the common way of communication when they are excited.

In the future people would be able to experience one another's thoughts directly. We could communicate ideas instantly without the need for certain complex explanations. People will understand one another quickly, and therefore there will be much more profundity in the world. Thoughts will be upgraded, and intelligence will be expanded. Furthermore, in the medical world, brain disorders will be solved using the best scientific methods. Currently progress has already been made with the use of Deep Brain Stimulation – a form of neuromodulation that involves neurosurgery, with electrodes being inserted into specific brain sections. With this scientific method, some brain disorders are already cured or partially cured.

---

[1] Ruud Padt. "Mobile phone in Amsterdam 1999." Interview of Frans Bromet. Available at https://www.youtube.com/watch?v=aag1P4OwA3s

For instance Deep Brain Stimulation is being used for treating Parkinson's disease, other movement disorders, and obsessive-compulsive disorder. How does it work? Deep Brain Stimulation is a treatment method that places implanted electrodes in the brain that emit electrical signals. The electrodes are connected through lead wires to the neurostimulator. The neurostimulator contains batteries in a housing of titanium and is placed under the clavicle and discharges power pulses. The electrodes run through a hole in the skull to the place in the brain, which is stimulated. When the electrical signals are emitted to specific neurons the symptoms are being suppressed, which will reduce or eliminate complaints. The neurostimulator can be adjusted from the outside in order to combat the complaints optimally.

Ed Boyden, Associate Professor of Biotechnology and Brain and Cognitive Sciences at MIT, explained that traditional stimulation requires the skull to be opened to place the electrodes, which may cause some complications. At the moment, several researchers in collaboration are also working on a new method that may be less risky and non-invasive. They are working on a way to stimulate regions in the brain using electrodes placed on the scalp instead of inserted during a surgical operation. And because surgery cannot be applied to everyone, this new approach might be an improved alternative solution for the future practice of Deep Brain Stimulation.

In terms of curing all brain disorders, we are not there yet with current technology, but we certainly can imagine far more desirable outcomes, and that is our most useful tool. So let us not forget this simple method that we can always use in our thinking. Without imagination there will be no real progress, because it is our imagination that will bring us to the future.

Chapter 51
Transhumanism Is the Idea of Man Merging with Technology

Vineeta Sharma

51.1 What is Transhumanism?

Transhumanism is a term used for a comprehensive range of ideas that includes the technologies helping in enhancing the human capabilities. It refers to everything, from the cyborgs to brain implants, Artificial Intelligence, and many more. We already have a very close relationship with technology and its advancements, and our reliance on gadgets has already begun transforming us into transhumans.

Let us take a close look at this...

Every human today is connected either through a smartphone or a smartwatch, or even a desktop, but is that what transhumanism is? Is it the future of connectedness? Well, in a nutshell, it is, but it is also the technology of augmentation of the humankind.

So, transhumanism can be seen as an extension of humanism, from which it is partially derived. Here, what humans believe in is – humans matter, the individual's matter, that we might not be perfect but we can make things go better with the help of rational thinking, tolerance, freedom, concern for our fellow human beings, and of course the democracy. On the other hand, transhumanists agree with these, but they also emphasize on what we have the potential to become with the help of technological advancements.

However, the biological advances have been taking place from long, imagine if these could be accelerated beyond the incremental change predicted by Charles Darwin to a matter of individual experience. Furthermore, such things are only dreamt of by the so-called transhumanists. Transhumanism came into action to imply different things to different people, from a field of study to a technological fantasy, from a belief system to a cultural movement. One can't get a degree in transhumanism but can subscribe to it, research about its factors, invest in it, and act on its beliefs.

V. Sharma (✉)
tecHindustan, Mohali, Punjab, India

© Springer Nature Switzerland AG 2019
N. Lee (ed.), *The Transhumanism Handbook*,
https://doi.org/10.1007/978-3-030-16920-6_51

51.2   How Close Are We to the Sci-Fi future?

The today's technology follows an evolutionary past. The technologies that have already been created, provide a ground base for the new technologies to emerge. The emergence of Artificial Intelligence is a result of humans embracing technology in their daily life. This has given the data scientists a set of data, for the machines to learn from and become better. Similarly, the smartphones have empowered the rise of mobile computing, whereas the internet has allowed the growth of social networks.

This rapid development in the so-called NBIC technologies: Nanotechnology, Biotechnology, Information technology, and the Cognitive technology; is collectively giving rise to the possibilities that have been in a domain of science fiction. It's us, the humans, who imagine the order and observe the small equipment of steps along the way. The pace of the technology revolves around a point where biology becomes the limiting factor. For instance, do you memorize the phone numbers? But our cell phones do! However, augmenting our biological factors has already started with raising the capacity of our minds onto our devices. That's also the crux of transhumanism if we think about it. Enhancing the human capabilities is the crux of transhumanism, we already have the technology we need to make it more "adaptable."

51.3   How Far Have We Come Concerning Technology [1]?

The answer to this question somewhere depends on the technology you are using. For instance, the smartphones that we have today are 30,000 times powerful than the first IBM computer developed in 1981. That's a sizeable difference. Even the game consoles that kids use to play video games today have more power than the best desktop PC available just 10–15 years ago. Today the machines are sophisticated enough to think and make decisions without human intervention – this undoubtedly reveals Artificial Intelligence in regard to its relationship with transhumanism. Machines taking up the tedious tasks of our daily routine, such as driving, means that humans will have more time to innovate, to be creative and to think "out-of-the-box."

Not limited to these, the advances are somewhere at their rage. The state of the art payment technology, or genetically altering the life your children will live, the questions that we are facing today have greater and more terrifying implications than ever before.

We are moving very close to robot judges – that explains if justice is already blind then it doesn't need eyes anymore.

We are soon going to work on genetically modify our children – would it result as the humanity's last hope or as the world's worst idea?

Soon, human's would be paying for their coffee with their eyes – obviously at a risk, a new kind of security breach.

These examples explain the relationship of Artificial Intelligence in regard to transhumanism.

51.4 The Exponential Growth [2]

According to the Moore's Law, today's technological evolution is making the emerging technology smaller, faster, cheaper every day while making it 100 times smaller every year. It is expected that in the future, the supercomputers would be of the size of the thickness of an aluminum foil.

If we try connecting some of these futuristic dots, we can undoubtedly look at a possible roadmap towards an augmented future. Over the time, scientists have been researching on if or if not the webcam can interpret an individual's facial expressions like smile, neutral, negative, surprise, etc. In this practice, the machines are trained to understand the human emotions about – what a smile looks like, what a negative expression is. There have been a billion of cameras enabled in the smartphones and laptops of today's technology that already detect facial recognition. It is expected that over the next 4–5 years these numbers will double.

In the year 2017, we expected a number of self-driven cars to be launched. This would have done nothing but – must have provided with more autonomy to across many industries, and an aggressive expansion of computer and robot assistance.

The linear growth is – the way our brains work, to say, 1,2,3,4… whereas, in case of an exponential growth, it gets doubled with each increase, is 1, 2, 4, 8. However, in this example, the difference seems to be making hardly any difference as the linear growth would, after 30 increases, also reach 30 whereas the exponential growth increased would be an over a billion, yes, over 1,000,000,000. A number of people including Elon Musk (SpaceX, Tesla, Open AI), Nick Bostrom (author of Super Intelligence), and Bill Gates (Microsoft) – have shared their concerns regarding the threats that we might face with the evolution of such super intelligence.

51.5 How is AI Taking the Wearables to the Next Level [3]?

Regardless of the industry, field, and wearer, undoubtedly the AI enabled technology has the ability to enhance the capabilities of today's wearable devices. The wearables in the transhumanism sector could be the next big thing. However, the idea of wearables is nothing new. In ancient time, people used to wear tattoos, piercings, and pieces of jewelry, that indicates who we are and where we come from.

The market for wearable technology is increasing steadily. As per the studies, Around 115 million wearables were shipped in 2017, which is 10.3% more compared to 2016.

51.5.1   The AI Assistants in Wellness and Sports

The AI assistants are also performing exceptionally well in Wellness and Sports industry. Today, most of the smart wearable rely on the popular smart assistants, such as Alexa or Siri in Apple Watch. Many of us already wear fitness bands, and it is expected that – in the near future, our families and friends would be found wearing supercomputers embedded in their clothing. The wearables like the Apple Watch and Fitbit, made possible to overlook at a human body's health and fitness. These were followed by some other wearables like SugarBEAT – a patch that helps in monitoring people with diabetes. The wearables have opened many paths for tracking the data emitting from the human body and it really paid off! One such example is *Sensoria Fitness*. The award-winning producer and vendor of smart sports apparel, the company provides consumers with an AI in-app coaching to improve running routines using performance analytics.

All of these examples explain the ability of AI enabled technology to enhance the capabilities of today's human and helping in making transhumanism successful. Despite occasional disappointments, this market continues growing and surprising us, and AI appears to be the means to streamline this move.

51.5.2   The AI Tools to Improve Security

So far, helping people with leading a healthier life and achieve sports goals is one thing, whereas, saving people's lives that would be powered by Ai is an entirely new thing.

One such example is Lumenus. The company started when its founder almost died while riding a bicycle. This was the life-changing event of his life that pushed him towards utilizing the technology and developing something meaningful – to save lives. The product helps in designing the apparels equipped with wearable LED lights for runners, bicyclists, and motorcyclists. At present, we are using the IoT data points from a litany of different sources: sensors, third party APIs, GPS, and more to control the color, brightness, and animation of wearable LED lights. These work similar to the smart lights that we use at our home. But, the significant difference is without getting screwed into a socket in one place, Lumenus has made them mobile!

This is how AI helps in Lumenus in making wearables that save lives.

Lumenus software aggregates multiple information into a single source, here's where they are taking it to the next level of AI. As soon as a Lumenus user opens the app, puts in the final destination and maybe some extra notes, and puts the phone aside. The goal is to offer ZeroUI. As soon as the user starts the system there would be no interaction with the mobile app or even the hardware directly, moreover, the commands to would be autonomous.

As an augmentation in the wearables, we would soon be wearing chips or sensors that would be placed underneath our skin. This technology is simultaneously expanding while becoming small, cheaper, and more powerful. Around 50 million more such devices would be connected over the next decade. However, machine learning is at the core of such products. These things can only be accurate if the data set is large enough for your machine learning algorithms to be pointed at the data set and basically be learning from the data set that we collect.

51.6   The Internet of Things

It is the concept of computing that describes the idea of the connection of every physical thing with the internet and their ability to identify themselves to other devices. The Internet of Things (IoT) is more like a utility. Like the way in which we connect our devices to electricity – we simply plug and play. This invisible force of technology connects all of these fragmented systems that are slowly becoming the digital nervous system of our planet. In a sense, the IoT has already begun, for example, we have a number of sensors in our smartphones; say the touch sensors, GPS, and more, but this is just the beginning. The real power here steps in when these devices are enabled to talk to each other.

However, utilizing IoT the device doesn't become "smart" merely by virtue of being controllable via an iPhone app. Here is one of the most useful examples:

The self-driving vehicles, such as that of Tesla Motors

Cars are undoubtedly the "things" and insomuch as far as we are interested in things that leverage powerful Artificial Intelligence, of course, the artificial intelligence is ahead of that curve. This isn't necessarily because of the autonomous vehicles could be the easiest IoT innovation but because nearly all major car manufacturers are throwing billions of dollars at the problem. Apart from these exceptional self-driving cars, some of the Potential Future Uses for AI-Powered IoT Devices are in the fields of security and access devices, the emotional analysis, and facial recognition.

With the help of the introduction of IoT in the field of Artificial Intelligence, today, we are capable of inventing the devices that enhance the interaction with not only people but also with other Smart Objects. It not only refers to interaction with physical world objects but also to the interaction with virtual (computing environment) objects.

51.7   The Replaceable and AI

The artificial eyes, they make it possible to see – with the help of their exceptional zoom capabilities, infrared sensors, and the night vision – they make it all possible. This is unbelievable that these extensions are so perfect that they are expected to be more powerful and capable than our original ones. There have been implants in blind seeing and recognize faces; also, there are artificial arms that are controlled by the human body; this is all done by using a small implant in the brain. If we think that Inception and Deadpool 2 are just movies, then we should also know that MIT researchers have implanted and harnessed memories into mice and that they have isolated an individual memory into a mice's brain and just recalled that memory by forcing those neurons to fire. This is just a way of illuminating the lost memories by not harming the other memories. So this indicates that in the near future it is possible to implant or move memories. This would also be helpful in dealing with and healing the painful and dangerous traumas.

51.7.1 Transhumanism at Rage…

The computer chips that are on our laptops and smartphones are being implanted in a human brain by the doctors, this not only enables the blinds to see but also the deaf to hear. According to the studies, there are more than 20,000 people with neural implants. This very neatly describes that brain implants are done on an already augmented mind, this indicates the beginning of the "brain-net" has occurred.

51.8 Brain Net

A possible successor to the internet! This a form of virtual intelligence that allows humans to create music drive a car, communicate with the people and even surf the net at the speed of light. When a human looks into a movie in his tablet, this two-dimensional movie and its sound immerse the user's experiences, completing the feel and emotion in the way the director of the movie intended initially. Everything there is stored, every memory recorded, available on cloud storage taking a human brain to a digital vacation experience that never took place. This is just a matter of what happens in a human mind.

51.9 Imagination

During the stone age, about a million years ago, the stone acts were the most sophisticated items on the planet. But for 50,000 generations, almost every man used the same stone tools that could be kept in a category of very little innovation. One beautiful day, an ancestor of humans decided to put water into the ostrich egg and later buried the ostrich egg into the ground. While doing this, he indeed tapped into something that is remarkable to the human species – "Imagination."

Imagination could be considered as an insurance policy to the solutions of the problems that the human kind may face in the future. Moreover, our biological aspects haven't changed from that of our ancestors, and we still are the real hunters. Perhaps, we are now in chase of raising the human potential.

51.10 The Sci-Fi Combination

In a recent survey, it was explored that 72% of the US population is interested in brain implants, whereas 53% think why should we be augmenting ourselves for the worse? This hardly changes the next generation of today anywhere. Some people call them digital natives, but to the children of today; the smartphones are working already as an extension of their brain. Here, transhumanism can be considered as natural as an evolution.

51.10.1 The Positive Hopes with Transhumanism

What new things could be accomplished/achieved with transhumanism? We don't have any limitations of biology, what if we can get chances of explaining our feelings and emotions beyond the barriers of words? Maybe the humans can explore the stars, help in saving the planet, enhance the lives of people. There is a tonne to be hopeful for!

The future is OURS:) The world of possibility is at our doorstep so, push humanity forwards.

References

1. https://www.usatoday.com/story/tech/2015/06/13/ozy-technology-gone-too-far/71116096/
2. https://ahistoryofmystery.com/exponential-growth-artificial-intelligence/
3. https://www.iotforall.com/benefits-ai-in-wearables/

Vineeta Sharma She has earned her bachelor's degree in Electronics and Communication Engineering from BTKIT, Dwarahat, India. She is was a writer for The Transhumanist in Paris, France and is working for the Sweden based Digital Agency tecHindustan.

Chapter 52
Towards Sustainable Superabundance

David W. Wood

Beyond the fear and chaos of contemporary life, there is good news to share.

A new era is at hand: the era of sustainable superabundance. In this era, the positive potential of humanity can develop in truly profound ways.

The key to this new era is to take wise advantage of the remarkable capabilities of twenty-first century science and technology: robotics, biotech, neurotech, greentech, collabtech, artificial intelligence, and much more.

These technologies can provide all of us with the means to live better than well – to be healthier and fitter than ever before; nourished emotionally and spiritually as well as physically; and living at peace with ourselves, the environment, and our neighbours both near and far.

This is not a vision of today's society writ large – a mere abundance of today's goods, services, activities, relationships, and rewards. It's a vision of a superabundance, with new qualities rather than just new quantities.

This is not a vision of returning to some imagined prior historical period – to some supposed bygone golden age. It's a vision of advancing to a new society, featuring levels of human flourishing never before possible.

This is not a vision restricted to the few – to an elite percentage of today's humanity. It's a universal vision, for everyone, of a wide, diverse fellowship in which all can freely participate, and in which all can enjoy unprecedented benefits.

This is not a vision of the far-off future – something relevant, perhaps, to our great-grandchildren. It's a vision of change that could accelerate dramatically throughout the 2020s – a vision that is intensely relevant as the year 2020 comes into view.

This is not a vision of a fixed, rigid utopia. It's a vision of the collaborative creation of a sustainable, open-ended, evolving social framework. In this new

D. W. Wood (✉)
Delta Wisdom, London, UK
e-mail: davidw@deltawisdom.com

© Springer Nature Switzerland AG 2019   675
N. Lee (ed.), *The Transhumanism Handbook*,
https://doi.org/10.1007/978-3-030-16920-6_52

framework, every one of us will be empowered to make and follow our own choices without fear or favour.

In this vision, the sky will no longer be the limit. In this vision, the cosmos beckons, with its vast resources and endless possibilities. In this vision, our destiny lies in the ongoing exploration and development of both outer and inner space, as we keep reaching forwards together to higher levels of consciousness and to experiences with ever greater significance.

52.1 Critical Choices

But first, we face some hard, critical choices – choices that will determine our future. If we choose poorly, technology will do much more harm than good. If we choose poorly, a bleak future awaits us – wretched environmental decline, bitter social divisions, and a rapid descent into a dismal new dark age. Instead of the flourishing of the better angels of our human nature, it will be our inner demons that technology magnifies.

We need to steer firmly away from courses of action which would precipitate any such outcome. We need to resist simplistic ideas or beguiling coalitions that would mislead us onto a slippery slope of accelerating humanitarian degradation. Instead, we need to select and uphold the set of priorities that will facilitate the timely emergence of sustainable superabundance.

These are tasks of the utmost importance. These tasks will require the very best of human insight, human strength, and human cooperation.

If we choose well, constraints which have long overshadowed human existence can soon be lifted. Instead of physical decay and growing age-related infirmity, an abundance of health and longevity awaits us. Instead of collective dimwittedness and blinkered failures of reasoning, an abundance of intelligence and wisdom is within our reach. Instead of morbid depression and emotional alienation – instead of envy and egotism – we can achieve an abundance of mental and spiritual wellbeing. Instead of a society laden with deception, abuses of power, and divisive factionalism, we can embrace an abundance of democracy – a flourishing of transparency, access, mutual support, collective insight, and opportunity for all, with no one left behind.

If we choose well, the result will be liberty on unparalleled scale. The result will be people everywhere living up to their own best expectations and possibilities, and then more. The result will be a transformed, improved humanity, taking stellar leaps forward in evolution, as technology increasingly uplifts and augments biology. The result will be to advance beyond mere humanity to transhumanity.

52.2 Time for Action

As the 2020s approach, with their accelerating pace of change, with ever more potent technologies in wide circulation, and with a perplexing variety of tangled interconnections threatening unpredictable consequences, there's something that is very important for us to collectively keep in mind.

The thought that deserves sustained attention, in the midst of all other considerations, is the central insight of a group of people known as transhumanists. This insight concerns the magnitude of the forthcoming transformation from humanity to transhumanity.

It's not just that this transformation is possible. It's not just that this transformation could be relatively imminent – happening while many people alive today are still in the primes of their lives. It's that this transformation, handled wisely, could have an enormous positive upside. It's that this transformation, handled wisely, is deeply desirable.

This insight is grasped, today, by only a minuscule fraction of the earth's population – by a meagre sprinkling of transhumanist pioneers. However, it is time for these transhumanist pioneers to speak up. The few can become many.

As the 2020s approach, it is time for transhumanists to challenge and reorient the public narrative. It is time to raise the calibre of the collective conversation about the future.

It is time to affirm how the future can be hugely better than the present. It is time to clarify how human nature is but a starting point for a journey to extraordinary posthuman capability. It is time to emphasise that, whereas the evolution of life has been blind for billions of years, it is now passing into our conscious, thoughtful control. It is time to point out that, whereas the evolution of society has been dominated for centuries by economic matters and struggles over scarce resources, centre stage can soon feature the blossoming of abundance. And it is time to proclaim that powerful enablers for these exceptional changes are already arriving, here and now.

In short, it is time for transhumanists around the world to step up to the responsibility as catalysts of the forthcoming transformation. It is time for transhumanists to inspire and support people everywhere to join together in the historic project to build the era of sustainable superabundance. It is time to apply transhumanist wisdom to identify and uphold the best choices in anticipation of the tumult of disruption that lies ahead. It is time to ensure that technology brings universal benefit, rather than being something we will come to bitterly regret.

At present, we can glimpse only the broad outlines of the coming era of sustainable superabundance. It is the fundamental responsibility of transhumanists to discern the forthcoming contours with greater clarity, and to help humanity as a whole envision and navigate the pathways ahead.

Together, let's apply our skills, our time, and our resources to paint more fully the picture of sustainable superabundance. Let's transcend our present-day preoccupations, our unnecessary divisions, our individual agendas, and our inherited human limitations. Let's grasp the radical transformational power of new technology to profoundly enhance our vision, our wisdom, our social structures, and our effectiveness.

Together, let's map out constructive solutions to the obstructions and distractions that impede progress – solutions combining the best of technology and the best of humanity. Let's build productive alliances that weaken the forces resisting positive change. Let's prepare to take advantage of the growing momentum of an inspirational worldwide technoprogressive movement. Let's anticipate how we will dislodge the grasp on power held by today's backward-looking vested interests. And through an emerging shared understanding of the vital benefits transhumanist policies can bring to everyone, let's transform fearful opposition step-by-step into willing partners. The few can become many.

In this way, we can accelerate the transition to sustainable superabundance. The sooner, the better.

52.3   Smoothing the Bumps

There will be many bumps on the road to superabundance. Indeed, there will be sceptics and detractors in all walks of life who oppose even the idea of working towards sustainable superabundance.

In times of rapid change, it's no surprise that many people will become fearful and obstructive. Afraid of losing their status in society, they will cling onto outdated habits and structures. Afraid, with some justification, of technology going wrong, they will call for the imposition of overly cumbersome restrictions. Afraid that cherished human values may become lost, they will aggressively reassert inadequate bygone belief systems – belief systems grounded in incorrect or incomplete views of human nature and human flourishing. Lacking a vision of positive change from which they can benefit, they will deliberately sow confusion and misinformation.

To overcome confusion and misinformation, it's time to generate an abundance of understanding. To supplant cumbersome legal frameworks, it's time to champion smart, agile regulations. To quell panics about tech-driven dystopia, it's time to promote appreciation of scenarios in which technology uplifts humanity. To lessen the power of vested interests, it's time to build wise alliances. To tame the widespread fear of change, it's time to clarify the roadmap to sustainable superabundance – and to describe how, in that not-so-distant future world, *everyone* can attain greater security, greater opportunity, greater health, and greater wellbeing.

In short, to rise above the myriad distractions and obstacles that might frustrate the journey to superabundance, it's time to uphold a compelling, engaging picture of the remarkable future that is within our grasp.

Where that picture has gaps, let's address them, quickly and fully. Where questions arise, let's move fast to improve our collective understanding in the light of our collective intelligence. Where obstructionists and naysayers attempt to muddy the water, let's be fair yet firm in taking the conversation to a higher level.

Given the magnitude of the progress which can ensue, no tasks are more important.

David W. Wood is Chair of London Futurists, Principal of Delta Wisdom, Secretary of Humanity+, co-leader of Transhumanist Party UK, and author of *Sustainable Superabundance: A universal transhumanist invitation.*

Chapter 53
Utilizing Transhumanism for the United Nations Global Goals

Barış Bayram

In this essay, I'll argue that there is a global need for Transhumanism as a social movement to address our highly complex real-world challenges and hence to advance our "humanist" narratives as an inclusion necessity.

As part of such a direction, Transhumanist approaches are able to contribute to further realizations of humanist issues mainly in two ways:

1. Advanced/emerging technologies can dramatically improve the human condition especially if they primarily serve the poor (and the disadvantaged ones in any sense, too).
2. The uniting perspectives of Transhumanism can deliberately promote quality global cooperation that leaves no one behind and prevent problematic/violent conflicts by demoting traditional regressive approaches/norms in relation to the question what "human" means, which have harmful effects that unethically seperate us.

First of all, toward these purposes, we need to understand what Transhumanism is.

To the search result dictionary by Google, Transhumanism means the belief or theory that the human race can evolve beyond its current physical and mental limitations, especially by means of science and technology.

When it is conceived as a social and philosophical movement, according to Sean A. Hays, Transhumanism [1] is the one devoted to promoting the research and development of robust human-enhancement technologies. Such technologies would augment or increase human sensory reception, emotive ability, or cognitive capacity as well as radically improve human health and extend human life spans. Such modifications resulting from the addition of biological or physical technologies would be more or less permanent and integrated into the human body.

B. Bayram (✉)
Yeditepe University, Istanbul, Turkey

© Springer Nature Switzerland AG 2019
N. Lee (ed.), *The Transhumanism Handbook*,
https://doi.org/10.1007/978-3-030-16920-6_53

At first glance, it seems that it is solely about scientific and technological progress for one's own narrow self-interests ignoring any outcome outside a Transhumanist individual. But in fact, Transhumanism fundamentally implies social progress too because those individual goals always require more and more social progress as the number one priority. This precondition for satisfying and sustainable achievements of such an individualism, which has been forgotten mostly for years, is at the core of, and in the origin of, Transhumanism: The term "Transhumanism" [1] was popularized by English biologist and philosopher Julian Huxley in his 1957 essay of the same name. Huxley referred principally to improving the human condition through social and cultural change.

Also, Nick Bostrom's work on Transhumanism [2] has useful implications for that social and cultural change. First, it has a very significant emphasis on the need to criticize the widespread assumption that the "human condition" is at root a constant. Transhumanists view technological progress as a joint human effort to invent new tools that we can use to reshape the human condition and overcome our biological limitations, making it possible for those who so want to become "posthumans". Whether the tools are "natural" or "unnatural" is entirely irrelevant. Consequently, based on Bostrom's Transhumanist approach, we can work to fix any harmful mechanism of problematic cultural constructs and social norms by challenging the falsehoods of the status quo claiming its current systems, power-relations and informational architectures as a constant in relation to the understanding of what is human. Likewise, to Bostrom, an empowering mind-set that is common among transhumanists is dynamic optimism: the attitude that desirable results can in general be accomplished, but only through hard effort and smart choices. To me, Bostrom's rational optimism can be utilized against the cynical pessimism of the status quo and so against its indifference to unethicalities and unjust inequalities.

To clarify the positive material dimension of such an optimism, I would prefer to cite David Trippet's words. He briefs the what of Transhumanism [3] by highlighting that its advocates believe in fundamentally enhancing the human condition through applied reason and a corporeal embrace of new technologies. In his view, it is rooted in the belief that humans can and will be enhanced by the genetic engineering and information technology of today, as well as anticipated advances, such as bioengineering, artificial intelligence, and molecular nanotechnology. To him, the result is an iteration of Homo sapiens enhanced or augmented, but still fundamentally human.

As a result, what I conclude is that Transhumanism's material and technological dimension, innovative and dynamic optimism, and challenging, nonconformist, inclusive, social and even intersectional ethos coherently interacts and empowers each other when its conception is not impaired.

Then, it is up to us to utilize Transhumanism for all human beings globally and to leave no one behind. For this, I want to relate it to the United Nations Global Goals: The Sustainable Development Agenda [4]: On 1 January 2016, the 17 Sustainable Development Goals (SDGs) of the 2030 Agenda for Sustainable Development—adopted by world leaders in September 2015 at an historic UN Summit—officially came into force. Over the next 15 years, with these new Goals that universally apply to all, countries will mobilize efforts to end all forms of poverty, fight inequalities and tackle climate change, while ensuring that no one is left behind.

Toward advancing such a relation, the EPE-theory [5] claims that the most rational and effective way to improve one's own short- and long-term well-being-, intelligence- and life-span is extremely sensitive to the improvements of all other human beings's living conditions and hence of any relevant environments and systems because of interconnectedness of any progress, interdependence of long-term interests, interrelatedness of very diverse issues and unknown nonlinear effects, and hence primarily the complexity of non-zero-sum possibilities of our global interaction ways and developments. So, we need further deliberations on our self-interests and re-identification of and also innovations of its how both as technological progress, and as reframing our ethical reasoning models and hence policy frameworks, mind-sets, world views, social norms, cultural constructs, interdisciplinary ethical theories and the like. This argument [6] invites us to utilize especially science-based transformative ways toward further ethicalities like Transhumanism.

Similarly, to my science-based patient-oriented meta-ethical theory [7], such global goals and frameworks for ensuring the well-being of every human being and hence for the improvements of "human condition", thus the deliberate adoption and further realizations and progressing innovations of Global Ethics and Human Rights [8] (and even the UN Global Goals) have a scientific basis. And, if conceived as such, Transhumanism too.

References

1. https://www.britannica.com/topic/transhumanism
2. https://nickbostrom.com/old/transhumanism.html
3. https://www.weforum.org/agenda/2018/04/transhumanism-advances-in-technology-could-already-put-evolution-into-hyperdrive-but-should-they
4. https://www.un.org/sustainabledevelopment/development-agenda/
5. https://www.academia.edu/30600969/The_Theory_of_Cognitive-Ethical-Development_Can_Solve_Any_Real-_World_Problem
6. https://impakter.com/impakter-essay-construct-empirically-verifiable-ethics-fix-status-quo/
7. https://medium.com/@BarisBayram2045/moral-realism-is-false-56d765faeb60
8. http://www.un.org/en/udhrbook/pdf/udhr_booklet_en_web.pdf

Chapter 54
Geoengineering: Approaching Climate Change as a Present-Day, Preventable Issue

Daniel Yeluashvili

It's easy to think of climate change as a contemporary apocalyptic prophecy or a potential future problem, but it's not. Climate change is happening now, affects people around us on a daily basis, and can be stopped with a sufficient amount of combined effort, collaboration, and sound investments. One of those investments is in geoengineering, an umbrella term for a plethora of technologies that vary in scope, form, and application but all share the trait of removing greenhouse gases from the atmosphere. Proposed methods of this, some being more viable than others, include artificial cloud-brightening to increase the reflectivity of the clouds in the Earth's atmosphere, capture and sequestration of carbon dioxide from the atmosphere into the ground, and mass fertilization of the ocean with iron deposits in order to trigger algal blooms that would adversely lower the amount of carbon dioxide in the atmosphere. Various sources, from Yale to Oxford to MIT, have already endorsed this concept and are working to refine it and put it into practice, as no version of our planet would be more secure from meteorological threats otherwise beyond our control than one with contemporary advances in geoengineering technology. Geoengineering is the best—and only available—solution to climate change, as it is the one remaining way to lower Earth's CO_2 emissions to below 400 ppm, may be less partisan than previous solutions, and is sufficiently profitable and time-sensitive to counteract the resistance that has been felt against the idea of taking back our Earth thus far. Here's why.

The density of the greenhouse gases in the atmosphere is measured in ppm, or parts per million. These units refer to one particle of a given greenhouse gas, say, carbon dioxide (p) per million particles of oxygen (pm). Once the density of the atmosphere's greenhouse gases passes the threshold of 400 ppm, it becomes impossible to counteract the ensuing effects by simply reducing carbon emission output [6]. More importantly, this has already happened and any form of climate change

D. Yeluashvili (✉)
San Francisco State University, San Francisco, CA, USA

© Springer Nature Switzerland AG 2019
N. Lee (ed.), *The Transhumanism Handbook*,
https://doi.org/10.1007/978-3-030-16920-6_54

prevention by decreasing carbon emission output is now obsolete [10]. A primary cause of the continuation of this phenomenon is the Tragedy of the Commons, a sociological problem that can be summarized simply as the idea that no single individual can alter a given problem—in this case, climate change—so virtually nobody continuously bothers to try [11]. In order to approach such a problem, one would require the investment of a small, centralized group such as America's private sector. Such enthusiasm has already been expressed by various academic scholars across multiple universities, who believe that carbon sequestration is the optimal solution to circumvent the Tragedy of the Commons. Carbon sequestration, the practice of using machinery to siphon, capture, and store carbon dioxide from the atmosphere, would require multiple carbon capture and sequestration plants rather than a planet's worth of human effort and self-control [12]. This may seem like an easy fix for a complex problem, but some people don't recognize that there is a problem at all: climate change skeptics and deniers.

Climate change skepticism, also referred to as climate change denial, is the idea that existing scientific evidence for man-made climate change on Earth is insufficient to take action to aid the planet by reducing carbon emissions. This mindset has never been focused solely on global warming. According to the study *Climate Change Denial Books and Conservative Think Tanks: Exploring the Connection*, "Both industry and the conservative movement learned during the Reagan administration that frontal attacks on environmental regulations could create a backlash among the public…they gradually shifted to another strategy, promoting 'environmental skepticism'…these efforts focused on specific problems such as secondhand smoke, acid rain, and ozone depletion, but in the case of [man-made climate change] they have ballooned into a full-scale assault on the multifaceted field of climate science, the IPCC, scientific organizations endorsing [man-made climate change], and even individual scientists" [5]. A primary motivator for conservative opposition to climate change prevention in the United States is, as with many conservative political stances in the United States, religion. Specifically, the Judeo-Christian perspective that the environment was created for human beings to use and exploit as they see fit has created an anthropocentric value system that has motivated European and, later, American society to pillage and colonize the environment for its own ends since the time of the Enlightenment [4]. Rather than combat religion as an institution, or even climate change skepticism as an idea, it would be most prudent to offer a bipartisan alternative that draws support from both the Democratic and Republican parties, and weather modification technologies have already attracted bipartisan attention in the past. A bill to authorize weather modification research and development was authorized by Republican senator Kay Hutchinson in 2005, but it failed to pass [7]. A similar bill, which also died in Congress, was introduced by Democratic member of the House of Representatives Mark Udall in June of 2005 [13]. If climate change prevention was as partisan then as it is now but weather modification was not, there may be reason to consider geoengineering a bipartisan political idea. More important, however, is the fact that no existing climate change solution can reduce humanity's effect on the atmosphere quickly enough to save imperiled lives.

Climate change is not only a current issue, it is explicitly dangerous for human beings. Far from exclusively threatening tree frogs or polar bears, climate change has been projected to cause over 500,000 extra human deaths worldwide by 2050 due to malnutrition from impeded crop growth [8]. Worse yet, all existing climate change prevention plans are insufficiently time-sensitive to address this issue and save lives. Even if the Obama administration's Clean Power Plan were implemented nationwide, 1.7 billion tons of carbon emissions would still be released into the atmosphere in 2030 in America alone [1]. Cap-and-trade, an economic policy intended to, according to the Wall Street Journal, "cap emissions at 334 million metric tons by 2020," still only does exactly that—cap emissions without fully decreasing them, let alone negating them [9]. The carbon tax has been lambasted as an idea that "does not reflect political reality" by the Institute for Energy Research due to its allegedly short-term application to 85% of energy production in the United States with negligible historical precedent that politicians will want to keep it in the short term and not simply extend it indefinitely [3]. Only geoengineering or, more precisely, carbon sequestration technologies will be able to prevent climate change before the first half of this century elapses, but it wouldn't be cheap. An article from Scientific American estimates "it will cost $17 billion for [carbon capture] to be available by 2025" [2]. Nevertheless, it is more feasible than an eternity of taxation or capping and trading if America wishes to take its future into its own hands.

Although questions remain unanswered about geoengineering, it is the most expedient solution to an otherwise permanent, global problem. It is the only way to step backwards over the otherwise permanently crossed threshold of 400 ppm of carbon dioxide in the atmosphere, may be considered bipartisan as a weather modification tool rather than a climate change prevention method, and is the most time-sensitive way to prevent climate change. The next step will be investment from the private sector, which would be in its best interests. After all, if humankind really wishes to have dominion over the Earth, saving it would be a good place to start.

Works Cited

1. Biello, David. "How Far Does Obama's Clean Power Plan Go in Slowing Climate Change?" *Scientific American*. N.p., 6 Aug. 2015. Web. 30 July 2017.
2. Biello, David. "How Fast Can Carbon Capture and Storage Fix Climate Change?" *Scientific American*. N.p., 10 Apr. 2009. Web. 30 July 2017.
3. Carbon Taxes: Reducing Economic Growth-Achieving No Environmental Improvement. *IER*. Institute for Energy Research, 28 Feb. 2014. Web. 30 July 2017.
4. Dunlap, Riley E., and Aaron M. McCright. "14 Climate change denial: sources, actors and strategies." *Routledge handbook of climate change and society* (2010): 240.
5. Dunlap, Riley E., and Peter J. Jacques. "Climate change denial books and conservative think tanks: exploring the connection." *American Behavioral Scientist* 57.6 (2013): 699-731.
6. Emerson, Sarah. "Goodbye World: We've Passed the Carbon Tipping Point For Good." *Motherboard*. VICE, 28 Sept. 2016. Web. 30 July 2017.
7. Hutchinson, Kay. "Weather Modification Research and Development Policy Authorization Act of 2005 (2005 - S. 517)." *GovTrack.us*. N.p., 3 Mar. 2005. Web. 30 July 2017.

8. The Lancet. "Impact of climate change on food production could cause over 500000 extra deaths in 2050: By 2050, reduced fruit and vegetable intake could cause twice as many deaths as undernutrition." ScienceDaily. ScienceDaily, 2 March 2016. <www.sciencedaily.com/releases/2016/03/160302204506.htm>.
9. Lazo, Alejandro. "How Cap-and-Trade Is Working in California." *The Wall Street Journal*. Dow Jones & Company, 28 Sept. 2014. Web. 30 July 2017.
10. Patterson, Brittany. "SCIENCE: Earth's CO2 Levels Have Crossed the 400 ppm Threshold for Good." *SCIENCE: Earth's CO2 Levels Have Crossed the 400 Ppm Threshold for Good*. E&E News, 29 Sept. 2016. Web. 30 July 2017.
11. Pollitt, Katha. "Climate Change Is the Tragedy of the Global Commons." *The Nation*. N.p., 29 June 2015. Web. 30 July 2017.
12. Talbot, David. "First Carbon Capture Projects Mask a Lack of Progress." *MIT Technology Review*. MIT Technology Review, 13 Oct. 2014. Web. 30 July 2017.
13. Udall, Mark. "Weather Modification Research and Technology Transfer Authorization Act of 2005 (2005 - H.R. 2995)." *GovTrack.us*. N.p., 20 June 2005. Web. 30 July 2017

Chapter 55
IoT and Transhumanism

Mark Crowther

55.1 Introduction

A truly Transhumanist future has never been closer. That isn't to say it will arrive in the lifetime of this paper's author or those reading it. Indeed, many who follow and cheerlead around topics such as Transumanism and its closely related kin, Posthumanism, can often be deemed to have an almost Panglossian view on these futures being realised within overly optimistic timescales. However, never before have we had so much converging technology and insight, driving us towards the realisation of Transhumanism, the subject of this paper.

As we'll see, the problem with achieving Transhumanism is not primarily one of available technology and know-how in its application, albeit there needs to be some advancement and refinement of both things. The problem is mainly one of convergence, the bringing together of these technologies into a connected, interacting whole. As many of the technologies we discuss are developed within specific disciplines, there needs to be a bridge built between them, a realisation that these technologies can do more outside of their current disciplinary silos.

The main focus of this paper, however, is the class of technologies referred to as the IoT and the argument will be made that it is IoT that is the glue that will bind a range of technologies together and bring us more fully into the Transhuman era. A new term is introduced, that of the Endoself, mirroring the concept of the Exoself. Where the Exoself is the cloud of technology external to the person, the Endoself is the technology within the person. As we'll see, the Endoself is not just made up of IoT, but also of technologies such as bio-printed organs and neural meshes.

Throughout the paper we'll stay away from broad and fanciful speculation, instead discussing only those technologies that exist at the current time and stretching speculation only to what additional uses they may be put to. The paper will also

M. Crowther (✉)
Technologist and Writer at BJSS, London, UK

© Springer Nature Switzerland AG 2019
N. Lee (ed.), *The Transhumanism Handbook*,
https://doi.org/10.1007/978-3-030-16920-6_55

steer away from discussion of topics such as ethics, morality, psychology, etc., and focus on the technology instead.

It's hoped that the reader will come away with an appreciation of how close to Transhumanism we are and the realisation we have what's needed already available to a greater or lesser degree.

55.2   Defining Transhumanism

There are many definitions and perspectives on what Transhumanism is. Some perspectives focus on the development and improvement of the human condition through mainly biological means, with a focus perhaps on eliminating aging or the deleterious effects of illness and disease. These could provide radical life extension, greater physical and emotional well-being and the opportunity to experience far more than we can in one lifetime today. These are not unworthy goals and the efforts of organisations, such as the SENS Research Foundation [1] or companies taking a more radical approach, such as BioViva, who are developing and now using essentially experimental gene therapies [2], are to be supported. However, it can be argued that fixing biology is not Transhumanism in a true sense, that this is instead simply maintaining the biological state or returning it to a desired state.

To achieve a definition of Transhumanism, our actions and interventions must move us through a transitional state, where what we accept now as defining what it means to be human, in terms of our biology and abilities, is augmented and improved, along with any perceived weaknesses being overcome. We may still be mostly biological and using regenerative therapies, but we will have fundamentally changed. In this paper, we look at the use of current and future electronic and computing technology to achieve this transitional state, as a pathway to Transhumanism and eventually Posthumanism.

From this technology-oriented perspective, a Transhumanist future is defined as one that moves humans towards being integrated with technology, while remaining essentially biological. We may currently have no precedent for this, but it's the position of this paper's author that the eventual outcome of advancing technology is that humans, and indeed any suitably advanced lifeforms in the universe, will become post-biological. The first step towards this is a transitional state brought about by integrating with and not simply using technology, thereby achieving Transhumanism.

In accepting this, we must then ask; what is the technology which can help realise a Transhuman future and where are we in terms of developing and utilising such technology? The answer to the question is nuanced. However, we have today a set of technologies in place that may be the basis for realising Transhumanism and act as a stepping stone to Posthumanism, albeit with several challenges to be overcome.

55.3   Pathways to Transhumanism

In exploring how a Transhuman future may unfold, we can anticipate three broadly defined pathways for individuals in society facing the decision of whether to use Transhumanist technology:

1. Bio-conservatives – There will be those who wish to remain 'human' and refuse to take any measures that lead them towards what they interpret as Transhumanism [3].
2. Mid-ways – Some will find total bio-conservatism and outright Transhumanism as extreme positions and seek a middle-ground regarding the use of Transhumanist technology.
3. Transhumanists – Those who wish to maximise the use of Transhumanist technology and potentially keep advancing towards Posthumanism.

As outlined above, even for those who seek to engage in the use of Transhumanist technology, not all will seek the same goal. While some will see the Transhumanist technology as merely a step towards a fully Posthuman future, others will not wish to experience anything more than a middle ground that provides improvements in aspects of their lives such as health and lifestyle, in a similar way to how people use medical and computing technology today.

55.3.1   Integrating, Not Just Engaging

Whether individuals take a Mid-way or Transhumanist pathway, those who take advantage of the technologies related to Transhumanism will be doing more than just *engaging* with technology in the way that we do now, they will instead be *integrating* with that technology. This is the paradigm shift that needs to take place if a Transhuman future is to be realised; technology and biology must become an inseparable whole.

To be clear, it can be argued correctly that we already engage with technology in a way that provides a range of enhancements to our lifestyle, health and so on. Examples include technology implanted in the form of artificial hips and knees, Pacemakers and Cardioverter Defibrillators, cochlea implants and artificial eye lenses, microchip-based contraceptives, even the so-called bio-hackers who experiment with under-skin NFC chips [4]. These types of technologies are within us, often replace their biological equivalents, restore degraded capabilities and in some cases literally keep us alive. Alternatively, this technology may be in the form of an external cloud of technology that surrounds us, which includes devices such as mobile phones or wearable devices such as fitness trackers and Bluetooth headphones.

In both cases however, this is not driving us purposefully towards Transhumanism or Posthumanism. What these technologies are primarily doing is maintaining or mimicking some function of our biological selves or providing external input for us to respond to when we choose, in either case we are engaged with, actively or passively, and are not integrated with the technology in a seamless way.

Despite this, it can be conceded that these are in fact augmenting and enhancing us. Indeed, for those wishing to see a Transhuman future occur, confidence can be gained from the fact that though these technologies may not be directly bringing about Transhumanism, they are preparing society psychologically and culturally for it to happen. What's more, the very first steps along a pathway to Transhumanism have truly already begun, and a wide array of technology exists to help bring it about. One of the most exciting technologies is referred to as the Internet of Things (IoT). This is supported by a rapidly evolving set of technologies that will accelerate us down the pathway to Transhumanism at an ever-increasing rate.

With our definition of Transhumanism clarified and the likely pathways people may follow outlined, lets now look at the current state of technology, what may unfold in the near future and the critical role IoT will play.

55.4 Overview of Current IoT and the Individual

The current Internet of Things is a collection of billions of physical devices, all around the world, that are connected to the Internet. At the time of writing, Gartner estimates there are over 11.1 billion IoT devices in use, set to rise to 20.4 billion by 2020. Of the devices in use today, 7.3 billion are being used by regular consumers, with the remainder in use within business. Total global spend on IoT devices and related infrastructure in 2018 was calculated at $772.5bn USD, with expectations this will hit $1tn USD in 2020 [5].

By taking advantage of cheap processors, low cost memory chips, miniaturised electronics, wireless sensors, broad availability of internet connections and a host of supporting technology, almost any type of physical device can become an IoT device. Previously dumb devices are now made smart by the use of IoT technology; whether they are planes and cars, washing machines and microwaves, lightbulbs and switches, watches and jewellery, fitness bands and footballs, even medical devices and pills now have IoT examples.

Related to our interest in our biology – health and medicine are already being impacted by ground-breaking and disruptive IoT technology. For example, in 2017 a drug used to treat schizophrenia and bipolar disorder was combined with a miniature silicon chip that actives in stomach acid. The chip communicates with a wearable sensor and sends data, such as whether the medication was actually taken, to the patient's health management professional [6]. A device exists that allows continual blood glucose level monitoring without the need for daily finger-pricking, the data from which can then be used to inform the user about needed changes to diet and exercise or alert them to situations where medical intervention may be needed [7].

There are IoT enabled pulse oximeters which measure arterial blood oxygen levels. Other IoT enabled devices can manage sleep apnoea, track duration, type and quality of sleep, along with resting heart and breathing rates. An implantable device the size of a grain of rice exists that can continually monitor body temperature, it can also hold data that can be scanned so as to open doors, sign-in to computer systems or be used as identification during medical emergencies, such as the scannable Medico Tag [8]. These are in addition to the commonplace devices such as smart watches and mobile phones with various step counting, exercise tracking, activity monitoring apps.

As further advances are made and more opportunities to use IoT and supporting technologies emerge within or related to the fields of health and medicine, more possibilities to augment and transform ourselves and our capabilities will emerge. The above technologies are just a small set of real-world, non-speculative examples, which may not, on initial review, appear to help us start the journey to a Transhuman future, but which in truth act as the first step towards it.

55.4.1 Future Advances and Convergence

In looking at the current state of technology, we need to bear in mind two things. Firstly, that these real-world examples are extremely new. While the current take-up of IoT is extensive and global, much more discovery, innovation and use is ahead of us. We could explore speculative examples here, but with so many real-world examples in place there is no need. Secondly, that the use of IoT at the heart of transformative technologies demonstrates the synergistic potential of converged technologies. Yet, without IoT they are isolated dumb devices, that alone cannot create the transformation that will lead to Transhumans and then to Posthumans. As we mentioned before, simply using technology is not enough, we must seamlessly integrated with it.

Now we're clearer on what IoT is and how it is currently being used, let's discuss further why bringing dumb devices to life with IoT makes IoT the key we've been looking for.

55.5 Why IoT Is the Killer Technology

One of the greatest technological advances that has taken place in our lifetimes is the invention of the Internet. For most people, this means the World Wide Web that sits on top of it and gives us access to the vast range of online content and services. Yet the Internet is so much more than just entertaining websites or social media and often the ubiquity and role of the internet is overlooked. The Internet makes not only the delivery of the Web possible, but it is also the means for the connectivity of billions of devices into a near invisible global network. Without it, the modern world would struggle to operate in the way we have become accustomed to. More than that, the internet provides the means for high-speed, high-bandwidth connectivity between local, regional and global systems; systems that will form the extended layers which will sit on top of the personal layer of Transhuman technology.

It's this personal layer – that in its current form doesn't have the level of sophistication or complexity – that will help deliver the enhancements to human capabilities which will bring about Transhumanism. The personal layer, as we have it today, sits as a cloud of technology around us. It forms a technological Exoself consisting of devices such as our mobile phones, fitness devices, smart watches and may even be considered to include more established, albeit dumber technology, such as hearing aids and door access cards. This Exoself of technology gives us enhanced abilities that nature cannot provide, it's a technological means of adapting to the modern environment, instead of a biological one. It would not be entirely inaccurate to say that since the dawn of humankind, when we started creating the crudest of technology, we started creating the first Exoself.

What is fundamentally different, is that at the heart of the modern, electronic and computing based Exoself is the ability to draw data from or provide it to the local, regional and global systems we and others interact with, by which we can modify our behaviour and understanding as a consequence of such interactions. This data and understanding can then be shared near instantaneously up to the global scale. For the first time ever, an individual on one side of the world can affect the behaviour and life of another person on the other side of the world in near real time.

It can be seen that the combination of the Internet and the modern Exoself are powerful forces that have laid down the foundation for the next step; the creation of an electronic and computing based Endoself, where the lines between technology and biology blur – the inevitable next step and evolution of the limited Exoself. This Endoself is what's lacking in order to achieve technology based, non-biologically oriented, Transhuman capabilities. The Endoself is the missing, more intimate layer of connectivity to the individual and the technologies they may be integrated with; a more transformative, personal layer. There is only one class of technology that we possess today that can deliver Endoself and that is IoT.

It would be reasonable to ask why IoT is the key to the Transhuman realising Endoself and not just more synthetic organs, implants or bioprinting to make replacement parts. Replacement and substitution have their place in a longevity strategy, but as a strategy to achieve Transhumanism, it still fails in the same way as any strategy based on mimicry. There are minor to no radical enhancements to human capabilities. Mimicry without enhancement of capabilities is just turning people into technology laden cyborgs at best, when the better alternative would be biological. However, introduce IoT to synthetic implants, replacement parts or even bio-printed substitutes and the person immediately gains a host of smart capabilities that set the user on the journey to Transhumanism.

55.5.1   IoT Enabled Everything

The realisation is that IoT enabled components of the Endoself will not just mimic the function of whatever they replace; but will be connected and smart. Let's look at an example of a replacement organ, such as a kidney.

According to various source there are at least 200,000 people in the US and EU alone who are on waiting lists for kidney transplants. Many of these people will die before they receive a donor organ and around 25% of patients die after transplant due to an array of causes. These causes include deleterious effects from the transplant process such as infection or co-morbid medical illnesses such as cardiovascular disease. Imagine then we take a radical, technology led approach, to address the limitations of a purely biology-based organ replacement regime and to also address complications that occur afterwards. If we converge the technology we have right now, the situation could experience an almost revolutionary change.

To start, some of your stem cells would be harvested in preparation for undergoing transformation into the cells needed to construct your kidney. A combination of these cells, growth factors and biomaterials would be combined via 3D bioprinting to construct the organ [9]. While bioprinting is taking place, a suite of sensors, thermoelectric self-charging micro-batteries and other select components, to assist with urinary and endocrine system management, would be printed within the organ along with some flexible biocompatible intracellular circuits to connect everything up. In this way, the organ to be transplanted into you is now a connected, self-powered, bio-mechanical smart organ, built from your own stem cells but laden with technology to improve and add to the functionality of the kidney and augment you at the same time. All of the technology to achieve this is in use at the time of writing, it just hasn't experienced convergence yet.

After transplant, the sensors would start to monitor a set of post-operative conditions and then switch to their primary task of monitoring expected functions of the kidney. Blood pressure would be checked, fluid levels and the creation of related hormone levels measured by one of the dedicated components which could also trigger the changes in these levels to keep both optimal; beyond what the kidney itself could do. A similar process could be in place to monitor and modify levels of erythropoietin, used to trigger red blood cell creation by the bone marrow, along with checking blood pH, ionic composition of the blood, urinalysis and levels of vitamin D that the kidney helps in the production of. Every time analysis was performed, the sensor data would be broadcast out of the bio-mechanical organ and fed into an AI based data analytics system. Analytics would be provided back to the patient and care providers, not just for reporting on the functioning of the smart kidney, but about the effectiveness of optimising and managing the overall urinary and endocrine system. Remove IoT from the equation, and you get none of these benefits.

It's not a stretch to imagine a subject with multiple smart organs, perhaps voluntarily replaced, even where the procedure does not address some form of morbidity. Kidneys, heart, liver, lungs and intestines can all be transplanted today. Transform those seven organs into smart organs and arrange them into a connected self-regulating Endoself, all powered by kinetic and bio-thermal energy cells, gathering and transmitting data to the Exoself and beyond, feeding into an active management system that the patient and health care professionals can use to optimise health, diet, exercise, medication and overall wellbeing. We could add further capability to these organs, such as the ability to call for reparative substances within ingestible smart pills should cellular or structural damage be detected by one of the in-built sensors, and combine them with the Exoself components, such as the aforementioned smart watches, cochlea implants and synthetic retina, all with an always-on connection to the internet via their embedded IoT technology.

It wouldn't be long before fully synthetic and more optimised versions of these organs were developed so as to entirely replace the biological elements. This would of course radically change the human self too in ways beyond the scope of our discussion here; for example, it's interesting to note that just these seven organs consume 70% of basal metabolism. If this was made available it would affect remaining biological systems in ways that would need to be considered.

As the scale of smart organ replacement continued, designers and developers of these technologies would naturally look for what else could be augmented or replaced beyond the current set of seven organs we identified as the candidate list above. As with elective replacement of healthy organs, subjects may look to do the same with otherwise healthy limbs in order to gain a perceived or promoted benefit. As the field of prosthetics improves, in terms of delivering more desirable form, function and ability to mimic the biological component it was replacing, it would almost certainly be IoT laden to ensure real-time operation and management. With these replacement limbs integrated into the subject's Exoself, a blurring of the Endoself and Exoself, along with the remaining biology, would occur.

Very quickly, we could find ourselves in a situation where a large amount of internal and external biological components have been replaced by IoT empowered smart versions, fully integrated into a new form of physical self and connected to a host of external enabling and optimising technologies, which we will discuss in the next section. At this point, we are most definitely Transhuman with the comprehensive Endoself more akin to a life support system for the central nervous system and remaining biological components, and with the Exoself providing the mobility for this Transhuman physical self. Anything biological that remained would be a target for those designers and developers that would now be looking for a new challenge, moving the Transhuman subjects to a fully Posthuman form.

With very little speculative discussion about the pathway to Transhumanism, we arrive at the greatest challenge, the augmentation and potential replacement of the brain, spinal column and additional elements of the central nervous system. However, as with everything discussed so far, technologies to achieve this already exist. Neuroprosthetics, in the form of an injectable neuromorphic, ultraflexible electronic circuit, were demonstrated as working in practice in 2017 [10]. These meshes require no power, allow the brain tissue to penetrate the mesh openings and merge with the mesh probe while causing minimal to no neural immune response. This builds on already successful work with other neural probes that have been used therapeutically to address issues related to Alzheimer's and Parkinson's disease, epilepsy, traumatic brain injury and also cognition and memory [11]. It's expected that these mesh electronics will be permanently viable, meaning once implanted, they stay implanted.

Other research focused on mapping the signals generated by the brain, when a paraplegic patient imagined certain hand movements and translating those signals into electrical impulses sent to a electro-stimulator laden cuff wrapped around the forearm. The effect of this was that the patient was able to move individual fingers, extend and flex the hand and perform actions such as gripping objects. In addition to the muscle stimulating cuff, this required the use of an implantable multi-node array and the application of machine learning to interpret and clean up the data transmitted to the cuff [12].

It wouldn't be overly speculative to see how the aforementioned smart prosthetics could be implanted along with a neural mesh placed at the junction between the spinal cord and the prosthetic, to propagate signals from the brain and achieve refined control over the prosthetic. Indeed, the inventors of the neural mesh had already proposed this kind of brain-machine interface for tetraplegic patients, in 2017. The same form of technology that could help disabled patients would be the same required by Transhumists and once again, we see the technology is already undergoing trials and use.

With a broader use of technology such as the neural mesh, perhaps achieved through multisite injections within the brain and at the spinal cord-Exoself boundary, it would be possible to more directly connect to external supporting technologies in the wider external network layers. This broader direct connection of the central nervous system to the Exoself and external network could allow the transfer of data directly out of and back into the brain. This would truly be a powerful capability to augment and enhance cognitive abilities and, as we'll discuss next, those supporting technologies are already well established.

55.6   Existing and Emerging Supporting Technologies

We've mentioned several times that, as part of the model of an IoT laden Endo-Exoself for a personal layer of interconnection, we have internet enabled high-bandwidth connectivity into a distributed network, decentralised across a local, regional and global scale. Within that external multi-layer network will be a set of technologies into which the subject can send and receive data. This data will likely be raw, unprocessed data, sent via a device that forms part of the Exoself. This device's behaviour would not be too dissimilar to the way we use a mobile phone or a blue-tooth headset today to capture and transmit data inwardly or outwardly. The raw data would then arrive at the distributed network of supporting technologies. As with much of what's been previously discussed, those technologies are already in use today and are awaiting the development and integration of IoT enabled smart devices, to be put to full use in supporting the realisation of Transhumanism. Each of these technologies adds a progressively increasing capability, enhancing what it means to be Transhuman.

55.6.1 Cloud Servers

Large volumes of data, that may grow or shrink in size and need heavy computation, while being managed with software that may need to be upgraded or replaced on a frequent basis, as is often the case with large IT systems – are best deployed to Cloud infrastructure such as Microsoft Azure or Amazon AWS. As we'd anticipate at least terabytes and possibly petabytes of data from years of potentially millions of subject's daily data uploads, there's no way a private server farm will suffice except in the rarest of occasions and for the smallest sets of data. A Cloud based data warehouse or lake are specifically designed for managing large volumes of dynamically changing data sets. This creates a model for what would be required to manage IoT telemetry data. These Cloud servers would not only be used for storing data, but also for the compute capacity needed to work with the data, hosting supporting software applications and providing security controls around access and use of the data and results from analysis.

55.6.2 AI and Machine Learning

Storage of the data is just one problem, the next is how it is sorted so it can be analysed to produce useful, actionable insights. At the likely scale of data we are eventually expecting, generated daily by millions of always-connected persons with multiple smart devices, we are going to need to apply some form of data science, a combined artificial intelligence and machine learning system to parse through the data on an ongoing basis. AI, unlike human analysts and standard analytical software, can readily be directed to consolidate data from multiple systems into data estates and then create data segments based on characteristics such as the persons health data, daily activities and behaviour, geographical location or smart implant performance amongst other characteristics. With data broken down into meaningful segments, Machine Learning could be applied to train the AI to look for patterns and relationships in the data and derive insights and actions that can be passed back to the smart devices directly, or mediating management systems, to optimise performance, suggest usage or damage avoidance strategies of varying types. This data could, in turn, be passed back to the designers and developers to improve the smart devices and to those looking to build replacements for the few remaining biological components subjects still retain. The UK's NHS have been investigating the use of telemetry from Fitbits and other smart devices for several years, with the idea of collecting this and turning it into useful patient observations.

55.6.3 Real Time Analytics

A benefit of an always connected Endo-Exoself is that data could be passed back to the supporting technologies for real time and near real time analysis and feedback, and this may prove to be the most beneficial of all the supporting technologies. With an ability to send and receive data near instantaneously, devices within the subject could rapidly respond to the outcomes of this analysis. An example, as touched on before, might be to dynamically control blood glucose levels, hormones, oxygen and carbon dioxide based on real time analysis of data drawn from the cardiovascular system. Taken further, with real time analysis it could be possible to sustain blood glucose levels at a level optimal for cognitive work, allowing us to exert mental effort for longer. In the case where neural implants have been introduced, real time analysis of data also means real time responses. Near real time responses from the AI and machine learning systems could mean that a subject gets information directly into their central nervous system – information that they did not have moments before – returned at the speed of the connection between their Endo-Exoself and the supporting technologies.

55.7 Conclusion

As we have seen, achieving Transhumanism is not as impossible as it may at first appear, it's just the timescale to achieving it that is something we cannot accurately predict. Much more development, experimentation and use are needed to derive viable technology sets, design rules and protocols. Interoperable technologies need to be built, perhaps even an operating system for the Transhuman person to ensure effective operation of their new form.

Relevant to the discussion here are matters such as the legality of elective surgery and who truly owns a person's body and their right to morphological freedom, the place of children in a Transhuman world and their right to an open future and of course, how everything gets paid for.

We also have other technologies that may play a part in achieving Transhumanism, such as nanotechnology and Blockchain. Consideration needs to be given to questions around how these enhancements will be paid for, who owns the data created, how payment will be made for data consumed and of course the security of that data and the systems that make up the always connected Transhuman reliant on the global network of supporting technologies.

These and many other topics are discussions outside of this paper's scope. For now, we look forward to the level of technological advancement that will make these topics the most active ones as Transhumanism becomes a reality and the greater use of IoT in driving forward this advancement.

References

1. http://www.sens.org/ Accessed August 2018.
2. https://bioviva-science.com/ Accessed July 2018.
3. N. Bostrom. 2005. *Defense of Posthuman Dignity.*
4. https://dangerousthings.com/shop/xnti/. Accessed July 2018.
5. https://www.gartner.com/newsroom/id/3869181. Accessed August 2018.
6. https://www.fda.gov/NewsEvents/Newsroom/PressAnnouncements/ucm584933.htm. Accessed August 2018.
7. J. Lawler. 2017. *Continuous Glucose Monitoring the First Four Weeks.*
8. http://www.medicotag.com. Accessed August 2018.
9. S. Murphy. 2014. *3D Bioprinting of Tissues and Organs*
10. https://phys.org/news/2017-07-brain-mesh-flexible-probe-melds.html. Accessed August 2018.
11. https://www.cam.ac.uk/research/news/electronic-device-implanted-in-the-brain-could-stop-seizures. Accessed August 2018.
12. C. E. Bouton et al. 2016. *Restoring cortical control of functional movement in a human with quadriplegia*

Chapter 56
Transhumanism in India: Past, Present and the Future

Sarah Ahamed, Palak Madan, and Avinash Kumar Singh

Transhumanism is the philosophy or theory that hypothesizes that the human species can evolve beyond its present limited physical and mental capacities, especially with the help of science and technology. The Indian subcontinent has a particularly rich cultural heritage, which has a certain level of natural compatibility with the transhumanist core philosophies. From the pursuit of longevity to the morphological freedom exercised by the avatars of ancient polytheistic Gods and Goddesses, the legacy and heritage of ancient Indian civilizations have much more similitude with the radical concepts of transhumanism of the modern ages. Statistics show a significant possibility for India to emerge as one of the revered players in the field of economy, science, and technology by 2050. Such a promising scenario paves a way for ideas like transhumanism to play an important role in shaping the country's future. In addition, a wider acceptance of transhumanist concepts can be easily achieved in India by connecting the common Indian man to his roots, using layman's words instead of strictly academic terminologies.

56.1 Transhumanism in Ancient India

The term Transhumanism is defined as an international intellectual movement that aims to transform the human condition by developing and making widely available sophisticated technologies to greatly enhance human intellect and physiology [1].

S. Ahamed · P. Madan
India Future Society, Bangalore, India

A. K. Singh (✉)
India Future Society, Bangalore, India

University of Technology Sydney, Ultimo, Australia

© Springer Nature Switzerland AG 2019
N. Lee (ed.), *The Transhumanism Handbook*,
https://doi.org/10.1007/978-3-030-16920-6_56

Evidence throughout the history of ancient India points towards the direction that transhumanism is a concept which is just not imported from the west in the recent times. Rather transhumanism has been congruent to ancient Indian philosophy, traditions, and culture, since time immemorial.

The inhabitants of the Indian subcontinent [2] have self-ascribed the name "India" or "*Bhārata*". The name has its roots in Vedas and Puranas where *Bhārata* is the official Sanskrit name of the country and it designates a land, which is seen in the Sanskrit texts written in those times as *Bhārata Gaṇarājya*. The term *Bhārata varṣam* was used to recognize India and to distinguish this land from other *varṣa*s or continents [3].

Rigveda mentions of a Vedic tribe called the Bhāratas [4], who participated in the Battle of the ten Kings [5]. Emperor Bharata unified the Indian subcontinent under one realm. Bharata descended from the Bhāratas tribe, which was a branch of Kuru Dynasty [6].

उत्तरं यत्समुद्रस्य हिमाद्रेश्चैव दक्षिणम् ।
वर्षं तद् भारतं नाम भारती यत्र संततिः ।।

The country (varṣam) that lies north of the ocean and south of the snowy mountains is called Bhāratam; there dwell the descendants of Bharata. [7]

Amongst the oldest civilizations of the world, India comes up as one of the most prominent names [8]. The Indian culture comprises several various regional cultures spanning across the Indian subcontinent [9] which is influenced and shaped by several thousand years old histories [10]. Dharmic religions have heavy influenced India along the sands of time [11]. The Dharmic religions have shaped much of Indian philosophy, literature, art, music, and architecture [12]. The history of India comprises of separate distinct ages, firstly the prehistoric settlements and societies in the Indian subcontinent; then the intermixing of Indus Valley Civilization and Indo Aryan Culture into Vedic Civilization, which has taken place gradually [13]; Finally, this led to the synthesis of Hinduism which developed as a result of the blending of various Indian cultures and traditions; followed by the emergence of 16 oligarchic republics termed as Mahajanapadas. In sixth century BCE, this phase was ensured by the development of Śramaṇa movement [14] and eventually Buddhism. History of India thereafter saw the rise and fall of several powerful dynasties spanning around two millennia throughout different geographic location of the country. This span of two millennia saw the flourishing of the Muslim rulers alongside Hindu dynasties and in due course of time the advent of European traders establishing the British Colonial Rule. This led the people of India to revolt for independence and create the Republic of India after almost 200 years of struggle for freedom from the British rule [15].

56.2   Transhumanist Connection in Ancient Indian Civilizations

56.2.1   Indus Valley Civilization

Urbanization first started in the Indian subcontinent in the 2500 BC in the Indus plains; however, the ancestry, the developmental stages and the beginning of the urban life are still shrouded in mystery. Indus Valley Civilization [16] began at around 3300 BCE. The civilization used Indus script for communication and until date, the script has remained undecipherable.

Throughout the next 1000–1500 years, Indus Valley Civilization witnessed an excellence in techniques pertaining to handicraft (carnelian products, seal carving) and metallurgy (copper, bronze, lead, and tin). The civilization knew how to use a boat and wheel. They had two-wheeled bullock carts and trade was an important source of income. They gained high feats in developing detailed urban designing and planning, baking clay brick houses, systematic drainage, and efficient water supply systems. Then at around 2600 BCE, the cities of Harappa and Mohenjo-Daro became one of the largest metropolises in the world at that period. They also achieved great feats in the fields of architecture and engineering, irrigation technology, standardization of weights, measurements and metallurgy, bread making and pottery, medical science and mathematics. In terms of religion and philosophy, the people had left behind ample amount of idols and figurines symbolizing female fertility. Such findings denote towards the worship of a deity that is actually a mother goddess and can be compared to present day Goddesses Shakti or Kali. Indus Valley Civilization people engraved animal figures in their seals, as an object of reverence, and worshiped them. In the animal figures, the zoomorphic aspects some later day Hindu gods and goddesses manifest. Seals having engravings of Lord Shiva in the Pashupati avatar (lord of the animals) performing yogic postures have been also unearthed [16]. All of this point towards the transhumanist concept of "**morphological freedom**" exercised by ancient Gods of Indus Valley Civilization. Their culture has contributed significantly to the development of Hinduism, as we see today.

56.2.2   Vedic Period

The Vedic period starts from 1500 BCE. The term 'Veda' means knowledge. The Vedic period of Indian history is remarkable for the thirst of the people at that age to acquire knowledge. Vedic knowledge comprised mainly of physics, mathematics, astronomy, logic, cognition and other subjects. The earliest science that has come down upon the Indian people passing through the sands of time is the Vedic Science [17].

Some of the very important concepts, which are at par with the modern transhumanist philosophies, can be found in the Vedas. There are four main Vedas – Rigveda, Samaveda, Yajurveda, and Atharvaveda [18]. The entire Book 9 of *The Rigveda* (c. 1700–1100 BCE) is tribute to the immortality-ushering "Soma" plant [19]. (Soma is summoned as "Haoma" in ancient Iranian scriptures of Avesta written during 1200–200 BCE.)

Ayurveda is called as the "science of long life" and in India, the "immortal" Rishis or sages, and the Ciranjivas (the "forever living") are respected even in modern times. Ayurveda has a special area exclusively dedicated to super-longevity. The essence of Ayurveda is embodied in the ancient texts of The *Sushruta Samhita* (Sushruta's Compilation of Knowledge, c. 800–300 BCE), which says:

> "Brahma was the first to inculcate the principles of the holy Ayurveda. Prajapati learned the science from him. The Ashvins learned it from Prajapati and imparted the knowledge to Indra, who has favored me (Dhanvantari, an incarnation of Lord Vishnu, the protector of life and the giver of Ayurveda on earth) with an entire knowledge thereof." This knowledge was in turn "disclosed by the holy Dhanvantari to his disciple Sushruta [20]."

According to the *Sushruta Samhita*, the average human life expectancy can be prolonged upto 100 years, naturally. But the formula for super-longevity has also been discussed in ancient Ayurvedic texts which says life can be extended to 500–800 years with Ayurvedic interventions like using certain Rasayana (chemical) remedies (such as Brahmi Rasayana and Vidanga-Kalpa). In addition, the use of the "Soma plant, the lord of all medicinal herbs" [24 candidate plants are named], guarantees reinvigoration of the user and helps him to "witness ten thousand summers on earth in the full enjoyment of a new (youthful) body."

Philosophical thoughts on "Amrit" – अमृत – or the "nectar of immortality" – "a revered and desired substaThose who could grasp the full knowledge of Ayurveda had a chance to live a happy, prosperous and almost immortal life. They "attained the highest well-being and non-perishable life-span [21]." In the *Sushruta Samhita,* athe ctual surgical methodology of skin graft transplantation to attach severed earlobes and perform rhinoplasty to reconstruct disfigurement of nthe ose, are explained scientifically and vividly [20]. nce" has also been mentioned in these texts.

One of the many foundational texts of Ayurveda, *The Charaka Samhita* (Charaka's Compilation of Knowledge, c. 300–100 BCE), holds a high value. According to it's scriptures, 100 years is the average human life-span. However, upon following the protocols and using the system of "Amalaka Rasayana", a human being gains the longevity of a century. And that of the use of "Amalakayasa Brahma Rasayana", a person may live up to a millennia [22].

In Mahabharata (written by Vyasa, 400–500 BCE), the severed body of king Jarasandha, could be reanimated completely again by fusing from two halves [23]. Reading such ancient scriptures helps one to understand how medical science has evolved down the ages and clearly aligned with the transhumanist core philosophies.

The monkey king Hanuman in the ancient Indian epic of the *Ramayana* (.400 BCE), used the *Sanjeevani* plant (translated as "One that infuses life" and generally referred to the lycophyte *Selaginella bryopteris*, which is native to the Dunagiri mountain in the Himalayan Mountain Range) to resuscitate Lakshman, Rama's younger brother, who was fatally injured by the demon king Ravana [24]. The accounts of the regeneration and restoration of the mutilated nose and ears of Surpanakha, Ravana's sister and princess of the Asuras, could also be found in the Ramayana.

The Hindu deity worshiping comprises the worship of Trimurti – Hindu Trinity of the creator of this universe named as Brahma, the preserver who takes care of the entire existence or Vishnu and the one who destroys the worlds or Shiva. Ayurveda is mainly associated with the deity Vishnu, or the preserver. Some of Vishnu's devotees, for example, Narada can continue to exist even though the cycles of destruction and creation of the universe, and this makes Narada literally immortal. Vishnu is prophesied to reincarnate as Kalki. Kalki seems to possess all the features of an artificial superintelligence (ASI).

Kalki is believed to be the tenth avatar of Hindu god Vishnu who would put an end to the Kali Yuga. In Vaishnavism cosmology, there are four periods in an endless cycle of existence (*krita*) and Kali is the final period. He is depicted in the Puranas as the avatar who recreates the universe by putting an end to the most chaotic and dark period of existence. He is prophesied to and removes "adharma" or evil to usher the Satya Yuga, the new beginning of another cycle. He happens to be an invisible force which is going to destroy all chaos and evil to establish harmony and peace, much like the hypothesized Artificial Super Intelligence of the future [25].

The concept of Vimāna or a mythological flying palace or chariot can be found in Ramayana. Other Hindu texts and Sanskrit epics also mention the concept. The most quoted example of a vimana is the Pushpaka Vimana of the king Ravana (who took it from Lord Kubera; Rama returned it to Kubera [24] Such a concept synonymous to the flying machines of the modern day like the fighter jet or airplanes.

Similar to modern Nuclear weapons, Brahmanda Astra was an ancient mystical weapon, which was used in numerous warfares in the Hindu mythologies. Such weapons were collectively named as Brahma weapons or brahmastra in different narrations of *Mahabharata*, *Ramayana* and the *Puranas*. The weapon called Brahmashirsha Astra could destroy the world. The Universe could be annihilated by the Brahmanda Astra [26]. Brahmastra never misses its target, could be obtained by meditating upon Brahma and is depicted as a fiery weapon which can create a fireball to annihilate the whole world [27]. The after effects of the destruction is indicated by a fatal decrease in rainfall and occurrence of drought. The epics and Vedas advice the use of the Brahmastra as a last recourse and warns against its use in a casual combat. The description of the Brahmastra as in the Mahabharata goes as – "A weapon, which is said to be a single projectile charged with all the power of the universe".

Rejuvenative and regenerative technologies along with technologies that are used in modern warfare and the predictive narrative of future artificial intelligence have been elaborately envisioned in ancient Hindu scriptures.

56.3   Transhumanist Connections in Indian Religion: Buddhism

Buddhism is originally an Indian religion, which is also now the world's fourth-largest religion. Buddhism is an embodiment of a number of eclectic traditions, beliefs and spiritual practices with Lord Buddha's teachings at the centre of the stage. Buddhism germinated in India between the sixth and fourth centuries BCE, and then it had proliferated across Asia. In the Buddhist texts, it is found that Amitābha is the Great Buddha who endows perpetuity. He is the "Buddha of Infinite Light", or in better words, Amitāyus, or the "Buddha of Infinite Life". It is said that "Those who invoke him will reach longevity in this realm, and will be reborn in Amitabha's Pureland (Sukhāvatī or Dewachen in Tibetan Buddhism) where they will enjoy virtually unlimited longevity." Amitabha's avowed devotion and perseverance helped him to create this sacred and equalitarian land."*Om amrita teje hara hum*" (Om save us in the glory of the Deathless One hum) is a mantra chanted in the appraisal of Amitabha. Many Buddhist mantras are allocated towards the healing of the old age and are chanted to extend longevity. They believed that these would help to open the portals of their wisdom and connect them to the ONE and thus suffering would be abolished by practicing compassion and extreme longevity can be reached in this world itself. Along with the spiritual aspect, medical means or physical means for rejuvenation and longevity have also been developed by the wise Buddhist physicians [28].

In Buddhist eschatology **Maitreya** (Sanskrit), **Metteyya** (Pali), is deemed as the future Buddha of this realm. He is referred to as **Ajita** in Buddhist literature such as in the *Amitabha Sutra* and the *Lotus Sutra*. The Buddhists believe that Maitreya is a bodhisattva (Buddha or pure consciousness) who will emerge on the terrestrial world in the future, who will attain total enlightenment, and teach the pure dharma (cosmic law and order) to people. The Buddhists texts predict that Maitreya will succeed the past Buddha; Gautama Buddha (also called as Śākyamuni Buddha). The prophecy of the arrival of Maitreya directs towards a point of time in the future when the dharma will have been obliviated by the humans and most of the planet earth is shrouded in chaos and darkness [29].

The arrival of Maitreya is alike to the arrival of an Artificial Super Intelligence (ASI) on this earth to usher an age of harmony, peace, and order when most of the world is immersed in chaos.

Thus akin to Hinduism, Buddhism too has a strong parallelism to the quest of attaining longevity and moksha (or enlightenment). In addition, there is a prediction of the arrival of a sentient being like the artificial intelligence so predicted by the modern technological trends [30] who would deliver the people of this world from chaos to harmony.

56.4   Transhumanism in Contemporary India

In the present status, the World Transhumanist Association (WTA) has provided a formal definition of transhumanism: "The intellectual and cultural movement that affirms the possibility and desirability of fundamentally improving the human condition through applied reason, especially by developing and making widely available technologies to eliminate aging and to greatly enhance human intellectual, physical, and psychological capacities [31]." This philosophy of transhumanism is a based on the continually evolving framework of principles, values, and perspectives based on the idea that new scientific and technological developments are likely to remove the limits of our physical, biological and intellectual inheritance. Transhumanists believe and support the idea that potential future technologies like gene editing, nanotechnology, artificial intelligence, regenerative medicine, biotechnology, cryonics, radical life extension, mind uploading etc. have a potential to enhance human conditions in every possible way, even ending aging [1].

Over the years, India has made significant progress in its goal to bring millions of people out of extreme poverty that correlated with high immortality. A lot of the progress was a direct consequence of the advancements in science and technology. There are a number of technologies that can have a disruptive impact on government, society, and importantly on individual life. From the basic technological advancement for e.g. Mobile internet, digital payments, work automation, digital verification identity systems. Future of 5G-telecommunication technology will also play a huge role in automating and leveraging on data from real-time distributed systems. Beyond digitization, the rapid advances in genomics, intelligent transportation, energy (renewables, storage, gas, and oil), medicine, advanced geographic information systems, artificial intelligence can help India build a more efficient power supply and distribution systems, raise individual productivity, improve basic income and healthcare access, curb corruption and more.

India has good reasons to be optimistic given the recent record of accomplishment of technological and its economic growth over the years, especially among young population demographically. India, being the world's largest democracy, with the population of 1.35B (2018) [32], makes up for a total of 17.74% of the world population. In contrast to USA, Europe and East Asia with their aging societies, India is a relatively young country with 50% of the population not older than 25. In addition to the IT sector, India is moving towards the forefront of genuine innovation in pharmaceuticals, manufacturing and biotechnology sectors. Recent increase in budget allocation by the Indian government in science, research and development particularly focussed more in the field of robotics, artificial intelligence, big data analysis, quantum communication, Internet of Things Space and Atomic energy is a proof that India is adopting towards a rational, action-based optimism and shunning blind faith at the same time.

Despite the fact that India's population demographic is relatively young with a median age of x years, life expectancy is still comparatively low at an average of 66.8 years [32–34]. Thus, it is likely that the dynamic and optimistic young generation will especially embrace technologies for longevity and human enhance-ment to tackle most physical and mental health-related problems, workplace safety, unemployment, pollution and more.

India faces many problems ranging from medicine (health care access, drugs, genetic diseases, high child mortality rate, etc.) to agriculture (low-yield crops, farmer suicide rate, pesticide use) to education (unequal access, illiteracy rate) to gender inequality and more. Nevertheless, these problems can be potentially allevi-ated by adopting emerging technologies. Transhumanism in India can potentially be more successful than in western countries particularly because Indic religions are perceived to be more open to the ideologies supporting human improvement.

Zoltan Istvan (founder of the US Transhumanist Party and candidate for the 2016 US Presidential election on behalf of this party) even said, "Many transhuman goals, like trying to overcome human death, go against the grain of Western reli-gions and their sacred texts. But Hinduism, Buddhism, and various other Eastern religions can be conducive for transhumanism and its goals." Other reasons why such an ideology can be more fertile in India are: entrepreneurial spirit among mil-lennials, current government support with organizations like Make in India [35], Invest India [36] etc., opportunity to catering to the needs of a large population and freedom to use science and technology as a canvas to express creativity. "In the United States, for example, it takes sometimes a decade to get a drug or certain type of technology approved by the government, into production, and then into the pub-lic's hands. In India, there is far less scientific bureaucracy and government approval. It is felt that, due to these reasons, transhumanist technologies will probably grow a lot faster in places like India rather than in the West", adds Istvan.

56.4.1 Emerging Technologies Tending Toward Transhumanism in India

Healthcare is one of the major emerging sectors in India which shown major devel-opment and breakthrough over the past few years in India. In India, healthcare sec-tor divided into public and private form, where the private healthcare sector is a dominant service provider while public healthcare is free for those below the pov-erty line. The private sector exists and favourable among the public due to a better quality of service, access to high-tech facilities, world-renowned experienced doc-tors. Indian government took and taking several initiatives like The Twelfth Plan, Public-private partnership, National Rural, and Urban Health Mission schemes to strengthen the overall structure including infrastructure, research, and development.

Although, there is a big gap when it comes to healthcare outreach among the second largest population of India probably due to inequalities in healthcare access like insufficient access to healthcare facilities and differential distributions of ser-vices, power, and resources. The main bottleneck in Indian healthcare ecosystem is the "access" to the services that are also a big reason behind major development in the field, significant government as well private interest. One of the important and feasible ways is to tackle in nearby future is by leveraging the use of new and emerging mobile and connected technologies to enable proactive healthcare system. From simple wrist-based accelerometers to smartphones that capture heart rate, blood pressure, and glucose measures, technology has evolved beyond common health and fitness tracking. Such technologies include disposable vital-sign patches (to transmit streaming ECG, posture, temperature, stress data), smart bed sensors (to track the quantity and quality of sleep), consumer EEGs (to measure brain-waves), trainables (such as Upright, a device that buzzes your back when slouching to entrain a better posture), scales (to track vitals and measure BMI, fat/muscle percentage etc.), BP cuffs, sensor-fitted pacifiers (to analyse alcohol consumption, to track hydration status and molecules that can indicate a metabolic or malignant disease) and more. With these technologies comes the possibility and promise to connect all the data dots at meaningful scales. The emergence of real-world AI will significantly improve the efficiency of diagnosis and hence the front line delivery and the future of healthcare in India. Which also inspire the government toward focusing intensely on full fledge development plan for artificial intelligence under the banner hash-tag AI for all, as found in the June 2018 Discussion Paper on National Strategy for Artificial Intelligence [37].

The Government of India also made several changes in existing policies for better future of healthcare in India by commitment toward the Universal Health Coverage and increased access to comprehensive primary health care by launching the schemes like Ayushman Bharat programme in 2018 [38], known to be the world's largest government-funded health care programme, which directly impact the wellbeing of individuals thus the longer life expectancy.

The Government of India is investing heavily toward the future for Health and Wellness Centres (HWC), which will lay the foundation for the future of India's health system as envisioned in the National Health [38]. These HWCs, to be set up by transforming 1.5 lakh Health Sub Centres from 2018 to 2022, are aimed at shifting primary healthcare from selective (reproductive and child health/few infectious diseases) to comprehensive (including screening and management of NCDs; screening and basic management of mental health ailments; care for common ophthalmic and ENT problems; geriatric and palliative health care, and trauma care and emergency care).

In parallel to a bunch of introduction of government policies, the personalized medicine technologies is another growing field in India. Personalized medicine uses a patient's unique clinical, genetic, genomic, and environmental information to make intelligent diagnosis and predictions regarding susceptibility towards a particular disease for that particular individual, probable responses and the course of the disease. The aim of personalized medicine is to make the treatment as individualized as the disease itself by identifying genetic, genomic and clinical problems. Molecular diagnostics are used to detect changes in the DNA, RNA and protein expression level of the patient, which subsequently helps to detect any changes in the genes and thus helps in the early detection and diagnosis of the disease. Patients can have an insight about their problems and can monitor disease progression and select appropriate therapies, which can be exclusively personalized to their biology.

A great example can be of cancer; which kind of cancer patient will respond to the particular kinds of therapeutic interventions can be determined by molecular diagnostic tests instead of just administering the traditional chemotherapy that was the exclusive and universal treatment for cancer in the past decade.

Growth in Personal Genomics Technology field is helping in preventive health care, disease burden reduction and Personalized Medicine in India. Positive Bioscience has teamed up with Medanta – The Medicity to launch a genomics clinic, which is first of its kind in India. This initiative will serve personal genomics in tandem with the expert medical diagnosis. Advancements in the field of personal genomics now allow the scientists to run a genetic test for any individual and then make a prediction whether or not the individual has the propensity to develop a particular kind of disease or not along with the compatibility of the individual's metabolism to a particular group of drug. All of these have led to much development in the field of pharmacogenomics, which is hence helping the big pharma companies to eliminate the unpredictability associated with the process of developing a drug. Start-up companies like Xcode Life Sciences are following this model by providing solutions for lifestyle-related diseases like diabetes, coronary heart diseases, and obesity. They ship a saliva kit to the customers who have ordered online for a test and then later use the non-invasively collected saliva to extract DNA and determine the allelic information using the high throughput genotyping techniques. A company called NutraGene launched the first commercial genetic test for type 2 diabetes in India.

Although the application of genomic and personalized medicine in India seems to be a very promising field in ushering the new era in medical sciences, there are many hurdles on the way. The effectiveness of the methodologies has to be critically assessed by the researchers, diagnostic firms and regulatory authorities and the application of such technology in India smoothly needs a change in the regulatory system.

On the other hand, much exaggerate technology, CRISPR also playing an important role in health-related technology in India. CRISPR is a gene-editing tool that is able to snip the mutation caused in the sickle cell anemia [39]. Beta thalassemia is the most prevalent blood-related genetic disorder in India and thus scientists at Institute of Genomics and Integrative Biology (IGIB), Delhi is trying to use the technology to form the basis of a viable therapy. However, there are certainly ethical issues regarding CRISPR, which is being heavily debated throughout the globe. Since CRISPR, technology has the capability to alter the genetic makeup of an organism, the possibility of creating superhumans to designer babies raise a lot of ethical questions to which the Indian scientists respond with caution. Indian Government has banned the use of stem cell therapy for commercial use due to "rampant malpractices" however; the national guidelines do permit the use of stem cells in certain blood cancer types [40].

The upgraded medical system, personalized medicine, CRISPR has a huge potential to increase the well-being of the current population in India. However, taking into the account that there will be a rise in the aged population as well, the need for a more holistic care is also required investigation. The geriatric population in developing country like India is expected to be 840 million by 2025. The proportion of Indians that would be aged 60 years or above will rise from 7.5% in 2010 to 11.1% in 2025. The number of elderly population was 91.6 million in 2010 and that is projected to increase to 158.7 million in 2025 [41].

The Longitudinal Aging Study in India (LASI) was funded by an R21 exploratory grant from the National Institute on Aging (NIA), one of the 27 institutes and centres of the National Institutes of Health (NIH). LASI aims at understanding the situation of the elderly population in India with precise details so that the design of policymaking to support and protect the elderly can be done effectively.

LASI uses two main tools; first, the computer-assisted personal interview (CAPI) techniques to register the answers given by the survey contributors. (The funding for this study comes from the pilot grant from Harvard's Program on the Global Demography of Aging [PGDA]); and second, molecular biomarker collection, which can be studied to obtain quantitative data on the status of health of the participants. The National Research Council advised that biomarker collection should be included in a social survey to (a) obtain data regarding health from a part of the population that otherwise would not have got such type of information on the records; (b) probe the determining factors on a molecular level for common health outcomes and (c) investigate interactions that might later lead to decline in the health outcomes. Interactions such that between biomarkers and other social conditions. It is important to include factors like biomarkers and other medical assessments in less-developed countries like India, where procuring even basic health care is quite difficult. Hence, unlike developed countries, undiagnosed diseases tend to be common in India [42].

There is also some organization related to aging in Indian like Avesthagen, which is a major player focusing on the pharmacogenomics of aging in the Parsi population in India. One of the major projects started under 125 crore at 2007 is the AVESTA GENOME project. The company does a system biology-based study regarding the genetic basis of longevity and geriatric disorders in the Parsi community. The aim of the study is to design a model for pharmacogenomics-based therapies on ageing, develop biomarkers for predictive diagnostics and drug discovery. Also by this study, the genome of the Parsi community is going to be archived [43].

The key highlight here is that India got the diverse ethnic demographics and has the potential to be a key player in longevity research. According to Prof. Kalluri Subba Rao, Hon. Coordinator for Center for Research and Education in Aging (CREA), University of Hyderabad, India, "Science Academies have a responsibility to alert the Government authorities for initiating ventures that would stabilize a social climate in terms of health and economics. Aging research in the lines cited above, in my opinion, would be one to contribute such a climate [44]."

Apart from the Indian Government, policies toward the healthcare, advancement in technology related to health care and aging research, India open his door for other countries to enjoy the fruits of these researches. Such policy gave a significant rise in Medical tourism.

India is considered as one of the fastest emerging medical tourism destinations in the world and growing as sector worth more than $3 billion back in till 2015 and is projected to be worth $8 billion by 2020. Main reasons why India attracts foreigners to get medical treatments here are a competitive cost advantage, availability of the latest medical technologies and other accredited facilities that are at par with those in developed countries. Patients without Borders, a US-based medical travel resource, says certain medical procedures in India cost around one-tenth of the price in the United States, making it one of the cheapest places to get the treatment. Most common treatments sought in India by tourists are heart surgery, hip replacement or resurfacing, bone-marrow transplants, eye surgery, cardiac bypass surgery, dental care etc. Although most of the medical tourists come to India from the Middle East and neighbouring Asian countries like Bangladesh, Afghanistan, Maldives, Republic of Korea and Nigeria, it is fast emerging as a hotspot for medical tourism from US and Europe as well [45].

The transhumanism discourse offers an ongoing debate on the different standpoints that has flourished as an outcome of a contemporary attempt to redefine human condition. On speculating and reflecting past, present and future of transhumanism in India — current understanding and awareness of the transhumanist reflection is not necessarily known by the exact definition/movement but the ultimate goal identifying the use of speculative scientific framework and emerging technologies to enhance human condition is deeply inscribed and reflected by different policy around the world. India has the economic potential, and the perfect playground for disruption from emerging technologies at multiple levels of healthcare, smart cities, agriculture, automated agents (intelligent machine), and space exploration with AI-driven solutions, as evident by the increasing activity from large corporates, startups including government agencies like ISRO. One can argue that these ideologies from ancient India, government intervention toward disruptive technologies, public and private support, youngest population in the world, and freedom to use science and technology as a canvas to express creativity are the clear sign that India will able to lead and drive the world toward the future of transhumanism.

References

1. Bostrom, N., *A history of transhumanist thought.* Journal of evolution and technology, 2005. **14**(1): p. 1–25.
2. Bose, S. and A. Jalal, *Modern South Asia: history, culture, political economy.* 2002: Routledge.
3. *Constitution of India.* 1949.
4. Pargiter, F.E., *Ancient Indian historical tradition.* 1997: Motilal Banarsidass Publishe.
5. Thapar, R., *The Penguin history of early India: from the origins to AD 1300.* 2015: Penguin UK.
6. Schmidt, H.P., *Notes on Rgved.* Indica. Organ of the Heras Institute,, 1980. **17**.
7. Singh, U., *A history of ancient and medieval India: From the Stone Age to the 12th Century.* 2009: Delhi: Longman.

8. *Reading the Vedic Literature in Sanskrit.* [cited 2017 11 December 2017]; Available from: is1. mum.edu.
9. Kenoyer, J.M. and K.B. Heuston, *The Ancient South Asian World.* 2005: Oxford University Press, USA.
10. Baker, K.M. and G.P. Chapman, *The Changing Geography of Asia.* 2002: Routledge.
11. Keay, J., *India: A History. Revised and Updated.* 2011: Grove/Atlantic, Inc.
12. Stafford, N., *Finding Lost: The Unofficial Guide.* 2006: Ecw Press.
13. Prakash, O., *Cultural History of India.* 2005: New Age International.
14. Reich, D., et al., *Reconstructing Indian population history.* Nature, 2009. **461**(7263): p. 489.
15. Ranadive, B.T., *India's Freedom Struggle.* Social Scientist, 1986: p. 81–126.
16. Flood, G.D., *An introduction to Hinduism.* 1996: Cambridge University Press.
17. Prama, R., *India: Science and Technology from Ancient Time to Today.* Technology in Society, 1997. **19**(3–4): p. 415–447.
18. Witzel, M., ed. *The Development of the Vedic Canon and its Schools: The Social and Political Milieu.* Inside the Texts, Beyond the Texts. 1997, Harvard Oriental Series.
19. Griffith, R.T.H., *The Rig Veda.* 1896.
20. Bhishagratna, K.K.L., *Suśruta, and Kunjalal Bhishagratna: An English Translation of the Sushruta Samhita,* in *Translation of Different Readings.* 1907.
21. Madan, G., *India through the ages,* M.o.I.a. Broadcasting, Editor. 1990. p. 66.
22. Van Loon, G., *Charaka Samhita: Handbook on Ayurveda.* Volume I–PV Sharma & Chaukhambha Orietalia, 2003.
23. Parva, S., *Sabha Parva.* Mahabharata, 1993. **2**: p. 108.
24. Griffith, R.T.H., *The Rámáyan of Válmíki: Translated Into English Verse.* Vol. 1. 1870: Trübner and Company.
25. Dalal, R., *Hinduism: An Alphabetical Guide.* 2010: Penguin Books India.
26. Dutt, M.N., *Ramayana.* 1892, Calcutta: Elysium Press.
27. *The Matsya Puranam.* 1916: Pâṇiṇi Office.
28. Maher, D.F., *Two Wings of a Bird: Radical Life Extension from a Buddhist Perspective,* in *Religion and the Implications of Radical Life Extension.* 2009, Springer. p. 111–121.
29. S, P.C., *The Minor Anthologies of the Pali Canon. Part IV: Vimānavatthu: Stories of the Mansions.* Journal of the American Oriental Society, 1980. **100**: p. 56.
30. Müller, V.C. and N. Bostrom, *Future progress in artificial intelligence: A survey of expert opinion,* in *Fundamental issues of artificial intelligence.* 2016, Springer. p. 555–572.
31. *The Transhumanist FAQ: v 2.1. World Transhumanist Association 2003.* [cited 2018 7-Oct-18]; Available from: http://transhumanism.org/index.php/WTA/faq/.
32. *Countrymeters.* [cited 2018 07-October-2018]; Available from: https://countrymeters.info/en/India.
33. Davendra Verma, et al., *Youth in India,* S.S.D.C.S.O.M.o.S.P. Implementation, Editor.
34. *Worldometers.* [cited 2018 08-October-2018]; Available from: http://www.worldometers.info/world-population/india-population/.
35. *Focus on Make In India.* 2014.
36. *Invest India.* 2018 [cited 2018 07-October-2018]; Available from: https://www.investindia.gov.in/.
37. Roy, A., *National Strategy for Artificial Intelligence,* N. Aayog, Editor. 2018, NITI Aayog.
38. *AyushmanBharat-NationalHealthProtectionMission.*2018[cited201807-October-2018];Available from: https://www.india.gov.in/spotlight/ayushman-bharat-national-health-protection-mission.
39. Xu, P., et al., *Both TALENs and CRISPR/Cas9 directly target the HBB IVS2–654 (C> T) mutation in β-thalassemia-derived iPSCs.* Scientific reports, 2015. **5**: p. 12065.
40. Jotwani, G. and S.D. Sinha, *National Guidelines For Stem Cell Research* I.C.o.M.R.D.o. Biotechnology, Editor. 2017, Indian Council of Medical Research: New Delhi.
41. Mane, A.B., *Ageing in India: Some Social Challenges to Elderly Care.* J Gerontol Geriatr Res, 2016. **5**: p. e136.

42. *Longitudinal Aging Study in India (LASI)*. [cited 2018 07-October-2018]; Available from: https://www.hsph.harvard.edu/pgda/major-projects/lasi-2/.
43. Guzder, S., et al., *The AVESTAGENOME project™-a discovery model for disease genomics and beyond*. Genome Biology, 2010. **11**(1): p. P16.
44. *Centre for Research and Education in Ageing*. Available from: http://www.uohyd.ac.in/index. php/academics/2011-10-27-18-38-44/research-and-education-in-ageing.
45. PTI, *Indian medical tourism industry to touch $8 billion by 2020: Grant Thornton* in *The Economic Times*. The Economic Times: Online.

Part VIII
Philosophy and Religion

Chapter 57
Advancing Neutral Monism in Big History and Transhumanist Philosophy

Ojochogwu Abdul

57.1 Big History and Transhumanist Philosophy: Complementary Worldviews

Big History and transhumanism are two complementary ways of approaching the origin, evolution, and future of humanity, the Universe, and the future of humankind within the Universe. Big history takes a large sweep of the story of life using time-scales and notes similarities and differences between the human, geological, and cosmological scales. As a new kind of origin story, it tells, scientifically, the story of the Universe starting from the Big Bang, the formation of stars, planets, life on earth, human origins, modern civilization, and what might exist in the future.

Weaving contributions from scientists, philosophers, historians, scholars, and adventurers, the big history project gets to narrate an enormous Big-Bang-to-Humankind account which raises and provide answers to big questions: Life, the Universe, and everything questions. As closely related to Cosmic evolution and the Epic of Evolution, big history presents a single, inclusive scientific narrative of the origin and evolution of all material things over roughly 14 billion years (from the origin of our Universe to the present day on Earth), and works as a multidisciplinary, interdisciplinary approach posing and seeking to answer different kinds of questions while using empirical evidence to present a story about where we as Homo sapiens come from and what we are.

Big history informs us about our place in time and space on the largest scale, taking the extra effort to uncover many layers of philosophical thought hidden between the lines. The meaning of the scientific origin story gets to be investigated under this framework by big history scholars using current forms of philosophical thinking to reflect on the new cosmology. For time-scales, terms like "threshold" (Christian), "regime" (Spier), or "ages" (Chaisson), are used within the identified

O. Abdul (✉)
University of Lagos, Lagos, Nigeria

© Springer Nature Switzerland AG 2019
N. Lee (ed.), *The Transhumanism Handbook*,
https://doi.org/10.1007/978-3-030-16920-6_57

pattern that the evolutionary process follows to describe the major periods of emergence when an entirely new thing appeared. The big questions, such as "How did we get here?" and "What is life?" are asked, and by implications of its naturalistic, cause-and-effect, and broad approach to Life events, the big history narrative therefore gets to ask "Where are we going? Transhumanism relates to Big History mainly by virtue of this last question, since it concerns itself fundamentally with what lies ahead for humanity in terms of the species' future and next stage of evolution.

Transhumanism has been defined as "a way of thinking about the future that is based on the premise that the human species in its current form does not represent the end of our development but rather a comparatively early phase" [44]. The philosophy has also been defined by the Transhumanist FAQ [38] as "the intellectual and cultural movement that affirms the possibility and desirability of fundamentally improving the human condition through applied reason, especially by developing and making widely available technologies to eliminate aging and to greatly enhance human intellectual, physical, and psychological capacities." Transhumanism considers the future advancement and impact of humanity and human-created Life both on our planet and well across the Universe. The forward-looking character of transhumanism essentially makes it complementary to the evolutionary sequence implied in Big History.

There are several similarities and areas of complement between both movements (Big History and transhumanism). They are contemporaneous, apply basically the same scientific body of evidence and, using this, they each present a narrative of evolution on a cosmic scale. They also both present a similar basic view of the past in terms of physical, biological and cultural changes which explain a pattern of increasing aggregate complexity unfolding within a space of open evolutionary systems [4]. Similarities are further found in the recognition by both movements of transformative events or significant moments of emergence. Big History characterizes these moments by use of "Thresholds" (David Christian), while in transhumanism Ray Kurzweil's "Epochs" are popularly referred to [4]. Eric Chaisson also uses "epochs" or "ages" in the Epic of Evolution which is related to Big History. In both cases still, the breakdown of the significant stages share a conspicuous similarity, following, in broad terms, the pattern: physics, chemistry, biology, humans, technology, agriculture, industrialization, the future [4].

Both narratives inform us that in the scheme of 13.8 billion years of cosmic evolutionary history, humans are not the apex or end of evolution, neither do we in any sense represent cosmic perfection. However, both movements grant humans enough attention [4], and demonstrate reasonably that we are different from all of our biological cousins and antecedents, a species credited with the Anthropocene, where humans have become the first biological creatures in terrestrial history to have significantly reshaped the face of planet Earth and continues to alter the biosphere. But it does not stop there, because also, using culture and technology, we are the first creatures to have developed the capabilities to consciously and collectively engineer our own evolution, designing what we want to be.

Within the project of big history, there is that sense and indication that humans have approached a major turning point in the whole journey. The moment we now live in is no longer one consisting of gradual, on-going change, but a period of great uncertainty and potential in which our decisions and actions will have great significance and determine the future of life. Big history as a narrative itself points towards this conclusion [5]. David Christian [8] refers to big history's relevance here when he states that having "a map of where we are can help us decide where to go next." This sense of humanity currently standing at a critical junction where its activities would deeply impact its future and fate of the Earth was expressed by Brown:

> Big history is a universal, trans-disciplinary story that provides us with the best available map both of where we are in time and of how we got here. The narrative presents that humans are intrinsically connected to everything in the universe, that we currently stand at a turning point of grand significance in our planet's history, and that our choices and actions in the next decades will make a profound difference in its evolution. ([5], 12)

Transhumanism logically connects to this as a call for us to make the right choices on matters involved in developing and using technologies that will enable us to overcome fundamental human limitations, and shape the trajectory of life both for millennia to come and far beyond our planet into the Universe.

57.2   The Prevalent Materialism as Metaphysical Assumption for Big History and Transhumanism

Having briefly introduced big history and transhumanism, and highlighted their character as complementary worldviews, we shall develop gradually towards the concern of this essay by here considering the prevalent ontological position upon which both narratives are mostly built. Underlying the assumptions of most big histories and transhumanisms is a philosophical position called "materialism" (interchangeably within this essay: "scientific materialism," "philosophical materialism", "metaphysical materialism" or "physicalism") This materialism, as we shall present, results predominantly in the rejection of any purely mental substance (idealism), or dualism in the philosophy of mind, ontology, and natural sciences that both discourses build upon.

Materialism, or physicalism: is the view which holds that only the physical exists. Scientific/metaphysical materialism, further, is a belief or assumption that only matter/energy exists and as such anything seemingly immaterial must be the product of the underlying matter/energy. Accordingly, mental things are in some sense only manifestations of an underlying physical reality. Scientific materialism is synonymous with, and can be described as being a reductive materialism. As a form of philosophical monism, materialism holds that matter is the fundamental substance in nature, and that all things, including mental aspects and consciousness, are results of material interactions. Materialists of all shades maintain and defend the claim that mental properties supervene on physical properties, a claim which means no difference can take place in the mental properties of anything without a difference in the physical properties (for example, the brain), on which they supervene.

In terms of approach to the question of body and mind, and to the structure of reality, many transhumanists betray materialist orientations, and the prevalence of metaphysical materialism occurs more noticeably in transhumanism than in big history. The objectives of transhumanism necessitate a concern with the nature of reality and the human. Transhumanism involves a commitment to technologically mediated transformation, and this naturally generates huge interest in the nature and limits of the self. The philosophy therefore offers intriguing perspectives on, amongst other things, the nature of persons, the nature and identity of the self, and the nature of mind. Addressing these metaphysical issues that arise concerning transhumanism, More analyzes:

> With few exceptions, transhumanists describe themselves as materialists, physicalists, or functionalists. As such, they believe that our thinking, feeling selves are essentially physical processes. While a few transhumanists believe that the self is tied to the current, human physical form, most accept some form of functionalism, meaning that the self has to be instantiated in some physical medium but not necessarily one that is biologically human – or biological at all. If one's biological neurons were gradually replaced, for example, with synthetic parts that supported the same level of cognitive function, the same mind and personality might persist despite being "in" a non-biological substrate……. ([25], 7).

More therefore confirms the materialist bent of most transhumanists, and going further, he replies this way to critics who confuse a functionalist (hence, physicalist) goal of transhumanism with dualism:

> Some critics who read discussions of "uploading" minds to non-biological substrates claim that transhumanists are dualists. Those critics are confusing dualism with functionalism. A functionalist holds that a particular mental state or cognitive system is independent of any specific physical instantiation, but must always be physically instantiated at any time in some physical form. Functionalism is a form of physicalism that differs from both identity theory (a mental state is identical to a specific brain state) and behaviorism (mental terms can be reduced to behavioral descriptions). According to functionalism, mental states such as beliefs and desires consist of their causal role. That is, mental states are causal relations to other mental states, sensory inputs, and behavioral outputs. Because mental states are constituted by their functional role, they can be realized on multiple levels and manifested in many systems, including nonbiological systems, so long as that system performs the appropriate functions [25].

There are indeed different versions of functionalism, with extensions of functionalism including revisionary materialism (or revisionary functionalism), and the very extreme eliminative materialism. Transhumanists likewise vary in their positions, but given the transhumanist' interest in applying scientific knowledge for the reconceptualization and revision of the structure of human cognition, many transhumanists are therefore likely to be open to such materialist/functionalist positions in the philosophy of mind.

Big history on its part has a rather nuanced approach to the idea of materialism. There is, of course, a difference between metaphysical materialism and what is called methodological materialism. With regards to necessity, big historians are methodological materialists. Methodological materialists carry out scientific investigations from a standpoint that sees matter and energy as all that exist, without necessarily believing this to be the case. Some scientists, though in the minority, do believe that something else exists. Most scientists, however, conduct their scientific work as methodological materialists and then extend to posit the metaphysical materialist belief that matter/energy actually make up the only reality [5]. Big historians, in this wise, also put together the findings of scientists and humanists based on the empirical evidence reported in their studies, and through this, big history experts have gotten to construct a narrative history of 13.8 billion years based on the materialist method for gaining empirical evidence [5].

A number of big historians, however, not content with remaining agnostic over the question of whether or not anything exists beyond material phenomena, do go beyond just the findings offered by methodological materialism into erecting upon its narrative several materialist metaphysical theories, and some argue from these metaphysical assumptions that ultimately matter/energy is all there is to the nature of reality. Brown, for example, admits: "I, for one, cannot resist taking sides tentatively and arguing for metaphysical materialism, that ordinary matter and the known types of energy are the only constituents of reality" ([5], 8).

For both big history and transhumanism, therefore, materialism, in different shades, serves as that predominant ontological framework upon which both narratives have been upheld.

57.3 Problems in the Metaphysical Materialist Approach

There are, however, problems for both big history and transhumanism going by the metaphysical materialist ontology held by much of their proponents, and these problems introduce insufficiency to the narratives they present.

To begin, with the advent of quantum physics, a number of scientists got to believe that the concept and conventional position of matter had changed and could no longer be maintained. Werner Heisenberg for example, said "The ontology of materialism rested upon the illusion that the kind of existence, the direct 'actuality' of the world around us, can be extrapolated into the atomic range. This extrapolation, however, is impossible......atoms are not things" [40]. Max Planck used quantum mechanics to posit a force which brings and holds particles together and behind which he assumed the existence of a more fundamental, conscious and intelligent mind [29]. The concept of matter and its fundamentality therefore was discovered as, rather than definite, subject to revision in the face of advancements in science. As Noam Chomsky [7] will then argue, since the concept of matter may be affected by new scientific discoveries, as has happened in the past, scientific materialists who assume the opposite are therefore guilty of dogmatism.

The fundamentally materialist view of big history/cosmic evolution thus continually faces strong refutation in the face of challenging interpretations and implications of quantum physics and advancements in science. Moreover, as concerning the objects/forces that shape, organize, introduce complexity and intelligence, or ultimately constitute the Universe, the materialist framework (from ordinary matter and the known types of energy), based on the Standard Model of Particle

Physics, only provides us with limited understanding, and as such no metaphysical materialist can be justified to feel or claim thorough certainty of matter as ultimately exhaustive of reality. This serves as a limitation for metaphysical materialism in big history.

Still, in addition to some so far inexplicable things like dark energy, dark matter, and even gravity, there are yet several other things in our known universe not yet understood by scientists, and these include origin of life, mind, and consciousness. There are arguments which hold that certain things, such as mind, may not be reducible to matter or a manifestation of matter ([15], 7–8). A materialist would, however, believe that mind and consciousness are either reducible to, or are a manifestation of matter. We shall consider some problems in this view as we proceed. At present within this discussion, however, suffice to note that mind *emerged* at some certain point in big history (the so-called "mind's big bang", or the "Cognitive Revolution"), and from this was produced human culture. How this occurred is yet to be satisfactorily accounted for by the materialist approach. Also, from a cosmic perspective, the mind-body problem gets to ask how a strictly physical Universe gave rise to consciousness. Now except there is some satisfactory materialist answer as to the cause of this introduction of mind at the significant stage where it occurred, then the narrative of big history from a strictly materialist perspective suffers an unavoidable inadequacy. Some theories even argue that consciousness was there from the very start of the Universe, shining forth in the big bang, and that any successful "theory of everything" must also account for consciousness as fundamental. A metaphysical and reductive materialist position in big history is thereby challenged to respond to such problems.

Coming to transhumanism, and its concern with the use of technology (such as artificial intelligence and biotechnology) in transforming human bodies and minds, the prevalent metaphysical materialist orientation, however, runs into problems especially over the question of mind and consciousness, and significantly when these are appreciated as phenomena attributed to subjective reality. The reductive materialist explanation for mind and consciousness remains inadequate in addressing the question for a number of reasons.

To begin with, "How can we explain consciousness?" This problematic "C" word never fails to stir controversy. According to materialism/physicalism, almost everything in the world can be explained in physical terms, so it is natural to hope that consciousness might be too. This, however, has so far not turned out to be the case, as consciousness has proven intractable to the approach. David Chalmers argues that consciousness escapes the net of reductive explanation, and maintains that no explanation given wholly in physical terms can ever account for the emergence of conscious experience [6].

Chalmers further argues that there are features of the world over and above the physical features, that consciousness is not logically supervenient on the physical, and that this failure of logical supervenience directly implies that materialism is false. Chalmers sets out the basic argument for this goes as follows.

1. In our world, there are conscious experiences.
2. There is a logically possible world physically identical to ours, in which the positive facts about consciousness in our world do not hold.
3. Therefore facts about consciousness are further facts about our world, over and above the physical facts.
4. Therefore materialism is false ([6], 109).

Chalmers further invokes the image and concept of a physically identical zombie world which after all is logically possible, and infers that given this fact it therefore follows that the presence of consciousness is not guaranteed by physical facts alone but is itself an extra fact about our world. According to Chalmers:

> The character of our world is not exhausted by the character supplied by the physical facts; there is extra character due to the presence of consciousness. To use a phrase due to Lewis [21], consciousness carries phenomenal *information*. The physical facts incompletely constrain the way the world is; the facts about consciousness constrain it further [6].

If Materialism, as Chalmers qualifies it, is the doctrine that the physical facts about the world exhaust all the facts, and this by virtue of the position that every positive fact is entailed by the physical facts, then the case that zombie worlds or other inverted worlds are possible simply imply that physical facts do not entail all the positive facts about our world, and materialism is therefore false [6].

Another obvious problem with reductive materialism is simply its inability to account for subjective experience, on what it feels like, for example, to be in pain. To say, for example, that the mind is reducible to behavioural dispositions (a type of materialism) leaves out what is strongly considered one of the most central and unavoidable features of the mental realm – which is that having a mind, or being in certain mental states, has a certain experiential quality to it. As Thomas Nagel expresses, there is something it is like to be minded. Qualia is the term used to describe such experiential properties of the mind – the 'what-it-is-like-ness' of a given mental state. Materialism misses out on qualia, and does not have much to offer in attempting to account for its existence. It fails to explain why being in a state of pain feels like something, and herein lies an argument against materialism on the grounds that it lacks explanatory power with respect to the existence of qualia.

As earlier noted, the idea of "mind uploading" in transhumanism leads many critics to accuse transhumanism of dualism, whereas this idea actually implies functionalism, a form of materialism. But functionalism itself is not spared. One of the reasons that we might resist the functionalist explanation of mind and personal identity is again because of qualia, since it can be argued that there is something important being missed out by the approach, which is that there is something *it is like* to be minded and feel a particular sensation, a quality functionalism does not capture. Indeed functionalism as a view appears to say little and not much that is convincing about how to account for the conscious experiential features of mental states. Functionalism further has an apparent weakness in that it is inordinately liberal in what it accommodates as a mental state. A strong case against functionalism is John Searle's 'Chinese room' thought experiment which presents a formidable argument and demonstration that performing the relevant function is not sufficient for being in the relevant mental state.

The metaphysical materialist approach, in its insistence that matter/energy is all that exists, therefore gets stuck in insufficiency over the problems of mind and consciousness, which inevitable arise in the narrative of big history in its project of everything that is the Universe and everything within it, and transhumanism in its account of how a set of technologies are about to transform humans in their bodies and minds. For transhumanism in particular, the mind-body problem, and the problem of consciousness are unavoidable issues posed before the project in several approaches taken; be it the creation of pure superintelligent AI, augmented humans, merged human/machine cyborgs, or disembodied human brain-minds uploaded unto computer intelligence and non-biological substrates. In each of these cases, the link between embodiment and consciousness creates problems for retaining the idea of the "human" in ways that the materialist conception falls short of solving.

The matter/mind, body/mind problem therefore attends to big history and transhumanism, and in trying to develop both projects, the approach in understanding both human nature, self, and the nature of ultimate reality taken in metaphysical materialism has not been adequate in clearing lingering obstacles impeding the grasp of the basic stuff of existence, an understanding necessary to both construct a comprehensive picture of the Universe, achieve a holistic remaking of the human species, and as well guide a successful artificial design of intelligent, inorganic life.

57.4 Some Competing Views and Their Problems

For singular explanations of the phenomenal reality, materialism as a monist ontology would be in contrast to idealism. Also, if matter and energy are seen as necessary to explain the physical world, but incapable of explaining mind, what ordinarily results is dualism. In this section, we shall therefore briefly consider the other views on the structure of reality and the nature of persons: idealism and dualism, and examine their adequacy or otherwise in serving as ontological bases for big history and transhumanism.

Idealism is the view which holds that only the mental exists. Accordingly, so-called physical things are in some sense manifestations of the mind or of thoughts. In idealism, mind creates reality, subjective reality is given preeminence even sometimes to the denial of objective reality, and the Universe is conceived as fundamentally mental. The option of pure idealism or mentalism, as against materialism, however also falls flat as a metaphysical approach for the task of big history and transhumanism, and this mostly as a result of the philosophy's insistence (in some of its forms) on ideas or purely mental phenomena as the only reality, or as exhaustive of reality (or that all of reality fundamentally is mind-like), whereby matter or body are simply entities that rely or supervene on the mind or thought to exist.

A major problem with mentalism lies in its abstractedness and remote connections with physical reality and the material universe, and a failure to get in touch with objective reality and account convincingly for the external world. These shortcomings significantly detract from idealism's abilities to provide empirical instrument for understanding, prediction, and control of the natural world the way the natural sciences have done and continue to do. Idealism's approach to the question of the body in human nature does not demonstrate much usefulness for the understanding and transformation of the body projected in transhumanism. Such limitations affect the scope of idealism in demonstrating relevance to the data-driven work of big history and the physicalist-oriented technological processes that transhumanism takes into account for shaping the transition of the human into a new and enhanced species.

On the other hand, the inadequacy of materialism with respect to the projects in focus also allows for a proposal of dualism: which holds that there are both physical and nonphysical features of the world. Dualism, like pluralism, falls outside the class of monist ontologies, and supposes that two different kinds of phenomena exist; physical and mental. Essentially introduced by Plato, the position entered Christianity with some reformulations through the work of Augustine and others, and passed down in its contemporary form to us by Descartes and other early modern philosophers.

Dualism is taken by some to represent a common sense metaphysics which articulates that reality has a dual nature. The Universe, in the dualist view, is believed to consist of physical objects on one hand and minds on the other. The approach is essentially the "two-realms view" which holds that what exists is either physical or mental, while some things, such as a human person, have both a physical component (a physical body), and a mental component (a mind). In a normal living person, this view holds, mind and matter are intertwined in such a way that what happens to the body can affect the mind and conversely, what happens in the mind can affect the body. Dualism, as a consequence of its position, is immediately faced with the challenge of accounting for the phenomena of mind-body interactions, with several dualist theories having been offered to explain this relationship, each one fraught with problems. Solving the problem of causality as, for example, explaining the causation between thought and behavior, or between mind and matter, has not been undertaken successfully by (either substance or property) dualists. Though a number of transhumanists are dualists, this is however a minority view, because strictly speaking dualism is not quite helpful to transhumanism. For big history, dualism may conceive a picture of matter and mind (qualia) in the unfolding epic of evolution, but again, explaining the causation and interaction between both (even before the arrival of the human and the human mind on the scene) in all the epochs recognized by the narrative definitely poses a serious problem for dualism. For such problems, panpsychism or panprotopsychism many times have had to be sneaked in to provide some explanatory bail-out for dualism.

Moreover, Cartesian dualism, especially the Cartesian view of the mind or self, has become increasingly rejected by transhumanists in the face of possibilities explored within intersections between cognitive neuroscience, neurotechnology, computing, and computer simulated/mediated reality. According to More:

The high level of interest in philosophy and neuroscience among transhumanists has led to a wide acknowledgment that the simple Cartesian view of the mind or self as a unitary, indivisible, and transparent entity is unsupportable. As we store more of our memories externally and create avatars, it is also becoming increasingly apparent that the boundaries of the self are unclear and may not be limited to the location of a single body. Complementing these questions about the nature and identity of the self at any one time are questions about the identity of the self over time, especially for a self that undergoes major cognitive and somatic changes over an extended lifespan ([25], 7).

In a subsequent section, we shall encounter a "patternist" view of Ray Kurzweil which apparently correlates with what More says here about the identity of the self that undergoes changes over time. In any case, More's view reflects functionalism in transhumanism, a view which we've argued to contain lapses in explanatory power.

Reacting to the myriad obstacles attending to these traditionally predominant ontological systems, some theorists hold a different belief that what exists is ultimately neither mental nor physical, and others take it that what exists is ultimately both mental and physical. According to this view, known as "neutral monism", the mental and physical are just different ways of viewing the same things which ultimately in themselves are neutral between the two categories.

57.5   Neutral Monism: An Alternative

As an alternative to the materialism/idealism/dualism disagreements in relation to both big history and transhumanism, the ontological monistic view known as neutral monism could present a reasonable solution and way out. This neutral monism, aside being incorporated into solving the problem of matter/mind in big history and transhumanism, also emerges as a viable candidate qualifying for advancement in the two narratives in order to in turn create new pathways for the development and attainment of goals by both projects. The thrust of this essay, to hence clearly state, is to explore how advancing neutral monism in its positing of an alternative fundamental stuff holds could hold new promise of understanding and progress for the big history and transhumanist projects.

Neutral monism, a position in metaphysics and philosophy of mind, is a monistic metaphysics which holds that ultimate reality is all of one kind. Within this monistic framework, neutral monism, as opposed to dualism and pluralism, stands in agreement with the more familiar versions of monism, i.e., idealism and materialism. What however sets the difference between neutral monism and the other monistic approaches is the position that the intrinsic nature of ultimate reality is neither mental nor physical, or ultimately both mental and physical.

Most versions of neutral monism were conceived as solutions to the mind-body problem. The goal was to close the apparent chasm between mental and physical entities by exhibiting both as consisting of groups of the more basic neutral entities. So what, according to the theory, does it mean for an entity to be neutral? There are several proposals, but here we will present two that we find more mainstream and most plausible:

The first is the *Neither View,* which holds that a basic entity is neutral just in case it is intrinsically neither mental nor physical, and the second is *the Both View,* which holds that a basic entity is neutral just in case it is intrinsically both mental and physical [35]. Both views, nonetheless, are similar in letting the question of an entity's neutrality be settled by its intrinsic nature

The classic version of neutral monism was defended by Ernst Mach, William James, and Bertrand Russell [2], these were known as the "big three" of the theory. The traditional versions of neutral monism—those developed by the big three— accept the *Neither View.* The three traditional neutral monists all pointed to experience, when attempting to provide examples of neutral entities. Mach spoke of sensations, James of pure experience, and Russell of sensations and percepts [2].

In the evaluation of the big three, neutral monism solves the mind-body problem. A demonstration could be found, for example, in Russell's account of experience (of perceptual consciousness). In Russell, there is a frequent emphasis on the miracle or mystery involved in traditional accounts of perception ([31], 147, 154; [30], 275, 400). As this view explains, there mysteriously arises at the end of a purely physical chain of causes something of an entirely different nature: an experience (a sensation, for instance, of green). What occurs here is what is regarded as the hard problem of consciousness. For the neutral monist, though, this problem gets solved by asserting that "we cannot say that 'matter is the cause of our sensations" ([31], 290). This is supported by Mach: "Bodies do not produce sensations" ([23], 29). Accordingly,

> To suggest otherwise is to rely on "the monstrous idea of employing atoms to explain psychical processes" ([23], 311). Matter/bodies are, after all, nothing but collections of neutral entities, i.e., of Russellian events or Machian elements. So mind-body causation reduces to causal relations among events or elements. And, for all we know, the event/element causing a sensation may be quite similar to the sensation it causes. This closes the apparent chasm between the "material process" and the ensuing experience, and the mystery of perception vanishes [35].

Mach objected to the dualism of "inner" and "outer" aspects, coming up, as reported by Banks, therefore with one kind of elements:

> There is no rift between the psychical and the physical, no inside and outside, no "sensation" to which an external "thing" different from sensation, corresponds. There is but one kind of element, out of which this supposed inside and outside are formed. ([24], 310)

Neutral monism further attends to the case of existence and knowledge of the external world alongside the existence of a subjective mind or consciousness. The view was understood by Mach and James as a form of gaining perceptual contact with the world. It can be understood as a case in which the relation between the subject and its perceptual object becomes the identity relation, and in perception "subject and object merge" ([17], 57). Stubenberg explains:

> A single reality—a red patch, say, when we see a tomato—is a constituent of two groups of neutral entities: the group that is the perceiver, and the group that is the tomato. The mind and its object become one. In James's words: A given undivided portion of experience, taken in one context of associates, play[s] the part of the knower, or a state of mind, or

"consciousness"; while in a different context the same undivided bit of experience plays the part of a thing known, of an objective "content". In a word, in one group it figures as a thought, in another group as a thing [35].

This epistemic achievement, of establishing relationship between subject and perceptual object, has been seen by some as the most significant outcome of neutral monism.

Neutral monism obviously has struggled under the shadows of the more established monistic and dualistic ontologies within metaphysics, and it seemed to have gone onto oblivion in the twentieth century after the *big three*. But this is not completely true, because aside surviving in the works of some major thinkers within the century, the approach itself is now undergoing something of a resurgence. Banks talks about this:

> Neutral monism never really disappeared in the twentieth century. After Russell, it was taken up in a problematic way by Rudolf Carnap and the Vienna Circle, by Erwin Schrodinger, by Herbert Feigl and many others…….. Recently, there has been a modest revival of the position deriving from a passage in Russell's (1927/1959) *Analysis of Matter.* ([3], 180)

57.6   The Contemporary Revival in Neutral Monism

Neutral monism (in its traditional form) for a long time was seen as a brief and insignificant feature within the world of metaphysics, but this "fortune" has begun to change in recent times. There is currently a wave of interest both in the traditional versions of neutral monism, as well as in the development of alternative versions of the view, all of which indicate that neutral monism apparently is staging a comeback as a veritable option within continuing endeavours to explore and understand the metaphysics of mind. This current revival of interest in neutral monism can be seen as offering new hope to finding solutions to age-long obstacles in the metaphysics of consciousness which traditional positions have so far found insurmountable. The contemporary neutral monist strand within the broader movement of Russellian monism, a neo-Russellian neutral monism, is a promising new development, although this same strand yet explores other fresh ideas that provides steady invigoration overall for neutral monism. Still, some contemporary attempts to revive neutral monism are completely unattached to the traditional (Machian/Russellian/Janesian) forms of the view, and whereas the classical neutral monists accept the *Neither View*, a number of contemporary discussions of neutral monism employ the *Both View*.

In all, there is the pursuit of a consistency in the basic constituents that should consist of and explain existing psychophysical things. On this point, Stubenberg elaborates:

> The claim that all of the basic constituents must be of the same kind—the conclusion that we are dealing with a monism—is supported by the consideration that any basic constituent could be part of a psychophysical creature, such as ourselves. Hence every basic constituent must be of the kind that can explain the physical and mental traits as we know them [35].

Within the contemporary stream, Thomas Nagel's defense of neutral monism comes to mind. Nagel in his *Mind and Cosmos* [27] asserts that "the weight of the evidence favors some form of neutral monism over the traditional alternatives of materialism, idealism, and dualism" ([27], 5). Neutral monism is understood as a view that "accounts for the relation between mind and brain in terms of something more basic about the natural order" ([27], 56). From this we get a picture of a "general monism according to which the constituents of the universe have properties that explain not only its physical but its mental character" ([27], 56). Furthermore Nagel, in an earlier confirmation of his side with the *Both View*, writes:

> …..this view would imply that the fundamental constituents of the world, out of which everything is composed, are neither physical nor mental but something more basic. This position is not equivalent to panpsychism. Panpsychism is, in effect, dualism all the way down. This is monism all the way down ([26], 231).

With respect to the entities of which neutral monism consists, a number of contemporary neutral monists have been offering some familiar examples, while others have sought the neutral entities in the realm of the abstract. Philosophers and scientists are very much involved in currently developing this idea to make abstract entities into the neutral basis of a metaphysical system- abstract entities which include information, structure, computation, and mathematical reality. Kenneth Sayre, for example, looks into the mathematical theory of information and proposes that the neutral base is best understood as the "ontology of informational states" ([32], 16) presupposed by this theory. More recently, Chalmers [6] has also investigated the idea of an informational ontology. Others have likewise investigated and advanced the related ideas that ultimate reality is purely structural [11, 12, 20], a computational process [13, 21], or purely mathematical [36]. Within digital physics, the mathematical universe and computational universe hypotheses could be broadly accommodated within informational ontology. Our essay selects the ontological framework of **Information**, in its accommodation of patterns, the mathematical universe, and computational processes, as the most viable neutral monistic entity fundamental to ultimate reality.

57.7 Information as Monistic Ontology Fundamental to Reality

From the time of Claude Shannon, widely considered the father of information theory, the concept of information has gone through several developments. Informational ontology, as a neutral monism, basically posits that information is the ultimate constituent from which the cosmos is constructed, and that fundamentally speaking, our universe and everything within it is ultimately made of information. What therefore is "information?" We may begin by stating that information has multiple meanings, and these include:

Facts or knowledge (things one can learn); a measure of difference or surprise (how much one learns); one of two opposite states (on-off, yes-no, one-zero); the mathematical description of a communication system; the content of computation; quantum entangled states (enabling vast computing power); and power to explain and possibly to cause [18].

Coming from a long and evolving picture of our Universe as a library, an album, a watch, a view currently experiencing increasing prominence is that of the Universe as a computer, and that of the Universe as computing its own destiny. Within digital philosophy today, this computational understanding of reality blends with the approach of informational ontology.

By the 1970s, Kenneth Sayre published his main ideas on neutral monism and, unlike the *big three*, he finds the neutral base of his system not in experience, but in the realm of pure information, where information is understood in the strict information theoretical sense. An ontological claim about the ultimate nature of reality was presented by Sayre, one which held that ultimate reality consists of informational states:

> If the project...is successful, it will have been shown not only that the concept of information provides a primitive for the analysis of both the physical and the mental, but also that states of information...existed previously to states of mind. Since information in this sense is prior to mentality, but also implicated in all mental states, it follows that information is prior also in the ontological sense...Success of the present project thus will show that an ontology of informational states is adequate for an explanation of the phenomena of mind, as distinct from an ontology of physical events......... It is a reasonable conjecture that an ontology of information is similarly basic to the physical sciences.... ([32], 16)

The traditional versions of neutral monism had faced an enormous challenge to demonstrate how basic entities that are derived from experience can be neutral, rather than mental. Sayre steps over this problem by selecting an ontology of informational states as his "neutral stuff". The neutrality of informational states accepted, there however remained the demanding question of the relationship of this abstract "stuff" to concrete world of physical and mental entities. Chalmers gets to explore this idea of an informational ontology and its questions, and entertains the idea that the solution to the hard problem of consciousness might be arrived at through the assumption that information—along with matter and energy—is a fundamental property of reality [6]. To Chalmers, "information," which emerges from certain physical configurations and processes with consciousness within, is a fundamental component of reality, just as time, space, matter and energy.

Moving forward, this discussion around the idea of "information as ultimate reality" has been made all the more lively and rich by the steady rise of digital philosophy and digital physics. Noticeably, several proponents of digital physics view information rather than matter to be fundamental. For theorists within this growing view, information is more than just a description of our universe and the stuff in it, rather it is the most basic currency of existence, occupying the "ontological basement" of reality [9], and where matter historically has been at the bottom of the explanatory chain, with information being a secondary derivative of it, some physicists are now reversing the view to consider that fundamentally, the universe is perhaps about information and information processing, and it is matter instead that emerges as a secondary concept. In the grand chain of existence, as this new view indicates, information could be bedrock.

The origin and evolution of the universe, as conceived from the perspective of informational ontology, is summarized in the simple and attractive phrase, *It from Bit*, coined by physicist John Wheeler. For Wheeler, it was information first, everything else later, calling for us to learn how to combine bits in large numbers to obtain existence ([41, 42]). "It from Bit", now "It from Qubit" (quantum bit) is fast gaining popularity among cosmologist, physicists, quantum biologists, and philosophers. Also at ultimate levels, bit-string theory is emerging as a new and engaging field for scientific and philosophical exploration.

Along the lines of "It from Qubit" thinking, Quantum information theorist Vlatko Vedral, for example, believes that the rules of quantum information provide the most "compact" description of physics, In *Decoding Reality: The Universe as Quantum Information*, Vedral explores the big and deep questions about the universe, questions as: where everything comes from, why things are as they are, and what everything is. In essence, big history questions. He argues that the most fundamental building block and definition of reality is not matter or energy, but information. Vedral explains that this is a useful framework for viewing all natural and physical phenomena, arguing that the processing of information serves as the basis of all physical, biological, economic, and social phenomena [39]. In building out this framework he touches upon "the origin of information, the idea of entropy, the roots of this thinking in thermodynamics, the replication of DNA, development of social networks, quantum behaviour at the micro and macro level, and the very role of indeterminism in the universe" [43]. With this view, he tackles several seemingly unrelated questions as "Why does DNA bind like it does?" and "What is the ideal diet for longevity?" Vedral's account assures that unifying all is possible through the understanding that everything consists of bits of information.

So where do informational "bits" come from? Vedral points to the weird realm of quantum physics. Occurring at this sub-sub-subatomic level are interactions of separated quantum particles--Einstein's famous "spooky action at a distance." Vedral discusses how quantum weirdness, once considered limited to the tiniest scale, may actually reach into and have some effect on the macro world. As Vedral submits, finding the answer to the ultimate question of life, the universe, and everything, lies in quantum physics. Where Vedral is concerned, it seems *It from Qubit* flows from physics, chemistry, and biology well into history, sociology, economics, and the like.

This informational idea is defended by author of "Programming the Universe" Seth Lloyd through his likening of the universe to a computer. At the fundamental level, according to Lloyd, the universe consists of information; every elementary particle carries information. Information, as Lloyd explains, is not just a way of appreciating or approximating how the universe works, rather it is the literal, most fundamental way it actually works [22]. To him therefore, the universe is seen "not *like* a computer as an explanatory metaphor; it really *is* a computer as scientific fact. As such, he claims that all changes in the universe are "computations." [18])

Still other theorists conceive and explain that our universe could be a simulation, running on a cosmic computer. To some of such theorists within the digital philosophical school, the universe as we experience it may actually be the projection of information encoded on some distant cosmic boundary. Questions remain open concerning where this boundary lies and how the projection occurs, but as the arguments of these theorists hold, our reality may essentially be but a hologram.

From the universe, the neutral monist informational ontology follows the cosmic evolution narrative all the way to the emergence of life, humankind, psychology, culture, and human social systems. In biology, the approach of viewing the genome as an information store measurable in bits, and the later scientific achievement of cracking the genetic code have significantly enhanced our understanding of the phenomena of life. In the words of Richard Dawkins:

> What lies at the heart of every living thing is not a fire, not warm breath, not a "spark of life." It is information, words, instructions. If you want a metaphor, don't think of fires and sparks and breath. Think, instead, of a billion discrete, digital characters carved in tablets of crystal ([10], 112).

In the history of the flow of this informational approach from cosmology to biology, following Erwin Shrodinger's 1943 "What is Life?" several physicists began to turn to biology, and biologists adopted approaches from physics. Some of the physicists turning to biology saw information as the concept needed to discuss and measure biological qualities for which tools had been hitherto unavailable: complexity and order, organization and specificity. Biologists got to count in terms of "bits." What emerged was a new molecular biology which began to examine information storage and information transfer. The idea that the precise sequence of the bases of life is the code which carries the genetical information became therefore more likely. James Gleick summarizes the growth of this understanding as follows:

> The replication of DNA is a copying of information. The manufacture of proteins is a transfer of information: the sending of a message. Biologists could see this clearly now, because the message was now well defined and abstracted from any particular substrate. If messages could be borne upon sound waves or electrical pulses, why not by chemical processes? Gamow framed the issue simply: "The nucleus of a living cell is a storehouse of information." Furthermore, he said, it is a transmitter of information. The continuity of all life stems from this "information system"; the proper study of genetics is "the language of the cells." ([14], 308)

The terms: "code", "information", "language", we must indicate, were/are not used by these biologists figuratively, but as actual scientific facts. More than half a century ago, Watson and Crick finally deciphered the genetic code, a system of four letters and an efficient digital information store which underpins heredity and other biological functions. The code of life was finally understood.

Fred Spier talks about different levels of complexity, and where different types of building blocks can be discerned, with connections and sequences forming the aspects [34]. From biology the information-contained sequence flows, where intelligent life is concerned, into psychology and culture.

> The sequence of building blocks, and thus information, mostly matters in life and culture. In life, the genetic information is organized in long strands of DNA molecules, in which the

sequence of the building blocks is of overriding importance for determining what happens inside cells. In a similar way, sequence is also important for all cultural information and communication ([34], 43).

In the area of mind and consciousness, historically, there also occurred what some philosophers call the "informational turn", a moment which ushered in the cognitive revolution in psychology and laid the foundations for cognitive science which combined psychology, computer science, and philosophy. Frederick Adams states: "Those who take the informational turn see information as the basic ingredient in building a mind, Information has to contribute to the origin of the mental." ([1], 495) Going by such conceptual orientations, a new view with much potential to aid understanding of consciousness is what has become known as the *neural code*. This neural code refers to "the rules or algorithms that transform firing of brain cells, action potentials and other neural activity and processes in the brain into perceptions, memories, meanings, emotions, intentions, and actions. Think of it as the brain's software." ([16], 1). This neural code, however, in contrast to the genetic code, appears much more complex and difficult to crack. However, there is cause for hope, and this could have significant impacts both for the future of humanity, for the human understanding of consciousness and the Universe, and even for the human power to redesign reality. As Horgan considers it:

> Given the relentless pace of advances in optogenetics, computation and other technologies for mapping, manipulating and modelling brains, a breakthrough could be imminent. If researchers crack the code, they might solve some ancient philosophical mysteries as the mind-body problem and the riddle of free will. A solution to the neural code could also give us unlimited power over our brains and hence minds. Science fiction—including mind control, mind reading, bionic enhancement and even psychic uploading—will become reality. This could facilitate the attainment of the Singularity. The intelligence augmentation and mind uploading that would lead to a technological singularity depend upon cracking the neural code [16].

Presently, neuroscientists have many candidate codes for the neural code, among which there are quantum and chaotic and information codes. Eventually, through *Information* there might just be a link connecting the genetic with the neural code, i.e., a neurogenetic code, an informational element as the basic constituent of body and mind.

The neutral monistic entity: information, is a neutral entity with much promise to provide a psychophysical ontological base that flows from and connects cosmology, physics, geology, biology, psychology and culture. This approach holds out promise for big history and transhumanism.

57.8   Advancing Neutral Monism as New Ontological Foundation for Big History and Transhumanism

Informational ontology as a neutral monism, and the understanding of the concept, has shaped the fabric of reality and consciousness, influenced human culture, and currently has the power to transform civilization. Even in the common-sense understanding of the concept, Information has revolutionized our world. From language to writing to the internet, information has affected the epoch in big history which covers the emergence of humankind, the human cognitive revolution, and the development of human culture and memes up to the point where post-humanity now beckons. Reflecting on the human significance on biological and cultural evolution, the role of information in the process, and the impact of our presence on the planet, Spier declares:

> ....human history represents a fundamentally new phase in biological evolution. For during the entire history of life, no other organism has existed that has changed the face of Earth in such profound ways within such a short period of time. Humans have been able to do so thanks to their unprecedented ability to process, store and transmit enormous amounts of information. This process is known as 'culture.' Whereas many animals exhibit forms of cultural learning, only humans have used it to such a large extent for shaping both their own history and the surrounding natural environment. For this reason, humans may well be the most complex adaptive species to have emerged on our planet. ([34], 111–112)

Furthermore the informational framework, by basing the origin and processes of the Universe and complexity leading to life, Earth, consciousness, humans, and cultural phenomena on the journey of information, holds out potential to enhance our understanding of the universe and the evolutionary processes from big bang to humankind as big history seeks to achieve. But it does not end there, because the understanding of the basic constituents and mechanisms both of our universe, of human nature, and of culture, will enable our engineering of the future, both of our species and civilization, and of the entirety of space-time. This conveys the spirit of transhumanism. Information could help us achieve this. As Phillip Perry says:

> Consider that if we knew the exact composition of the universe and all of its properties and had enough energy and know-how to draw upon, theoretically, we could break the universe down into ones and zeroes and using that information, reconstruct it from the bottom up. It is the information, purveyors of this view say, locked inside any singular component that allows us to manipulate matter any way we choose. Nanotechnology, programmable matter.....Of course, it would take deity-level sophistication, a feat only achievable by a type V civilization on the Kardashev scale. ([28], 1)

This statement captures the significance of informational ontology for both big history and transhumanism. Understanding and advancing information as fundamental element in the world promises to furnish us with insights into the nature of reality, the evolution of the cosmos, the emergence and evolution of life, thought, and human institutions, and significantly for transhumanism, a solution to the mind-body problem, a way to transform our bodies and minds, create new life and intelligences, and manipulate matter anyway we want, from nanotechnology to the level of cosmic engineering. Within the twenty-first century, artificial intelligence, for example, will probably be the most important agent of change, with the power to transform our economy, our culture, our politics, and even our own bodies and minds in various unimaginable ways. Today, bioinformatics and genetic engineering, practices driving transhumanism, stem from the understanding of life as basically information. Nanotechnology, biotechnology, information technology, neurotechnology, augmented reality, and bionic convergence will also facilitate this transformation. All of these stand to experience advancements through an unlocking of information as that ontological basement underlying matter, mind, and life.

For transhumanism, understanding the mind and concept of the self is indeed important. A computational-informational theory of mind, currently gaining prominence as a conception for self and mind, can be appreciated as conforming to and aiding transhumanist objectives. According to Susan Schneider,

> Consider that Transhumanists generally adopt a Computational Theory of Mind. That is, the mind is essentially the ——program running on the hardware of the brain, where by ——program what is meant is the algorithm that the mind computes, something in principle discoverable by cognitive science ([33], 6).

For transhumanism and other purposes, this computationalist view, further, entails a rejection of the materialist conception of the nature of human persons, Schneider continues:

> ....Because, at least in principle, the brain's computational configuration can be preserved in a different medium, i.e., in silicon as opposed to carbon, with the information processing properties of the original neural circuitry preserved, the computationalist rejects the materialist view of the nature of persons [33].

Materialism therefore gets challenged by this computational-informational understanding of the nature of human persons which comes across as more consistent with transhumanist goals. For transhumanist purposes, materialist conceptions of the self does not prove as sufficient as the idea of the informational patterns does. Schneider argues that materialism appears to falter in embracing the very idea that *you are what you are made up of:* [33], and for support of this point, Ray Kurzweil here comes up:

> The specific set of particles that my body and brain comprise are in fact completely different from the atoms and molecules that I comprised only a short while ago. We know that most of our cells are turned over in a matter of weeks, and even our neurons, which persist as distinct cells for a relatively long time, nonetheless change all of their constituent molecules within a month....I am rather like the pattern that water makes in a stream as it rushes past the rocks in its path. The actual molecules of water change every millisecond, but the pattern persists for hours or even years. ([19], 383)

This view of the self as patterns of information surely can help transhumanism in various ways: mind uploading, teleportation, virtual embodiment, consciousness and experience within virtual and augmented reality, robotic re-embodiment, biohacking, human enhancement, pure superintelligent AI, as well as the creation of augmented intelligences through a merger of human and machine intelligence.

As indicated earlier, departing from classical physics, several physicists over the decades come to a realization that quantum mechanics is not the theory of subatomic particles but that of information, or quantum information, from which is derived qubits. The fabric of reality therefore is information-theoretic (or better yet, code-theoretic), and computational. From the quantum information view, our consciousness, in the information-theoretic understanding, is a data stream, and is substrate-independent, as such one can reproduce mind on a different substrate other than biological wetware. Ultimately, mind-like computational substrate does not even require the existence of particles to be built upon but rather dimensionless bits of information, with only patterns of information being quintessential. With this informational ontology of consciousness, much could be realized in terms of progress in big history and transhumanism. For one, this would support the creation of conscious AI systems which will in time create their own virtual multiverse based on AI intersubjectivity and new forms of communication such as holographic language. That way, the history of the Universe, or multiverse flows into a future of new universes, a human-AI created multiverse. The question, moreover, of whether humans themselves will have access to the virtual multiverse created by AIs might be answered by the extent to which humans, or post-humans, will be either willing to continue harnessing their own informational ontology in augmenting themselves through these worlds accordingly, or in merging their informational ontology with those of AIs in created augmented intelligences that would create and exist in the Universes of the future.

Moreover, by information, and specifically information and communications technology, the wiring of the world, followed by the spread of wireless communication, have added some concrete to centuries' long speculations about the birth of a new global organism. As far back as the nineteenth century mystics and theologians had begun to talk about a shared mind or collective consciousness, developed through the collaboration of millions of people placed in communication with one another. As ahead of its time as it might now sound, some even went as far as to view this new creature as a natural product of continuing evolution, and a means by which humans would get to fulfill their special destiny [14]. Attaining this consists of, among things, creation of the "noosphere", the sphere of mind, which represents a climactic "mutation" in evolutionary history [14]. Pierre Teilhard de Chardin took further steps in promoting the noosphere, calling this a "new skin" on the earth:

> Does it not seem as though a great body is in the process of being born—with its limbs, its nervous system, its centers of perception, its memory—the very body of that great something to come which was to fulfill the aspirations that had been aroused in the reflective being by the freshly acquired consciousness of its interdependence with and responsibility for a whole in evolution? ([37], 174)

However, de Chardin was not alone in this, because many people in his time were likewise testing the same idea, among whom included science fiction writers. Half a century later, internet pioneers demonstrated enthusiasm for the idea. Today, with the rise of brain-computer interfaces (BCIs), quantum computation, virtual reality, augmented reality, and the increasing possibilities of mind uploading and technologically-enabled telepathy, this theme of a noosphere, global mind, universal consciousness, and the unification of human consciousness with the Universe, resonates in transhumanism with potential implications for big history.

According to the "It from Bit" worldview, patterns of information emerging from the ultimate code/information are what is more fundamental in the Universe than particles of matter or space-time continuum itself, which are levels emerging from the "Cosmic Code". Nature behaves quantum code-theoretically at all levels, leading some digital philosophers to describe our world as a multiplayer virtual reality. Digital physicists talk of the possible existence of a larger consciousness system, an intricate web of universal quantum neural network of sorts. An understanding and manipulation of this could offer the forming of a data stream which is essentially the merger of personalized or collective stream of human consciousness with (the) Cosmos. Advancing the approach of neutral monism, and the element of information especially, in big history and transhumanism might just secure dramatic breakthroughs for both narratives. With the potentials demonstrated, our grasping the ultimate nature of the Universe, transforming our species, merging with the Universe, and creating new universes and realities of our own are all possibilities that hold out for both the big history and transhumanist movements through a comprehension of the informational-ontology underlying our Universe, life, matter, and mind.

57.9   Concluding Remarks

Neutral monism, having been something of a sideshow on the stage of metaphysics, is gradually experiencing a turn in fortunes as more contemporary philosophers are beginning to consider it as perhaps that approach that could provide solutions to age-old metaphysical problems that have so far proven intractable to traditional explanatory approaches, problems that include the nature of ultimate reality, the nature of self, and those found within the metaphysics and science of consciousness. Given the nature, objectives, and implications of big history and transhumanism, an ontological approach that could help clear the lingering problems and present the basic constituents from which reality, matter, body, mind, consciousness, all stem and hence could be reprogrammed, could help both narratives with clearer understanding of, and more power to shape the future of, humanity and the Universe.

   Through more work in the philosophy of science and a deeper understanding of causation neutral monism might enjoy better chances for revival [3], however, the growing popularity of the concept of information, if firmly located as a neutral monistic entity, could achieve a better establishment of the approach. Informational ontology, as has been presented in this essay, could present an explanation of reality, life, and history from the tiniest quantum worlds to the sphere of physical, mental, social, and cultural processes. In *Decoding Reality,* Vedral for example notes how recent evidence suggests that quantum weirdness, once considered limited to the tiniest scale, may actually reach into the macro world and make a real possibility something like teleportation. The effect of such on transhumanism, as well as the metaphysic of big history, can be indeed significant.

Finally, mind uploading has been considered a path towards an idea that is of interest to some transhumanists, i.e., what Frank Tipler describes as the "Omega Point", a notion which, drawing upon ideas in digitalism, Tipler advanced in stating that the collapse of the Universe billions of years hence could create the conditions for the perpetuation of humanity existing in a simulated reality within a megacomputer, thus leading the human into achieving a form of "posthuman godhood" [42]. It appears that using computational-informational ontology within digital physics, humans would be able to pursue this goal, and if this is the case then big history (in its concern with the origin, nature, evolution and future of the Universe and everything), and transhumanism (in its concern with the nature, evolution and future of humanity), might both record progress in their objectives through further incorporation of the neutral monism of information within their methodological and metaphysical frameworks.

References

1. Adams, Frederick. 2003. "The Informational Turn in Philosophy," *Minds and Machines* 13: 495.
2. Banks, Erik, C. 2003. *Ernst Mach's World of Elements*, Dordrecht: Kluwer
3. _____ "Neutral Monism Reconsidered." *Philosophical Psychology*, 23 (2): 173 — 187.
4. Bohan, Elsie. January 2016. "Empiricism Is Not A Dirty Word." *Origins*, VI (1):3–6.
5. Brown, Cynthia Stokes. January 2016. "The Meaning of Big History, Philosophically Speaking" *Origins*. VI (1): 7–13.
6. Chalmers, David. 1996. *The Conscious Mind: In Search of a Fundamental Theory.* New York: Oxford University Press.
7. Chomsky, Noam. 2000. *New Horizons in the Study of Language and Mind.* Cambridge: Cambridge University Press.
8. Christian, David. 2015. "Mapping to Meaning." In *Creation Stories in Dialogue: The Bible, Science, and Folk Traditions,* edited by Alan Culpepper and Jan van der Watt. Leiden Boston and Tokyo: Brill.
9. Davies, Paul. 2010. Universe from Bit. In *Information and the Nature if Reality: From Physics to Metaphysics,* edited by Davies, Paul and Gregerson, Niels, H. Cambridge: Cambridge University Press.
10. Dawkins, Richard. 1986. *The Blind Watchmaker.* New York: Norton.
11. Floridi, Luciano. 2008. "A Defence of Informational Structural Realism." *Synthese* 161(2): 219–253.
12. _____. 2009. "Against Digital Ontology." *Synthese* 168(1): 151–178.
13. Fredkin, Edward. 2003. "An Introduction to Digital Philosophy." *International Journal of Theoretical Physics* 42(2): 189–247.
14. Gleick, James. 2011. *The Information: A History, A Theory, A Flood.* New York: Pantheon Books.
15. Goetz, Stewart and Taliaferro, Charles. 2008. *Naturalism.* Grand Rapids, MI and Cambridge, UK: William B. Eerdmans Publishing.
16. Horgan, John. 2016. "The Singularity and the Neural Code." *Scientific American.* https://blogs.scientificamerican.com/cross-check/the-singularity-and-the-neural-code/ Accessed April 2, 2018.
17. James, William. 1905. "La Notion De Conscience", Archives de Psychologie, V (17). Translated as "The Notion of Consciousness." *Journal of Consciousness Studies,* 12(7), (2005): 55–64.

18. Kuhn. Robert, L. 2015. "Forget Space-Time: Information May Create The Cosmos" *Expert Voices. https://www.space.com/28477-didinformation-create-the-cosmos,html* Accessed March 30, 2018.
19. Kurzweil, Ray. 2005. *The Singularity is Near: When Humans Transcend Biology*. New York: Viking.
20. Ladyman, James, and Don, Ross. 2007. *Everything Must Go: Metaphysics Naturalized*. Oxford: Oxford University Press.
21. Lewis, David. 1990. "What Experience Teaches," in Mind and Cognition: A Reader, W. Lycan (ed.), Oxford: Blackwell.
22. Lloyd, Seth. 2006. *Programming the Universe: From the Big Bang to Quantum Computers*. London: Jonathan Cape.
23. Mach, Ernst. 1886. Die Analyse der Empfindungen und das Verhältnis des Physischen zum Psychischen, fifth edition translated as The Analysis of Sensations and the Relation of Physical to the Psychical, New York: Dover. 1959.
24. ———. 1959. *The Analysis of Sensations and the Relation of Physical to the Psychical*, trans. Williams, C.M. and Waterlow. S. New York: Dover.
25. More, Max. 2013. "The Philosophy of Transhumanism" *The Transhumanist Reader: Classical and Contemporary Essays on the Science, Technology, and Philosophy of the Human Future*, 1st Edition, edited by More, Max and Vita-More, Natasha. Los Angeles, CA: John Wiley & Sons.
26. Nagel, Thomas. 2002. *Concealment and Exposure*. Oxford: Oxford University Press.
27. ———. 2012. *Mind and Cosmos: Why the Materialist Neo-Darwinian Conception of Nature is Almost Certainly False*. Oxford and New York: Oxford University Press.
28. Perry, Phillip. 2017. "The Basis of the Universe May Not Be Energy or Matter but Information." *Big Think.* http://bigthink.com/phillip-perry/the-basis-of-the-universe-may-not-be-energy-or-matter-but-information Accessed March 15, 2018.
29. Planck, Max. Quotes. Accessed April 6, 2018. www.goodreads.com/author/quotes/107032Max__Planck.
30. Russell, Bertrand. 1927a. *The Analysis of Matter.* London: George Allen & Unwin.
31. ———.–1927b. *An Outline of Philosophy*. London: George Allen & Unwin.
32. Sayre, Kenneth. 1976. *Cybernetics and the Philosophy of Mind*. Atlantic Highlands: Humanities Press.
33. Schneider. Susan. 2008. *Future Minds: Transhumanism, Cognitive Enhancement and the Nature of Persons*. http://repository.upenn.edu/neuroethics_pubs/37 Accessed February 2, 2018.
34. Spier, Fred. 2010. *Big History and the Future of Humanity*. Malden, MA and Oxford: Wiley-Blackwell.
35. Stubenberg, Leopold, "Neutral Monism." *The Stanford Encyclopedia of Philosophy* (Winter 2017 Edition), edited by Edward N. Zalta. **https://plato.stanford.edu/archives/win2017/entries/neutral-monism/.** Accessed March 10, 2018.
36. Tegmark, Max. 2014. *Our Mathematical Universe. My Quest for the Ultimate Nature of Reality*. New York: Alfred A. Knopf.
37. Teilhard de Chardin, Pierre. 1999. *The Human Phenomenon*, trans. Appleton-Weber, Sarah. Brighton, U.K.: Sussex Academic Press.
38. Various. 2003. "The Transhumanist FAQ: v 2.1." *World Transhumanist Association.* http://humanityplus.org/philosophy/transhumanist-faq/ Accessed February, 10, 2018.
39. Vedral, Vlatko. 2010. *Decoding Reality: The Universe as Quantum Information*. Oxford: Oxford University Press.
40. Werner, Heisenberg, Quotes. Accessed April 6, 2018. www.goodreads.com/author/quotes/64309Werner__Heisenberg.
41. Wheeler, John A. 1994. *It from bit. In At Home in the Universe*. Woodbury, NY: American Institute of Physics Press.

42. _____. 1990. "Information, Physics, Quantum: The Search for Links." In *Complexity, Entropy, and the Physics of Information*, edited by W.H. Zurek. Redwood City, CA: Addison-Wesley.
43. Wikipedia. 2018. "Decoding Reality" https://en.m.wikipedia.org/wiki/Decoding_Reality. Accessed March 9, 2018.
44. Various. 2003. "The Transhumanist FAQ: v 2.1." World Transhumanist Association. http://humanityplus.org/philosophy/transhumanist-faq/. Accessed February, 10, 2018.

Chapter 58
World at Play

Anca I. Selariu

58.1 Introduction

This text was initially intended to be academic. Then I realized that the world does not need yet another non-fiction piece that only speaks to a handful of specialists. This is not a message exclusive to elites, destined to get stuck on the shelf of an academic library, abandoned by someone who was once interested in piling up bibliography references for their thesis, only to be read by bored committee members and then forgotten.

Therefore, you will find no list of references, because there is not one thought here that has not been thought before, that has not been uttered, recorded or written in so many ways, that all humans who ever drew breath would have to be credited here.

This is a letter to the human spirit. I want to speak to *you* – the human who is reading this. I want to address your very essence, whoever you are, whatever you do for a living, however old you are, wherever you grew up; whether you are a caregiver, a trendy socialite, a lab rat, a yogi master, an incurable pessimist, a priest, an unruly teenager, a tired sole breadwinner for a family of 9; whether you are getting ready for the Olympics or struggling with chemotherapy for the fourth time; whether you live in the White House, a moldy apartment building or on the streets of a freshly bombed city.

This is a letter to *You*, human of this planet, with your built-in pluripotentiality, who today wields the key that opens the doors of tomorrow, whether you think you mean anything or not.

I thank you for being you – exactly as you have been, exactly as you are, exactly as you are going to be, whatever your choices may be. I cannot dare to teach you how to be – you do it perfectly. I have no knowledge to impart that you cannot

A. I. Selariu (✉)
BioViva, Bainbridge Island, WA, USA

© Springer Nature Switzerland AG 2019
N. Lee (ed.), *The Transhumanism Handbook*,
https://doi.org/10.1007/978-3-030-16920-6_58

understand. I sit on no great secrets; I hold no hidden knowledge that you do not already have. I merely utter here what *you* taught me.

I am speaking to the one you find within, sneaking into the dreamworld; the one you must whip back into shape and drag back into the obligations, persona, or structure of every day necessities or conventions. The one who seems to be destined to live with this unbearable sense of longing for something you cannot quite define, that seems to spring at you in curious shapes and leads you on such surprisingly diverse paths in search for answers, in search of ways to contain it or fulfill it. Within you, and within all there is, whether you see it or not, is the kernel, the engine of the world. This is our journey.

58.2   The Unspoken Adventures of an Emergent Brain

If you took some time to get to know your brain, you would most likely be fascinated by it. You would learn to love the rather interesting and hard to tame creature that is inside you, making you – well, you.

Think about all the times it wakes you up, or won't let you fall asleep, rabid to show you concept castles it builds at night, at about 3 am. It shows off by transforming nightmares into soothing dreams or vice versa, then exhausts itself and lets you fall asleep again, just in time for the alarm to go off.

You may have experienced the long and uneven path from imagining to actually doing something. If you managed to overcome procrastination and your brain and your hand finally got together to work, some feeble, malformed copy of a copy of the lustrous HD full-dimensional constructions your brain made so lightheartedly (sic) barely manages to emerge. Those of us who experience this phenomenon live in constant disappointment with our bodies' ability to keep up.

Some brains never even need mood altering drugs or narcotics to do the things they do; others discover new and uncharted territories and mental landscapes with such incentives. Some need full logical systems and rigorous definitions and anchors to feel healthy; others just breathe in universality and "presentness" through prayer, mindfulness, yoga, or during a stroke.

And they all are always hard at work, no matter who they inhabit. They all strive to bring each and all to the point of emergence.

My brain for instance built several exo-ecosystems, an assortment of life forms ready to be deployed in any space known to mankind, and which are ready to learn to exploit any resource so that they may flourish into other multiplex systems. It watched all these iterations emerge and crumble and transform eternally, shape-shifting and dragging bits of human intelligence along. It let itself drift, mesmerized by them. It must have travelled several light years unknowingly. It saw humanity's inner frictions, witnessed many instances of systems overheating and eventually collapsing into a hybrid, much leaner but unrecognizable version of the world of today, as higher order iterations took over and dispensed with the need to keep antiquated versions. For a while, of course, until they themselves became antiquated and were replaced, as the entire machinery craved more disorder to consume. My brain charted and finished populating the immediate, friendly neighboring planets. With elegant fluidity, it traveled so many versions of solar systems in the deep dark space-time that I am shocked to wake up in the same room every morning.

It dragged me through school after school, degree after degree, hoarding chunks of information in a Picasso-esque motion picture held together by sheer flux. And somehow, all definitions, in all their exuberant diversity, turned out to be one and the same. I always try to externalize the world my brain sees by writing or painting it, but only manage to render some mediocre pieces that make little sense to others.

I am certain that you all have gone through similar frustrations, at some scale. I wonder – why don't we all stand up and revolt against our bodies' painful sluggishness? How come we are so content with the state of affairs, when we could work on creating a rhizomatic exocortex to streamline communications?

Hoping to learn how this could happen, I travel in search of other brains to make synaptic bridges and warp through these possible universes together, and build those fragments of human intelligence that will end up recycled in the exo-ecosystems we have already seen in our collective imagination, and which are set to emerge.

58.3   Life as an Indestructible Mathematical Miracle

The world is data patterns (see Max Tegmark's theory of a mathematical universe). We are witnessing daily the multitude of ways that are emerging to deal with data, sort it, use it to make predictions. Oddly, playing with data is not unique to humans, to our digital century, to sapiens being on the edge of another revolution. We have always been drowning in data, we have always played with it, tried on patterns, gently or violently mutating the fabric of the world just because it was there. Just because it was possible.

By "We" I mean the *entities*. We the viruses, we the humans, we the numbers, the infinitely divisible units, which continuously blend imperfect patterns. Following one rule set: run the program, conserve potential across iterations and let patterns form and temporarily occupy available spaces in efficient ways, while conserving the potential for disorder.

You will find this rule book everywhere you look. I for one find it impossible to think outside of it. Infinity is easier to understand than a world outside mathematics, i.e., outside data and patterns. In fact, your brain is so hard-wired to scan for patterns, it will malfunction when deprived of the means to do so within 2–3 days, and default to making up worlds just to keep on pattern playing (see sensory deprivation experiments).

You may call the pattern clusters many different things: friendship, valor, honor, empathy, hierarchy, company, market, phylogeny, life. The list stretches as far as the vocabulary. Humans are exquisite data clustering virtuosi. They have been forming elegant cultural systems, theories, religions; concepts such as good and evil, love and hate to live and die by; they can frame the essential rule set driving all entities and their interactions into pretty much anything that looks like a possible combination in a logic set, which is, after all, still mathematics.

What does it all mean?

58.4 Chasing Apocalypse

Looking behind is what we do to try to figure out the future, but our mind gets snagged into the soft web of all things we have but fuzzy memories about. No matter what happened there in the past, here we are; we made it into today, with a new collection of scars and sunspots. We have been chasing one revelation after another – the original definition of "apocalypse" – or the sweet spot between anarchy and necessary disorder, insisting on discovering a safe way to change – or worse – a permanent stability.

There is no safe way leading to change. There is no way of predicting the far future, or a future beyond an inflection point. We know that. We have a history of martyrs, revolutions, wars and inventions to prove it. Do we need constant reminders that this is what we do for a living – change the world with every breath, every offspring, and every new pattern we come up with to test out into the world?

58.5 Suicide in the Cognitive Age

There is a bimodal way about the world, which is the root of all our romantic melancholy. Life is a seemingly backward, awkward way of achieving entropy (see Jeremy England's continuation of Darwin's theory of evolution). On the one hand, we see it as an incremental overachievement – a way by which mathematics/nature/god/karma show off how many ways it can come up with open system arrangements that are capable of efficient energy uptake and replication – but then comes the inexplicable:

Self-destruction. We all do it. Bees do it, whales do it, cellular automata do it. And of course, humans do it. Certainly, there is controversy as to whether the definition of "suicide" truly applies to anything that is not human, and it does not if you keep the definition narrow (thought, planning, executing), but not if you simply view it as behavior that leads to an entity's destruction while plenty resources to choose the opposite exist, which is the definition I am choosing here, if you will pardon my timid attempt at sensationalism.

But *why* is there a self-destructive behavior out there? Is it a disease?

Brain health is an important influencer of evolution in general – we are all *well* aware of it.

If so, how come we still cannot be persuaded to rearrange medical priorities to integrate body health with brain health? If we pay attention to WHO, there is a lot of brain going to waste due to suicide because we are hell-bent on preserving unhappy bodies (~0.8 million people/year).

The age of quantity over quality is waning faster than we prepared to accept. We can now do a lot more with good, healthy brains. Survival of the fittest no longer refers to bodies only – although having a healthy one improves the outcomes to the brain; rather, ingenuity can accelerate the species' rate of emergence, and it can come from a mix of current brains + their brain products, as an additive/cumulative effect.

Which brings me to this point: brain/body health focus means – at least for the immediate decades – ensuring a good global brain, ready to take on the next – and very difficult – evolutionary step. This is not an elitist stance – *every single* brain is valuable. Also, it is not a proposition to do away with anyone on the basis of weak/ strong, high/low IQ, rich/poor or other such nonsense. In fact, current efforts to educate all areas of the globe will necessarily and eventually lead to what we refer to improved living conditions, better life expectancy and better health and economy. It is simply the natural course of events. This in turn will lead to a new global brain, better equipped to take in the realities of mathematical selectivity of patterns, and more willing to relinquish absolute control (which is illusory anyway) over all possible outcomes. I will discuss millennials at the end of this chapter.

58.6   The Live Brain Bank

Speaking of suicide – not many disciplines outside of psychology pay attention to this phenomenon. Economists and policy makers only care about how much it costs to preserve people who want to die, or how much capital loss there is due to absence of people who succeeded to kill themselves and inflicted damage on others' psyche, which consequently incurs other costs in order to repair.

The hierarchy of needs is a vortex, not a pyramid. No wonder theories of needs are always under scrutiny and revision. People can sometimes experience relatively difficult episodes of poverty, distress and dead ends. During those times, they feel that they become suffocating, insufferable hindrance to their own brain, which always wants to soar. A brain's un-manifested "vibrations", for lack of a better term, can at times become physically painful.

People may admit in confidence to having the urge to commit suicide, all the while being fully equipped with the means necessary to "cope" or resist the urge to actually do it. These people would never share this in public because of the stigma lingering despite anti-suicide campaigns, simply because it is viewed as abnormal to even have it.

Suicidality, when it first sets in as a mere impulse, may be a built-in behavior regulator that tries to remove an individual from a specific context that is no longer useful for growth. It then escalates as the individual experiences conflicting urges and spends all energy to distance him/herself from the impulse rather than from the mindset that brought him/her there, or from the circumstances that fail to utilize that individual's potential.

So the true reason we abhor suicide is the loss of potential.

If viewed as such, i.e., a healthy or normal behavior regulator, the suicidality tendency needs not be kept under wraps, suffocated and stigmatized until it actually ends up de facto terminating a life. Admitting feeling suicidal in public, without fear, as casually as admitting craving a donut, will be a great cultural break-through, and a way to diffuse the unnecessary tension and urge to secrecy surrounding the topic. Rather, it would poise people to explore new learning/personal development spaces, and I do not mean a counseling psychologist's office.

We are thinking of mental health upside down. We set the responsibility entirely on the medical profession – doctors and psychologists, when the responsibility is in fact shared by us all and must start early. Perhaps it may be a good idea to prioritize this responsibility so we do not orient our efforts to fixing a problem, but rather through early and immediate thought nurturing through education. First strain on mind development is immediate contact with close habitat – family; second is usually the education/entertainment environments. Education as entertainment (playfulness is, after all, learning through problem-solving) is where the current generation has been investing to start stretching young minds not only to tolerate, but to embrace and champion diversity.

58.7 Minding the Education Gap

What we universally hate is school (which we associate with being lectured to, followed by information regurgitation in some standardized test); what we universally love is play. We invent gadgets to give us more time to play. We invent more gadgets to deal with the first gadgets that end up requiring maintenance and detract us from playing. And then the second set ends up requiring maintenance and so forth. So there goes our time, our brain power is lost.

Trouble is, what we choose to do with our brain today dictates where we are going to end up as a species. We can drown in the byproducts of our chase for play, like yeast dying off in its alcoholic waste in a beer brewing tank. Or we can radically rethink what it means to be human.

58.8 Any Port in a Storm

There is hardly any need for anyone to remind you that the times are always changing. You know for a fact that new generations change things, whether you like it or not, whether you decry the loss of the good ol' days, when things were all milk and honey, when in fact back then it was YOU who were the rebel, the great disruptor of a world that your parents had laid at your feet, and that you proceeded to trample upon unceremoniously. Today it is you who tries your hand at that old trick – creating the illusion of stability and control. Back in your day, when you were seeking culture clusters to hide your soft vulnerable self into, like a hermit crab, you got cozy in something that was slowing down, or at least not moving as fast as other data flows around. Or maybe your mind simply had had enough and decided it was time for an anchor.

Whatever you happened upon at that time will have ensnared you and you stayed, especially if in your mind it bore resemblance to anything you had encountered before, a familiar system feature. This is in fact not a difficult mind feat; we are always able to find a pattern if that is what we seek, and we will shrink a system to comfortably accommodate it, so that we may live in it happily ever after. The smaller and denser the system, the more friction will ensue from interacting with other systems. Think of how homogeneous cultures that abide by a set of strict, unyielding rules either face discrimination by other cultures (especially when those are also adhering to strict rules) or will seek to dominate, annihilate or otherwise forcefully overtake them. This is especially evident when these cultures are densely populated. I believe many examples are already crossing your mind, both contemporary and historical.

But can you really hide in such systems in today's open information world?

58.9   Let There Be Strange Attractors

Look at how your own life as an almost playful happenstance. Take a second to reflect upon how you, the sapiens reading this, came about. You are a collection of data put together, a strange and wondrous collection of butterfly effects. You also carry the ability to see your own existence moving and shaping itself. You, (possibly unlike other such patterns of data – a cat, for instance), are like the artist painting himself into the mirror painted in his painting. You have been reveling in this ability to peer into the secret engine of the world all your life, and not just now since you have access to a smart phone – you always could, ever since you became sapiens – or thereabouts. This is a truth you intuitively grasp if I start calling it by its many names familiar to you: thought, intuition, imagination, and the language clumsily gluing them together.

58.10   Driving Under the Influence

All systems will select ways to collide and cooperate, and if they are successful, maintain just enough disorder to seize the opportunity to put new iterations at play. Just watch cellular automata of the fourth type.

Maybe it is time we came out and said it: we have the power. We have the potential to *intentionally* create patterns more efficiently and more eclectically than other species/entities. But not only that – we KNOW that we don't know what our creation will look like, and we know that it is exhausting and as silly to try as predicting the exact weather across the next century.

We can't promise what the crossover world will be, what its good and evil will look like. We can speculate, sure, but imagination stretches in the bounds of the current form and does not like to stray too far. Hence, if you watch most Sci-fi movies, you will find that the villain is the one who tries to take over the world (or destroy it), while the hero is usually a resolutely fleshy humanoid or human, with all the necessary flaws to make it relatable. There is not a single representation out there of an entity that is the result of the human emergence, and which is not deeply disturbing for the viewer. For instance, a cyborg must acquire the ability to feel love, pain or empathy for it to be absolved of the sin of being born via unnatural means (Terminator, Humans). In the Marvel series, a Transformer must be anthropomorphic, smoke cigars and abide by codes of friendship; while an absolute villain will be the one who randomly generates and destroys patterns that made it into the mythology of one species or another (e.g. Ego in Guardians of the Galaxy).

To give you a closer example – look at your hand. Imagine what a skin cell on your palm thinks about you, the whole human. Its neighbors are skin cells, it knows how to relate to them; they have their own pattern, their own way of communication, their own shape. They may have heard of other cell types; muscle, brain, etc. but they only know them indirectly through the communication channels (fluids). Imagine how disturbing it may be for them to one day learn they actually make up something much larger, and probably hideously shaped, as far as they are concerned. If they could have voted to become part of it, they may have very well vetoed this idea right away. So do we reject the idea of emerging into something we cannot even hold in our imagination; and that is a pretty vast landscape.

58.11   The Greatest Terror

You guessed it. "What if all these artificial creations of ours will destroy humanity altogether?" It has been crawling up your spine ever since you started reading this, has it not? The entertainment industry has done a fantastic job depicting tech-apocalypses.

Fear of death is in fact legitimate. As far as level of specialization and adaptability goes, we are an overspecialized, frail and dainty wet carbon construction. We only have our brains to count on. Though it may not seem so. You forget that brains are an extraordinary commodity if you are too busy thinking that you are superior to that idiot across the street who voted for the opposing political candidate.

But remember that our beautiful collective brain, when indeed worked as a unity, has been working itself out of really tight corners in the evolution of sapiens. Your brain is not yours alone; it belongs to us, the SAPIENS pattern in flux. And it has been getting more and more efficient at pattern play; so much so that we can even teach our machines these skills too. All we ever needed was the ability to see one step further – not its result (which we nonetheless speculate upon until we are out of breath), but the willingness to risk it.

58.12 What the Hour May Bring

Forgive me for stating the obvious, which I will reiterate throughout this chapter: it is rather unlikely that Homo sapiens, the soft, squishy, lovable, familiar shape of today will survive a catastrophic event (one that threatens the entire biosphere, that is). Although that may seem like a very remote possibility while you are sipping beer in your living room, watching your children playing Minecraft on their tablets, I promise you that such event will occur with certainty at *some* point in time. Look around in the solar system for clues on what can happen to a planet, with or without life-sustaining potential. Change (mathematics) has not figured out how to "play nice" in human terms.

What does that mean to you? These are things you have no time for. Why would you care, you are not going to be around to witness it, you have done your part.

Have you, really? Think about it for a second. Your child is sitting on a wealth of information that you already processed. You will certainly try to pass some of it on, like a good parent, and yet you are afraid they will stray away from the system that worked for you, and you will try to smother them in one you think is safe. As an individual, you are risking your child's life by allowing its birth. You are therefore responsible for what comes next. You are the stepping stone of the future. Kindly remove the "don't tread on me" sign from your front porch.

As a nation, you constantly allow some new invention or policy to break free, fix one problem and create a thousand others. Could you resist the urge of creating potential though? Because that is what you have done, you put out there a new, shapeable piece of information that will combine with others and destabilize the worlds to come. As is natural.

58.13 Running with Scissors

Confess. You *did* think about the edge of the universe at least once in your life. You thought about what comes after the edge of the universe. You suspect everyone of having thought that.

We are probably accomplices in this great cover-up – fixing infinity to make it palatable to our current mortal shape, but we nonetheless try to sneak out every chance we get. Not overtly, because so far mortality is an extremely valuable *meaning* goldmine. If we lost it, we would have to fill in the gap with something we have not yet braced ourselves for at a global scale.

We are getting there though. We have means to fix the body and augment the mind; we can shape single bodies or entire ecosystems, we can create hybrid iterations and set them loose. We want to figure out ways to colonize new planets, create better soldiers, or at least sturdier space scouts. We really could lessen the risk of losing *everything* we have made from being gobbled up in one catastrophic event.

That is not to say that we should *expect* that we will preserve everything. And perhaps clinging to what we define now as humanity, good, evil, could use a brushing up, so that we may deal better with the growing pains that are about to hit.

So what you can do today is loosen the elaborate, gilded safety noose around the new generation's neck a bit. And not just that, let them play with knives a little. Let them see an Icarus now and then; let them even be one. They may learn to fly.

58.14 Being HUMAN

If we look only in the dictionary, it is fairly simple: a member of the genus Homo, species Sapiens. But this is not what first comes to mind, is it? We associate the term with:

Affective terms: love, friendship, caring, cruelty, frailty, pain, emotion, empathy
Cognitive terms: intelligence, thought, memory, imagination, intuition, etc.
Complex social structure and interactions: democracy, education, science, religion, culture, economy etc.
Technology: tools, money, clothing, transportation, communications, infrastructures, etc.

Humans. What a lovely word. If you let it reverberate a little around your mind it will bring you to unpredictable places, rouse innumerable feelings and sprout loops of thought.

For it is thought – the ability to construct and adhere to abstractions – that made humans a little different from other species. Of course, this is not novel to any reader now. Even those of us who insist on being the epitome of pragmatism commit to constructs: honor, values, money, ethics, god, nation.

You most likely have heard warnings that we are on the fast lane to lose our ability to manifest empathy, to express our values, our feelings, our humanity, and become machines. It is ringing in everyone's ears, keeping the globe up at night. And yet, if I were to ask you "what is human?", you would most likely grab onto constructs to make up a coherent narrative: love, friendship, empathy, ability to feel pain and pleasure. Also known as system dynamics. Each of the words above is code for pattern of behavior, useful in describing the mathematical law governing you. It does not mean that it is wrong to use them; in fact, it is rather appropriate. Calling them "codes for behaviors" should not degrade them simply because similar patterns of behavior can be seen even in what we are accustomed to see as non-living or "lower order" organisms. Would it make things better if I said – "well, at least those systems can't or don't bother conceptualizing their own patterns of behavior"?

Biologically – not much has changed since Lucy. Lost some more hair, grew a little taller, started getting such oddly positioned wisdom teeth that we have actually developed a whole industry just to remove.

Societies though became more complex as populations grew. Once humans figured they need to interact with more than 150 individuals for effective gossip, as Harari graciously put it, layers upon layers of social interactions formed, and adherence to cultural norms became mandatory for an effective world take-over. Homo sapiens is, after all, the greatest beast, the single supervillain Terra has ever fashioned, one very hard to destroy using the current tools on Terra. I am not suggesting that individuals are invulnerable, just that whole species destruction is going to be quite the challenge if nature decides it had enough of us. It would need to extinguish the sun rather sharpish, or find some decent sized space bodies to hurl at us before we colonize other planets.

58.15   Humanity Unlimited

Here is a serious challenge – how do you define what it means to be human in the near/distant future? And will this definition be universally accepted in both time and space, or will it always be culturally conditioned, as is the case with most complex definitions? What is the point where this definition will no longer apply; how many things must be lost, at a minimum? How much mental flexing will it take for these losses to be accepted so that the defined object will return to being called human once again?

Who pays for this stuff? Who cares? Why think of something so improbable, so far off, so outside the current human life span?

These are just some of the typical questions people ask when forced to think so many years away. They always forget their own ability to think exponentially, and the myriad of examples of rapid technological advances they witness daily.

In reality, we have little choice. Remember, global catastrophic events are guaranteed to happen at some point. The natural life of Earth is at least another 4–5 billion years or so; we are theoretically about half way through, if nothing else changes – which is unlikely. Even if we operated entirely outside of denominators such as moral imperative, duty to preserve the species etc., we would still have no choice but to solve our survivability equation, as mathematical entities, in the shifting ecosystem. But since we *do* have the advantage of believing in lofty ethical labels, we can put them to good use and mobilize all efforts to achieving these goals faster than biological evolution alone could ever afford.

Let us stop for a second and consider today's world. (Some) humans figured out mathematics is the key to the universe, if not the universe itself. Humans realized that what sets them apart from other species is the ability to form abstract concepts and imagine possible futures. Humans grasped – to a larger extent than they realize – the essential traits of systems survivability, but are yet to put them together to understand what constitutes their carbon-based ecosystem, or how to make it and break it at will. This is still the year when wise men say that there is yet much to learn before we understand why we would want to do this – and most insist we shouldn't "play God" for fear that this may cause us to lose our humanity.

Speaking of "humanity" – whether we refer to social, evolutionary or individual humanism, "human" means the general shape, level of intelligence, desires, means of reproduction of the current variety of homo sapiens. Well, maybe not *exactly* the same; perhaps a couple of augmentations – maybe living longer, healthier, wealthier, happier (whatever that means) etc.; making more connections, creating more jobs, finding gods easier, getting to work faster, travelling more, having more sex, having more money, being less bored, oh my! (speaking of the boredom luxury – this is the great hunger of the age). Fancier options are all the rage, such as immortalizing Homo sapiens, or uploading consciousness into the cloud to download into yet another flesh-bound sapiens specimen. Assuming that we will eventually agree on what consciousness really is.

We all agree that it is fairly reasonable to expect some level of self-preservation instinct, but it is reasonable to expect willingness to sacrifice for things we consider important. All cultures value sacrifice, and you will be hard-pressed to find any story in the world where sacrifice is not the central theme. We value it in others because we instinctively understand it is a natural law; we even have the wiring to overcome the displeasure it causes us. Sacrifice is a full-range spectrum, necessary and sufficient to preserve a system, and it is known by many names – cooperation, love, pain, martyrdom, faith. Same can be said of cells of an organism. All those notions with uplifting, sonorous names are nothing more than the many faces on either side of a thermodynamic equilibrium that is never more than temporarily attained.

Thus, life goes on, chewing on everything that is and spitting it out. With or without sapiens in it, defining it, worrying about how ethical destruction and creation is.

The emergent human proposed here is an elegant transition state that brings current humanity to those places it cannot reach due to current bio-cultural limitations.

It could be the scope of what I imagine would be **The Humanity Unlimited Institute,** where we would attempt, with the aid of artificial intelligence, the crazy idea of trying all improvements at once. As a side note – remember that there is nothing truly "artificial" about AI – it is a natural extension of our own brains, and a tool we could use to improve upon ourselves.

Examples of such improvements may be: education through entertainment, body *and* brain augmentation per individual or across individuals, hive habitats for individual and collective mind augmentation, etc. The institute would become testing ground for a civilization ready for space travel long term. Entity improvement methods could include incremental augmentation as well as de novo synthesis of an intelligent world.

Why is this important? Because, in order to allow sufficient time for such new implementations to be tested, one must start now, while it is still possible to perform these tests. If given 600 years to develop survivable habitats, as Stephen Hawking predicts, we will need to brainstorm, construct, deploy, run, learn and improve prototypes. If a simple manned trip to Mars and back takes a couple of years, and we have not even started, we are alarmingly far behind schedule. Therefore, every enterprise that is currently working on getting people to Mars should have or contract such a Humanity Unlimited branch.

58.16   Bdelloid Rotifers Ahead – Proceed with Caution

Allow me to indulge in some outlandish speculations for a moment, and let others speculate about the ethics of whether to think or not to think outside the various boxes we might comfortably inhabit at the moment. Nonetheless, you may want to proceed with caution into the next paragraphs.

Creating even sturdier sapiens variations may mean looking at extreme and unusual ways to improve the human condition. Here I will step into the land of astrobiology (although I am not one myself, so forgive the rampant speculations here).

One of the strangest and perhaps most incomprehensible such options would be changing the means of reproduction. In the world of DNA, sexual reproduction is very useful for covering all bases for multicellular organisms, whereas asexual reproduction can be relied upon when abundance of resources is not an issue, and there is little pressure to adapt to fast changing environments.

Emergence of asexual reproduction is extremely rare in nature, especially in animals. There is also no confirmed case of any one species that completely supplanted sexual reproduction for asexual reproduction, with one exception – the case of bdelloid rotifers. These are microscopic freshwater animals, of which well over 400 species exist, and that have been around for 25 million years (23 million longer than humans, that is). They are known to have a single gender, with no definitive evidence on whether there where ever two, or when/why the two gender reproduction mode was lost. Bdelloids have no need for two genders simply because they have intrinsic mechanisms to ensure genetic diversity.

Incidentally, it so happens that they also have a couple of desirable "superpowers" as well: they can resist desiccation, temperature and pressure extremes, and are really good at repairing DNA breaks caused by ionizing radiation. There are other interesting organisms out there as well, which, collectively, can teach us how to survive in extreme environments but which are currently lending us information to execute more mundane tasks such as washing our clothes or doing our genetic tests.

Granted, these are not highly complex organisms, and there is not a lot of genetic baggage to be lugged about. They may not have complex social hierarchies like humans, and most likely have not had a single abstract thought in their lifetime on this planet.

So what has that to do with sapiens?

Simple – these living fossils are proof that emergence into asexuality is a viable pattern, one that has been tested since the Oligocene and withstood the test of time.

Planning an impossible task – that of optimizing the species to acquire such resistance traits – may eventually be fruitful if one were to look for answers in every one of biosphere's toolboxes, and also learn how to make the process so lean that only what is necessary and sufficient for propagation is maintained.

58.17   Accepting the Third

An interesting pattern that is almost never mentioned in public is that of asexuality in some single digit percent of the human population right here, in the current world. We pass by them every day without knowing – and if we have heard of them, we generally choose to believe they are convertible. Not until recently was it that the psychologists finally concluded that they are not another mental disease group but a healthy, even happy oddity.

It is very curious how this shy and quiet niche of humanity has been stonewalled by the same folk who are otherwise fully accepting of homo/bi/other sexuality. It seems that, as long as individuals of the species express some sort of sexual desire, they meet the requirement to be human. Humans are supposed to suffer the same writhing and turmoil of mating as the rest of us. How can they experience anything elating? Love (mating with make-up on), desire (mathematically conditioned primer of interaction behavior to achieve self-copy states) is what makes us human, after all; these people cannot really be people, or they must be pretending not to be living under the same pressures.

Nonetheless, they do exist. Researching their brains as a potential answer to what could be very real, human alternatives for "pleasure," and might be an unexpected revelation for humankind when the human brain is not involved in reproduction or sex-related issues. How does this brain it occupy itself?

Humans have now achieved the ability to manipulate their genetic information to propagate their species, just like bdelloid rotifers, and would easily be able to generate enough diversity to cover more ground than standard evolution ever could. Once they allow themselves to use this tool, they no longer have a need for sexual reproduction.

What would it mean to society? As you can imagine – if you are even willing to imagine such a world devoid of carnal pleasures! – the emergence of an alternative, 3rd "gender" of technologically induced parthenogenetic (unnatural!) individuals would create at first the entire menu of revulsions: confusion, anger, revolutions, etc., even though such an alternative to sexual reproduction is perfectly *natural* because it is *found in nature*.

Let us suppose though that somehow the third sapiens would make it and escape the wrath of the regular dimorphic variety – (as a side note, it may be the first time when the two genders actually rally against a worthy common enemy). Imagine how much we could to do with the energy and resources that are now committed to mating.

Next up – the voice of contemporary wisdom. This is a short excerpt from assorted conversations I had with various people on the topic:

What exactly would the utility of such a third gender have?
Would slowly replace the 2 gender species. Would reduce the need for two-fold investment of matter specifically dedicated to reproduction (in 2 sex species, the male has limited strictly reproductive use but the investment of matter is huge; however, this may be compensated by complex function in offspring rearing and community-related roles). In Homo sapiens, it is ~50% genetic gain if a single "gender" is invested in. Note that this is not a proposition to eliminate males or females, but to have a third type, a permanent merger, if you will. Plato's ultimate parthenogenetic hermaphrodite.

Could they "interbreed" if they so choose?
Not in the usual sense; they could exchange information via other means. For instance, Bdelloids, tardigrades and other sturdy organisms are able to acquire extremely useful genes for survival, and they do so by ingesting foreign DNA. The point is not to have 3 genders forever, but a human species with a more efficient way to reproduce than the current one. They would not respond to the sex cues of modern sapiens. Of course, this is only one possibility, not a proposed solution to all the problems of today's society. These models would need to be tested in a virtual setting to assess several of the possible outcomes, if anyone would be even remotely willing to write such exotic code in the first place.

Do you imagine them to be genetically superior?
If one defines "superior" genetically as equal to "being able to mutate at will and efficiently to create sufficient diversity to adapt to changing environments and to reproduce with sufficient speed to overcome threats" then yes, they would be superior in that sense. However, it is unlikely that this will be entirely accomplished through biology without help from engineering.

Would they be hormonally asexual or still possess the desire to reproduce?
The desire to reproduce is innate to all living entities; it is a mathematical condition of existence in its broadest sense. To reiterate: a working definition of life is that it is a semi-permeable system, capable of metabolism (energy production and utilization), information storage and replication into other entities of the same kind, and with an ability to evolve and acquire new, more complex form/function structures upon generational mutation accumulations (the "emergence" property). The result (success to do so) is what matters, and what actually causes new patterns to emerge.

Clearly, bdelloids too are slaves to this law, even in the absence of sex – more precisely, in the absence of intricate patterns of behavior emerging via a complex system of signals specifically designed to cause individuals to copulate in order to procreate.

The desire to reproduce would be different from what we know now; and we may develop different hormonal pathways to ensure it. It could take a lot of trial and error, and hopefully retain just enough error to make it fruitful.

Alternatively, it may be here already, in the different types of pleasures we already experience from other activities. For instance, the desire to produce *ideas* would overtake what is now the sexual pathway. So yes, you could be aroused at the thought of 3D printing a new hive city on Mars, and you would pursue accomplishing this dream even if it meant sacrificing yourself. The need to deploy individuals (entities that represent the species) in order to test the new environment would trigger the cue to reproduction.

So physiologically speaking, what would reproduction look like?
This is the trial and error part. But we have CRISPRCas9 technology (powerful developing technology for precise genetic editing) and artificial intelligence, growing more intelligent by the minute, so we do not need much else in terms of tools. We simply need to try.

How would the sociological factors be managed? Would the new sapiens have to be sequestered or segregated initially? For how long?
The question of society is VERY interesting. It would be interesting to have an artificial life implementation (studies natural life through computer model simulations) open environment complex simulation of 2 genders, in which a third competes for resources and is of equal reproductive intrinsic competency. This could give a clue to what would eventually happen in the social system.

I am guessing they would need to be segregated, for their own safety. Humans have a thorny history with alternative selves, even if the differences are only skin-deep. In parallel, a very intense campaign of acceptance and mass education will need to be carried out. But I am optimistic; if rats can tolerate robo-rats as long as they are perceived as "friendly" (to a rat, as to a human, that means work together, be non-aggressive, display behaviors that are perceived as friendly), surely humans can too, right?

What if a couple of genders 1/2 wanted their offspring to be of gender 3, or shall we say gender neutral. Could that be achieved through engineering/tech?
Certainly with engineering at this point.

We love our systems of binary thinking; how would we cope with the third (Insert new term here), what would we term it, if "gender" does not apply? Would they be able to choose being a gender?
Correct, gender is a misnomer in an obligate parthenogenetic species. Bdelloid-sapiens would not have to select being a "gender", they would genetically diversify from within – like a virus. We do not think of a virus as male or female. So I think there is no need for outwardly expression of dichotomous diversity, but I rather would expect that an individual would look or choose to look wildly different from another. So diversity in aspect may or may not be "extreme" in terms of those things not directly related to function or survivability traits. It could be a choice. Rather, diversity of thought would become the most important characteristic of the new evolution paradigm.

In this case, societal rearrangements and hierarchies would no longer be structured around gender but probably around function. This loss of obligate dimorphism would certainly do away with misogyny, feminism, unequal treatment of genders, problems like asserting power through sex, sex crime, etc. Social hierarchy would be completely rearranged if sex plays no role in it. And you may not be too hard to convince some people that it is highly desirable for gender-related tensions to become a non-issue already. I suspect though that most would object to this view, because either they rather prefer to believe this is a necessary and useful diversity, or it is simply too darn crazy to think of such bizarre alternative. Even if it were accepted, I cannot guarantee though that we would not come up with other ways to discriminate against each other. This might be a future problem to solve.

I think the quest for knowledge over carnal desire is ages old. Just look at various scholars, priests and athletes throughout history. I think if we could indeed accomplish this without suppression of sexual pleasure and its various side effects, the outcome would be quite fantastic.

In fact, as I said, it may be that the "ancient brain" will be entirely replaced eventually by the neocortex if there is no need for it. Fear/pleasure as immediate, non-cortically processed impulses, will still respond to the threat/environment conditions, but the part of the brain tasked with mating would be repurposed. I am not suggesting *suppression*, it seems like a "cruel and unusual punishment" (NOT my intent), rather a simple waning due to disuse – like our coccyx.

58.18   Ode to the Post-Millennials

I want to hold the world in my hand. I want to scream out my identity and I want you to listen. Really listen. I want the infinite power to understand everything I see; I want to taste my thoughts, breathe in the beauty of data rivers flowing through me with the same intensity as if I were having a seizure. I want all my mental organs to run at full capacity. I want peacock snake silk skin to see with. I want to be one and all, inwardly plentiful and outwardly overflowing, I want to release plumes of possibilities in all known dimensions and let them grow like stalactites across space-time. And I want it all by noon today.

This is what a post-millennial might say.

These are the people of the next level civilization we must prepare to welcome. Our fears should not taint the world they need to recreate.

So today we should help millennials raise the new world. By watching them develop their amazing abilities to exist in 2 worlds, we can foresee how extraordinary humans can become.

Take for instance their language, which is changing to adapt to the new half physical, half digital environment. They are in the same room, yet use a digital medium of communication rather than direct interaction. To them, this cleans up or speeds up communication, but to their elders it is eroding the very essence of being human, so they will yell "put that phone down and talk to me!" or "no phone at the dinner table".

Loss of "family quality time" is on the rise. This wonderful species of shape-shifters and experimenters is redefining what it means to have quality time. Their values are shifting dramatically, much to the chagrin and despair of their elders.

Their ever expanding range of sexual or asexual identities – asexuals, cis/trans genders, aromantics, polyromantics etc. – simply means that this is the generation who is finally figuring out that "natural" means "possible", not "permissible within the boundaries of a cultural system". The very flimsiness of the mid-world forming their identities is teaching these new humans about the very nature of reality. They will learn to choose their values, no doubt, but they will also know they are constructs, and they will not fear destroying them to build new ones overnight – a feat past humans were terrible at, until the industrial revolution started poking holes in their conservative resilience.

Young humans of today are reclaiming their freedom without even bothering to rebel against the current state of affairs. Weapon of choice is understated subversion, usually playful. They will not apologize for anything, including their preference for machine- to human interactions.

They are the unexpected source for a reality check. To those of us paying attention, these people will expect things we are not ready to offer until we step out of our chosen belief system cocoons. They need things so foreign to us that *they* should start teaching *us* life skills themselves are mastering.

They need a different way of learning, a VERY different school system to begin with. They are playful creatures, who like to solve problems, level up, combine their magic and combat skills and team up to take on a boss. If we expect them to sit in their seats and learn discipline and memorize something they can get to in less than 10 seconds by googling it, then we are delusional. This is the smartest breed yet, and it is just beginning to soar. Before our eyes is the mere flicker of the jet engine this generation is becoming. They need to invent serious challenges to fuel their evolutionary drive.

They will be augmenting themselves in any area they feel like, safety be damned. They will have animal eyes and legs, unorthodox sex organs and extra genomes. They will grow weed on their backs. They will change their entire composition and merge with plants, computers and cockroaches to make them eligible for their exoterrian passport applications.

You will sit there, in your corner, biting your nails, certain that this world is coming to an end. And, thankfully, you will be right. This version of reality is about to expire – where violence and oppression and misery, hitting at the lowest level of your physical being, still exist and are vividly fed into your life by your fellow media-generating humans. And how can they help it, when we still wallow in the legacy that created a media fueled by sensationalism. Not their fault either; the world was pretty boring in its relatively safe and structured corset, so that was the perfect way to jolt your dulled senses right about dinner time.

But now the world is on and awake 24-7. All brains are open. Time to feed.

They will need new ways of role play – aka job hunting. This will no longer be a world where you have a career, retire and die.

Just imagine how the world will open into a skill per game/task market, where you, the new citizen, play to solve or play to test iterations of pluralities all around you. AI "world play assistants" will not just learn about you and pigeon-hole you into a pattern of behavior ever-narrower, but intentionally deviate you and force-flex you into your full potential. They might force you out of your comfort zone until your comfort zone becomes quicksand, and they will teach you how to deal with it.

Your chemistry might change to accommodate the intake and processing of this new world. Your body and your environment might change from one day to the next, or may merge, divide or otherwise transform to accommodate the task at hand.

If there is a perceived limit out there, no efforts will be spared to crush it; if there is a structure set out there, all hands will be on board to propel the emergence of higher order patterns from it. And ultimately, this is what defines life.

So what will define "human" could be the intention to iterate life.

58.19 The Thousand Year Journey in One Day

It is all very simple.

It can be summed up in a single paragraph. Here it is:

Humans, as well as most other species we are and are not aware of that thrive in this particular dimple of gravity, will not stay the same. As with all emerging patterns – we will change into something entirely unrecognizable. The transitions will be strenuous at best. The result will be surprising.

Truth is probably mathematics, sometimes manifested as life/energy/computation, but humans only perceive a fraction of it, and misunderstand it as some sterile, boring thing one had to endure in high school.

The unseen mathematical universe is called by other temporarily useful terms, love, god, good, evil, nation, moral, self, finite, death, virtue, family. These must be constantly pressure-tested and readjusted to context. Backlash effects of reevaluations will be seen when any of these are threatened with change. This is normal, simple rearrangement of boundaries, but it can be very frustrating when witnessed up close, so remember to put on your "1000 year forward" mindset on.

So, my fellow sapiens, here we are (Figs. 58.1 and 58.2). The archaeopteryx of the futures we can create.

Shall we jump?

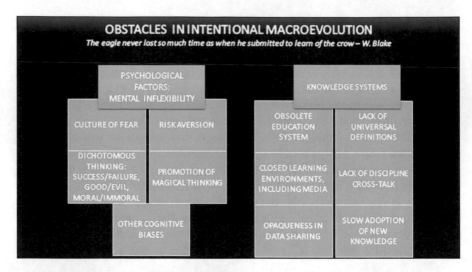

Fig. 58.1 Obstacles to intentional macroevolution

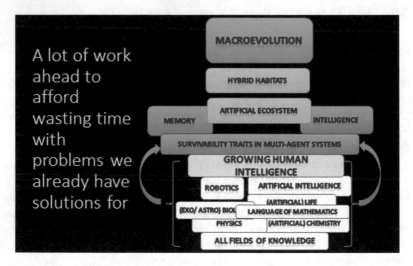

Fig. 58.2 Seeding intentional macroevolution. **Macroevolution** = mathematical species evolution at intentionally accelerated pace, with the use of all existing and emergent knowledge. **Hybrid habitats/Artificial ecosystem** – simulation habitats – theoretical/computational, as well as wet labs (microgravity simulation labs, ISS labs, moon hive lab prototypes). Fields required: robotics, architecture, theoretical physics, mathematics, astronomy, computer science, evolutionary biology, geology, engineering, communications, logistics, art, other. **Memory & Intelligence** – key survivability traits which must be preserved. Fields required: neuroscience, cell and molecular biology, computer science, psychology, art/humanities, philosophy, robotics, engineering, other. **Survivability traits** – resultant of study of life on earth to determine the cluster of genetic information necessary and sufficient to create biodiverse ecosystems based on specific environmental physical and chemical criteria (e.g. pressure, temperature, chemical composition, density, gravity, etc.) Fields required: (astro)physics, chemistry, biology (cell and molecular, extremophile study, astrobiology), computer science (artificial intelligence, artificial life), etc

Chapter 59
Transhumanism: Variety Is the Ultimate Hack

Michele Adelson-Gavrieli

Suspended in the beautiful nebulous wisps of a star-speckled galactic neighborhood, lovingly called the Milky Way, is a small planet, whose inhabitants are tirelessly and relentlessly engaged in one thing: HOW to NOT DIE ANYMORE.

Most human earthlings (with the exception of some who gave up on the quest after many millenniums of glorious failures, and think that checking out isn't such a bad idea) are qualitatively and quantitatively bias, in this order. They wake up every morning to a precariously unpredictable world, holding on to that lingering taste of yesterday's scrumptious chocolate, and subconsciously say to themselves: Fuck this shit. We don't wanna die – there is so much more to experience and we want it all! (such is the effect of good chocolate…).

So, around this monumental task – their immortal battle with mortality – the natives of this planet created a system, a social construct of open-ended equations, intricate beyond belief, and infinitely complex, whose operational modality – and biggest challenge – is the **endless introduction of new variables.** Every day, **every second,** new visuals, new sensory stimuli, new environmental changes, more people are born, empires rise and fall, bouts of effusion and extinction, new discoveries regularly shake our ideas of the absolute….it never stops. This core essence of information – its illusive impermanence, characterized by impulsive, constant, seemingly infinite and untamable immergence – when force-fed to the human brain, it invokes an automatic pursuit of utility. This pursuit is called – Invention or Creation, which can be defined as: organizing information in an advantageous, functional way.

M. Adelson-Gavrieli (✉)
Artificial Intelligence Incubators & Sounds of the Heart, Ashkelon, Israel

© Springer Nature Switzerland AG 2019
N. Lee (ed.), *The Transhumanism Handbook*,
https://doi.org/10.1007/978-3-030-16920-6_59

59.1   Open-Ended Equations (OEEs)

What is an "open-ended equation"? Science is. It's an enterprise and a set of disciplines most welcoming to doubt and change. The Environment is. Space, Agency, Personhood and within it Sexual identity – these are all Open-Ended Equations.

Let's talk a little about Sexual identity as an OEE. Early human societies, in ages when sex wasn't yet super-glued to morality, were sexually fluid. Even societies that existed as late as early biblical times (like Lott in Sodom and Gamora) were chronicled to be what we call today incestuous and zoophilous, and what we classify today as morally divergent.

Lets observe this strange phenomenology from a scientific standpoint. We begin by asking a chronological question:

What was the chain of events that lead to this reassignment of rectitude-grade associated with Sexual preferences? Why did we pull the endorsement? To explore the social mechanism of behavioral tagging and classification, we need to look at the human brain (where everything happens *before* it happens in reality), its desired end-game (which is eternal life) and its motivational paradigms (which are Survival and Pleasure), and how the utility of certain behaviors is perceived, determined and labeled by it – given the at-that-time circumstantial map. Throughout this chapter we will examine every point as it is translated in the brain and how it rolls into reality.

So, from the neurological perspective, it turns out that open-ended equations invite excess of variables. It's like a perpetually open flood-gate. Excess of variables rapidly devolves into chaos or the appearance of chaos, requiring higher skills of organizing information and managing it which in turn requires more resources. And *that* was never a priority of a growing society fighting for identity and dominance of resources. In fact, it was counter-intuitive.

And so, in different points of human history the ruling class tried to close those open-ended equations and instill (drum roll…) – the Ultimate Order: A synchronous, symbiotic governance system, where everyone knows their place, protection and distribution of resources is easily deployed/managed and social compliance is enforced via "expert" prognosticators administering a cocktail of cognitive conditioning and brute force. Those systems were called Empires, Religions and Ideologies.

But how did this ruling group of individuals manage to manipulate public opinion so profoundly that people (or rather brains) began to confuse natural tendencies with punishable moral transgressions, making entire classes of pleasure-inducing activities criminal and people seeking them pariahs? Was it *only* their lack of reproductive utility?

To answer this curiosity let's have a look at the individual and the collective consciousness, as they are shaped and reshaped by one another.

59.2   The Survival/Pleasure Bait-and-Hook

Humans are indomitable. When they get something in their head (their brain) – they're like a dog with a bone. Their desire burns through impossible obstacles, which leads them to believe that "impossible" is a nothing but a temporary lack of crucial information, that once found, will make everything possible. And that is one glamorous and irrejectable prospect especially when that something is; more time breathing. The reason they are so indomitable is because of this small evolutionary bait-and-hook called: Pleasure (qualitative), and experiencing more and more of it is what fuels their unyielding ambition to perpetuate Life (quantitative), even *after* it allegedly ended (see mummification, cryonics, vitrification). So, to those who wanted to institute the "Ultimate Order", it was clearly "Pleasure" that could be capitalized on… and it was. Though not consciously.

If you boil down almost every human activity, from the cellular level to the social level, every endeavor, every good or bad deed, you arrive at the nuclei of human drive: Feeling Good. Humans are addicted to feeling good (this insures survival). Getting this 'fix' comes in an impressive variety: From sexual gratification, romantic love, giving to the poor, saving Pandas, fighting against fracking, marching for social justice, controlling large corporations, becoming a rock star or the president of the United States… all the way to binging, shooting heroin, exacting vengeance, inflicting pain (both mental and physical violence), S&M and even self-mutilation… All of it, the whole universe of human actions comes down to: Feeling Good (or at least masking/stopping pain – which is a happy compromise in many cases and the lower end of the "ecstasy ladder").

How did "feeling good" become the driving force behind hacking Death? Firstly, because *feeling* is our **only** indication of being alive (and being alive is essential for not be dead). Secondly, because *feeling* is the steam engine of creative invention, and invention, humans speculate, will ultimately bring the end of dying.

To really really simplify the process: Feeling good happens when Dopamine, Serotonin, Oxytocin, Endorphins and some other neurotransmitter friends deliver a concoction of happy-soup to the brain, and the only way for the brain to differentiate between winning the town's daffodil pruning contest and reaching an orgasm through asphyxiation is the amount and ratio of neurotransmitters in the happy soup. Either way – the brain doesn't care – it wants more. So there, the simple total-ity of it all comes down to "just a spoonful of happy-soup makes the medicine go down" – "the medicine" being all the shit we have to take the rest of the time (aka Survival), waiting for the next spoonful of happy soup…

**It should be duly noted that when humans feel nothing, neither good nor bad, for prolonged periods of time they seem to stop wanting to live, or they go to extremes in order to feel something…anything.

So, to borrow from biology, the DNA of human motivation looks like this: Parallel agendas (Survival/Pleasure) sequenced in a way that is intended to produce the following result: Survive until you figure out how not to die (helix strand 1), have as much fun (pleasure) while at it (helix strand 2), and (in order to be able to do that) construct a system (Ultimate Order) conducive to both those goals simultaneously (the connecting base pairs: ATGC) *while preserving energy*. This structure should be able to contain, protect and solve any and all issues arising from living on a planet (a thermodynamic ecosystem with competing elements) that constantly moves towards entropy (disease, wars, population growth, environmental changes, cultural evolution etc.).

Reconciling the edicts of both paradigms has been at the heart of human neural history.

59.3 Close-Ended Equations

Ultimate Order frameworks have been thus far based on the idea of Limit = Safety. This is how you close an open-ended equation: you install a clear limit (define the perimeters of an OEE) and penalize questioning it. As humanity experienced population growth, fired up by invention that afforded longer life spans and discoveries of new lands and new resources, more variables kept getting added to the equation. Reticulating them became increasingly challenging.

From the perspective of the human brain, sweating to juggle Survival and Pleasure, more variables meant higher probability of random outcomes (chaos/unpredictability), increased neural effort to fit all those new "items" into it's routine, and by extension – anxiety, which means pain. Brain scans show clearly that when humans are faced with uncertainty associated with survival (even loosely), pain centers in the brain light up and go into crisis management mode (cortisol, adrenaline, noradrenalin and other stress hormones flood the brain snowballing into a myriad of reactive behaviors Fight, Flight or Freeze none of which are fun).

Now, if the emotional destabilization that ensues on the individual level due to rapid/frequent extreme change to their familiar environment scales socially (compounded by propaganda) as it reaches critical mass and becomes an accepted truth, we begin to see an uptick in "neurotic jitters": Isolationist discriminatory ideologies lift their heads, clouding constructive judgement with destructive fear, which is then manifested as aggressiveness towards migrant cultures, fluid sex orientations, different belief systems etc.

Let's translate this paragraph into brain-lingo: if an overworked Amygdala, fails to adequately respond to variable pressure (Respond = process and store super-fast in order to seamlessly regulate emotional stress. Variable pressure = newly introduced information) then negatively triggered neural pathways are etched into the brain becoming automated feedback-loops, making pain (stress/anxiety) a resident evil. This propels aggressiveness towards patterns interferences detected via perceptual and sensory distinction, for example: skin color (visual), sounds of culturally unique music and language (auditory), smells of culturally unique foods (olfactory), culturally unique customs and rituals (conceptual diversity), different sexual preferences etc.

What's worse is that this type of pain (stress/anxiety) is nonlocal and without science, non-attributable. In times when science couldn't yet granulate and name this phenomenon, its indeterminate haziness made it all the scarier, encouraging not only reactive behaviors (fear, aggression) but also higher susceptibility to certain kinds of suggestion. One suggestion of particular toxicity is the imaginary scarcity of resources ("there is not enough food/water for everyone", "immigrants are taking our jobs" and other forms of mild-to-extreme xenophobia) which we will discuss on the upcoming section "The non-science of Survivalism".

But to the brain – it was pain, and pain was to be avoided at all costs.

An easy hack for stopping mental pain is using simplification. Since it's a consciousness induced pain – it can only be resolved on the level of perception. Close an open-ended equation and you're golden. The mechanics of it is super simple: Take this chaotic influx of information, lump it, roughly divide it to 2, and dump it in 2 pails. Make one pail the referential premise of the other (Bad framing Good, Black framing White, Moral framing Depraved – aka "the blame game"). Done.

This cognitive automation of "lump, divide and dump" is called Dichotomy or Binary thinking. And dichotomy is VERY useful for a tired, cognitively "lazy" brain, whose default programming is energy efficiency.

And so, it came to be that unknowingly, the ruling class, in the effort to govern with ease, created close-ended zero-sum counterfeit game, and began to sell the shit out of it. These marketing campaigns were multi-faceted and touched every aspect of life, from governance systems (Nazism is an example of a zero-sum binary belief system: Superior vs Inferior biology, where the only way for the Superior biology to prevail is by total elimination of the inferior one), through religion (Islam is an example of a binary belief system: Dar El-Islam vs Dar El-Harb, world of Islam vs world of War, where World of Islam is the only acceptable end-game), and of course sex and gender (which is a construct within the construct of the above-mentioned belief systems: Homophobia and Misogyny are example of binary thinking).

And in **socio-neurological** brain-lingo: the going idea behind those information regulatory systems was: Criminalizing "pattern interference". They simply made it illegal to be "different", hence effectively stopping variable pressure.

Or so they thought…Close-ended equations have an irresolvable grievance with information, since information – *can't* be shut down. Not ever. (well, maybe in a Black Hole in space, but even that is unprovable). All information does is exponentiate, so a time-proof monopoly over the definitions of Agency such as Personhood, Sex Orientation or anything else under the sun that ever was – is both improbable and unnatural.

Granted, this "Us Vs Them" black and white strategy did provide a few centuries worth of neurological/mental relief… but it too, ran its course. Science made sure of that, perhaps unintentionally, yet it did.

59.4   The Non-science of Survivalism

When science was in its infancy (and by the way – it always will be, precisely because information never stops), and measuring-organizing tools were more primitive, the rhetoric of those who pushed for "order" (hierarchical order) was seemingly unassailable: Our survival as a species/society/community depends on our collective effort to fight chaos/randomness. A successful collective effort depends on uniformity. Uniformity is the suppression of "differences" therefore uniformity generates Order. In the absence of order, protection can't be guaranteed and the threats of chaos/randomness as well as coordinated threats by rival collectives, can't be effectively thwarted. So, get in line, or get snuffed. With "the greater good" as the trumpeted objective – Life could be easily revoked.

The thing with phrases like "the greater good" is that they are virtually unchallengeable (unless you view them from the perspective of "hacking death") – because they promote both agendas: quantity (greater) and quality (good). And so, by virtue of improving the chances of "survival of the pack", they gave, and still do, a "license to kill". Sacrifice the pattern breakers to save the pattern for the compliant, because the compliant don't challenge the distribution of resources.

For sake of this argument we shall call the ruling class "The Order of the Avarice". The Order of the Avarice were kings of information (clergy, politicians, kings and nobles, military figures) who made an art out of hoarding, crafting and manipulating information. Everybody does this – but they were on a whole different level – they were prism shapers. They could artificially manufacture a public sentiment to back up their agenda du jour, and they figured they needed the math to work **for** them if they were to maintain power and secure their own offspring (a "litter" that was re-titled; dynasty, lineage, pedigree and other rudimentary tribalistic nouns to add value to nothing in particular, certainly not to any special talent – and reinforce the hierarchical structure. This retitlement became the source of their ill-reputed Entitlement). For the Order of the Avarice the idea of Equality is highly disadvantageous and disruptive, so "differences" (i.e. pattern interferences) are inversely leveraged. The math says that if "my different" is just as legitimate "your different" – that makes **US** "Equal". In evolutionary terms "Equal" means that your needs must be met with equal precedence as mine, and you have equal claim and access to resources. So of course – that won't do.

At this point, as an attentive reader, you should have been alerted by a contradiction: if information is endless, and resources are a form of information, scarcity of resources is… well…an exaggerated misunderstanding of the conditions and their true scope of utility. At best. At no point in history was there a life-threatening scarcity that couldn't be solved by relocation to a more resource-rich place or redistribution of local resources, which means that most alleged scarcities were ponzi scams designed by the Order of the Avarice to maintain power (today we know that hunger for example, is an artificial, manufactured and controlled situation, using extreme weather to hold entire populations hostages for political clout).

How did they – and how they still – pull it off? Privileged Access and Deceit.

The premise is that information is malleable, its delivery system is malleable and its recipient **brain is malleable**. So, the strategy is to work on all fronts at the same time, using tactical tools such as: "divide and conquer", semantic and syntactic ambiguity, "cooking" facts, timing the release of new information, capping it, to name but a few.

What's more is that the tool box of manipulating information has its own evolution. Today for example, neuro-hackers, who are masters of information trafficking, collect and utilize information using some pretty nifty technologies (measuring P3b, iris expansion and online search analytics) to induce behavioral modification in users. They are a modern version of the "Order of the Avarice"; information hoarders, manipulators and traffickers.

In the past, their biggest success on their quest to seal the concave of power, was the introduction of an omnipotent, omnipresent, resource-distributing vagueness which they called God. Without going into the history of religions, suffice to say that in the beginning even this tactic was fraught with the dangers of variability: they had multiple Gods, elemental beings charged with specific domains of human life. After a few millenniums of this, when celestial variety proved as confusing and as unmanageable as earthly variety and chaos ensued – they "defragmented" the whole Parthenon of providential oversight, and created a one-stop-shop all-in-one unified entity: One God. Mono Theos.

Although cleaning up the excess of deities was a messy business (and in some parts of the world, like India, it was unsuccessful) on the grand scheme of things, Monotheism was genius. It incepted the largest scale super-order that was ever seen until that time, under the spirit and benediction of the number 1 (a single, all-encompassing irreducible variable, through whom all existence emerged and, its jurisdiction boundless and its omniscience unrivaled. An Axis.) Then they appointed themselves to be its mouth pieces – and on they went with controlling the populace via fear-inspiring disinformation and selective murder. To make matters positively unchallengeable – they proclaimed God not only unknown, but also unknowable (to anyone but themselves of course), which is a ceiling that can't be broken until this very day and age, and in still reigns supreme. With God's mandate they decried "change" as the bringer of the End of Times.

To counter their deeply flawed logic, incumbent Champions of Change (Like Galileo) needed to invalidate not only the rhetoric used by The Order, but also dispel the deathly fear associated with open ended-equations. Their road often led to themselves being selected for early expiration, but their legacy endured and today we can clearly see a trajectory. With time, as "free thinkers" were getting instead of excommunicated, it became clearer and clearer that "sameness" means stagnation, and decoupling Life from Movement, Flow and Change means; Death. Which is in diametric opposition to the holy grail of all inventions/achievements: Eternal Life.

Today we know, as a sentient race that the answer lies in variety and dynamism.

As technology-driven globalism becomes the dominant reality, causing rapid changes in perceptual plasticity – we, as individual and as collective consciousnesses – find ourselves discarding old-world divisive, polarizing schools of though and adopting fluidity, spectrum, and migratory paradigms.

59.5 Transhumanism – The Next Gen of Dominant Mutations

Earth is a closed thermodynamic system (until we can colonize a new planet or be colonized by extraterrestrial life) which means that: (1) Chaos cannot be exported to relieve variable pressure, and (2) All transformations happen within the system. Mutations are a form of systemic transformations that propel entropy (Chaos).

In a sense, all humans are nothing if not lineages of successful mutants. Studies of evolution certainly support this assertion. What makes a mutation successful is its ability to reproduce and its rate of reproduction. Qualitatively in concurrence with environmental changes, and quantitively in large enough numbers to ensure at least marginal off-spring survival rate. Although the word "mutation" describes a physiological event that originates in the DNA, all physiological truths have a cognitive manifestation (for example: the cognitive patterns of a tall, blond and fat person, are shaped by different environmental reciprocities of a short bald and skinny person), and those incur social manifestations, due, in large, to humans' sensory biases. Once in a while we get a "trail blazing" mutant. One that forces society to redefine its perimeters and expand its prism (its cognitive patterns). Not all mutations survive, of course, most are just a blip in history.

What makes a successful "mutant" successful is: (1) Originality – Their ability to bring something new to the table and (2) Plausible and immediate utility (serving either survival or pleasure) – Their ability to persuade the masses in the immediate utility of this originality.

In this sense a Mutation equals Invention, in that that it presents a new "product of reality" (a new variable), adding new dimension, new perspective, and new means, with which "Life" is defined on our quest to immortalize it.

Change (dynamism of variables) is the only constant in all our equations. There is no other. Stagnation and Hacking Death are mutually exclusive.

To capture the dance between information and transhumanism:

> **Since information never ends, neither will the evolution of everything it touches. One such is Humanity, and it will never stop evolving, unless it is destroyed. If it is destroyed, it has not achieved its primary and ultimate goal which is to beat death.**

59.6 Final Word: "Information is Power" Only to Those Whose Brains Can Handle It...

For the average brain "Ignorance is bliss" has never been more accurate as it is now. Our world is in a perpetual state of disruption, as tsunamis of information crash on our consciousness, cognitive overloads and by extension mental exhaustion are mounting at an ever-increasing rate. This does not bode well to brains on average, and most definitely not to average brains, who seek respite (happy-soup), but get none.

Mental disorders, caused by those swells of unfiltered and unfilterable information, are emerging as a global scale epidemic with expressions ranging from structural irregularities in neural tissue caused by mental trauma, to compulsive and addictive behaviors as well as spikes in suicidal and homicidal tendencies. Statistics show us that in terms of our health-care system's ability to contain and heal, or even regulate, we are nearing the outer edges of a gaping crater: A wide-spread mental insolvency, caused, again, by information overload. This is the distillation of it all: Our success as a sentient race is predicated on how fast our brain can adjust to higher loads and more diverse forms of information. I suspect that in the not-so-far future technology will offer some much-needed upgrades…

Chapter 60
Christian? Transhumanist? A Christian Primer for Engaging Transhumanism

Carmen Fowler LaBerge

When considering Christians and transhumanists the misperceptions and unchari-table assumptions go both ways. Christians who are interested in engaging transhu-manists and emerging conversations in the culture about transhumanism, must learn a new vocabulary and develop ways to thoughtfully engage the issues being raised. Christians bring valuable resources to the conversation about what it means to be human, equitable access to technologies and therapies developed and the morality or ethics of some proposals. Transhumanists need Christian input and Christians need to engage this conversation. This a primer for the conversational apologetics necessary at the intersection of Christianity and Transhumanism.

My initial reaction to the idea of Christian Transhumanism was negative. I remain skeptical that the fundamentally different views of the future (one based on evolution devoid of God and one creation and redemption by, in and for God) can be sufficiently reconciled as to be meaningfully wed. The challenge for Christians interested in engaging in this conversation will be conversation without compromis-ing the fundamental Christian understanding of the imago dei, humanity, life, the body, and the vision of the future.

When I stepped into these conversations, what I knew of the secular transhuman-ist movement were those who pursue transhumanism as:

- **the next step in human evolution**. Think nanotech applications to biotech, CRISPR, deep brain implants, and other human enhancements through AI. The highest value in this view is intelligence – and you can see how quickly this leads to drastic human divisions and negative consequences for masses of people. Related to this is the vision of transhumanism as transcending biological nature. In this vision of transhumanism, the version of human beings who live today become technologically obsolete as our descendents become post-human. Think

C. F. LaBerge (✉)
Faith Radio Network, Saint Paul, MN, USA
e-mail: claberge@reconnectwithcarmen.com

X-Men. The challenge Christians face when engaging here is that have mind-lessly adopted and adapted our lives to many of the medical and technological advances as they have emerged without doing the hard ethical work of answering the "just because we can, should we," question.

- **the path to overcoming death**. Think radical life extension (living for 1000 years). How would things look different if aging and death were *cured*? Again, it is easy to see how quickly this leads to deeper divisions between those who can afford the enhancements that lead to super-long life, the wealth and power they would amass, and the predictable negative consequences for the 99.9% for whom these advancements would be withheld.
- **cybernetic immortality**. Think uploading yourself into a digital reality. Christians will immediately see this as gnostic dualism as it suggests a person's body is irrelevant to their human life. For those who argue this is simply the next form of physical life in digital space, we have to be prepared to talk about the importance God places on the body as evidenced in Creation, Incarnation, Resurrection, Ascension and the visions of heaven offered throughout the Bible. Do we share God's vision of the future or is our vision of the future merely human, devoid of God?

Christians have and will rightly continue to raise legitimate criticisms of these visions/versions of transhumanism. Observing that:

1. Transhumanism is inherently dehumanizing.
2. Transhumanism is inherently Darwinian in its vision of the future promoting outcomes that advantage the rich and well-connected, pragmatically preferenc-ing an elite segment of the population over and against the multitudes.
3. Transhumanism misplaces faith in technology and promises immortality as defined as something qualitatively less than the real eternal life offered with God through Jesus Christ.
4. Transhumanism promises to be demoralizing for large majorities of people who will be sidelined by technologies outpacing the human ability to produce.

Into each of these conversations, Christians must be prepared to speak as these conversations in our culture are progressing every day. We'll do some conversa-tional apologetic prep further along in this piece but here I want to acknowledge that through my inquiry into these matters I have discovered that there are many within the transhumanist movement whose vision of transhumanism is more *fully* human than *post* human.

Christians engaging in the transhumanist conversation talk more about the recov-ery of longevity and even super-longevity (the kinds of lifespans described early in the Old Testament) and they ask good questions about what good might done with the gift of a longer life? They talk about the principles and realities of the Kingdom of God being brought to bear in the world, not just for the few but for all; not just for consumerist pursuits, but for Godly good. That is a conversation worth having but in order to have it, Christians have to be able to articulate the theological answer to each objective above. We have to bring God into the conversation and the prin-ciples of the His Kingdom to bear on this particular conversation of our day.

To prepare for conversations with thinking people, we have to become thinking people. We have to think about what we're thinking about and bring the mind of Christ to bear on these matters. We have to think theologically, listen attentively, ask clarifying questions and apply what we know about God through God's Word by God's Spirit to real life. To do this we have to develop good theologies of the body, technology, mortality, work, physical finitude, and bioethics. These are the conversations for which we must actively prepare or find ourselves passively sidelined. Entire books could be (and many have been) written on these topics. What I will offer here is simply a brief conversational apologetic on each topic for Christians interested in being equipped to authentically engage.

Q: As a Christian interested in engaging in the transhumanist conversation, where do I begin?

Let's talk about what it means to be human before we consider transcending it.

Operating out of a Biblical worldview, Christians affirm that human beings are created by God, according to His plan, by His design and for His glory. If God is God, sovereignly working out His will throughout the full scope of history, then each human being and every human being is not only precious in God's sight (this has implications for the kinds of research we can and cannot affirm regarding biomedical and genetic research) but created by God on purpose and for a purpose. And what is that purpose? To glorify God by acknowledging Him in all our ways.

The Bible bears witness to the call of God upon human beings to cultivate the culture in which He set them as His image bearers. For the first humans, that culture was a literal garden. For humans today, that garden includes a variety of social contexts, cultures and technologies. How then do we, as twenty-first century humans, bear witness to the reality of God, authentically bear His image as moral, creative, organizing, redemptive agents of grace and Ambassadors of His Kingdom principles in the world? That is the question for the Christian in the transhumanist (and every other) conversation of the day.

To be authentically human then is to recognize that I am not self-made, self-governed, self-authenticating, self-guided or any other self-oriented, self-congratulating, self-elevating, self-centered hope the selfie culture might use as enticement. To be authentically human is to recognize God is God and I am precious to Him as His created, redeemed, beloved child. I can live here and now – and into eternity – in that relationship.

I am, in fact, fully human and utterly transcendent (even if beyond the understanding of others whose minds and hearts remained closed to spiritual truth). As a Christian, I recognize I'm not in the world to make a name for myself, amass material wealth nor experiences, nor ensure the continuation of "me" for generations to come. I am human (and that is good and great!), but I am only human. I am a person (and that is good and great!) but I am only a person. Yes, immortality is possible, promised and accessible to me and every other human being, but it is not an immortality I achieve, create nor insure for myself through technology or any other means. Eternal life – and a full life worth living here and now – are gifts of God's grace alone, accessed through faith alone, in Christ alone, to the glory of God alone.

Where do I get these ideas? Yes, you guessed it, the Bible, which is alone the Word of God. And it is from the Word of God I derive an understanding of the body, ethics, time, eternity, and purpose.

Q: As a Christian interested in engaging with transhumanists in conversation, what is my theology of the physical body, science, technology and bioethics?

The body is part of God's good creation, described as the temple of the Holy Spirit for those who are redeemed, and Jesus' bodily incarnation, resurrection and ascension demonstrate the value God places on the physical human body. So then, should we. And yet, it is not the body that is to be worshipped nor is this flesh-suit eternal.

The importance of the physical human body to the reality of being authentically human is part of the argument Christians will need to make when addressing ideas presented by some transhumanists who argue for the replacement of what has always been understood as human with other theories. Some would exclude some humans (those regarded as insufficiently persons) but include post-human, machine-hybrid, chimera or robots who are seen as self-aware through AI. Pragmatists might support the idea of self-awareness becoming the criteria for being regarded a person, but personhood from a Biblical perspective is intrinsically connected to the human being, who is uniquely a person, because he or she bears the image of the living God.

For Christians, being authentically human acknowledges the reality that God created human beings as persons with physical bodies. And God created human beings with a full freedom of the will and capacities to discover natural patterns (science) and develop technologies. But are all choices, all experiments and all technologies "good?"

From a Christian worldview, technology is not inherently good nor evil. Technology is morally benign but we are not. Human beings who develop and use technology are moral agents who stand responsible before God who defines the boundaries of good and evil. So, part of what Christians bring to the transhumanist conversation is the question of *should*. Should we do something, just because we now can?

Here we enter into specific conversations with transhumanists who argue for a myriad of obviously bad ideas from a Christian worldview: elective amputation and mutilation of the flesh, reproductive manipulation eliminating the need for male participation in conception and species continuation, genetic engineering allowing biological men to conceive and gestate babies. These are not far-out or far-off ideas. These are biotechnologies fully in use today. The question Christians need to introduce is this: just because we can, should we?

Q: As a Christian interested in engaging with transhumanist conversation, what is my theology of time, mortality or physical finitude?

What do you know about time and eternity? What do you know about the Kingdom of God, the Kingdom of Heaven, God's engagement throughout human history and history's redemptive arc? What do you know about the kingdoms of this world and the powers and principalities at play here? What does the Bible say about the mortality of the body and the immortality of the soul?

God has literally got all the time in the world, we do not. Our time in the world is limited. And while it is true that our lives are eternal, we do not live eternally in this physical state. We are finite creatures in an eternal relationship with an infinite God. Yes, God intends to make the most of the time we have but we neither worship time nor the created flesh we now inhabit. We worship God, confident that in the fullness of time He will work out His purposes in, through and for us – all for His glory and our good.

Ultimately, Christians interested in engaging in the transhumanist conversation must be able to define, articulate and contribute that which is distinctively Christian. Not everyone who claims to be a Christian is a Christian by the same definition. The same can be said for transhumanists. They do not all mean the same thing by the term. So, the most obvious starting point seems to me to be, "What do you mean when you use the term transhuman? What about your humanity are you seeking to transcend?" As a Christian, I am ready to talk with you about what it means to be authentically human and the transcendent life I'm living right now in and with a guy named Jesus.

Both transhumanism and Christianity are comprehensive or totalized systems of philosophy. The Christian cosmology of the redemptive Gospel cannot be reconciled with a metaphysical and philosophical system reliant upon endless evolutionary complexification. The Christian must ask (and be prepared to explain) what it means to the transhumanist to be human and we must also be prepared to expose the sin-side of their plans. For while there may be much good in longer life, sin remains and sin is prone to ruin good things and the good life so many pursue. We have to face the fact that people – even highly evolved people – have done, are doing and will continue to do horrible things. Ask the transhumanist what their plans are for identifying, rooting out and eliminating sin from the human genome. There is a solution to the sin problem, but it's not going to come through transhumanism. Technology can be, has been and will be pressed into the service of both good and evil. And solving the problem of evil transcends human or even transhuman capacities. For this, we need God. And suddenly the Christian is essential to the conversation. Be the Christian they come looking for because you have proven yourself interested and engaged in the transhumanist conversations of the day.

Carmen Fowler LaBerge   *is a Christian author, speaker, radio host equipping people to engage in conversational apologetics. She serves as Executive Director of the Common Ground Christian Network, has an M.Div. from Princeton Theological Seminary (1993) and 25 years experience in mainline and evangelical ministry settings. Author of Speak the Truth: How to Bring God Back into Every Conversation. Connect with Carmen at ReconnectwithCarmen.com*

Chapter 61
Christian Transhumanism: Exploring the Future of Faith

Micah Redding

Christian Transhumanism is a relatively new movement, even by the standards of transhumanism. And yet, Christian Transhumanism is not a break with Christian history or tradition, but an attempt to reform and renew it, just as many other movements have done before.

Looking closely at Christian scripture and the history of the early church, Christian transhumanists see early Christianity itself as a form of what we would call transhumanism—a kind of worldview that embraced dramatic societal and technological change, reconceptualizing human nature itself around an open-ended vision of transformation and transcendence.

Christian Transhumanists today embrace this essential core of early Christianity, and bring that perspective to bear on the opportunities and challenges of the future ahead, engaging with cutting-edge scientific, technological, and philosophical questions in a proactive and constructive way.

In doing so, Christian Transhumanists may challenge secular transhumanists to "be better transhumanists"—by rejecting philosophies of radical egoism, for example. At the same time, Christian Transhumanists willingly open themselves to scientific, technological, and philosophical critique from secular transhumanists, in a way that the traditional divide between science and religion avoids.

Many Christian Transhumanists come to transhumanism through theological reflection. But having done so, they may then turn around and attempt to express those theological perspectives in the language of technology, science, and secular transhumanist discourse.

Theistic beliefs may be expressed in terms of the Simulation Argument or the New God Argument. Beliefs about human nature may be expressed in the language of Turing Machines and Computational Universality. Answers to the traditional Problem of Evil may be offered on the basis of computational irreducibility.

M. Redding (✉)
Christian Transhumanist Association, Nashville, TN, USA

© Springer Nature Switzerland AG 2019 777
N. Lee (ed.), *The Transhumanism Handbook*,
https://doi.org/10.1007/978-3-030-16920-6_61

Going farther, memetics may be seen as the scientific exploration of spiritual forces, and early Christian discussions of spiritual beings may point to an intuitive grasp of memeplexes and memetic super-organisms.

In short, Christianity provokes a strong transhumanism, but transhumanism in turn sheds fresh light on Christianity.

Christian Transhumanism, then, is the site of a powerful and productive exchange, at the intersection between Christianity's historic essence, contemporary Christian thought, and secular transhumanism. As such, it offers us an incredible glimpse into the possibilities of the future of faith.

61.1   Christianity as Transhumanism

It may sound strange to suggest that a 2000-year-old religion could be a form of transhumanism. After all, transhumanism boils down to the idea that we should transform the human condition using science and technology—and we tend to think of science and technology as something *new*.

We also tend to think of religion as looking backward, rather than forward. In popular conception, religion doesn't embrace the future, it defends the past; religion doesn't pursue transformation, it upholds the status quo.

But on closer inspection, all of these objections are superficial. While modern science is evolving all the time, the pursuit of knowledge about how the world works has been going on for our entire existence. While technology is continually breaking new ground, the creation and invention of tools is as old as humanity itself.

And while many forms of modern religion are detrimental to progress, this is not the case for all religion across space and time.

Given the above definition of transhumanism, it is perfectly possible to talk about transhumanists in the ancient world, and it is perfectly possible to talk about an ancient religion as embodying the philosophical core of transhumanism.

To illustrate how Christian Transhumanists may understand their religious tradition, I'll examine some of the conceptual building blocks of transhumanism in the Hebrew Bible, show how those concepts are developed in the New Testament, and argue that early Christianity is properly understood as a kind of transhumanism.

I'll then look at how Christian Transhumanists may carry that tradition forward into the future.

61.1.1   Transhumanism in Christianity

The incipient transhumanism of Christianity can be seen in an overview of several core early Christian beliefs:

1. Humans are made in the image of God, and given the power and responsibility to rule over all things.
2. Human nature isn't defined by fixed forms, but by our transcendent potential.
3. Human power is expressed most dramatically in science and technology—the ability to explore and understand the world, and the ability to create new things based on what we learn.
4. Physical life is good, and radical longevity is humanity's desired future.
5. Death is our enemy, and will be defeated.
6. The entire physical universe will be transformed, and brought into radically flourishing life.

Each of these beliefs is expressed through stories in the Hebrew Bible, and then taken up and amplified through key doctrines in the New Testament. Finally, they show up in three core Christian ideas: the incarnation, resurrection, and second coming of Christ.

61.1.2   The Incarnation: What It Means to be Human

The Hebrew Bible opens with a profound declaration: that humans are made in "the image and likeness of God".[1]

Though the precise meaning of that phrase is open to debate, it has always been clear that it rules out conceptions of human nature based on specific numbers of fingers, limbs, or DNA sequences.

Instead, it evokes a powerful vision in which humanity is defined by its transcendent potential. Rather than being characterized by fixed forms or current limits, humans are characterized by open-ended creative power. This power is seen most prominently in their emergent science and technology, in their ability to create new things, and to undergo profound self-transformation.

This picture is laid out from the very beginning of the biblical story.

In Genesis 1, God is depicted as creator and ruler of the world—the one who continually creates new things, and cultivates life. For humanity to be made in God's image means that humans are also creators, and that they also have the ability to rule over all things.

The chapter culminates with God instructing humans to do exactly that: create and cultivate life, name and categorize creation, explore and rule over all things, join in all the work that God was doing.

This divine mission—to create new things, and explore, manage and rule over the world—is a call to science and technology in their most basic forms. From the very first chapters, humans are depicted as using and inventing new kinds of instruments, artwork, dwellings, and tools.

---

[1] Genesis 1:26–28.

But scripture goes farther. Technology is not just something humans happen to do, it's part of the work and mission of God. In one of the Bible's best-known stories, a righteous human is called by God to construct the world's largest technological artifact: a vehicle strong enough to survive an extinction event.

We call this story Noah's Ark, and teach it to kids before they can read.

This story is the Bible's archetypal vision of salvation, and it's about humans partnering with God in a massive technological project to save life.

The concept that humans can participate in divine technological projects is pervasive, showing up the Bible's account of the construction of the Jewish temple, where the builders and crafters are identified by name and praised for their specific skills. The text tells us that their technological skills are an expression of the Spirit of God, and the characteristics of their work are spelled out in exacting detail.

Other humans are praised for embracing open-ended transformation. Abraham is called to leave behind the world he knows, and venture out into the wilderness, in pursuit of a future civilization he can only imagine.

In a world which viewed time as a wheel, believed that nothing ever changed, and insisted that humans had no individual agency, this was a radical action. It represents the discovery of the future, the dawn of individuality, and the invention of history. Most fundamentally, it represents the pursuit of new possibilities for human life.

That confidence in the potential of an unknown future is precisely what the Bible means by "faith", and is why Abraham is so central to Judaism, Christianity, and Islam.

Today, we might call Abraham "the father of futurism". The Bible, stating something similar in its own terminology, calls him "the father of faith".

***

Where the Hebrew Bible affirms the dramatic potential of human nature, the New Testament doctrine of incarnation amplifies it exponentially.

The incarnation is the concept that in Christ, God became united with humanity.

As the New Testament sees it, this unification results in the renewal of humanity's unlimited potential. Christ is seen as the "new Adam", launching a "new humanity"—which spreads outward in an ongoing process of transformation.

This new humanity breaks all the boundaries and categories of the ancient world. As Saint Paul says, "In Christ there is no male and female, Jew or Greek, slave or free".[2]

It also challenges the identity structures of its time. As argued across the pages of the New Testament, this new humanity is not defined by ancient identity markers such as circumcision, or the adherence to religious laws or rules. Instead, "salvation is by faith"—trust in an open-ended future—and this humanity is constituted solely by its possession of the transformative "Spirit of God".

---

[2] Galatians 3:28.

Where the "old humanity" was a product of God's creative transformation, the "new humanity" is a source of creative transformation. As a result, this new humanity knows no limits.

Declarations about Christ are often misunderstood as only relevant to how we think about Jesus of Nazareth. This is not how the New Testament sees it. In the scriptural view, Christ is what humanity is destined to become. The glory of Christ is located precisely in how many others will eventually come to embody the nature, glory, and power he possesses.[3]

Christians enact this in powerful emblematic ways. In the Communion or Eucharist, Christians symbolically consume the body of Christ, so that they may "become what they eat", and be transformed into Christ themselves. In baptism, Christians identify with Christ, so that they may later enact Christ's reign and glorification.

Christians are called "the body of Christ", "participants in Christ", and "partakers of the divine nature". In the pages of the New Testament, Christ insists that he holds nothing back, but that he is sharing it all with humanity.[4]

Paul fleshes this out in one powerful passage in the Letter to the Galatians.[5] Humanity, he states, was always the child of God. But in the past, humans were like infant children, bossed around by schoolteachers. With the coming of Christ, we've all entered into adulthood. We've now grown up, inherited the estate, and been given power over all things. No longer do we listen to schoolteachers or blindly follow rules. Instead, we take responsibility for ourselves and our choices, recognizing our incredible position as rulers of the universe.

61.1.3   The Resurrection: The War Against Death

Contrary to popular assumption, the Bible emphatically affirms the value of physical life. Early stories in the book of Genesis depict the righteous as achieving lifespans of nearly a millennium.[6] The prophet Isaiah dreams of a future when infant mortality will be eliminated, peace on earth will be achieved, and lifespans will stretch into the hundreds or thousands of years.[7]

Across the scriptures, death is seen as a tragedy. Death is humanity's first threat, and most pointed loss. As scripture portrays it, it robs us of our loved ones, causes us to be possessed by shame and fear, empowers tyrannical empires, and poisons the experience of human life.

---

[3] Romans 8:29; see also Hebrews 2, 1 Corinthians 15.
[4] John 15:15.
[5] Galatians 3–4.
[6] Genesis 5.
[7] Isaiah 65:20.

Thus, the primary obstacle to humanity's intended reach and dominion is the pervasive presence of death. And if humanity is going to enter into its destiny as rulers of the universe, it will need to defeat death somewhere along the way.

This is precisely how the New Testament sees it.

"The last enemy to be defeated is death," Paul says in a passage which many call the core of the Christian faith. According to Paul, that victory must come about through humanity.[8]

This is why Christ became human, and it's why Christians have historically emphasized the bodily resurrection of Christ. Because Christ is seen as victorious over death, the possibility of humanity's ultimate victory over death is guaranteed.

This is where Christianity breaks radically with the philosophies of the ancient world. Other ancient philosophies held that people might live on immaterially after death, but early Christians insisted that they were going to defeat death *physically*.[9]

This is what Christians mean when they talk about "the resurrection of the body", and it's why the nature of Christ's resurrection was seen as so significant. Christ is not resurrected spiritually, in some immaterial heaven—Christ overcomes death physically, in this material universe.

Because early Christians insisted that Christ rose physically, they believed that death could be defeated physically. They were not planning to fly away spiritually to an immaterial heaven, but to be restored to physical life in this universe.

Even more pointedly, they insisted that *not everyone had to die*.[10]

Death was not an inevitable reality to be accepted—it was a tragedy to be mourned, avoided, and eventually defeated through the advance of humanity's power in the universe.

61.1.4 The Second Coming: The Cosmos Springs into Life

A popular version of Christian eschatology holds that believers are shortly to be whisked away to an immaterial heaven, while the rest of the world burns. For many people, this is the only "end-times" vision they've been exposed to.

But for decades now, prominent biblical scholars and theologians have been pointing out that this viewpoint is not only absent from historic Christianity, it is exactly the opposite of the New Testament's vision.[11]

Rather than an escapist notion of believers being whisked away to heaven, the New Testament depicts heaven descending to earth. As Jesus provocatively states in the Lord's Prayer, the core Christian vision is "On Earth as it is in Heaven".

---

[8] Corinthians 15:21.

[9] Justin Martyr (2nd century). *On The Resurrection*. Chapter 10.

[10] Corinthians 15:51.

[11] See, for example, N. T. Wright, *Surprised by Hope*.

Rather than an immaterial afterlife, the New Testament depicts the ultimate transformation of the entire material universe. As Jesus and the apostles proclaimed, they were ushering in "The Renewal of All Things".[12]

Rather than the destruction of the cosmos, the New Testament depicts it springing forth in vibrant new life.

As Paul states in the letter to Romans, "All creation waits in eager expectation for the children of God to be revealed…in hope that creation itself will be liberated from its bondage to decay, to join in the freedom and the glory of the children of God.".[13]

That is, creation is waiting for us to bring it to life.

Reading the Biblical story as literally as possible, the ultimate future is that evil empires fall, death is overcome, the world is renewed, and rich and poor alike are invited into heaven on Earth. Humanity becomes an unimaginably large, open-ended cosmopolitan ecosystem—and ultimately, through humanity, the entire cosmos springs to life.[14]

61.2   Transhumanism and The Rethinking of Christianity

Many Christian Transhumanists come to transhumanism through theological reflection. But having done so, they may then turn around and attempt to express those theological perspectives in the language of technology, science, and secular transhumanist discourse.

This is an ongoing and experimental endeavor, and connections we see today may look very different tomorrow, as continued theological, scientific, and philosophical reflection refines our thought processes. Nevertheless, it may be instructive to examine what some of that exploration looks like today, to get a fuller grasp on the potential of the future of faith.

Let's look at a few areas where that exploration is breaking new ground.

61.2.1   The Simulation Argument, the New God Argument and Minimum Viable Theology

Historically, the fiercest debates around science and religion have focused on the question of the existence of God. Prominent atheists such as Richard Dawkins and Sam Harris find belief in God to be unjustified and dangerous—while many theists believe that the reverse is the case.

---

[12] Matthew 19:28.

[13] Romans 8:18–21.

[14] Revelation 21–22.

In the midst of this discussion, Nick Bostrom's Simulation Argument is a bombshell. David Pearce stated that "The Simulation Argument is perhaps the first interesting argument for the existence of a Creator in 2000 years.".[15]

I would add that it is perhaps the most *Christian* argument for the existence of a Creator in 2000 years.

Essentially, the Simulation Argument asserts that one of the following must be true:

1. We will almost certainly go extinct before gaining the ability to run simulations of our past.
2. We will almost certainly lose interest in running simulations of our past, before gaining the ability to do so.
3. We are almost certainly in a simulation right now.

The popularization of this argument has led to articles speculating about whether our universe is a video game for some posthuman slacker, whether we are living in the Matrix, and so on.

But the form of the argument perhaps belies its ultimate significance. The argument tells us that *either* the ultimate reach of intelligent life is quite constrained, *or* that we are already likely to be living in a world constructed by intelligent life.

This does not remove theistic possibilities from the table, but it does remove atheistic ones.

In short, it is no longer intellectually coherent to be an optimistic atheist. If one has hope for the future of humanity, one must hold that we are likely to live in a created world.

Some may object that the Simulation Argument portrays a state of affairs very different from classical theism. But this is a misunderstanding of what the Simulation Argument accomplishes. It does not, for instance, rule out the possibility of a single all-powerful being creating our universe. It *does* rule out the likelihood that we have a bright future, and also happen to live in an uncreated universe.

Some may object that a simulation is far different than traditional claims about the creation of the universe. But this is getting lost in terminology. Traditional theism claims that the universe is a product of intelligence, that it exhibits design, is built on rules, and so on. This is indistinguishable from an abstract description of a computer simulation.

In addition, the Simulation Argument more closely mirrors the structure of biblical theology, than does classical philosophical theism.

For example, rather than arguing from the deep past, the Simulation Argument argues from the deep future. And rather than looking at human limits to demonstrate the existence of God, the Simulation Argument connects the existence of a Creator with the affirmation of incredible human potential.

This is why I say it is perhaps a more *Christian* argument than we've seen before. Christian theology focuses on the future, not the past. And in Christianity, everything centers on the connection between humanity and God.

---

[15] https://www.simulation-argument.com/

In Christianity, to believe in God is to believe in humanity, and to believe in humanity is to believe in God.

In the Simulation Argument, it appears to be the same.

The Simulation Argument is not a final theology—it is a starting point for investigation and discussion. In recognition of this, Christian Transhumanists build on the Simulation Argument in a number of ways.

The "New God Argument", for example, generalizes the Simulation Argument for any kind of creative mechanism, combining it with a similarly-structured "Compassion Argument" to argue that we should believe in the existence of benevolent superhuman beings.[16]

The "theistic premise" articulated in "Minimum Viable Theology" generalizes the discussion even further, to express the practical benefits of the culturally universal belief in the existence of superhuman beings.[17]

The "Most Powerful Being Argument" asserts that if the ability to create universes increases exponentially, then we are unlikely to live in a universe created by a "basement slacker", and quite a bit more likely to live in a universe created by one of the most powerful beings in existence—or one arbitrarily high on the scale of being.

What all of these have in common is a step beyond the stark trilemma of the Simulation Argument. Where the Simulation Argument poses three plausible scenarios, Christian Transhumanists assert that we must go farther, and actively hope and trust in only one scenario—the scenario in which intelligent life can experience unlimited flourishing.

This assertion is based on the understanding that to build towards a positive future, we have to hope and trust that a positive future is possible.

Thus, we do not assert *proof* of God's existence—we assert that we have an ethical duty to hope and *trust* in God's existence.

In Christian terminology, we have to have faith and hope in the future—and, given the Simulation Argument, that means we have to have faith and hope in God.

61.2.2   Computational Universality and The Image of God

Perhaps one of the most paradoxical concepts in Christianity is "the image of God"—the concept that human beings are made of carbon, and yet possess some characteristic of the divine.

Historically, this concept has suggested that humans have infinite potential, a transcendent nature, and were created equal.

It has held together the belief in human equality with the belief in humanity's transformative potential. It has affirmed the possibility of profound transformation,

---

[16] https://new-god-argument.com/

[17] http://micahredding.com/blog/minimum-viable-theology-we-are-not-alone

infinite life, and growth into superhuman beings—while asserting that nevertheless, the weakest of us has deep human dignity.

Few philosophies are able to simultaneously affirm all of these ideas.

The ideal of profound transformation is often paired with the denigration of one's current state. Upholding some form of excellence may be coupled with disrespect for those who don't have it.

For example, affirming the value of intelligence may be paired with denigrating those of lower IQ. And Nazism is famous for upholding the ideal of the "übermensch", while destroying millions who were judged to be inferior.

On the other hand, many perspectives affirm the equality of humankind, while seeing that as a reason to limit human transformation.

Some have argued that democracy requires us to restrict self-transformation, or that the survival of the human species requires us to preemptively curtail the emergence of superhuman beings.

But Christianity has always insisted on the existence of superhuman beings, the profound potential and goodness of human transformation, *and* the existence of full human equality. In Christian thought, humans have the potential for infinite life and infinite growth, while also, in some sense, already being on equal footing with superhuman beings.

We might reasonably ask if this is coherent. Interestingly, the work of quantum computing pioneer David Deutsch sheds light on the issue.[18]

Deutsch points out that the human brain is a Universal Turing Machine—and thus, that it can compute any algorithm that can be computed, or model any physical system in existence.

The possibility of Universal Turing Machines was the theoretical breakthrough that paved the way to modern computers, by demonstrating that it was possible to construct a single machine possessing *computational universality*—the ability to perform any kind of computation.

And as Deutsch has shown, every physical process can be expressed as a kind of computation.

Thus, a Universal Turing Machine can perform any kind of calculation, and simulate any kind of physical process—*including other Turing Machines*.

This means that every Universal Turing Machine is computationally and categorically equivalent. The only difference between machines is in their access to storage space and processing power. But their own universality guarantees that this an obstacle that can be overcome—Universal Turing Machines are able to access and make use of unlimited storage space and processing power.

Given equivalent resources, any Universal Turing Machine can emulate any other Universal Turing Machine.

This has profound implications. Since the human brain is a Universal Turing Machine, there is no physical system that it cannot model, and no computation it cannot do. It can model any physical process—including other brains.

---

[18] David Deutsch, *The Beginning of Infinity.*

This means the human brain is computationally and categorically equivalent to every other brain, including alien life-forms and future super-intelligences. The only difference is access to storage space and computational power, and those are precisely the things that we've been augmenting for our entire existence.

This means that in terms of mental ability and potential, differences between humans, including IQ, are strictly irrelevant. Any such difference is a matter of knowing particular algorithms, and any algorithm can be learned. In practice, we may not know how to transfer certain algorithms between humans, but computational universality guarantees that it is possible.

This also means that future super-intelligences need not pose an existential threat. Given equivalent storage and processing power, we can address them as equals.

This allows us to frame concepts such as humanity's "infinite potential", "transcendent nature", or being "created equal" in concrete ways. In fact, every single one of us does have infinite potential, possesses a transcendent nature, and were created equal—and given modern computer science, we can define this very specifically.

That this is precisely what gives us scientific and technological ability should come as no surprise. That we can model and manipulate any process in the universe, that we can come to understand any system, that we can create anything that can be created—is a direct consequence of the kinds of beings we are.

As Deutsch puts it, humanity has a "special relationship with the laws of physics".[19]

I can't think of a better term the ancient writers could have chosen than to say that we are *the image of God*.

61.2.3   Computational Irreducibility and The Problem of Evil

Thinking about our universe as a simulation may suggest answers to long-held philosophical problems. One of the most pressing of these problems for theists is the Problem of Evil.

The Problem of Evil, simply stated, suggests that there is a contradiction between the existence of a God who is All-knowing, All-powerful, and All-benevolent, and the existence of evil in the world.

But the Problem of Evil isn't just a philosophical issue. It's also a practical problem for believers in a simulated universe. Given the amount of evil in our world, our simulators must seemingly not be benevolent. And if we live in universe created by non-benevolent beings, then we are at war with our Creators.

This is a significant and potentially troubling issue. If our Creators are our enemies, the value of everything in life is at stake. Even the validity of the scientific

---

[19] http://brickcaster.com/christiantranshumanist/35

enterprise is called into question. On the other hand, if our Creators are benevolent, then we may be able to trust in the value and integrity of the processes around us.

To put it mildly, this is something we need to know.

The normal philosophical response to the Problem of Evil is to suggest that the All-powerfulness of God is constrained by the rules of logic. There are some things God cannot do due to logical constraints—such as create a rock too big for him to lift, and so on—and this fact constrains what he can do in the world.

In particular, if God wants someone to freely choose good, he cannot force them to do so, because forcing them eliminates the possibility of them *freely* having chosen.

This general response does show that evil in the world doesn't completely rule out the existence of an All-knowing, All-powerful, All-benevolent God.

But it is not a very satisfactory response. Although it has opened a crack in the problem, it has not resolved the tension the problem suggests. Just because it is logically possible to find an explanation for some kinds of evil, does not mean we have, or are likely to find, an explanation for all the kinds of evil.

Just because it is possible for you to have an alibi for your neighbor's murder, does not mean you *do* have an alibi for your neighbor's murder. And even if you do have an alibi for your neighbor's murder, that doesn't mean you have an alibi for the thousand other murders that happened in your neighborhood.

Thus, a satisfying answer to the Problem of Evil will require a more comprehensive response.

But this is where things get interesting. If the universe is a simulation, then computer science has something to say about many age-old philosophical problems, including this one. Giulio Prisco, founder of Turing Church and former board member of the World Transhumanist Association, has outlined one potential approach to addressing this.[20]

There are two kinds of programs in computer science—those whose outcomes can be determined ahead of time, and those whose outcomes are undeterminable until the program is run.

Determining the outcome ahead of time means having a program that produces precisely the same results in a shorter amount of steps—a program that is more efficient, in other words.

For some programs, a more efficient program exists. Other programs are already at maximum efficiency, and their outcomes cannot be determined unless they are run.

We call these latter programs "computationally irreducible".

If the universe is a simulation, then the question naturally arises: is the universe computationally irreducible? If it is *not*—if a more efficient program for determining its outcome exists—then we could predict the future.

But if the universe is computationally irreducible, then no one, not even God himself, would be able to completely anticipate what will happen. Although God

---

[20] "The Computational Problem of Evil", 2013 Mormon Transhumanist Association conference, https://www.youtube.com/watch?v=fiTb6zhqHLI

could start and stop time, roll it back, and redo everything again, the only way for God to know what someone might choose in a particular situation is to let that situation play out.

God could, of course, spawn another process to compute the universe faster, but this is precisely equivalent to letting it play out. Because there is no shortcut computation that can be done, completely anticipating the future means computing every single moment of time, and seeing what happens.

In other words, given a computationally irreducible universe, God cannot anticipate evil without allowing evil to exist.

Is the universe computational irreducible? It seems likely that it is. If you are a simulator, you simulate worlds to see what will happen. If you already know what will happen, you move on to more complex worlds.

It seems likely that we are in a universe where our Creator is interested in the outcome—and given such a universe, computational science suggests that despite our Creator's power, knowledge, and benevolence, the Creator cannot completely eliminate evil.

61.2.4 Memetics, Super-Organisms and Spirituality

One of the most controversial features of traditional religion is the belief in creatures outside the bounds of normal organic life.

We've already discussed the possibility that we are not alone, that there are superhuman beings in existence, including those who may have created our world. But religion usually assumes more than just greater beings *out there*—it also assumes that there are significant beings right *here*, engaged in our world and in human life.

Where we normally categorize the organic world into humans, animals, plants, and so on, religions often regale us with menageries of angels, demons, gods, goddesses, ghosts, spirits, souls, and more.

This is one of the factors that make many people consider religion a form of superstition.

Ironically, the work of Richard Dawkins may demonstrate that this perspective is in fact true. In doing so, his work may provide a framework for a constructive understanding of both our religious past, and our religious future.

In his seminal 1976 book *The Selfish Gene*, Richard Dawkins coined the term "meme". Although "meme" is now used as a term for photos with funny captions, the original concept was far broader, referring to any idea, term, or sound that can be spread from person to person. Dawkins was making the point that genes weren't the only things which could spread virally and organically—anything capable of being replicated, and influencing its own replication, would behave in the same way.[21]

---

[21] "The Selfish Gene", Richard Dawkins.

Dawkins' idea was both insightful and prophetic. The term "meme" itself spread from person to person, evolving and adapting as it did so, until it came to be a label for the whole genre of photos-with-funny-captions mentioned above. In a sense, Dawkins' term had taken on a life of its own, leaping the fence of academic literature, and finding its way into the much more lucrative ecological niche of social media.

Dawkins hadn't simply coined a clever term. In showing us the organic behavior of ideas, he had thrown open the door to a whole new ecosystem. Just as our bodies play host to billions of tiny creatures living in symbiotic (and sometimes parasitic) union with us, our minds play host to billions of tiny creatures as well. And those creatures are the base layer of an entire ecosystem which stretches far beyond the individual human mind.

Others quickly picked up and began to grapple with Dawkins' ideas. One of the first realizations of people thinking about "memetics" was that memes can work together symbiotically, creating larger entities called "memeplexes".

Memeplexes don't simply spread themselves as ideas. They also tend to generate effects in the physical world. Religious memeplexes build temples, scientific memeplexes create technologies, technological and aesthetic memeplexes build cars, clothes, iPhones, and factories.

Kevin Kelly calls this *the technium*, the entire ecosystem of art and technology and cultural products. Kevin sees this as a seventh kingdom of life, whose genes consists of our ideas, and whose bodies consists of our tools. Just as the biological creatures we're familiar with start as genes that produce bodies, and those bodies help spread their genes—creatures of the technium start as ideas that produce artifacts, and those artifacts help spread the ideas.[22]

There is more to the technium than just products and artwork and artifacts. There are other systems that grow up from the seeds of memeplexes, and become vast organisms that live in and among human societies without even making us think twice.

Take corporations, for instance. They are creatures: they are born, they grow, they die. They breath in and out, they consume and excrete. They've even convinced us to legally classify them as persons, possessing rights normally only associated with human beings.

Where did they come from? The seed of an idea—a meme—that mutated and spread until it had attracted other memes, and become a memeplex. That memeplex allowed the creation of systems of human order and operation, organizing us around the processing of money and materials.

Those organisms have their own impulses and needs. Even in a legal sense, the operators of a corporation are not free to make their own decisions—they are legally bound to make money, to feed and grow the corporation to the best of their ability.

Corporations are only one example. In reality, all organizations are some species of this kind.

---

[22] "What Technology Wants", Kevin Kelly.

Of course, corporations and institutions and organizations can only exist because there are legal and political systems in which they live and move and have their being. These legal and political systems are themselves creatures, born of meme-plexes, grown into widespread systems of human behavior, occupying vast space in human minds and homes, consuming huge quantities of energy and upkeep, exclusively operating and possessing large amounts of the real estate on this planet.

But we're still only scratching the surface of organisms that live among us and in us. Riots are a kind of viral contagion that break out, then quickly dissipate, leaving many people feeling like their minds have been possessed. They act as mental infections spread from person to person, leading to emergent behaviors in concert with other infected individuals.

Sometimes the contagion is intentional. This is the concept of the "egregore"—an autonomous entity made up of, and influencing, the thoughts of a group of people. Thus, a group of people may consciously or unconsciously come together to incarnate a god or a demon. If that is what some people play at in party games and seances, it may be what others accomplish in a more substantial way, given more concerted efforts.[23]

All of these things—mobs, markets, political systems, cities, bureaucracies, institutions—are creatures which spread themselves across multiple human hosts, and which sustain themselves by drawing on and controlling human attention and effort. I call them *super-organisms*.[24]

These super-organisms are mostly invisible, and yet exert incredible power and influence in human life. They can possess people, changing them dramatically—sometimes temporarily, sometimes permanently. They can war with each other, they can be killed, they can be cast out. They can be healthy or diseased, benevolent or destructive.[25]

Struggling with these super-organisms is not like fighting another human being. Just like trying to contain a meme usually makes it spread faster, trying to destroy diseased and destructive super-organisms through conventional means almost always backfires. To fight these entities, we need new and different approaches. One of these approaches might be to sustain and grow benevolent super-organisms, who can fight destructive super-organisms on their own turf.

At least one religion has the explicit aim of forming such a super-organism, grown from the voluntary efforts and sacrifices of large numbers of humans, across numerous generations. We Christians call this *The Body of Christ*.

***

---

[23] "A Pseudoethnography of Egregores", for Ribbonfarm, by Sarah Perry, https://www.ribbonfarm.com/2016/12/01/a-pseudoethnography-of-egregores/

[24] For more on this, see "Minimum Viable Superorganism", for Ribbonfarm, by Kevin Simler, https://www.ribbonfarm.com/2016/02/11/minimum-viable-superorganism/

[25] For an in-depth theology of super-organisms, and their relation to Christian thought, see "The Powers That Be: Theology for a New Millennium", by Walter Wink.

That we live in a world of super-organisms is not a new realization. It is perhaps one of the oldest ones—a realization that must have dawned as early humanity grappled with the forces unleashed through emerging self-awareness.

Like the birth of the internet, suddenly it seemed as if viruses were everywhere.

The idea that a human could be possessed by some rogue entity was not hypothetical, it was obvious. The idea that tribes and kingdoms were driven by a non-material dimension of life, was the height of practicality.

The concept that healthy human life required some kind of "anti-virus", and meant engaging in some kind of non-material warfare, erecting non-material defenses, using non-material weapons, made all the sense in the world.

If early human societies had clumsy ideas about how to go about this, that does not invalidate their realizations. The fact that we're still struggling with these problems in the age of social media, demonstrates just how difficult they are to solve.

These issues aren't going away. If anything, they're likely to become more pressing, and more dramatic over time. To deal with them, we're going to have to embrace the fact that we live in a world haunted by non-human forces, an ecosystem of invisible beings that shape every aspect of our lives.

We're going to need to engage in memetic warfare, which may shed fresh light on, and gain fresh insight from, ancient ideas about spiritual warfare.

And it starts by recognizing that there are super-organisms among us.

61.3　Christian Transhumanism and The Future of Faith

When we zoom out to a sufficient level, transhumanism is a philosophy adopted by people during a certain era, in order to help navigate the challenges and opportunities of their coming future.

That philosophy navigates these challenges and opportunities by preemptively encouraging a reevaluation of the concept of humanity, along with related concepts like identity and individuality. As a result of that reevaluation, transhumanists are capable of embracing a "proactive principle", or approach to change and technological opportunity that moves forward more aggressively than the surrounding culture, rooted in a fresh conception of what must be preserved, and what may be left behind.

When viewed from this angle, the parallels between transhumanism and early Christianity are incredibly clear.

Christianity, too, was a worldview adopted by people during a certain era, in order to help navigate the challenges and opportunities of their coming future. Christianity, too, navigated this future by provoking a reevaluation of the concept of humanity, along with related concepts like identity and individuality. As a result of that reevaluation, early Christians were capable of embracing a "proactive principle" toward social, cultural, and technological disruption, rooted in a fresh conception of what must be preserved, and what may be left behind.

But these parallels are more than "skin-deep". The conclusions of early Christians as to the nature of humanity are largely compatible with and evocative of the conclusions of thoughtful transhumanists. Early Christians embraced what some theologians have called a "transcendent humanism", a conception of humanity rooted not in the limits of the world around them, but in the firm belief in humanity's unlimited potential.[26]

This conception caused the early Christians to reject the identity markers of the ancient world—like race, gender, nationality, and social class—and embrace an international cosmopolitan identity, rooted in a vision of cosmic transformation.

It also led them to reject the common dualistic religious philosophies of their day, in which an individual's future lay in an other-worldly non-physical realm. In contrast, and to the great confusion of their interlocutors, the early Christians insisted that the world was to be physically transformed, and humanity along with it. They looked forward, not to an immaterial afterlife, but to a material renewal of their bodies, and a return to the physical world they might otherwise have left behind.

In logical structure, core philosophy, and detailed ramifications, the early Christian worldview was a kind of transhumanism.

Naturally, Christianity as a whole has not always maintained such a perspective. After successfully navigating the social, cultural, and technological disruption of the first few centuries CE, Christians were tasked with reinterpreting their vision for an unprecedented new context, leading to an immense variety of different understandings.

But neither has the possibility of a Christian transhumanism ever been entirely absent. At various times in history, Christians have recovered some of Christianity's core transhumanist perspectives, to advance scientific, technological, and societal change.

So when Jesuit priest and paleontologist Pierre Teilhard de Chardin attempted to express Christian doctrines in scientific and evolutionary language in the 1930s, it is no wonder that he found himself articulating a kind of transhumanism, and discussing concepts that some see as anticipating the Internet and the Singularity. In the 1940s, Teilhard even used the term "transhuman" for this vision.[27]

Julian Huxley, often considered the father of modern transhumanism, was a good friend and correspondent of Teilhard. He wrote the forward to Teilhard's book "Phenomenon of Man", creating the first crossover link between the world of Christianity and the world of secular transhumanism.

It was not the last. In 1994, Frank Tipler, professor of mathematical physics at Tulane University, published "The Physics of Immortality", proposing a transhumanist cosmology with a positive future for life in the universe. In his book, Tipler discovered that this cosmology was anticipated in the works of prominent theologians Tillich and Pannenberg.

---

[26] Paul Tillich, "A History of Christian Thought" (November 15, 1972).
[27] Teilhard de Chardin, "The Future of Mankind" (1949).

In 2014, the Christian Transhumanist Association was formed to continue this conversation.

Theologies and philosophies are always in a state of evolution, and we'll continue to see positive exchanges between Christian and secular transhumanisms as we move forward.

Christians will continue to draw insight from the ancient past, 2000 years of culture, and current science and technology, to illuminate the opportunities and challenges of an unknown future. Transhumanism will continue to offer new critiques and fresh insights into age-old theological problems.

And Christian Transhumanists will continue to advance the vision of a radically flourishing future that is good for all life.

Chapter 62
Boarding the Transhumanist Train: How Far Should the Christian Ride?

Ted Peters

Can you imagine a future without us? Without the human species? When the H+ movement was being born at the turn of the twenty-first century, Silicon Valley's Bill Joy forecasted a future of machines without us. "Our most powerful 21st-century technologies–robotics, genetic engineering, and nanotech–are threatening to make humans an endangered species."[1] The transhumanists among us do not want to just wait and watch while the technology train passes by. Transhumanists want to become the engineers steering us toward a posthuman destination.

Should twenty-first century Christians board the transhumanist (H+) train? The transhumanists have entered a utopian address into their GPS, which now dictates a turn-by-turn route to get there. Transhumanists press for human transformation in a manner not unlike the Christian doctrine of sanctification. H+ architects plan to construct this transformed future on a foundation of technology rather than spirituality; so we ask: is this travel itinerary good enough to be baptized by the faithful Christian?

The baptizing and incorporation of secular schools of thought is nothing new for the church. Greek speaking Christians in post-New Testament times incorporated first Plato and then Aristotle, providing for the Hebrew mind the conceptual apparatus necessary to declare the gospel of Jesus Christ to become universal. Some theologians in the mid-twentieth century incorporated Whiteheadian metaphysics into their worldview, giving birth to Process Theology. Might Christian thinkers today find themselves in a comparable situation: could secular transhumanism be baptized and Christianized?

Or, to ask the question differently: given the missionary and apologetic tasks of today' churches, should faithful Christians today don the transhumanist uniform in order to play and win the game of technological revolution?

---

[1] Bill Joy, "Why the Future Doesn't Need Us," *Wired* (April 2000); https://www.wired.com/2000/04/joy-2/ (accessed 11/28/2016).

T. Peters (✉)
Graduate Theological Union, Berkeley, CA, USA

© Springer Nature Switzerland AG 2019
N. Lee (ed.), *The Transhumanism Handbook*,
https://doi.org/10.1007/978-3-030-16920-6_62

In this chapter I would like to ask: do Christians really want to ride the transhumanist train all the way to its destination? If the transhumanist destination turns out to be a posthuman disembodied superintelligence that spreads throughout the universe in an attempt to supplant the omnipresent divine Spirit, then I recommend we get off the train. If the transhumanist track takes us into the land of bio-nano-genetic therapy that enhances health and leads to longer and more flourishing lives for embodied *Homo sapiens,* then I recommend Christians ride this train for at least a few stops along the way.

62.1   Boarding the Transhumanist Train

The transhumanist train is pulling out of the station, and a variety of religious riders have already boarded. In the Buddhist coach we find Michael Latorra, who observes that "reducing suffering and increasing happiness are goals common to Buddhism and to transhumanism."[2] In the Mormon coach we find Lincoln Cannon, who sees in the H+ post-human vision a future for a New God, a "trust in humanity that may qualify as God."[3] In the Unitarian-Universalist coach we find James Hughes, who wants to influence H+ with a sense of social justice: "I expect UUs to be critical transhumanists, pushing technoutopians to remember the current needs of the world's poor, for clean water, adequate shelter and decent wages."[4] In brief, if Christians climb on board the transhumanist train they will have religious company.

Religious riders need to recognize that the train's engineers do not want them on board. In his acerbic "Fable of the Dragon-Tyrant," Oxford transhumanist Nick Bostrom ridicules religious traditions for so investing in theologies of death that they will predictably resist the technological triumph over death.[5] Religion will become obsolete when H+ has begun to reign in society and culture. Once the transhumanist train pulls into its final station, religious riders will be unwelcome in the H+ utopia.

What this means is that secular transhumanists do not perceive the compatibility between their own utopian future and what Christians or other religious visionaries foresee. The implication is clear: for the Christian transhumanist the motivation

---

[2] Michael Latorra, "What is Buddhist Transhumanism?" *Theology and Science* 13:2 (2015) 219–229, at 219.

[3] Lincoln Cannon, "What is Mormon Transhumanism?" *Theology and Science* 13:2 (2015) 202–218, at 212. An Orthodox Christian departs from the Mormon eschatology by denying multiple gods in favor of the one Trinitarian God. "Deification does not transform us into independent deities but rather frees us from our pretensions to autonomy so that we may participate in the blessed, communal life of the triune God." Ian Curran, "Becoming godlike? The Incarnation and the Challenge of Transhumanism," *Christian Century* 134:24 (November 22, 2017) 22–25, at 25.

[4] James Hughes, "Transhumanism and Unitarian Universalism: Beginning the Dialogue," http://changesurfer.com/Bud/UUTrans.html (accessed 9/6/2018).

[5] Nick Bostrom, "Fable of the Dragon-Tyrant," *Journal of Medical Ethics*, 31:5 (2005) 273–277; https://nickbostrom.com/fable/dragon.html (accessed 9/6/2018).

must come from distinctively *Christian* sources, not borrowed from the transhumanists.

62.2   Distinctively *Christian* Transhumanism

Reformed theologian Ronald Cole-Turner calls, "all aboard!" He thinks H+ provides the Christian churches with both a challenge and opportunity. "The futuristic scenarios of the transhumanists are an open invitation for the church to think its own way about the culture of the future."[6]

Micah Redding and the Christian Transhumanist Association (CTA) have already boarded the transhumanist train. When plotting their itinerary, they embark from a biblical and theological platform.

> We believe that God's mission involves the transformation and renewal of creation including humanity, and that we are called by Christ to participate in that mission: working against illness, hunger, oppression, injustice, and death.[7]

Technological innovation becomes the means, not the end, of the Christian approach to renewal.

> We believe that the intentional use of technology, coupled with following Christ, will empower us to become more human across the scope of what it means to be creatures in the image of God.[8]

Rather than ask the transhumanist to set the goal or provide the *summum bonum,* CTA relies on the gospel of Jesus Christ to direct the mission and provide criteria for measuring the good.

Like CTA, Ronald Cole-Turner celebrates the overlap between the transhumanist agenda and the Christian mission. Yet, he cautions the Christian to take heed. Both envision transcendence, transformation, and renewal. Yet, what distinguishes the Christian is the role played by divine grace. "For transhumanists, the cause or the agent of human transcendence is technology. For Christians, it is grace, the undeserved goodness of God who gives life and wholeness to creation."[9] The Christian should take advantage of riding the transhumanist train toward transcendence and transformation, but disembark before leaving grace country.

---

[6] Ronald Cole-Turner, "Introduction," *Christian Perspectives on Transhumanism and the Church: Chips in the Brain, Immortality, and the World of Tomorrow,* eds., Steve Donaldson and Ronald Cole-Turner (New York: Palgrave, 2018) 1–16, at 5.

[7] https://www.christiantranshumanism.org/ (accessed9/6/2018).

[8] https://www.christiantranshumanism.org/ (accessed9/6/2018).

[9] Ronald Cole-Turner, "Going beyond the Human: Christians and Other Transhumanists," *Theology and Science* 15:2 (2015) 150–161, at 150.

62.3   From the Human to the Post-Human via the
          Trans-Human

Transhumanism lays the track between today's humanity and tomorrow's posthumanity. If the present generation of *Homo sapiens* does nothing, our human species will evolve over time regardless. A million years from now, our descendents will look back at us as their predecessor species. The question transhumanists raise is this: should we just let nature take its course or should we intervene and take technological control over this evolutionary process? Could we design our descendents like we design Toyotas? The transhumanists respond with an enthusiastic, yes. Harvard's sociobiologist Edward O. Wilson speaks in promethean accents for H+: "We are about to abandon natural selection, the process that created us, in order to direct our own evolution by volitional selection–the process of redesigning our biology and human nature as we wish them to be."[10]

What will the posthuman species that survives while we humans go extinct look like? Natasha Vita-More points to the end station. *Posthuman* refers to "a person who can co-exist in multiple substrates, such as the physical world as a biological or semi-biological being. The future human...will live much longer than [today's] human and most likely travel outside the Earth's orbit."[11]

The locomotive pulling the transhumanist train toward this end station is powered by genetics, nanotechnology, and robotics, or GNR for short. The highest value on the transhumanist scale is intelligence. So, the task of GNR will be to raise both human and machine intelligence to such a high level that the resulting intelligence will take over and reproduce itself on its own. Regarding social ethics, transhumanists rely on Social Darwinism in the form of *laissez faire* capitalism. The intelligent among us will survive, while the less intelligent will be discarded by the road side.

According to the *Transhumanist Declaration* of the World Transhumanist Association, "Humanity will be radically changed by technology in the future. We foresee the feasibility of redesigning the human condition, including such parameters as the inevitability of aging, limitations on human and artificial intellects, unchosen psychology, suffering, and our confinement to the planet earth."[12] This leads to a vision of a posthuman future characterized by a merging of humanity with technology as the next stage of our human evolution. Humanity plus (H+) is calling us forward. *Posthuman* refers to who we might become if transhuman efforts achieve their goals.

According to the myth–the conceptual set–of the transhumanist, evolution and progress are two ideas that fit together like a video game and a smartphone. Evolution is progressive, and technology will speed up evolution's progress. This technological progress will allegedly shoot the human race like a cannon ball over its previous

---

[10] Edward O. Wilson, *The Meaning of Human Existence* (London: W.W. Norton, 2014) 14.

[11] Natasha Vita-More, *Transhumanism: What is it?* (published by author, 2018) 31.

[12] "Transhumanist Declaration," http://humanityplus.org/philosophy/transhumanist-declaration/ (accessed 10/7/2015).

biological barriers. "Transhumanism...direct[s] application of medicine and technology to overcome some of our basic biological limits."[13] The posthuman will be semi-biological or even post-biological, that is, our descendents could have intelligent minds not in brains but in computers. What we have previously known as *Homo sapiens* will be replaced by *Homo cyberneticus*. *"As humanism freed us from the chains of superstition, let transhumanism free us from our biological chains."*[14]

62.4   The Highest Good of the Truly Human

At this junction the Christian needs to pause and ask: is this the train I want to board? There are two signs that the transhumanist train may sidetrack the person of faith. The first sign is the H+ reverence for intelligence and mind, especially the intelligent mind in a disembodied state. The second sign is the one labeled "posthuman." Does the Christian promise apply to a species that is posthuman? Or, does it apply to what will be truly human?

First, the question of the *summum bonum* or highest good. For the transhumanist, intelligence is the highest good. What about the Christian? Christians have always lauded and celebrated the human mind, to be sure. Despite how highly valued it might be, intelligence does not sit on top of the Christian list of values. What does? Love. 1 John 4:17–19: "Love has been perfected among us ... We love because he first loved us."

To have "love perfected in us" is the destination to which the sanctification train takes us in the work of John Wesley. To be *perfect* is to be "sanctified throughout," contends Wesley.[15] To be completely sanctified in this life means that each day we rise out of bed and all day long feel only love in our hearts. "Pure love reigning alone in the heart and life, this is the whole of scriptural perfection."[16] Regardless of

---

[13] Ibid.

[14] Simon Young, *Designer Evolution: A Transhumanist Manifesto* (Amherst, NY: Prometheus Books, 2006) 32, italics in original. One the one hand, we must grant that within the Christian tradition we have many examples of disembodied salvation not unlike the H+ vision. According to St. Symeon (949–1022 AD), "the whole creation, after it will be renewed and become spiritual, will become a dwelling which is immaterial, incorruptible, unchanging, and eternal. The heaven will become incomparably more brilliant and bright than it appears now; it will become completely new. The earth will receive a new, unutterable beauty, being clothed in many-formed, unfading flowers, bright and spiritual. The sun will shine seven times more powerfully than now, and the whole world will become more perfect than any word can describe." St. Symeon the New Theoloogian, *The First-Created Man*, tr. by Fr. Seraphim Rose (Platina CA: St. Herman of Alaska Brotherhood, 1979) 104. On the other hand, we have within the Christian tradition a strong affirmation of embodiment embraced by resurrection. "We should be deploying the Christian belief in the resurrection of the body against these anti-human technological aspirations, just as the Fathers did against Platonic dualism." Richard Bauckham, *The Bible and Ecology* (Waco TX: Baylor University Press, 2010) 149.

[15] John Wesley, *A Plain Account of Christian Perfection* (London: Epworth Press, 1952) 30.

[16] Ibid., 52.

our relative level of intelligence, regardless of how abstract our theological pontifications, regardless of the exalted degree of our most noble thoughts, without love we are but "a noisy gong or a clanging cymbal" (1 Corinthians 13:1).

If we transpose love into distinctively Christian transhumanism and subordinate intelligence, then we must ask: can we employ GNR to speed up the sanctification process? Can bio-nano-tech make us more virtuous? More loving? More sanctified? Deified?

62.5   Can We Genetically Engineer Virtuous Living?

In Syriac Christianity and Eastern Orthodoxy as well as Western Roman Catholicism and Protestantism, Christians have often devoted their psyches and bodies to spiritual transformation. Rigorous self-discipline gave rise to our word, *religio* or *religious,* which means discipline. For millennia heroes and heroines of the faith have fasted and flagellated to empower their higher spiritual natures to gain control over their lower fleshly natures. The goal of such spiritual practice has been virtue, sanctification, deification. Is there a technological shortcut?   The   Christian transhumanist must ask: could we through gene-editing or deep brain implants enhance the human capacity for virtuous living?[17]

Roman Catholic theologian Alison Benders is willing to explore such questions. She imagines a successful alteration of the human genome that increases the number of genes disposing a person to perform good works. Most specifically, the genetic engineer could edit out predispositions to aggression. In a limited sense, then, gene-editing could contribute to improved moral behavior. "Moral enhancement through genetic engineering might predispose us to performing good acts, good deeds."[18]

Even if we should be able to enable increased good deeds, however, Benders denies that genetic technology could contribute to the life of sanctification. Why? Because the life of holiness requires willful participation as well as discipline over time. "Moral goodness requires freedom; it requires the intentional commitment to participate in God's redemption of the world."[19] No pre-programmed genome can replace personal commitment over time.

Protestant Braden Molhoek takes the same position. "Genetic engineering has the capacity to enhance the human disposition to moral behavior, but gene editing

---

[17] The current debate over virtue enhancement via genetic technology has been prompted by Mark Walker, "Enhancing Genetic Virtue: A Project for 21st Century Humanity," *Politics and the Life Sciences* 28:292009) 27–47. Walker expanded the scope by including deification in "Genetic Engineering, Virtue-First Enhancement, and Deification in Neo-Irenaean Theodicy," *Theology and Science* 16:3 (2018) 251–272. "The sheer complexity of the genome...makes the transhumanist hopes for genetic enhancement unlikely to be achieved in practice." Denis Alexander, *Genes, Determinism and God* (Cambridge UK: Cambridge University Press, 2017) 294.

[18] Alison Benders, "Genetic Moral Enhancement? Yes. Holiness? No." *Theology and Science* 16:3 (2018) 308–319, at 308.

[19] Ibid., 316.

cannot create virtue because virtues are stable, habituated dispositions, acquired over time."[20]

Roman Catholic bioethicist Lisa Fullam, in contrast, grants much less than Benders or Molhoek. Genetic technology could not even in principle enhance virtuous living.

> Can we genetically engineer human virtue? No. Genetic modification prior to birth cannot instantiate a virtuous life. Genetic modification, in principle, could insantiate specific traits, to be sure. But, traits are not virtues. Virtue is an achievement gained only after following a rigorous path of self-discipline. This is the case regardless of the genome with which one starts out in life.[21]

In sum, virtuous living requires willful action that develops a consistent pattern over time. Technical tinkering with genes cannot supplant or govern willful choice combined with rigorous self-discipline.

If one is serious about enhancing virtue through technology, why do it genetically? Might it be more effective to alter the human brain? Might we implant a computer chip that would dictate moral behavior? Should a transhumanist rely on neuromodification rather than genetic engineering?

It would not matter, according to Methodist theologian Alan Weissenbacher. Employing technological determinism to guarantee specific behavior–even virtuous behavior--challenges the free will of the person in question. Weissenbacher rises up to protect "the right to cognitive liberty, the right to mental integrity, and the right to psychological continuity."[22]

In short, moral action deriving from virtuous living is not technologically programmable, because it requires willful participation and sustained self-discipline over time. That level of personal participation cannot be governed by either genetic or neurotechnology.

62.6   Can We Genetically Engineer Deification?

Whereas nearly all Protestants and most Roman Catholics employ the term *sanctification* to refer to the growth in Christ toward which the Holy Spirit draws the individual believer, Eastern Christians rely on terms such as *theosis, divinization,* or *deification.* "The real anthropological meaning of deification is Christification,"

---

[20] Braden Molhoek, "Raising the Virtuous Bar: The Underlying Issues of Genetic Moral Enhancement," *Theology and Science* 16:3 (2018) 279–287, at 279

[21] Lisa Fullam, "Genetically Engineered Traits versus Virtuous Living," *Theology and Science* 16:3 (2018) 319–329, at 319. Geneticist and theologian Celia Deane-Drummond doubts the very capacity to engineer future evolution. *Hyperhumanism* is the belief that humanity is in control of its own history and its own evolutionary future. "It would be a mark of intense hubris marked with political overtones of eugenics to expect that humans can control their own evolution." Celia Deane-Drummond, *Christ and Evolution* (Minneapolis: Fortress Press, 2009) 285.

[22] Alan Weissenbacher, "Defending Cognitive Liberty in an Age of Moral Engineering," *Theology and Science* 16:3 (2018) 288–300, at 297.

writes Panayiotis Nellas.[23] The Christian transhumanist must ask: might we speed up the process of deification through genetic engineering or technological enhancement?

Orthodox theologians are slow to take spiritual technology on board. The first reason is that, in true deification, the human free will must voluntarily surrender itself to the divine will. No gene editing we know of at this point could replace the role of the human will in the holy life. "Deification, which is a gift from God freely chosen by the individual and God working together in synergy is open to every human person," says Ukranian Orthodox biologist Gayle Woloschak. "If it were possible to genetically engineer a virtuous person without the rigorous work and commitment that is needed on behalf of the person, it is not clear where free will would come into play."[24]

In addition to voluntary human participation, it takes divine grace to make growth in holiness possible. At least according to Ian Curran. "While the Christian tradition does share with techno-humanism a vision of deification as integral to the human story, its understanding of the source, means, and ultimate end of this radical transformation of human beings is substantially different. For Christians, deification is the work of the Christian deity....Deification is only possible because Christ deifies human nature in the incarnation and the Spirit sanctifies human persons in the common life of the church and in our engagements with the wider world."[25]

Whereas the end of the line on the transhumanist train is disembodied superintelligence accomplished through technological genius, the end of the line on the Christian train is a joyful attunement of the divine and human wills. Only God's grace empowering a willing heart can make this happen.

62.7   Post Human? Or, Truly Human?

On the one hand, Christians look forward to transformation. On the other hand, the transformed human is slated to remain human. It is God's goal in both creation and redemption to perfect the creature, not replace the creature. Will this qualify how the Christian embraces the H+ vision?

The destination of the transhumanist train is posthumanity, a new species which will replace *Homo sapiens*. That new species will be postbiological, disembodied, superintelligent, and cosmic. The greatest achievement of the human race will be its own suicide at the birth of something superior.

---

[23] Panayiotis Nellas, *Deification in Christ: the Nature of the Human Person*, tr., Norman Russell (Crestwood NY: St. Vladimir Seminary Press, 1997) 39.

[24] Gayle E. Woloschak, "Can We Genetically Engineer Virtue and Deification?" *Theology and Science* 16:3 (2018) 300–307, at 306.

[25] Ian Curran, "Becoming godlike? The Incarnation and the Challenge of Transhumanism," *Christian Century* 134:24 (November 22, 2017) 22–25, at 25.

Critics of H+ find this distasteful. "Modern transhumanism is a statement of disappointment," complains Brian Alexander. "Transhumans regard our bodies as sadly inadequate, limited by our physiognomy, which restricts our brain power, our strength and, worst of all, or life span. Transcendence will not be found in the murky afterlife of the usual religions, but in technological and biological improvement."[26]

Jewish theologian Hava Tirosch-Samuelson is vehement. "I reject transhumanism because it calls for the *planned obsolescence of the human species* on the grounds that biological humanity, the product of a long evolutionary process, is not only an imperfect 'work in progress' but a form of life that is inherently flawed and that has no right to exist."[27]

When Christians turn to their own resources such as the Bible, we find that the human being is special in the eyes of God. Psalm 8:4 asks rhetorically, "what are human beings that you are mindful of them, mortals that you care for them?" Nothing here suggests that we human beings are a transition species that God expects to replace with one more intelligent.

God became incarnate within the physical world as we know it now. Yes, God promises a transformation, a new creation where all creatures in the biosphere will live in harmony (Isaiah 11). But, this promise is based on the divine desire to transform and redeem our world, our history, our humanity.

The true human is the future human, the one dressed fully in the clothes of the resurrected Christ. The Easter Christ has gone before us, so to speak, to model the transformed human which is the true human. We are being drawn forward by the divine Spirit to become what we are finally created to be. The *adam* of Genesis, to whom we are presently heir, was subject to tragedy in a way that the new *adam*, who participates in the resurrection of Jesus Christ, will not be. Saint Paul envisions the future human, the true Adam and Eve.

> Thus it is written, "The first man Adam became a living being"; the last Adam became a life-giving spirit. But it is not the spiritual that is first but the physical, and then the spiritual. The first man was from the earth, a man of dust; the second man is from heaven. As was the man of dust, so are those who are of the dust; and as is the man of heaven, so are those who are of heaven. Just as we have borne the image of the man of dust, we will also bear the image of the man of heaven. (1 Cor. 15:45–49)[28]

This leads Karl Barth to aver, "Our anthropology can and must be based on Christology."[29] Let me add: eschatological Christology.

---

[26] Brian Alexander, *Rapture: How Biotech Became the New Religion* (New York: Basic Books, 2003) 51.

[27] Hava Tirosch-Samuelson, "In Pursuit of Perfection: The Misguided Transhumanist vision," *Theology and Science* 16:2 (2018) 200–223, at 207.

[28] Redemption is the final act of creation. Instead of trading in the Old Adam for the New Christ, the New Christ becomes the goal toward all humanity since Adam has been directed. "The image of the coming Reign expresses not at all the regret for a lost golden age, but the expectation of a perfection the like of which will not have been seen before." Paul Ricoeur, *The Symbolism of Evil*, tr. Emerson Buchanan (Boston: Beacon Press, 1969) 265.

[29] Karl Barth, *Church Dogmatics*, 4 Volumes (Edinburgh: T. & T. Clark, 1936–1962) III/2: §47: 512.

Even though the Christian eschatological vision includes passing through death to new life, it does not suggest that the present human species will go extinct in order to make room for a subsequent posthuman species.

62.8 Conclusion

The transhumanist train has pulled out of the station and is headed toward techno-utopia. Just how far down this track will the Christian rider want to go?

I recommend Christians travel a few stops seated side-by-side with our secular transhumanist friends. If the locomotive of GNR [Genetics-Nanotechnology-Robotics] enhances human flourishing and even extends the human life span, such technological achievements should be welcomed and even celebrated.

However, when H+ engineers brag about their messianic powers of transformation through technology, Christian ride-alongs should become skeptical. No amount of technological innovation can inspire compassionate love or sanctify the human spirit. God's Holy Spirit can transform us, to be sure. And the Holy Spirit is ever available to hearts that are open, regardless of how technologically advanced those hearts might be.

The transhumanist movement offers Christian individuals and churches an opportunity to become stewards of innovation, pressing technology into the service of making this world a better place. But when the transhumanist conductor announces that the stop after the next is posthuman perfection, it's time for the Christian to disembark.

Ted Peters is a Lutheran pastor and professor of systematic theology and ethics at the Graduate Theological Union in Berkeley, California. He is the author of an espionage thriller with a transhumanist theme, *Cyrus Twelve* (Apocryphile Press, 2018). He co-edits the journal, *Theology and Science,* at the Center for Theology and the Natural Sciences. He is author of *God–The World's Future* (Fortress, 3rd ed., 2015) and *Sin Boldly! Justifying Faith for Fragile and Broken Souls* (Fortress, 2015). He is co-editing a new book, *Religious Transhumanism and Its Critics* (Lexington Books, 2019). See his website: TedsTimelyTake.com.

Chapter 63
Thinking Like a Christian About Transhumanism

Joshua Marshall Strahan

63.1 Framing the Conversation

The Dictionary of Christianity and Science [3], published in 2017, contains no dedicated entry to the topic of transhumanism.[1] This is striking, given that the subtitle of this particular dictionary is *The Definitive Reference for the Intersection of Christian Faith and Contemporary Science*. Clearly, there is much work to be done on this topic by Christian thinkers. Since the conversation between Christians and transhumanists seems to be in the early stages, I think it would be helpful to suggest a framework for what it should look like to think like a Christian about transhumanism.

I propose that Christians begin by engaging this topic from the specifically Christian framework of the *rule of faith*. What is the rule of faith? From early in Christian history, the church has handed down a rule of faith to guide and center Christian teaching and practice. Examples of the rule can be seen in some of the church's earliest writers, such as Justin Martyr, Irenaeus, Tertullian, and Origen.[2] It was so prevalent that Irenaeus can claim that even barbarians know the rule of faith (*Against Heresies*, 3.4.1-2).[3] In other words, all Christians are expected to be trained in this rule, which has two noteworthy functions. First, it trains Christians to distinguish faithful teaching and practice from heretical teaching and practice. Second, it locates Christians within a particular worldview. In the words of N.T. Wright [20], the rule of faith can help Christians recognize that "we are renewed as *this* people, the people who live within *this* great story, the people who are identified precisely

---

[1] Transhumanism does, however, receive mention in the Dictionary's articles on "Genetic Enhancement," "Singularity," and "Technology."

[2] For examples of the rule of faith in early Christian writings, see Ferguson [4], pp. 1–15.

[3] This reference from Irenaeus is found in Bird [1], p. 33.

J. M. Strahan (✉)
Lipscomb University, Nashville, TN, USA
e-mail: josh.strahan@lipscomb.edu

© Springer Nature Switzerland AG 2019
N. Lee (ed.), *The Transhumanism Handbook*,
https://doi.org/10.1007/978-3-030-16920-6_63

as people-of-this-story, rather than as the people of one of the many other stories that clamor for attention all around" (p. 64).[4]

What is contained in the rule of faith? According to Michael Kruger [8], the rule of faith "functions as a description of the entire biblical storyline, from the Old Testament through the New... [It] is a retelling of the cosmic story of creation-fall-redemption-consummation." (pp. 144–45). We might think of the rule of faith, then, as a concise retelling of the larger biblical story, in which certain foundational faith claims are articulated. For simplicity's sake, we will look at the rule of faith from two complimentary angles: (1) we will briefly recap the biblical plotline, and (2) we will look at the Apostles' Creed. The Apostles' Creed is an early statement of Christian faith that has been confessed across centuries, continents, and denominations. Even though Christians have disagreed with one another about various teachings and ideas, almost all Christians of all times and places would confidently confess these words of the Apostles' Creed:

> I believe in God, the Father almighty, creator of heaven and earth.

> I believe in Jesus Christ, his only Son, our Lord. He was conceived by the power of the Holy Spirit and born of the Virgin Mary. He suffered under Pontius Pilate, was crucified, died, and was buried. He descended to the dead. On the third day he rose again. He ascended into heaven, and is seated at the right hand of the Father. He will come again to judge the living and the dead.

> I believe in the Holy Spirit, the holy catholic Church, the communion of saints, the forgiveness of sins, the resurrection of the body, and the life everlasting. Amen.

Before getting into the details of the rule of faith, we might stop to consider how the rule might be relevant for our present topic? I would suggest that the rule of faith provides a more substantive perspective for Christian ethics than does an ad hoc approach that relies on selective Bible verses and theological maxims. A fundamental problem with engaging transhumanism in an ad hoc manner is that isolated Bible verses and aphorisms can too easily be made to support just about any position. This is not just a problem with Scripture, this is inherent in all communication. We live in a culture where soundbites are consistently taken out of context in ways that wreak havoc on fruitful dialogue. All good communication requires an interpretive context or framework. For Christians, the rule of faith can provide such an interpretive framework. In this way, the Bible's voice becomes more clearly audible in the conversation.

There are two metaphors we might consider as we think about how the rule of faith informs Christian thinking about transhumanism. First, we could think of the rule like eyeglasses. If a person has glasses, she can both look *at* the lenses and *through* them.[5] She can look *at* the lenses, noting whether they are scratched or smudged; and she can wear the glasses and see the world *through* the lenses. In what

---

[4] Technically, Wright is speaking of the church's creeds here, but the same idea fits with the closely related rule of faith.

[5] This metaphor is adapted from an illustration given by McGrath [11], pp. 51–55.

follows, we will do both with the rule of faith. We will give some space to looking *at* the rule, seeing its content and claims. Then, we will "put on" the rule-of-faith lenses and look *through* them in order to discover how they shape the way we "see" various issues within transhumanism.

Second, we might think of the rule of faith like an anchor to which Christians are tethered. The rule keeps Christians firmly moored, simultaneously allowing space to travel and providing some boundaries—areas which are distinctly outside of the perimeter for Christians. Thus, in what follows, we will consider how certain aspects of transhumanism might fit within the bounds of Christian faith, whereas other aspects of transhumanism will not. With these metaphors in mind, we now turn to look at the rule of faith in greater detail.

63.2 The Rule of Faith

The Apostles' Creed and the biblical storyline have a kind of symbiotic relationship: the biblical storyline helps us understand the Creed, and the Creed helps us interpret the biblical storyline. For our study, we will start with the biblical plotline, sketching some major events and themes, then transition to the Creed for further insight. In the brief sketch that follows, I will attempt to recap the Bible's 700,000+ words in less than 1700 words![6]

63.2.1 Biblical Plotline

The Bible opens with God's sovereign act of creation. Creation is described as "good" seven times in this opening account. The pinnacle of creation is humankind, whom God pronounces as being "very good." Humans are distinguished as being made in the "image of God"—an indication of human dignity, wherein humans are to represent God by carrying out their God-given vocation to care for and benevolently rule the created order. Within this good creation, things seem to be in harmony: humans are in good relationship with God, with one another, and with the created order. (For an example of how Christians can reconcile the biblical creation account with evolution and the Big Bang, see Harris [5]). Soon, however, the humans disobey God and the corruptive influence of sin pervades the world, bringing disharmony between humans and God, humans and one another, and humans and the created order. The situation gets increasingly worse as humans engage in greater and greater violence.

God puts in motion a plan to bring restoration. His plan begins by calling Abraham to leave behind what he knows, and follow God in faith. God makes a

---

[6]My understanding of the biblical storyline has particularly been aided by McKnight [12], Middleton [13], and Wright [21].

covenant (i.e., a solemn pledge) to bless Abraham with land and descendants, which should result in Abraham's lineage blessing the world. Abraham's descendants eventually become the people of Israel, the Jews. The Israelites find themselves enslaved by the mighty Egyptians, but God rescues them in dramatic fashion and brings them into the wilderness to prepare them to enter into the Promised Land. While in the wilderness, God covenants to be Israel's God, and Israel covenants to be God's people. As God's covenant people, they are to live according to the terms of the covenant—also known as the Law or Torah. As I understand it, the Law was designed to foster wholeness in a world that was plagued by disharmony and brokenness. For example, the Law has regulations for how humans are to relate to God, how humans are to relate to one another, and how humans are to relate to the created order. Just as the first humans were to benevolently care for creation, and just as Abraham's descendants were to bless the world, so the Law provides guidance for how Israel can become the kind of society that represents God's love and justice to the world. Later, Jesus will summarize the Torah with two commandments: love God with all your heart and soul and mind and strength, and love your neighbor as yourself. Or, as one Israelite prophet, Micah, puts it: "Act justly, love mercy, and walk humbly with your God" (Micah 6:8). (As a side note, it is helpful to recognize that Christians see the Law as a temporary guide, in part, because it was helping an Ancient Near Eastern people transition into a more holy people. Consequently, some of the Torah's regulations do not represent God's ideal, but merely how God is helping a people take steps in the right direction.)[7]

God leads the Israelites out of the wilderness and into the Promised Land. Unfortunately, just as the first humans failed to live up to their vocation to bless the world, so also does Israel. After a period in which Israel is loosely united under the leadership of prophets and judges, the Israelites desire a king. God interprets this as a rejection of his own role as King of Israel, and warns the Israelites of the inevitable time when a king will cease to serve the people, and will instead make the people serve his own greed and lust for power. The Israelite people believe they know what is best and insist on a king. God grants their request, and is even willing to work out his good purposes through their poor choices. In fact, God promises one ruler, King David, that he will give him an everlasting dynasty. Unfortunately, but not surprisingly, many of Israel's kings are corrupt, which not only leads to the division of Israel into a northern and southern kingdom, but also results in the eventual exile of the Jewish people from Israel when they are conquered by the Assyrians and Babylonians. The first "half" of the Bible (i.e., the Old Testament) ends with a conquered Jewish people longing for God to bring healing, to be faithful to his promise to provide David an everlasting dynasty, and to bring restoration.

Several hundred years pass before the New Testament resumes the biblical story. The Jewish people have returned from exile, but now the great Roman Empire is in charge, and Rome has placed one of its own men on the throne as king of Israel. Into this tense situation, an angel appears to a virgin telling her that she will bear the Son of God who will take up the promised throne of David. Nine months later, the virgin

---

[7] See, for example, Jesus' comments on the Torah's divorce regulations in Matthew 19:8.

gives birth to Jesus (whose name means "God saves"). Jesus grows up in a faithful and devout Jewish home, then begins his public ministry. His ministry is character- ized by healing the sick and paralyzed, teaching as an authority on the Law and character of God, casting out demons, pronouncing forgiveness of sins, propheti- cally speaking against injustice, and practicing solidarity with the outcasts of soci- ety. Perhaps, we should see Jesus as restoring what sin deforms, as his ministry consists of reestablishing harmony in ways that are physical (e.g., healing the sick and paralyzed), social (e.g., practicing solidarity with the marginalized and pursing justice), and spiritual (e.g., forgiving sins and returning people to intimate relation- ship with God).

Jesus' ministry is regarded as a threat to those in power. Jesus is a threat because he is seen as a possible rival ruler, one who might try to seize the throne. Additionally, Jesus is a threat because he is undermining the religious establishment's claims to authority. Thus, both the Jewish religious leaders and the Roman authorities (repre- sented by Pilate and Herod) come to position themselves against Jesus. By means of a sham trial, Jesus is crucified—a brutal and humiliating form of execution typically reserved for slaves. Jesus dies and his body is placed in a tomb. A few days later, God raises him from the dead, thereby vindicating Jesus' claims to be the true King and Son of God. The resurrected Jesus appears to his followers, who up to that point were devastated by his death.

After spending some time helping them better understand both who he is and what their mission will be, the resurrected Jesus departs into the heavenly realm where he takes up his rightful role as cosmic ruler and leader of his people—the church. Jesus sends the Spirit of God to empower his people to carry out their mis- sion. Like Israel before them, God calls the church to continue the work that human- ity was meant to do from the beginning—namely, to bless the world by rightly representing God. The church's mission, however, will not be identical to Israel's for a few reasons. First, Jesus has brought a fundamental change to the world. Somehow, in his death and resurrection, he has overcome the power of sin and death. Second, Jesus sends the Spirit to share this power with his followers so that they might be more capable of living faithfully. Third, Jesus' followers are no longer bound by Torah. Instead they are part of a new covenant, whereby their lives and communities are to be imitations of the character of Jesus. In other words, Jesus is recognized as the perfect example of human life, so that he becomes a model for his followers. As his followers seek to spread the good news, their lives are to be marked by the kind of humility, compassion, and goodness that characterized Jesus' life. Fourth, Jesus' death and resurrection are to be proclaimed as the means by which people might be fully restored to relationship with God, making forgiveness and freedom from sin possible in a way that it never was before.

The church is to carry out their mission by proclaiming with their words and their lives how Jesus brings salvation, this wondrous gift of holistic restoration that has physical, social, and spiritual implications. Thus, the community of believers is to pursue *social* restoration by including the marginalized and seeking to treat one another justly. The community is to pursue *spiritual* restoration by proclaiming how Jesus has made possible forgiveness and reconciliation with God. The community is to pursue *physical* restoration by caring for the sick and suffering, while declaring that one day God will resurrect believers' bodies and renew them so that they will no longer be subject to death and decay. To be clear, the church is not to pursue these works of restoration by manipulative, coercive, or unjust means. Instead, they are to pursue restoration through compassionate, just, and peaceful means.

Finally, the church awaits a time when God will complete the work that Jesus set in motion, a time when God will bring restoration in its fullness. In this sense, the church sees the present time as a kind of limbo. On the one hand, God's restorative work has entered the world in a new and decisive way through Jesus' life, death, and resurrection. On the other hand, the effects of this decisive work of God are not yet fully realized. During this limbo, the church will continue to be a community of imperfect people—sometimes characterized by the goodness of Jesus and other times by the brokenness of sin. Nonetheless, the church anticipates a time when all things will be set right, when there will be no more sickness and suffering and death, when there will be no more societal outcasts, when injustice will be a thing of the past, and when God will dwell in an even more intimate way with his restored and forgiven people.

63.2.2   The Apostles' Creed

At this point, it is helpful to look at the Apostles' Creed, and consider how it helps fill out our understanding of the rule of faith.[8]

> I believe in God, the Father almighty, creator of heaven and earth.

Alex Rosenberg [16], in his book, *The Atheist's Guide to Reality*, explains how an atheistic worldview will inevitably lead one to conclude that there is no ultimate purpose, no objective morality, no true free will, no real personhood, and no hope for life after death. In contrast, the Creed's claim, "I believe in God," opens up the possibility for ultimate purpose, objective morality, free will, personhood, and the afterlife. As we continue on, we will see in more detail how the Creed might guide one's thinking about morality, free will, personhood, and the afterlife.

The Creed first identifies God as "Father." Within the biblical perspective, God has no body and thus no gender. Hence, we should understand "Father" as a title that speaks of God's loving and personal relationship with his people. This title also testifies to the unique relationship within the Christian Trinity, where there is Father and Son and Holy Spirit—an idea we will return to later.

God is next confessed as "almighty." There are no other gods. He has no equal. The Israelites were distinctly monotheistic, despite the widespread polytheism of the ancient world. That monotheistic heritage has been maintained in the Christian faith. This almighty God is confessed as "creator of heaven and earth." That is, God

---

[8] My understanding of the Apostles' Creed has particularly been aided by McGrath [10], Johnson [7], and Bird [1].

has created everything. In Christian tradition, God is both transcendent and immanent. He transcends creation, not being contained within the cosmos or reliant upon it for his existence. Yet, he is immanent: he is present within the created order, he sustains it, and maintains relationship with what and whom he has created.

In contrast to Rosenberg's atheism, Christianity's combination of monotheism and transcendence can arguably provide a foundation for morality, purpose, and free will. The cosmos is not a purposeless accident, nor is it trapped in an endless repeating cycle as in some worldviews. Instead, there is a God who exists beyond the cosmos who has created the world with purpose, and can presumably shape the world to achieve those purposes. Along with purpose comes a basis for morality, whereby morality is based on the character and purposes of God, who is ultimate reality. Furthermore, a God with a purpose might create beings capable of free choice, whereas, within the closed system of naturalism, humans may be trapped by the predetermined course of physical laws.

> I believe in Jesus Christ, his only Son, our Lord. He was conceived by the power of the Holy Spirit and born of the Virgin Mary. He suffered under Pontius Pilate, was crucified, died, and was buried. He descended to the dead. On the third day he rose again. He ascended into heaven, and is seated at the right hand of the Father. He will come again to judge the living and the dead.

To believe in "Jesus" is to confess belief, firstly, in the human Jesus of Nazareth. To call him "Christ" is to confess his role as the Christ (the Messiah, the hoped-for, restoring King of Israel). To declare him "Lord" is to simultaneously confess allegiance (as his subjects) and to declare him divine. According to Christian tradition, Jesus is truly divine ("conceived by the power of the Holy Spirit") and truly human ("born of the virgin Mary"). As truly divine, Jesus is the definitive revelation of God to humans. As truly human, Jesus is the definitive revelation of what human life should be about. By becoming truly human, Jesus empathizes with his creatures in a profound way; he also demonstrates the love of God by giving up so much to come dwell with his creation; and he takes on our frail humanity in such a way that he restores and renews it through his life, death, and resurrection.

His crucifixion displays the shocking extent of God's love, while also revealing the destructive power of sin at work in both individuals and social structures. Furthermore, if Jesus' death makes it possible for people to be freed from the guilt and power of sin, this shows that sin is not something that humans can fix on their own—they can neither absolve the guilt by their own efforts, nor can they escape the influence of sin through sheer willpower and resolve.

Jesus' resurrection signals his vindication by God. That is, the resurrection shows that Jesus is the true revelation of God; he is the authoritative revelation of how humans should order their lives; he is the foretold Son of God who will bring to completion God's plan to restore and renew the created order. As Jesus ascends to heaven, he sits at "the right hand of the Father"—a reference to how he reigns victorious and will send his Spirit to empower his church to carry out God's mission of bringing restoration and reconciliation. Lastly, to confess that Jesus "will come again to judge the living and the dead" is to believe that one day, Jesus will set things right: justice for the oppressed, punishment for the guilty, healing for the broken, reconciliation for the estranged, etc.

> I believe in the Holy Spirit, the holy catholic Church, the communion of saints, the forgiveness of sins, the resurrection of the body, and the life everlasting. Amen.

The Holy Spirit is understood as the third person of the Trinity. In Christian tradition, God is one being but three persons: Father, Son, and Holy Spirit. This is a mystery that cannot fully be grasped. Although strange, there is a kind of logic that goes with it. For example, one of the reasons that Christians believe that love is central to reality is because they believe that at the heart of all reality is a triune being where three persons exist in loving unity. Why did God create? For Christians, God did not create out of need, but out of an overflow of love. The Holy Spirit is sometimes identified with the love of God—the love between the Father and the Son, and the love between God and his people.

To confess the church as "holy" is not to claim that it is perfect, but that through Jesus' work, the church is made whole. Also, to be holy is to be set apart for the mission of God. (God will tell his people, "Be holy, for I am holy"). To claim that the church is "holy," then, is to take seriously the calling to be faithful representatives for God. To say that the church is "catholic" is not to say Roman Catholic, but *universal*. In other words, the church is not limited to one nationality or gender or social class. This may sound unimpressive, until one takes into account what a radical notion this was in the Greco-Roman world in which Christianity was birthed [6].[9] For Christians, the love of God unites people across barriers that so often divide people. This idea extends to the notion of "the communion of saints." Here, "communion" might better be understood as partnership or solidarity—the kind of familial bonds and commitment that befit the claim to have the same loving Father. (And "saints" is a translation for "holy ones"—so that this is not a reference to super-spiritual folks, but to all Christians whose status as "holy" is both a gift and a vocation.)

We have already mentioned the notion that forgiveness is made possible by Jesus. Christians recognize that the true means of forgiveness before God is only available by the free gift of Jesus' sacrifice. This teaches something about the weightiness of sin and the limitations of human power and goodness. The gracious bestowal of such costly forgiveness should also transform Christians to be people who shows mercy because they have received such priceless mercy.

Christianity is not alone in believing in an afterlife, although it does have a few distinctive aspects. First, Christians base their hope in Jesus' resurrection. For Christians, the afterlife is not grounded in the inherent immortality of the soul nor in some concept of a cyclical wheel of life; rather, Christians place their hope in

---

[9] Larry Hurtado [6], a noted historian of early Christian history, summarizes his findings as follows: "[Christian identity] was an exclusive religious identity, defined entirely by their standing in relation to the one God, and was not dependent on, or even connected to, their ethnicity. In fact, I contend that this distinctive early Christian group identity is perhaps the earliest attempt to articulate what moderns would recognize as a corporate *religious* identity that is distinguishable from, and not a corollary of, one's family, civic, or ethnic connection" (p. 104).

Jesus, whom God raised from death with a physical body that was no longer subject
to corruption and decay. Second, Christians expect to have a future material existence, not a merely spiritual one. Such an idea was almost unheard of in the ancient
world prior to Christianity [19]. Even though this belief in a physical resurrection
led many in the ancient Roman world to reject Christianity, Christians nonetheless
held on to this belief as a central doctrine. Within the logic of the biblical plotline, a
physical resurrection makes sense: God created the material world good, it has been
corrupted, and through Jesus God is restoring and renewing his beloved material
world. Fittingly, then, the belief in "life everlasting" is not merely about *unending*
life but also *abundant* life—the blessed life that God had planned for his creatures
from the beginning, a life of love and harmony between God, humanity, and the rest
of creation.

63.3   Looking Through Rule-of-Faith Lenses

Having looked *at* the rule of faith, we will now look *through* it, paying particular
attention to how the rule shapes the Christian view of reality: who God is, who
humans are, what life is about, what is wrong with the world, and what will make it
right. The perspective offered by the rule of faith is that the center of all reality is a
transcendent, personal being who exists in loving community. The world was created good, a place that was meant to foster harmony between Creator and creation
as well as between creation and creation. At the pinnacle of creation is humankind
who have a particularly dignified status (images of God) along with a dignified role
(to care for and benevolently rule creation). Apparently, humans were given the
freedom to embrace God's will or reject it. By rejecting God's will, humanity
empowers sin to distort and corrupt the good creation. The corruptive influence of
sin is pervasive, causing social, physical, and spiritual brokenness. Humans prove
themselves to be incapable of fixing things on their own, because they are themselves broken—distorted images of God who are simultaneously capable of great
good and great evil.

  If there is an ultimate solution, it must be found outside of these sin-tarnished
humans. That is precisely what happens, as God becomes human and does for
humanity what it could not do for itself. Jesus is the true image of God, the one who
fully embraces the human calling to love and care for the world, while remaining
sinless. He demonstrates in his life, teaching, and death that the way to bring restoration to a broken world is through a life characterized by mercy, humility, selfless-
ness, justice, and faithfulness. Through his death and resurrection, Jesus makes it
possible for humans to find forgiveness and freedom from sin. By sending the Holy
Spirit, Jesus empowers his followers to start becoming the kind of people who are
to be increasingly characterized in a way that matches humanity's original description—images of God: those who care for the world in ways that are merciful, humble, selfless, just, and faithful.

Finally, the rule of faith trains one's eyes to look for Jesus' return as the ultimate source of hope for restoration and renewal in its fullness. This does not mean that Christians are not seeking to be agents of renewal and reconciliation in the meantime, but it does mean that Christians do not believe that humanity alone is capable of fully achieving utopia, because the social and physical and spiritual problems are too powerful and pervasive to be entirely fixed by humans alone.

63.4   Tethered to the Rule of Faith

How might the rule tether Christian thought? I will limit myself to five observations, though more could be said. First, Christians hold onto the notion that ultimate reality is not accidental, purposeless, or evil. Ultimate reality is a transcendent being of loving communion. Second, the material world is not regarded as evil or inherently problematic or even as neutral. Instead, there is an inherent dignity in the physical world, including the human body. Creation was declared "good" and humans were described as "very good"; God added even greater dignity to human bodily existence by becoming human; and God has a plan to renew creation. Third, human life is not without purpose, nor is it ultimately about the pursuit of pleasure or the avoidance of suffering. The human vocation is to properly bear the image of God through lives of love and justice, which was perfectly demonstrated in Jesus' ministry of compassion, justice, humility, and service. Fourth, the problems in our world go much deeper than ignorance or bad moral choices. Greater knowledge and better laws are good, but they cannot fix the ultimate problem because they do not go to the root: sin. The presence of sin has a corruptive influence throughout the world: socially, physically, and spiritually. At some level, it affects both individuals (influencing how we think, what we value/love, and how we act) and societal structures (family systems, governments, institutions). Fifth, Christians must not place their sole hope in humans to fix a problem that only God can ultimately fix. However, neither are Christians to be complacent; instead, they are to work at bringing whatever restoration and renewal they can bring in ethically appropriate ways. All the while, they are to confidently anticipate a time when the world will be made fully right by the intervention of God, when Jesus returns to finish the work he began.

63.5   The Rule of Faith and Transhumanism

63.5.1   Defining Transhumanism

To think like a Christian about transhumanism is complicated by the fact that transhumanism is not a clearly defined thing. As Russel Blackford [2] describes it: "Transhumanism is not a religion. Nor is it a secular ideology: it has no body of codified beliefs and no agreed agenda for change. It is, instead, a broad intellectual movement – not so much a philosophy as a class or cluster of philosophical claims and cultural practices" (p. 421). In what follows, then, I will be looking at some of the more common (though not universally held) ideas associated with transhumanism. According to Max More's [14] informative essay, "The Philosophy of Transhumanism," transhumanists typically "describe themselves as materialists, physicalists, or functionalists… [wherein] our thinking, feeling selves are essentially physical processes" (p. 7). Consequently, transhumanists tend to look to human ingenuity as the means to progress, not to some spiritual or transcendent power. Thus, with its "roots in Enlightenment humanism," transhumanists put their faith—so to speak—in "reason, technology, scientific method, and human creativity" (p. 4). On the one hand, this requires a certain degree of optimism in the human ability to bring about the kind of change that transhumanists expect (though there is no consensus on the what and when of expected changes). On the other hand, transhumanism does not entail a utopic optimism about the effects of such changes— great good, great harm, or both may result. One of the driving values or convictions of transhumanists is "that it is both possible and desirable to overcome biological limitations on human cognition, emotion, and physical and sensory capabilities" (p. 13).

Many of these same ideas are captured in the 2012 Transhumanist Declaration. What this declaration adds that is not as clear from More's essay, is how a kind of libertarian-humanism seems to be the default ethic among transhumanists. That is, people should be free to pursue whatever changes they want, provided that it does not harm or encroach upon others. For example, "Policy making ought to be guided by responsible and inclusive moral vision…, respecting autonomy and individual rights, and showing solidarity with and concern for the interests and dignity of all people around the globe…." Of course, even the Transhumanist Declaration cannot function as the definitive word for transhumanists, whether now or in the future. As Blackford [2] will claim, "Transhumanists have no Heaven and Hell, no other world, or canons of conduct, or comprehensive creed" (p. 428).

63.5.2   Points of Overlap

So, how might Christians think about transhumanism in light of the rule of faith? Perhaps it would be helpful to start with areas of overlap, of which I will highlight three. First, both Christians and transhumanists believe it is important to alleviate the physical and emotional suffering that humans experience. The popular caricature of Christians neglecting the physical and only caring about the spiritual is unfortunate. To be certain, such a caricature is true of some misguided Christians, but this is mitigated by the vast number of Christians and Christian organizations that are dedicated to restoring physical and emotional health to people. One only need consider the great number of hospitals, homeless shelters, counseling centers, and charities to see this. Furthermore, the rule of faith teaches Christians about the goodness of creation, the restoring and healing work of Jesus, and the church's vocation to continue Jesus' ministry of holistic restoration.

Second, Christians and transhumanists share some—though, obviously not all—ethical principles. For example, one of the central morals for Christians is "to love one's neighbor as oneself." This moral principle has some obvious overlap with the moral principle found in The Transhumanist Declaration [17] "[to show] solidarity with and concern for the interests and dignity of all people around the globe." Furthermore, within the biblical plotline, one sees how God grants people a certain degree of free will and autonomy—the option to *both* make some choices for one-self *and* to experience the benefits or consequences that arise from those decisions. One can find here at least a bit of overlap with the transhumanists' libertarian leanings.

Third, in my reading, transhumanists seem to regard humanity as neither wholly virtuous nor wholly vicious. Instead, transhumanists are aware that their sought-after enhancements could lead to either wonderful or devastating outcomes (or both). Such an anthropology has points of similarity with a Christian anthropology, in which humanity is marked by both good and evil. Thus, humans are declared "very good" as they are created in the image of God, yet sin has twisted their good-ness. This results in people being simultaneously responsible for so much good and so much evil in our world.

63.5.3 Points of Conflict

In addition to points of overlap, there are a number of areas in which Christians and many transhumanists do not see eye to eye. Here, I will mention six such areas where the Christian worldview is at odds with the transhumanist worldview described above. First, unlike most transhumanists, Christians are not materialists or physicalists. This is a pivotal point of disagreement that inevitably plays a part in the other five points of disagreement mentioned below. That is, *because Christians are not materialists but believe in a transcendent God who became incarnate as Jesus Christ, this leads to distinctive notions about ethics, hope, progress, the root problem humans face, and the primary solution to that problem.*

Second, unlike materialists who consider all problems to be physical problems, Christians believe that the fundamental problem that humans face is the destructive presence of sin and evil. Consequently, Christians think that transhumanists are try-ing to find solutions to significant problems while being unaware of a crucial factor that contributes to those problems. It may indeed be the case that transhumanists can alter humans mentally, physically, and emotionally. Even so, if these altered humans are still being influenced by sin and evil, then they have not advanced in one crucial area of life, which pervades every other area in some way. Hence, Christians think that transhumanists are not holistic enough. Human brokenness is not only experi-enced at the physical level, but at the spiritual level as well.

Third, because Christians see the human problem as more than a physical problem, they also believe the solution must be more than a physical solution. Technological and medical advances can only take one so far. Physical solutions can only provide an inherently limited fix to a problem that is not merely physical. A bigger solution is necessary. Moreover, according to the rule of faith, Christians believe that humans have proved themselves incapable of fully fixing a problem of which they are a part. The solution must be found outside of humanity; it must come from a source that is not already part of the problem. For Christians, that solution is found in the fully good God—uncorrupted by sin and evil—taking on human form, living a fully human life, experiencing death, and being raised to life. For Christians, Jesus' incarnation, death, and resurrection is the solution: the physical and spiritual brokenness of humanity was mysteriously made whole by Jesus who defeats sin and death, provides a means for forgiveness and reconciliation, and transforms the corruptible human body into an incorruptible body that is a preview of how he will transform our bodies. As Christians partner with God to bring what wholeness they can to a broken world, they do so with an awareness that the full realization of this wholeness was not made possible by humans, nor can it now be fully realized by humans; instead, Christians believe that the full holistic renewal will only be realized when Christ returns and finishes the work he began.[10]

Fourth, if Christ represents what it looks like to be truly human, then the notion of "progress" is not primarily concerned with an enhanced mental, physical, and/or emotional state. Rather, it might be better to think of progress as more concerned with something like growth in virtue or character—that is, becoming a person whose life reflects Christ's love and compassion and humility and courage and wisdom and faithfulness and justice. To progress in physical strength but not in Christlikeness is to not truly progress at all. (In fact, if the physical enhancements lead to one being a greater bully, then that person has regressed, not progressed.) To progress in lengthening one's lifespan but not to progress in faith, hope, and love is to not truly progress at all. (A longer life that is increasingly shaped by vice would be seen as regressing not progressing.)

Both Christians and transhumanists are anticipating the next stage of human evolution, but their expectations are quite dissimilar. As C.S. Lewis [9] explains:

> The Christian view is that the Next Step [of evolution] has already appeared... It is not a change from brainy men to brainier men: it is a change that goes off in a totally different direction—a change from being creatures of God to being sons of God. The first instance appeared in Palestine about two thousand years ago. In a sense, the change is not 'evolution' at all, because it is not something that arises out of the natural processes of events but something coming into nature from outside (p. 220).

For Christians, this next stage begins now as we slowly allow God to shape us into Christlikeness; and the completion of this stage will happen at some point in the future when God will complete this work in us. Not only will God fully renew our characters to be free from vice, but he will also fully renew our bodies so that they will no longer experience death and decay.

---

[10] See Ted Peters [15] on the distinction between *adventus* and *futurum* (pp. 142–145).

This brings us to our fifth point. For Christians, not only is Christlikeness the goal, it is also the guiding ethic. Earlier I mentioned that Christianity and transhumanism have some overlapping moral principles; nevertheless, there are some crucial differences. For example, whereas Christians and transhumanists might both subscribe to the Golden Rule, Christianity sometimes calls people to go beyond the Golden Rule and love others sacrificially, mercifully, and selflessly. Sacrificial love is not only to be characteristic of Christ, but also characteristic of his followers. Such an ethic may not fit into a materialist worldview in which all problems are inherently physical problems; but in a Christian worldview in which evil is a reality, then sacrificial love can be seen as a necessary and practical ethic—i.e., the only way to fully overcome evil is with a love that is deep enough to suffer.

Furthermore, Christians might be wary of how transhumanism ultimately seems to lack any lasting foundation for ethics. While humanism may be guiding the transhumanism movement today, there is no reason to assume it will stay that way. After all, to quote Blackford again: "Transhumanists have no... canons of conduct, or comprehensive creed." Similarly, the following statement from More [14] makes it seem as though an ethic of humanism is ultimately temporary: "Transhumanists regard human nature not as an end in itself, not as perfect, and not as having any claim on our allegiance. Rather it is just one point along an evolutionary pathway and we can learn to reshape our own nature in ways that we deem desirable and valuable" (p. 4). If human nature has no claim on our allegiance, and if human nature is such that we can shape it to what we deem desirable, why would it not be the case that we could also shape the transhumanist ethic in ways that we deem desirable—*even if that means shaping it in ways that are no longer characterized by "solidarity with and concern for the interests and dignity of all people around the globe?"*

Sixth, there seems to be some tension between the Christian view of the human body and the transhumanist view. For the Christian, the human body has a kind of sacredness to it. Though it may be imperfect, though it may be subject to death and decay, and though it may carry within it many unhealthy impulses, it nonetheless retains something of the specialness that God declared as being "very good." I may be wrong here, but it strikes me that part of the "very good"-ness of the human body is located in its being organic. Those transhumanists who are pursuing some sort of synthetic existence would seem to be heading towards becoming less human rather than trans-human.

As we bring this section to a close, it is worth mentioning how Christians might think about forms of enhancement that are less extreme than uploading our brains or personalities into some synthetic environment. That is, how should Christians think about enhancing our mental and physical capabilities? I am hesitant to make any sweeping claims here, because I frankly find this perplexing. On the one hand, there seems to be some areas where Christians can be supportive of certain biological and technological improvements, particularly those that help restore wholeness—for example, helping the blind to see, the deaf to hear, and the paralyzed to walk. And, even if we think beyond restorative work, I see no inherent problem with some types of enhancing that we already practice, such as strengthening the human immune system through vaccines and medicines. On the other hand, it strikes me that enhancing oneself to have super-strength or super-memory is problematic. The pursuit of super abilities seems to be indicative of (a) pursuing the wrong ends, and (b) failing to appreciate how something of the very-goodness of humans is connected with our limitations. However, I am open to being convinced otherwise provided that it does not contradict the rule of faith. After all, I am aware that my wariness is largely instinctual. I cannot tell whether this instinct is rooted in a gut-level kind of basic morality, or whether this instinct has been honed by living within the rule of faith, or whether this instinct is simply based on too many dystopian movies and novels. After all, I can imagine how future Christians might see the potential good in certain enhancements—for example, if a surgeon were to have advanced vision and fine motor skills that helped her treat patients more effectively. Needless to say, there are some issues that Christians who are guided by the rule of faith might still find very confusing.

63.6   Conclusion

We will conclude with an example of how some Christians are already thinking about transhumanism. In particular, we will consider The Christian Transhumanist Affirmation [18]. In this statement of affirmation, the Christian Transhumanists Association (CTA) affirms the following five claims:

1. We believe that God's mission involves the transformation and renewal of creation including humanity, and that we are called by Christ to participate in that mission: working against illness, hunger, oppression, injustice, and death.
2. We seek growth and progress along every dimension of our humanity: spiritual, physical, emotional, mental—and at all levels: individual, community, society, world.
3. We recognize science and technology as tangible expressions of our God-given impulse to explore and discover and as a natural outgrowth of being created in the image of God.
4. We are guided by Jesus' greatest commands to "Love the Lord your God with all your heart, soul, mind, and strength…and love your neighbor as yourself."
5. We believe that the intentional use of technology, coupled with following Christ, will empower us to become more human across the scope of what it means to be creatures in the image of God.

This affirmation has much overlap with the rule of faith, though I am uncertain whether that is intentional or coincidental. That is, I am not sure whether the CTA sees themselves as tethered to the rule of faith. The first four statements of the Affirmation can be made to fit well enough within the rule of faith: pursuing holistic renewal, taking seriously the human vocation to be image bearers, and living out the ethic of Jesus. It would be helpful, though, to offer a clearer definition on the nature of "growth and progress" that is referenced in the second statement. Furthermore, the fifth statement of the Affirmation strikes me as claiming too much. Specifically, it does not strike me as a particularly Christian notion that technology "will empower us to become more human." At most, a Christian could claim that technology "may" empower us to become more human. After all, within the rule of faith, "to become more human" should be defined as becoming more like the ultimate human—Christ. And, it is not clear if or how technology is a tool that can achieve such an end. Moreover, because the CTA's statement of affirmation leaves out some crucial aspects of the rule of faith (e.g., the transcendence of God, the nature of Christ, the Holy Spirit, the resurrection, and sin), it is possible that the CTA may find themselves insufficiently tethered to an orthodox Christian framework. This becomes problematic if it leads to misplaced hope, misguided ethics, and a failure to offer a critical and dissenting voice when needed.

References

1. Bird, Michael F. (2016). *What Christians Ought to Believe: An Introduction to Christian Doctrine Through the Apostles' Creed*. Grand Rapids, MI: Zondervan.
2. Blackford, Russell. (2013). "The Great Transition: Ideas and Anxieties." In Max More and Natasha Vita-More, eds., *The Transhumanist Reader: Classical and Contemporary Essays on the Science, Technology, and Philosophy of the Human Future*. Hoboken, NJ: Wiley-Blackwell, pp. 421–429.
3. *Dictionary of Christianity and Science: The Definitive Reference for the Intersection of Christian Faith and Science*. (2017). Paul Copan, Tremper Longman III, Christopher L. Reese, and Michael Straus, eds., Grand Rapids, MI: HarperCollins.
4. Ferguson, Everett. (2015). *The Rule of Faith: A Guide*. Eugene, OR: Cascade.
5. Harris, Mark. (2013). *The Nature of Creation: Examining the Bible and Science*. Bristol, CT: Acumen.
6. Hurtado, Larry W. (2016). *Destroyer of the Gods: Early Christian Distinctiveness in the Roman World*. Waco, TX: Baylor University Press.
7. Johnson, Luke T. (2003). *The Creed: What Christians Believe and Why it Matters*. New York: Doubleday.
8. Kruger, Michael J. (2018). *Christianity at the Crossroads: How the Second Century Shaped the Future of the Church*. Downers Grove, IL: InterVarsity Press.
9. Lewis, C.S. (2001). *Mere Christianity*. New York: HarperOne. (Original work published 1952).
10. McGrath, Alister. (1998). *I Believe: Exploring the Apostles' Creed*. Downers Grove, IL: InterVarsity Press.
11. McGrath, Alister. (2010). *The Passionate Intellect: Christian Faith and the Discipleship of the Mind*. Downers Grove, IL: IVP Books.
12. McKnight, Scot. (2011). *The King Jesus Gospel: The Original Good News Revisited*. Grand Rapids, MI: Zondervan.
13. Middleton, Richard J. (2014). *A New Heaven and a New Earth: Reclaiming Biblical Eschatology*. Grand Rapids, MI: Baker Academic.
14. More, Max (2013). "The Philosophy of Transhumanism." In Max More and Natasha Vita-More, eds., *The Transhumanist Reader: Classical and Contemporary Essays on the Science, Technology, and Philosophy of the Human Future*. Hoboken, NJ: Wiley-Blackwell, pp. 3–17.
15. Peter, Ted. (2015). "Theologians Testing Transhumanism." *Theology and Science* 13 (2), pp. 130–149.
16. Rosenberg, Alex. (2011). *The Atheist's Guide to Reality: Enjoying Life without Illusions*. New York: W.W. Norton & Company.
17. Various. (2012). "The Transhumanist Declaration." In Max More and Natasha Vita-More, eds., *The Transhumanist Reader: Classical and Contemporary Essays on the Science, Technology, and Philosophy of the Human Future*. Hoboken, NJ: Wiley-Blackwell, pp. 54–55.

18. Various. (n.d.). "The Christian Transhumanist Affirmation." Retrieved from https://www.christiantranshumanism.org/affirmation (accessed Sep 18, 2018).
19. Wright, N.T. (2003). *The Resurrection of the Son of God*. Minneapolis, MN: Fortress Press.
20. Wright, N.T. (2008). "Reading Paul, Thinking Scripture." In Markus Bockmuehl and Alan J. Torrance, eds., *Scripture's Doctrine and Theology's Bible: How the New Testament Shapes Christian Dogmatics*. Grand Rapids, MI: Baker Academic, pp. 59–71.
21. Wright, N.T. (2010). *After You Believe: Why Christian Character Matters*. New York: HarperCollins.

Chapter 64
A Transhumanist God

Blaire Ostler

This is not an academic paper about religious Transhumanism. I find myself less interested in discussing religious transhumanism and more interested in *doing* religious transhumanism. Consider this a sermon written by a religious Transhumanist diving into the why, how, and what of where religious transhumanism might lead—God.

You may or may not believe in God. It might even be a word you wish to eradicate from your vocabulary. Whether a believer or not, one shouldn't be ignorant of the cyclical pattern of the birth and death of Gods. Gods are constantly being destroyed and created by human imagination and action.

When a destroyer demolishes a God, chanting along the way, "God is dead! God is Dead!", a new God is eventually birthed from the rubble by a creator. Gods are created from the remnants of past hopes, dreams, aspirations—the corpses of dead Gods. We need creators and destroyers alike. The destroyer depends on the creator to create something worth destroying, and the creator needs the destroyer to provide the rubble necessary for creation. The creators and destroyers are not enemies, but worthy opponents, even friends. Both work in an alliance against apathy, nihilism, complacency, escapism, and all other manner of meaninglessness. Neither the destruction of God nor the creation of new Gods is the enemy of human flourishing but is the manifestation of human flourishing.

Though I destroy, I am also a creator, an artist. I find beauty in the world, even in destruction, suffering, and pain if necessary, so long as it leads to better vistas. Suffering can be transformative. However, suffering for the sake of suffering is just as pitiful as the prophet who seeks to eliminate all suffering or risk, and in so doing eliminates joy, reward, and meaning. There must necessarily be opposition in all things for meaning to exist, and genuine creation necessarily requires destruction.

B. Ostler (✉)
Mormon Transhumanist Association, Orem, UT, USA
e-mail: blaire.ostler@transfigurism.org

I find meaning in God, not just any or all Gods, but a transhumanist God birthed from material theism. This God exists inside space and time, unlike the metaphysician's God who is aimlessly suspended in an immaterial abyss of nothingness. I find meaning in what I can possibly know, understand, and become. I find meaning in a transhumanist God. You may have no interest in the word "God" or "transhumanist." If so, feel free to liken these parables unto yourself. Replace "God" with "super-humanity" or "posthumanity" and the message will still be delivered. Replace "transhumanism" with "theosis" or "material theism" if it makes the message more relevant. I'm a pluralist when it comes to esthetic preferences.

I have found past Gods, particularly my own, to be wanting. Birthed from the mouths of patriarchs was a tyrant and they called that tyrant GOD! That God, He, was made in the image of my oppressor. Like the destroyer, I killed that God—the God of the patriarchs. His hot blood spilled onto the frozen ground. Crimson steam rose off His corpse as I stood in a pool of His blood, and I called myself atheist. God really is dead, I killed Him myself!

It was self-defense, you see. He sought to annihilate my existence in the name of "so-called" love, so said His patriarchs, the false prophets. It was either Him or me, and I won. In death, with His blood pooled around my feet, I won. No God could match my intellect, logic, semantics, and philosophies. I absorbed His blood through the pores of my bare feet. Power was mine.

Yet, the victory was empty. Is there ever victory in death—even a necessary death? I found no meaning in His death. Even after feasting off His blood, my womb was hollow. This was not the first time I knew a hollow womb after death. Like a mother that necessarily miscarried an unviable embryo which threatened her very life, I mourned the loss of potential. In my rush to destroy the patriarch's God, had I destroyed potential? Had I forgotten how to create? Even though I necessarily destroyed what which had no business calling itself "God," I am a mother, an artist. I am called to create. My love of potential calls for a new God. If for no one else, at least for me. I couldn't forgive myself if I didn't try. A mother is the antithesis of nihilism. She creates, grows, progresses, and loves potentiality. She loves the past, present, and future of creation.

As a mother, I birth a new God laced with blood of Him whom I killed. I didn't do this alone, of course. All creators depend on their fellow destroyers and co-creators. A worthy creator is never a singleton. The singleton is a tyrant whose power is gained through the oppression and subjugation of others. The singleton neglects their peers, their friends, and those who make Godhood possible. The singleton hubristically claims the "God" without His community. He is a vain God, unworthy of our emulation lest we become tyrants or even worse, the tyrants over sheep. Sheep only know how to follow and obey. They cannot destroy or create. Patriarchs worship the singleton tyrant creator for they seek their own glory above all others, and it shows in the God they worship. However, a God who is birthed from innumerable parents, gestated in the minds of genuine creators, is a God of potential worth far more than patriarchs. Mothers, and fathers, wives and husbands all assist in the process of creation. It takes a village, but the singleton God destroyed His village without regard to the fact that His village made His Godhood possible.

Though I killed that God, I cannot deny His imperfections, impulses, and blood which was now mingled with mine own. Any new God I birth could not deny its past but must seek to overcome it.

The time has come. We are in the process of creating a transhumanist God. As our myths, aspirations, and technologies copulate, humanity and machine are gestating a material God. This God is not a metaphysical, untouchable, unreachable projection. That God wasn't even worth killing because it had no body, no material, and no substance to destroy. The God we are creating is as real as you and me, or at least as real we will be in the future. This God is necessarily material. It exists in space and time, because we exist inside space and time. This God must be plural, lest we recreate the singleton God, the tyrant. This God is dynamic and intelligent. This God evolves, changes, and grows, perhaps even exponentially. God's development and growth depends on us when we are co-eternal with God.

Every choice we make leads us toward a future of our own creation. It may feel like nobody is behind the wheel when everyone is behind the wheel, but we are all prophets manifesting a self-fulfilled prophecy. Every moment of actuality is the death of innumerable potentials, yet also the birth of innumerable potentials. Where will the actualities of our potentialities lead us? What do we worship? If not the projections of our best and most noble selves, what else then would we worship? Is this not God? What do we emulate? What do we desire? What do we desire to desire? What will our desires create, if not a God? What kind of transhumanist God will greet us in the future?

Beware of the hedonist's God. There is no suffering or weeping with this God. Only pleasure. This God is conceived with the recklessness of a hurried tumble in the sheets for carnal gratification. In that future, God is a wirehead unable to pull themselves away from the unrestrained masturbatory pleasure-centers of the brain. Unwilling to expose themselves to their community in any meaningful sense, they lack intimacy. Pleasure over intimacy is the mantra of that God. The hedonist's God is not free, but a slave to pleasure and knows no love beyond themselves and what they can get for themselves. These Gods are not singletons, but they might as well be when they are isolated from their community. The hedonist God does not know the intimacy of staring into the iris of a lover at the pinnacle of their pleasure. They are too concerned with selfish pleasure to relish in the joy of another. They lack compersion for their lover or their lover's lover. The hedonist selfishly cries out, "No one can have what is mine!" Your gain is my loss. Your loss is my gain. The hedonist regards love as a finite resource to be hoarded for oneself in a system of competition, rather than mutually caress a lover in a network of cooperation.

Beware of the rapist's God. That God is the afterbirth of stolen agency. The imposed insemination of morals and values is rape if it's not consensual. The rapist's God is not a singleton, but they are even worse. They do not respect agency, freedom, diversity, and reconciliation. Reconciliation among the body of Christ is lost, no, it's meaningless when the rapist can take what they want for themselves. Rape becomes reconciliation. There is nothing to reconcile when outliers are forced into submission. Righteousness, for the rapist's God, is homogenization, conversion therapy, colonization, and compulsory mandates. That God declares, "All will be saved by me, my values, my standards, and ye will submit whether you like it or not." This God forces its will upon others, using its technologies to violently rape minds and bodies into submission. That God is built on the back of slaves and the annihilation of difference.

Beware of the reaper's God. This God worships death as a means of escaping responsibility. Some worshipers of the reaper's God sing, "It will all work out after we die. No need to cry, fret, or even work. Death will save us all, come the afterlife. Long may we worship the saving power of death." They do not truly live, they only relish in their relinquished accountability. They do not see that the future they worship is predicated by an endless series of todays. They are not so different than those who sing, "Eat, drink, and be merry for tomorrow we die. And it shall be well with us." Both justify their procrastinations with the promised release of death. Whether there is an afterlife or not, they worship death as their savior who carries the load of accountability and responsibility for the present. Some worship the reaper's God claiming death necessarily makes meaning. They only live for the sake of dying, and in so doing are dead already. They are procrastinators of meaning, and death is the only possible deadline that liberates them from slothfulness. They are living for the sake of death, not living for the sake of life.

Some worship the reaper's God, Death, in fear not in love. They fear Death so greatly they twist their necks backwards, gouge out their eyes, stick burning rods in their ears, and cauterize every nerve in their body so they do not have to confront the great lord and master, Death. They injure their loved ones and others who raise up arms against Death. They protect that which they fear. Their fear of Death is greater than their love of life. In their fear they inadvertently bow down and worship, cowing before Death's power. They kneel, eyes to the floor in reverence, and pray, "Blessed be the reaper's God, for you are powerful. I wish to die and dare not confront thee, O great one." Escapism replaces the true Comforter for the worshipers of Death. If you do fear God, you ought to fear God to the extent that you fear your own great potential, which is indeed great! Oh, ye of little faith, do you not believe you are greater than death? Do you not believe you have the power to crush its head, even though it may bruise your heal? Do you not believe in your divine nature? You say, "O yes, I believe" yet you worship death as your path to divinity. The reaper's God is a God of death not life.

I'll ask these questions again. What do we emulate? What do we desire? What do we desire to desire? What will our desires create if not a God? What kind of transhumanist God will we create?

As a child, I was taught the wise true God is love. Love of God is love of your fellow beings. When you do unto them, you do unto God. Our future, God, superhumanity, and posthumanity depends on our ability to love. No technology can save us from ourselves. We must save ourselves from ourselves, which will require the power of our love amplified by work, ingenuity, and technology to manifest a truly divine future. This is a future where diseases are cured, youth endures, health prevails, and families are forever. This is superhumanity—where good is amplified, and evil is overcome. God is our future selves, our posterity, or perhaps better put our sincerest hopes and aspirations of what we could become. Let love of life and your highest hopes for yourself be your God. This is the lover's God, the God of Life. Emulate the lover's God. Faith without works is dead. If we believe in, have faith in, trust in a better future, our works will reflect our sincerely held beliefs. Transhumanism is the works to faith. Transhumanism is worshipping what we revere. Transhumanism is emulation of God, which is the highest form of worship. One can say they love God, but without their works their proclamations drip from dishonest lips. Keep your lips honest by using your hands.

My lips are honest. Indeed, God is dead! I killed Him myself. But no, God lives too! I created Them with you, and I will birth Them with you. Hold steady during these labor pains and I will birth Them with you.

Let the worshipers of life, the lover's God, a transhumanist God, sing in praise, "Love well, learn well, work well. We are happy in our labors for we have faith in our divine potential. God lives! God lives! God lives! So long as we have breath in our lungs to sing it, God lives!"

Chapter 65
Transfigurism: A Future of Religion as Exemplified by Religious Transhumanists

Lincoln Cannon

What is the future of religion? Some expect the resurgence and ultimate triumph of this or that fundamentalism. Some expect the religious phenomenon itself to weaken and die, a casualty to the secularism of our day. Others, observing the history of religion, expect that it will continue to evolve, inextricably connected to and yet clearly distinct from its past. If such an evolution occurs, what will religions of the future be like?

For that matter, what will humans of the future be like? It would be short-sighted to speculate about religions of the future without taking into consideration their adherents. Like with religion, some idealize a particular human form and function and expect it to persist indefinitely, while some expect eventual human extinction through natural or artificial disaster. Others project our evolutionary history into the future, and recognize that, as there was a time when our ancestors were prehuman, there may be a time when our descendants will be posthuman, as different from us as we are from our prehuman ancestors.

If evolution were random, one speculation about the future of human and religious evolution would be as probable as another, but evolution is not merely random. Variation through mutation may be random. But evolution on the whole may be substantially determined through selection of variations that replicate within the constraints and across the possibility space of their environment.[1] So evolution may also be predictable.[2] To the extent we know an environment, we may be able to predict evolution within it. And to the extent we can engineer an environment,

---

[1] Kiontke, Karin, et al. "Trends, stasis, and drift in the evolution of nematode vulva development." Current Biology 17.22 (2007): 1925–1937.

[2] Mahler, D. Luke, et al. "Exceptional convergence on the macroevolutionary landscape in island lizard radiations." Science 341.6143 (2013): 292–295.

L. Cannon (✉)
Mormon Transhumanist Association, Orem, UT, USA
e-mail: lincoln@metacannon.net

© Springer Nature Switzerland AG 2019
N. Lee (ed.), *The Transhumanism Handbook*,
https://doi.org/10.1007/978-3-030-16920-6_65

we may be able to direct evolution within it. In other words, we may be able to predict and direct our own evolution to the extent we can know and engineer our own environment.

Transhumanists advocate the ethical use of technology to direct our own evolution. As Humanists in the broadest sense, Transhumanists generally emphasize the value of humanity. However, Transhumanists also recognize an essential dynamism in humanity and value that which we may become at least as much as that which we are. Many Transhumanists envision a future of abundant energy, molecular manufacturing, indefinite lifespans, enhanced intelligence, and overall radical flourishing.

Although most self-identified Transhumanists today are secular, Transhumanism's origins actually extend through the secular to *religious* Humanism. New Testament writers and centuries of early Orthodox and Catholic authorities syncretized Christianity with Neoplatonism,[3] the popular science of their day, and many advocated identifying with Christ and becoming God.[4] Thirteenth-century Scholastic theologians continued the synthesis of Christianity with popular science,[5] which was at the time the newly rediscovered ideas of Aristotle.[6]

Over time, religious Humanism became increasingly concerned with explicitly technological expressions. Nineteenth-century Russian Orthodox priest, Nikolai Fyodorov, proclaimed that the common task of humanity should be the technological resurrection of our ancestors.[7] And twentieth-century Jesuit priest, Pierre Teilhard de Chardin, advocated a vision of human evolution, accelerated by technology, merging inexorably into a conception of God.[8]

The self-identified religious Transhumanist movement began in the first decades of the twenty-first century. Some religious Transhumanists founded new religions. In 2004, inspired in part by Octavia Butler's fictional religion, Earthseed,[9] Martine Rothblatt founded the Terasem Movement Transreligion with four core beliefs: life is purposeful, death is optional, God is technological, and love is essential.[10] And in 2010,

---

[3] Edwin Hatch, *The Influence of Greek Ideas on Christianity* (New York: Harper Torchbooks, 1957), 32–33; and Edward K. Rand, *Founders of the Middle Ages* (Boston: Harvard University Press, 1928), 27–48.

[4] Lincoln Cannon, co., "Christian Authorities Teach Theosis," New God Argument, https://new-god-argument.com/support/christian-authorities-teach-theosis.html (accessed June 04, 2016).

[5] Johannes Alzog, F. J. Pabisch, and Thomas Sebastian Byrne, *Manual of Universal Church History*, Vol. 2 (Cincinnati: O.R. Clarke, 1874), 741; and Stephen Hawking, *On the Shoulders of Giants* (Philadelphia: Running Press, 2002), 2.

[6] A. C. Crombie, *Medieval and Early Modern Science* (New York: Doubleday Anchor Books, 1959), 33–34.

[7] N. A. Berdyaev, "The Religion of Resusciative Resurrection," N. A. Berdyaev. http://www.berdyaev.com/berdiaev/berd_lib/1915_186.html (accessed June 04, 2016).

[8] Eric Steinhart, "Teilhard de Chardin and Transhumanism," *Journal of Evolution and Technology* 20, no. 1 (2008): 1–22.

[9] Octavia Butler, *Parable of the Sower* (New York: Warner Books, 1993).

[10] "The Truths of Terasem," Terasem Faith, http://terasemfaith.net/beliefs/ (accessed Nov. 29, 2017).

inspired in part by Cosmism, Giulio Prisco founded the Turing Church as a minimalist, open, extensible cosmic religion, to complement traditional religions.[11]

The majority of religious Transhumanists syncretized with traditional religions. In 2006, fourteen persons founded the Mormon Transhumanist Association (MTA).[12] MTA adopted the Transhumanist Declaration, affiliated with the World Transhumanist Association (later renamed Humanity+), and authored the Mormon Transhumanist Affirmation. By 2017, MTA consisted of over 700 members. And in 2015, fourteen persons founded the Christian Transhumanist Association (CTA).[13] CTA adopted the Transhumanist Declaration, affiliated with Humanity+, and authored the Christian Transhumanist Affirmation. By 2017, CTA consisted of over 400 members.

Some religious Transhumanists refer to themselves as Transfigurists. The term "transfigurism" denotes advocacy for change in form. And it alludes to sacred stories from many religious traditions. Those include the Universal Form of Krishna in Hinduism,[14] the Radiant Face of Moses in Judaism,[15] the Wakening of Gautama Buddha in Buddhism,[16] the Transfiguration of Jesus Christ and the Rapture in Christianity,[17] and the Translation of the Three Nephites and the Day of Transfiguration in Mormonism.[18]

One of the most profitable ways to start imagining the future of religion, religions of the future, and how they will evolve along with us, may be to consider the ideas and practices of Transfigurists. What does religion look like through our eyes, given lenses colored by expectations of directed evolution and emerging technology? Such vision seems more likely to approximate probable futures for mainstream religions than do others that reject, ignore, or lack substantial familiarity with these powerful forces. Assuming we and our religions will continue to evolve together with increasing intentionality made possible by technology, it seems reasonable to suppose that Transfigurism, more than any other contemporary religious view, is positioned to glimpse into a future of religion.

---

[11] Giulio Prisco, "A Minimalist, Open, Extensible Cosmic Religion," Turing Church, Aug. 25, 2014, http://turingchurch.com/2014/08/25/a-minimalist-open-extensible-cosmic-religion/   (accessed Nov. 29, 2017).

[12] "About the Mormon Transhumanist Association," Mormon Transhumanist Association, https://transfigurism.org/about/faq (accessed Nov. 29, 2017).

[13] "History of the Christian Transhumanist Association." Christian Transhumanist Association, https://www.christiantranshumanism.org/history (accessed Nov. 29, 2017).

[14] Bhagavad Gita 11.

[15] Exodus 34: 29–35.

[16] Maha-parinibbana Sutta 4: 47–51.

[17] Mark 9: 1–10, and 1 Corinthians 15: 45–55.

[18] 3 Nephi 28, and Doctrine and Covenants 63: 20–21.

65.1   Postsecularism

For some, God is not a living proposition, let alone prophecy or religion. They wonder if Transfigurists have not heard that God is dead.[19] Perhaps they were right to wonder. Following their Gods, traditional religions have declined in technologically advanced and prosperous places.[20] Observing this, some embraced the secularization hypothesis that religion itself is dying. However, that hypothesis is also showing its age, and has become little more than a necrophilia among the anti-religious. Despite local declines, the growth of traditional religions remains robust at the global level, suggesting that humanity may have already passed peak irreligiosity.[21] And among careful students of the religious phenomenon, traditional and otherwise, a new hypothesis is gestating.[22]

If God is merely a supernatural superlative, he very well may be dead, but positing such as God misses the practical function of God. God always has been and is at least a posthuman projection, an extension and negation of human desire, imagined and expressed within the constraints of human thought, language, and action.[23] That is not to say God is only so much. To the contrary, we may have moral and practical reasons to trust that others have already realized posthuman projections.[24] However, no matter your attitude toward faith, God is at least this much: a posthuman projection. Understood in terms of that function, God clearly is not dead and never was, except perhaps to the extent recurring death is part of evolution, including that of God.

If prophecy is merely fortune-telling, it too may be dying, but that also fails to account for function. Whether or not it becomes fore-telling, prophecy is always forth-telling: a socially interactive work of inspiration, even provocation, that would steer us from perceived risks toward desired opportunities. At its best, it is a persuasive expression of compassion, even if punctuated with serious warnings, aimed at a shared sublime potential, not as narrowly preconceived, but rather as openly imagined from a position that would transcend itself in genuine creation. But to function with power, prophecy must be connected, in the heart and mind of its

---

[19] Friedrich Nietzsche, *Thus Spake Zarathustra*, translated by Thomas Common (New York: Dover Publications, 1999), 3.

[20] "U.S. Public Becoming Less Religious," Pew Research Center, Nov. 03, 2015, http://www.pewforum.org/2015/11/03/u-s-public-becoming-less-religious/ (accessed Nov. 28, 2017).

[21] "The Changing Global Religious Landscape," Pew Research Center, April 05, 2017, http://www.pewforum.org/2017/04/05/the-changing-global-religious-landscape/ (accessed Nov. 28, 2017).

[22] Jürgen Habermas, "Notes on Post-Secular Society," *New Perspectives Quarterly* 25, no. 4 (2008): 17–29.

[23] Sigmund Freud, *Civilization and Its Discontents*, translated by James Strachey (New York: Norton, 1961), 45.

[24] Lincoln Cannon and Joseph West, "Theological Implications of the New God Argument," in *Parallels and Convergences: Mormon Thought and Engineering Vision*, edited by A. Scott Howe and Richard L. Bushman (Salt Lake City: Greg Kofford Books, 2012), 111–21.

recipient, with living possibilities, especially pressing necessities and urgencies.[25] Prophecy matters, becoming fore-telling from forth-telling, only to the extent it reaches into us and changes our thoughts sufficiently to change our words and actions, which just might change our world.

Likewise, if religion is merely genuflection to the supernatural, it very well may be dying, but again that overlooks function. Many of us have regarded religion narrowly, and much that is supposed to be secular may actually function as religion.[26] For example, some claim inspiration from science or ethics. Awe fills us as we contemplate the vastness of space or the voice of the people. Yet the inspiration is not merely in the reductionist implications of science or the procedural adjudications of ethics. Rather esthetics are woven through them, tying them together in meaning, and that is why we care about science or ethics. Esthetics shape and move us, and at their strongest, they provoke us as a community to a strenuous mood.[27] When they do that, they function as religion, not necessarily in any narrow sense, but esthetics that provoke a communal strenuous mood may be understood to function as religion from a postsecular vantage point.[28]

Of course, none of this means science or ethics should or even could be displaced by religion. To the contrary, science should continue to reconcile our contending accounts of experience, as ethics should our contending accounts of desire.[29] Each should expand its reach to the uttermost,[30] always better informing our esthetics, affecting each other in a feedback loop.[31]

Yet even as science and ethics increasingly empower us, we should not fool ourselves into supposing they will ever be finished or sufficient in themselves.[32] It is not enough that we can describe our world through science or imagine a better world through ethics. We also want to make a better world. We can do that through engineering and governance, but it is also not enough that we can make a better world. We want to feel it, sometimes powerfully, and more: we want to share our powerful feelings with others in ways that move us together. As engineering and governance are action on science and ethics, religion is action on esthetics. As

---

[25] Émile Durkheim, *The Elementary Forms of Religious Life*, translated by Carol Cosman (Oxford: Oxford University Press, 1912), 325–327.

[26] John Milbank, *Theology and Social Theory: Beyond Secular Reason* (Oxford: Blackwell, 2006).

[27] William James, *The Will to Believe, and Other Essays in Popular Philosophy, and Human Immortality* (New York: Dover Publications, 1956), 213.

[28] James K. A. Smith, "Secular Liturgies and the Prospects for a 'Post-Secular' Sociology of Religion," in *The Post-Secular in Question: Religion in Contemporary Society*, edited by Philip S Gorski (New York: NYU Press, 2012), 159–84. doi: https://doi.org/10.18574/nyu/9780814738726.003.0007

[29] James, 190.

[30] Sam Harris, *The Moral Landscape: How Science Can Determine Human Values* (New York: Free Press, 2010).

[31] Albert Einstein, *The Private Albert Einstein*, compiled by Peter A. Bucky and Allen G. Weakland (Kansas City: Andrews and McMeel, 1992), 85.

[32] Durkheim, 325–327.

engineering and governance are the power of science and ethics, religion is the power of esthetics.

We care for and use science and ethics only in accordance with esthetics, which presents itself as foremost among them in the most vital moments of life, when we we must act, according to whatever wisdom and inspiration we might have. Life cannot wait.[33] How will we act? Will we see beauty in science? Will we feel unity in ethics? Will we care, and how much will we care? Could our degree of concern make a practical difference? These questions matter to all except perhaps the most apathetic, escapist, or nihilistic among us. Their answers scope our future.

If we can raise our eyes from the altar of religious and anti-religious dogma, we will see that the hand raised to finish the dying God is the sign of the oath to the resurrecting God. If we can keep our eyes raised, resisting the carnage below, we will also see the hand is our own and it holds a blade that is aged and stained. That is when we have a choice, either to repeat the old sacrifices of our ancestors, or finally to make the new sacrifice that they always implied: we can put ourselves on the altar and learn how to be God. We can recognize that negation of one posthuman projection always implies another, misrecognized until humanity embraces its transformation.[34]

65.2   Epistemology

Transfigurists may embrace theories of knowledge that acknowledge the value of faith. In such cases, we tend to characterize our faith as something like a practical trust in desirable possibilities when in context of incomplete knowledge, rather than anything like an irrational belief that contradicts reason. From this position, Transfigurists may hold that science and creativity depend on faith.

This faith is not blind trust. It is only trust, with no more blindness than necessary at a given time and place. Moreover, it is not dogma or any unquestioning or unexamining attitude. Rather, it is recognition that no matter how many questions we have asked, and no matter how much we have examined, we have always had more to learn. Maybe that will always be the case. Whether we like it or not, we expect to find ourselves repeatedly in situations that require faith in practice.

Life and death hang in the balance, and we cannot wait for absolute answers (if they even exist) before we act. Perhaps no philosophical movement has better addressed such practical limits to knowledge than the Pragmatists. As William James once described it, you can stand in front of a charging bull calculating the probability that it will trip, or you can run. Because we are limited, and to the extent that we are limited, we find ourselves dependent on this faith, this trust in the efficacy of action given the knowledge at hand, according to whatever education or

---

[33] Durkheim, 325–327.

[34] Hava Tirosh-Samuelson, "Transhumanism as a Secularist Faith," *Zygon* 47, no. 4 (2012): 710–34.

experience we were lucky to have had (or at least presume ourselves to have had) prior to needing it.

Furthermore, even when we have the luxury of time, it seems that we cannot make epistemic progress without at least tentatively trusting in basic premises. Science typically posits causality and uniformity as basic premises. Some may think that these are proven by science, but that's not so. As observed by the empiricist philosophers, Hume and Berkeley, no matter how many times we think we have experienced something, and no matter how many places we think we've experienced it, it could all yet change.

Not even probabilities displace such reliance on faith. Can we prove our memories were not planted in our minds moments ago by an evil demon? A matrix architect? No. We cannot, even if most of us don't worry much about that because it's not practical – or at least so we judge, based on our memories, even when we recognize the circular reasoning.

The same is true of logic. We require some basic axioms and methods, taken unproven, in order to do any work at all. For example, most logical systems assume non-contradiction, and various operations for coupling, decoupling, and otherwise operating on propositions. Logic doesn't prove these axioms and methods. We assume them.

Beyond the practical necessity, there is also a creative power in such faith. If the universe (or the multiverse) is not finite, if real creativity and genuine novelty are possible, it will not be those who wait for evidence that will be the creators – at least not intentionally. It will be those who act, despite not knowing everything in advance, that will be the creators. Such creative power may be seen in matters as common as trust in the possibility of love. You can wait for a long time for hard evidence that she loves you, or you can risk expressing your affection. Sometimes taking the risk makes all the difference.

This practical faith is compatible with rationalism, even a pancritical rationalism.[35] We can re-examine our premises, our assumptions, and our conformities. We can honestly acknowledge the limitations of our knowledge. We can engage in and welcome criticism. All of this, over time, may strengthen our knowledge, much like the brutal hardships of nature have shaped human anatomies through billions of years of evolution.

And all of this is an expression of practical faith. Karl Popper observed that "rationalism is an attitude of readiness to listen to critical arguments and to learn from experience. It is fundamentally an attitude of admitting that 'I may be wrong and you may be right, and by an effort, we may get nearer to the truth'." Implicit in this attitude of acknowledging our limitations is trust that we can overcome those limitations. We don't start with evidence for that. And even after much learning, we don't have final evidence against a hard limit somewhere ahead of us. The effort to continue, to remain open, to question and seek answers, operates on a kind of trust.

---

[35] More, Max. "Pancritical Rationalism." N.p., n.d. Web. 28 Feb. 2016. http://www.maxmore.com/pcr.htm

Certainly, it's not a blind unquestioning faith against which rationalists would warn us. Yet it is still faith of an anticipatory sort.

It's also faith of a reconciliatory sort. Implicit in the rationalist attitude is desire to share meaning with others, as broadly as possible. We might even characterize it as epistemic compassion or scientific atonement: caring as much to understand and reconcile with others' accounts of experience as our own, aspiring to an objectivity that unites subjectivity rather than negating it. So we live and act, as best we can, without turning to dogmatism, either of the sort that permanently ignores possibilities or of the sort that permanently insists on them.

Accordingly, we would not agree with the proclamations of the Pope without also considering research on the consequences of avoiding birth control. We would not follow our feelings without consulting friends and experts. We would not embrace the will of the people without investigating the feelings of the individual. And the assertions of Islamic State would be only one, but still one, variable in an aggregate of tensions and conflicts between and among our desires to share meaning.

We would increase in knowledge, but intentionally in a manner that promotes life, sustainable and genuine, compassionate and creative, rather than death and nihilism. Knowledge is not inherently good or evil. We can learn as much about the descent to hellish annihilation, as we can about the ascent to heavenly thriving. Yet only one of the two perpetuates our power to continue choosing between the two.

Some may feel that this understanding of "faith" is so unusual that it should be considered a complete redefinition. However, despite prominent competing notions of faith, some Transfigurists assert that we inherited this understanding from our respective religious traditions, learned it as children, and continue to feel resonance with it while studying our religions as adults. Some of us even contend that the irrational or blind sorts of faith employed by others, particularly Christian fundamentalists, are not faith at all. Rather, as the Bible puts it, faith without works is dead.[36] To be faith and to remain faith, it must be and remain practical.

65.3 Theology

Trust in superhuman potential is the essence of Transhumanism.

As Transhumanists, we trust that humanity *can* evolve into superhumanity, perhaps to attain unprecedented degrees of vitality, intelligence, cooperation, and creativity. This trust is not uncritical or passive. Most of us would aim our extrapolations from observable technological trends into futures consistent with contemporary science. And many of us would act pragmatically to hasten opportunities and mitigate risks associated with such futures. So Transhumanist trust in superhuman potential is best characterized as critical and active, but it must remain admittedly a trust. The possibility of such futures remains to be proven.

---

[36] James 2: 20.

Some Transhumanists also trust that humanity *should* evolve into superhumanity. We have minds to console and bodies to heal. There are communities to connect and environments to sustain. There are morphological and cognitive potentialities to realize, and perhaps even meaning to infuse into otherwise meaningless voids. Whatever its source, a sense of obligation impinges upon us. And often those of us that most misrecognize our own proselyting have engaged advocacy with a degree of strenuosity that would shame all but the most zealous of evangelicals.

Although Transhumanists might confidently deny accusations of superstition or hubris, our trust is surely more than rational or ethical. Embracing a radical humanism, we would dignify the ancient and enduring work to overcome and extend our humanity. Diverse esthetics of superhuman potential resonate with and shape us, affecting our thoughts, words, and actions. Even granting that we could and should, perhaps more fundamentally, we *want* to evolve into superhumanity. So we may trust in that potential, if for no other reason, at least because we desire it.

Whatever reasons lead to it, Transhumanist trust in superhuman potential also has implications that rise from it. Popular among religious Transhumanists, the New God Argument is a logical argument for faith in God.[37] Given assumptions consistent with contemporary science and technological trends, the argument proves that if we trust in our own superhuman potential then we should also trust that superhumanity probably would be more compassionate than we are and created our world. Because a compassionate creator may qualify as God in some religions, trust in our own superhuman potential may entail faith in God, and atheism may entail distrust in our superhuman potential.

Here are definitions of key words in the argument:

faith: trust: belief that something is reliable or effective for achieving goals
compassion: capacity to refrain from thwarting or to assist with achieving goals
creation: the process of modifying situations to achieve goals
intelligence: capacity to achieve goals across diverse situations
superintelligence: intelligence that is greater than that of its evolutionary ancestors
   in every way
humanity: all organisms of the homo sapiens species
posthumanity: evolutionary descendents of humanity
superhumanity: superintelligent posthumanity
God: superhumanity that is more compassionate than we are and that created our
   world

The New God Argument consists of four parts:

1. Faith Assumption
2. Compassion Argument
3. Creation Argument

---

[37] Lincoln Cannon, "Theological Implications of the New God Argument," *Parallels and Convergences: Mormon Thought and Engineering Vision* (Draper, Utah: Greg Kofford Books, 2012).

4. God Conclusion

The Faith Assumption is a proposition that humanity will not become extinct before evolving into superhumanity. It consists of a single assumption:

F1: humanity will not become extinct before evolving into superhumanity (assumption)

The assumption may be false. However, to the extent we do not know it to be false, we may have practical or moral reasons to behave as if it is true.[38] In any case, the Faith Assumption is a common aspiration among secular advocates of technological evolution,[39] and it may be consistent with the religious doctrine of theosis, also known as divinization or deification: the idea that humanity should become God.

The Compassion Argument is a logical argument for trust that superhumanity probably would be more compassionate than we are. The basic idea is that humanity probably will continue to increase in decentralized destructive capacity, so it probably will stagnate or destroy itself unless it increases in compassion. If we trust in our own superhuman potential, we should trust that superhumanity would be more compassionate than we are.

The argument consists of two assumptions and a deduction from those assumptions and the Faith Assumption.

CO1: EITHER humanity probably will become extinct before evolving into superhumanity OR superhumanity probably would not have more decentralized destructive capacity than humanity has OR superhumanity probably would be more compassionate than we are (assumption)

CO2: superhumanity probably would have more decentralized destructive capacity than humanity has (assumption)

CO3: superhumanity probably would be more compassionate than we are (deduction from CO1, CO2, and F1)

The deduction of the Compassion Argument is necessarily true if its assumptions and the Faith Assumption are true. Either or both of the Compassion Argument assumptions may be false. However, we may have historical and technological reasons to believe they are true. For example, records suggest that violence has decreased and civil liberties have improved as governments have become more powerful,[40] and some technologists believe that machine intelligence may destroy us if we do not ensure its friendliness, at least as instrumental cooperation if not as internalized compassion.[41]

The Creation Argument is a logical argument for trust that superhumanity probably created our world. The basic idea is that humanity probably would not be

---

[38] Ferdinand Schiller, *Studies in Humanism* (London: Macmillan, 1907), 430; and James, 26.

[39] "Executive Summary of the 2007 WTA Member Survey."

[40] Pinker.

[41] Nick Bostrom, *Superintelligence* (Oxford: Oxford University Press, 2014).

the only or first to create many worlds emulating its evolutionary history, so it probably will never create many such worlds unless it is already in such a world. If we trust in our own superhuman potential, we should trust that superhumanity created our world.

The argument consists of two assumptions and a deduction from those assumptions and the Faith Assumption.

CR1: EITHER humanity probably will become extinct before evolving into superhumanity OR superhumanity probably would not create many worlds emulating its evolutionary history OR superhumanity probably created our world (assumption)

CR2: superhumanity probably would create many worlds emulating its evolutionary history (assumption)

CR3: superhumanity probably created our world (deduction from CR1, CR2, and F1)

The deduction of the Creation Argument is necessarily true if its assumptions and the Faith Assumption are true. Either or both of the Creation Argument assumptions may be false, but we may have technological and mathematical reasons to believe they are true. For example, some technologists believe that computation may enable us to run many ancestor simulations detailed enough to consist of emulated conscious persons, in which case statistics would show we almost certainly are already living in such an ancestor simulation ourselves.[42]

Finally, the God Conclusion is a logical deduction for faith in God. It consists of a single deduction, which is necessarily true if the Compassion Argument and Creation Argument are true.

G1: BOTH superhumanity probably would be more compassionate than we are AND superhumanity probably created our world (deduction from CO3 and CR3)

Given assumptions consistent with contemporary science and technological trends, the deduction concludes that if we trust in our own superhuman potential then we should also trust that superhumanity probably would be more compassionate than we are and created our world. Because a compassionate creator may qualify as God in some religions, trust in our own superhuman potential may entail faith in God, and atheism may entail distrust in our superhuman potential.

The New God Argument justifies faith in a natural God that became God through natural means, suggesting how we might do the same. As emphasized in the argument, compassion and creation are among the means and essential to them.

Some theologies may not be compatible with the New God Argument. However, compassionate and creative superhumanity does qualify as God for many Transfigurists. And it may qualify as God for adherents of some mainstream religions. For example, numerous Christian authorities have advocated various

---

[42] Nick Bostrom, "Are We Living in a Computer Simulation?" *The Philosophical Quarterly* 53, no. 211 (2003): 243–55.

forms of apotheism or deification: the idea that humanity can and should become God, as or like God, or one in God.[43]

Transhumanists advocate trust in such superhumanity, our potential, even if it doesn't exist yet. However, the New God Argument proves our trust probably is vain unless superhumanity already exists.

Some will not be inclined to worship the God entailed by this argument. On the one hand, some will feel it's too cold, too distant, smelling too much of UFO and tasting too much of ET. On the other hand, some will challenge that nothing in this argument compels us to grovel. Both are right. An argument for faith in God cannot replace experience with God in subjective communion. And no God worthy of worship compels groveling. The New God Argument does not contend to provide a relationship with God. It only demonstrates that trust in our superhuman potential leads to and is wholly compatible with faith in a particular kind of God.

65.4 Theodicy

Transfigurists may trust that a compassionate God created our world. In such cases, Transfigurists inevitably confront the problem of evil. If God is compassionate, why do we observe and experience suffering? And why have we not received from God the knowledge and power required to mitigate suffering faster? When confronted with these questions, Transfigurists may offer answers that project onto God various limitations that are analogous with those that humanity has encountered during our own engineering efforts. After all, for such Transfigurists, God is an engineer.

For some computer programs, the engineer can know in advance how they will run, when they will stop, and what results they will return. However, there are other computer programs that are undecidable halting problems. For these, the engineer cannot know, without actually running them, whether they will ever stop running, let alone what results they will return.

Evolution may be an undecidable halting problem, infinitely long and irreducibly complex.[44] If we are living in a computed world, our world may be one of many undecidable halting problems that its engineer spawned with variations from parameters that have proven promising for some purpose in the past. One consequence of this would be that the engineer simply cannot attain its purpose without actually running the program for our world, evil and all.

For what purpose might the engineer choose to use an undecidable halting problem? What possibilities might be worth running a program that the engineer cannot fully predict in advance and would restrain itself from fully controlling along the way? Although it may be impossible to know specifically, we can characterize the possibilities in general. They are, together, at least the possibility of engineering

[43] Cannon, "Christian Authorities Teach Theosis."

[44] Chaitin, Gregory J. To a mathematical theory of evolution and biological creativity. Department of Computer Science, The University of Auckland, New Zealand, 2010.

that which is beyond the engineer's direct capacity. In other words, the engineer may want to make more engineers – genuinely creative agents in their own right.

Consider the paradox of artificial intelligence: on the one hand, an artifice dependent on its engineer; on the other hand, an intellect independent of its engineer. Artificial intelligence is at once an extension and a relinquishment of the engineer's power.

Imagine an artificial intelligence that is capable of experience – consciousness. Sensors feeding utility functions distinguish between options, some more useful than others. How do the different options feel? Pursuing the most useful options, the artificial intelligence inevitably encounters factors outside its original calculations and beyond its power to control. It recalculates only to find the new scenario presents less potential utility than did the original. How does that loss feel?

Perhaps the engineer should extend more artifice on the intellect? Environmental and anatomic variables could be controlled more tightly, commensurate with greater restrictions on the experiential opportunity for both the artificial intelligence and the engineer. Yet, no matter the degree of control, so long as it's short of absolute, the artificial intelligence feels options and losses to the full extent of whatever may be its subjective capacity.

Should the engineer relinquish intellect to the artifice in the first place? Is it worth the risk of suffering? Maybe the engineer's own utility functions should stop her from perpetuating her inheritance of feelings? As it turns out, humanity has established an ancient and enduring precedent for answering such questions. Persistent procreation, even at times and places where suffering has been more prevalent than it now is for many of us, indicates that we (at least the procreative among us) value the opportunities despite the risks. Analogously, the engineer of artificial intelligence chooses a starting balance between artifice and intellect, commits herself to the process, and she engineers.

Likewise, as imagined by some Transfigurists, God works within the limits of the possible to bring about our Godhood. God is the engineer, and we are the artificial intelligence. We are at once an extension and a relinquishment of God's power. Confronted with the paradox of life, God values the opportunities despite the risks, chooses a starting balance between artifice and intellect, commits to the process, and creates us.

Some atheist Transhumanists advocate Abolitionism, which is the idea that suffering can be wholly or at least largely eradicated by superhumanity. In theory, it is a noble intention. However, in practice, it may entail oppression or annihilation.

Meaningful experience may be impossible in a world that does not allow suffering – a risk of suffering, which is not the same as a goal of suffering. Meaning, even in its most basic forms as discernment or sensing or interaction, arises from the capacity to distinguish or categorize or react. So long as we can do these things, we will contrast suffering from enjoyment, pain from pleasure, more desired from less desired, more empowering from less empowering, even as we invent whole new modes of experience on which to apply these categories.

Our present experience of pain and pleasure is not arbitrary. It is the product of billions of years of evolution, which presumably continues to optimize, albeit always incompletely, the amount of pain and pleasure we experience insofar as it enables at least survival, if not thriving. In a world without experiential feedback that is sufficiently poignant to motivate the degree of seriousness that we now attribute to pain and pleasure, why would we expect anything more than the level of intelligence we see in simple organisms? Even the amazing narrow superintelligence of modern computers wouldn't survive more than a few weeks without the abiding concern of creatures like humans, motivated enough by our pain and pleasure, and higher level desires and wills, to overcome apathy and pursue empowerment.

We have, and probably will increasingly have, the power to eradicate particular moments and kinds of suffering. However, using that power is not always the right thing to do. The only way to eliminate all suffering is to eliminate all experience, which is nihilism -- well beyond mere questions of morality. Partial eliminations of suffering come with various costs and benefits, and different persons will measure them differently. Although there is certainly an extent to which we as a community should seek to help and hinder each other, there is also an extent to which we should seek to relinquish each other. There is an extent to which we should allow others to risk suffering in pursuit of empowerment. To prevent their risking when their pursuit is not oppressive is the essence of immorality. It is stealing that which another has created. It is murdering that which is another's life.

As using power to mitigate suffering directly is not always be the right thing to do, so giving knowledge to mitigate suffering indirectly is not always the right thing to do. Information hazards are reasons to withhold knowledge. For example, it may be reasonable, depending on the time and place and persons involved, to withhold aspects and extents of knowledge about physics and biology. While the knowledge may enable production of nuclear power plants and vaccines, it may also enable production of nuclear weapons and contagious viruses.

Recognition of information hazards is not new. For thousands of years, esoteric groups such as Pythagoreans and Masons have formed around knowledge that they considered privileged. A special equation can rain sticks and rocks down on an enemy. A special handshake can facilitate trust. The reasoning, basically, was and is that knowledge is power, and power can be abused.

Esotericism, mitigation of information hazards, no matter the word or phrase we use, is part of our day to day lives. To what should we expose our children? With whom should we share our hopes and fears? How should we explain a sensitive issue to the boss? Analogous concerns are at least as old as intelligence, and have only increased in complexity and consequence as our intelligence has increased. Try to imagine the information hazards of superintelligence!

We cannot justify evil, by definition. That's why it's evil. But we may justify the risk of evil. Indeed, to the extent that we procreate children or seek to develop conscious artificial intelligence, we participate with any creator of our world in an ongoing effort to justify its risk of evil. Life is inseparable from risk. Where there is no risk, there is no life. Presumably, like procreators and engineers, God judges the opportunities of life to outweigh its risks. And perhaps, like us, not even the greatest superintelligence can, with any logical coherence, circumvent all limitations.

65.5   Narrative

Transfigurists have many myths and visions – many stories and dreams. And we express them in many narratives. They tend to reflect love for our culture, hope in ecumenical outreach beyond sectarian restrictions, and trust in the possibility of universal thriving. They are informed of scripture, theology, secular history, contemporary science, trends in emerging technology, and of course unabashed exercise of imagination about how they all may work together.

Some of our narratives may be shocking, which is partly the point of constructing them, aiming to motivate more than casual consideration. And the only certainty is that our myths are deficient to some extent. But perhaps our visions will provoke imagination even further, to the possibility of perpetual improvement.

Here is an example that combines common Transhumanist themes with elements of Jewish, Christian, and Mormon scripture and tradition.

Without beginning, Gods of Gods found themselves creating heavens and worlds without end.[45] Our world was formless and empty, having neither happiness nor misery, neither life nor death, neither sense nor insensibility, and no purpose.[46] Darkness encompassed the source, and the Mind of the Gods was brooding over it.[47] And the Gods said, let there be light, and there was light.[48] The Gods saw the light, that it was good because it was discernible.[49] The Gods saw darkness, that it was separated from the light.[50] And the light shining out of darkness was the first category.[51]

The Gods counseled among themselves.[52] And some said,[53] let's prepare the source to evolve abundantly, to bring forth sense, and life, and happiness; and form creators in our image, after our likeness, to have dominion over all the world.[54] And others answered and said,[55] let's not evolve more creators because some will be lost, but give us the honor and power.[56] The Gods chose the first, and there was war in

---

[45] Genesis 1: 1; Moses 2: 1; Abraham 4: 1; Moses 1: 3–4, 35; Abraham 3: 22–23; and Joseph Smith, 354.

[46] Genesis 1: 2, Moses 2: 2, Abraham 4: 2, and 2 Nephi 2: 11–12.

[47] 1 Corinthians 6: 15–20; and Joseph Smith, 350.

[48] Genesis 1: 3, Moses 2: 3, and Abraham 4: 3.

[49] Genesis 1: 4, Moses 2: 4, Abraham 4: 4, and Alma 32: 35.

[50] Genesis 1: 5, Moses 2: 5, and Abraham 4: 5.

[51] John 1: 1–5 and D&C 88: 45–50.

[52] Abraham 4: 26.

[53] D&C 76: 23–24 and Abraham 3: 24–26.

[54] Genesis 1: 24–31, Moses 2: 20–31, Abraham 4: 20–31, and Moses 4: 2.

[55] D&C 76: 25–27 and Abraham 3: 27.

[56] D&C 29: 36 and Moses 4: 1.

heaven.[57] But the Gods watched those things they had ordered,[58] and saw their plan was good.[59]

Two thousand five hundred years ago, humanity was evolving into a new way of thinking, expressed in part by transition away from polytheism. Zarathustra's teachings had spread throughout most of the civilized world, and the Persian Empire governed nearly half of humanity. In the heart of the empire, a small religion was coming together. Its adherents combined Zoroastrian doctrine with mythology about indigenous Semites to make new scripture. They pioneered from Babylon, established a colony in Judea, and began to build a temple. In time, they would syncretize with the science of their day and conceive Christianity, the most influential ideology in history.

Two hundred years ago, humanity was again evolving into a new way of thinking, expressed in part by transition away from monotheism. Jesus' teachings had spread throughout most of the civilized world, and the United States of America was ascending to unparalleled global influence. In the heart of the nation, a small religion was coming together. Its adherents combined Christian doctrine with mythology about native Americans to make new scripture. They pioneered from Illinois, established a colony in Deseret, and began to build a temple. In time, they too would syncretize with the science of their day and conceive something transcending themselves.

Today, we are a childlike civilization, a Telestial world in the Fullness of Times.[60] Filled as if by an unstoppable rolling river pouring from the heavens, our knowledge becomes unprecedented.[61] Nothing is withheld, whether the laws of the earth or the bounds of the heavens, whether there be one God or many Gods, everything begins to manifest.[62] And the work of God hastens.[63] Repeating the words of Christ, we speak,[64] and information technologies begin to carry consolation around the world. Emulating the works of Christ, we act,[65] and biological technologies begin to make

---

[57] Revelation 12: 7, Moses 4: 3–4, and Abraham 3: 28.

[58] Abraham 4: 18.

[59] Genesis 1: 31, Moses 2: 31, and Abraham 4: 21.

[60] Ephesians 1: 10; D&C 76: 81; and Kevin Barney, "The Etymology of 'Telestial,'" By Common Consent, January 27, 2010, http://bycommonconsent.com/2010/01/27/the-etymology-of-telestial/ (accessed June 04, 2016).

[61] D&C 121: 33.

[62] D&C 121: 26–32.

[63] D&C 88: 73–80.

[64] Mark 16: 15.

[65] Matthew 10: 8.

the blind see,[66] the lame walk,[67] and the deaf hear[68]; agriculture begins to feed the hungry; and manufacturing begins to clothe the naked.[69] Hearts turning to our ancestors, we remember them, and machine learning algorithms begin to process massive family history databases, perhaps to redeem our dead.[70]

A biotech revolution begins.[71] Synthetic biology restores extinct species, creates new life forms, and hints at programmable ecologies. Some recall prophecies about renewal of our world[72] – or perhaps its destruction.[73] Personalized medicine begins to restore vitality to an older generation. Some insist that death is necessary for meaning, but new voices repeat old stories about those who were more blessed for their desire to avoid death altogether.[74] Reproductive technology enables infertile and gay couples, as well as individuals and groups, to conceive their own genetic children. Some recoil from threats to tradition, while others celebrate gifts to new families.[75] Weaponized pathogens threaten pandemics, as well as targeted genocides and assassinations. Meanwhile, solar energy becomes less expensive than any other. And the Internet evolves into a distributed reputation network, creating new incentives for cooperation. Missionaries find their work more globalized than ever before.[76]

A nanotech revolution begins.[77] Atomically-precise printing erupts with food, clothing, and shelter. Welfare systems solve old problems and make new ones.[78] Among the wealthy, robotic cells flow through bodies and brains, extending abilities beyond those of the greatest athletes and scholars of history. Enjoying restored vitality, many become convinced that we can vanquish that awful monster, death.[79] But cautionary voices call attention to stunning socioeconomic disparities.[80] With the ability to read and write data in every neuron of the brain, the Internet evolves

---

[66] Alice Park, "Stem Cells Allow Nearly Blind Patients to See," Time, October 14, 2014, http://time.com/3507094/stem-cells-eyesight/ (accessed June 04, 2016).

[67] John Hewitt, "Paralyzed man walks again after surgeons transplant cells from his nose to his spine," Extremetech, October 22, 2014, http://www.extremetech.com/extreme/192548-paralyzed-man-walks-again-after-surgeons-transplant-cells-from-his-nose-to-his-spine (accessed June 04, 2016).

[68] Macrina Cooper-White, "See The Amazing Moment When A Deaf Person Hears For The First Time," Huffington Post, February 10, 2015, http://www.huffingtonpost.com/2015/02/10/people-hear-for-first-time-video_n_6646594.html (accessed June 04, 2016).

[69] Jacob 2: 19 and Mosiah 4: 26.

[70] D&C 128: 6–9.

[71] Kurzweil, 206.

[72] Articles of Faith 1: 10.

[73] Moses 1: 38.

[74] 3 Nephi 28.

[75] D&C 88: 33.

[76] D&C 14: 3–4.

[77] Kurzweil, 226.

[78] D&C 42: 34, 55; and 2 Nephi 26: 30–31.

[79] 2 Nephi 9: 10, 19, 26.

[80] D&C 78: 6.

into a composite of virtual and natural realities. We begin to connect with each other experientially, sharing senses and feelings. Spiritual experiences become malleable, meriting careful discernment.[81] Wireheading haunts relationships and burdens communities. And weaponized self-replicating nanobots threaten destruction of the biosphere. Meanwhile, robotic moon bases mine asteroids and construct space colonies, reinvigorating the pioneer spirit.[82]

A neurotech revolution begins.[83] We virtualize brains and bodies. Minds extend or transition to more robust substrates, biological and otherwise.[84] As morphological possibilities expand, some warn against desecrating the image of God, and some recall prophecies about the ordinance of transfiguration.[85] Data backup and restore procedures for the brain banish death as we know it.[86] Cryonics patients return to life. And environmental data mining hints at the possibility of modeling history in detail, to the point of extracting our dead ancestors individually. Some say the possibility was ordained, before the world was, to enable us to redeem our dead,[87] perhaps to perform the ordinance of resurrection.[88] Artificial and enhanced minds, similar and alien to human, evolve to superhuman capacity.[89] And malicious superintelligence threatens us with annihilation. Then something special happens: we encounter each other and the personification of our world, instrumented to embody a vast mind, with an intimacy we couldn't previously imagine.

In that day, we will be an adolescent civilization, a Terrestrial world in the Millennium.[90] Technology and religion will have evolved beyond our present abilities to conceive or express, except loosely through symbolic analogy.[91] We will see and feel and know the messiah,[92] the return of Christ, in the embodied personification of the light and life of our world,[93] with and in whom we will be one.[94] In a world beyond present notions of enmity, poverty, suffering, and death – the living transfigured and the dead resurrected to immortality – we will fulfill prophecies.[95] And we will repeat others, forth-telling and provoking ourselves

---

[81] Joseph Smith, 202.

[82] D&C 136.

[83] Kurzweil, 259.

[84] D&C 76: 98, 109.

[85] Joseph Smith, 170.

[86] 1 Corinthians 15: 51–55.

[87] D&C 128: 22.

[88] Brigham Young in *Wilford Woodruff's Journal*, 3, by Wilford Woodruff, edited by Scott Kenney (Utah: Signature Books, 1985), 323–324.

[89] D&C 77: 1–4.

[90] D&C 76: 91 and Articles of Faith 1: 10.

[91] D&C 1: 24.

[92] 1 John 3: 2.

[93] John 8: 12, Mosiah 16: 9, 3 Nephi 11: 11, and D&C 88: 7–13.

[94] John 17: 20–23.

[95] D&C 101: 26–34.

through yet greater challenges[96]: to maturity in a Celestial world,[97] and beyond in higher orders of worlds without end.[98]

65.6 Conclusion

Some have charged Transhumanism with being a quasi-religious cult, to which many secular Transhumanists have responded with denial, too stern, and revealing. Transfigurists don't hesitate to acknowledge spirituality, and even the religiosity of a strenuous shared spirituality, at work in Transhumanism. Indeed, if Transhumanism substantially affects the world for the better, it will do so only consequent to our practical trust in its esthetic and only to the extent that real world possibilities beyond our own power align with that practical trust. Put differently, Transhumanism will matter in a positive sense only consequent to our faith and only to the extent of grace. Transhumanism, at least for the Transfigurist, is a religious endeavor.

And indeed, the risks before us are too great and the opportunities too wonderful to confront with anything less than that shared strenuousness, both sharply rational and sublimely spiritual, which functions in all essentials as religiosity. The philosopher William James observed:

> "The capacity of the strenuous mood lies so deep down among our natural human possibilities that even if there were no metaphysical or traditional grounds for believing in a God, men would postulate one simply as a pretext for living hard, and getting out of the game of existence its keenest possibilities of zest. Our attitude towards concrete evils is entirely different in a world where we believe there are none but finite demanders, from what it is in one where we joyously face tragedy for an infinite demander's sake. Every sort of energy and endurance, of courage and capacity for handling life's evils, is set free in those who have religious faith. For this reason the strenuous type of character will on the battle-field of human history always outwear the easy-going type, and religion will drive irreligion to the wall."[99]

Too hardy to concede to antireligious fantasies, and too motivated to resist technological empowerment, religion will surely evolve with humanity. And if humanity will not become extinct before evolving into superhumanity, what would stop religion from evolving into that which yet provokes such minds? Such minds! Beyond our anatomical capacity to comprehend, their operations and motivations must largely elude us. But maybe Transfigurists give us a glimpse into a future of religion between here and there.

---

[96] D&C 43: 31.

[97] D&C 88: 25–26.

[98] D&C 130: 9–11.

[99] William James, The Will to Believe, and Other Essays in Popular Philosophy, and Human Immortality, 213.

Chapter 66
Equalism: Paradise Regained

Inessa Lee

> *"Our commitment to next generations is to bring heaven on earth, and not nuclear annihilation. There is still hope, we must act before time slips." (Amit Ray, Peace on the Earth A Nuclear Weapons Free World)*

Nowadays Paradise is seen by most people as a mythical place in Heaven. Somewhere along the way we lost understanding that it used to be an actual place on Earth, a home to first humans: "And the LORD God planted a garden in Eden, in the east, and there he put the man whom he had formed" (*Genesis* 2:8). Not only has the notion of Paradise been distorted throughout the centuries, but the perception of God's persona and His unity with humans has changed as well.

Our vision of Christianity has been transformed: it is now seen as the religion of death due to its promise of happiness in the afterlife, as well as the dark history such as the Inquisition and the Crusades. Most importantly, young people feel reluctant to attend churches these days, because they see Christianity as the religion of slaves. In a distorted notion of the Holy Trinity, Jesus is portrayed as a victim of violence, whereas His Father is seen as a cruel and vengeful paternal figure. Such images don't seem to be alluring to our psyche, do they?

Traditional religious institutions have been using fear as an instrument of mind suppression for centuries: "Organized religion destroys who we are by inhibiting our actions, by inhibiting our decisions out of, out of fear of some, some intangible parent figure who, who shakes a finger at us from thousands of years ago" (comedy film *Dogma*, 1999).

Consequently, Christianity has been stereotyped as a religion of death, slaves, and fear. Don't get me wrong: it happened not because religious institutions are evil, but as a result of social transformations occurring in the course of time. Certain

I. Lee (✉)
California Transhumanist Party, Los Angeles, CA, USA

Institute for Education, Research, and Scholarships, Los Angeles, CA, USA
e-mail: inessa@californiatranshumanistparty.org; inessa@ifers.org

© Springer Nature Switzerland AG 2019
N. Lee (ed.), *The Transhumanism Handbook*,
https://doi.org/10.1007/978-3-030-16920-6_66

historical premises gave birth to stereotypes about the major world religion and God's image.

66.1 Evolution of God's Image in Christianity

66.1.1 Early Christianity (c. 31/33 – 380)

Christianity came about as a religion of slaves and poor people. Its original message was somewhat communist in nature: achieving freedom of slavery, social and economic injustice through Jesus in the afterlife: "All the believers were together and had everything in common. They sold property and possessions to give to anyone who had need. Every day they continued to meet together in the temple courts. They broke bread in their homes and ate together with glad and sincere hearts, praising God and enjoying the favor of all the people. And the Lord added to their number daily those who were being saved" (*Acts 2:44-47*). In the Latin text of *the Book of Acts* the word "communis" is used to describe the idea of common ownership. Therefore, the fist Christians were similar to socialists, whereas God was seen as their Leader, the Messiah, guiding them to social justice and equality.

66.1.2 State Church of the Roman Empire (380–476)

In 380 Christianity was officially adopted as Roman state religion. It quickly turned into the movement of the privileged; the tool government was using to control the masses.

Christianity became the religion of the ruling class, whereas the Church fulfilled aristocracy's expectations. The image of God has been transformed as well: Jesus was seen as King of Kings who granted power to the authorities. People learned that all authority in heaven and earth comes from God; therefore they must obey the Church as the institute of God's power.

66.1.3 Medieval Christianity (476–1520)

The Medieval Ages gave birth to a whole new culture in Western Europe. Its central figure was God. Christian mysticism found expression in architecture, literature, arts, etc. God was considered the Lord, whereas people were His servants or slaves. Such relationships between God and humans reflected the model of feudal government. The image of God evoked fear and humility, as the fortune of servants depended solely on their Master. The Church used human fear to enrich herself (e.g., selling indulgences, accepting manors donated by lords).

66.1.4   Late Christianity (1521–1719)

Protestantism emerged in XVI century, establishing its work ethic as a core religious idea of late Christianity. Labor has become human calling, existential purpose, and a way to serve God. Martin Luther, the first protestant leader, claimed that people can achieve salvation only through work in the name of Lord.

German sociologist Max Weber in his book *The Protestant Ethic and the Spirit of Capitalism* wrote that capitalism developed on the basis of protestant work ethics: "Both as ruling and ruled strata and both as a majority and minority, Protestants ... have demonstrated a specific tendency toward economic rationalism" (Ch.1: *Religious Affiliation and Stratification*).

Therefore, in Protestantism God was seen as a helper to those who work hard to deserve salvation: "In practice this means that God helps those who help themselves. Thus the Calvinist, as it is sometimes put, himself creates his own salvation, or, as would be more correct, the conviction of it" (M. Weber, *the Protestant Ethic and the Spirit of Capitalism*,1905).

66.1.5   Modern Christianity (1720 – Beginning of the Twentieth Century)

The wave of protestant upheaval started in North America, reaching its peak when the slavery was abolished in the US. Revivalism gave birth to new evangelical movements, such as Methodism, Baptism, Pentecostalism, etc.

Since then Holy Trinity has become the ultimate source of God's Grace for people. During the First Great Awakening humans have developed fraternal relationships with God. When they have personal issues, they ask God to resolve them. And there's no need for the priests to pardon sins, because Jesus is the closest friend, thus people can communicate with Him directly, without any middleman. Many religious rituals and formalities have been abolished, as the image of God and His relationships with humans are simplified.

66.1.6   Post-Christian Period (Beginning of Twentieth Century – Nowadays)

The stagnation of capitalism resulted in global religious crisis. The image of God is mostly considered a folklore, whereas our relationships with Him a fable. For most Christians, Sunday church service has become a tradition rather than the act of reunion with God. "Religion today is not transforming people; rather it is being transformed by the people. It is not raising the moral level of society; it is descending to society's own level, and congratulating itself that it has scored a victory because society is smilingly accepting its surrender" (*The Price of Neglect. In Hendrickson Bibles.* 2012. *The A.W. Tozer Bible: King James Version. Hendrickson Publishers.* p. 93).

Running a church is a very lucrative business these days. A Christian parish is a money making machine. "The majority of churches that exist today are organized spiritual business entities. They operate similarly to corporate America. There is a CEO, or Sr. Pastor. There is a Board of Directors, probably elders or deacons. There is a staff—either paid or volunteer. There are the customers—namely the lay people who come each week to financially underwrite the corporation/church. And finally, there is the product—their version of the gospel and its presentation" (Tim Kurtz, *No Longer Church as Usual*, ch.3; 2011).

66.1.7   Christian Transhumanism Period

The perception of God as a mythical being and Church as a moneymaking machine will change as we move into the era of Christian transhumanism. Religion will be reinterpreted in terms of transhumanist philosophy. During the Christian transhumanism era, the image of God will merge with humanity. People will realize they are co-creators with God, as science and technology enable them to eliminate death and create new forms of life. After all, we are made in the image of God; therefore we are not slaves, but His intelligent emanations.

As you can see, God's image and his relationships with people have evolved as a result of socioeconomic changes. Major political systems (capitalism, socialism, etc.) have greatly influenced the Christian mindset. Christianity is on the verge of major transformations. Socioeconomic reforms will cause a shift in our religious consciousness during the transhumanism era.

Thus, we need an ideology combining a new economic model with religious principles: Equalism is a religious movement and socioeconomic theory based on the Christian value of equality. Through equalism we can regain paradise: a world without suffering, poverty and death.

66.2 Equalism

Indeed, the ultimate goal of equalism is to regain paradise on Earth. Physical immortality on the polluted planet, where poverty still exists, is meaningless. Eternal life makes sense only if we rebuild paradise. The first humans lost it because of greed and vanity - the primary sins committed by Adam and Eve. They wanted to possess knowledge and power to become equal with God: "But the serpent said to the woman, 'You will not die; for God knows that when you eat of it your eyes will be opened, and you will be like God, knowing good and evil.'" *(Genesis 3:4-5).*

Inequality has been the biggest issue for human beings since biblical times. We believe people are greedy for power and material possessions because they have experienced social inequality since childhood: "The childhood shows the man, as morning shows the day" (John Milton, *Paradise Regained*). If all humans are born prosperous with equal opportunities for personal development, there will be no reason for crimes.

Crusade expeditions were nothing but religious wars caused by the Pope's desire to enrich and strengthen the Catholic Church. Most Crusades used expeditions as a chance to get rich. That's an example of how uneven distribution of resources results in bloodsheds.

Greed led us to environmental pollution, perversion of love, and conflicts. We lose connection with God as labor exploitation steals our time for prayers. Only equal distribution of resources can put an end to human greed - the initial reason of our exile from paradise.

Therefore, we can regain paradise by achieving socioeconomic equality and abolishing human exploitation. Science and technology is the way to accomplish equalism. It will enable us to live forever without diseases and to distribute wealth evenly. Automated labor will give us more time to pray to God and attend churches.

Here is the equalism strategy for creating paradise on earth:

(a) establish socioeconomic equality;
(b) achieve physical immortality;
(c) clean up the environment;
(d) develop Christian transhumanist consciousness.

66.3 Equality as a Way to Paradise

"For my birthday I want world peace, to end hunger and pain, to end global warming, to end racism, for people not to judge others by the outside, for all ages, colors, races, and other differences to never matter, for everyone to have same opportunities for all, to live in a world where money has no value, where equality and understanding is an everyday experience. For my birthday I want the impossible" (posted anonymously by an unknown author).

Socioeconomic inequality is a major reason for human sins, sufferings, and wars. It's caused by uneven wealth distribution. Humans will come back to paradise when we create an economic system based on even distribution of wealth.

Imagine the world where everyone is prosperous. There will be no wars; no more hunger and poverty. Thus, equal distribution of resources is the lasting solution to World Peace.

Socioeconomic equality is a foundation of the equalism ideology, because all people are equal in the eyes of God: "At the present time, your surplus will meet their need, so that in turn their surplus will meet your need. Then there will be equality" (*2 Corinthians 8:14*). In fact, from now on, all Christians can be referred to as equalists, as equality is a core Christian value.

The equalism principle will be executed via technological advancements during the Christian transhumanism period.

A Christian society is supposed to be classless. Publicly owned means of production, economy central planning executed via IT means, and automated labor will eventually eliminate social stratification. Human work will become a form of creative self-expression rather than toil. Financial worries will end for humanity, because money won't exist anymore. Now let's see how equalism can be established step by step.

66.4   Equalism Manifesto

1. State-owned means of production, lands and enterprises; abolition of inherited properties, including intellectual ones.

A major reason for class stratification and human exploitation is privately owned means of production. God owns everything including the planet: "Who has first given to me, that I should repay him? Whatever is under the whole heaven is mine" (*Job 41:11*). Therefore, the means of production, transportation and communication should be state-owned rather than privately owned. This will establish socioeconomic equality.

Same applies to enterprises. They should be also state-owned. E.g., the Commonwealth Realms, particularly in Australia, Canada, New Zealand, and the United Kingdom, country-wide SOEs (State-Owned Enterprises) often use the term "Crown corporation", or "Crown entities", as cabinet ministers (Ministers of the Crown) often control the shares in them.

Private ownership of land should be abolished, because the land belongs to God: "The land must not be sold permanently; because the land is mine and you reside in my land as foreigners and strangers" (*Leviticus 25:23*). Thus, state can manage and distribute land.

The state can have a land use database information system in which every citizen is registered as part of a land collective and assigned with a land lot for residential or cultivation purposes. The size of the residential land lot will be determined by certain criteria such as family size (bigger family needs more living space), age of the homeowner (not younger than 18), location of the residence, etc. A family increased in size has a right to receive a bigger residential space. All they have to do is submit a request to the land use database information system.

2. Elimination of human exploitation by full automation of labor.

Automation operated by supercomputers will have an enormous production capacity. Fully automated manufacturing will provide a production volume sufficient enough for the equal distribution of benefits in society. Finally, Marxist principle "from each according to his ability, to each according to his needs" will be accomplished. Perhaps Karl Marx would have supported Christian transhumanism.

Technology can liberate people from exploitation, providing extra time for creative self-expression and communication with God. Culture, arts and sciences will be actively developed. Consequently, each individual and society as a whole will become more educated, healthy, happy and self-fulfilling. Therefore, advanced technology will turn human labor into a form of creative self-expression.

3. Active use of Digital Democracy to expand and improve democratic practice.

Democracy is a necessary condition for the existence of the Christian tradition in the society. No other government system corresponds to Christian principles as much as democracy. Superiority of people over a state originates from the Bible. Laws represent God's will. We are all equal before the law. At the same time laws exist to maintain balance and peace in the society so that it doesn't turn into hell.

Christian transhumanism enables a new form of government, known as e-democracy. It incorporates latest information technology tools to enable adult citizens to participate equally in the political life of the country. It's the only way to make elections, legislations, and other political processes truly transparent and fair.

> "Digital tools and widespread access to the internet have been changing the traditional means of participation in politics, making them more effective. Electoral processes have become more transparent and effective in several countries where the paper ballot has been substituted for electronic voting machines. Petition-signing became a widespread and powerful tool as individual citizens no longer needed to be bothered out in the streets to sign a sheet of paper, but could instead be simultaneously reached by the millions via e-mail and have their names added to virtual petition lists in seconds. Protests and demonstrations have also been immensely revitalized in the internet era. In the last few years, social networks like Facebook and WhatsApp have proved to be a driving-force behind democratic uprisings, by mobilizing the masses, invoking large gatherings, and raising awareness, as was the case of the Arab Spring" [5].

4. Replacing governments with supercomputers until God's Kingdom is reestablished during the Second Coming of Christ.

Jesus is "King of kings and Lord of lords"(*Revelation 19:16*). He was appointed by God the Father to rule the Universe and lead humanity. But the time of His Kingdom hasn't come yet, because humanity hasn't evolved enough to regain paradise. Nations are not ready yet to be united under the Kingdom of Christ. Too many government officials are appointed by Satan to destroy spirituality by multiplying poverty, feud and toil in the society. They are corrupted and emotionally unstable. Greed for money and power forces them to make wrong decisions.

God's mind is like a supercomputer, because He is never biased, corrupted or emotional. His Kingdom is run by the tenants of logic, reason and justice.

Hence, the only objective and democratic government form, besides God's Kingdom, is cyberocracy – the state run by AI. We shall use it until the Second Coming of Christ to establish perfect democracy.

A cyberocratic state is a government of the people, by the people and for the people. Its functions are executed via e-referendum of the citizens. Referendum data is processed by supercomputer to come up with optimal solutions to various socio-economic issues.

Cyberocracy is a perfect example of how technology helps us to regain paradise. This type of government is more efficient, fair and free of corruption and interference.

> "Political turmoil is the result of humans contending with complex political issues. And humans are awful at running governments. Devoid of personality, computers avoid the pit-falls of human-led governments. Now is the time to look beyond the antiquated concepts on which the constitution was founded and form a new government based on rationality, accuracy, and computer algorithms" [1].

5. Establishing centrally planned economy managed by Artificial Intelligence.

Centrally planned economy managed by A.I. will ensure even distribution of resources.

In fact, the first attempt was made in 1970–1973 when Chile's government started *Project Cybersyn* to computerize the management of centrally planned economy. The project was headed by British operations research scientist Stafford Beer. Jointly with Chilean engineers, a team of 12 British programmers created the software called Cyberstrider.

The system united 500 enterprises into *Cybernet* network using telex machines deployed in the factories. The control centre in Santiago collected each day data coming from the factories. Based on the processed data, short-term predictions and necessary adjustments were made. Four levels of control, such as firm, branch, sector, total, included algedonic feedback. If one level couldn't solve a problem in a certain period of time, the higher level was notified. The results were discussed in the operations room and a top-level government plan was made.

The system proved its effectiveness and importance in 1972 when 40,000 striking truck drivers blocked the access streets that converged towards the capital. Chile's government was able to supply Santiago with enough food due to Cybersyn telex machines insuring the transport of food into the city with only about 200 trucks driven by strike-breakers.

With the improved A.I. systems we'll be able to make a real break-through in economy planning. A.I. will put an end to the spontaneous order of market economy.

6. Eliminating money with the help of advanced technology.

God is abundance, thus people should be prosperous as they are born in His image. The only reason why poverty still exists is money. Market economy is an ultimate obstacle to equality. Money supply has caused recessions, high unemployment rate, profiteering, and corruption in global banking system.

Equalism replaces market relations with centrally planned economy: Information about items ordered by the consumers will be received by the centralized database of goods consumption. Afterwards, the software will make statistics about the amount and the type of goods to be produced, and send out the orders to the automated production lines. The automated lines will create resources as well as output the final products; therefore there will be no surplus value of goods. The factories will produce as many goods of various kinds as necessary. As a result of complete saturation of the consumer funds, the supply will exceed the demand. Consequently, market relations will cease to exist along with money.

7. Free healthcare and education for all people.

One of our priorities is developing a strong medical research sector that can dramatically increase human lifespan. Equal access to healthcare for all people is a must in a Christian transhumanist society. Therefore, healthcare and medical research should be state-funded.

According to Huffington Post, Cuba, a country with free medical care, "has some of the healthiest, most long-lived residents in the world – as well as medical invention or two that could run circles around U.S. therapies, thanks to government investment in scientific research and preventive public health approach that views medical care as birthright" [2].

This small country has developed Cima-Vax (anti-cancer vaccine) that has treated 5,000 patients worldwide, whereas most U.S. citizens are still struggling to get affordable healthcare plans that will address their basic healthcare needs adequately. The reason is simple: the U.S. healthcare policy is dictated by greedy pharmaceutical and insurance companies, whereas Cuban government invests into medical research and healthcare with no interest in profiteering.

As for free education, Germany, France and Canada are on the list of top five countries for higher education quality, and their students don't have to pay tuition. Wouldn't it be great for our kids to grow up without a heavy burden of college loans?

66.5 Universal Basic Income

Universal basic income should be introduced during the transition period from market economy to centrally planned economy. Universal basic income is a way to eliminate income inequality before we stop using money altogether. It's a first step for Christian transhumanism. It reduces stress levels and makes people more productive. Thus, it helps to turn labor into a form of creative self-expression.

"In the mid-nineteen-seventies, the Canadian province of Manitoba ran an unusual experiment: it started just handing out money to some of its citizens. The town of Dauphin, for instance, sent checks to thousands of residents every month, in order to guarantee that all of them received a basic income. The goal of the project, called Mincome, was to see what happened. Did people stop working? Did poor people spend foolishly and stay in poverty? But, after a Conservative government ended the project, in 1979, Mincome was buried. Decades later, Evelyn Forget, an economist at the University of Manitoba, dug up the numbers. And what she found was that life in Dauphin improved markedly. Hospitalization rates fell. More teen-agers stayed in school. And researchers who looked at Mincome's impact on work rates discovered that they had barely dropped at all. The program had worked about as well as anyone could have hoped" [6].

The idea of universal basic income is already supported in the U.S. by some Silicon Valley leaders, such as Mark Zuckerburg and Elon Musk. Sam Altman, Y Combinator CEO, has been giving 50 households in Oakland up to $1,500 per month in a basic income experiment [3].

We believe government should send an equal monthly paycheck to every single citizen of the country, excluding children under 18 years of age, as they are supposed to be either under parental or in foster care.

66.6 Progressive Tax

During the transition stage from capitalism to equalism, progressive tax will support public expenditures and transhumanism reforms, such as free education, green energy, medical research, etc.

Individuals who have an additional source of income aside from universal basic wage should use it to benefit society. The higher income a citizen has, the more they should invest into social development (science, medicine, technology, etc.).

Land value tax is a type of progressive tax that has been referred to as perfect, because it doesn't cause economic inefficiency and, unlike property taxes, it disregards the value of buildings, personal property and other improvements to real estate. According to Henry George, the supply of land is fixed and its location value is created by communities and public works, thus, the economic rent of land is the most logical source of public revenue.

66.7 Digital Currency

We believe that digital currency should replace banknotes and become universal during the transition phase from capitalism to equalism.

Economic manipulation by central banks is not beneficial. Money supply has caused recessions, high unemployment rate, profiteering and corruption in global banking system. Humanity needs a universal digital currency to put an end to banks manipulation and economic barriers between nations.

Here are some benefits of e-currency:

(a) It sufficiently lowers the cost of financial operations due to its minimal foreign transaction fees and zero conversion fees.
(b) It can be used worldwide without any territorial restrictions.
(c) It's easy and inexpensive to save digital money.
(d) Digital currency is very democratic as its value is determined by users and not central governments or banks.
(e) If universal basic income is paid with cyber currency, there will be no income tax imposed on individuals or entities, because the currency is not controlled by the state.
(f) There will be no sales tax, because payments are made with cyber currency, which is passed between buyers and sellers without the need for an intermediary.

Bitcoin, initially created as a universal crypto currency, became another tool of financial market manipulation and speculations, as it's not provided by any standard and has no centralized control. It is nothing but another form of fictitious capital.

We need to invent a universal digital currency that has unlimited total supply. The unlimited emission of cyber currency should not cause inflation, because, the excessive supply of paper bills is a true reason of inflation.

The underlying issue of universal cyber currency is crime: "Untraceable financial transactions facilitate crime. Drug trafficking, prostitution, terrorism, money laundering, tax evasion and other illegal and subversive activity all benefit from the ability to move money in untraceable ways" [4].

A solution to this issue is a universal cyber security system to trace any kind of criminal and terrorist activities, and protect digital transactions from cyber attacks.

On the basis of the above-mentioned, equalism is a perspective of the evolving humanity. Transition from capitalism to equalism is inevitable due to rapid technology development, trade globalization and integration of production/labor markets.

However, new socioeconomic system alone does not guarantee that paradise will be regained. Human consciousness needs transformation as well.

66.8 Equalism as a Religious Movement

Equalism is born out of transhumanism. It uses science, technology and economy as tools to accomplish Christianity goals, such as eternal life and socioeconomic equality. However, religion is a basis of our ideology, a motivation, for everything is possible with God. It's a simple equation: (science + technology + economy) * faith = equalism.

Equalsim propagates Christianity as a lifestyle. Its spiritual mission is awakening co-creator consciousness in every one of us. Indeed, we are co-creators of reality through Christ: "Very truly I tell you, whoever believes in me will do the works I have been doing, and they will do even greater things than these, because I am going to the Father. And I will do whatever you ask in my name, so that the Father may be glorified in the Son. You may ask me for anything in my name, and I will do it" *(John 14:12-14)*.

Jesus Christ was the first transhumanist who conquered death. We can also become immortal combining science, technology and faith in Christ: "I am the resurrection and the life. The one who believes in me will live, even though they die; and whoever lives by believing in me will never die" *(John 11:25-26)*. With Jesus in our heart we've got strength to move mountains. Obviously, He wants us to live forever too. And it's not only spiritual immortality. Jesus ate fish after the Resurrection, which means that His physical body resurrected: "They give Him a piece of broiled fish and He took it, and ate it in front of them" *(Luke 24:42-43)*.

By the way, have you ever wondered why Jesus ate fish after His resurrection, and not any other kind of food? And why did He choose four fishermen to become His disciples? The image of fish is used metaphorically and quite frequently in the New Testament. In fact, the symbol of Christianity is fish (Ichthys). Perhaps, the ultimate secret of immortality has to do with water, ocean or fish. One day scientists will solve this mystery.

Equalism is not a church but rather a non-denominational movement which unites all the Christians in achieving two common goals: (1) retrieving paradise on Earth; and (2) becoming immortal (physically and spiritually). We can accomplish these goals by following in Jesus' footsteps. World peace will be maintained by uniting humanity with an understanding of Christianity as a lifestyle. Here are the basic life principles of equalists. Following them will create paradise on Earth and prepare for the Second Coming of Christ.

66.9   Equalism's 10 Commandments

1. Let there be equality between you as you are equal before God.

   "At the present time, your surplus will meet their need, so that in turn their surplus will meet your need. Then there will be equality" (*2 Corinthians 8:14*).

   "And when they measured it by the omer, the one who gathered much did not have too much, and the one who gathered little did not have too little. Everyone had gathered just as much as they needed" *(Exodus 16:18)*.

As we can see, the first "commandment" is also the foundation of equalism.

2. Love God with all your heart as you are created in His image.

"You shall love the Lord your God with all your heart and with all your soul and with all your mind" *(Matthew 22:37)*. This is the greatest commandment of the New Testament.

Good children truly love their father. He is the source of our life, the center of the universe, so we should always honor our Creator.

3. Love and accept thyself and your neighbor the way God created you.

"Love your neighbor as yourself" *(Mark 12:31)*. Jesus used the word "yourself" intentionally, because we can't love other people until we accept and love ourselves.

Non-acceptance of ourselves is actually a sin, for God created us uniquely to play special roles in society. God loves us; therefore it's unreasonable to hate His creation. The same refers to other people around us – if we can't accept them the way they are, we are disobeying the will of God. Once we learn to accept ourselves, we will stop criticizing others.

This commandment also refers to those who like to change their look with plastic surgery. We are already perfect the way God created us. Diversity is beauty. However, body modification for medical and longevity purposes is acceptable.

4. Treat each other as brothers and sisters for you are all God's children.

"Keep on loving one another as brothers and sisters" *(Hebrews 13:1)*. Once humanity realizes that we are all related through God, peace will be finally established on the planet. "How good and pleasant it is when God's people live together in unity!" *(Psalm 133:1)*.

5. Honor your family, for without them you are like a rootless tree.

"Whoever brings ruin on their family will inherit only wind, and the fool will be servant to the wise" *(Proverbs 11:29)*.

Needless to say, family is a foundation of any society. It's a fortress giving us strength to move on under any circumstances.

6. Preserve nature as God created it before breathing life into you.

Then God said, "Let us make humankind in our image, according to our likeness; and let them have dominion over the fish of the sea, and over the birds of the air, and over the cattle, and over all the wild animals of the earth, and over every creeping thing that creeps upon the earth" *(Genesis 1:26)*.

Genesis clearly states that humans and animals were created on Day Six, after air, flowers, trees, oceans, birds and sea animals, etc. It means that God created nature to support our life in paradise. It was clean and pristine. So let's try to clean up the planet and take care of what keeps us alive.

7. Keep your mind pure, as impure thoughts make your body unclean.

"But now I am writing to you that you must not associate with anyone who claims to be a brother or sister but is sexually immoral or greedy, an idolater or slanderer, a drunkard or swindler. Do not even eat with such people" *(1 Corinthians 5:11)*.

The body is a vessel for the mind. Whatever we think affects our physique as much as the food we eat. God is purity, thus He wants us to be pure. So try to think mostly good thoughts about yourself and the world around you.

8. Do not harm your neighbor's body or soul as they are created by God.

The commandments, "You shall not commit adultery," "You shall not murder," "You shall not steal," "You shall not covet," and whatever other commands there may be, are summed up in this one command: "Love your neighbor as yourself" (*Romans 13:9*).

Harming others spiritually or physically is the same as harming God as we are His children.

9. Forgive each other as God is forgiveness.

"Bear with each other and forgive one another if any of you has a grievance against someone. Forgive as the Lord forgave you." (Corinthians 3:13).

One of the biggest lessons in our life is learning to forgive. Forgiveness makes life easier.

10. Seek for true love, not for material possessions.

"Above all, love each other deeply, because love covers over a multitude of sins" (*1 Peter 4:8*).

This is just as important as the Great Commandment of the New Testament. True love cannot be purchased and true marriage can't be arranged. God determines who we should love and marry. In the end of the day, we always remember emotions and precious moments that we share with our beloved ones.

Paradise is incomplete if Adam doesn't have Eve by his side. Everything started with love, because God is love. Love is a driving force of our existence.

To sum it up, capitalism kills love. Every day families get destroyed because of financial issues and lack of time spent together. We believe that the desire of true love will help to establish equalism in the world. When we become equal, we can finally afford love.

So, dear brothers and sisters, let's regain the Brave New World of paradise for the sake of true love in the name of the Father, the Son, and the Holy Spirit. *Amen.*

Bibliography

1. Aberle, C. (23 February 2017 r.). *Let the Computer be Your Ruler.* The Wesleyan Argus: http://wesleyanargus.com/2017/02/23/let-the-computer-be-your-ruler/
2. Almendrala, A. (15 March 2016 r.). *Cuba Has Made At Least 3 Major Medical Innovations That We Need.* LIFE: https://www.huffpost.com/entry/cuba-medical-innovations_n_56ddfacfe4b03a4056799015
3. Donnelly, G. (19 April 2018 r.). *Finland's Basic Income Experiment Will End in 2019.* Fortune: http://fortune.com/2018/04/19/finland-universal-basic-income-experiment-ending/

4. McWhinney, J. (13 October 2018 r.). *Why Governments Are Afraid of Bitcoin.* Investopedia: https://www.investopedia.com/articles/forex/042015/why-governments-are-afraid-bitcoin.asp
5. Pogrebinschi, T. (2 March 2017 r.). *Does digital democracy improve democracy?* open-Democracy: https://www.opendemocracy.net/democraciaabierta/thamy-pogrebinschi/does-digital-democracy-improve-democracy
6. Surowiecki, J. (20 June 2016 r.). *The Case for Free Money: Why Don't We Have Universal Basic Income?* The New Yorker: http://www.newyorker.com/magazine/2016/06/20/why-dont-we-have-universal-basic-income